1795.

D0081057

TURFGRASS:
SCIENCE
AND
CULTURE

TURFGRASS: SCIENCE AND CULTURE

James B. Beard

Michigan State University

Prentice-Hall, Inc., Englewood Cliffs, N.J.

© 1973 by Prentice-Hall, Inc.
Englewood Cliffs, N.J.

All rights reserved. No part of this
book may be reproduced in any form
or by any means without permission
in writing from the publisher.

10 9 8 7 6 5 4 3 2 1

ISBN: 0-13-933002-X

Library of Congress Catalog Card Number 79-168621

Printed in the United States of America

PRENTICE-HALL INTERNATIONAL, INC., *London*
PRENTICE-HALL OF AUSTRALIA, PTY. LTD., *Sydney*
PRENTICE-HALL OF CANADA, LTD., *Toronto*
PRENTICE-HALL OF INDIA PRIVATE LIMITED, *New Delhi*
PRENTICE-HALL OF JAPAN, INC., *Tokyo*

To My Wife
Harriet

Contents

Preface

This book has been written as a comprehensive basic text and reference source on turfgrass culture for use throughout the world. It is designed for turfgrass undergraduate and technical training students, the practicing professional turfmen, and interested individuals in allied professions such as ornamental horticulture. The stimulus for writing this book has come from the rapid expansion in the turfgrass industry during the past two decades, and an increasing demand for professional turfgrass managers who are equipped with advanced technical and science-oriented knowledge. There have been elementary manuals and a few books written on several phases of turfgrass culture. However, a textbook was needed that presents the fundamental facts and concepts on which the culture of various types of turf is based.

Emphasis is placed on the scientific principles of turfgrass culture that will provide the reader with a sound basis for formulating decisions and methods of operation. Factual statements are documented by reference citations whenever possible. If uncertainty or controversy exists concerning the topic under discussion, the author indicates that this is the case. The science of turfgrass culture is comparatively young, and the pool of knowledge is far from adequate. In areas of on-going research, the conclusions are tentative and the concepts controversial.

Much of the research information concerning turfgrass culture is widely scattered in turfgrass conference proceedings, trade publications, scientific journals, and field day programs, and therefore is not generally available. Since this literature is not accessible to most individuals, it has been compiled and utilized in substantiating the basic concepts of turfgrass culture. Over 12,000 references were reviewed during the preparation of the text. A list of the more pertinent references is cited at the end of each chapter for those wishing to study a particular area in greater detail. Figures and tables have been selected to illustrate the principles discussed. In writing this book, the author has assumed that the reader possesses an elementary knowledge of botany, chemistry, and soil science.

The book is divided into three parts. Part I involves a detailed treatise on the turfgrasses including their anatomy, growth, development, characteristics, adaptation, cultural requirements, use, and cultivars. A new approach to the ecology of turfgrass communities is also presented. Part II covers the interrelationships of atmospheric and soil environmental factors influencing turfgrass growth and development. The light, temperature, water, air, and traffic chapters are quite unique in their emphasis. The basic principles and practices of turfgrass culture and establishment are discussed in Part III. Specific cultural systems are not covered to avoid a restrictive, provincial approach. The discussion of specific cultural systems is left to the discretion of each instructor, who can best meet the student's needs in a specific climatic situation.

Arrangement of the text in three parts permits flexibility in course organization. It may be used in a 1, 2, or 3 course sequence, depending on the detail desired. Portions of the text can be omitted as appropriate if used for only one course. A separate turfgrass soils and fertilization course can also be taught utilizing the soil, traffic, fertilization, and cultivation chapters. The light, temperature, water, and air chapters could provide the basis for an advanced course in turfgrass culture.

I wish to express appreciation for the generous assistance and helpful critical reviews of individual chapters by my colleagues E. O. Burt, J. D. Butler, R. E. Chapin, W. H. Daniel, R. R. Davis, R. M. Endo, R. E. Engel, C. R. Funk, V. A. Gibeault, W. W. Huffine, M. Kabalin, R. A. Keen, J. W. King, B. Langvad, C. W. Laughlin, D. P. Martin, A. J. Miller, R. W. Miller, C. R. Olien, K. T. Payne, D. Penner, A. M. Radko, R. E. Rieke, R. E. Schmidt, C. R. Skogley, S. N. Stephenson, A. J. Turgeon, J. M. Vargas, H. Vos, W. E. Wallner, J. R. Watson, C. G. Wilson, and V. B. Youngner. Individuals who so courteously provided illustrations or photos are mentioned in the appropriate legends. Establishment of the O. J. Noer Memorial Turfgrass Library at Michigan State University greatly facilitated the literature review for this book. Finally, the assistance of my wife, Harriet, in typing, proofing, and assembling the manuscript is gratefully acknowledged.

JAMES B. BEARD

TURFGRASS:
SCIENCE
AND
CULTURE

chapter 1

Introduction

Turfs were developed by modern man in order to enhance his environment. The more technologically advanced the civilization, the more widely turfs are used. Turfs are important in human activities from the (a) functional, (b) recreational, and (c) ornamental standpoint.

Functional. A turf has numerous, important functional purposes as well as being attractive. Turfs control wind and water erosion of soil and are essential in eliminating dust and mud problems on areas surrounding homes, factories, schools, and businesses. More recently, the importance of turfs in climate control has been recognized. Turfs reduce glare, noise, air pollution, heat buildup, and visual pollution problems. Quality turfs are also of economic importance through increased property values and providing commercial appeal. A well-groomed lawn surrounding a factory or business is an asset in conveying a favorable impression to the employees and general public. Roadside turfs are important in highway safety because they function as a stabilized zone for emergency stopping of vehicles that stray from the roadway. Turfs are utilized for soil and dust stabilization on airfields in order to prolong the operating life of engines. Smaller airstrips also utilize turf on the runways.

Recreational. Many outdoor sports and recreational activities utilize turf, including baseball, cricket, croquet, field hockey, football, golf, hiking, lawn bowling, lawn tennis, lacrosse, polo, racing, rugby, shooting, skiing, and soccer. Turfs provide a cushioning effect that reduces injuries to the participants, particularly in the more active sports such as football, rugby, and soccer (72). The enjoyment and

1

benefits to physical health derived from recreational and leisure activities on turf are a vital part of modern man's activities.

Ornamental. A turf provides beauty and attractiveness for human activities. Cities can be very dismal without green turfs surrounding homes and businesses, in parks, and beside boulevards. The clean, cool, natural greenness of turfs provides a pleasant environment in which to live and work. Such aesthetic values are of increasing importance to the mental health of modern man because of the rapid life style and increasing urbanization.

History of Turfgrass Development

Reference to grass and gardens is found in the biblical literature. Lawns were an integral part of the Persian pleasure garden carpets and later of the Arabian gardens (58). Low growing, flowering plants were the basic constituents of these garden lawns. Subsequently, the Greeks and then the Romans adapted the Persian lawn gardens to their cultures.

The culture of mowed lawns is a relatively recent development in the history of man. References to "lawn gardens" are found in the English literature of medieval times (37, 84). Lawns of this period were composed of low growing grasses interplanted with flowers similar to the vegetation found in a meadow. Some gardens of the thirteenth century had turfs composed of grass monostands (1). Turfed seats were a feature of this period (37). The thirteenth-century literature also contains references to the sport of lawn bowling that utilized turf. Cricket, in the elementary form of club-ball, was played on turfs during the latter portion of the thirteenth century. The bowling green was probably the forerunner of modern, fine turfs. Outdoor sports such as lawn bowling, cricket, soccer, and golf played a vital role in the development of modern, high quality turfs.

More elaborate gardens and bowling greens or "allies" were developed during the sixteenth century (42). A form of soccer was played on public greens during this period. Lawns became more common in Great Britain, Germany, France, Netherlands, Austria, and other locations in northern Europe during the sixteenth and seventeenth centuries. Most towns and villages had what was called a green, common, or heath that was turfed and served as a medieval park and recreational area. The game of golf flourished initially on the natural upland downs and seaside turfs that were composed largely of *Agrostis* and *Festuca* species. "Mowing" was accomplished by the close grazing of sheep.

In 1665, John Rea published recommendations on preparing the planting bed, selecting the sod, sod harvesting, and sod transplanting (81). His instructions were as follows:

> "The next work, is to prepare the places intended for grass, and to provide turfs for them. First, level the ground, and consider the thickness of the turfs, which when layed, must be three inches lower than the upper edge of the rails, and the allies four inches, so the grass will be an inch higher, remembering still from the

rails to fetch your measures, and level, to keep the whole work in order; and if the ground under the turfs be not barren of it self, it should be covered some thickness with hungry sand to make it so, that the grass grow not too rank. The best turfs for this purpose are had in the most hungry Common, and where the grass is thick and short, prick down a line eight to ten feet long, and with a spade cut the turfs thereby, then shift the line a foot or fifteen inches further, and so proceed until you have cut so far as you desire, then cross the line to the same breadth, that the turfs may be square, and cut them thereby; then with a straight bitted spade, or turving-iron (which many for that purpose provide) and a short cord tied to it near the bit, and the other end to the middle of a strong staff, whereby on thrusting the spade forward under the turfs, and another by the staff pulling backward, they will easily be flaid and taken up, but not too many at a time for drying, but as they are layed; which must be done by a line, and a long level, placing them close together, and beating them down with a mallet; having covered the quarter, or place intended, let it be well watered, and beaten all over with a heavy broad beater: lastly, cut away by a line what is superfluous, that the sides may be straight and eaven, or in what work you shall please to fancy."

Turfs were cultured for use in lawn gardens, flower gardens, pleasure gardens, and greens during the seventeenth and eighteenth centuries. The great houses built during this period were designed with large lawns. "Grass walks" were a vital component of the elaborate gardens developed during this period (28, 29, 53, 68). In John Evelyn's instructions for the gardener, the "grass walks" and bowling greens were to be mowed and rolled every 15 days (29). The use of a "turf beater or rammer" during moist seasons was suggested for smoothing uneven turfs. The culture of turf in cemeteries was also receiving attention during the eighteenth century (37). Numerous garden books of this period contained instructions on lawn care (68). The authors emphasized the value of mowing, rolling, edging, weeding, and obtaining good seed.

Edwin Budding of Stroud, Gloucestershire, England, invented and first patented a mowing machine for turf in 1830, with manufacture of the mower beginning in 1832. Books on gardening devoted even more attention to turfgrass cultural practices during the early nineteenth century. It was suggested that the grass walks and plats be mowed as often as there was the least hold for the scythe, that the cut leaves be removed, and that the edges of the walks be cut occasionally (63). One of the most complete discussions of turfgrass establishment and cultural practices is found in McDonald's "Complete Dictionary of Practical Gardening" (66). He indicated that sodding is preferred over seeding; that September through April is the best time to sod or seed; that the seed should be free of weeds and be a deep-rooted, prostrate, permanent, heat resistant type; and that poling be done regularly to break and scatter wormcasts. Much of the early knowledge of lawn culture evolved from practices originating in England. During this period, turfgrass culture was an art attained through experience, observation, and trial-and-error methods.

Turfgrass research initiated. The first investigations of turfgrasses and their culture were initiated in the United States. Turfgrass species and mixture evaluation studies were initiated around 1880 at the Michigan Agricultural Experiment Station by the noted botanist W. J. Beal (10, 11, 12). Preliminary evaluations of grasses for lawn use were begun at New Haven, Connecticut, in 1886 (47). A more formal turfgrass experimental garden was established by J. B. Olcott in association with the Connecticut Agricultural Experiment Station at South Manchester, Connecticut, in 1890 (Fig. 1-1). The objective was to describe, evaluate, and select improved turfgrasses for use in lawns (78, 79). Turfgrass experiments were also initiated in 1890 at Kingston, Rhode Island, by the Rhode Island Agricultural Experiment Station (52). Additional studies, initiated in 1905, involved comparisons of turfgrass cultivars, grass mixtures, and the influence of different fertilizers for use on lawns, golf courses, and polo grounds (21). The Rhode Island turfgrass experiments were further expanded in 1928 (31).

Serious problems were encountered in the establishment and growing of turf on sand at the National Links near Southhampton, Long Island, in 1908. United States Department of Agriculture (USDA) scientists C. V. Piper and R. A. Oakley were called upon for assistance. Subsequently, numerous experiments were initiated on golf courses in the area and much valuable information concerning turfgrass culture for golf courses was developed (70). The Arlington Turf Garden was established by the USDA at Arlington, Virginia, in the spring of 1916 (Fig. 1-2). Initially, the experimental plots were under the direction of Piper and Oakley. The Green Section was established in 1920 under the auspices of the United States Golf Association (USGA). A cooperative agreement was drawn up between the

Figure 1-1. View of the Olcott Turf Garden at South Manchester, Connecticut, as it appeared about 1910. (Photo courtesy of USGA Green Section—79).

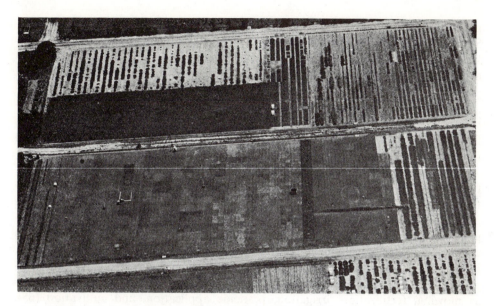

Figure 1-2. Aerial view of the Arlington Turf Garden as it existed in August, 1927 (Photo courtesy of USGA Green Section—70).

Green Section and the USDA in 1921 for jointly financing the turfgrass research at Arlington (39). The pioneering investigations at Arlington emphasized turfgrass cultivar evaluation, fertilization, and disease control (14). The turfgrass experimental plots were moved from Arlington to the USDA Plant Industry Station at Beltsville, Maryland, in 1942.

Stimulated by the interest and encouragement of Piper and Oakley, turfgrass experimental plots were initiated at the Agricultural Experiment Stations in the following states: California in 1921 (40), Florida in 1922 (27, 89), Massachusetts in 1923 (26), Kansas in 1924 (20, 98), New Jersey in 1925 (73, 89), Nebraska in 1925 (33), Ohio in 1927 (96), and Pennsylvania in 1929 (74). In 1928, the USGA Green Section established 15 demonstration turfgrass gardens on golf courses in different sections of the country (69). The USGA Green Section also established experimental turfgrass gardens north of Chicago, at Everett, Illinois, and on the grounds of Leland Stanford University at Palo Alto, California.

The successful efforts of the United States in initiating turfgrass research stimulated the Royal Canadian Golf Association to request federal funds for support of turfgrass research (67). As a result, turfgrass experimental plots were established at the Central Experimental Farm near Ottawa, Ontario in 1924. The newly formed Board of Greenkeeping Research in Great Britain established a turfgrass research station on the St. Ives estate near Bingley, Yorkshire, England in 1929 (24, 25). The original board was reorganized in 1951 and has since been known as the Sports Turf Research Institute. Studies concerning the fertility requirements of bermudagrass turf were initiated in South Africa in 1929 (38). More detailed turfgrass investigations were started in 1933 at the Frankenwald Botanical Research Station in South Africa. The New Zealand Golf Association established a Green-keeping Research Committee in 1935 (22). The committee was reorganized in 1949 as the New Zealand Institute for Turf Culture (93). During this period, similar programs of turfgrass research were being developed in Australia and India.

Most of the early turfgrass research was initiated through the efforts and mone-tary support of the golf interests in English-speaking countries. The United States fostered turfgrass research initially and has continued to provide leadership in the development of knowledge concerning the science of turfgrass culture.

Advances in turfgrass culture. Information obtained during the early history of turfgrass research contributed significantly to the understanding of turfgrass cul-tural practices. However, most of the major advances in the science of turfgrass culture have occurred since 1946. The 1950's was a decade of revolution in turf-grass science. Major achievements were (a) the release of improved turfgrass cultivars; (b) the development of many effective pesticides for turfgrass weed, insect, and disease control; (c) the formulation of fertilizers specifically for turf-grasses; and (d) major advances in the mechanization of such cultural practices as mowing, fertilization, irrigation, pesticide application, and cultivation (65). These advances were made possible through the combined efforts of the State and Federal Agricultural Experiment Station turfgrass researchers, private research conducted by firms associated with the turfgrass industry, and individual profes-sional turfmen. The 1950's was also the decade when the demand and use of turf expanded beyond the expectations of most individuals associated with the turf-grass industry. Turfgrass culture, in its many forms, has continued to grow rapidly during the 1960's and has become a major industry in the United States.

Modern day turfgrass culture is becoming more and more scientific. A certain degree of art is still retained and will remain, especially in the maintenance of greens. The increasing demand for professional turfgrass managers who have college training indicates the importance of science in modern turfgrass culture. A further evidence is the expansion of turfgrass research and educational programs (Fig. 1-3). Most states in the United States now have turfgrass research programs and similar efforts are being developed in other countries such as Germany, Japan, Netherlands, and Sweden. One needs to know when and why, as well as how, various turfgrass cultural practices are utilized. A considerable amount of scientific knowledge has accumulated concerning the culture of turfgrasses; yet many ques-tions and problems still exist.

Value and Extent of the Turfgrass Industry

The extent and value of the turfgrass industry varies greatly throughout the world. Turf is most widely used in the more highly developed industrial regions of the world such as North America, England, Europe, New Zealand, Japan, and Aus-tralia. No accurate information is available on the acreage and value of the world turfgrass industry.

Figure 1-3. General view of the bentgrass experimental plots at Michigan State University, East Lansing, Michigan.

One of the problems in determining the value of the turfgrass industry is that a turf, with the exception of sod, is not a marketable commodity. Since a large segment of the turfgrass industry involves the maintenance of established turfs that are not bought and sold through commercial channels, no concerted effort has been made to determine the total value of the industry.

The turfgrass industry can be evaluated in three basic ways: (a) the initial capital investment required to develop and establish turfs, (b) the annual cost for maintenance of turfs, and (c) the turfgrass acreage. There are also benefits that cannot be measured in terms of monetary value or acreage. It is difficult to quantitatively measure the value of beautification, contributions to human physical and mental health, dust suppression, noise abatement, injury prevention, and soil erosion control that is achieved through the use of turfgrasses.

Some attempts have been made since 1960 to determine the value of the turfgrass industry (Table 1-1). The Pennsylvania and Texas reports are based on statistically representative survey methods while the other five state values are comparatively rough estimates. The estimates appear to be rather conservative. The extent of the turfgrass acreage in the United States has been estimated at more than 20 million acres. Based on the scattered state estimates, the total annual turfgrass maintenance expenditure in the United States was estimated at $4,325,794,086 in 1965 (76). The $4.3 billion total can be broken down into 11 categories of speciality turfgrass use (Table 1-2). The distribution of expenditures among these turfgrass use specialities probably varies among countries throughout the world.

Table 1-1

ESTIMATE OF THE TOTAL ANNUAL TURFGRASS MAINTENANCE COSTS FOR SEVEN STATES

State	Total annual turfgrass maintenance cost	Year estimate was made	Reference
Florida	$120,700,000	1960	44
Michigan	260,000,000	1969	13
New Jersey	95,000,000	1960	5
New York	142,000,000	1961	6
Oklahoma	72,020,900	1963	45
Pennsylvania	565,261,896	1966	17
Texas	211,568,126	1964	43

Table 1-2

ESTIMATED PERCENTAGE DISTRIBUTION OF EXPENDITURES IN 1965
FOR TURFGRASS MAINTENANCE BY TURFGRASS SPECIALITY IN
THE UNITED STATES [AFTER NUTTER (76)]

Turfgrass use category	Percent of total
Lawns—residential	69.4
Roadsides	10.9
Cemeteries	8.4
Golf courses	5.5
Parks—municipal	1.4
Schools—public	0.9
Airfields	0.8
Lawns—commerical	0.6
Churches	0.6
Colleges and universities	0.4
Miscellaneous	1.1

Labor, equipment, and water are the major expenditures in turfgrass maintenance (43).

The available estimates indicate that the turfgrass industry is a major component of the American economy. The growth rate of the industry has been especially rapid during the 1950's and 60's. People want the benefits of lawns, parks, playgrounds, and athletic fields and are willing to support them financially. With the anticipated continued expansion in population and urban development plus increased leisure time available for outdoor recreational activities and lawn care, the value and importance of the turfgrass industry will continue to grow in countries with prosperous economies (87).

Turfgrass Terminology

Turfgrass culture is a unique phase of plant culture defined as the science and practice of establishing and maintaining turfgrasses for specialized purposes such as lawns, recreational areas, roadsides, and airfields (4, 57). **Turfgrass management**

is a broader term that encompasses labor supervision, record keeping, budgeting, and cost accounting as well as the turfgrass cultural phases.

Sod, sward, and turf are used synonymously by many individuals. Each term has a more specific usage that should be recognized. The word **turf** originated from the Sanskrit word **darbha** that denoted a tuft of grass. More recently the word has evolved from Middle English. **Turf** refers to a covering of vegetation plus the matted, upper stratum of earth filled with roots and/or rhizomes. Turfs are commonly cut close and uniform in character. The term **turfgrass** is preferred to turf since the latter term has been associated with horse racing for many years.

Sod refers to plugs, blocks, squares, or strips of turfgrass plus the adhering soil that are used for vegetative planting. The term sod has become closely associated with the sod production industry in the United States. A **sward** is the grassy surface of a turf that may be composed of one or more species. **Green** refers to a smooth, grassy area maintained for games of golf, bowling, or other sports. Ground that is covered with fine textured grass and is kept closely mowed is called a **lawn**.

Evaluating Turfgrass Quality

Turfgrass evaluation is a complex, difficult problem. Turfgrass quality is a relative term that varies with the type of turf, the time of year, the individual making the evaluation, and the purpose for which the turf is to be used. What is acceptable turfgrass quality for one purpose or individual may be inferior for another. The degree of detail involved in evaluating turfgrass quality varies with the needs and objectives of the individual making the evaluations. For example, the turfgrass quality observations of a homeowner would be rather simple in comparison to the detailed criteria used by a turfgrass researcher.

Turfgrass quality is difficult to measure quantitatively since it is a composite of many characteristics and factors. The characteristics of high quality turf have been established over the years by the personal preference and needs of the user. There are six basic **components of turfgrass quality**: (a) uniformity, (b) density, (c) texture, (d) growth habit, (e) smoothness, and (f) color. The relative importance of these components varies depending on the purpose for which the turf is used. For example, smoothness is very important on putting and bowling greens, but is much less important for a home lawn.

Uniformity. A high quality turf should be extremely uniform in appearance. The presence of bare areas, weeds, blemishes due to insect or disease injury, or an irregular growth habit lowers the level of turfgrass quality. A visual estimate is the most representative measure available for the evaluation of uniformity.

Density. Density is one of the more important components of turfgrass quality (83). Visual quality ratings are positively correlated with density. Turfgrass density can be measured quantitatively by counting the number of shoots or leaves per unit area (3, 8, 9, 18, 55, 61, 62, 83). Shoot counts are preferred because of the greater ease and accuracy with which the counts can be made. A high turfgrass shoot density or plant population is also desired because of the increased competition to invading weeds. Attempts have been made to measure density using a portable, transistorized densitometer. Measurements with the densitometer should be taken soon after mowing. As yet, the technique has not been perfected sufficiently to be utilized as a standard method for determining turfgrass shoot density.

Shoot densities vary among turfgrass species and cultivars (Table 1-3). The

Table 1-3

COMPARATIVE DENSITIES OF TEN TURFGRASSES MOWED AT 1.5 IN. COUNTS
MADE DURING OCTOBER AT EAST LANSING, MICHIGAN

Relative density	Shoots per square decimeter	Turfgrass cultivar
High	>200	Toronto creeping bentgrass
		Pennlawn red fescue
		Prato Kentucky bluegrass
		Merion Kentucky bluegrass
Medium	100 to 200	Park Kentucky bluegrass
		Kenblue Kentucky bluegrass
		Norlea perennial ryegrass
Low	<100	Meadow fescue
		Kentucky 31 tall fescue
		Timothy

shoot density of any one turfgrass cultivar also varies widely depending on the cultural regime, environment, and time of year (88). The rate of seeding and degree of initial establishment may also be important factors in determining the turfgrass shoot density of the noncreeping turfgrass species. Adequate soil moisture, close mowing, and nitrogen fertilization generally increase the shoot density (83). One of the highest turfgrass shoot densities occurs on bentgrass greens with over 1700 shoots per square decimeter being recorded. This is equivalent to 164.8 billion bentgrass shoots per acre.

Texture. Turfgrass texture is a function of the width of individual leaves. A medium-fine to medium texture, ranging from 1.5 to 3 mm in width, is generally preferred for most turfgrass uses. Comparative leaf texture measurements must be made with leaves of the same age or stage of development and should be at the same location on the individual leaves. Turfgrasses vary greatly in leaf texture (Table 1-4). Since considerable variability occurs within any one turfgrass species, exceptions can be found to the general rankings given in Table 1-4. Cultural practices such as cutting height, nutritional level, and topdressing will affect texture. For example, the leaf texture of creeping bentgrass and annual bluegrass can be reduced 50 percent by lowering the height of cut from 1.5 to 0.3 in. Leaf texture also varies with stand density and environmental stress. In the formulation of turfgrass seed blends or mixtures, it is desirable to utilize cultivars and species of similar leaf texture in order to achieve a uniform appearance.

Growth habit. An upright or vertical positioning of turfgrass leaves is preferred on greens. The trueness of ball roll can be altered by the presence of **grain**, which is the tendency of turfgrass leaves and stems to grow horizontally in one or more

Table 1-4

A GENERAL COMPARISON OF THE AVERAGE LEAF TEXTURE OF TWENTY-
FIVE TURFGRASSES WHEN CUT AT 1.5 IN.

Texture category	Leaf width (mm)	Turfgrass species or cultivar
Very fine	<1	Sheep fescue
		Pennlawn red fescue
		Chewings fescue
		Velvet bentgrass
Fine	1–2	Emerald zoysiagrass
		Rough bluegrass
		Tiffine bermudagrass
		Norlea perennial ryegrass
Medium	2–3	Astoria colonial bentgrass
		Toronto creeping bentgrass
		U-3 bermudagrass
		Merion Kentucky bluegrass
		Annual bluegrass
		Meyer zoysiagrass
		Buffalograss
Coarse	3–4	Meadow fescue
		Italian ryegrass
		Redtop
		Timothy
		Centipedegrass
Very coarse	>4	Kentucky 31 tall fescue
		Kikuyugrass
		St. Augustinegrass
		Bahiagrass
		Carpetgrass

directions rather than vertically (22). Considerable variation in graininess exists among the turfgrass cultivars. Arlington creeping bentgrass is noted for its swirling tendency.

Turfgrass species vary in stem growth habit from a very erect, vertical type to a low growing, prostrate type. Creeping bentgrass and bermudagrass are low growing, prostrate types that tolerate a short cutting height. This contrasts with the ryegrasses that have an erect, vertical type of growth. Bentgrass species have a low, prostrate type of stem growth but can still provide a dense, upright positioning of the leaves if properly maintained.

Growth habit variations also occur within species. For example, Park and Delta Kentucky bluegrasses are characterized by an erect, vertical type of leaf and stem growth, while Merion and Cougar have a prostrate orientation and a short internode length. The latter two characteristics result in a moderately low growth habit that is more tolerant of close mowing. The variations in internode length among bermudagrass cultivars are illustrated in Fig. 1-4.

Smoothness. Smoothness is a component of turfgrass quality that is particularly important on greens. Greens must be free of obstructions or depressions that might deflect a rolling ball. Leaf growth habit and smoothness can be evaluated by (a)

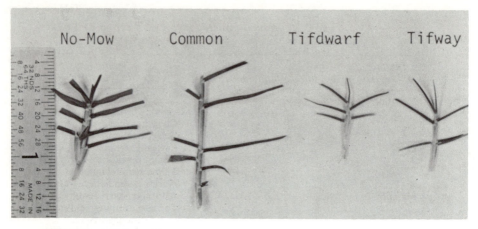

Figure 1-4. Variations in internode lengths among four bermudagrass cultivars. (Photo courtesy of G. C. McBee, Texas A & M University, College Station, Texas).

visual estimates or (b) the trueness and distance of ball roll. No quick, accurate method of quantitatively measuring this component of turfgrass quality has been perfected. Several experimental techniques for measuring the degree of friction of a turfgrass surface have been developed for use in critical investigations (34, 50, 71). One technique that uses an inclined plane principle is shown in Fig. 1-5. The Michigan Inclined Plane Technique is limited to comparisons of greens surfaces having similar, uniform slopes. It is very sensitive to the effects of wind velocity and direction.

Color. Color is an evaluation of the spectral composition of light by visual sense. As radiant energy contacts the turf, certain wavelengths are absorbed and transmitted while other wavelengths are reflected. Reflected wavelengths in the range from 380 to 760 mμ are perceived by the human eye as the color of the turf. Color is one of the best indicators available concerning the general condition of a turf. A loss of green color may indicate the development of nutritional, disease, insect, nematode, excessive water, wilt, or other environmental stress problems. Turfgrass color is basically a personal preference, with a dark green color being preferred by most individuals. Turfgrass color is extremely difficult to measure since it is a composite of green, tan, and brown leaves; stems; and clippings rather than a single, uniform green color.

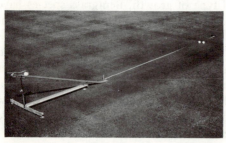

A **chlorophyll index** measured as chlorophyll content in milligrams per square decimeter of area has been used (60, 62). This index is effective in measuring certain nutritional responses but does not respond to the effects of mowing, irrigation, or shading.

Figure 1-5. An inclined plane apparatus used in the measurement of the friction on greens in terms of its effect on ball roll.

A photocell has been used to quantitatively measure the light reflection from a turf (62, 80). The light reflection, shoot density, chlorophyll content, and visual rating methods of turfgrass evaluation correlate with one another when turfgrass responses to water and nitrogen are being evaluated. Light reflection evaluates certain aspects of turfgrass quality but fails to measure texture, uniformity, and smoothness. More recently a portable single beam spectrophotometer has been used to objectively evaluate turfgrass color (16, 94). This method measures a two wavelength reflectance ratio that is correlated with visual ratings of turfgrass color. These methods of color evaluation are correlated with turfgrass color but do not represent the true color of the turf as a whole. Perhaps the accurate measurement of turfgrass color is not critical since color acceptability is based on personal preference and does not affect the major components of quality.

Color charts have been used as reference points in the determination of the actual turfgrass color. This technique is commonly used in soils work. It can also be used for turfgrass color determination but is more difficult due to the variegated nature of turfgrass color.

A wide range of colors exists among turfgrass species and cultivars as illustrated in Table 1-5. Color is a consideration in mixing turfgrass species or blending cul-

Table 1-5

A Comparison of Colors Among Eleven Colonial and Creeping Bentgrass Cultivars as Established Using the Nickerson Color Fan [after Butler (19)]

Turfgrass color	Munsell color notations	Bentgrass cultivar
Dark yellowish green	2.5G4/6	Congressional
Dark yellowish green	2.5G4/6	Toronto
Dark yellowish green	2.5G4/6	C-52
Strong yellow green	5GY6/8	Cohansey
Strong green	5G4/7	Seaside
Strong green	7.5G5/8	Penncross
Dark green	7.5G3/6	Washington
Dark green	7.5G3/6	Evansville
Moderate olive green	7.5GY4/4	Nimisila
Moderate olive green	7.5GY4/4	Arlington
Strong bluish green	5BG5/8	Pennlu

tivars. For example, the light, yellow green color of rough bluegrass is not compatible with the color of the commonly used Kentucky bluegrasses, while the color of Pennlawn red fescue blends quite well. The ability to retain a green color late into the fall and through the winter period is also an important characteristic to be considered in selecting improved cultivars.

Other measurements. The **yield** of dry matter collected from a turf during a given period does not necessarily measure turfgrass quality (95). Yield measures the growth rate of the shoots and the response to nutritional or environmental factors but gives no reliable indication of leaf texture and shoot density (85). A turf may have a high yield but very poor shoot density, leaf texture, and rooting. The growth rate does give some indication of the ability of a turfgrass to resist the invasion of weeds as well as the potential ability to recover from injury. Actually, a moderate growth rate is preferred if shoot density, color, and uniformity are not sacrificed. Turfs having a moderate shoot growth rate are in a more healthy state than a rapidly growing turf because the additional shoot growth is achieved at the expense of the plant carbohydrate reserves, root system, and recuperative potential.

Verdure has been proposed as a better measure of turfgrass quality than yield (59). It is defined as the quantity of growing turf per unit area remaining after mowing. A certain amount of verdure and mat is required on a golf green in order to have the proper resiliency for approach shots. Verdure is correlated with the shoot density component of turfgrass quality but gives no measure of the other five components.

Sod strength is one of the most important characteristics in sod production. A sod crop cannot be harvested until root, rhizome, and/or stolon development is sufficient to permit handling and transplanting. The criteria for adequate sod strength that is utilized by the layman normally involves lifting the sod piece off the ground by one end and determining whether tearing occurs. A more quantitative technique has been developed for use in sod production experiments (82). The Michigan Sod Strength Test involves the application of force at a uniform rate to a sod piece until tearing occurs (Fig. 1-6). This technique permits comparisons of sod strength measurements among species, cultivars, and individual cultural practices.

Other quantitative measurements such as rooting depth, quantity of roots, botanical composition, plant height, plant weight, erosion, and amount of bare ground provide additional information regarding the quality and vigor of a turf. ***Visual estimates.*** The complex nature of turfgrass quality has prevented the development of an adequate quantitative measure that does not involve an empirical approach to some degree (23). Each of the six components of quality varies with the turfgrass species, cultivar, and cultural system. Any one component of quality is not necessarily representative of the overall quality. For example, just because a turf has a high shoot density does not ensure that it will rank high in overall quality (32, 83). Also, the relative importance of each component of quality varies with the particular use for which the turf was established. Because of the unique nature of turfgrass quality, the visual estimate is the most practical, rapid, and representative method of evaluation now available.

Figure 1-6. The Michigan sod strength apparatus utilized for the quantitative measurement of sod strength in sod production investigations.

Individuals have attempted to measure overall quality by totaling the values obtained from measurement of each individual component of turfgrass quality (75). Weighted coefficients have also been assigned to the components based on the estimated importance of each component (41, 97). The value of the coefficient assigned to each component varies depending on the purpose for which the turf is to be utilized and the environment in which the turf is to be grown. This method also requires, for some components, an empirical or observational approach.

The visual estimate or rating should be done by several competent observers on an independent basis. The ratings are made on one of several arbitrary scales and the scores averaged. A common scoring system involves a range from 1 to 9 with 1 ranked as the best possible quality and 9 the poorest or vice versa. The subjectivity of visual estimation is recognized. The reliability of this approach is limited by the skill and, to a degree, the art of the estimator. Visual estimates are quantitative within an experiment but comparisons of estimates among locations, years, and estimators are limited to a relative basis. Even with its limitations, it appears that visual observations will continue to be used in turfgrass quality evaluations until a more objective method is devised.

Botanical composition. At times it is necessary to determine the botanical composition of turfgrass polystands. This may be done quantitatively using variations of the point-quadrat or line-transect methods as well as by actual counts of the different species made on a randomly selected area (30, 56, 61, 77, 91, 92). These methods are time consuming and costly. Studies have shown that visual estimates of botanical composition are usually sufficient when the species involved have distinctly different vegetative characteristics (46, 56, 64).

Controls of turfgrass quality. The components of turfgrass quality are (a) the result of the hereditary potentialities of the turfgrass plant, (b) controlled by the surrounding environment, and (c) expressed through internal developmental, physiological, and biochemical processes. These hereditary and environmental factors are termed the **controls of turfgrass quality** and may be grouped into two general categories.

The first group of controls encompasses the inherited characteristics of a turfgrass cultivar including (a) disease, insect, and nematode susceptibility; (b) wear tolerance; (c) heat, cold, drought, flood, shade, wilt, and salinity tolerance; and (d) the recuperative potential. The professional turfman must consider these inherited controls in order to select the cultivar that is best adapted to the environment, soil, use, and intensity of culture under which it is to be grown. These inherited characteristics vary widely in terms of the degree of tolerance or resistance within and among turfgrass species. This range of inherited characteristics provides the sources of germ plasm from which improved turfgrass cultivars are developed.

The second group of controls of turfgrass quality involves the atmospheric, soil, and biotic components of the environment which influence the growth and survival of a turfgrass cultivar. Included among these environmental factors are light, temperature, water, nutrient level, soil aeration, and traffic. Many of these can be controlled at least partially through turfgrass culture. A major portion of this book is devoted to a detailed discussion of the controls of turfgrass quality.

References

1. ALBERTUS, M. 1867. De Vegetabilibus Lilori VII. Critical edition began by Ernest Meyer and completed by Karl Jessen. G. Reimer, Berlin, Germany. pp. 1–752.
2. ALLEN, W. W. 1961. Factors limiting turf quality. USGA Journal. 14(3): 30–31.
3. ANONYMOUS. 1935. Plant count excites research interest. The Greenkeepers' Reporter. 3(1): 9–22.
4. ANONYMOUS. 1956. Turf is big business. New Jersey Agriculture. 38(1): 8–11.
5. ANONYMOUS. 1960. Grass—a $95,000,000 crop. Report from Rutgers, the State University. 12(2): 4.
6. ANONYMOUS. 1961. Turfgrass is a big crop in New York. New York Turfgrass Association Bulletin. 67(2): 260.
7. ANONYMOUS. 1962. Summary of terms. Crop Science. 2(1): 85–87.
8. BARROWS, E. 1935. Studying types of putting green turf. The Greenkeepers' Reporter. 3(3): 1–7.
9. BARROWS, E. M., and L. J. FESER. 1934. The plant count. The Greenkeepers' Reporter. 2(8–9): 4–8.
10. BEAL, W. J. 1892. Some selections of grasses promising for field and lawn. Proceedings of the Society for the Promotion of Agricultural Science. 13: 135–137.
11. ————. 1893. Mixtures of grasses for lawns. Proceedings of the Society for the Promotion of Agricultural Science. 14: 28–33.
12. ————. 1898. Lawn-grass mixtures as purchased in the markets compared with a few of the best. Proceedings of the Society for the Promotion of Agricultural Science. 19: 59–63.
13. BEARD, J. B. 1969. Sod and turf—Michigan's $350 million carpet. Michigan Science in Action. pp. 1–24.
14. BENGEYFIELD, W. H. 1966. The USGA Green Section—Past, present, and future. Proceedings of the 20th Annual Northwest Turfgrass Conference. pp. 52–57.
15. BEUTEL, J., and F. ROEWEKAMP. 1954. Turfgrass survey of Los Angeles County, California. Southern California Golf Association. pp. 1–12.
16. BIRTH, G. S., and G. R. MCVEY. 1968. Measuring the color of growing turf with a reflectance spectrophotometer. Agronomy Journal. 60: 640–643.
17. BOSTER, D. O. 1966. Pennsylvania turfgrass survey, 1966. Pennsylvania Crop Reporting Service, P.D.A. CRS-42. pp. 1–38.
18. BROWN, D. 1954. Methods of surveying and measuring vegetation. Commonwealth Bureau of Pastures and Field Crops. Bulletin 42. Commonwealth Agricultural Bureau, Farnham Royal, Bucks, England.
19. BUTLER, J. 1963. Some characteristics of the more commonly grown creeping bentgrasses. Illinois Turfgrass Conference Proceedings. pp. 23–25.
20. CALL, L. E. 1926. Turf and lawn grass experiments. Director's Report of the Kansas Agricultural Experiment Station for 1924–1926. pp. 47–48.
21. CARD, F. W., M. A. BLAKE, and H. L. BARNES. 1907. Lawn experiment. 19th Annual Report of the Rhode Island Agricultural Experiment Station for 1905–1906, Part II. pp. 162–166.
22. CORKILL, L. 1967. Turf improvement—the future. New Zealand Institute of Turf Culture Newsletter No. 52: 3–4.
23. DANIEL, W. H., and N. R. Goetze. 1957. Subjective versus objective measurements of turf quality. 1957 Agronomy Abstracts. p. 78.
24. DAWSON, R. B. 1929. St. Ives Research Station, its surroundings and historical associations. The Journal of the Board of Greenkeeping Research. 1(1): 9–11.
25. ————. 1929. Some greenkeeping problems and practices, the programme of the research

station and the factors influencing growth of turf. The Journal of the Board of Greenkeeping Research. 1(1): 12–23.

26. DICKINSON, L. S. 1928. Turf experiments at the Massachusetts Agricultural College. Bulletin of USGA Green Section. 8(12): 248–249.

27. ENLOW, C. R. 1928. Turf studies at the Florida Experiment Station. Bulletin of USGA Green Section. 8(12): 246–247.

28. EVELYN, J. 1669. Kalendarium Hortense. 3rd ed. John Martin and James Allestry. London, England. pp. 1–33.

29. ———. 1932. Directions for the Gardiner at Says-Court. Edited by Geoffrey Veynes for the Nonesuch Press, England. pp. 1–109.

30. FOOTE, L. E., J. SHEPHERD, J. A. JACKOBS, and E. GOULD. 1964. Instrumentation, methods and techniques for research in roadside development. 23rd Short Course on Roadside Development. pp. 63–78.

31. GARNER, E. S. 1928. Turf experiments at Rhode Island Experiment Station. Bulletin of USGA Green Section. 8(12): 254–255.

32. GILL, W. J., W. R. THOMPSON, and C. Y. WARD. 1967. Species and methods for overseeding bermudagrass greens. The Golf Superintendent. 35: 10–17.

33. GRAU, F. V. 1928. Turf experiments at Nebraska College of Agriculture. Bulletin of USGA Green Section. 8(12): 253–254.

34. ———. 1933. Drift and speed of putted ball on bents as determined by mechanical putter. Bulletin of USGA Green Section. 13(3): 74–81.

35. ———. 1954. Turf quality. Proceedings of the Arizona Turfgrass Conference. pp. 20–24.

36. ———. 1954. Techniques and turf quality. Proceedings of the Midwest Regional Turf Conference. pp. 14–19.

37. HALDANE, E. S. 1934. Scots Gardens in Old Times (1200–1800). Alexander Maclehose & Co., London, England. pp. 1–244.

38. HALL, T. D. 1948. Introduction. Experiments with *Cynodon Dactylon* and Other Species at the South African Turf Research Station. pp. V–VII.

39 HARDT, F. M. 1939. Introducing "Turf Culture." Turf Culture. 1(1): 1–5.

40. HARING, C. M. 1922. Grass and forage plant investigations. Report of the College of Agriculture and the Agricultural Experiment Station of the University of California for 1921. p. 70.

41. HARPER, J. C. 1951. Tests of fairway grasses. 20th Annual Pennsylvania State College Turf Conference. pp. 71–75.

42. HILL, T. 1568. The Profitable Art of Gardening. Thomas Marshe, London, England. pp. 1–192.

43. HOLT, E. C., W. W. Allen, and M. H. Ferguson. 1964. Turfgrass maintenance costs in Texas. Texas Agricultural Experiment Station Bulletin B-1027. pp. 1–19.

44. HORN, G. C. 1962. Establishment and maintenance of large turf areas. Soil and Crops Science Society of Florida Proceedings. 21: 114–119.

45. HUFFINE, W. W. 1963. The value of turf in Oklahoma. Turfgrass Research, Oklahoma State University. p. 1.

46. HUNT, O. J. 1964. An evaluation of the visual weight estimation method for determining botanical composition of forage plots. Agronomy Journal. 56: 73–76.

47. JOHNSON, S. W. 1888. The station forage garden. 11th Annual Report of the Connecticut Agricultural Experiment Station for 1887. pp. 163–176.

48. JUSKA, F. V. 1961. Estimated acreage in established turf in the United States. Proceedings of the 2nd Annual Lawn and Turf Conference—University of Missouri. pp. 39–40.

49. ———. 1961. The nitrogen variable in testing Kentucky bluegrass varieties for turf. Agronomy Journal. 53(6): 409–410.

50. KAWAMURA, N. 1965. Machines for managing golf courses. Turf Research Bulletin, Kansai Golf Union Green Section Research Center. 9(5): 1–21.

51. KIMBALL, M. H. 1954. Turf is big business. Proceedings of the Northern California Turfgrass Conference of 1954. pp. 37–39.

52. KINNEY, L. F. 1891. Trial of lawn grasses. 3rd Annual Report of the Rhode Island State Agricultural School and Experiment Station for 1890. Part II. p. 156.

53. LANGLEY, B. 1728. New Principles of Gardening. A. Bettesworth, et al., London, England. pp. 1–207.

54. LANGVAD, B. 1968. Ball bouncing and ball-rolling as a function of mowing height and kind of soil have been studied at Weibullsholm. Weibulls Grastips. 10: 355–357.

55. LARSON, A. H. 1934. A study of some turf grasses. The Greenkeepers' Reporter. 2(8–9): 1–3.

56. LEASURE, J. K. 1949. Determining the species composition of swards. Agronomy Journal. 41: 204–206.

57. LEONARD, W. H., R. M. LOVE, and M. E. HEATH. 1968. Crop terminology today. Crop Science. 8: 257–261.

58. LOUDON, J. C. 1878. An Encyclopedia of Gardening. Longmans, Green, & Co., London, England. pp. 1–1278.

59. MADISON, J. H. 1962. Turfgrass ecology. Effects of mowing, irrigation, and nitrogen treatments of Agrostis palustris Huds., 'Seaside' and Agrostis tenuis Sibth. 'Highland' on population, yield, rooting and cover. Agronomy Journal. 54: 407–412.

60. MADISON, J. H., and A. H. ANDERSEN. 1963. A chlorophyll index to measure turfgrass response. Agronomy Journal. 55(5): 461–464.

61. MAHDI, Z., and V. T. STOUTEMYER. 1953. A method of measurement of populations in dense turf. Agronomy Journal. 45(10): 514–515.

62. MANTELL, A., and G. STANHILL. 1966. Comparison of methods for evaluating the response of lawngrass to irrigation and nitrogen treatments. Agronomy Journal. 58: 465–468.

63. MARSHALL, C. 1805. Introduction to the Knowledge and Practice of Gardening. Bye and Law, Clerkenwell, England. pp. 1–420.

64. MARTEN, G. C. 1964. Visual estimation of botanical composition in simple legume-grass mixtures. Agronomy Journal. 56: 549–552.

65. MASCARO, T. 1965. 20 years of progress in turfgrass maintenance. Proceedings of the 20th Annual Texas Turfgrass Conference. pp. 11–12.

66. McDONALD, A. 1807. A Complete Dictionary of Practical Gardening. R. TAYLOR and Co., London, England. Vols. I and II.

67. McROSTIE, G. P. 1928. Turf studies at the Central Experimental Farm, Ottawa. Bulletin of USGA Green Section. 8(12): 250–251.

68. MILLER, A. 1745. The Gardeners Kalendar. J. Rivington, London, England. pp. 1–341.

69. MONTEITH, J. 1928. Demonstration turf gardens on golf courses. Bulletin of USGA Green Section. 8(12): 239–243.

70. ———. 1928. The Arlington turf garden. Bulletin of USGA Green Section. 8(12): 244–245.

71. ———. 1929. Testing turf with a mechanical putter. Bulletin of USGA Green Section. 9(1): 3–6.

72. MORROW, R. 1967. Grass for player safety. Turf-Grass Times. 3(1): 2–16.

73. MUSGRAVE, G. W. 1926. Report of the Department of Agronomy, Miscellaneous. 46th Annual Report of the New Jersey State Agricultural Experiment Station for 1925. p. 275.

74. MUSSER, H. B. 1929. Unprojected Research. 42nd Annual Report of the Pennsylvania Agricultural Experiment Station. Bulletin 243. pp. 33–34.

75. NORTH, H. F. A., and T. E. ODLAND. 1934. Putting green grasses and their management. Rhode Island Agricultural Experiment Station Bulletin 245. pp. 1–44.

76. NUTTER, G. C. 1965. Turf-grass is a $4 billion dollar industry. Turf-Grass Times. 1(1): 1–22.

77. NUTTER, G. C., B. B. SUMRELL, and R. W. WHITE. 1957. The double quadrat method of measuring lateral growth of turf grasses. 1957 Agronomy Abstracts. p. 78.

78. OLCOTT, J. B. 1891. Grass-gardening. 14th Annual Report of the Connecticut Agricultural Experiment Station for 1890. pp. 162–174.

79. PIPER, C. V. 1921. The first turf garden in America. USGA Green Section. 1: 22–23.

80. POWELL, A. J., R. E. BLASER, and R. E. SCHMIDT. 1967. Physiological and color aspects of turfgrasses with fall and winter nitrogen. Agronomy Journal. 59: 303–307.

81. REA, J. 1665. Flora, Ceres and Pomona. Richard Marriot, London, England. pp. 1–239.

82. RIEKE, P. E., J. B. BEARD, and C. M. HANSEN. 1968. A technique to measure sod strength for use in sod production studies. 1968 Agronomy Abstracts. p. 60.

83. ROBERTS, E. C. 1965. A new measurement of turfgrass response and vigor. The Golf Course Reporter. 33(8): 10–20.

84. ROHDE, E. S. 1928. The garden. II. Lawns. Nineteenth Century and After. 104: 200–209.

85. ROTHWELL, V. T., and J. G. KEMP. 1966. A grass clipper for laboratory use. Canadian Journal of Plant Science. 46: 97–98.

86. RYDSTROM, A. G. 1964. The economic value of turfgrass in Colorado. Proceedings of the 11th Annual Rocky Mountain Regional Turfgrass Conference. pp. 18–21.

87. SCHERY, R. W. 1962. Turf is big business. Proceedings of the 1962 Wisconsin Turfgrass Conference. pp. 1–7.

88. SPAULDING, S., and V. B. YOUNGNER. 1968. Evaluation of bluegrass cultivars for turf use by various methods. 1968 Agronomy Abstracts. p. 64.

89. SPRAGUE, H. B. 1928. Turf experiments at the New Jersey state station. Bulletin of USGA Green Section. 8(12): 251–252.

90. STOKES, W. E. 1922. Lawn grass studies. Annual Report of the University of Florida Agricultural Experiment Station for 1922. p. 38R.

91. TINNEY, F. W., O. S. AAMODT, and H. L. AHLGREN. 1937. Preliminary reports of a study on methods used in botanical analysis of pasture swards. Journal of the American Society of Agronomy. 29: 835–840.

92. VAN KEURAN, R. W., and H. L. AHLGREN. 1957. A statistical study of several methods used in determining the botanical composition of a sward. I. A study of established pastures. Agronomy Journal. 49: 532–536.

93. WALKER, C. 1957. 7th report on greenkeeping research. The New Zealand Institute for Turf Culture. p. 1.

94. WESTFALL, R. T., and G. R. McVEY. 1967. Spectrophotometric determination of turf color under field conditions. 1967 Agronomy Abstracts. p. 55.

95. WHITE, R. W., and G. C. NUTTER. 1957. Yield versus qualitative evaluations for inter and intra variety comparisons in turf research. 1957 Agronomy Abstracts. p. 78.

96. WILLIAMS, C. G. 1928. Lawn and golf course problems being studied. 46th Annual Report of the Ohio Agricultural Experiment Station for 1926–1927. Bulletin 417. pp. 18–20.

97. WRIGHT, L. N., and H. B. MUSSER. 1950. Tests of new strains of grasses for fairways and greens. 19th Annual Pennsylvania State College Turf Conference. pp. 15–25.

98. ZAHNLEY, J. W. 1928. Experiments with turfgrasses in Kansas. Bulletin of USGA Green Section. 8(12): 249–250.

part I

THE TURFGRASSES

A basic pool of knowledge concerning the turfgrass plant is required in order to make sound decisions regarding the proper utilization and culture of turfgrasses. Of primary concern is an understanding of the (a) growth and developmental processes, (b) environmental and soil adaptation, (c) specific cultural requirements, and (d) proper use. Only through a fundamental understanding of these characteristics can the appropriate turfgrass cultivar or species be selected for a particular environmental situation and use.

The gross appearance of a turf is a green mass of vegetation. A more detailed investigation reveals that the turf is composed of individual plants that possess certain distinct characteristics. Actually, the basic unit involved in the culture and utilization of turfs is the turfgrass plant. Most grass species utilized for turfs are characterized by a low growth habit and frequently have a prostrate, creeping tendency. This characteristic is very important since it enables the species to tolerate the close, frequent defoliation commonly practiced on turfs. Other important characteristics include a high shoot density and a medium fine leaf texture.

Fossil records indicate that the grasses became a dominant life form on earth considerably before the advent of man. Based on the limited fossil records available, it has been concluded that the *Gramineae* emerged as a distinct class of angiosperms some 70 million years ago during the late Cretaceous period of the Mesozoic era. The evolution of grass species and the development of grasslands assumed an important role in the earth's vegetation during the Miocene epoch. Evolution of the herbivorous grazing mammals also occurred during this period. As a result, the herbivorous mammals exploited the grasslands by adapting to a different type of feeding. As the herbivorous mammals continued to evolve toward close grazing, natural selection resulted in grasses that were structurally adapted to survive under close grazing. This relationship between the evolution of grasses and grazing mammals developed over a period of 40 million years. The result was grasses having short basal internodes and a prostrate, creeping growth habit that is well adapted for turfgrass use.

Grass species that originated and persist in a specific region are referred to as **native grasses**. Through the activities of man these native grasses have been distributed to areas distant from their region of origination. Grasses from other regions that become established, adapt to a given region, and persist for a long time are termed **naturalized grasses**. The turfgrass species now in common use originated from a relatively few locations on the earth. They are now widely distributed throughout the world, however, and have become naturalized grasses. For example, the majority of the turfgrass species utilized in North America are not native grasses but have become naturalized through continued use over a long period of time.

The geographical distribution and utilization of turfgrasses are influenced primarily by the temperature and precipitation patterns of a given region. An idealized diagram of the major climatic types affecting turfgrass adaptation is presented in Fig. I-1. The bluegrasses, fescues, ryegrasses, and bentgrasses are the most commonly used turfgrasses within the cool humid and cool subhumid climatic regions and are also used in the cool semiarid regions when irrigated. The most commonly used turfgrasses in the warm humid and warm subhumid regions are bermudagrass, zoysiagrass, St. Augustinegrass, bahiagrass, and centipedegrass. These species are also adapted to warm semiarid conditions but usually require irrigation. Kentucky bluegrass, ryegrass, tall fescue, bentgrass, zoysiagrass, and bermudagrass are the most frequently utilized turfgrasses in the transitional zone between the cool and warm regions. In the semiarid and subhumid regions where irrigation is

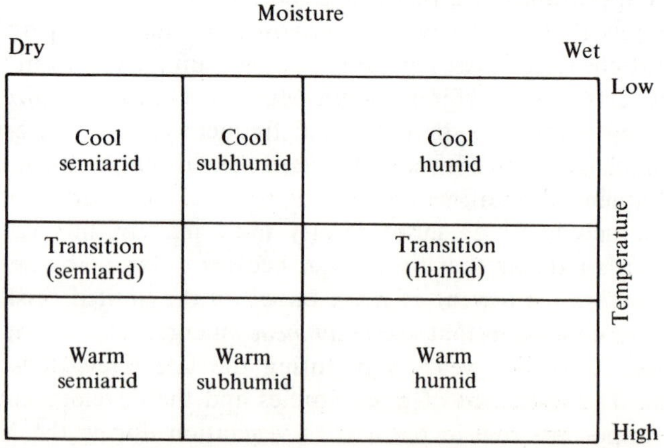

Moisture

Dry			Wet
			Low
Cool semiarid	Cool subhumid	Cool humid	
Transition (semiarid)		Transition (humid)	Temperature
Warm semiarid	Warm subhumid	Warm humid	
			High

Figure I-1. An idealized diagram of the major climatic zones of turfgrass adaptation.

not practiced, the wheatgrasses are commonly used in the cool areas while buffalo-grass is used in the warm regions.

Turfgrasses are basically **herbaceous**; that is, they are flowering plants whose aboveground stems do not contain woody tissue that persists. Herbaceous grasses that normally live for only one growing season or year are called **annuals.** Crab-grass, barnyardgrass, goosegrass and Italian ryegrass are typical annuals. **Biennials** are plants that complete their normal life term in 2 years. Herbaceous grasses that continue to live for more than 2 years are called **perennials**. A true perennial turf-grass will live indefinitely if not subjected to abnormal stresses. Bermudagrass and Kentucky bluegrass behave as perennials when grown in their normal regions of adaptation. Turfgrasses that die within 2 to 4 years are termed **short-lived perennials**. Certain perennial ryegrass and tall fescue cultivars are short-lived perennials when grown in climates characterized by low soil temperatures during the winter. These cultivars would behave as true, long-lived perennials in more moderate climates.

Taxonomy. The grass family is one of the most important in the entire plant king-dom. The *Poaceae* encompasses approximately 600 genera and 5000 species that have worldwide distribution. Of these 5000 species, less than 30 are of major impor-tance in turfgrass culture.

Specific units of classification or **taxa** have been established. This is best illus-trated by summarizing the taxa used in classifying Merion Kentucky bluegrass:
KINGDOM-*Plantae*, plant kingdom.
  DIVISION-*Embryophyta*, embryo plants.
    SUBDIVISION-*Phanaerogama*, seed plants.
      BRANCH-*Angiospermae*, seeds enclosed in an ovary.
        CLASS-*Monocotyledoneae*, monocotyledons.
          SUBCLASS-*Glumiflorae*, having chaffy flowers.
            ORDER-*Poales*, grasses and sedges.
              FAMILY-*Poaceae*, grass family.
                TRIBE-*Festuceae*, fescue tribe.
                  GENUS-*Poa*, bluegrasses.
                    SPECIES-*Pratensis*, having rhizomes.
                      CULTIVAR-*Merion.*
Trinomial-*Poa pratensis* L. Merion.

The *Glumiflorae* are subdivided into three families. Included are (a) the *Juncaceae* or rush family, (b) the *Cyperaceae* or sedge family, and (c) the *Poaceae*, (*Gramineae*) or grass family. There are two subfamilies of the *Poaceae*, the *Festucoideae* and *Panicoideae*. The two subfamilies are grouped into as many as twenty-five tribes. The tribal classification is based on the structure of the spikelets. Most of the major turfgrasses are classified in one of six tribes as follows:

Poaceae: Stems jointed; leaves always present and two ranked; fruit a grain or caryopsis, seldom angled.
    I. *Festucoideae*. Spikelets one to several flowered, sterile florets (if any) above the fertile florets in the spikelet; articulation usually above the persistent glumes and between the florets, the glumes remaining

on the plant when the seed is shed; spikelets laterally compressed, the glumes and leaves usually folded lengthwise.

A. *Chlorideae.* Inflorescence consists of one or more spikes, spikes one-sided.
   1. (*Buchloe*) buffalograsses.
   2. (*Cynodon*) bermudagrasses.
   3. (*Bouteloua*) gramas.

B. *Hordeae.* Inflorescence consists of one or more spikes that are symmetrical; articulation above the glumes or the rachis sometimes disarticulating.
   1. (*Agropyron*) wheatgrasses.
   2. (*Lolium*) ryegrasses.

C. *Zoysieae.* Inflorescence consists of one or more spikes that are symmetrical; articulation below the glumes, the rachis not disarticulating.
   1. (*Zoysia*) zoysiagrasses.

D. *Agrostideae.* Inflorescence a panicle or occasionally a raceme, sometimes narrow and spike-like; spikelets usually perfect and one-flowered; articulation usually above the glumes, the glumes not reduced.
   1. (*Agrostis*) bentgrasses and redtop.
   2. (*Ammophila*) beachgrasses.
   3. (*Phleum*) timothys.

E. *Festuceae.* Inflorescence a raceme or panicle, sometimes narrow and spike-like; spikelets with two or more florets; glumes shorter than the lower most lemma.
   1. (*Bromus*) bromegrasses.
   2. (*Festuca*) fescues.
   3. (*Poa*) bluegrasses.

II. *Panicoideae.* Spikelets with one perfect terminal floret and one sterile floret below it, the sterile floret usually represented by an empty lemma; articulation below the glumes, the whole spikelet falling when the seed is shed; spikelets dorsally compressed, the glumes and lemma flat.

F. *Paniceae.* Spikelets essentially uniform, but not paired; glumes herbaceous or membranous, not indurate.
   1. (*Axonopus*) carpetgrasses.
   2. (*Paspalum*) bahiagrasses.
   3. (*Pennisetum*) kikuyugrasses.
   4. (*Stenotaphrum*) St. Augustinegrasses.

chapter 2

Growth and Development

Introduction

A turfgrass plant passes through various growth and developmental phases during its life history. **Growth** involves an irreversible increase in size that is quantitative in nature. There are a number of growth indices including (a) length, (b) area, (c) volume, and (d) weight. In contrast, **development** involves changes in the form, structure, and general state of complexity of the plant that are primarily qualitative. The processes involved in development are cell division, differentiation, and morphogenesis. Both growth and development involve complex, interrelated physiological processes. From the standpoint of tissue and organ differentiation these processes may also be subdivided into **vegetative growth and development** involving the growth of plants from a seedling to the mature plant stage. This is contrasted to **reproductive growth and development** that involves initiation of reproductive processes in mature plants and includes the formation of flowers and seeds. The former is of primary concern in turfgrass culture. Knowledge of the basic patterns of turfgrass growth and development is essential in order to understand the responses of turfgrasses to various cultural practices and environmental influences.

Life Cycle

Typically a turfgrass plant increases in size through growth processes and develops new organs in an orderly manner throughout its life cycle (38). All the potential for growth and development of the entire plant is contained within the embryo.

When a mature turfgrass seed is placed under suitable environmental conditions, germination occurs and a seedling is produced from the embryo. Initially these processes involve the elongation of existing embryonic cells. This is followed by cell division, cell enlargement, and, at appropriate intervals, differentiation into specific structural organs and forms. The principal regions of growth are in the meristematic tissues of the shoot and root apices, as well as in the nodes of rhizomes and stolons. During the vegetative growth and development there is no basic alteration in the structural design and development of the turfgrass plant. There is an increase in plant size and in the number of specific plant organs. These organs originate in a repetitious manner that is well defined.

The developmental pattern can be changed to the reproductive phase at a certain stage of maturity or can be initiated by certain environmental conditions. The induction, initiation, development, and growth of flowers and seeds occur during the reproductive phase. Thus, with the formation of a mature seed and a viable embryo the life history is completed. Grasses maintained under turfgrass conditions may not go through the reproductive phase.

Growth Pattern

The growth rate of a root, stem, leaf, or individual plant follows a distinct pattern from the time of growth initiation to maturity. The typical sigmoid growth curve involves relatively slow initial and terminal growth rates and an extended intermediate period of rapid growth. After maturity, a senescing leaf, root, or shoot may actually decrease in weight. The rate and order of growth of individual grass tissues are partially regulated by plant growth hormones.

Differentiation

Differentiation occurs as the cells enlarge to full size. New external organs and tissues are developed as the result of cell differentiation. Growth and differentiation are regulated through morphogenesis in such a way that the resultant mature plant assumes a form distinctive for each turfgrass species. The type of tissue developing from any one cell or small group of cells in the apical meristem is influenced by its location relative to the other cells and by chemical stimuli received from associated mature cells. Differentiation into specific forms, structures, and tissues can be altered by exterior environmental influences as well as by interior stimuli.

The Embryo

The embryo is a relatively simple structure capable of forming a very complex turfgrass plant. It usually is formed from the fertilized egg but may develop by apomixis in certain species such as Kentucky bluegrass. The embryo is adjacent to one face of the endosperm and contains the primordia from which the first leaves, the primary root, and several adventitious roots are formed (6, 7, 63).

The general structure of a typical grass embryo is shown in Fig. 2-1. At one end of the embryonic axis is the **radicle** or primary root that terminates with a root cap. The nonvascular sheath of parenchymatous tissues surrounding the radicle and root cap is called the **coleorhiza**. At the opposite end of the embryonic axis is the **plumule** or embryonic shoot that has two to three leaf primordia. Surrounding the plumule is a sheath of tissue called the **coleoptile**. The coleoptile is attached to the scutellar node in most *Festucoid* grasses, while in the *Panicoid, Chloridoid,* and *Eragrostoid* grasses the coleoptile is attached above the scutellar node. The latter grasses have a distinct internode present between the scutellar node and plumule that is called the **mesocotyl**. Attached to one side of the scutellar node is a shield-like structure called the **scutellum** that

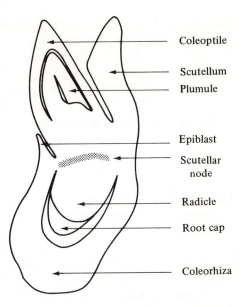

Coleoptile

Scutellum
Plumule

Epiblast

Scutellar node

Radicle

Root cap

Coleorhiza

Figure 2-1. Diagram of the general structure of a Festucoid type grass embryo in median longitudinal section.

is connected to the endosperm. It functions in enzyme secretion and absorption of carbohydrates from the endosperm. The **endosperm** is formed by the fusion of two polar nuclei of the embryo sack and a sperm nucleus (6). It provides the carbohydrate supply for the embryo and young seedling. A single endosperm is sometimes associated with more than one embryo (6). The flange of nonvascular tissue located just above the coleorhiza and opposite the scutellar node is called the **epiblast**. It arises late in embryonic development as an outgrowth of the coleorhiza (22). A vascular strand interconnects the plumule and the radicle with the upper part of the scutellum.

Grass embryos are of two basic types (100). The true *Festucoid* type is characterized by a small embryo, an epiblast, no cleft between the scutellum and coleorhiza, *Festucoid* vascularization, and, when viewed in transverse section, a primary leaf with relatively few vascular bundles and margins that do not overlap. The true *Panicoid* type has a rather large embryo, no epiblast, a distinct cleft between the scutellum and coleorhiza, *Panicoid* vascularization, and, when viewed in transverse section, a primary leaf with numerous vascular bundles and overlapping margins.

Germination. When a mature turfgrass seed is placed under suitable conditions, germination occurs resulting in the formation of a seedling containing all the essential structures for growth. The initial processes observed are water imbibition followed by enzyme activation, mitosis, and cell elongation. This results in the enlargement of the coleorhiza and coleoptile. The coleorhiza pushes through the wall of the caryopsis and forms a tuft of anchoring hairs that are quite similar to root hairs.

The primary root then develops endogenously through the side of the coleorhiza at right angles to the longitudinal axis of the embryo and penetrates downward through the soil. Shortly after this the coleoptile and mesocotyl, if present, elongate and emerge above the soil (111). The first leaf then elongates through a pore at the apex of the coleoptile and emerges into the light where new leaf initials develop from the apical meristem. Chlorophyll is synthesized and photosynthesis is initiated to support continued growth of the turfgrass seedling.

The Root

The roots of a turfgrass plant function in absorbing water and nutrients as well as anchorage. The primary or seminal roots developed from the embryo serve as the complete root system of a seedling (139). Adventitious roots are produced from the lower nodes of the stem within 2 to 3 weeks after germination (89, 111). The adventitious root system eventually replaces the seminal root system. The seminal root system of turfgrasses is generally active for 6 to 8 weeks (121).

Adventitious roots are those not originating from the root pole of the embryo. They may arise endogenously in connection with lateral shoot development or independently. The primordia of adventitious roots are initiated by divisions of parenchymatous cells. Adventitious roots may form from aerial plant parts, underground stems, or older roots (38). In turfgrass species having a tufted, noncreeping growth habit, the adventitious roots usually originate from the basal nodes of the main axis and from tillers near ground level. Adventitious roots may also develop from the nodes of stolons and rhizomes of perennial turfgrass species. In mature turfs, most of the new roots develop from the nodes of the youngest tillers on the outer edge of the original shoot or from the nodes of creeping stems (58, 121).

Root anatomy. A longitudinal section through a typical grass root reveals distinct zones of cellular activity (Fig. 2-2). At the distal end of the root is an assemblage of living parenchyma cells termed the **root cap**. The root cap functions in protecting the root meristem from the abrasive action of soil. As the outer cells of the root cap are sloughed off by abrasive action, they are replaced by cell division from the adjacent meristematic tissues. Behind the root cap is a zone of active cell division called the **meristematic region**. It is frequently no more than 1 mm in length. Immediately above the meristematic region is a **region of cell elongation**. Cell enlargement is primarily responsible for the extension of roots through the soil. Root hair and vascular tissue formation occurs primarily in the **region of differentiation** located just behind the region of cell elongation. The root hairs are particularly important in the absorption of water and nutrients.

Cell division of the root meristem is dependent on a continuing supply of metabolites from mature basal tissues. During the process of cell division and enlargement, various cells within the root differentiate into distinct tissues (111). Variability in root anatomy is evident among the grass species. The cross-sectional anatomy of a typical turfgrass root through the root hair zone is illustrated in

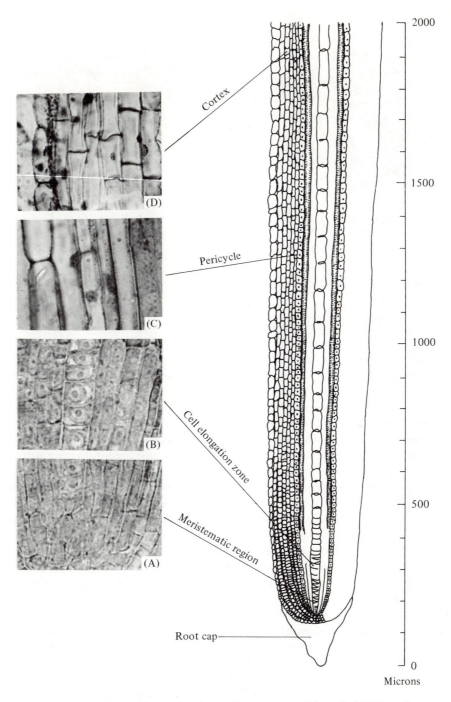

Figure 2-2. Diagram of a median longitudinal section of the apical 2000 μ of a normal colonial bentgrass root. Camera lucida drawing at 100× and photomicrographs at 1290×. (After Callahan and Engel—27).

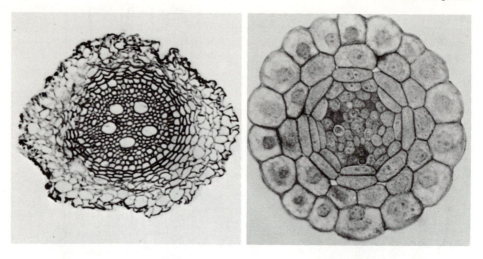

Figure 2-3. Cross section of a mature Kentucky bluegrass root (left) and a young creeping bentgrass root (right). (Photos courtesy of C. M. Harrison, Michigan State University, East Lansing, Michigan and R. M. Endo, University of California at Riverside, Riverside, California).

Fig. 2-3. Surrounding the root is a closely packed, thin-walled layer of cells called the **epidermis**. In many grasses the epidermis persists and becomes suberized. The **root hair** is a slender, elongated extension of an epidermal cell that lives for a relatively limited time. Root hair development in cool season turfgrasses occurs only from specialized epidermal cells called tricoblasts while root hairs can originate from any epidermal cell of a warm season species.

Just beneath the epidermis is a subepidermal cortical layer of differentiated cells termed the **exodermis**. The exodermis varies in thickness from one to three cell layers and is generally suberized. The **cortex** is a thick layer of homogeneous, relatively simple cells located beneath the exodermis. The root cortex of grasses is composed primarily of parenchyma cells with intercellular spaces frequently present. The cortex often breaks down in older roots forming cavities. Within and adjacent to the cortex is a specialized layer of cells called the **endodermis**. Casparian strips occur on the inner surface of the radial walls of endodermal cells. The **pericycle** is a layer of thin-walled parenchyma cells located beneath the endodermis. It may be one or two cell layers thick but is not necessarily continuous. The pericycle is disrupted by the protoxylem elements in certain grass species. Lateral roots usually originate in the pericycle. The vascular system of the grass root is contained within the pericycle. The number of xylem vessels in grass roots varies with the species. For example, perennial ryegrass normally has one vessel; timothy, five to ten; and orchardgrass, ten or more (130). In larger, well-developed roots, exarch xylem groups alternate with groups of phloem in a circular arrangement around a central pith. Fewer xylem and phloem groups and less pith are observed in smaller roots. A protoxylem element is present in very small roots but no pith. *Root system.* The root system of turfgrasses is typically multibranching and fibrous in nature. Root branching originates endogenously from fully developed root tissue. The roots of most turfgrasses are not major organs for the storage of carbohydrates (1, 9, 124).

A major portion of the root system of turfs mowed regularly at less than 2 in. is located in the upper 18 in. of the soil (34, 49, 64, 75, 117, 118, 120). The size and extent of the root system vary considerably among turfgrass species and cultivars (25, 26, 34, 64). Most warm season turfgrass species tend to have a larger root diameter and greater rooting depth than most cool season species that tend to be relatively fine rooted and more limited in potential depth. Bermudagrass, St. Augustinegrass, centipedegrass, and bahiagrass roots have been observed at soil depths of 5 to 7 ft (34, 64). In contrast, roots of Kentucky bluegrass and bentgrass maintained at a 2-in. cutting height are seldom observed at depths below 1.5 to 2 ft (55, 75).

Environmental and biotic influences. The first prerequisite for a deep, extensive root system is a healthy, actively growing turfgrass plant that is growing under conditions of adequate soil water availability and nutrition. The degree of rooting can vary depending on certain specific environmental and cultural influences. From the environmental standpoint, factors that can restrict rooting include (a) excessively high soil temperatures, (b) acidic soils with a pH below 5.0, (c) a lack of soil oxygen caused by compaction or waterlogged conditions, or (d) the presence of toxic pesticides or salts. A waterlogged or compacted soil condition is probably one of the major problems limiting the rooting of turfs especially when utilized for recreational purposes. Turfgrass roots are quite sensitive to herbicide injury (14, 27, 36, 101). This injury is not readily observed because tillering and shoot growth are not as easily affected. Cultural factors that can cause a reduction in the root system include (a) cutting at a low height or excessive frequency, (b) an excessive nitrogen level, (c) a potassium deficiency, and/or (d) an excessive thatch accumulation (55, 95).

Root longevity. The longevity of root life varies with the turfgrass species from less than 6 months to almost 2 years. Longevity also depends on the season of the year when the roots were initiated. Generally, roots produced in the fall or winter live longer than roots produced in the spring or summer (47). Root death and replacement is a continuing occurrence in certain turfgrass species, while in other species this occurs at a specific time of the year (121). Kentucky bluegrass, Canada bluegrass, crested wheatgrass, and orchardgrass retain a major portion of the functioning root system for more than 1 year and are perennial rooting types (121). In contrast, perennial ryegrass (58), colonial bentgrass, redtop, rough bluegrass, meadow fescue, and timothy replace most of the root system each year. Thus, the root system of these perennial species is termed annual in nature (121).

The root systems of many turfgrass species exhibit a seasonal periodicity of growth and death (43, 47). A major portion of the root initiation and growth occurs during the spring and early summer (9, 21, 46, 47, 59, 117, 121, 131). Root death and disintegration of cool season turfgrasses are most extensive during mid-summer heat stress periods (58, 117, 131). Roots associated with flowering tillers usually die during or just after floral development (110). Evidence of root deterio-ration includes an initial softening and finally sloughing of the cortex (58). Slough-ing occurs initially at the proximal region of the root below the crown. New root initiation of cool season turfgrasses is minimal when heat stress occurs (11, 121). For example, creeping bentgrass root initiation ceases at soil temperatures above 75°F, while the maturation rate of existing roots is increased (123).

Some root growth is initiated from certain species during the cooler fall and winter period (46, 98). Root growth of such cool season species as Kentucky bluegrass, timothy, and redtop may occur during the winter when soil temperatures are above freezing (52, 121). Root growth of warm season turfgrass species, such as bermudagrass and zoysiagrass, ceases when winter dormancy occurs and is most vigorous during the summer (131).

The Shoot

The development of a shoot is considerably more complex than that of a root. The **shoot** is a relatively broad concept that encompasses both the stem and leaf. A vegetative shoot consists of a short central stem with leaves or branches borne alternately at successive nodes. The stem internodes of the primary shoot of most turfgrasses do not elongate during vegetative growth (41). Thus, the stem is com-posed of a series of nodes separated by unelongated internodes. In certain turf-grasses, such as bermudagrass, zoysiagrass, and St. Augustinegrass, two to three leaves arise from each node on the lateral shoots (35).

The turfgrass shoot may originate directly from the embryo of a seed or from a vegetative bud in the leaf axil of an older shoot, or it may terminate a rhizome or stolon (38). The first embryonic shoot or plumule develops on the shoot meristem of the embryo on the side distant from the scutellum. The simple shoot developed from the embryonic meristem is termed the **primary shoot**. Shoots originating from vegetative buds in the axils of the leaves or from the nodes of older stems, rhizomes, or stolons are called **lateral shoots**. The **compound shoot system** is composed of the primary shoot plus one or more lateral shoots. The basic unit structure of a shoot is the **phytomer**. It is composed of an internode with a leaf plus a portion of the node at the upper end and a bud plus a portion of the node at the lower end.

Young stems develop from the basal nodes in two primary ways. One involves the erect growth of young vegetative stems upward within the enveloping basal leaf sheath that is termed **intravaginal growth**. This results in a tufted or bunch-type habit of growth. The persistent bases of the stem form a crown that initiates new shoots each year from within the old sheaths and beside the old stems. The second, **extravaginal growth,** is applied to young vegetative stems that grow outside the basal leaf sheath by penetrating through the sheath. Shoots having this type of growth habit originate independently rather than in clusters and result in a spread-ing or creeping habit of growth. Red fescue has an extravaginal type of growth, while intravaginal growth is exemplified by sheep fescue. Branches arising from the middle and upper nodes are commonly intravaginal, while those borne at the lower nodes are often extravaginal.

Tillers. Primary lateral shoots that arise intravaginally from the stem with limited elongation are termed **tillers** (Fig. 2-4). The tillers arise from vegetative buds in the axils of the leaf sheaths. Tiller primordia arise from subhypodermal cell layers of the meristem (107). The physiological control of tillering involves the concentration of auxin produced by the apical meristem and the carbohydrate level. The apical meristem and expanding leaf

Figure 2-4. A Kentucky bluegrass plant showing a primary shoot in the center and several tillers. (Photo courtesy of L. E. Moser, Ohio State University, Columbus, Ohio).

synthesize auxins in sufficiently high concentrations that tiller development is inhibited. Tillering can be stimulated by mechanical removal of the auxin-synthesizing region of the apex or the use of chemical agents to counteract the auxin (71).

The node at which tillering begins depends on the species and environment (92). Tillering of Kentucky bluegrass is observed primarily in the axils of the sixth to ninth leaf sheaths (38). Tillering of Italian ryegrass begins at a lower node than on perennial ryegrass (31). Tiller development in any axil does not usually occur until the leaf above it is fully expanded (92). Grass plants must develop to a certain minimum level of maturity before tillering is possible (111). Tillering of timothy occurs only after the first five leaves are fully developed (66). Also, the meristematic bud in any one axil of certain species such as Kentucky bluegrass is capable of initiating tillers for only a limited period of time (38). The rate of tillering is greater in turfgrass plants having a high photosynthetic capability and high levels of reserve carbohydrates (4, 8, 76, 92). Thus, tillering is generally enhanced by low temperatures, short day lengths, and high light intensities. Moderately close mowing also stimulates tillering (2, 9, 38, 54, 71). Procumbent stem development increases under short days and cool temperatures (5, 38, 78, 129).

Individual tillers of perennial grasses have a limited life span. For example, vegetative tillers of timothy live for approximately 1 year (66). The shoot density of turfs is maintained through a continuing process whereby new tillers replace mature, dying tillers (38). There is no distinct period of annual shoot death unless a lethal environmental stress occurs. Reproductive development of a tiller will cause early death (66). Tillering occurs throughout the growing season and even during the winter at temperatures above 32°F (66, 127). The rate of tillering is most frequent in the spring and fall, however, and is at a minimum during periods of extremely high or low temperatures or during drought stress (38, 45, 46, 67, 68, 74, 79, 105, 126, 127, 132).

A tufted, noncreeping growth habit results when tillering is the only type of lateral shoot development operative in a species. Typical examples include the ryegrasses and sheep fescue. A single sheep fescue plant can spread 4 in. per year by tillering (53).

Rhizomes and stolons. Secondary lateral shoots that arise extravaginally with considerable aboveground, horizontal stem elongation are called **stolons**. The plagiotropic growth of stolons results from the development of a positive geotropic response in the stolon (91). Since the stolons grow indeterminately at the soil surface and are exposed to light, they develop new shoots and roots quite readily from the nodes. Typical stoloniferous turfgrasses include creeping bentgrass, rough bluegrass, and zoysiagrass.

A **rhizome** is an extravaginal, secondary lateral shoot that elongates underground. Rhizomes are distinguished from aerial stems by their response to gravity. However, the initial downward growth of Kentucky bluegrass rhizomes is independent of gravity. Typical rhizomatous turfgrasses include Kentucky bluegrass, red fescue, and redtop (Fig. 2-5). The rhizomes of redtop are short and stocky, while those of Kentucky bluegrass are quite long and slender (40). Kentucky bluegrass rhizomes exhibit a circumnutational or boring type of motion when penetrating dense soils rather than straight intrusive growth. The mode of rhizome branching is quite variable (40). It occurs in a successive sequence starting at the terminal end. Injury to the terminal bud of a rhizome results in branching (37). New rhizomes of Kentucky bluegrass arise from nodes at the base of the aboveground shoot and from the nodes of older rhizomes. The rhizome bud of Kentucky bluegrass first appears as a swelling at the base of a leaf primordium. In contrast, new rhizomes of Canada bluegrass usually develop from nodes of other rhizomes beneath the surface of the soil.

A turfgrass seedling must reach a certain size and leaf area before rhizome development occurs. Lateral shoots that form rhizomes eventually terminate in aboveground shoots. When the tip of a rhizome is exposed to light or the lower carbon dioxide concentration of the atmosphere, linear growth is inhibited and the internodes adjacent to the tip turn upward with chlorophyll being formed in the leaf scales at the tip. Emergence of the terminal growing point of a rhizome above the soil is also stimulated by excessively high temperatures, excessive nitrogen levels, or short day lengths (20). The emergence and formation of shoots at the terminal end of a rhizome cause the parent plant to initiate a second set of new rhizomes. Newly initiated rhizomes from both the parent and daughter plants grow outward from the original mother plant. This outward orientation of new rhizome growth does not occur if the rhizome connecting the mother plant is severed prior to emergence of the rhizome tip above the soil. When a rhizome is severed, the tip turns up

Figure 2-5. Closeup of the crown of a Kentucky bluegrass plant showing lateral shoot development in the form of tillers and rhizomes. (Photo courtesy of O. M. Scott and Sons, Marysville, Ohio).

rapidly and becomes a shoot. Leafy shoots also develop from the nodes on each side of the severed point on the rhizome.

There is an intergradation of rhizomes and stolons in certain species such as bermudagrass. Both stoloniferous and rhizomatous stem types are capable of initiating shoot and root growth from the meristematic regions of each node. Branching of lateral stolons and rhizomes also occurs from the axils of scale buds in the nodes (7, 38). Branching and secondary lateral shoot development occurs more frequently from nodes associated with short internodes. These types of lateral shoot development become a form of asexual reproduction. When the stolons or rhizomes that connect with the maternal shoot are severed, they become propagules.

Rhizome and stolon development can occur throughout the year as long as soil temperatures are above 32°F (40). Maximum rates occur under favorable environmental conditions such as spring, late summer, and fall and is minimal during periods of drought and heat or low temperature stress (21, 40, 52). A large proportion of the rhizome tips turn upward to form aboveground shoots in early spring (40). The rate of spread of a turfgrass plant is a function of the length of secondary, lateral shoot growth and frequency of new rhizome or stolon development (40). Rhizome growth is favored by long day lengths, high light intensities, and lower levels of nitrogen nutrition.

Following both winter and summer dormancy, shoot regrowth is initiated from meristematic regions on the rhizomes, stolons, and crowns of shoots (90). The growth and development of shoots from rhizomes, stolons, and crowns of older shoots reoccur on a cyclic basis in perennial turfgrasses unless the shoot develops into a flowering culm. The shoot will usually die if it flowers.

Rhizomes and stolons of turfgrasses (72) are much larger in diameter than the roots and can function in the storage of carbohydrates (62). Rhizome and stolon growth is greater in turfs having a high photosynthetic potential and substantial carbohydrate reserves (72). Mowing enhances horizontal growth of stolons and rhizomes (72). Leaf, rhizome, and tiller development in terms of length and number can be significantly reduced by certain herbicides (48).

The Stem Apex

The apical meristem of the turfgrass shoot is comprised of the apical dome plus a series of alternating leaf primordia (Fig. 2-6). This stem apex is commonly referred to as the growing point. At the extreme tip of the stem apex is a one or two-layered **tunica** that surrounds an irregularly arranged central core of cells called the **corpus**. The outer layer of cells of the tunica is called the **dermatogen** and the inner layer is called the **hypodermis** (107). Two tunica layers are most common in the *Festucoideae*, while a single layer is most common in the *Panicoideae* (23).

A varying number of alternating leaf primordia appear as ridges located below and to either side of the apical dome (38, 41, 107). The number of leaf primordia

Figure 2-6. Closeup of the stem apex of perennial ryegrass showing the bulbous apical dome and two leaf primordia. (Photo courtesy of J. P. Cooper, Welsh Plant Breeding Station, Aberystwyth, Wales—29).

and length of the shoot apex varies among turfgrass species (106). For example, Italian ryegrass has a relatively long stem apex with approximately fifteen to twenty leaf primordia (41, 42). In contrast, annual bluegrass, bentgrass, perennial ryegrass, and timothy have a somewhat shorter stem apex with approximately five to ten leaf primordia. One of the shortest stem apexes is found on bermudagrass with 2 to 3 primordia. Most cool season grasses have a stem apex of between 0.5 and 1.0 mm in length (41). The stem apex of Kentucky bluegrass is reported to be between 1 and 2 mm in length (7).

The stem apex of most commonly used perennial turfgrasses is located at or near the ground level (41) (Fig. 2-7). Internode elongation does not normally occur in primary shoots of turfgrasses. Thus, the growing point is seldom subjected to removal during mowing. Certain erect growing species such as Canada bluegrass and western wheatgrass may have the growing point elevated to a sufficient height during a portion of the growing season that it is subject to removal by mowing (19, 39). When this occurs,

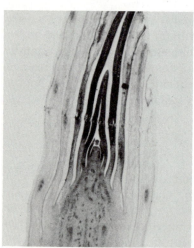

Figure 2-7. Longitudinal section through the stem apex of Kentucky bluegrass. (Photo courtesy of C. M. Harrison, Michigan State University, East Lansing, Michigan).

new shoots can develop only from the vegetative buds of lower growing stems. Species having an elevated stem apex are not adapted to close, frequent mowing.

Stem internode elongation is usually limited to modified stems such as stolons, rhizomes, and culms. Most of the creeping perennial turfgrasses possess intercalary meristems on the rhizomes and stolons. During early development of the stem apex, the leaf primordia form immediately above each other with no distinct separation by internodes. Internodes are formed later as a result of cell division in the axils between the leaf primordia. It is intercalary meristem activity involving primarily cell elongation that results in the growth of long stolons and rhizomes. Intercalary meristems are also active during the reproductive phase of growth and development. While the inflorescence is forming,

internode elongation is initiated by the intercalary meristems. The result is an elongated
culm with an inflorescence at the apex (38).
Stem anatomy. The epidermis and ground
tissue organization of elongated stems is similar to that of the leaf blade (Fig. 2-8). The
epidermis consists of a one cell thick, outer
layer organized in rows of long and short cells
with stomata located at intervals. Beneath the
epidermis is the cortex tissue. The parenchyma
cells of young stems vary in thickness and
contain chlorophyll. One or two layers of
lignified sclerenchyma cells develop below the
epidermis as the stem matures.

Differentiation of the vascular tissues also
occurs during development of the stem. The
vascular bundles of grass stems consist of an
orderly arrangement of xylem and phloem
elements. Inside and flanking the outer
phloem are the metaxylem vessels. To the
center and inside are located several protoxylem elements. The number of xylem and
phloem elements varies with the size of the
vascular bundle. The vascular bundle is
enclosed in a sheath of sclerenchymatous
fibers. Numerous types of vascular bundle
arrangements occur within grasses but are

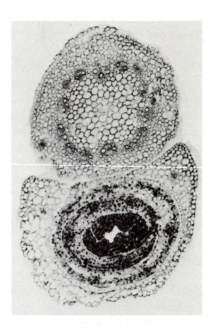

Figure 2-8. Cross section of a bentgrass
stem (above) growing adjacent to a stem
(below) having a higher cross section
which illustrates the leaves rolled in the
bud shoot. (Photo courtesy of C. M.
Harrison, Michigan State University,
East Lansing, Michigan).

usually arranged in one or two concentric rings around the central pith tissues.
The stem matures shortly after it elongates to form a culm. After stem growth ceases,
the cortical tissues and the protoxylem elements of the culm become lignified and
the central pith tissue breaks down.

Crown. The portion of the turfgrass plant that includes the stem apex, the unelongated internodes, and the lower nodes from which adventitious roots are initiated is
termed the **crown**. The crown tissues are the most vital portion of the turfgrass plant
since the adventitious roots, the lateral shoots, and the leaves are initiated from
this region (111). Should death of the root system, the leaf tissue, or both occur,
the plant can still survive as long as the crown tissues remain viable.

The junction of stem and root vascular systems occurs in the crown. The root
vascular system usually extends up to the scutellar node or vascular plate. Above
is a poorly defined transitional region that extends from the vascular plate to the
third internode. Vascular bundles typical of stem tissue are observed at this point.
Adventitious roots arise from nodes below the vascular plate. Leaves and lateral
shoots arise from axillary buds at successive nodes above the vascular plate.

The Leaf

The turfgrass **leaf** is borne on a stem and consists of a flattened upper blade and a lower basal sheath that encircles the stem. Both the blade and sheath function in photosynthesis and respiration. The blade is long, narrow, parallel veined, and nonpetiolate (111). The sheath is shaped like a hollow cylinder that is split down one side. The leaf blade width ranges from less than 1 mm for red fescue to widths exceeding 4 mm for St. Augustinegrass.

Leaf primordia are initiated by the periclinal division of cells in the hypodermis and dermatogen of the stem apex located at ground level (107, 111). Continued periclinal divisions occur on a horizontal plane extending partially around the axis and give rise to a series of alternate distichously arranged ridges (38, 41, 107). The blade develops as a result of marginal growth from periclinal divisions and apical growth from a rib meristem (Fig. 2-9). The leaf primordia forms a ridge, then a collar, and eventually a hood shape (111). Initially, meristematic activity occurs in the tip and the margins of the leaf primordium (41). Subsequently the meristematic activity shifts toward the base resulting in the formation of a basal intercalary meristem.

The youngest expanding turfgrass leaf assumes a protective hood-like shape over the stem apex (107, 111). No distinct boundary is evident between the leaf sheath and leaf blade. A boundary is established approximately when the ligule develops from the adaxial protoderm at the apex of the sheath. The auricles develop laterally from the boundary region at the same time. During this period of basal meristematic activity, a vegetative bud develops from the intercalary meristem on the opposite side from which the leaf originated.

The grass leaf matures from the tip downward with elongation being primarily basipital (41, 111). Blade elongation continues from basal intercalary meristem growth located just above the ligule until approximately the time the ligule and leaf blade are fully exposed (12, 65). Unfolding of the grass leaf is caused by greater enlargement of cells on the upper portion of the mesophyll than of the remainder of the mesophyll. The unfolding process of grass leaves is stimulated by light with the red (660 mμ) portion of the spectrum being the most active (133). The sheath is the last segment of the leaf to develop with most of the growth occurring from intercalary meristem activity located just below the ligule (111).

The leaf completes elongation when the blade fully emerges from the enclosed sheath. Only those leaves that are not fully extended at the time of mowing are capable of further growth (65). The ultimate leaf length is primarily a function of cell size and to a lesser degree of cell number, while leaf width is a function of cell number (16, 17, 44, 122). Only a limited number of externally visible leaves on an individual shoot grow at any one time and the number is independent of the total number of leaves formed by the shoot (65). For example, only the leaf blade and next oldest sheath of Kentucky bluegrass elongate at any one time (37). The youngest, partially expanded leaves function in support and protection of the apical

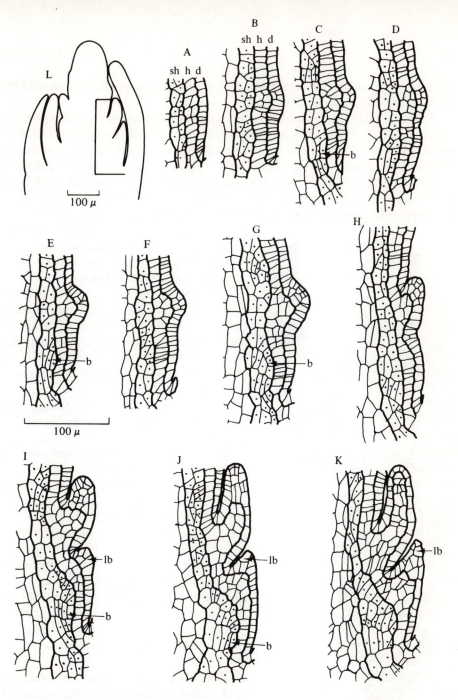

Figure 2-9. Radial longisections in plane of leaves of *Agropyron repens* showing stages in initiation of primordium on side of apex. b, cells which foreshadow bud in axil of leaf lower down; lb, base of primordium growing around from other side of axis. d, h, sh, dermatogen, hypodermal and subhypodermal layers. (Dots employed to indicate subhypodermal layer are purely conventional and in later stages are not intended to define absolutely the inner limits of this layer). L, longisection of whole apex showing position and extent of tissue included in K. (Photo courtesy of Sharman—107).

39

meristem. The intercalary meristems of the leaf are also protected and held erect by the more mature leaves.

The total quantity of shoot growth formed per day above the cutting height is a function of the rate of tillering, leaf appearance, growth, and death. The rate of appearance and elongation of successive leaves is closely integrated but can vary depending on genetic and environmental influences (78). The number of externally visible leaves of a turfgrass plant that are actively growing is constant with a given environment, cultural intensity, and species or cultivar. Seasonally, the rate of leaf appearance is most rapid in the spring and early summer and declines during periods of heat or low temperature stress (38, 65, 79, 92, 126). In general, the length of life of a turfgrass leaf is relatively short when compared to many dicotyledonous species (126). The distribution of carbohydrates is altered as the fully expanded, older leaves located on the outer edge of the turfgrass plant mature. Senescent leaves may become a negative drain on the plant carbohydrate balance (4). Death of a turfgrass leaf is a gradual process that occurs from the tip of the blade to the base (38). The rate of leaf death in a given environment is about equal to rate of leaf appearance (38, 65).

The emerging turfgrass leaf develops within the sheath of the next oldest leaf. Thus, the new leaf must elongate to a greater height above the soil surface than the previous leaf in order for it to be exposed to the light. As this process continues, the newly formed leaves eventually emerge above the cutting height and are completely removed. This means that only the leaf sheaths and older leaves remain and any new, actively photosynthesizing leaves must arise from newly formed lateral shoots. Species or cultivars possessing short leaf sheaths and internodes are more tolerant of close mowing.

Leaf anatomy. The leaf surface is characterized by a series of distinct parallel veins, with smaller ones usually alternating with large veins (111). A large midrib may or may not be present. A transverse section of a turfgrass leaf reveals three distinct types of tissues (Fig. 2-10). The epidermal tissue consists of a pro-

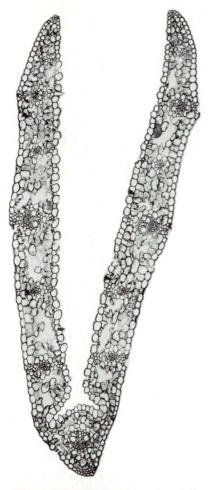

Figure 2-10. Cross section of a fully developed Kentucky bluegrass leaf. (Photo courtesy of R. M. Endo, University of California at Riverside, Riverside, California).

tective layer of epidermal cells located on both the upper and lower leaf surface. The epidermal cells are generally elongated and arranged in longitudinal rows of various combinations depending on the vein arrangement (111). The lower epidermis of grass leaves generally has a thicker cutical than the upper epidermis (73).

The modified subsidiary epidermal cells are relatively short and include the silica, cork, fiber, bulliform, and guard cells. Silica cells are characterized by a high silica content, while the cork cells have suberized cork-like walls. Turfgrass species having distinct silica cells include bahiagrass and zoysiagrass. Short epidermal hairs may arise from the cork and silica cells. Frequently located above and below the parallel vascular bundles are the fiber cells. The stiff, upright growth of zoysiagrass leaves is attributed to the numerous groups of fiber cells contained in the leaves (69).

Bulliform cells are frequently located in longitudinal rows on each side and parallel to the main midrib of certain species such as Kentucky bluegrass or they may be arranged in parallel rows throughout the upper surface of the leaf as with creeping bentgrass (73). Bulliform cells are large, thin-walled types that have a relatively high water content. The bulliform cells lose water and collapse quite readily when an internal water stress occurs. This response is thought to effect the rolling or folding characteristic of turfgrass leaves.

With the exception of the stomata, the epidermal cells of the grass leaves have a very compact structure. This structure combined with a wax-like layer of cutin on the outer epidermal wall, functions in preventing excessive water loss as well as in mechanical support of the leaf. A high degree of suberization, lignification, and silicification is typical of turfgrass leaves. The silica content of the epidermal walls is quite high in certain turfgrass species such as *Zoysia*.

The stomata of the epidermis occur in well-defined bands that vary in arrangement with the turfgrass species (111). Through differentiation of a specialized guard mother cell, two guard cells are formed. When fully developed the guard cells are elongated, somewhat enlarged at the end, and constricted in the middle. The shape and variable cell wall thickness of the guard cells result in an opening or pore between the two cells when they are fully turgid. The plant-atmospheric exchange of water vapor, oxygen, and carbon dioxide occurs through the stomatal openings. Opposite the stoma and on each side of the guard cell are specialized cells called subsidiary or accessory cells. They are thought to be functionally associated with the guard cells and are usually triangular in shape. A majority of the stomata are located on the upper epidermis of most turfgrass species (109, 111).

Within the upper and lower epidermis of the leaf and surrounding the vascular bundles are the ground tissues. This ground tissue, the mesophyll, is primarily a specialized photosynthetic tissue. The mesophyll of most turfgrasses is seldom differentiated into palisade and spongy parenchyma cells, but it is composed of thin-walled chlorenchyma cells that contain chloroplasts and colorless parenchyma cells (18, 111). Interspersed through the loose cellular arrangement of the mesophyll tissues are intercellular spaces through which carbon dioxide and oxygen diffuse

(42, 109). Measurements of the cross-sectional area of a perennial ryegrass blade show that it consists of 65% mesophyll and intercellular space, 32% epidermal and supporting tissues, and 3% vascular tissue (111).

The vascular bundles of a leaf contain the same type of xylem and phloem structure as the stem. Provascular strands of the midrib are observed developing at about the time that internode elongation occurs (111). This is followed by the development of lateral bundles in the intermediate and smaller veins of the leaf primordia. The smallest veins develop near the tip and margins and extend down the leaf to the base where they fuse with adjacent vascular bundles. Most of the veins in a blade are made up of a single vascular bundle. An exception is the midrib that may contain several vascular bundles.

The vascular bundle is enclosed in one or two sheaths (73, 109). The outer sheath is made up of large, thin-walled cells (18). The inner sheath, when present, consists of small, thick-walled cells (111). The xylem and phloem elements in larger veins are of the normal type and arrangement described for the stems. In the smaller veins, only one to a few tracheids in the xylem and a similar number of phloem elements may be observed (111). The network of vascular tissues in the roots, stems, and leaves functions primarily in the conduction of water and nutrient elements from the soil to the leaves and in the translocation of photosynthetic products from the leaves to other nonphotosynthetic organs of the plant.

The organizational leaf anatomy of the turfgrasses is of two basic types. The *Festucoid* type is the least specialized with a relatively simple epidermis, a well-developed inner bundle sheath, an outer bundle sheath that is not clearly differentiated, and homogeneous irregularly arranged chlorenchyma tissue with considerable intercellular space. In the *Panicoid* type there is more diversification of the epidermal cells; the inner bundle sheath is usually absent; the parenchyma sheath is well developed; and the chlorenchyma tissues are usually arranged radially around the vascular bundles.

Root-Shoot Growth Relationships

The growth of the roots and shoots of a turfgrass plant is closely interrelated and strongly influenced by the environment. The roots depend on the shoots as the source of carbohydrates, while the shoots are largely dependent on the root system for the uptake of water and nutrients. Both the root and shoot systems require carbohydrates for growth. The meristematic growth regions of the shoot are nearest the photosynthetic tissues and assume priority in utilization of the available carbohydrates when the quantity of carbohydrates is less than that required for root and shoot growth (77, 130, 134). Thus, shoot growth occurs at the expense of root growth and may even cause death of a substantial portion of the root system.

The ratio of the root weight to shoot weight is called the **root-shoot ratio**. This ratio is frequently used to express the interrelationship between the roots and shoots. A high root-shoot ratio is preferred under turfgrass conditions. Factors that stimulate shoot growth and reduce the root-shoot ratio include (a) temperatures above the optimum for root growth, (b) close cutting, (c) excessive nitrogen nutrition, and (d) low light intensities (13, 50, 55, 77, 105, 134).

Recuperative Potential

Turfs are periodically injured by environmental stresses, turfgrass pests, and the activities of man. Injury may involve thinning, severe defoliation, kill of the above-ground tissue, or the loss of turf by divots. **Recuperative potential** is the ability of turfgrasses to recover from injury, usually through vegetative growth from crown meristematic buds and the nodes of creeping stems. This is an important characteristic in the overall performance and persistence of turfgrasses.

Turfgrasses vary widely in recuperative potential (80, 143) (Table 2-1). Species that have a good recuperative potential generally possess vigorous rhizome and/or stolon development. A poor recuperative potential is associated with turfgrasses having a bunch-type growth habit. The rate of recuperation is a function of the growth rate. Two species may have the same recuperative potential but substantially different recovery rates. For example, the recuperative rate of the improved bermudagrasses is superior to the zoysiagrasses but the recuperative potential is fairly comparable. The rate of recovery will also vary with the severity of injury, especially to the rhizomes and stolons.

Table 2-1

THE RELATIVE RECUPERATIVE POTENTIAL
OF FOURTEEN TURFGRASSES

Recuperative potential	Turfgrass
Excellent	Bermudagrass
	Zoysiagrass
Good	St. Augustinegrass
	Kentucky bluegrass
	Creeping bentgrass
Intermediate	Red fescue
	Carpetgrass
	Colonial bentgrass
Poor	Centipedegrass
	Meadow fescue
	Bahiagrass
	Tall fescue
Very Poor	Perennial ryegrass
	Timothy

Inflorescence Growth and Development

A transition occurs from the vegetative to the reproductive growth and development phase when the turfgrass plant reaches a certain stage of maturity or is stimulated by a specific environment. Floral induction of turfgrasses is controlled primarily by the day length and low temperature exposure (5, 28, 30, 94). Nitrogen fertilization enhances vegetative growth and impairs floral development of turfs (56, 99). Seed head formation in bentgrass stolon nurseries has been prevented by using maleic hydrazide. Floral growth and development of most turfgrasses occurs in late spring or early summer while floral induction occurs in the fall (28, 41, 94). Certain species such as annual bluegrass and bermudagrass can initiate flowers with varying degrees of intensity throughout the growing season.

The first response observed is a rapid elongation of the shoot apex followed by lateral bud primordial development in the axils of leaf initials (29, 37, 41, 107). Subsequently, the lateral bud and growing point primordia differentiate into branches of the inflorescence and spikelets similar to primordial growth of lateral vegetative shoots (41). Elongation of the culm internodes occurs from basal intercalary meristems. Development of the inflorescence is completed at the time of emergence from the enclosing leaf sheaths (38). Further growth and maturation of the inflorescence occurs in a determinant manner with the terminal and middle spikelets maturing first and the basal spikelets last (41). Flowering and pollination proceed to the eventual formation of a seed and enclosed embryo.

Assimilation

The turfgrass plant increases in size and usually increases in dry matter content during growth. Carbohydrates are utilized in synthesis of the cell walls and protoplasm (Fig. 2-11). The protoplasm is composed largely of proteinaceous materials, while polysaccharides are the primary constituents of cell walls. Light is the energy source for photosynthesis in which simple carbohydrates are synthesized. The primary site of photosynthesis is in the mesophyll tissue of green leaves. The light energy used in photosynthesis is absorbed by chlorophyll contained in saucer-shaped chloroplasts. The end product of photosynthesis is a hexose sugar that is readily translocated throughout the plant from the chloroplasts. The efficiency of photosynthetic utilization of light is lower in plants that fix CO_2 by the C_4-dicarboxylic acid path than for the Calvin cycle plants. The photosynthetic rate increases with increased levels of nitrogen nutrition (103).

The carbohydrates produced by photosynthesis are utilized in the assimilation of more complex compounds. Included are (a) polysaccharides such as starch and fructosans that serve as reserve storage compounds; (b) cellulose and lignins that are utilized primarily as structural components in cell walls (137); and (c) proteins that are required in such vital cell constituents as the nucleotides involved in the transfer of hereditary characteristics, the enzymes for catalyzing most metabolic reactions, and the other protoplasmic constituents.

Carbohydrates are also utilized as a source of energy for the maintenance and growth of plant tissues. Energy is constantly required for the vital processes involved in the support of living cells. During respiration, oxygen is absorbed,

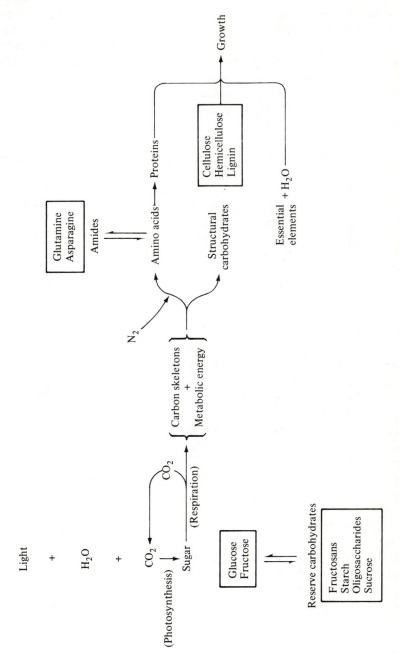

Figure 2-11. Schematic diagram of the general assimilation processes and sinks required for normal growth of turfgrasses.

carbohydrates oxidized, carbon dioxide and water formed, and chemical energy converted to free energy. Living cells capture a portion of the free energy and utilize it in support of many vital life processes. The enzymes and reactions involved in respiration are contained within protoplasmic organelles called mitochondria. Respiration is a continuing process common to all living tissues including the leaves, stems, crowns, stolons, rhizomes, and roots of turfgrasses. Respiration utilizes about 17% of the carbon fixed by photosynthesis under normal turfgrass growing conditions (57). The respiration rate increases as the temperature or level of nitrogen nutrition increases (3, 15, 57, 97, 102).

Carbohydrate reserves. Carbohydrates that accumulate in the more permanent organs of the plant in nonstructural forms and are available for subsequent utilization in assimilation processes are termed **carbohydrate reserves**. Carbohydrate accumulation occurs when carbohydrates are being synthesized at a rate in excess of the level required for assimilation. Fructosans and oligosaccharides containing monoglucopolyfructans are the primary storage carbohydrates found in cool season turfgrasses (9, 51, 57, 82, 83, 84, 85, 86, 88, 94, 96, 112, 113, 114, 115, 116, 119, 124, 125, 135, 136, 138, 141, 144). Starch and sucrose are the primary storage carbohydrates in most warm season turfgrasses (13, 24, 32, 33, 62, 82, 114, 140, 141). The fructosans are 2–6 linked fructose polymers that have a low degree of polymerization compared to starch (51). Sucrose also occurs in certain cool season turfgrass species (1, 9, 13, 33, 62, 88, 96, 115, 119, 124, 125, 135, 136, 138, 140). The primary carbohydrate reserve in turfgrass seeds is starch (88).

The accumulation of storage carbohydrates is greatest during periods of minimal shoot growth and high light intensity (3, 102, 128). A high rate of carbohydrate accumulation also occurs in turfgrasses during the late fall hardening period just prior to the onset of winter dormancy (21, 33, 74, 93, 97, 102, 128, 141, 144). The depletion of carbohydrate reserves is usually correlated with periods of active shoot growth (33, 81, 97). Thus, optimum temperatures, close mowing, irrigation, or a high nitrogen fertility level stimulates shoot growth and causes a drain on the carbohydrate reserve (1, 3, 8, 13, 15, 50, 61, 70, 81, 86, 87, 94, 104, 119, 128, 144, 145). There is a diurnal variation in nonstructural carbohydrates of turfgrasses. Higher levels occur during the latter portion of the daylight period.

A pool of reserve carbohydrates is desired at all times, since it can serve as an immediate source of carbohydrates for regrowth and recovery should the turf be injured or thinned by turfgrass pests, environmental stress, chemicals, or human activities. The potential level of carbohydrate reserves in turfgrasses is limited because of the frequent, close mowing that is practiced. The stems (1, 2, 3, 9, 84, 113, 119, 124, 125, 135, 141, 144, 145), rhizomes (38, 93, 94, 141), and stolons (72) are the primary storage tissues for carbohydrates. Glucose, fructose, and sucrose constitute a high proportion of the total soluble carbohydrate content of root and leaf tissue where the storage carbohydrate level is relatively low compared to the stems, rhizomes, and stolons (1, 3, 9, 93, 94, 97, 102, 112, 113, 116, 124, 125, 135, 136, 141, 144).

References

1. ADEGBOLA, A. A., and C. M. McKell. 1966. Effect of nitrogen fertilization on the carbohydrate content of coastal bermudagrass [*Cynodon dactylon*, (L.), Pers.]. Agronomy Journal. 58: 60–64.

2. ALBERDA, T. 1957. The effects of cutting, light intensity, and night temperature on growth and soluble carbohydrate content of *Lolium perenne* L. Plant and Soil. 8: 199–230.

3. ——. 1965. The influence of temperature, light intensity and nitrate accumulation on dry-matter production and chemical composition of *Lolium perenne* L. Netherlands Journal of Agricultural Science. 13: 335–360.

4. ALBURQUERQUE, H. E. 1967. Leaf area, and age, and carbohydrate reserves in the regrowth of tall fescue (*Festuca arundinacea* Schreb.) tillers. Ph.D. Thesis. Virginia Polytechnic Institute. pp. 1–278.

5. ALLARD, H. A., and M. W. Evans. 1941. Growth and flowering of some tame and wild grasses in response to different photoperiods. Journal of Agricultural Research. 62: 193–228.

6. ANDERSON, A. M. 1927. Development of the female gametophyte and caryopsis of *Poa pratensis* and *Poa compressa*. Journal of Agricultural Research. 34: 1001–1018.

7. ARBER, A. 1934. The Gramineae. Cambridge University Press. London, England. pp. 1–480.

8. AUDA, H., R. E. Blaser, and R. H. Brown. 1966. Tillering and carbohydrate contents of orchardgrass as influenced by environmental factors. Crop Science. 6: 139–143.

9. BAKER, H. K. 1957. Studies on the root development of herbage plants. III. The influence of cutting treatments on the root, stubble, and herbage production of a perennial ryegrass sward. Journal of the British Grassland Society. 12: 197–207.

10. BARNARD, C. 1964. "Form and structure" in Grasses and Grasslands. Ed., C. Barnard. Macmillan & Co. Ltd. London, England. pp. 47–72.

11. BEARD, J. B., and W. H. Daniel. 1966. Relationship of creeping bentgrass (*Agrostis palustris* Huds.) root growth to environmental factors in the field. Agronomy Journal. 58: 337–339.

12. BEGG, J. E., and M. J. Wright. 1962. Growth and development of leaves from intercalary meristems in *Phalaris arundinacea* L. Nature. 194: 1097–1098.

13. BENEDICT, H. M., and G. B. Brown. 1944. The growth and carbohydrate responses of *Agropyron smithii* and *Bouteloua gracilis* to changes in nitrogen supply. Plant Physiology. 19: 481–494.

14. BINGHAM, S. W. 1967. Influence of herbicides on root development of bermudagrass. Weeds. 15: 363–365.

15. BLASER, R. E., and R. E. Schmidt. 1964. Effect of nitrogen on organic food reserves and some physiological responses of bentgrass and bermudagrass grown under various temperatures. 1964 Agronomy Abstracts. p. 99.

16. BORRILL, M. 1959. Inflorescence initiation and leaf size in some *Graminae*. Annals of Botany, N. S. 23: 217–227.

17. ——. 1961. The developmental anatomy of leaves in *Lolium temulentum*. Annals of Botany, N. S. 25: 1–11.

18. BRABROV, R. A. 1955. The leaf structure of *Poa annua* with observations on its smog sensitivity in Los Angeles County. American Journal of Botany. 42: 467–474.

19. BRANSON, F. A. 1953. Two new factors affecting resistance of grasses to grazing. Journal of Range Management. 6: 165–171.

20. BROWN, E. M. 1939. Some effects of temperature on the growth and chemical composition of certain pasture grasses. Missouri Agricultural Experiment Station Research Bulletin 299. pp. 1–76.

21. ———. 1943. Seasonal variations in the growth and chemical composition of Kentucky bluegrass. Missouri Agricultural Experiment Station Research Bulletin 360. pp. 1–56.

22. BROWN, W. V. 1959. The epiblast and coleoptile of the grass embryo. Bulletin of the Torrey Botanical Club. 86: 13–16.

23. BROWN, W. V., C. Heimsch, and W. H. P. Emery. 1957. The organization of the grass shoot apex and systematics. American Journal of Botany. 44: 590–595.

24. BURRIS, J. S., R. H. Brown, and R. E. Blaser. 1967. Evaluation of reserve carbohydrates in Midland bermudagrass (*Cynodon dactylon* L.). Crop Science. 7: 22–24.

25. BURTON, G. W. 1943. A comparison of the first year's root production of seven southern grasses established from seed. Journal of the American Society of Agronomy. 35: 192–196.

26. BURTON, G. W., E. H. DeVane, and R. L. Carter. 1954. Root penetration, distribution, and activity in southern grasses measured by yields, drought symptoms and P^{32} uptake. Agronomy Journal. 46: 229–233.

27. CALLAHAN, L. M., and R. E. Engel. 1965. Tissue abnormalities induced in roots of colonial bentgrass by phenoxyalkylcarboxylic acid herbicides. Weeds. 13: 336–338.

28. COOPER, J. P. 1950. Day-length and head formation in the ryegrasses. Journal of the British Grassland Society. 5: 105–112.

29. ———. 1951. Studies on growth and development in *Lolium*. II. Pattern of bud development of the shoot apex and its ecological significance. Journal of Ecology. 39: 228–270.

30. COOPER, J. P., and D. M. Calder. 1964. The inductive requirements for flowering of some temperate grasses. Journal of the British Grassland Society. 19: 6–14.

31. COOPER, J. P., and K. J. R. Edwards. 1961. The genetic control of leaf development in *Lolium*. 1. Assessment of genetic variation. Heredity. 16: 63–82.

32. DeCUGNAC, A. 1931. Recherches sur les glusides des Graminees. Ann. Sci. Naturelles. 13: 1–129.

33. DODD, J. D., and H. H. Hopkins. 1958. Yield and carbohydrate content of blue grama grass as affected by clipping. Transactions of the Kansas Academy of Science. 61: 280–287.

34. DOSS, B. D., D. A. Ashley, and O. L. Bennett. 1960. Effect of soil moisture regime on root distribution of warm season forage species. Agronomy Journal. 52: 569–572.

35. DUELL, R. W. 1961. Bermudagrass has multiple-leaved nodes. Crop Science. 1: 230–231.

36. ENGEL, R. E., and L. M. Callahan. 1967. Merion Kentucky bluegrass response to soil residue of preemergence herbicides. Weeds. 15: 128–130.

37. ETTER, A. G. 1951. How Kentucky bluegrass grows. Annals of the Missouri Botanical Garden. 38(1): 293–376.

38. EVANS, M. W. 1949. Vegetative growth, development, and production in Kentucky bluegrass. Ohio Agricultural Experiment Station Research Bulletin 681. pp. 1–39.

39. EVANS, M. W., and J. E. Ely. 1933. Usefulness of Kentucky bluegrass and Canada bluegrass in turfs as affected by their habits of growth. Bulletin of USGA Green Section. 13: 140–143.

40. ———. 1935. The rhizomes of certain species of grasses. Journal of the American Society of Agronomy. 27: 791–797.

41. EVANS, M. W., and F. O. Grover. 1940. Developmental morphology of the growing point of the shoot and the inflorescence in grasses. Journal of Agricultural Research. 61: 481–520.

42. EVANS, P. S. 1964. A comparison of some aspects of the anatomy and morphology of Italian ryegrass (*Lolium multiflorum* Lam.) and perennial ryegrass (*L. perenne* L.). New Zealand Journal of Botany. 2: 120–130.

43. FITTS, O. B. 1925. A preliminary study of the root growth of fine grasses under turf conditions. USGA Green Section Bulletin. 5: 58–62.

44. FORDE, B. J. 1966. Effect of various environments on the anatomy and growth of perennial ryegrass and cocksfoot. 1. Leaf growth. New Zealand Journal of Botany. 4: 455–468.

45. FRAZIER, S. L. 1960. Turfgrass seedling development under measured environment and management conditions. M.S. Thesis. Purdue University. pp. 1–61.

46. GARWOOD, E. A. 1958. The seasonal initiation of new roots in a perennial ryegrass/white clover sward. Experiments in Progress. 10: 51.

47. ———. 1967. Seasonal variation in appearance and growth of grass roots. Journal of the British Grassland Society. 22: 121–130.

48. GASKIN, T. A. 1964. Effect of pre-emergence crabgrass herbicides on rhizome development in Kentucky bluegrass. Agronomy Journal. 56: 340–342.

49. GIST, G. R., and R. M. SMITH. 1948. Root development of several common forage grasses to a depth of eighteen inches. Journal of the American Society of Agronomy. 40: 1036–1042.

50. GRABER, L. F. 1931. Food reserves in relation to other factors limiting the growth of grasses. Plant Physiology. 6: 43–71.

51. GROTELUESCHEN, R. D., and D. Smith. 1968. Carbohydrates in grasses. III. Estimations of the degree of polymerization of the fructosans in the stem bases of timothy and bromegrass near seed maturity. Crop Science. 8: 210–212.

52. HANSON, A. A., and F. V. Juska. 1961. Winter root activity in Kentucky bluegrass (*Poa pratensis* L.). Agronomy Journal. 53: 372–374.

53. HARBERD, D. J. 1962. Some observations on natural clones in *Festuca ovina*. New Phytologist. 61: 85–100.

54. HARRISON, C. M. 1931. Effect of cutting and fertilizer applications on grass development. Plant Physiology. 6: 669–684.

55. HODGES, T. K. 1960. Factors influencing the rooting depth of *Poa pratensis*. M.S. Thesis. University of California at Davis. pp. 1–53.

56. HOSHIZAKI, T. 1953. Nitrogen uptake and temperature response of U-3 bermuda grass. M.S. Thesis. University of California at Los Angeles. pp. 1–70.

57. IYAMA, J., Y. Murata, and T. Homma. 1964. Studies on the photosynthesis of forage crops. II. Influence of the different temperature levels on diurnal changes in the photosynthesis of forage crops under constant conditions. Proceedings of the Crop Science Society of Japan. 33: 25–28.

58. JACQUES, W. A. 1956. Root development in some common New Zealand pasture plants. IX. The root replacement pattern in perennial ryegrass (*Lolium perenne*). New Zealand Journal of Science and Technology. Sec. A. 38: 160–165.

59. JACQUES, W. A., and R. H. Schwass. 1956. Root development in some common New Zealand pasture plants. VII. Seasonal root replacement in perennial ryegrass (*Lolium perenne*), Italian ryegrass (*L. multiflorum*) and tall fescue (*Festuca arundinacea*). New Zealand Journal of Science and Technology. Sec. A. 37: 569–583.

60. JONES, K. 1956. Species differentiation in *Agrostis*. II. The significance of chromosome pairing in the tetraploid hybrids of *Agrostis canina* subsp. *montana* Hartm., *A. tenuis* Sibth, and *A. stolonifera* L. Journal of Genetics. 54: 377–393.

61. JORDAN, E. E. 1959. The effect of environmental factors on the carbohydrate and nutrient levels of creeping bentgrass (*Agrostis palustris*). M.S. Thesis. Purdue University. pp. 1–63.

62. JULANDER, O. 1945. Drought resistance in range and pasture grasses. Plant Physiology. 20: 573–599.

63. KENNEDY, P. B. 1899. The structure of the caryopsis of grasses with reference to their morphology and classification. USDA Bulletin 19. pp. 1–44.

64. LAIRD, A. S. 1930. A study of the root systems of some important sod-forming grasses. Florida Agricultural Experiment Station Bulletin 211. pp. 1–27.

65. LANGER, R. H. M. 1954. A study of leaf growth in timothy (*Phleum pratense*). Journal of the British Grassland Society. 9: 275–284.

66. ———. 1956. Growth and nutrition of timothy (*Phleum pratense*). 1. The life history of individual tillers. Annals of Applied Biology. 44: 166–187.

67. ———. 1958. A study of growth in swards of timothy and meadow fescue. I. Uninterrupted growth. Journal of Agricultural Science. 51: 347–352.

68. ———. 1959. A study of growth in swards of timothy and meadow fescue. II. The effects of cutting treatment. Journal of Agricultural Science. 52: 273–281.

69. LeCroy, W. C. 1963. Characterizing zoysia by field and anatomical studies. Ph.D. Thesis. Purdue University. pp. 1–82.

70. Leonard, R. T. 1967. The influence of nitrogen fertilizer treatments on the carbohydrate reserves of timothy (*Phleum pratense* L.). M.S. Thesis. University of Rhode Island. pp. 1–57.

71. Leopold, A. C. 1949. The control of tillering in grasses by auxin. American Journal of Botany. 36: 437–440.

72. Leukel, W. A., and J. M. Coleman. 1930. Growth behavior and maintenance of organic foods in bahia grass. Florida Agricultural Experiment Station Technical Bulletin 219. pp. 1–56.

73. Lewton-Brain, L. 1904. On the anatomy of the leaves of British grasses. Transactions of the Linnean Society of London. 6: 315–360.

74. Lopex, R. R., A. G. Matches, and J. D. Baldridge. 1967. Vegetative development and organic reserves of tall fescue under conditions of accumulated growth. Crop Science. 7: 409–412.

75. Madison, J. H., and R. M. Hagan. 1962. Extraction of soil moisture by Merion bluegrass (*Poa pratensis* L. 'Merion') turf, as affected by irrigation frequency, mowing height, and other cultural practices. Agronomy Journal. 54: 157–160.

76. Mitchell, K. J. 1953. Influence of light and temperature on the growth of ryegrass (*Lolium* spp.). II. The control of lateral bud development. Physiologia Plantarum. 6: 425–443.

77. ———. 1954. Influence of light and temperature on growth of ryegrass (*Lolium* spp.). III. Pattern and rate of tissue formation. Physiologia Plantarum. 7: 51–65.

78. ———. 1954. Growth of pasture species. I. Short rotation and perennial ryegrass. New Zealand Journal of Science and Technology. Sec. A. 36: 191–206.

79. Mitchell, K. J., and R. Lucanus. 1960. Growth of pasture species in controlled environment. II. Growth at low temperatures. New Zealand Journal of Agricultural Research. 3: 647–655.

80. Morrish, R. H., and C. M. Harrison. 1948. The establishment and comparative wear resistance of various grasses and grass-legume mixtures to vehicular traffic. Journal of the American Society of Agronomy. 40: 168–179.

81. McKee, W. H., R. H. Brown, and R. E. Blaser. 1967. Effect of clipping and nitrogen fertilization on yield and stands of tall fescue. Crop Science. 7: 567–570.

82. McKell, C. M., and V. B. Youngner. 1967. Carbohydrate reserves of a warm season and a cool season perennial grass in relation to temperature regimes. 1967 Agronomy Abstracts. p. 42.

83. Norman, A. G. 1936. The composition of forage crops. I. Rye grass (Western Wolths). Biochemical Journal. 30: 1354–1362.

84. Norman, A. G., and H. L. Richardson. 1937. The composition of forage crops. II. Rye grass (Western Wolths) changes in herbage and soil during growth. Biochemical Journal. 31: 1556–1566.

85. Norman, A. G., C. P. Wilsie, and W. G. Gaessler. 1941. The fructosan content of some grasses adapted to Iowa. Iowa State College Journal of Science. 15: 301–305.

86. Nowakowski, T. Z. 1962. Effects of nitrogen fertilization on total nitrogen, soluble nitrogen, and soluble carbohydrate contents of grass. Journal of Agricultural Science. 59: 387–392.

87. Nowakowski, T. Z., R. K. Cunningham, and K. F. Nielsen. 1965. Nitrogen fractions and soluble carbohydrates in Italian ryegrass. I. Effects of soil temperature, form and level of nitrogen. Journal of the Science of Food and Agriculture. 16: 124–134.

88. Okajima, H., and D. Smith. 1964. Available carbohydrate fractions in the stem bases and seed of timothy, smooth bromegrass, and several other northern grasses. Crop Science. 4: 317–320.

89. OLMSTED, C. E. 1941. Growth and development in range grasses. I. Early development of *Bouteloua curtipendula* in relation to water supply. Botanical Gazette. 102: 499–519.

90. ———. 1942. Growth and development of range grasses. II. Early development of *Bouteloua curtipendula* as affected by drought periods. Botanical Gazette. 103: 531–542.

91. PALMER, J. H. 1956. The nature of the growth response to sunlight shown by certain stoloniferous and prostrate tropical plants. New Phytologist. 55: 346–355.

92. PATEL, A. S., and J. P. Cooper. 1961. The influence of several changes in light energy on leaf and tiller development in ryegrass, timothy, and meadow fescue. Journal of the British Grassland Society. 16: 299–308.

93. PETERSON, M. L. 1946. Physiological response of Kentucky bluegrass to different management treatments. Ph.D. Thesis. Iowa State College. pp. 1–87.

94. PETERSON, M. L., and W. E. Loomis. 1949. Effects of photoperiod and temperature on growth and flowering of Kentucky bluegrass. Plant Physiology. 24: 31–43.

95. PHILLIPPE, P. M. 1942. Effects of some essential elements on the growth and development of Kentucky bluegrass (*Poa pratensis*). Ph.D. Thesis. Ohio State University. pp. 1–57.

96. PHILLIPS, T. G., J. T. Sullivan, M. E. Loughlin, and W. G. Sprague. 1954. Chemical composition of some forage crops. I. Changes with plant maturity. Agronomy Journal. 46: 361–369.

97. POWELL, A. J., R. E. Blaser, and R. E. Schmidt. 1967. Physiological and color aspects of turfgrasses with fall and winter nitrogen. Agronomy Journal. 59: 303–307.

98. ———. 1967. Effect of nitrogen on winter root growth of bentgrass. Agronomy Journal. 59: 529–530.

99. PRINE, G. M., and G. W. Burton. 1956. The effect of nitrogen rate and clipping frequency upon the yield, protein content and certain morphological characteristics of Coastal bermudagrass [*Cynodon dactylon*, (L.) Pers.]. Agronomy Journal. 48: 296–301.

100. REEDER, J. R. 1957. The embryo in grass systematics. American Journal of Botany. 44: 756–768.

101. ROBERTS, E. C., F. E. Markland, and H. M. Pellett. 1966. Effects of bluegrass stand and watering regime on control of crabgrass with preemergence herbicides. Weeds. 14: 157–161.

102. SCHMIDT, R. E. 1965. Some physiological responses of two grasses as influenced by temperature, light and nitrogen fertilization. Ph.D. Thesis. Virginia Polytechnic Institute. pp. 1–116.

103. SCHMIDT, R. E., and R. E. Blaser. 1967. Effect of temperature, light, and nitrogen on growth and metabolism of 'Cohansey' bentgrass (*Agrostis palustris* Huds.). Crop Science. 7: 447–451.

104. ———. 1969. Effect of temperature, light, and nitrogen on growth and metabolism of 'Tifgreen' bermudagrass (*Cynodon* spp.). Crop Science. 9: 5–9.

105. SCHWASS, R. H., and W. A. Jacques. 1956. Root development in some common New Zealand pasture plants. VIII. The relationship of top growth to root growth in perennial ryegrass (*Lolium perenne*), Italian ryegrass (*L. multiflorum*) and tall fescue (*Festuca arundinacea*). New Zealand Journal of Science and Technology. Sec. A. 38: 109–119.

106. SHARMAN, B. C. 1942. Shoot apex in grasses and cereals. Nature. 149: 82–83.

107. ———. 1945. Leaf and bud initiation in the *Gramineae*. Botanical Gazette. 106: 269–289.

108. ———. 1947. The biology and developmental morphology of the shoot apex in the *Gramineae*. New Phytologist. 46: 20–34.

109. SOPER, K. 1956. The anatomy of the vegetative shoot of *Paspalum dilatatum* Poir. New Zealand Journal of Science and Technology. Sec. A. 37: 600–605.

110. ———. 1958. Effects of flowering on the root system and summer survival of ryegrasses. New Zealand Journal of Agricultural Research. 1: 329–340.

111. SOPER, K., and K. J. Mitchell. 1956. The developmental anatomy of perennial ryegrass (*Lolium perenne* L.). New Zealand Journal of Science and Technology. Sec. A. 37: 484–504.

112. SMITH, D. 1967. Carbohydrates in grasses. II. Sugar and fructosan composition of the stem

bases of bromegrass and timothy at several growth stages and in different plant parts at anthesis. Crop Science. 7: 62–67.

113. ———. 1968. Carbohydrates in grasses. IV. Influence of temperature on the sugar and fructosan composition of timothy plant parts at anthesis. Crop Science. 8: 331–334.

114. ———. 1968. Classification of several native North American grasses as starch or fructosan accumulators in relation to taxonomy. Journal of the British Grassland Society. 23(4): 306–309.

115. ———. 1969. Removing and analyzing total nonstructural carbohydrates from plant tissue. Wisconsin College of Agricultural and Life Sciences Research Report 41. pp. 1–11.

116. SMITH, D., and R. D. Grotelueschen. 1966. Carbohydrates in grasses. I. Sugar and fructosan composition of the stem bases of several northern-adapted grasses at seed maturity. Crop Science. 6: 263–266.

117. SPRAGUE, H. B. 1933. Root development of perennial grasses and its relation to soil conditions. Soil Science. 36: 189–209.

118. SPRAGUE, H. B., and G. W. Burton. 1937. Annual bluegrass (*Poa annua* L.) and its requirements for growth. New Jersey Agricultural Experiment Station Bulletin 630. pp. 1–24.

119. SPRAGUE, V. G., and J. T. SULLIVAN. 1950. Reserve carbohydrates in orchard grass clipped periodically. Plant Physiology. 25: 92–102.

120. STEVENSON, I. M., and W. J. White. 1941. Root fiber production of some perennial grasses. Scientific Agriculture. 22: 108–118.

121. STUCKEY, I. H. 1941. Seasonal growth of grass roots. American Journal of Botany. 28: 486–491.

122. ———. 1942. Some effects of photoperiod on leaf growth. American Journal of Botany. 29: 92–97.

123. ———. 1942. Influence of soil temperature on the development of colonial bentgrass. Plant Physiology. 17: 116–122.

124. SULLIVAN, J. T., and V. G. SPRAGUE. 1943. Composition of the roots and stubble of perennial ryegrass following partial defoliation. Plant Physiology. 18: 656–670.

125. ———. 1949. The effect of temperature on the growth and composition of the stubble and roots of perennial ryegrass. Plant Physiology. 24: 706–719.

126. TAYLOR, T. H., and W. C. Templeton, Jr. 1966. Tiller and leaf behavior of orchardgrass (*Dactylis glomerata* L.) in a broadcast planting. Agronomy Journal. 58: 189–192.

127. TEMPLETON, W. C., G. O. Mott, and R. J. Bula. 1961. Some effects of temperature and light on growth and flowering of tall fescue, *Festuca arundinacea* Schreb. I. Vegetative development. Crop Science. 1: 216–219.

128. THOMPSON, W. R., and C. Y. Ward. 1966. Effects of potassium nutrition and clipping heights on total available carbohydrates in Tifgreen bermudagrass (*Cynodon* spp.). 1966 Agronomy Abstracts. p. 37.

129. TROUGHTON, A. 1955. Cessation of tillering in young grass plants. Nature. 176: 514.

130. ———. 1956. Studies on the growth of young grass plants with special reference to the relationship between the shoot and root systems. Journal of the British Grassland Society. 11: 56–65.

131. UENO, M., and K. Yoshihara. 1967. Spring and summer root growth of some temperate-region grasses and summer root growth of tropical grasses. Journal of the British Grassland Society. 22: 148–152.

132. VAARTNOU, H. 1967. Responses of five genotypes of *Agrostis* L. to variations in environment. Ph.D. Thesis. Oregon State University. pp. 1–149.

133. VIRGIN, H. I. 1962. Light induced unfolding of the grass leaf. Physiologia Plantarum. 15: 380–389.

134. VOSE, P. B. 1962. Nutritional response and shoot/root ratio as factors in the composition and yield of genotypes of perennial ryegrass, *Lolium perenne* L. Annals of Botany, N. S. 26: 425–437.

135. WAITE, R., and J. Boyd. 1953. The water-soluble carbohydrates of grasses. I. Changes occurring during the normal life-cycle. Journal of the Science of Food and Agriculture. 4: 197–204.

136. ———. 1953. The water-soluble carbohydrates of grasses. II. Grasses cut at grazing height several times during the growing season. Journal of the Science of Food and Agriculture. 4: 257–261.

137. WAITE, R., and A. R. N. GORROD. 1959. The structural carbohydrates of grasses. Journal of the Science of Food and Agriculture. 10: 308–317.

138. ———. 1959. The comprehensive analysis of grasses. Journal of the Science of Food and Agriculture. 10: 317–326.

139. WEAVER, J. E., and E. Zink. 1945. Extent and longevity of the seminal roots of certain grasses. Plant Physiology. 20: 359–379.

140. WEBSTER, J. E., G. Shyrock, and P. Cox. 1963. The carbohydrate composition of two species of grama grasses. Oklahoma Agricultural Experiment Station Technical Bulletin T-104. pp. 1–16.

141. WEINMAN, H., and L. Reinhold. 1946. Reserve carbohydrates in South African grasses. Journal of South African Botany. 12: 57–73.

142. WIND, G. P. 1955. Flow of water through plant roots. Netherlands Journal of Agricultural Science. 3: 259–264.

143. YOUNGNER, V. B. 1961. Accelerated wear tests of turfgrass. Agronomy Journal. 53: 217–218.

144. ZANONI, L. J. 1967. Factors which influence the levels of soluble carbohydrate reserves of cool season turfgrasses. M.S. Thesis. University of Massachusetts. pp. 1–99.

145. ZANONI, L. J., L. F. Mickelson, W. G. Colby, and M. Drake. 1969. Factors affecting carbohydrate reserves of cool season turfgrasses. Agronomy Journal. 61: 195–198.

chapter 3

Cool Season Turfgrasses

Introduction

Turfgrasses having a temperature optimum of 60 to 75°F are referred to as **cool season turfgrasses**. These species are widely distributed throughout the cool humid, cool subhumid, and cool semiarid climates and also extend into the transitional zone (96). Over twenty cool season turfgrass species are utilized throughout the world.

Very few of the commonly used cool season turfgrass species originated from the large, natural grasslands of the world. Most were forest-margin species indigenous to Eurasia. The major reason most cool season turfgrasses originated from the fringe forest species of Eurasia is attributed to the type of grazing animal that was the dominant biotic factor in the evolution of those species (97). The *Bovidae* family, which includes domesticated cattle, goats, and sheep, arose in Eurasia during the late Pliocene and Pleistocene epochs. These species of mammals were well adapted to close grazing of grasses. As a result, certain species of *Bovidae* played a very important role in the evolution of the structural features of grasses in Eurasia. The structural adaptations of significance in turfgrass use include (a) the habit of forming short basal internodes; (b) a mode of branching by basal tillers, stolons, and rhizomes; and (c) the capability of developing lateral shoots from basal buds when the main shoot is removed. The growth of grass leaves from basal meristems was also a factor in the adaptation to close grazing. Very few species of *Bovidae* reached North America prior to their domestication. Therefore, most indigenous grasses on the North American continent did not develop a growth habit adapted to continuous, close grazing and are usually not desirable for use under closely mowed turfgrass conditions.

Other factors contributing to the Eurasian concentration of indigenous cool season turfgrasses were the (a) taxonomic affinities at the tribal level, (b) ecological factors including fertile soils and substantial, well-distributed rainfall, and (c) development of Eurasia as one of the early centers of agriculture. The Eurasian region is a major source of improved plant introductions of the cool season turfgrass species.

The plant description, adaptation, use, and culture of the various turfgrasses are discussed in the following sections. Exceptions and wide variations in the plant description, adaptation, and culture of a species can be found among the large pool of germplasm available in the world. Therefore, the comments in this text apply to the species and cultivars that are most widely utilized for turfgrass purposes. In the cultivar sections, comparisons are made among cultivars of that species. When referring to the turfgrass quality, shoot density, growth rate, tolerance, resistance, or susceptibility of a species or cultivar, the following relative scale will be used: superior, excellent, good, intermediate (moderate or medium), fair, poor, and very poor.

The Bluegrasses (*Poa* L.)

The bluegrasses are the most important and widely utilized of the cool season turfgrass species. The *Poa* genus comprises over 200 species that are widely distributed throughout the cool humid and transitional climates. The most distinguishing characteristic used in the vegetative identification of the *Poa* genus is the boat-shaped leaf tip.

The growth habit of the bluegrasses ranges from an erect, bunch type to a prostrate, stoloniferous type to an extremely dense sod former having a vigorous rhizome system. Most of the *Poa* species utilized for turfgrass purposes are perennials that respond to fertility and irrigation. The turf-type *Poa* species are best adapted to moist, fertile, fine textured soils having a pH of 6.0 to 7.0. With the exception of annual bluegrass, the best sustained growth and quality is achieved at a cutting height of 1 to 2 in. The seven species of *Poa* that are utilized to varying degrees in turfgrass culture are *P. pratensis*, *P. trivialis*, *P. compressa*, *P. annua*, *P. glaucantha*, *P. nemoralis*, and *P. bulbosa*.

Kentucky Bluegrass (*Poa pratensis* L.)

Kentucky bluegrass is a native of Eurasia that has been introduced for turfgrass use throughout the cool humid climates of the world (41). In certain regions it is called Junegrass and smoothstalk meadowgrass. Kentucky bluegrass is the most important and widely utilized of the cool season turfgrass species. It forms a turf of high quality under proper culture. Considerable variability exists within the species in terms of color, texture, shoot density, growth habit, adaptation, cultural requirements, and disease resistance.

Plant Description

Vernation folded; *sheaths* slightly compressed, glabrous, somewhat keeled, split with overlapping, hyaline margins; *ligule* membranous, 0.2–1 mm long, truncate; *collar* medium broad, divided, glabrous, yellowish green; *auricles* absent; *blades* V-shaped to flat, 2–4 mm wide, soft, usually glabrous, keeled below, parallel sided, abruptly boat-shaped apex, transparent lines on each side of midrib, finely scabrous margins; *stems* slightly compressed, erect, relatively slender with long, slender, multibranched rhizomes; *inflorescence* an erect, open, pyramidal panicle with slender, spreading lower branches usually occurring in whorls of five.

Kentucky bluegrass forms a medium textured, green to dark green turf of good shoot density (Fig. 3-1). The leaves are smooth, soft, and shiny with a distinct boat-shaped tip. The aggressive sod forming nature of Kentucky bluegrass is attributed primarily to vigorous rhizome development (68). A Kentucky bluegrass plant can produce from 20 to 60 ft of rhizomes from an original shoot in a 5-month period from mid-June to mid-November (150). The rhizomes are capable of initiating shoots and roots from each node. The extensive root system is concentrated primarily in the upper 6 to 10 in. of the soil profile. Some roots may penetrate to depths of 16 to 24 in. under mowed conditions (187). The root system persists as a perennial (215).

Kentucky bluegrass is propagated primarily by seed, although it may be established vegetatively from rhizomes. It is classed as a facultative apomictic and is highly polyploid, with chromosome numbers ranging from 28 to 150 having been reported (1, 18, 30, 36, 44, 92, 137, 174, 175, 182). Chromosome numbers below 40 and above 90 are infrequent. The establishment rate is slower than for the ryegrasses and fescues (56, 188). The recuperative potential is quite good.

Figure 3-1. Closeup of a Kenblue Kentucky bluegrass turf cut at 1.5 inches.

Adaptation. Kentucky bluegrass is a long-lived perennial that is widely adapted throughout the cool humid and transitional climates of the world (202). It can also be effectively utilized in the cool semiarid and arid regions if irrigated. There is a substantial reduction in shoot growth during extended periods of water and temperature stress. Summer dormancy may eventually occur with the aboveground foliage becoming brown and inactive (34). Kentucky bluegrass is capable of surviving extended drought periods and can initiate new shoot growth from the nodes of underground rhizomes and older crowns when moisture conditions become favorable. Growth can be initiated and an adequate turfgrass cover formed within 2 to 3 weeks after termination of a drought.

The low temperature hardiness, fall color retention, and spring green-up rate are quite good (213). Full sunlight or only slight shading are preferred. Most Kentucky bluegrass cultivars do not persist under shaded conditions in the cool, humid climates, partly due to powdery mildew thinning. In contrast, Kentucky bluegrass persists under partial shade in the warm humid regions. The wear tolerance is medium to good (171).

Even though Kentucky bluegrass is widely distributed throughout the cool humid regions, the range in soil adaptation is limited. Moist, well-drained, fertile, medium textured soils with a pH of 6 to 7 are preferred (70, 145, 177, 210, 213, 217). It does not tolerate extremes in acidity or alkalinity and forms a poor sod on acidic, infertile soils (84, 98). It tolerates wet, moderately waterlogged soil conditions and high phosphorus levels (128).

Use. Kentucky bluegrass is widely utilized for general-purpose turfs in the cool, humid and transitional climates. It is commonly used on lawns, parks, cemeteries, institutional grounds, fairways, tees and roughs, roadsides, airfields, athletic fields, and other comparable general-purpose lawn areas. Good recuperative potential plus vigorous rhizome development result in Kentucky bluegrass being well adapted for use on athletic fields and other heavily utilized playfields. The vigorous, dense rhizome and root system is a major factor in its wide use for commercial sod production in the cool humid climates.

Kentucky bluegrass is frequently used in mixtures with red fescue (67, 124). The establishment rate of Kentucky bluegrass is slower than for red fescue. Red fescue functions as a quicker establishing companion species that does not compete excessively with Kentucky bluegrass during establishment. Subsequently, the red fescue will become dominant in the shaded, droughty environments, particularly if maintained at a low intensity of culture; whereas, Kentucky bluegrass will be dominant under full sunlight, a high intensity of culture, and moist soil conditions (51, 52, 129, 130, 200). Because the recuperative potential is quite good, Kentucky bluegrass is well adapted for use on tees, fairways, and athletic fields that are subjected to frequent damage from the action of cleats or divots. The underground rhizome system is capable of reinitiating shoot growth quite rapidly.

Culture. Kentucky bluegrass requires a medium to medium high intensity of culture. An excellent quality turf is formed when properly maintained. A cutting height of 1 to 2 in. is preferred (20, 49, 52, 115, 217, 236). More diminutive, low growing Kentucky bluegrass cultivars are being developed that tolerate lower cutting heights. Most cultivars are not capable of maintaining a permanent, high quality turf if cut at heights lower than 0.7 in. The nitrogen fertility requirement varies with the specific cultivar. The range is from 0.4 to 0.7 lb of actual nitrogen per 1000 sq ft per growing month for Park, Delta, and Kenblue to as high as 1.0 to 1.3 lb per 1000 sq ft per growing month for Merion (60, 126).

Kentucky bluegrass responds to irrigation during periods of moisture stress.

Table 3-1

CHARACTERISTICS OF THIRTY KENTUCKY BLUEGRASS CULTIVARS

Cultivar	Released By	Date	Turfgrass characteristics	Adaptation	Other comments
Adorno	Gebr. van Engelen, Netherlands	1954	Medium dark green color; fine texture; medium shoot density; semierect growth habit	Very poor low temperature hardiness; fairly well adapted to droughty soils	Moderate resistance to rust; very susceptible to *Helminthosporium* spp.
Arboretum (75, 76, 80)	Missouri Botanical Garden, St. Louis, Mo., U.S.	early 1950's	Medium green color; medium texture; low shoot density; erect growth habit	Fairly good heat and drought tolerance; good spring green-up rate; poor tolerance to close mowing	Moderately susceptible to stripe smut; very susceptible to *Helminthosporium* spp.
Arista (ZWB) (82)	Gebr. van Engelen, Netherlands	1965	Medium dark green color; medium texture; medium shoot density; low growth habit	Medium establishment rate	Moderately susceptible to *Helminthosporium* spp. and stripe smut
A-10	Warren's Turf Nursery, Ill., U.S.	1967	Dark green color; medium fine texture; medium shoot density; moderately low growth habit	Medium good tolerance to low temperature and heat stress; established vegetatively	Good resistance to stem rust, stripe smut, and *Helminthosporium* spp.; moderate susceptibility to powdery mildew
A-20	Warren's Turf Nursery, Ill., U.S.	1969	Dark green color; medium coarse texture; medium high shoot density; low growth habit	Good low temperature color retention and spring green-up rate; good tolerance to close mowing; established vegetatively	Excellent resistance to *Helminthosporium* spp.; good resistance to powdery mildew, stripe smut, stem rust, and *Fusarium* blight
A-34	Warren's Turf Nursery, Ill., U.S.	1968	Light green color; medium texture; high shoot density	Medium good shade tolerance	Good resistance to stem rust and powdery mildew; moderate resistance to *Helminthosporium* spp. and stripe smut

Table 3-1 (Cont.)

					Good resistance to *Helminthosporium* spp.
Baron	Barenbrug Zaadhandel N.V., Netherlands	1970	Dark green color; medium coarse texture; high shoot density; low growth habit; medium slow vertical shoot growth rate	Good low temperature color retention and spring green-up rate; rapid establishment rate	Good resistance to *Helminthosporium* spp.
Birka	Weibullsholm Plant Breeding Institution, Sweden	1968	Dark green color; medium fine texture; high shoot density	Excellent low temperature color retention	Good resistance to *Helminthosporium* spp. and powdery mildew
Campus (78, 80, 82, 114)	Gebr. van Engelen, Netherlands	1965	Medium dark green color; medium coarse texture; medium shoot density; moderately low growth habit	Poor low temperature hardiness and heat tolerance; good low temperature color retention and spring green-up rate; poor recuperative potential	Moderately susceptible to *Helminthosporium* spp., leaf rust, and stripe smut
Celle	Weibullsholm Plant Breeding Institution, Sweden	1969	Medium green color; medium coarse texture; very high shoot density; moderately low growth habit	Good low temperature hardiness and establishment rate	Good resistance to *Helminthosporium* spp. and powdery mildew
Cougar (75, 80, 147, 190, 191)	Washington AES, U.S.	1964	Blue-green color; medium coarse texture; medium shoot density; low growth habit	Excellent drought tolerance; poor low temperature color retention; good establishment rate; medium tolerance to close mowing	Very susceptible to *Helminthosporium* spp.; moderately susceptible to stripe smut and powdery mildew; moderately tolerant of leaf rust
Delft (82)	Cebeco, Netherlands	1958	Blue-green color; medium texture; medium shoot density; moderately low growing; moderate vertical growth rate	Medium establishment rate	Good resistance to stripe smut; moderately susceptible to *Helminthosporium* spp.; very susceptible to rust

Table 3-1 (Cont.)

Cultivar	Released By	Date	Turfgrass characteristics	Adaptation	Other comments
Delta (32, 51, 75, 76, 78, 80, 82, 87, 110, 147, 190)	Central Experimental Farm, Ottawa, Ont., Can.	1938	Medium green color; medium fine texture; medium low shoot density; erect growth habit; rapid vertical growth rate; poor sod strength	Medium drought and low temperature tolerance; excellent rate of establishment and spring green-up; poor tolerance to close mowing	Good resistance to stem rust, leaf rust, and powdery mildew; moderately resistant to stripe smut; very susceptible to *Helminthosporium* spp.
Fylking (78, 82, 90, 111, 147)	Swedish Seed Association, Sweden	1936	Dark green color; medium fine texture; high shoot density; low growth habit; fairly slow vertical growth rate; prone to thatching; good sod strength	Good low temperature hardiness and drought tolerance; good low temperature color retention; good establishment rate; good tolerance to close mowing	Good resistance to *Helminthosporium* spp. and stripe smut; moderate resistance to stem rust; moderately susceptible to powdery mildew and *Fusarium* blight
Geary (75, 76, 80)	Ed Geary, Ore., U.S.	1929	Medium green color; medium texture; medium low shoot density; erect growth habit; rapid vertical growth rate and rhizome development	Good establishment rate	Very susceptible to *Helminthosporium* spp.; moderately susceptible to powdery mildew
Golf (115)	Hammenhog Seed Co., Sweden	1958	Medium green color; medium coarse texture; medium shoot density; medium low growth habit	Poor establishment rate; intermediate recuperative potential	Moderately susceptible to *Helminthosporium* spp.
Kenblue (32)	Kentucky AES, U.S.	1967	Medium green color; medium texture; low shoot density; erect growth habit; rapid vertical growth rate	Medium poor low temperature hardiness; good low temperature color retention and spring green-up rate; good establishment rate	Very resistant to stem rust; moderately susceptible to powdery mildew; very susceptible to *Helminthosporium* spp.; requires a medium low intensity of culture

Table 3-1 (Cont.)

Cultivar	Origin	Year	Morphological characteristics	Adaptation	Disease resistance
Merion (B-27) (9, 20, 31, 51, 52, 64, 75, 76, 77, 78, 80, 82, 87, 123, 124, 126, 147, 151, 172, 178, 183, 190, 218, 228, 232)	USGA Green Section and CRD-ARS, U.S.	1947	Dark green color; medium coarse texture; high shoot density; low growth habit; moderately slow vertical growth rate; vigorous tillering; good sod strength; prone to thatching	Good low temperature hardiness and drought tolerance; poor spring green-up rate and low temperature color retention, turns purplish; medium tolerance to close mowing; slow establishment rate; good recuperative potential	Excellent *Helminthosporium* spp. resistance; very susceptible to dollar spot, stem rust, leaf rust, powdery mildew, stripe smut, and *Fusarium* blight; requires a high level of nitrogen nutrition
Newport (32, 75, 78, 80, 82, 87, 126, 158, 190, 238)	Washington and Oregon AES and SCS, U.S.	1958	Dark green color; medium coarse texture; medium high shoot density; moderately low growth habit; moderately rapid vertical growth rate; poor sod strength	Excellent low temperature color retention; poor drought and low temperature tolerance; medium poor establishment rate and recuperative potential	Immune to stem rust; moderately good resistance to powdery mildew; moderately susceptible to *Helminthosporium* spp., dollar spot, leaf rust, and stripe smut; vigorous seed head formation causes stemminess
NuDwarf (75, 78, 82)	R. H. Rasmussen, Neb., U.S.	1963	Medium green color; medium fine texture; medium high shoot density; medium low growth habit; vigorous	Good drought tolerance; rapid establishment rate	Very susceptible to *Helminthosporium* spp.
Nugget	Alaska AES and CRD-ARS, U.S.	1965	Dark green color; medium texture; high shoot density; leafy, low growth habit; excellent sod strength	Excellent low temperature hardiness; poor low temperature color retention and spring green-up rate	Good resistance to *Helminthosporium* spp. and powdery mildew
Park (50, 51, 75, 76, 78, 80, 158)	Minnesota AES, U.S.	1957	Medium dark green color; medium leaf texture; medium low shoot density; erect growth habit; rapid vertical growth rate	Medium low temperature hardiness; medium good drought tolerance; good spring green-up rate; excellent establishment rate	Good resistance to dollar spot, stripe smut, stem rust, and leaf rust; very susceptible to *Helminthosporium* spp.; intermediate susceptibility to powdery mildew

Table 3-1 (Cont.)

Cultivar	Released By	Date	Turfgrass characteristics	Adaptation	Other comments
Pennstar (K5-47) (48, 75, 80, 87, 208)	Pennsylvania AES, U.S.	1967	Dark green color; medium texture; high shoot density; low growth habit	Good low temperature hardiness and color retention; rapid spring green-up rate; good recuperative potential and establishment rate	Good resistance to *Helminthosporium* spp. and stripe smut; medium resistance to stem rust; moderately susceptible to dollar spot and powdery mildew
Prato (32, 82, 114)	D. J. van der Have, Netherlands	1959	Medium green color; medium fine texture; very high shoot density; medium low growth habit; moderate vertical growth rate	Medium establishment rate	Moderate resistance to powdery mildew; moderately susceptible to *Helminthosporium* spp.; very susceptible to stripe smut and stem rust
Primo (75, 78, 82, 142)	Weibullsholm Plant Breeding Institution, Sweden	1949	Light green color; medium fine texture; medium shoot density; semierect growth habit; moderately slow vertical growth rate	Rapid establishment rate and recuperative potential	Moderately resistant to powdery mildew; very susceptible to *Helminthosporium* spp. and stem rust
Sodco	Indiana AES, U.S.	1967	Dark green color; medium coarse texture; medium shoot density; very low growth habit; very slow vertical shoot growth rate; good sod strength	Medium shade tolerance; slow establishment rate	Moderately good resistance to stem rust and *Helminthosporium* spp.
Sydsport	Weibullsholm Plant Breeding Institution, Sweden	1967	Dark green color; medium texture; high shoot density; low growth habit; intermediate vertical growth rate	Medium spring green-up rate; good recuperative potential and establishment rate	Good resistance to powdery mildew and *Helminthosporium* spp.

Table 3-1 (CONT.)

S-21 (75, 76, 80)	Jacklin Seed Co., Washington, U.S.	1953	Dark green color; coarse texture; low shoot density; erect growth habit; rapid vertical growth rate	Rapid establishment rate with vigorous rhizome development	Very susceptible to *Helminthosporium* spp.; moderate resistance to stem rust
Troy (75, 82, 110, 125)	Montana AES and CRD-ARS, U.S.	1955	Medium green color; coarse texture; low shoot density; erect growth habit; rapid vertical growth rate	Poor tolerance to heat stress; best adapted to cool climates	Very susceptible to powdery mildew, *Helminthosporium* spp., and stripe smut
Windsor (75, 78, 172, 183)	O. M. Scott Co., Ohio, U.S.	1962	Medium dark green color; medium fine texture; medium high shoot density; moderately low growth habit; medium vertical growth rate	Medium good low temperature hardiness; responds to high levels of nitrogen nutrition; good spring green-up rate	Moderate susceptibly to *Helminthosporium* spp., rust, and powdery mildew; susceptible to stripe smut

Thatch can become a serious problem, particularly with the vigorous growing cultivars maintained at a high intensity of culture. Kentucky bluegrass is fairly tolerant of the commonly used herbicides (1, 58, 86). The most serious disease over the years has been caused by *Helminthosporium* (93). Other diseases that may cause serious injury include rust, stripe smut, *Fusarium* blight, powdery mildew, *Fusarium* patch, *Pythium* blight, dollar spot, brown patch, *Ophiobolus* patch, and *Typhula* blight.

Cultivars. A number of cultivars of Kentucky bluegrass have been selected and developed for turfgrass use (Table 3-1). Merion, Newport, Park, Prato, and Windsor have been the most widely used cultivars. Some of the newer cultivars receiving attention are A-20, Fylking, Nugget, Pennstar, Sodco, and Sydsport. Due to the apomictic character of Kentucky bluegrass, most of the cultivars have originated through plant selection rather than from controlled hybridization and selection. Considerable variability in leaf texture, color, shoot density, growth habit, rhizome development, and disease resistance exists among the available cultivars (183). Characteristics desired in future cultivar development include improved disease resistance, shade adaptation, and tolerance to close mowing.

Rough Bluegrass (*Poa trivialis* L.)

Rough bluegrass is a native of northern Europe. It is sometimes referred to as roughstalk bluegrass, roughstalk meadowgrass, and rough meadowgrass as well as by the scientific name. The soil and temperature adaptation of rough bluegrass is somewhat similar to Kentucky bluegrass but the color, growth habit, and cultural requirements are quite different.

Plant Description

> *Vernation* folded; *sheaths* compressed, retrorsely scabrous, keeled below, frequently purplish, split part way; *ligule* membranous, 4–6 mm long, acute, entire, or sometimes ciliate; *collar* conspicuous, broad, glabrous, divided; *auricles* absent; *blades* flat or V-shaped, 1–4 mm wide, yellowish-green, soft, glossy and keeled below, strongly scabrous, tapering from base to narrowly boat-shaped apex, transparent lines on each side of midrib, margins scabrous; *stems* compressed, erect to somewhat decumbent at base with creeping, thin, leafy stolons rooting at the nodes; *inflorescence* an erect, oblong, open panicle with slender, spreading branches occurring in whorls of about five.

Rough bluegrass forms a fine textured, yellowish-green turf of high shoot density. The plants are more diminutive with considerably smaller stems than for Kentucky bluegrass (Fig. 3-2). The growth habit is prostrate with slender, creeping stolons. Rough bluegrass does not form a tight, vigorous sod due to a lack of rhizomes. The root system is quite fibrous, relatively shallow, and annual in nature (215, 226). It is readily distinguished from Kentucky bluegrass by the shiny, greenish-yellow leaves and the aboveground, creeping stolons. Propagation is primarily

Figure 3-2. A comparison of Kentucky bluegrass (left) and rough bluegrass (right) turfs cut at 1.5 inches.

from seed. It is principally cross-fertilized, with chromosome numbers of 14 and 15 having been reported (1, 74, 92).

Adaptation. Rough bluegrass is a long-lived perennial, adapted throughout the cool, humid climates of the world. It grows best in cool, moist, shaded environments (213). The low temperature hardiness and color retention are excellent. However, the tolerance to drought and heat stress is poor (42). A rough bluegrass turf can be permanently thinned or lost during extended periods of water stress, especially on sandy soils. Rough bluegrass will not persist in turfs subjected to continuous traffic because of very poor wear tolerance. It is well adapted to wet, imperfectly drained, fine textured soils. Best turfgrass formation occurs on fertile soils having a pH of 6 to 7.

Use. Rough bluegrass can form a very dense, uniform turf of fairly high quality under certain limited environmental conditions and cultural levels. Its use is usually restricted to cool, moist, shaded environments that are not subjected to concentrated traffic. Rough bluegrass is quite aggressive and tends to dominate polystands under these conditions.

Rough bluegrass has been used in polystands with red fescue and Kentucky bluegrass. The greenish-yellow color is objectionable when combined with the two darker green species. Rough bluegrass can become a weedy species, particularly on wet soils that are not subjected to water stress. The stoloniferous growth results in the formation of distinct, light green patches that invade Kentucky bluegrass turfs similar to creeping bentgrass. Because of these problems the use of rough bluegrass in turfgrass seed mixtures has been declining, even for moist, shaded sites. Rough bluegrass has been successfully utilized for winter overseeding of dormant warm season turfs in the warm humid regions.

Culture. Rough bluegrass tolerates cutting heights as low as 0.5 to 1.0 in. The nitrogen requirement is 0.5 to 1.0 lb per 1000 sq ft per growing month. Rough bluegrass responds more to irrigation than any other cultural practice. It will persist, become aggressive, and even invade other turfgrass species under wet soil conditions, particularly if mowed regularly at less than 1 in. The thatching tendency is medium to low. The most common diseases associated with rough bluegrass are caused by the *Helminthosporium* species. Other diseases that cause occassional problems include rust, stripe smut, brown patch, *Ophiobolus* patch, *Fusarium* patch, and *Typhula* blight. It is more prone to 2,4-D injury than Kentucky bluegrass. The use of rough bluegrass for turfgrass purposes is quite limited and is one of the reasons no improved turf-type cultivars have been developed.

Canada Bluegrass (*Poa compressa* L.)

Canada bluegrass is a native of western Eurasia that is widely distributed throughout the cooler portions of the cool humid climates.

Plant Description

Vernation folded; *sheaths* strongly compressed, strongly keeled, glabrous, split with overlapping, hyaline margins; *ligule* membranous, 0.5–1.5 mm, truncate, entire; *collar* narrow to medium broad, divided, glabrous; *auricles* absent; *blades* V-shaped to flat, 1–3 mm wide, glabrous, bluish to grayish-green, glaucous, keeled below, tapering to a boat-shaped apex, transparent lines on each side of midrib, margins slightly scabrous and hyaline; *stems* strongly compressed, slender, stiff, solitary, erect, or ascending from a decumbent base with extensively creeping, wiry rhizomes that are sparsely branched; *inflorescence* a narrow, oblong to ovate, stiff, contracted panicle with short, appressed branches.

Canada bluegrass forms a rather open, low density turf of inferior quality (213). The leaves are medium in texture with a distinctive bluish-green color. The basal stem internodes are elongated as much as 0.8 in. As a result, the stem apex is elevated well above the soil and a major portion of the leaves are removed by mowing (70). This causes an objectionable turfgrass appearance because of the stiff, stemmy nature. Canada bluegrass has numerous short, creeping rhizomes but does not form the tight, vigorous sod typical of Kentucky bluegrass. The root system is fairly deep and persists as a perennial (215). Canada bluegrass can be distinguished from Kentucky bluegrass by the compressed stems, longer ligule, slate-blue color, lower shoot density, shorter leaf blades, and retention of seed stalks. Propagation is primarily by seed. It is largely apomictic with chromosome numbers of 35, 42, 49, and 56 having been reported (1, 92).

Adaptation. Canada bluegrass is a long-lived perennial adapted primarily to the cooler portions of the cool humid climates. The drought and shade tolerance is better than for most Kentucky bluegrass cultivars (223). The wear tolerance is fairly good (171). It is well adapted to infertile, droughty soils where Kentucky bluegrass does not persist (70). It will tolerate a variety of soil types ranging from fine textured, imperfectly drained clays to well drained, gravelly soils. Canada bluegrass is more tolerant of acidic soil conditions than Kentucky bluegrass, with acceptable growth occurring at pH's of 5.5 to 6.5.

Use and culture. Canada bluegrass does not form a turf of particularly good shoot density or quality. Thus, its use is restricted to low quality, minimum maintenance situations. It is usually seeded in mixtures with sheep fescue. Best growth occurs at cutting heights of 3 to 4 in. at a relatively low mowing frequency typical of roadsides and other functional, nonuse turfgrass areas. It does not maintain an acceptable shoot density at lower cutting heights because the growing point is elevated well above the soil surface (70, 98). Although tolerant of infertile soils, it will respond to fertilization. The nitrogen requirement ranges from 0.2 to 0.6 lb per 1000 sq ft per growth month. Canada bluegrass usually becomes dominant on acidic, droughty, infertile soils when planted in mixtures with Kentucky bluegrass; whereas Kentucky bluegrass dominates on fertile, moist soils having a pH above 6 (99, 223). Diseases that can be a problem include the *Helminthosporium* spp., rust, stripe smut, brown patch, *Ophiobolus* patch, and *Pythium* blight.

Due to the extremely limited use, only a few Canada bluegrass cultivars have been developed. **Canon** was developed by the Ontario Agricultural College of Canada and released in 1965. It is leafier and more low temperature hardy than the common commercial strains.

Annual Bluegrass (*Poa annua* L.)

Annual bluegrass is a native of Europe that is widely distributed throughout the civilized regions of the world. It is sometimes referred to as annual meadowgrass, wintergrass, or by its scientific name. Annual bluegrass is generally classified as a turfgrass weed and is rarely included in seed mixtures for turfgrass use. Nevertheless, it will frequently invade, persist, and become a major component of irrigated, close cut, intensely fertilized turfs. Annual bluegrass becomes the dominant species under these conditions and the cultural program is frequently adjusted to meet the specific requirements of this species. For this reason annual bluegrass is included here as well as in the chapter on pests.

Plant Description

Vernation folded; *sheaths* distinctly compressed, glabrous, whitish at base, keeled, split with overlapping, hyaline margins; *ligule* membranous, 1–3 mm long, thin, white, acute, entire; *collar* conspicuous, medium broad, glabrous, divided by the midrib; *auricles* absent; *blades* V-shaped, 2–3 mm wide, usually light green, glabrous, soft, boat-shaped apex, transparent lines on each side of midrib, parallel sided, flexuous transversely, margins slightly scabrous and hyaline; *stems* flat, erect to spreading sometimes rooting at the nodes and forming stolons; *inflorescence* a small, pyramidal, open panicle, branches few and spreading.

Annual bluegrass forms a very fine textured turf of high shoot density, uniformity, and quality under the proper cultural, environmental, and soil conditions. The leaves are generally shorter, broader, softer, and lighter green in color than those of Kentucky bluegrass. The color is usually light green to greenish-yellow. It is a low growing, diminutive type plant that is well adapted to close mowing. The rooting depth is generally comparable to that of Kentucky bluegrass and colonial bentgrass (212). However, it is capable of surviving and growing with a very shallow root system.

Historically, annual bluegrass has been classified as a tufted, bunch type annual, but many plants persist under close cut, irrigated situations as perennial, prostrate, creeping types that root at the nodes. This variability in annual bluegrass has been attributed to its origination as a hybrid between *P. infirma* H. B. K., an annual, and *P. supina* Schrad., a perennial (109). The annual types have an upright bunch-type growth habit, produce few adventitious roots, tillers, and shoots, and are quite prolific seed producers with the seed possessing a dormancy factor (109) (Fig. 3-3). *P. annua* var *annua* (L.) Timm is the subspecies classification given to this plant type. In contrast, the perennial types have a prostrate stoloniferous growth habit, produce numerous shoots, tillers and adventitious roots, and are more restricted in seeding potential although those seeds produced have no dormancy factor (17, 109, 235). The perennial type is classified as *P. annua* var *reptans* (Hausskn.) Timm.

Propagation and dissemination are primarily by seed. It is cross-pollinated and has a somatic chromosome number of 28 (24, 135, 136, 154). Annual bluegrass is a prolific seed producer even at a cutting height of 0.25 in. (235). One plant produced 360 seeds between May and August in western British Columbia (198). It also has the unique capability to ripen viable seeds on panicles excised from the plant only 1 or 2 days after pollination (136). The seed heads can be very objectionable during the peak flowering period and can drastically reduce the

Figure 3-3. The upright, non-creeping growth habit of an annual type of annual bluegrass (left) compared to the prostrate, stoloniferous growth habit of a perennial type (right).

turfgrass quality of annual bluegrass greens (Fig. 3-4). The vegetative recuperative potential is poor but it can reestablish readily from the many seeds that are widely distributed in soils where it has been grown.

Adaptation. The heat, drought, and low temperature tolerances are quite poor, particularly for *P. annua* var *annua* (112, 212). *P. annua* var *reptans* is capable of persisting as a perennial under moderate environmental conditions where the plant is not subjected to severe cold, heat, or drought stress. Turfs containing major portions of annual bluegrass are

Figure 3-4. Seed head formation of annual bluegrass on a green mowed daily at 0.25 inch.

frequently subject to severe injury and thinning during periods of environmental stress. This is the major reason that annual bluegrass is considered a weed in turfs. In the warm humid regions, it behaves more as a winter annual (166, 212), whereas in the cooler portions of the cool humid climates it behaves as a summer annual. It is also capable of escaping extended periods of drought by behaving as either a summer or winter annual depending on when the drought stress occurs. Annual bluegrass is well adapted to moist, shaded environments. Sensitivity to smog injury ranks highest of the cool season turfgrasses (33, 121).

Annual bluegrass grows best on moist, fine textured, fertile soils having a pH of 5.5 to 6.5 and a high phosphorus content (127, 212, 235). It persists on coarse textured, droughty soils if irrigated frequently. The tolerance to compacted soil conditions is quite good (134). Annual bluegrass does not tolerate extended periods of submersion or waterlogged soil conditions, particularly if this occurs in conjunction with high temperatures. The tolerance to high salt levels is also poor.

Use. Although not seeded specifically for turfgrass use, annual bluegrass invades close cut, irrigated, intensively cultured turfs within a period of 3 to 5 years and may become a dominant component. Greens, fairways, tees, athletic fields, and similar intensively cultured turfs are most readily invaded. Annual bluegrass is less likely to invade turfs maintained at a cutting height of 1.5 in. or higher and irrigated on a relatively infrequent basis.

Culture. Annual bluegrass responds to a high intensity of culture. It is most aggressive and competitive at a cutting height of 1.0 in. or less (115, 235). Annual bluegrass can form a high quality turf at cutting heights as low as 0.2 in. and is easy to mow. The nitrogen requirement ranges from 0.5 to 1.0 lb per 1000 sq ft per growing month. This species thrives under moist soil conditions and when frequently irrigated. It tends to thatch, especially under higher cutting heights, high nitrogen fertility, and irrigated conditions. Annual bluegrass is subject to serious injury by dollar spot, brown patch, *Pythium* blight, *Fusarium* blight, *Fusarium* patch, *Typhula* blight, *Helminthosporium* spp., red thread, *Ophiobolus* patch, and stripe smut (235). In certain areas the turfgrass weevil has caused serious damage.

A wide range of variability in growth habit, color, leaf texture, and tolerance to environmental stress exists within annual bluegrass. No improved cultivars have been developed even though it is present as a major component of many close cut, irrigated, intensively cultured turfs.

Upland Bluegrass (*Poa glaucantha* Gaudin)

Upland bluegrass is a noncreeping, bunch-type, cool season turfgrass. The plants are mostly glaucous with the stems distinctly compressed. It is adapted primarily to the cooler portions of the cool humid climates and especially to moist soil conditions. Upland bluegrass has shown promise for erosion control and for general-purpose, low quality turfs in the Pacific northwest regions of North America.

Draylar is a cultivar of upland bluegrass that was released cooperatively by the Washington AES and the Soil Conservation Service in 1951. The original selection was made from plant material collected in Turkey in 1935 and was subsequently tested under the number P-410. It is highly apomictic and has a 2n chromosome number of 50. Draylar has fine, wiry stems that are decumbent at the base. It resembles Canada bluegrass in growth habit but has a darker green color. It is well adapted to infertile soils and is sometimes used as a substitute for Canada bluegrass in low quality, minimum use turfs.

Wood Bluegrass (*Poa nemoralis* L.)

Wood bluegrass is a native of Europe that has also been called wood meadow-grass. The leaves are dark green, narrow, and limp. Wood bluegrass has a prostrate growth habit, but is basically a bunch type since only occasional short, creeping rhizomes are produced. This slow growing species is not aggressive (213). It is a long-lived perennial adapted primarily to the cooler portions of the cool humid regions. The shade tolerance is excellent. Chromosome numbers of 28 and 42 have been reported (1). It is susceptible to rust and brown patch. It is utilized to a limited extent in Europe in turfgrass mixtures for shaded environments. In North America, rough bluegrass is superior to wood bluegrass for shaded lawn turfs (98, 213, 226). No improved cultivars are available.

Bulbous Bluegrass (*Poa bulbosa* L.)

Bulbous bluegrass is native to the southern portions of Eurasia (88). It is a perennial adapted to the cool humid, transitional, and warm humid regions of the world (204). Bulbous bluegrass forms a rather open, bunch-type, fine textured turf having a bright green color (89, 222). Shoot growth seldom exceeds 3 in. in height (227).

Bulbous bluegrass is best adapted to climates having a dry summer and a mild winter that is devoid of low temperature extremes and has a uniform rainfall distribution. Under these conditions, most of the growth occurs during the fall, winter, and early spring periods (222). It enters a summer dormancy period when exposed to high temperatures, even if the soil moisture is adequate (146). The shade tolerance is fairly good but the best growth occurs in full sunlight. Bulbous bluegrass is adapted to a wide range of soil types including acidic, wet, droughty, and infertile conditions.

The tolerance to cutting heights of less than 1 in. is not good due to the erect, elongated stems (166). Propagation is primarily by the bulblets produced above-ground in the panicle or by true bulbs formed underground at the base of the stem (131, 227). Chromosome numbers of 28 and 45 have been reported (1). The use of bulbous bluegrass for turfgrass purposes has been quite limited. At one time it was suggested as a winter companion species in high cut, dormant warm season turfs such as bermudagrass and zoysiagrass (131, 222, 227). It has not been utilized for this purpose since other species such as red fescue, bentgrass, and ryegrass are more effective and desirable.

The Bentgrasses and Redtop (*Agrostis* L.)

The genus *Agrostis* comprises over 100 species including the bentgrasses and redtop (104). The common name bentgrass is applied to all turfgrass species within the genus with the exception of redtop. The growth habit varies from a bunch type with limited stoloniferous growth to an extensive stolon system (192). Due to the prostrate growth habit, bentgrass is the most tolerant cool season turfgrass to continuous, close mowing at heights as low as 0.2 in. Bentgrass forms an extremely fine textured, dense, uniform, high quality turf when closely mowed. Some species are annuals but most are perennials, including all those utilized for turfgrass purposes. The *Agrostis* species are primarily adapted to the cool, humid and transitional climates. Most perennials have excellent low temperature hardiness. The spring green-up rate is slower than for Kentucky bluegrass (209). Best growth occurs on moist, fertile soils having a pH of 5.5 to 6.5.

The three turf-type bentgrass species commonly used are (a) *A. palustris*-creeping bentgrass, (b) *A. tenuis*-colonial bentgrass, and (c) *A. canina*-velvet bentgrass. Velvet bentgrass is not used as extensively as the other two. A wide range of divergent types and intermediate forms exists among creeping, colonial, and velvet bentgrass with the chromosome number ranging from 14 to 42 (216, 221).

Creeping Bentgrass (*Agrostis palustris* Huds.)

Creeping bentgrass is native to Eurasia but has been distributed throughout the world for use in close cut, fine textured turfs. The common name creeping bentgrass is derived from the vigorous creeping stolons that develop at the surface of the ground and initiate new roots and shoots from the nodes. Considerable variation exists in the literature concerning creeping bentgrass, with some authors referring to it as *Agrostis stolonifera* var. *palustris* (Farwell) (18, 28, 192, 194).

Plant Description

Vernation rolled; *sheaths* round, glabrous, split with overlapping, hyaline margins; *ligule* membranous, 1–2 mm long, acute to oblong, may be notched; *collar* narrow to medium broad; *auricles* absent; *blades* flat, 2–3 mm wide, glabrous to minutely scabrous above, below, and on margins, prominent veins above, acuminate apex; *stems* erect or ascending from a spreading decumbent base with long, slender stolons rooting at the nodes; *inflorescence* a narrow, dense, contracted pancile that is a pale purple color, branches clustered.

When mowed closely, creeping bentgrass forms a fine textured turf with superior shoot density, uniformity, and turfgrass quality. The finer textured cultivars tend to be compact and low growing, while the coarser textured are usually more upright and open (108). The turfgrass color varies among the cultivars from greenish-yellow to dark green to blue-green. Creeping bentgrass is one of the most vigorous cool season turfgrasses in terms of stolon growth (Fig. 3-5). It is capable of spreading quite rapidly by means of long, leafy, creeping stolons that initiate roots and shoots from the nodes. The root system is dense, fibrous, medium to shallow in depth, and annual in nature. The creeping bentgrasses can be subdivided into two types based on the propagation method. One includes those cultivars propagated primarily by stolons; the other, those cultivars propagated primarily by seed. Creeping bentgrass is cross-pollinated with chromosome numbers of 28 and 56 having been reported (44, 92). The establishment rate is medium slow but the recuperative potential is fairly good (188).

Adaptation. Creeping bentgrass is a long-lived perennial that is utilized in most of

Figure 3-5. The typical stoloniferous growth habit of Toronto creeping bentgrass.

the cool humid climatic regions of the world. It is being introduced into the transitional zone and the cooler portions of the warm humid regions for use on greens. Creeping bentgrass is one of the most hardy cool season turfgrasses to temperature extremes (23). The spring green-up rate is slow, while fall low temperature discoloration occurs earlier than for Kentucky bluegrass and red fescue (158). Creeping bentgrass usually persists during midsummer heat stress but shoot growth is seriously impaired and death of the root system may occur (45, 209). Proper drainage, irrigation, and disease control are particularly important in maintaining bentgrass at high soil temperatures. Creeping bentgrass tolerates partial shading but grows best in full sunlight. The wear tolerance is poor.

Creeping bentgrass tolerates a wide range of soil types but is best adapted to fertile, fine textured soils of moderate acidity and good water holding capacity (177, 213). A soil pH of 5.5 to 6.5 is preferred (27, 145). Creeping bentgrass is superior to most cool season turfgrasses in tolerating saline soil conditions and flooding. However, the tolerance to compacted soils is quite poor.

Use. Creeping bentgrass produces one of the most beautiful, fine textured turfs known when mowed closely. It is the outstanding cool season species available for use on putting and bowling greens maintained at a cutting height of 0.2 to 0.3 in. (Fig. 3-6). It is also utilized on high quality, intensively cultured fairways, tees, lawns, and similar ornamental turfs. It is seldom used in polystands with erect growing, cool season turfgrasses such as Kentucky bluegrass due to an aggressive, patchy, creeping habit of growth (51, 177). A certain amount of creeping bentgrass is utilized for winter overseeding of dormant warm season species such as bermudagrass. When used for this purpose, it is usually seeded as one component of a mixture containing several other rapid establishing cool season turfgrasses.

Culture. Creeping bentgrass forms the best quality turf at a cutting height of 0.7 in. or less (145, 209). At higher heights the stoloniferous growth habit results in excessive thatch formation, scalping, and a subsequent decline in turfgrass quality (124). Certain cultivars even thatch on greens mowed daily at 0.25 in. The thatching rate can be minimized through regular topdressing. Vertical mowing promotes juvenile shoot development and rooting at the nodes on stolons. The nitrogen fertility requirement varies from (a) 0.8 to 1.4 lb per 1000 sq ft per growing month on greens and (b) 0.5 to 1.0 lb on higher cut turfs (60). Creeping bentgrass turfs usually respond to irrigation (94, 224). Supplemental irrigation is required for maintaining a quality bentgrass turf on droughty, coarse textured soils.

The creeping bentgrasses are susceptible to a wide range of diseases including dollar spot, brown patch, *Helminthosporium* spp., *Fusarium* blight, *Fusarium* patch, *Pythium* blight, red thread, stripe smut, and *Typhula* blight (85). A preventative fungicide program is frequently practiced in regions where disease problems are anticipated. Creeping bentgrass is more prone to herbicide injury than Kentucky bluegrass (59, 207). Root and leaf injury frequently result from 2,4-D and 2,4,5-TP applications (40). Such cultivars as Toronto, Congressional, and C-52 are quite prone to thinning and stolon injury by 2,4-D, while Cohansey, Washington, and Arlington are more tolerant (2).

Vegetatively propagated cultivars. There are a number of vegetatively propagated cultivars of creeping bentgrass available

Figure 3-6. Closeup of a Toronto creeping bentgrass turf cut at 0.25 inch.

Table 3-2

CHARACTERISTICS OF NINETEEN CREEPING BENTGRASS CULTIVARS THAT ARE ESTABLISHED VEGETATIVELY AND USED PRIMARILY ON GREENS

Cultivar	Released By	Date	Turfgrass characteristics	Adaptation	Other comments
Arlington (C-1) (5, 6, 10, 63, 65, 80, 106, 108)	USGA Green Section and CRD-ARS, U.S.	(1928*)	Moderate olive-green color; fine texture; high shoot density; erect growth habit; tends to swirl; slow shoot growth rate; deep root system	Good tolerance to heat stress and wear; slow establishment rate; best adapted to droughty soils; not aggressive; used on tees; frequently blended with Congressional for greens	Moderately resistant to dollar spot and *Helminthosporium* spp.; very susceptible to brown patch; requires a medium intensity of culture
Cohansey (C-7) (5, 6, 63, 81, 106, 162, 214)	U.S.	1946 (1935*)	Distinctive yellowish-green color; medium fine texture; medium high shoot density; erect growth habit; less prone to thatching; aggressive	Good hardiness to heat and low temperature stress; excellent tolerance to winter desiccation; good rate of spring green-up and establishment; widely used on greens	Moderately resistant to brown patch and *Helminthosporium* spp.; very susceptible to dollar spot, copper spot, and *Typhula* blight
Collins (C-27) (5, 63)	USGA Green Section and CRD-ARS, U.S.	(1937*)	Very dark green color; very fine texture; intermediate shoot density; erect growth habit; very slow shoot growth rate	Poor establishment rate and recuperative potential; lack of aggressiveness results in weed encroachment	Moderately resistant to brown patch; susceptible to dollar spot and *Helminthosporium* spp.
Columbia (4, 184)	U.S.	(1916*)	Medium green color; very fine texture; low shoot density; tends to be grainy	Poor establishment rate and recuperative potential	Susceptible to brown patch and dollar spot
Congressional (C-19) (5, 63, 122, 162)	U.S.	(1936*)	Very dark green color; very fine texture; very high shoot density; slow shoot growth rate; less prone to thatching; not aggressive	Good hardiness to heat and low temperature stress; good low temperature color retention and spring green-up rate; frequently blended with Arlington for greens	Good resistance to *Typhula* blight; moderate resistance to dollar spot; very susceptible to brown patch

Table 3-2 (CONT.)

C-52 (Old Orchard ®) (6, 63, 106, 144)	Old Orchard Grass Nursery, U.S.	(1934*)	Medium green color; medium fine texture; medium high shoot density	Adapted to a wide range of soil types; intermediate heat tolerance; good spring green-up rate	Moderately resistant to dollar spot and brown patch; very susceptible to copper spot; requires medium intensity of culture
Dahlgren (C-115) (5, 63, 65)	USGA Green Section and CRD-ARS, U.S.	(1946†*)	Distinct yellow-green color; coarse texture; low shoot density; erect growth habit; rapid shoot growth rate	Adapted to intermediate heights of cut; good establishment rate and recuperative potential; used on tees, fairways, and lawns	Moderately resistant to dollar spot; requires a low intensity of culture
Evansville (81, 85)	Indiana AES, U.S.	1963	Dark green color; very fine texture; very high shoot density; rapid shoot growth rate; very prone to thatching	Good establishment rate and recuperative potential	Excellent resistance to *Helminthosporium* spp.; moderately resistant to *Pythium* blight and dollar spot; very susceptible to *Typhula* blight, brown patch, copper spot, and stripe smut
Metropolitan (C-51) (4, 5, 108)	USGA Green Section and CRD-ARS, U.S.	(1917*)	Dark blue-green color; medium coarse texture; low shoot density; very slow shoot growth rate; tends to be puffy and very grainy	Good drought tolerance; poor establishment rate and recuperative potential	Good resistance to dollar spot; very susceptible to brown patch and *Helminthosporium* spp.; requires a high intensity of culture
Nimisilia (15)	Bill Lyons, Ohio, U.S.	1950	Moderate olive-green color; fine texture; high shoot density; erect growth habit; rapid shoot growth rate; prone to thatching	Good low temperature color retention and spring green-up rate; excellent establishment rate and recuperative potential	Moderately susceptible to dollar spot
Norbeck (C-36) (5, 105, 107)	USGA Green Section and CRD-ARS, U.S.	(1937†*)	Medium green color; medium fine texture; medium high shoot density; rapid shoot growth rate; prone to thatching	Has a high calcium requirement; good establishment rate and recuperative potential	Susceptible to dollar spot

Table 3-2 (CONT.)

Cultivar	Released By	Date	Turfgrass characteristics	Adaptation	Other comments
Northland (71, 140)	J. R. Watson, Toro Mfg., Minn., U.S.	1955	Blue-green color; medium fine texture; medium high shoot density; rapid shoot growth rate at cool temperatures	Excellent low temperature hardiness; good low temperature color retention and spring green-up rate; very poor heat tolerance	Excellent resistance to *Typhula* blight; medium resistance to brown patch and dollar spot; susceptible to *Helminthosporium* spp.
Pennlu (13, 85, 101, 122)	Pennsylvania AES, U.S.	1954	Dark blue-green color; fine texture; high shoot density; rapid shoot growth rate; prone to thatching	Good low temperature hardiness and shade tolerance; good establishment rate and recuperative potential	Fairly good resistance to brown patch and dollar spot; very susceptible to stripe smut
Pennpar	Pennsylvania AES, U.S.	1967	Medium dark green color; medium fine texture; high shoot density; intermediate shoot growth rate	Good establishment rate and recuperative potential	Good resistance to dollar spot, brown patch, and *Helminthosporium* spp.; susceptible to *Typhula* blight; requires a medium high intensity of culture
Springfield	Kansas AES, U.S.	1968	Distinctive blue-green color; fine texture; high shoot density	Good establishment rate and recuperative potential	Very susceptible to dollar spot and brown patch
Toronto (C-15) (5, 6, 55, 63, 106)	U.S.	(1936*)	Medium dark green color; fine texture; high shoot density; vigorous shoot growth rate; prone to thatching; shallow root system	Excellent low temperature hardiness; good low temperature color retention and spring green-up rate; poor tolerance to winter desiccation; widely used on greens	Moderately resistant to copper spot; very susceptible to dollar spot and brown patch; requires a high intensity of culture
Vermont (4, 165, 167, 168, 169, 170)	U.S.	(1917*)	Medium green color; coarse texture; low shoot density; semi-erect growth habit	Poor recuperative potential and establishment rate	Susceptible to dollar spot, *Fusarium* patch, and *Typhula* blight

Table 3-2 (CONT.)

Virginia (4, 184, 226)	U.S.	(1919*)	Medium light green color; medium coarse texture; low shoot density; rapid shoot growth rate; extremely grainy	Rapid establishment rate and recuperative potential	Very susceptible to brown patch; susceptible to dollar spot
Washington (4, 5, 54, 85)	USGA Green Section and CRD-ARS, U.S.	(1919*)	Light green color; medium fine texture; medium low shoot density; slow shoot growth rate	Good tolerance to heat stress; poor tolerance to winter desiccation; poor low temperature color retention and spring green-up rate, turns purplish; poor recuperative potential	Moderately resistant to brown patch; susceptible to dollar spot; susceptible to *Fusarium* patch, stripe smut, and *Typhula* blight

* Year selected.
† Not officially released.

(Table 3-2). All have been developed in the United States. The most commonly used cultivars are Cohansey, C-52, Northland, Toronto, and the Arlington-Congressional blend. Collins, Columbia, Dahlgren, Metropolitan, Norbeck, Vermont, and Virginia are no longer available. The rates of turfgrass establishment from stolons and seed are comparable. Vegetatively propagated cultivars are preferred where turfgrass uniformity is to be maintained for an extended period of time. This uniformity is important in the execution of cultural practices because it produces a similar response throughout the green. Spot patching of damaged areas can also be done with a minimum of discernable variation if the greens and nursery contain a single, uniform, vegetative bentgrass cultivar. The seed yield of the vegetatively propagated bentgrasses is comparatively low (185). When purchasing or propagating bentgrass stolons, it is important that the stolons were grown under conditions where the uniformity of the strain was ensured. This means that (a) the stolons were not permitted to form seed, (b) the area was isolated, and (c) the soil was sterilized prior to planting so that no off-type bentgrasses can be introduced.

Seeded cultivars. The seeded creeping bentgrass cultivars possess a certain degree of heterogeneity. A new stand of seeded bentgrass is relatively uniform in turfgrass appearance. The young plants have not had time to develop individuality. Over a period of time they tend to segregate into distinct patches under turfgrass conditions (108). It occurs more quickly in Seaside than in Penncross (Table 3-3). Segregation into various strains complicates cultural practices. Should a segregated strain of bentgrass on a portion of a green become grainy while a strain on another area is thin, brushing of the grainy strain probably could not be accomplished without seriously injuring the thin strain. The seeded cultivars have been widely utilized, primarily because they are more economical to establish. In addition, the establishment procedure is much easier on extensive areas such as fairways. A further advantage of the seeded cultivars is that a thinned or damaged green can be thickened or reestablished by seeding without seriously disturbing the playing surface.

Colonial Bentgrass (*Agrostis tenuis* Sibth.)

Colonial bentgrass is native to Europe but has been introduced for turfgrass use throughout the cool humid climates of the world and has become naturalized in New Zealand and in the Pacific Northwest and New England regions of North America (149). It is one of the two most widely utilized *Agrostis* species. This species has been called Brown, Browntop, New Zealand, Northwest, Prince Edward Island, and Rhode Island bentgrass. These are naturalized strains of colonial bentgrass that were named for the region in which the seed was grown. No significant differences have been found among the so-called strains. Colonial bentgrass was referred to as *Agrostis capillaris* Huds. or as *Agrostis vulgaris* With. in some of the older literature (194).

Table 3-3

CHARACTERISTICS OF THREE CREEPING BENTGRASS CULTIVARS COMMONLY ESTABLISHED FROM SEED

Cultivar	Released		Turfgrass characteristics	Adaptation	Other comments
	By	Date			
Penncross (80)	Pennsylvania AES, U.S.	1954	Medium dark green color; medium fine texture; high shoot density; vigorous shoot growth rate; less prone to segregation into mottled patches	Aggressiveness results in good recuperative potential and proneness to thatch; widely used on greens; rapid establishment rate	Moderate resistance to dollar spot; susceptible to stripe smut
Seaside (Cocoos) (55, 63, 86, 162, 237)	Lyman Carrier, Ore., U.S.	1923	Medium light green color; medium texture; medium shoot density; prone to segregation into mottled patches	Excellent tolerance to winter desiccation and saline soils; intermediate low temperature hardiness; poor low temperature color retention; good spring green-up rate; widely used on fairways, tees, lawns, and greens	Moderately susceptible to dollar spot; very susceptible to brown patch, *Helminthosporium* leaf spot, and *Typhula* blight; tolerates a medium to low intensity of culture
Smaragd	Weibullsholm Plant Breeding Institution, Sweden	1965	Dark green color; medium fine texture; high shoot density	Good low temperature hardiness	Good resistance to *Typhula* blight

Plant Description

Vernation rolled; *sheaths* round, glabrous, split with overlapping, hyaline margins; *ligule* membranous, 0.4–1.2 mm long, truncate; *collar* conspicuous, narrow to medium broad, may be divided, glabrous, light green; *auricles* absent; *blades* flat, 1–3 mm wide, moderately scabrous above and on margins, prominent veins above, acuminate apex; *stems* erect, slender, tufted, with stolons or rhizomes absent to weak and short; *inflorescence* an open, loose, delicate, oblong to ovate panicle, branches slender and clustered.

Figure 3-7. The mottled, patchy nature of an eight-year-old Astoria colonial bentgrass turf cut at 0.25 inch.

Colonial bentgrass forms an upright, fine textured, dense turf under close mowing. The stems and leaves are delicate, fine in texture, and rather low growing with the lower internodes being quite short. The low growth habit results in good tolerance to close mowing. Segregation into off-type clones is likely to occur as the turf matures because of the heterogeneity of certain colonial bentgrass cultivars (Fig. 3-7). Thus, the overall uniformity and turfgrass quality frequently declines. The color ranges from a greenish-yellow to a medium dark green. The root system is fibrous, relatively shallow, and annual in nature (215). The creeping tendency is minimal since the rhizome and stolon growth of colonial bentgrass is either lacking or quite short (192). Colonial bentgrass is propagated primarily by seed. It is cross-pollinated with a reported chromosome number of 28 (19, 24, 29, 74, 92, 118, 120). The establishment rate is fairly good (56) but the recuperative potential is fair to poor.

Adaptation. Colonial bentgrass is a long-lived perennial utilized throughout the cool humid regions of the world. The low temperature hardiness is quite good, although inferior to creeping bentgrass (23, 71, 140). The spring green-up rate is comparatively slow (98). The tolerance to heat and water stress is relatively poor. Colonial bentgrass has medium shade tolerance. The wear tolerance is poor.

Colonial bentgrass is adapted to a wide range of soil types but does best on fertile, moist, fine textured soils having a pH of 5.5 to 6.5 (84, 98, 124, 210, 220). It has the capability to utilize nitrogen at a low soil pH and persists on acidic soils (177, 210).

Use. Colonial bentgrass is utilized in polystands with other cool season turfgrasses on fairways, tees, and high quality, fine textured lawn turfs. It is aggressive and eventually becomes dominant when seeded in mixtures with erect growing, cool season turfgrasses such as Kentucky bluegrass (50, 123, 124, 177, 186). Dominance occurs more rapidly if the colonial bentgrass contains creeping types and is mowed at a height of less than 1 in., irrigated frequently, and intensely fertilized (50). Colonial bentgrass is also utilized to a limited extent on greens, although to a lesser extent than creeping bentgrass (6, 170).

Culture. Colonial bentgrass requires a relatively intense level of culture to produce a high quality turf. The preferred cutting height is between 0.3 and 0.8 in. (209, 213). It tends to thatch excessively at higher heights of cut. Thatching can be a problem even at lower cutting heights if not topdressed at intervals. It is more tolerant of deviations from the optimum fertility level than Kentucky bluegrass but responds to a high nutritional level (177, 210). The nitrogen fertility require- ment is 0.5 to 1.0 lb per 1000 sq ft per growing month. The irrigation requirement is less than for the creeping bentgrasses.

Colonial bentgrass is extremely susceptible to a large number of diseases includ- ing dollar spot, brown patch, red thread, *Ophiobolus* patch, *Pythium* blight, *Typhula* blight, *Fusarium* patch, stripe smut, and *Helminthosporium* spp. (184). It is more prone to herbicide injury than most cool season turfgrasses (173). Serious root injury frequently occurs following an application of 2,4-D or 2,4,5-TP. Certain preemergence herbicides such as DCPA can also cause damage to the root system and reduce the shoot density (58, 86, 207).

Cultivars. For many years the only cultivars of colonial bentgrass commercially available were Astoria and Highland. More recently several new colonial bentgrass cultivars have been released (Table 3-4). A noncreeping colonial bentgrass cultivar is needed that (a) blends well in polystands with other cool season turfgrasses and (b) will not become dominant.

Velvet Bentgrass (*Agrostis canina* L.)

Velvet bentgrass forms one of the finest textured turfs of exquisite beauty avail- able for use on close cut greens and other fine textured lawn turfs (54, 84, 170, 184, 213). It is a native of Europe that has become naturalized in New England. Suc- cessful velvet bentgrass culture is generally restricted to certain specific soil and climatic conditions.

Plant Description

Vernation rolled; *sheaths* round, glabrous, split with overlapping, hyaline mar- gins; *ligule* membranous, 0.4–0.8 mm long, acute to oblong; *collar* medium broad; *auricles* absent; *blades* less than 1 mm wide, flat, soft, glabrous above and below, veins prominent above, margins scabrous and hyaline, acuminate apex; *stems* erect, often tufted and decumbent at base with creeping, scaly stolons; *inflorescense* a reddish, pyramidal-oblong, copiously flowered panicle with the the branches open at anthesis, sometimes contracting later.

Velvet bentgrass forms a turf of extremely fine texture with almost needle-like leaves (Fig. 3-8). It has an erect growth habit and forms a soft, velvety turf of very high shoot density and uniformity (54, 213). The rate of spread by creeping stolons

Table 3-4

CHARACTERISTICS OF FIVE COLONIAL BENTGRASS CULTIVARS

Cultivar	Released By	Date	Turfgrass characteristics	Adaptation	Other comments
Astoria (23, 55, 63)	Oregon AES, U.S.	1936	Medium green color; medium fine texture; medium shoot density; slow shoot growth rate; not aggressive; tends to segregate into mottled patches	Medium poor low temperature hardiness; good early spring green-up rate; used on lawns, fairways, and tees	Moderately resistant to dollar spot, copper spot, and Typhula blight; very susceptible to brown patch
Boral	Weibullsholm Plant Breeding Institution, Sweden	1965	Medium dark green color; medium fine texture; high shoot density; low growth habit	Good low temperature hardiness; poor low temperature color retention; used on lawns and fairways	Good Fusarium patch resistance
Exeter (RI No. 5) (65, 81)	Rhode Island AES, U.S.	1963	Bright medium green color; fine texture; high shoot density; leafy, erect growth habit	Good low temperature hardiness and spring green-up rate; used on lawns and fairways	Good resistance to dollar spot; best at medium cutting heights of 0.5 to 1 in.
Highland (65, 144, 165, 167, 168)	Oregon AES, U.S.	1934	Dark bluish-green color; medium fine texture; intermediate shoot density; erect growth habit	Fairly good drought tolerance; adapted to infertile soils; used on lawns, fairways, and tees	Moderately resistant to dollar spot; susceptible to brown patch; becomes puffy at higher cutting heights
Holfior	D. J. van der Have, Netherlands	1940	Medium dark green color; medium fine texture; high shoot density; semi-erect growth habit; less prone to thatching	Adapted to infertile, acidic soils of coarse texture; medium poor recuperative potential; used on lawns and fairways	Susceptible to Ophiobolus patch and Fusarium blight; best at medium cutting heights of 0.5 to 1 in.

is more aggressive than for colonial
bentgrass but not as vigorous as for
creeping bentgrass. The shoot growth rate
is rather slow (184). Root production is
excellent but the root decomposition rate
is quite slow (25). Propagation is either by
seed or vegetatively by stolons. A chromosome number of 14 has been reported
(19, 119). The establishment rate and
recuperative potential are relatively poor.
Adaptation. Velvet bentgrass is a long-
lived perennial that is utilized primarily
in the cooler portions of the cool humid
regions. It has good heat and low temperature hardiness and is fairly drought

Figure 3-8. Close-up of a fine textured Kingstown velvet bentgrass turf cut at 0.25 inch.

resistant compared to the other *Agrostis* species. However, the soft, succulent tissue
is prone to wilt. The shade tolerance is quite good (197).

Velvet bentgrass is quite tolerant of acidic, infertile soil conditions. It is not
adapted to poorly aerated, imperfectly drained soils. Best growth is made on coarse
textured, well-drained soils. A soil pH of 5 to 6 is preferred. Velvet bentgrass is
more tolerant of acidic soils than the other bentgrass species (196).

Use and culture. Velvet bentgrass is utilized primarily on closely mowed putting
greens, bowling greens, and fine textured ornamental lawn turfs. The use of velvet
bentgrass has been limited to certain specific soil and climatic regions such as the
New England area. It is sometimes used in polystands for shaded areas and other
fine textured lawn and park areas. However, it tends to dominate polystands under
acidic, well-drained soil conditions.

Velvet bentgrass produces the highest quality turf under close, frequent mowing of 0.2 to 0.4 in. The tendency to become grainy is minimal. Since there is a
strong thatching tendency, it must be topdressed at frequent intervals but at a light
rate. The nitrogen requirement is from 0.5 to 1.0 lb per 1000 sq ft per growing
month. Velvet bentgrass is prone to iron chlorosis problems. It is generally quite
susceptible to copper spot but has fairly good dollar spot resistance (54, 55). Other
diseases which may be a problem include brown patch, red thread, *Fusarium*
patch, *Pythium* blight, *Ophiobolus* patch, *Helminthosporium* spp. and *Typhula* blight.
Cultivars. Only a limited number of velvet bentgrass cultivars have been developed
(Table 3-5). Acme, Kernwood, Piper, and Raritan are not now available. Kingstown is available for establishment from seed.

South German Mixed Bentgrass (*Agrostis* L.)

A mixture of *Agrostis* species originating from native stands in southern Germany that has been widely utilized for fine turf is known as **South German Mixed
Bentgrass** (103). The names German bentgrass, German mixed bentgrass and

Table 3-5

CHARACTERISTICS OF FIVE VELVET BENTGRASS CULTIVARS

Cultivar	Released By	Date	Turfgrass characteristics	Other comments
Acme (4, 184)	USGA Green Section and ARS-USDA, U.S.	(1919*)	Light green color; medium fine texture; intermediate shoot density	Poor recuperative potential; somewhat tolerant of heat stress; very susceptible to copper spot
Kernwood (184)	Not officially released, U.S.	—	Medium dark green color; fine texture; high shoot density	Good low temperature color retention and spring green-up rate; very susceptible to copper spot
Kingstown (16, 80)	Rhode Island AES, U.S.	1963	Dark green color; very fine texture; very high shoot density; vigorous	Good resistance to dollar spot; susceptible to copper spot
Piper	USGA Green Section and ARS-USDA, U.S.	1929	Medium green color; very fine texture; high shoot density	Intermediate recuperative potential; very susceptible to copper spot
Raritan (63, 211)	New Jersey AES, U.S.	—	Medium dark green color; fine texture; very high shoot density	Moderately resistant to dollar spot and brown patch; very susceptible to copper spot

* Year selected.

German creeping bentgrass have also been used. The seed has been harvested commercially in a number of European locations including Holland, Belgium, England, and Germany. It is usually composed of from 75 to 85% colonial bentgrass, 10 to 20% velvet bentgrass, and 1 to 5% creeping bentgrass (194). South German Mixed bentgrass provided an adequate turf for greens during the period when it was used (165, 167, 168, 170). Established greens of South German Mixed bentgrass are quite mottled in appearance because of the heterogeneous nature of the seed source. Distinct patches of various textures, colors, and growth habits become quite evident. Many of the improved velvet and creeping bentgrasses such as Arlington, Cohansey, Congressional, Toronto, and Washington were selected from old South German Mixed bentgrass greens. These improved bentgrass cultivars have replaced the old South German Mixed bentgrass for use on greens.

Redtop (*Agrostis alba* L.)

Redtop is an *Agrostis* species that is native to Europe. It is no longer widely used in quality turfs although it was utilized extensively prior to 1945. The name redtop is derived from the color of the seed head.

Plant Description

Vernation rolled; *sheaths* round, glabrous, split with overlapping, hyaline margins; *ligule* membranous, 2–5 mm long, acute to obtuse, may be lacerated; *collar* conspicuous, medium broad, divided, spiral, glabrous, light green; *auricles* absent; *blades* flat, 3–5 mm wide, scabrous, veins prominent above, margins scabrous and hyaline, acuminate apex; *stems* erect, often tufted and decumbent at base with creeping, scaly rhizomes; *inflorescence* a reddish, pyramidal oblong, copiously flowered panicle with the branches open at anthesis, sometimes contracting laterally.

Redtop forms a stemmy, coarse textured, open turf of low shoot density (74, 145). The leaf texture is wider than for most of the turf-type *Agrostis* species. The ligule is longer than any of the other turf-type *Agrostis* species. Redtop possesses short rhizomes but does not form a particularly tight sod. The root system is regenerated annually (215). Redtop is propagated almost entirely by seed. It is cross-fertilized with chromosome numbers of 28 and 42 having been reported (74, 92, 181). The establishment rate is medium-fair, being slightly more rapid than for the bentgrasses.

Adaptation. Redtop is one of the most widely adapted turfgrass species (213). It may behave as either a long-lived or short-lived perennial depending on the intensity of culture, soil, and environmental conditions. A majority of the redtop plants from a given seed source behave as short-lived perennials that die out due to low temperature kill and other adversities (51). A few scattered plants persist as individual, gray-green tufts that become weeds in a quality turf. Redtop is adapted primarily to the cool humid climates but is occassionally utilized in the transitional and warm humid regions. It does not tolerate shaded conditions or high temperatures. Redtop turfs frequently turn brown and stemmy under heat stress. The wear tolerance is poor (93).

Redtop is one of the best adapted cool season turfgrasses for use on wet, imperfectly drained, fine textured, acidic soils of low fertility. It tolerates a wide range of soil types and moisture conditions. It even persists on droughty, coarse textured soils. Redtop is adapted to soils of a somewhat higher pH than the other commonly used *Agrostis* species (84, 98, 220).

Use. The use of redtop is restricted primarily to low quality turfs. The wide tolerance to soil pH, soil texture, and climate enables redtop to be commonly used in seed mixtures for roadsides, ditches, and waterways for erosion prevention. Redtop has also been used in quality mixtures for lawn turfs. With the development of improved Kentucky bluegrasses, fescues, and bentgrasses, however, it no longer contributes significantly to quality lawn turfs and its use for this purpose has rapidly declined (124).

For years redtop was widely accepted as a valuable temporary component of turfgrass mixtures. However, recent studies indicate that redtop is not particularly rapid in establishment compared to the improved ryegrasses or fescues and, thus, has no value as a temporary component of quality turfgrass polystands (188). It has been used for emergency overseeding of greens as well as for winter overseeding of warm season turfs (166). Even this use is limited since red fescue is usually preferred for the former use and several species are superior for the latter.

Culture. Redtop is best adapted to a medium to low intensity of culture. It does not persist under close, frequent mowing (209). The shoot density of a redtop turf can be maintained for a longer period of time if mowed at a cutting height of 1.5 to 2 in. (51). The nitrogen fertility requirement is 0.5 to 1.0 lb per 1000 sq ft per growing month. Although adapted to acidic, infertile soils, redtop responds to higher soil pH and fertility conditions. It grows best on sandy sites if irrigated. Redtop is fairly susceptible to *Fusarium* patch injury (46). Other potential disease problems include *Helminthosporium* spp., *Pythium* blight, red thread, stripe smut, dollar spot, and brown patch. Only a few turf-type redtop cultivars have been developed, due to the rather limited, nonspecific use. None of them are utilized to any extent.

The Fescues (*Festuca* L.)

The fescues compose a large genus of over 100 species that are members of the *Festuceae* tribe (193). Considerable variation in longevity, leaf texture, and growth habit occurs among the *Festuca* species. The annual species are generally considered weedy types, while a number of the perennials possess excellent characteristics for turfgrass use. The fescues are adapted to the cool humid regions of the world and tolerate droughty, infertile, acid soils with a pH of 5.5 to 6.5. The *Festuca* genus contains some of the most wear tolerant cool season turfgrasses. The seven species or subspecies of *Festuca* utilized for turfgrass purposes are (a) *F. rubra*-red fescue, (b) *F. rubra* var. *commutata*-chewings fescue, (c) *F. arundinacea*-tall fescue, (d) *F. elatior*-meadow fescue, (e) *F. ovina*-sheep fescue, (f) *F. ovina* var. *duriuscula*-hard fescue, and (g) *F. capillata*-hair fescue. Tall fescue and meadow fescue are coarse textured, while the others have a very fine leaf texture (Fig. 3-9).

Red Fescue (*Festuca rubra* L.)

Red fescue is the species of *Festuca* that is most widely used for turfgrass purposes. It is sometimes referred to as creeping red fescue. A number of red fescue cultivars have been developed for turfgrass use throughout the cool humid regions. Red fescue is a native of Europe where it was originally described long before being adapted for turfgrass use.

Plant Description

Vernation folded; *sheaths* oval to round, glabrous to finely pubescent, veins prominent, lower sheaths reddish, glossy, thin, and disintegrating with age, open part way; *ligule* membranous, 0.2–0.5 mm long, truncate; *collar* narrow, indistinct, glabrous; *auricles* absent or merely enlarged margins; *blades* folded or involute, 0.5–1.5 mm wide, usually bristle-like, deeply ridged above, bluntly keeled, glabrous, acute apex, margins glabrous; *stems* oval to round, erect to ascending from a decumbent base, extravaginal type growth, usually with slender, creeping rhizomes; *inflorescence* a narrow, lanceolate oblong, contracted panicle, branches rough, usually in pairs, unequal, reddish when ripe.

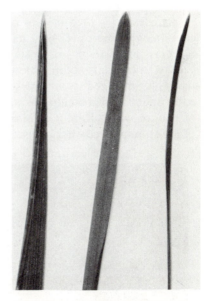

Red fescue forms a very fine textured turf of high shoot density, uniformity, and quality (133). The turfgrass color is medium to dark green. The vertical shoot growth rate is slower than for most cool season turfgrasses (229). The root system is fibrous and extremely dense. Red fescue can be distinguished from chewings and sheep fescues by the extravaginal type of shoot development that results in a creeping growth habit. The degree of rhizome development and sod formation is less than for Kentucky bluegrasses (128). Red fescue is propagated primarily by seed. The establishment rate is fairly good, being faster than Kentucky bluegrass but somewhat slower than ryegrass (56). It is largely cross-pollinated with

Figure 3-9. The comparative leaf textures of Kentucky 31 tall fescue (left), Merion Kentucky bluegrass (center), and Pennlawn red fescue (right).

chromosome numbers of 14, 42, 56, and 70 having been reported (24, 43, 74, 92, 117). The recuperative potential is medium good.

Adaptation. Red fescue is a long-lived perennial that is widely distributed throughout the cool humid regions of North America, Eurasia, North Africa, and Australia. The adaptation range is not as wide as for Kentucky bluegrass and bentgrass. It is not as low temperature hardy as creeping bentgrass and Kentucky bluegrass. Red fescue does not persist in the warm humid regions due to a lack of heat tolerance (213).

Red fescue is superior to most cool season turfgrasses in shade adaptation (213). It is capable of more rapid leaf growth than the other turfgrass species when grown under a reduced light intensity (229). However, the quality and shoot density of a red fescue turf grown under shaded conditions will not be as good as when grown in full sunlight. The water use rate of red fescue is much lower and the drought tolerance superior to Kentucky bluegrass and creeping bentgrass. The wear tolerance is medium. It is well adapted to dry, sandy soils having a pH of 5.5 to 6.5 (98, 124, 128). Red fescue does not tolerate wet, poorly drained soils or soil salinity.

Use. Red fescue is one of the three most widely utilized turfgrass species in the cool humid regions. It is the best cool season turfgrass available for use under dry, shaded conditions and is widely used on lawns, parks, cemeteries, institutional grounds, fairways, roughs, roadsides, airfields, and other general use turfgrass areas (Fig. 3-10). In Europe, it is used in polystands with bentgrass on putting and bowling greens.

Red fescue is used in seed mixtures with Kentucky bluegrass throughout the cool humid climates (124). It has a more rapid establishment rate than Kentucky bluegrass but does not compete excessively during the establishment period (129, 188). Red fescue is compatible with Kentucky bluegrass (67, 124). Where a Kentucky bluegrass-red fescue polystand is established, the red fescue usually predominates under shaded conditions; on droughty, sandy soils; and a minimum intensity of culture, while Kentucky bluegrass becomes more dominant under conditions of moist, imperfectly drained soils; intensive culture, and full sunlight (52, 77, 129, 130, 200).

Figure 3-10. Close-up of a fine textured Penn-lawn red fescue turf cut at 1.5 inches.

Red fescue does not perform well on sports fields and tees because of the weak rhizome system and slow recuperative rate. The amount of red fescue utilized in commercial sod production has also been minimal. The potential exists within the available genetic material for developing a more vigorous, rhizomatous, sod forming cultivar that would be better adapted for use in sod production.

Red fescue is used for winter overseeding of dormant bermudagrass in the warm humid regions (203). The fall and spring transition periods are quite good in comparison to ryegrass and rough bluegrass. It may also be overseeded into damaged greens in order to provide a quick cover and playable surface until the bentgrass recovers.

Culture. Red fescue requires a low intensity of culture, including minimal levels of nitrogen and water. It forms a high quality turf under proper culture. Cutting heights of 1.0 to 2.5 in. are utilized with the higher heights preferred under shaded conditions (52). The turfgrass quality of red fescue fairways cut between 0.5 and 1 in. is excellent. The putting surface of a red fescue-bentgrass turf cut at 0.3 in. is quite adequate.

The nitrogen requirement of 0.2 to 0.5 lb per 1000 sq ft per growing month is quite low compared to most turfgrass species (52, 60, 208). A decline in turfgrass quality occurs if excessive amounts of nitrogen are applied to a red fescue turf (52). Similarly, the quality deteriorates if irrigated excessively (52, 94, 224). The intolerance to high soil moisture levels is further emphasized by the fact that red fescue is one of the least tolerant turfgrasses to submergence. Red fescue tolerates high soil phosphorus levels (128). It thatches but the problem is not as serious as with bentgrass and Kentucky bluegrass. Once formed however, a red fescue thatch has a very slow decomposition rate due to the high lignin content of the leaf sheath. Red fescue is less tolerant of the commonly used herbicides than Kentucky bluegrass (58, 86, 173, 207). The most serious disease of red fescue is caused by the *Helminthosporium* spp. with red thread also being a serious problem in certain regions. It is more prone to *Fusarium* patch and *Typhula* blight injury than Kentucky bluegrass (46). Other diseases that occasionally cause injury are *Ophiobolus* patch, *Pythium* blight, powdery mildew, and dollar spot.

Cultivars. Only a limited number of red fescue cultivars had been developed specifically for turfgrass use prior to 1960 (91, 233). Included were Duraturf, Illahee, Olds, Pennlawn, and Rainier. Other early cultivars that are of no significance in modern turfgrass culture are Clatsop, Steinach, and Trinity. A number of new red fescue cultivars have been released more recently (Table 3-6). Pennlawn has been the most widely used cultivar. There is great potential for improving the red fescue cultivars in terms of a vigorous creeping habit, overall quality, disease resistance, heat tolerance, and low temperature hardiness.

Chewings Fescue (*Festuca rubra* var. *commutata* Gaud.)

Chewings fescue is a cool season turfgrass originating from Europe. It has also been called *Festuca fallax* (Thuill.) and *Festuca rubra* var. *fallax* (Hack.). Chewings is quite similar to red fescue with one major exception: It has a noncreeping, bunch-type growth habit.

Plant Description

Vernation folded; *sheaths* distinctly compressed, glabrous, or very short pubescence, soon splitting, lower sheaths red at base; *ligule* membranous, 0.2–0.4 mm

Table 3-6

CHARACTERISTICS OF TEN RED FESCUE CULTIVARS

Cultivar	Released By	Date	Turfgrass characteristics	Adaptation	Other comments
Arctared	Alaska AES and CRD-ARS, U.S.	1965	Dark green color; fine texture; medium high shoot density	Excellent low temperature hardiness; low temperature discoloration results in a distinct reddish color	Rapid establishment rate
Boreal	Alberta, Canada.	1966	Medium dark green color; fine texture; medium shoot density; deep root system	Good low temperature hardiness	Rapid establishment rate
Duraturf	Canada Dept. Agriculture, Ottawa, Ont., Can.	1943	Medium dark green color; fine texture; medium shoot density; rapid vertical shoot growth rate	Intermediate low temperature hardiness	Very susceptible to *Helminthosporium* blight
Illahee (8, 53)	Oregon AES and CRD-ARS, U.S.	1950	Medium green color; fine texture; medium high shoot density; moderately slow shoot growth rate	Poor low temperature hardiness	Very susceptible to *Helminthosporium* blight
Olds (53, 62)	School of Agriculture, Olds, Alberta, Can.	1931	Medium green color; fine texture; medium shoot density	Moderately good low temperature hardiness; moderately drought tolerant	Susceptible to *Helminthosporium* blight and red thread
Pennlawn (14, 38, 50 53, 66, 208)	Pennsylvania AES, U.S.	1954	Medium dark green color; fine texture; high shoot density; fairly rapid shoot growth rate	Intermediate tolerance to drought; medium good low temperature hardiness; intermediate tolerance to close mowing	Moderately susceptible to red thread; more responsive to high nitrogen fertility levels

Table 3-6 (CONT.)

Rainier (176)	Oregon AES and CRD-ARS, U.S.	1944	Dark green color; fine texture; medium high shoot density; fairly rapid vertical shoot growth rate	Moderately poor low temperature hardiness	Very susceptible to red thread and *Helminthosporium* blight
Reptans (53)	Weibullsholm Plant Breeding Institution, Sweden	1943	Medium green color; fine texture; medium low shoot density; very rapid vertical shoot growth rate	Good drought tolerance; fairly good tolerance to saline soils	Moderately resistant to *Helminthosporium* blight
Ruby	D. J. van der Have, Netherlands	1964	Dark green color; medium fine texture; medium shoot density; erect growth habit; vigorous rhizome development	Improved shade adaptation	Susceptible to red thread; fairly rapid establishment rate
S.59 (112, 114, 115)	Welsh Plant Breeding Station, Aberystwyth, G.B.	1932	Medium green color; very fine texture; very high shoot density	Intermediate tolerance to close mowing	Susceptible to dollar spot and red thread

Table 3-7

CHARACTERISTICS OF TEN CHEWINGS FESCUE CULTIVARS

Cultivar	Released By	Date	Turfgrass characteristics	Adaptation	Other comments
Barfalla	Barenbrug Zaadhandel N.V., Netherlands	1968	Medium green color; very fine texture; high shoot density; low growth habit	Good tolerance to close mowing	Moderately resistant to red thread
Cascade	Oregon AES, U.S.	1964	Dark green color; very fine texture; medium high shoot density; good establishment vigor	Well adapted to droughty, coarse textured soils	Susceptible to *Helmintho-sporium* blight
Dawson (Fr. 10) (11)	Sports Turf Research Institute, G.B.	1957	Dark green color; very fine texture; high shoot density; low growth habit	Good tolerance to close mowing	Moderately resistant to red thread
Erika (143)	Weibullsholm Plant Breeding Institution, Sweden	1965	Light green color; fine texture; high shoot density	Intermediate tolerance to close mowing	Moderately resistant to *Fusarium* patch
Golfrood (112, 113, 114)	D. J. van der Have, Netherlands	1940	Distinct light green color; very fine texture; high shoot density; semi-erect growth habit	Good low temperature color retention; good tolerance to saline soil conditions	Susceptible to red thread and dollar spot
Highlight (114)	Gebr. van Engelen, Netherlands	1959	Medium green color; very fine texture; high shoot density	Poor drought tolerance; good tolerance to close mowing	Good resistance to red thread; moderately susceptible to *Helmintho-sporium* blight
Jamestown (208)	Rhode Island AES U.S.	1967	Medium dark green color; fine texture; high shoot density; low growth habit; medium slow vertical shoot growth rate	Good tolerance to close mowing	Fairly resistant to red thread and *Fusarium* patch

Table 3-7 (Cont.)

Oasis	Gebr. van Engelen, Netherlands	1960	Light green color; very fine texture; high shoot density; low growth habit	Good low temperature color retention and spring green-up rate	Moderately resistant to *Helminthosporium* blight; very susceptible to red thread
Polar (53, 143)	Weibullsholm Plant Breeding Institution, Sweden	1963	Medium dark green color; very fine texture; high shoot density; moderately slow shoot growth rate	Good tolerance to close mowing	Susceptible to *Helminthosporium* blight
Wintergreen	Michigan AES, U.S.	1968	Dark green color; very fine texture; very high shoot density; minimal creeping habit	Good low temperature color retention and spring green-up rate	Prefers a low intensity of culture

long; *collar* broad; *auricles* absent; *blades* folded or involute, 1–2 mm wide, bristle-like, glabrous, stiff to rigid, bluntly keeled, acuminate apex, glabrous margins; *stems* oval, distinctly erect, mostly intravaginal growth; *inflorescence* an erect, linear to lanceolate, narrow, contracted panicle.

Chewings fescue forms a fine textured, erect growing, high density turf. It tends to become tufted under unfavorable soil or cultural conditions due to the bunch type growth habit. The degree of tillering is extensive (188). Chewings fescue can be distinguished from red fescue by its intravaginal shoot growth and bunch-type growth habit. It is propagated by seed and has a fairly good establishment rate. A chromosome number of 42 has been reported (157). The recuperative potential is medium poor.

Adaptation. Chewings fescue is a long-lived perennial adapted to the same cool humid climates as red fescue. It does not tolerate extremes in temperature. The low temperature color retention is poor (114). The shade adaptation and drought tolerance is good (148, 226). The wear tolerance is better than for most cool season turfgrasses (171). It is well adapted to acidic, infertile soils containing a high sand content.

Use and culture. Chewings fescue is used in seed mixtures with Kentucky bluegrass for lawns and general-purpose turfs throughout cool humid climates (67). It is also used in polystands with bentgrass on close cut greens, particularly in Europe (148). The cultural requirements are similar to those previously discussed for red fescue, with minimal fertility and irrigation levels being particularly important. A cutting height of 1.5 to 2 in. is preferred, although chewings fescue tolerates shorter heights. The nitrogen requirement is 0.2 to 0.5 lb per 1000 sq ft per growing month. Chewings fescue can accumulate an extensive thatch, particularly at a soil pH below 5.0 (61).

Cultivars. A number of chewings fescue cultivars have been developed (Table 3-7). Most originated as private selections from European countries. Major improvements needed in these cultivars are *Helminthosporium* blight resistance and improved heat and low temperature hardiness.

Sheep Fescue (*Festuca ovina* L.)

Sheep fescue is a tufted, perennial turfgrass that is indigenous to North America and Eurasia. It forms a relatively low quality turf and has not been widely utilized for turfgrass purposes.

Plant Description

Vernation folded; *sheaths* round, glabrous to dense, short pubescence, dull, split with overlapping margins; *ligule* membranous, 0.5 mm long, sometimes absent, truncate; *collar* narrow, indistinct, glabrous; *auricles* absent or merely enlarged at margins; *blades* strongly folded or involute, 0.5–1 mm wide, firm, bristle-like, deeply ridged above, glabrous below, grayish to bluish-green, acute apex, margins glabrous; *stems* erect to ascending, tufted, rounded, stiff, intra-vaginal type growth; *inflorescence* a narrow, contracted panicle, sometimes almost spike-like, branches ascending.

Sheep fescue has a very fine leaf tex-ture and tends to be quite tufted in growth habit due to the intravaginal type of shoot growth (Fig. 3-11). It seldom forms a turf of uniform shoot density and appearance. Sheep fescue is characterized by a distinct blue-green color with the leaves being stiffer than for the red fescues. The rooting depth is considerably less than for chewings fescue when cut closely (71). Sheep fescue is propagated by seed and is largely cross-pollinated with chromosome num-bers of 14, 42, 56, and 70 having been reported (43, 74, 92).

Figure 3-11. The clumpy nature of a 12-year-old sheep fescue turf.

Adaptation. Sheep fescue is a long-lived perennial adapted to the same cool humid climates as red fescue. The heat tolerance is quite poor. Sheep fescue is extremely drought resistant and grows best on sandy or gravelly soils (213). The wear toler-ance is fairly good (171). Acidic, coarse textured soils of low fertility are preferred.

Use and culture. Sheep fescue is used sparingly for turfgrass purposes with seed supplies being quite limited. The commercial seed is grown primarily in Europe. Sheep fescue is most commonly used as a low quality turf for roadsides, roughs, and other nonuse areas in the cooler portions of the cool humid climates. It is usually seeded in mixtures with Canada bluegrass but this use has declined consid-erably in recent years.

The intensity of culture is lower than for red fescue. The tolerance to cutting heights of less than 0.5 in. is poor (90, 209). Sheep fescue grows best under unirri-gated conditions and a nominal soil fertility level with no supplemental nitrogen fertilization. The tough leaves are somewhat difficult to mow. It is susceptible to red thread (113), *Fusarium* patch, powdery mildew, brown patch, and stripe smut.

Career, a cultivar of sheep fescue, is a private selection developed by Gebr. Van Engelen of the Netherlands. It is more greenish in color than common sheep fescue. Career is fine textured, tufted, and low growing. The drought resistance and shade tolerance are excellent.

Hard Fescue (*Festuca ovina* var. *duriuscula* L. Koch)

Hard fescue originated in Europe. It is a perennial, cool season species pos-sessing intravaginal-type shoot growth. The leaf blades are more grayish-green, wider, and tougher than sheep fescue. It has a high shoot density, a somewhat tufted growth habit, and an extensive root system. The adaptation and use of hard fescue are comparable to sheep fescue. The drought tolerance is less than for sheep fescue but better than red fescue. Hard fescue is adapted to shaded conditions and tolerates a higher soil moisture level than sheep fescue (209). It does not tolerate mowing at less than 1 in. (209). It has a somewhat higher nitrogen fertility requirement than sheep fescue. Hard fescue is less susceptible to disease including the *Helminthosporium* spp. Propagation is by seed. The primary use is on roadsides, ditch banks, and other minimum maintenance, low quality, nonuse areas.

Durar is a cultivar of hard fescue that was released cooperatively in 1949 by the Idaho, Oregon, and Washington Agricultural Experiment Stations and the Soil Conservation Service. It is more uniform, drought resistant, and shade tolerant than chewings fescue (205). Durar is well adapted to the cool subhumid and semi-arid regions where close mowing is not essential.

Biljart (C-26) is a cultivar released by N.V.H. Mommersteeg of the Netherlands in 1963. It forms a deep green turf of very fine texture, high shoot density, and slow vertical shoot growth rate. The growth habit is medium low and somewhat tufted with distinctly stiff leaves. The drought resistance is excellent. The resistance to red thread and *Helminthosporium* spp. is very good. Biljart is the best turf-type hard fescue cultivar available.

Hair Fescue (*Festuca capillata* Lam.)

Hair fescue originated in Europe. It is a low growing, densely tufted species with leaves that are extremely narrow or hair-like. The drought resistance and wear tolerance are excellent. The tolerance to wet, fine textured soils is poor. Hair fescue is preferred to sheep fescue in England but is seldom planted alone. It is used to a limited extent as a constituent in European seed mixtures for greens and high quality turfs. It is established by seed but supplies are limited. No cultivars are available.

Tall Fescue (*Festuca arundinacea* Schreb.)

Tall fescue is a cool season turfgrass that originated in Europe. It is a coarse textured, bunch-type species that is adapted to a wide range of soil and climatic conditions.

Plant Description

Vernation rolled; *sheaths* round, glabrous, or occasionally scabrous, split with overlapping, hyaline margins, reddish at base; *ligule* membranous, 0.2–0.8 mm wide, truncate; *collar* conspicuous, broad, divided, usually pubescent on margins, yellow-green; *auricles* small, narrow, celia present; *blades* flat, stiff, 5–10 mm wide, scabrous above at least near the apex, dull with veins prominent above, glossy, keeled, and glabrous below at least at base, midrib prominent, acuminate apex, margins scabrous and hyaline; *stems* round, erect, stout, tufted; *inflorescence* an erect, or nodding, lanceolate to ovate, somewhat contracted panicle, axis and branches scabrous, spikelets four to five flowered.

Tall fescue forms a turf of very low shoot density with a leaf texture that is coarser than any of the commonly used cool season turfgrasses. It does not blend well with the other commonly used cool season turfgrasses such as Kentucky bluegrass, red fescue, and ryegrass. Tall fescue has a wider leaf width than meadow fescue as well as more prominent ribbing and a darker green color on the upper leaf surface. Although possessing short rhizomes, it has a bunch-type growth habit with very weak sod forming characteristics (195). Most new shoots originate from the crown rather than from the nodes of rhizomes (152). The root system is extensive, coarse, and deeper than most cool season turfgrasses. Propagation is by seed. The establishment rate is quite good, ranking better than Kentucky bluegrass but somewhat less than ryegrass (188). A chromosome number of 42 has been reported (43, 48, 117, 180). Pollination is primarily by cross-fertilization (24, 74, 92). The recuperative potential is medium poor.

Adaptation. Tall fescue is a long-lived perennial when grown in the transitional zone between the cool humid and warm humid regions. In the cooler portions of the cool humid regions, tall fescue is prone to low temperature injury and tends to behave like a short-lived perennial. Because of the medium poor low temperature hardiness, it gradually thins out until only scattered, coarse textured plants remain that eventually become weeds. Tall fescue is fairly heat tolerant compared to most cool season turfgrasses. Although leaf growth is restricted, tall fescue is capable of retaining adequate color and appearance during periods of temporary heat stress (144). Tall fescue is one of the most drought and wear tolerant cool season turfgrasses (171, 236). The shade tolerance is intermediate when grown in the transitional zone.

Tall fescue is adapted to a wide range of soil types but is best adapted to fertile, moist, fine textured soils that are high in organic matter. It tolerates soils of low fertility but responds to fertilization. Soils with a pH of 5.5 to 6.5 are preferred with pH's from 4.7 to 8.5 tolerated. The tolerance to alkaline and saline soil conditions is better than for most cool season turfgrasses, especially if irrigated. Tall fescue tolerates wet soil conditions as well as extended periods of submersion and is frequently used in drainage ways.

Use. Tall fescue is well adapted for use in the transitional zone between the cool humid and warm humid regions on areas subjected to intensive traffic. The coarse leaf texture is sometimes objectionable where a high quality turf is desired. Tall fescue has been used on sports fields; playgrounds; roadsides; waterways; airfields; and minimum use, low quality turfgrass areas. It is very effective in slope stabilization along roadsides because of the rapid establishment rate, deep root system, and ability to become established on poor soils.

Polystands of Kentucky bluegrass and tall fescue result in a higher quality turf than tall fescue monostands. The tall fescue component in seed mixtures with other cool season turfgrasses should be more than 70% by weight. A high seeding rate is also important. Low temperature stress in the cooler portions of the cool humid climates frequently causes tall fescue to thin out to the extent that scattered, unsightly clumps develop (50, 123, 124, 144, 163) (Fig. 3-12). Tall fescue is some-

Figure 3-12. Scattered clumps of tall fescue in a Kentucky bluegrass turf.

times overseeded annually on intensely used sports fields in the cooler portions of the cool humid regions where perennial, cool season turfgrasses such as Kentucky bluegrass are not able to persist due to the excessive traffic. The rapid establishment rate and superior wear tolerance increase the likelihood that the tall fescue turf will persist throughout the playing season.

Tall fescue is planted for general grounds use in association with the more open, coarse textured bermudagrass cultivars in certain areas of the warm humid regions. A polystand of tall fescue and Pensacola bahiagrass has also been used in the warm humid regions for sports fields and playgrounds (234).

Culture. The best leaf texture and overall turfgrass performance of tall fescue is usually achieved at a cutting height of 1.7 to 2.2 in. It does not maintain an adequate shoot density at cutting heights of less than 1.2 in. and thus is not satisfactory for use on closely mowed turfs such as fairways. The mowing qualities are quite good. The nitrogen requirement is 0.4 to 1.0 lb per 1000 sq ft per growing month. A high nitrogen fertility level increases the low temperature injury problem of tall fescue in the cooler portions of the cool humid regions (163). There has been no thatch problem associated with tall fescue turfs. Although drought tolerant, tall fescue does respond to judicious irrigation. It is somewhat resistant to crown rust and *Helminthosporium* spp. (93, 138, 139) but is susceptible to brown patch (3), *Fusarium* patch, *Typhula* blight, and *Ophiobolus* patch.

Cultivars. Five tall fescue cultivars have been developed for turfgrass use (Table 3-8). Alta and Kentucky 31 have been the most widely used cultivars for turfgrass purposes. Tall fescue possesses a number of very desirable characteristics for turfgrass purposes but considerable improvement is needed in the leaf texture, rhizome development, and low temperature hardiness. Intergeneric hybridization of tall fescue with both perennial ryegrass and Italian ryegrass has succeeded in producing hybrids but no improved turf-type cultivars have resulted (37, 38, 47, 116, 156, 231).

Meadow Fescue (*Festuca elatior* L.)

Meadow fescue is a cool season, bunch-type turfgrass that originated in the temperate regions of Eurasia. It has sometimes been called English bluegrass and is utilized throughout Europe where it is known as *Festuca pratensis* Huds. (219). Meadow fescue is adapted to a wide range of soil and climatic conditions similar to tall fescue. Botanically, it resembles tall fescue.

Table 3-8

CHARACTERISTICS OF FIVE TURF-TYPE TALL FESCUE CULTIVARS

Cultivar	Released By	Date	Turfgrass characteristics	Adaptation
Alta	Oregon AES and CRD-ARS, U.S.	1940	Coarse texture; medium low shoot density; moderately upright growth habit	Improved tolerance to drought stress; adapted to a wide range of soil types
Goar (10)	California AES and SCS, U.S.	1946	Improved texture; medium low shoot density; upright, bunch-type growth habit; slower establishment rate	Superior tolerance to heat stress; poor low temperature hardiness
Kenmont	Montana AES, U.S.	1963	Coarse texture; improved shoot density; semierect growth habit	Improved low temperature hardiness
Kentucky 31 (78, 123, 124)	Kentucky AES (natural selection), U.S.	1940	Coarse texture; medium shoot density; slightly prostrate growth habit	Somewhat improved heat and low temperature hardiness; adapted to a wide range of soil types
Kenwell (78)	Kentucky AES and CRD-ARS, U.S.	1965	Somewhat improved texture and shoot density; slower vertical shoot growth rate	The reduced aggressiveness results in improved compatibility in polystands

Plant Description

Vernation rolled; *sheaths* round, glabrous, split with overlapping hyaline margins, reddish at base; *ligule* membranous, 0.2–0.6 mm wide, truncate to obtuse, whitish-green; *collar* continuous, broad, conspicuously glabrous, pale yellow to yellow-green, thickened, frequently flexuous; *auricles* small, blunt, cilia absent; *blades* flat, 3–8 mm wide, glabrous to scabrous and dull above, glabrous and glossy below, veins prominent above, acuminate apex, scabrous margins; *stems* erect, tufted with occasional short, stout rhizomes; *inflorescence* an erect or nodding, slender, somewhat contracted panicle.

Meadow fescue forms a bright green turf that has better uniformity and a finer texture than tall fescue (Fig. 3-13). The leaf texture approaches that of the coarser textured Kentucky bluegrass cultivars. Meadow fescue is a tufted, noncreeping, deep rooted species with the root system regenerated annually (215). Although possessing only a few short rhizomes, it forms a fairly uniform sod that is not as bunchy as tall fescue. The tillering and shoot density are superior to that of tall fescue. Propagation is by seed. The establishment rate is quite good, being more rapid than Kentucky bluegrass but somewhat slower than tall fescue and ryegrass (189). It is cross-fertilized with chromosome numbers of 14, 28, 42, and 70 having been reported (24, 43, 44, 69, 74, 92, 117, 181).

Adaptation. Meadow fescue can be a long- or short-lived perennial, depending on the degree of low temperature stress. It is not as persistent as tall fescue. Meadow fescue is adapted to the cool humid regions of the world and extends somewhat into the cooler portions of the warm humid regions. It is prone to direct low temperature kill in the cooler portions of the cool humid regions. It is more heat and drought tolerant than timothy but less tolerant than tall fescue. The shade adaptation is fairly good in the temperate regions (213). The wear tolerance is fairly good

Figure 3-13. A comparison of Kentucky 31 tall fescue (left) and meadow fescue (right) turfs cut at 1.5 inches.

in comparison to the other cool season turfgrasses (236). Meadow fescue is best adapted to fertile, moist soils but grows on sandy sites if a moderate amount of water is applied. It is well adapted to wet or waterlogged soil conditions.

Use and culture. In Eurasia, meadow fescue is frequently included in seed mixtures for use on general-purpose turfs, especially on fertile, moist, shaded sites. Its use for turfgrass purposes has been very minimal in the United States. Meadow fescue is more compatible in polystands with Kentucky bluegrass and ryegrass than tall fescue.

The cultural requirements are similar to those for tall fescue. A mowing height of 1.5 to 2 in. is preferred. It does not tolerate cutting heights of less than 1 in. The vertical shoot growth rate is less than for tall fescue. The nitrogen requirement is 0.4 to 1.0 lb per 1000 sq ft per growing month. It responds to irrigation, especially on sandy, droughty soils. Meadow fescue is quite susceptible to crown rust and *Helminthosporium* spp. (93, 138, 139). Other diseases that cause occassional problems include stripe smut, brown patch, *Fusarium* patch, and *Pythium* blight.

Only a few cultivars of meadow fescue such as Ensign, Mimer, and Trader have been developed and none have been of any significance for turfgrass use. **Ensign**, a synthetic cultivar released in 1944 by the Canada Department of Agriculture of Ottawa, Ontario, is leafy and rust susceptible. **Mimer**, developed by the Weibullsholm Plant Breeding Institution of Sweden, and **Trader**, a synthetic cultivar developed by the Canada Department of Agriculture and released in 1963, have improved leaf rust resistance.

The Ryegrasses (*Lolium* L.)

Ryegrass is the common name for those species included in the genus *Lolium*. Ryegrass should not be confused with cereal rye (*Secale cereale* L.) that is sometimes planted for rapid stabilization of roadside slopes and nonuse areas. *L. multiforum* (Italian ryegrass) and *L. perenne* (perennial ryegrass) are the only species of *Lolium* that are utilized to any extent in turfgrass culture. Common or domestic ryegrass is a mixture of many intermediate types resulting from the field cross pollination of Italian and perennial ryegrasses. The two species are closely related morphologically and the cultural requirements are very similar. They have the most rapid establishment rate of the commonly used cool season turfgrasses.

Perennial Ryegrass (*Lolium perenne* L.)

Perennial ryegrass is the *Lolium* species most widely used for turfgrass purposes and is thought to be one of the first cultivated grasses. It is sometimes referred to as English ryegrass. It is native to the temperate regions of Asia and North Africa.

Plant Description

Vernation folded; *sheaths* somewhat compressed, glabrous, loose, lower sheaths reddish at base, split with overlapping margins; *ligule* membranous, 0.5–1.5 mm long, truncate; *collar* conspicuous, narrow to medium broad, divided, glabrous; *auricles* small to moderate size, claw-like, soft; *blades* flat, 2–5 mm wide, glabrous, glossy below, dull with prominent veins above, keeled, acute apex, margins usually scabrous; *stems* compressed, erect to somewhat decumbent at base, tufted; *inflorescence* erect, spike-like, long, narrow, flat spikes with awnless spikelets positioned edgewise to the rachis.

Perennial ryegrass forms a medium textured turf of good shoot density and uniformity. The leaves are a glossy, light yellow-green color on the under side. These typical leaf characteristics of the older cultivars result in poor compatability when used in Kentucky bluegrass-red fescue polystands (145). A few improved perennial ryegrass cultivars are fairly compatible with Kentucky bluegrass because of the darker green color, slower vertical shoot growth rate, and improved mowing quality. The leaf texture is quite compatible. Perennial ryegrass is difficult to mow because of the extremely tough, fibrous vascular bundles in the leaves (Fig. 3-14). The shredded, mutilated leaf tips associated with most ryegrass cultivars cause a general reduction in turfgrass quality. Perennial ryegrass is a noncreeping, bunch-type species that forms a uniform sod if properly established and maintained. The turf is frequently thin and stemmy during late spring when numerous reproductive tillers are formed. The fibrous root system is annual in nature (215). Propagation is by seed. Perennial ryegrass is a large seeded turfgrass that possesses a rapid rate of seed germination, establishment, and vertical leaf extension (67, 112, 188, 220). It is cross-fertilized with a chromosome number of 14 reported (24, 43, 69, 74, 92, 181). Perennial ryegrass can be distinguished from Italian ryegrass by the vernation; the former is folded in the bud shoot and the latter is rolled. The recuperative potential is quite poor.

Adaptation. Perennial ryegrass is generally considered to be a short-lived perennial. It can persist indefinitely, however, if not subjected to extremes in high or low temperature stress. It is best adapted to cool, moist regions that have mild winters and

Figure 3-14. The poor mowing quality of perennial ryegrass (right), even when cut with a well adjusted, sharp reel mower, in comparison to a Kentucky bluegrass turf (left).

cool, moist summers. It does not tolerate climatic extremes of cold, heat, or drought (42). Perennial ryegrass has the poorest low temperature hardiness of the perennial cool season turfgrasses and is readily killed by low temperatures in the cooler portions of the cool humid climates (220). A few perennial ryegrass cultivars are available that have improved low temperature hardiness and, thus, are more perennial in nature. The drought tolerance is medium poor as is the recuperative potential. The adaptation to partial shading is good. The wear tolerance is fairly good compared to the other cool season turfgrasses (236).

Perennial ryegrass is adapted to a wide range of soil types. Best growth occurs on neutral to slightly acidic soils of medium to high fertility. However, an adequate turfgrass cover can be achieved on relatively infertile soils. The tolerance to wet soil conditions is good as long as there is good surface drainage. The tolerance to saline soil conditions is medium.

Use. Perennial ryegrass is a common constituent of seed mixtures used on home lawns, parks, cemeteries, institutional grounds, fairways, roughs, roadsides, airfields, and other general use turfgrass areas. It is traditionally utilized where rapid establishment and soil stabilization are desired such as (a) on slopes having a high erosion potential, (b) seeding at a time of year when the probability of successful establishment is low, and (c) when planting during droughty periods on sites that cannot be irrigated. It is seldom planted alone except as a short term, temporary vegetative cover. The use of perennial ryegrass in North America has been limited primarily to a temporary component of general-purpose turfgrass mixtures. Generally, ryegrass should compose no more than 20 to 25% of the seed mixture, on a seed number basis. A higher ryegrass content in the seed mixture results in excessive competition with the desirable turfgrass species, such as Kentucky bluegrass, and impairs the establishment of these species (26, 50, 67). Whenever possible, perennial ryegrass cultivars having a rapid vertical shoot growth rate should be withheld from quality turfgrass seed mixtures unless rapid establishment is needed (67, 220).

The traditional viewpoint concerning perennial ryegrass is expected to change with the development of diminutive cultivars having a slow vertical shoot growth rate. Improved, compatible perennial ryegrass cultivars have good potential for use in polystands with Kentucky bluegrass, especially on sports turfs in the warmer portions of the cool humid regions (Fig. 3-15).

Perennial ryegrass is utilized as a major constituent of general-purpose turfs in certain regions of Europe characterized by moderate temperatures and

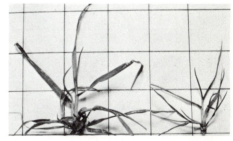

Figure 3-15. The diminutive size of a Manhattan perennial ryegrass (right) plant in comparison to Italian ryegrass (left) at 4 weeks of age. (Photo courtesy of J. M. Duich, Pennsylvania State University, University Park, Pennsylvania).

Table 3-9

CHARACTERISTICS OF SEVEN TURF-TYPE PERENNIAL RYEGRASS CULTIVARS

Cultivar	Released By	Date	Turfgrass characteristics	Adaptation	Other comments
Linn (78)	Oregon AES, U.S.	1961	Bright medium green color; low shoot density; stemmy, erect growth habit; very rapid vertical shoot growth rate	Poor tolerance to temperature extremes	Poor mowing quality
Manhattan (79)	New Jersey AES, U.S.	1967	Bright, moderately dark green color; medium fine texture; medium high shoot density; leafy, diminutive growth habit; profuse tillering; rooting at nodes; slow vertical shoot growth rate	Medium good tolerance to heat and shaded conditions	Medium mowing quality; moderate resistance to brown patch and Fusarium patch; moderately susceptible to rust, red thread, and dollar spot
NK-100 (83)	Northrup, King & Co., U.S.	1962	Bright, medium green color; medium fine texture; medium shoot density; leafy, diminutive growth habit; fairly rapid vertical shoot growth rate	Medium tolerance to drought, heat, and low temperature stress	Moderate susceptibility to Fusarium patch; susceptible to red thread
Norlea (77, 83, 155)	Central Exp. Sta., Ottawa, Ont., Can.	1958	Dark green color; medium texture and shoot density; semierect growth habit; medium slow vertical shoot growth rate	Medium good low temperature hardiness; medium poor tolerance to heat stress	Medium mowing quality; fair Fusarium patch resistance; susceptible to leaf rust
Pelo (79, 112)	D. J. van der Have, Netherlands	1962	Bright medium green color; medium fine texture; medium shoot density; extensive tillering; fairly rapid vertical shoot growth rate	Medium hardiness to heat and low temperature stress	Susceptible to red thread; moderately susceptible to rust; moderately resistant to Fusarium patch

Table 3-9 (Cont.)

| Pennfine | Pennsylvania AES, U.S. | 1969 | Bright, medium dark green color; medium fine texture; medium high shoot density; leafy, diminutive growth habit | Medium good tolerance to heat and low temperature stress | Medium good mowing quality; good rust resistance |
| S.23 (112, 114) | Welsh Plant Breeding Station, Aberystwyth, G. B. | 1933 | Bright medium green color; medium texture; medium shoot density; leafy, prostrate growth habit | Poor low temperature hardiness; good low temperature color retention | Poor mowing quality; susceptible to red thread; tolerates moderately low cutting heights |

minimal soil moisture stress. It is widely used in England as a constituent of winter sports turfs for soccer, football, and hockey fields. It is quite wear tolerant during the period of winter play.

Perennial ryegrass is also utilized for the winter overseeding of dormant warm season turfs such as bermudagrass. It establishes rapidly and provides a uniform, green surface for the winter period when seeded at high rates. The spring transition period is poor due to the abrupt, early loss of the ryegrass (203).

Culture. Perennial ryegrass requires a medium to medium low intensity of culture. A cutting height of 1.5 to 2 in. is preferred (77). The tolerance to cutting heights of less than 0.9 in. is poor (113, 115). It is quite difficult to mow even with a properly adjusted, sharp reel mower because of the tough, fibrous vascular bundles in the leaves (145). The nitrogen fertility requirement ranges from 0.4 to 1 lb per 1000 sq ft per growing month. Higher fertility levels decrease the tolerance to environmental stress. Irrigation is required to ensure the survival of ryegrass during prolonged drought periods. Thatch has not been a problem with this species.

Pythium blight is a warm weather seedling disease of perennial ryegrass that is a serious problem in the successful winter overseeding of dormant warm season species. Rust, *Fusarium* patch (83), brown patch (83), red thread, stripe smut, *Typhula* blight, and *Helminthosporium* spp. can also be injurious to perennial ryegrass.

Cultivars. Most of the turf-type perennial ryegrass cultivars have been developed since 1960 (Table 3-9). Although some improvements have been made, there are still major improvements that must be made before perennial ryegrass will be widely used as a major component of permanent turfgrass polystands. Characteristics needing further improvement are (a) mowing quality; (b) a more diminutive, leafy growth habit with a slow vertical shoot growth rate; and (c) improved tolerance to heat, drought, and low temperature stress.

Italian Ryegrass (*Lolium multiflorum* Lam.)

Italian ryegrass is native to the Mediterranean regions of southern Europe, North Africa, and Asia Minor. It is frequently referred to as annual ryegrass. The plant characteristics, adaptation, use, and cultural requirements of Italian ryegrass are quite similar to those for perennial ryegrass.

Plant Description

Vernation rolled; *sheaths* round, glabrous, lower sheaths yellowish-green to reddish at base, split with tightly overlapping, hyaline margins; *ligule* membranous, 1–2 mm long, obtuse, entire; *collar* broad, may be divided; *auricles* prominent, narrow, claw-like; *blades* flat, 3–7 mm wide, glabrous, dull above, glossy below, veins prominent, acuminate apex, margins glabrous and hyaline; *stems* round, erect to somewhat spreading, robust, tufted; *inflorescence* erect, spikelike, having long, narrow flat spikes with awned spikelets positioned edgewise to the rachis, alternating in two rows.

The leaves are lighter green and coarser in texture than perennial ryegrass (164). The light green, shiny leaves cause Italian ryegrass to be less acceptable for use in polystands with Kentucky bluegrass and red fescue. The shoot density, uniformity, and overall turfgrass quality of an Italian ryegrass turf is not as good as that for perennial ryegrass. Tillering is initiated at a lower node on the crown than for perennial ryegrass. This results in a larger total leaf area when in the seedling stage (45, 164, 189). The growth habit is an upright, bunch type with no rhizomes or stolons (Fig. 3-16). The depth and number of roots produced by Italian ryegrass is less than for perennial ryegrass. Italian ryegrass has a very rapid establishment rate (188) with propagation being by seed. It is cross-fertilized

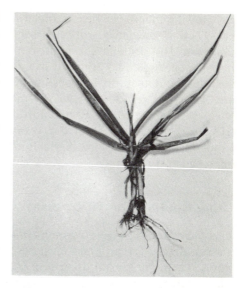

Figure 3-16. The upright, noncreeping growth habit of Italian ryegrass.

with a chromosome number of 14 reported (24, 69, 74, 92, 181). The recuperative potential is very poor. *Adaptation.* Italian ryegrass usually acts as an annual. Under moderate temperatures, however, it may behave as a biennial or even a short-lived perennial. The adaptation is comparable to that for perennial ryegrass. Italian ryegrass is the least low temperature hardy of the cool season turfgrasses. The tolerance to moisture and heat stress is even poorer than for perennial ryegrass (153, 213). It is best adapted to fertile, wet soils having a pH of 6.0 to 7.0. It forms an adequate turf under low fertility conditions. The tolerance to wet soil conditions is good but the submersion tolerance is poor.

Use and culture. Italian ryegrass is included in seed mixtures used for general turf-grass purposes where quick establishment is desired. The excessive competition to the permanent turfgrasses, and the maximum percentages used in seed mixtures are similar to those previously discussed for perennial ryegrass. Italian ryegrass is even more competitive than perennial ryegrass (77, 83). It can be effectively used as a temporary vegetative cover. It forms a uniform, green cover quite rapidly when seeded in late spring or summer and can be plowed in late summer in preparation for permanent turfgrass establishment. Italian ryegrass is used for winter overseeding of dormant warm season turfs in the warm humid regions. The early spring death of Italian ryegrass results in poor transition to the warm season turfgrasses (203). The overseeded ryegrass is also quite prone to injury from *Pythium* blight.

The cultural requirements are similar to those for perennial ryegrass. A cutting height of 1.5 to 2 in. is preferred. The mowing quality is almost as poor as perennial ryegrass. The nitrogen fertility requirement ranges from 0.4 to 1 lb per 1000 sq ft per growing month. Higher nitrogen fertility levels increase the low temperature kill. There is no thatch problem associated with Italian ryegrass.

Cultivars. The demand for improved turf-type cultivars has been limited since Italian ryegrass is used primarily as a temporary cover. Several cultivars have been developed but they do not possess any outstanding characteristics for turfgrass use. **Astor** was released in 1964 by the Oregon Agricultural Experiment Station. It is well adapted to the Pacific northwest region of the United States. **Gulf** was released cooperatively from the Texas Agricultural Experiment Station and the Crops Research Division, Agricultural Research Service in 1958. It possesses improved rust resistance.

Other Cool Season Turfgrasses

The *Poa, Festuca, Agrostis,* and *Lolium* genera encompass the majority of the grasses utilized for turfgrass purposes throughout the cool humid climates of the world. There are several other genera however, which have limited and very specific turfgrass uses. Included in this group are beachgrass, smooth brome, timothy, orchardgrass, and crested dog's-tail.

Beachgrass (*Ammophila* Host.)

The beachgrasses are primarily associated with sandy coastal areas. The genus name is derived from the Greek *ammos* meaning sand and *philos* meaning loving. Two species of *Ammophila* propagated for the purpose of sand stabilization are: American beachgrass (*Ammophila breviligulata* Fernald) and European beachgrass [*Ammophila arenaria* (L.) Link]. The two species are similar in morphology, adaptation, and use. American beachgrass is indigenous to North America, while European beachgrass is native to the coastal regions of northern Europe.

Plant Description

Vernation rolled; *sheaths* round, glabrous to slightly scabrous, split with overlapping, hyaline margins, numerous, tough, broad, overlapping sheaths at the base of stems; *ligule* membranous, 1–3 mm long, obtuse, possibly short cilia; *collar* broad, may be divided; *auricles* absent; *blades* involute, 5–10 mm wide, firm, stiff, scabrous above, glabrous and glossy below, margins glabrous and hyaline; *stems* round, erect, stiff with deep, strong, extensively creeping rhizomes; *inflorescence* pale, dense, nearly cylindrical, and spike-like.

Beachgrass is tough, coarse textured, and erect growing. It is deep rooted and possesses scaly, hard, creeping, underground rhizomes that spread rapidly. The tough, coarse stems have excellent resistance to the abrasive action of windblown sand particles (141). This is one of the key characteristics that enables beachgrass to establish, survive, and function in sand stabilization. Even young stems can resist the effects of wind driven sand. Beachgrass is propagated almost entirely by vegetative clones since viable seed is seldom produced. It initiates stems quite rapidly from underground buds and forms large clumps.

Adaptation. Beachgrass is a perennial, cool season grass adapted to the cool humid marine climates of coastal regions. It is very drought tolerant and grows best on unstable beach sand of low fertility. Beachgrass thrives on sand deposition and makes rapid growth through sand deposits of 2 ft or more per year (141).

Use and culture. Beachgrass is unexcelled for the specific purpose of sand stabilization (Fig. 3-17). It has excellent capabilities in the initial stabilization of shifting dunes and produces a rapid accumulation of organic matter from the leafy foliage. Mechanical barriers may be needed to assist establishment on sites subjected to severe drifting of sand. Mowing and fertilization are not normally practiced where beachgrass is used for dune stabilization. However, establishment is significantly enhanced by the application of fertilizer and lime.

American beachgrass is preferred for sand dune stabilization because it is more permanent in nature and more heat tolerant (35, 161). It is widely used on the shores of the Great Lakes, along the Atlantic coast of North America as far south as North Carolina, and along the Pacific coast. **European beachgrass** has been used along the Pacific coast of North America and the northern European coastline.

Figure 3-17. A clonal planting of beachgrass utilized in the stabilization of wind-blown sand at Provincetown, Massachusetts. (Photo courtesy of J. Zak, University of Massachusetts, Amherst, Mass.).

It is particularly effective in controlling shifting sand dunes along coastal regions. It will die out when the sand ceases to move, however, and thus is not suitable for permanent sand dune control. European beachgrass resembles American beachgrass except for a larger ligule on the former. There are no cultivars of American or European beachgrass available.

Smooth Brome (*Bromus inermis* Leyss.)

Smooth brome is native to Europe, Siberia, and China. It is now widely distributed throughout the temperate regions of the world. Because of a coarse leaf texture, smooth brome has only limited turfgrass use for erosion control on minimal use areas such as roadsides.

Plant Description

Vernation rolled; *sheaths* round, glabrous, white with pink veins, closed to near the top; *ligule* membranous, less than 1 mm long, obtuse; *collar* narrow to medium broad, divided, glabrous; *auricles* absent; *blades* flat, 8–12 mm wide, glabrous or minutely scabrous, keeled below, margins glabrous to scabrous; *stems* stout, erect with vigorous, creeping rhizomes; *inflorescence* an erect, dense, contracted panicle, branches whorled.

Smooth brome forms an upright, coarse textured, open sod of quite low turfgrass quality. It spreads vegetatively by vigorous, fleshy rhizomes that form a firm sod. The root system is extensive. Smooth brome is propagated mainly by seed. It is primarily cross-fertilized, with chromosome numbers of 28, 42, 56, and 70 having been reported (24, 74, 92). The seeds of smooth brome are large, flat, light, and chaff-like. The recuperative potential is medium good.

Adaptation. Smooth brome is a long-lived perennial that is widely distributed throughout the cool humid and transitional regions of the world. It has fairly good drought resistance with good heat and low temperature hardiness. Smooth brome is capable of becoming dormant during extended dry periods common to the semiarid regions and can initiate new growth when adequate moisture becomes available. The tolerance to traffic is poor (171). Smooth brome is best adapted to deep, well-drained, fertile, fine textured soils but also does well on coarse textured sandy soils if additional nitrogen is supplied. It is somewhat tolerant of alkaline soil conditions (98) and tolerates short periods of flooding with the associated deposition of silt.

Use and culture. Smooth brome is most effectively used on deep, fertile prairie soils. Due to the coarse texture and low shoot density, smooth brome has only limited turfgrass use for erosion control on roadsides and similar low quality, minimal use areas. It is included as a constituent of roadside seed mixtures used in the Great Plains of the United States. The tolerance to frequent, close mowing is very poor (95).

Cultivars. Smooth brome is usually differentiated into northern and southern types. The northern types originated from Siberia and are more low temperature hardy but less heat and drought tolerant. They are used primarily in the cooler portions of the cool humid regions. The rhizome development is less vigorous than for the southern types. In contrast, the southern types originated from central Europe and are more tolerant of heat and drought stress.

Timothy (*Phleum* L.)

Timothy is of European origin and was called Herd's grass at one time. There are two *Phleum* species of concern in turfgrass culture. *P. pratense* L. is the more upright growing type. It has been widely used for forage and to a limited extent on roadsides and functional turfgrass areas. *P. nodosum* L. is a lower growing type that is sometimes referred to as wild timothy or, more recently, turf timothy. The latter species is better adapted for turfgrass uses, while *P. pratense* has been utilized only sparingly for turfgrass purposes due to poor tolerance to close, frequent mowing.

Plant Description

Vernation rolled; *sheaths* round, glabrous, split with overlapping, hyaline margins; *ligule* membranous, prominent, 3–6 mm long, obtuse, distinctly notched on either side; *collar* broad, conspicuous, usually divided, glabrous, light green; *auricles* absent; *blades* flat, 4–8 mm wide, glabrous near base and scabrous near apex, slightly keeled, venation distinct, long acuminate apex, margins scabrous; *stems* erect, tufted, stout, swollen or bulb-like at base; *inflorescence* a densely flowered, cylindrical, compact, spike-like panicle.

Timothy tends to behave as a bunch type with poor sod forming qualities. The leaves frequently have a grayish-green appearance. Improved cultivars of *P. nodosum* are more prostrate growing and have a slower vertical shoot growth rate than *P. pratense*. Most cultivars of *P. pratense* form a poor quality, open, coarse textured turf when mowed at a height of 2 in. or less (213). The shallow, fibrous root system is replaced annually. Timothy differs from most turfgrasses in that the lower one or two internodes of the stem remain relatively short and become swollen into an ovid body referred to as a **haplocorm**. These corms and the associated culms behave as winter annuals, forming in early summer and dying within a year. The corms serve as storage organs for carbohydrates. Timothy is propagated by seed with a fairly rapid establishment rate (114). It is cross-fertilized with a 2n chromosome number of 14 for *P. nodosum* and 42 for *P. pratense* (24, 74, 92, 181).

Adaptation. Timothy is adapted to the cool humid climates where it may persist as a perennial. Timothy behaves as a short-lived perennial, however, if mowed short or subjected to heat and drought stress (95). The low temperature hardiness is good but the tolerance to heat and drought stress is inferior to that of tall fescue and meadow fescue. It is extremely slow to recover from the effects of midsummer heat and water stress. The recuperative potential of the turf-type timothys (*P. nodosum*) is fairly good (114). The wear tolerance is very poor.

Timothy tolerates soils having a relatively wide fertility range but responds to high fertility levels. It is best adapted to moist, fine textured soils having a pH of 6 to 7 (98). Acidic soils adversely affect timothy.

Use and culture. The use of *P. pratense* has been limited to low quality turfgrass areas such as roadsides and similar nonuse areas. Even then it only persists on well-drained, moist, fine textured soils that are not subjected to severe moisture stress. The newer turf-type timothy cultivars (*P. nodosum*) are being used on close cut sports turfs, particularly in northern Europe under cool, moist conditions. Turf-type timothys tolerate cutting heights as low as 0.6 in. Timothy has a fairly high nitrogen fertility requirement of 0.5 to 1.0 lb per 1000 sq ft per growing month. It responds to irrigation under droughty conditions. Thatch is not a problem.

Cultivars. Most *P. pratense* cultivars were developed for agricultural purposes and do not tolerate close mowing. There are improved selections of *P. nodosum*, however, that have a more diminutive plant size and the low growth habit needed for turfgrass uses (Table 3-10).

Orchardgrass (*Dactylis glomerata* L.)

Orchardgrass is a native of Europe where it is known as cocksfoot. The leaf texture is quite coarse with leaves folded in the bud shoot and sheaths distinctly compressed. It forms an open sod of low shoot density. Orchardgrass is basically a bunch-type grass since it has neither rhizomes nor stolons. Propagation is by seed with pollination being primarily by cross-fertilization (24, 74). Orchardgrass is adapted to the cool humid regions of the world. The low temperature hardiness is relatively poor, ranking less than smooth brome, timothy, and Kentucky bluegrass. Because of somewhat greater heat tolerance, it is used primarily in the warmer portions of the cool humid climates and to a certain extent in the transitional zone. Orchardgrass has a rapid early spring green-up rate and good low temperature color retention. The drought tolerance is greater than for timothy but not as good as for smooth brome. The shade adaptation is quite good but the wear tolerance is quite poor (171).

Orchardgrass is adapted to a wide range of soil types including infertile, acidic soils. However, it does respond to fertility and grows best on moist, fine textured soils of medium fertility. It tolerates cool, wet, imperfectly drained soils. Since orchardgrass forms a coarse textured, weak sod, its use is quite limited for turfgrass purposes. Typical use situations include roadsides and other low quality, minimal maintenance areas.

Crested Dog's-Tail (*Cynosurus cristatus* L.)

Crested dog's-tail is a perennial species that is native to Europe. It has a very fine leaf texture and a distinctly tufted growth habit similar to perennial ryegrass. This species is quite drought resistant and is adapted to a relatively wide range

Table 3-10

CHARACTERISTICS OF FOUR TURF-TYPE TIMOTHY CULTIVARS

Cultivar	Release By	Date	Phleum species	Turfgrass characteristics	Adaptation
Evergreen (114)	Weibullsholm Plant Breeding Institution, Sweden	—	*nodosum*	Medium green color; medium texture; medium high shoot density	Good low temperature hardiness; lacks heat and drought tolerance; prefers poorly drained, fine textured soils; tolerates close mowing
Heidimij (112, 114)	D. J. van der Have, Netherlands	1935	*pratense*	Medium light green color; medium coarse texture; prostrate growth habit; aggressive	Good wear tolerance; prefers medium to coarse textured soils
King (112, 113, 114)	D. J. van der Have, Netherlands	1961	*pratense*	Medium dark green color; coarse texture; low growth habit	Good low temperature color retention
S.50 (112, 113)	Welsh Plant Breeding Station, Aberystwyth, G. B.	1932	*nodosum*	Light green color; medium texture; medium high shoot density; prostrate growth habit; rooting at the nodes	Good wear tolerance; tolerates close mowing; resistant to rust

of soil types. The low temperature hardiness is quite poor. Crested dog's-tail tolerates moderately close mowing but forms rather hard stalks that are objectionable in appearance. The wear tolerance is good. Propagation is by seed with the establishment rate being quite slow. Seed heads can be a problem and are difficult to mow. Crested dog's-tail is utilized to a limited extent in England on cricket outfields and football grounds. It is not utilized for putting and bowling greens because of poor tolerance to close, frequent mowing. The recuperative potential is poor.

The Wheatgrasses (*Agropyron* Gaertn.)

The wheatgrasses are classified among the approximately 150 species in the *Agropyron* genus. They are (a) easily established, (b) adapted to a wide range of soil types, (c) extremely drought resistant, and (d) adapted to extremes in climate. Because of these characteristics, the wheatgrasses are quite valuable for soil stabilization in the cool subhumid and semiarid regions. The species most commonly used are *A. cristatum*-fairway wheatgrass and *A. smithii*-western wheatgrass. Other less frequently used species include intermediate wheatgrass [*Agropyron intermedium* (Host) Beauv.] and slender wheatgrass [*Agropyron trachycaulum* (Link) Malte]. A widely known species of this genus that was introduced from Eurasia is quackgrass [*Agropyron repens* (L.) Beauv.]. In certain regions it is one of the most troublesome perennial weedy grasses in turfs mowed at heights above 1 in. The rhizome system of quackgrass is deep and extensive.

Fairway Wheatgrass [*Agropyron cristatum* (L.) Gaertn.]

Fairway wheatgrass is native to the cold, dry plains regions of Russia and Siberia. In the past it has been confused with crested wheatgrass [*Agropyron desertorum* (Fisch.) Schult.], which is more variable, taller growing, more open in shoot density, coarser stemmed, and more rapid in vertical shoot growth. The adaptation, use, and culture of crested wheatgrass for turfgrass purposes are similar to the following description for fairway wheatgrass.

Plant Description

Vernation rolled; *sheaths* nearly round, glabrous, veins distinct, open; *ligule* membranous, 0.5–1.5 mm long, obtuse, short ciliate margin; *collar* medium narrow, distinct, divided; *auricles* claw-like; *blades* nearly flat, 2–5 mm wide, pubescent, veins prominent above, midrib prominent below, scabrous above, acuminate apex, margins weakly scabrous; *stems* round, erect to somewhat decumbent at base; *inflorescence* a spike with spikelets widely spreading.

Fairway wheatgrass is shorter, denser, finer stemmed, and slower growing than most wheatgrasses. The color is a bright green with an abundance of basal and stem leaves produced. The growth habit is an upright, bunch type that is less tufted than other wheatgrasses (132). The root system is extensive, deep, fibrous, and perennial in nature (215). Roots have been observed at depths of 8 to 10 ft. Fairway wheatgrass can be distinguished from crested wheatgrass by the pubescence present on practically all plant parts, whereas sparse pubescence is evident on only the upper surface of crested wheatgrass leaves. Fairway wheatgrass is a prolific seed producer with propagation being by seed. It is primarily cross-fertilized (74, 92). The germination and establishment rates are very good. Adequate stands can be obtained even when the seed is drilled into poorly prepared, weedy seedbeds.

Adaptation. Fairway wheatgrass is a perennial, cool season species that is particularly well adapted to the cool, dry plains and mountainous regions. The fibrous root system grows laterally and close to the soil surface. This enables fairway wheatgrass to compete readily with weeds for the available soil moisture. It possesses excellent drought resistance, good low temperature hardiness, and tolerance to frequent defoliation. Fairway wheatgrass initiates growth very early in the spring. The tolerance to high temperatures is poor. Fairway wheatgrass has the capability of becoming dormant during hot, dry summer periods. Growth is resumed when more favorable moisture conditions occur. The tolerance to prolonged flooding and wet soils is poor. Fairway wheatgrass is adapted to fertile soils having textures ranging from sandy loam to clay. The tolerance to alkaline soils is not as good as that for western wheatgrass.

Use and culture. The ease of establishment and outstanding tolerance to moisture stress make fairway wheatgrass one of the best species available for use in revegetating plains with low rainfall. It is used very effectively for controlling wind and water erosion and can successfully compete with coarse weeds. The primary turfgrass use of fairway wheatgrass is on roadsides and other minimal use turfgrass areas in the cool subhumid and semiarid regions (57) (Fig. 3-18). It also forms

Figure 3-18. A stand of fairway wheatgrass on a roadside in the semiarid plains region of western Nebraska in August. (Photo courtesy of A. E. Dudeck, University of Nebraska, Lincoln, Nebraska).

an adequate turf under unirrigated conditions in the cool semiarid and subhumid regions where it is used on general purpose turfgrass areas as well as on play-fields, fairways, tees, and roughs (132).

Fairway wheatgrass requires a medium to low intensity of culture. It tolerates frequent mowing at heights of 1.5 to 2.5 in. The nitrogen requirement is 0.2 to 0.7 lb per 1000 sq ft per growing month. Irrigation should be minimal, otherwise the turf thins out like red fescue. There is no thatch problem. No improved turf-type cultivars of fairway wheatgrass are available.

Western Wheatgrass (*Agropyron smithii* Rydb.)

Western wheatgrass is indigenous to the semiarid plains regions of North America. The adaptation is somewhat wider than for fairway wheatgrass, particularly the distribution into warmer and wetter climates.

Plant Description

Vernation rolled; *sheaths* round, glabrous split with hyaline margins; *ligule* membranous, 0.1–0.5 mm long, truncate, short ciliate margin; *collar* medium broad, glabrous, divided; *auricles* narrow, 1–2 mm long, claw-like, glabrous, generally colored; *blades* generally flat, stiff, 2–5 mm wide, veins prominent above, midrib indistinct, scabrous, or pubescent above, glabrous below, glaucous blue-green color, acuminate apex, sharply barbed or toothed; *stems* round, erect, with creeping rhizomes; *inflorescence* an erect spike with spikelets rather closely imbricate.

The growth habit is erect with stiff leaves prominently ribbed lengthwise and very rough on the upper surface. The leaves roll up quite tightly and assume a wire-like form during periods of moisture stress. The turf has a characteristic bluish-green color. It has vigorous, strongly creeping rhizomes that form a tight, heavy sod. The rhizomes proliferate to a depth of 3 in. below the soil surface. The root system extends to a depth of 8 ft or more depending upon the moisture conditions. Propagation is primarily by seed although vegetative establishment is sometimes practiced, especially in waterways. Western wheatgrass is largely cross-fertilized with chromosome numbers of 42 and 56 having been reported (92). *Adaptation.* Western wheatgrass is a long-lived perennial adapted to the cool sub-humid, semiarid, and transitional regions. It possesses good drought resistance and low temperature hardiness (225). The spring green-up rate is fairly good. It enters a dormancy period during midsummer droughts and initiates growth again when moisture becomes available. Western wheatgrass is adapted to a wide range of soil types but grows best on fine textured soils where an adequate supply of water is available. It has moderate tolerance to alkaline soil conditions. *Use and culture.* Because of the vigorous rhizomes, drought resistance, low temperature hardiness, and wide adaptation to soil and climatic conditions, western wheatgrass is used in seed mixtures for the revegetation of subsoils and droughty sites, particularly along roadsides and waterways. Its greatest value is in soil erosion control. Western wheatgrass is also utilized to a limited extent in the cool sub-humid and semiarid regions for unirrigated general-purpose turfs. Because of the slow establishment rate from seed, western wheatgrass is usually planted in mixtures rather than alone. It requires a medium to low intensity of culture comparable to that for fairway wheatgrass. The resistance to grasshoppers is better than for smooth brome. No improved turf-type cultivars of western wheatgrass are available.

The Legumes

Although not related to the cool season grasses, the legumes are discussed briefly since they are involved in certain specialized uses associated with general-purpose turfgrass areas. The legumes are classified in the *Leguminosae* family and comprise nearly 500 genera and over 11,000 species. Less than a half dozen species are of any concern for turfgrass use. The legumes are dicotyledons characterized by monocarpellary fruit that contains only a single row of seeds and dehisces along both ribs. Typically, legume leaves are arranged alternately, have large stipules, and are usually either pinnately or palmately compound.

Most herbaceous legumes have a taproot with symbiotic bacteria forming nodules capable of fixing free nitrogen from the atmosphere. This fixed nitrogen becomes available to the legume plant and, to a certain degree, to grasses growing in association with the legumes.

The legumes are seriously injured by 2,4-D, 2,4,5-TP, and related broadleaf herbicides. These herbicides are commonly used along roadsides but must be withheld from areas where legumes are desired as a permanent component of the plant community. Many perennial legumes used for turfgrass purposes are quite susceptible to injury from insects and, therefore, tend to behave as short-lived perennials. With the exception of crownvetch, the use of legumes in turfgrass seed mixtures is minimal and declining.

Crownvetch (*Coronilla varia* L.)

Crownvetch is a perennial, herbaceous legume that is native to Europe but is not a true vetch. It has a creeping, semivining growth habit that forms a solid green mat of foliage approximately 18 to 24 in. high. The vigorous growth habit is quite competitive and crowds out most competing weedy species. The leaves are pinnate with six to thirteen pairs of leaflets. The stems are leafy, hollow, and weak, which impairs the climbing tendency. Crownvetch is a relatively self-sterile species that seeds profusely (102). The root system is extensive, multibranched, and deep (179). In addition, numerous fleshy horizontal stems greatly enhance the rate of

Figure 3-19. The stabilization of a steep roadside slope in eastern Pennsylvania with Penngift crownvetch.

establishment and spread since new roots and shoots are initiated from the nodes. Individual creeping stems have been observed to grow 10 ft or more in length with a single plant spreading over an area of 75 to 100 ft within a 3- to 4-year period.

Crownvetch is well adapted to the cool humid regions and is also utilized in the cooler portions of the warm humid climates. The drought and low temperature hardiness are excellent (179). Crownvetch forms a good cover and deep root system even on infertile, acidic soils, such as those on steep roadside cuts. It is best adapted to well-drained soils having a pH of 6.5 to 7.0 and high phosphorus, potassium, calcium, and magnesium levels (160). Crownvetch is not adapted to wet, imperfectly drained soils (100).

Crownvetch is highly valued for soil improvement, erosion control, and similar uses along roadsides and other steep slopes that cannot be mowed (179, 199) (Fig. 3-19). An additional benefit is the pleasing appearance, particularly when in flower. The blossoms vary in color from a whitish-pink to a purplish-pink.

The establishment rate is rather slow. Propagation can be by seed or vegetative means. Seeding at a rate of 20 lb per acre is the common method for establishing crownvetch. Dehulling and seed scarification are required to minimize the hard seed problem and ensure an adequate stand of crownvetch from seed. The seed must also be inoculated prior to planting. Crownvetch seedlings grow slowly and may require as much as two full growing seasons to achieve an adequate cover (179, 201). For this reason, crownvetch is usually seeded in association with a companion species. Red fescue and ryegrass are the preferred companion grasses to plant with crownvetch (179). The companion species provides initial cover and slope stabilization until the crownvetch becomes established and eventually predominates. The crownvetch seedlings are subject to frost heaving if seeded in late fall, especially if a mulch is not used (199). Establishment can be achieved in 2 years by the vegetative planting of crowns on 18- to 24-in. centers. Successful vegetative establishment is unlikely should a drought occur during this period. This method is quite costly and time consuming.

Three cultivars of crownvetch are available (Table 3-11). Penngift is the most widely used of the three, particularly as a vegetative cover for slope stabilization on roadsides.

Table 3-11

CHARACTERISTICS OF THREE CROWNVETCH CULTIVARS UTILIZED IN ROADSIDE SLOPE STABILIZATION

Cultivar	Released		Turfgrass characteristics	Adaptation
	By	Date		
Chemung (57)	SCS, U.S.	1961	Tall growing with large leaves and coarse stems; medium good establishment rate	Better adapted to infertile soils
Emerald (57)	Iowa AES and SCS, U.S.	1962	Tall growing with large leaves and coarse stems; poor establishment rate	Prefers moderately fertile soils
Penngift (73, 159, 179, 201)	Pennsylvania AES (natural selection), U.S.	1954	Moderately tall growing with medium sized leaves and stems; intermediate rate of establishment; widely used for roadside slope stabilization	Good drought tolerance and low temperature hardiness

Alfalfa (*Medicago* L.)

Alfalfa is sometimes used as a legume component in roadside seed mixtures. It originated in southwest Asia and is also called lucerne in certain countries. This herbaceous, perennial legume has pinnately trifoliolate leaves arranged alternately on the stem. Alfalfa is adapted to a wide range of climatic conditions and is utilized in the cool humid, transitional, and warm humid climates. The drought resistance is good due to a deep taproot (26). The heat tolerance is also good but the low temperature hardiness is poor. Frost heaving is a common cause of winter damage. Alfalfa is adapted to a wide range of soils but is best utilized on deep, fertile loams having a pH of 5.5 to 7.5 and an open, porous subsoil. Roots have been known to penetrate soils to a depth of 25 to 30 ft. Alfalfa usually behaves as a short-lived perennial under roadside conditions where it is most commonly used.

Lespedeza (*Lespedeza* Michx.)

Lespedeza is an upright growing legume used on roadsides and other minimal use areas for erosion control and soil improvement. It is adapted to the warm humid and transitional regions and acidic soils of low fertility. The heat and drought tolerance are good. It lacks low temperature hardiness. Lespedeza grows on a wide range of soil types except for extremely sandy or wet soils. It does not tolerate soil pH's above 7. Propagation is by seed. The flowers are normally self-fertilized. Seed inoculation is usually needed only for initial establishment.

There are three *Lespedeza* species of Asiatic origin used on roadsides. Two are summer annuals that voluntarily reestablish from seed each year. **Common** lespedeza [*Lespedeza striata* (Thumb.) Hook. and Arn.] is adapted to warmer temperatures. Kobe, a cultivar of common lespedeza, has a poor reseeding capability. **Korean** lespedeza (*Lespedeza stipulacea* Maxim.) is the most extensively used *Lespedeza* species. It is well adapted to the cooler portions of the warm humid and transitional regions. The drought tolerance is quite good.

Sericea lespedeza [*Lespedeza cuneata* (Dumont) G. Don] is an herbaceous, perennial type, introduced from Japan. It provides better soil stabilization during the winter than the annuals. Sericea lespedeza is best adapted to the transitional climates.

The True Clovers (*Trifolium* L.)

The true clovers originated in southwestern Asia Minor and southeastern Europe. The species within this genus are the most widely distributed legumes. The clovers prefer a cool humid climate and do best on fertile, imperfectly drained soils. There are two species of *Trifolium*, (a) white clover (*Trifolium repens* L.) and (b) alsike clover (*Trifolium hybridum* L.), that are sometimes utilized in turfgrass mixtures, primarily for roadsides and other minimal use areas.

White clover is a native of southeastern Europe that is now one of the most widely distributed legumes. It is best adapted to moist, fine textured soils and cool humid climates. This decumbent, perennial legume spreads by means of stolons. It frequently behaves as a winter annual in warm humid climates. The leaves are composed of three sessile leaflets borne on a petiole that develops from the crown and at the nodes of stolons. The rooting depth is comparatively shallow. White clover is cross-pollinated, with propagation being primarily by seed.

Prior to 1940, white clover was frequently included in seed mixtures planted on ornamental lawns. The main attribute of white clover was the associated nitrogen-fixing capability that provided a certain amount of nitrogen for use by the turfgrasses growing in association with the white clover. This attribute is no longer needed in modern turfgrass culture. White clover is generally considered to be a weed in quality turfs. Distinct white to pinkish flowers are produced even at cutting heights of less than 1 in. These flowers detract from the turfgrass quality. White clover persists at mowing heights as low as 0.3 in. It is generally a greater problem on wet soils, during years of excessive rainfall, or under high potassium fertilization (98).

Alsike clover is an upright, perennial legume adapted primarily to the cool humid regions. The leaves are composed of three leaflets borne on numerous tillers originating from the crown. The heat and drought tolerance are quite poor. It is best adapted to cool, wet, imperfectly drained soils. The flooding tolerance is good. Alsike clover tolerates a wide soil pH range and responds to fertile soil conditions. Propagation is by seed with voluntary reestablishment from seed occurring on roadsides and minimal use areas where it is utilized.

References

1. AKERBERY, E. 1942. Cytogenetic studies in *Poa pratensis* and its hybrid with *Poa alpina*. Hereditas. 28: 1–126.

2. ALBRECHT, H. R. 1947. Strain differences in tolerance to 2,4-D in creeping bent grasses. Journal of the American Society of Agronomy. 39: 163–165.

3. ALLISON, J. L., H. S. SHERWIN, I. FORBES, and R. F. WAGNER. 1949. *Rhizoctonia soloni*, a destructive pathogen of Alta fescue, smooth brome grass and birdsfoot trefoil. Phytopathology. 39: 1.

4. ANONYMOUS. 1924. Named strains of creeping bent. Bulletin of USGA Green Section. 4(10): 240.

5. ANONYMOUS. 1943. Identity of creeping bent strains planted on experimental greens. Timely Turf Topics. December. p. 3.

6. ANONYMOUS. 1944. Four-year summary of ratings of creeping bents on experimental greens. Timely Turf Topics. June. pp. 2–5.

7. ANONYMOUS. 1948. Grass, The Yearbook of Agriculture, USDA. pp. 1–892.

8. ANONYMOUS. 1951. Illahee red fescue (Reg. No. 2) registration. Agronomy Journal. 43: 237.

9. ANONYMOUS. 1952. Registration of Merion Kentucky bluegrass (Reg. No. 1). Agronomy Journal. 44: 155.

10. ANONYMOUS. 1952. Turf performance evaluations. Southern California Turf Culture. 2: 3–4.

11. ANONYMOUS. 1957. St. Ives creeping red fescue (Fr. 10). Parks and Sports Grounds. 22(7): 422.

12. ANONYMOUS. 1958. Registration of the creeping bentgrass variety, Penncross (Reg. No. 1). Agronomy Journal. 50: 399.

13. ANONYMOUS. 1958. Registration of the creeping bentgrass variety, Pennlu (Reg. No. 2). Agronomy Journal. 50: 399.

14. ANONYMOUS. 1958. Pennlawn red fescue (Reg. No. 3), registration. Agronomy Journal. 50: 400.

15. ANONYMOUS. 1958. Nimisila—a new bentgrass. Golf Course Reporter. 26(3): 26–28.

16. ANONYMOUS. 1964. Kingston bent. Massachusetts Turf Bulletin. 2(8): 7.

17. ARBER, A. 1934. The Gramineae. Cambridge University Press, London, England. pp. 1–480.

18. ARMSTRONG, J. M. 1937. A cytological study of the genus *Poa* L. Canadian Journal of Research. 15: 281–297.

19. ASTON, J. L., and A. B. BRADSHAW. 1963. Natural variation in *Agrostis stolonifera* L. (creeping bent) and the value of this grass in turf. Journal of the Sports Turf Research Institute. 11(39): 7–18.

20. BEACH, G. A. 1963. Management practices in the care of lawns. 10th Annual Rocky Mountain Regional Turfgrass Conference Proceedings. pp. 42–45.

21. BEARD, J. B. 1965. Bentgrass (*Agrostis* spp.) varietal tolerance to ice cover injury. Agronomy Journal. 57: 513.

22. ———. 1966. Relationship of creeping bentgrass (*Agrostis palustris* Huds.) root growth to environmental factors in the field. Agronomy Journal. 58: 337–339.

23. ———. 1966. Direct low temperature injury of nineteen turfgrasses. Quarterly Bulletin of the Michigan Agricultural Experiment Station. 48(3): 377–383.

24. BEDDOWS, A. R. 1931. Seed setting and flowering in various grasses. Welch Plant Breeding Station Series H. No. 12. pp. 5–99.

25. BELL, R. S., and J. A. DEFRANCE. 1944. Influence of fertilizers on the accumulation of roots from closely clipped bent grasses and on the quality of the turf. Soil Science. 58: 17–24.

26. BELL, R. S., and J. C. F. TEDROW. 1945. The control of wind erosion by the establishment of turf under airport conditions. Rhode Island Agricultural Experiment Station Bulletin No. 295. pp. 1–22.

27. BENGTSON, J. W., and F. F. DAVIS. 1939. Experiments with fertilizers on bent turf. Turf Culture. 1: 192–213.

28. BRADSHAW, A. D. 1958. Studies of variation in bent grass species, I. Hybridization between *Agrostis tenuis* and *A. stolonifera*. Journal of the Sports Turf Research Institute. 9(34): 422–429.

29. ———. 1959. Population differentiation in *Agrostis tenuis* Sibth. 1. Morphological differentiation. New Phytologist. 58: 208–227.

30. BRITTINGHAM, W. H. 1943. Type of seed formation as indicated by the nature and extent of variation in Kentucky bluegrass, and its practical implications. Journal of Agricultural Research. 67: 225–264.

31. BRITTON, M. P. 1963. Stripe smut damage to Kentucky bluegrass lawns. Illinois Turfgrass Conference Proceedings. pp. 1–2.

32. BRITTON, M. P., and J. D. BUTLER. 1965. Resistance of seven Kentucky bluegrass varieties to stem rust. Plant Disease Reporter. 49(8): 708–710.

33. BROBROV, R. A. 1955. The leaf structure of *Poa annua* with observations on its smog sensitivity in Los Angeles County. American Journal of Botany. 42: 467–474.

34. BROWN, E. M. 1943. Seasonal variations in the growth and chemical composition of Kentucky bluegrass. Missouri Agricultural Experiment Station Research Bulletin 360. pp. 1–56.

35. Brown, R. L., and A. L. Hafenrichter. 1948. Factors influencing the production and use of beachgrass and dunegrass clones for erosion control. I. Effect of date of planting. Journal of the American Society of Agronomy. 40: 512–521.

36. Brown, W. L. 1939. Chromosome complements of five species of *Poa* with an analysis of variation in *Poa pratensis*. American Journal of Botany. 26: 717–723.

37. Buckner, R. C. 1960. Cross-compatibility of annual and perennial ryegrasses with tall fescue. Agronomy Journal. 52: 409–410.

38. Buckner, R. C., H. D. Hill, and P. B. Burrus, Jr. 1961. Some characteristics of perennial and annual ryegrass × tall fescue hybrids and of the amphidiploid progenies of annual ryegrass × tall fescue. Crop Science. 1: 75–80.

39. Butler, J. 1963. Some characteristics of the more commonly grown creeping bentgrasses. Illinois Turfgrass Conference Proceedings. pp. 23–25.

40. Callahan, L. M., and R. E. Engel. 1964. The control of chickweed and clover by phenoxy propionics and dicamba compounds in bentgrass. Northeastern Weed Control Conference Proceedings. 18: 535–539.

41. Carrier, L., and K. S. Bort. 1916. The history of Kentucky bluegrass and white clover in the United States. Journal of the American Society of Agronomy. 8: 256–266.

42. Carroll, J. C. 1943. Effects of drought, temperature, and nitrogen on turf grasses. Plant Physiology. 18: 19–36.

43. Church, G. L. 1929. Meiotic phenomena in certain *Gramineae*. I. *Festuceae, Aveneae, Agrostideae, Chlorideae*, and *Phalarideae*. Botanical Gazette. 87: 608–629.

44. ———. 1936. Cytological studies in the *Gramineae*. American Journal of Botany. 23: 12–15.

45. Cooper, J. P., and K. J. R. Edwards. 1961. The genetic control of leaf development in *Lolium*. 1. Assessment of genetic variation. Heredity. 16: 63–82.

46. Cormack, M. W., and J. B. Lebeau. 1959. Snow mold infection of alfalfa, grasses, and winter wheat by several fungi under artificial conditions. Canadian Journal of Botany. 37: 685–693.

47. Crowder, L. V. 1953. Interspecific and intergeneric hybrids of *Festuca* and *Lolium*. Journal of Heredity. 44: 195–203.

48. ———. 1953. A survey of meiotic chromosome behavior in tall fescue grass. American Journal of Botany. 40: 348–354.

49. Darrow, R. A. 1939. Effects of soil temperature, pH, and nitrogen nutrition on the development of *Poa pratensis*. Botanical Gazette. 101: 109–127.

50. Davis, R. R. 1958. The effect of other species and mowing heights on persistence of lawn grasses. Agronomy Journal. 50: 671–673.

51. ———. 1961. Turfgrass mixtures—influence of mowing height and nitrogen. Proceedings of the 1961 Midwest Regional Turf Conference. pp. 27–29.

52. ———. 1967. Population changes in Kentucky bluegrass-red fescue mixtures. 1967 Agronomy Abstracts. p. 51.

53. Davis, W. E. P. 1964. Performance of red fescues from the northwest. Proceedings of the 18th Annual Northwest Turfgrass Conference. pp. 15–18.

54. DeFrance, J. A. 1938. The effect of different fertilizer ratios on colonial, creeping, and velvet bent grass. Proceedings of the American Society for Horticultural Science. 36: 773–780.

55. DeFrance, J. A., T. E. Odland, and R. S. Bell. 1952. Improvement of velvet bentgrass by selection. Agronomy Journal. 44: 376–378.

56. DeFrance, J. A., and J. A. Simmons. 1951. Relative period of emergence and initial growth of turf grasses and their adaptability under field conditions. Proceedings of the American Society for Horticultural Science. 57: 439–442.

57. Dudeck, A. E., and J. O. Young. 1968. Establishment and use of turf and other ground covers. 1967 Annual Report of the Nebraska Highway Research Project 1. Progress Report 62. pp. 1–25.

58. DUICH, J. M., B. R. FLEMING, and A. E. DUDECK. 1961. The effect of preemergence chemicals on crabgrass and bluegrass, fescue, and bentgrass turf. Northeastern Weed Control Conference Proceedings. 15: 268–275.

59. DUICH, J. M., B. R. FLEMING, and F. SIRIANNI. 1964. The use of several phenoxy compounds for weed control on bentgrass. Northeastern Weed Control Conference Proceedings. 18: 530–534.

60. DUICH, J. M., and H. B. MUSSER. 1960. Response of Kentucky bluegrass, creeping red fescue and bentgrass to nitrogen fertilizers. Pennsylvania Agricultural Experiment Station Progress Report 214. pp. 1–20.

61. EDMOND, D. B., and S. T. J. COLES. 1958. Some long-term effects of fertilizers on a mown turf of browntop and chewing's fescue. New Zealand Journal of Agricultural Research. 1: 665–674.

62. ELLIOTT, C. R., E. C. STACEY, and W. J. DORAN. 1961. Creeping red fescue. Canada Department of Agriculture, Research Branch Publication 1122. pp. 1–15.

63. ENGEL, R. E. 1954. Performance of bentgrasses cut at one-quarter inch. Golf Course Reporter. 22(2): 20–22.

64. ———. 1955. Merion bluegrass resists disease, close cutting. New Jersey Agriculture. 37(2): 10–11.

65. ———. 1966. A comparison of colonial and creeping bentgrasses for $\frac{1}{2}$- and $\frac{3}{4}$-inch turf. 1966 Report on Turfgrass Research at Rutgers University. Bulletin No. 816. pp. 45–47.

66. ENGEL, R. E., and C. R. FUNK. 1966. Performance of red fescue types. 1966 Report on Turfgrass Research at Rutgers University. Bulletin No. 816. pp. 68–71.

67. ERDMANN, M. H., and C. M. HARRISON. 1947. The influence of domestic ryegrass and redtop upon the growth of Kentucky bluegrass and chewing's fescue in lawn and turf mixtures. Journal of the American Society of Agronomy. 39: 682–689.

68. ETTER, A. G. 1951. How Kentucky bluegrass grows. Annals of the Missouri Botanical Garden. 38: 293–375.

69. EVANS, G. 1926. Chromosome complements in grasses. Nature. 118: 841.

70. EVANS, M. W., and J. E. ELY. 1933. Usefulness of Kentucky bluegrass and Canada bluegrass in turfs as affected by their habits of growth. Bulletin of USGA Green Section. 13: 140–143.

71. EVANS, T. W. 1931. The root development of New Zealand browntop, chewing's fescue and fine-leaved sheep's fescue under putting green conditions. Journal of the Board of Greenkeeping Research. 2(5): 119–124.

72. FERGUSON, A. C. 1967. Some observations on winter injury in turfgrass experiments at the University of Manitoba. Summary of the 18th, Annual RCGA National Turfgrass Conference. pp. 7–12.

73. FOOTE, L. E., and A. G. JOHNSON. 1965. Seedling vigor of five strains of crownvetch. Journal of Soil and Water Conservation. 20: 220–221.

74. FRYXELL, P. A. 1957. Mode of reproduction in higher plants. Botanical Review. 23(3): 135–233.

75. FUNK, C. R. 1965. Kentucky bluegrass for New Jersey lawns. Rutgers Short Course in Turf Management. pp. 1–12.

76. FUNK, C. R., and R. E. ENGEL. 1963. Performance of Kentucky bluegrass varieties and various sources of common Kentucky bluegrass. Rutgers Short Course in Turf Management. pp. 21–28.

77. ———. 1963. Effect of cutting height and fertilizer on species composition and turf quality ratings of various turfgrass mixtures. Rutgers Short Course in Turf Management. pp. 47–56.

78. ———. 1967. Performance of Kentucky bluegrass, perennial ryegrass and tall fescue varieties. 1967 Report on Turfgrass Research at Rutgers University. Bulletin No. 818. pp. 67–71.

79. ———. 1968. Manhattan perennial ryegrass for turf. Turf-Grass Times. 4(1): 1–9.

80. FUNK, C. R., R. E. ENGEL, and P. M. HALISKY. 1966. Performance of Kentucky bluegrass

varieties as influenced by fertility level and cutting height. 1966 Report on Turfgrass Research at Rutgers University. Bulletin No. 816. pp. 7–21.

81. ———. 1966. Performance of bentgrass varieties and selections. 1966 Report on Turfgrass Research at Rutgers University. Bulletin No. 816. pp. 41–44.

82. ———. 1967. Summer survival of turfgrass species as influenced by variety, fertility level and disease incidence. 1967 Report on Turfgrass Research at Rutgers University. Bulletin No. 818. pp. 71–77.

83. FUNK, C. R., R. E. ENGEL, and H. W. INDYK. 1966. Ryegrass in New Jersey. 1966 Report on Turfgrass Research at Rutgers University. Bulletin No. 816. pp. 59–67.

84. GARNER, E. S., and S. C. DAMON. 1929. The persistence of certain lawn grasses as affected by fertilization and competition. Rhode Island Agricultural Experiment Station Bulletin 217. pp. 1–22.

85. GASKIN, T. A. 1965. Varietal reaction of creeping bentgrass to stripe smut. Plant Disease Reporter. 49(3): 268.

86. GIBEAULT, V. A., and C. R. SKOGLEY. 1967. Effects of DMPA (Zytron) on colonial bentgrass, Kentucky bluegrass and red fescue root growth. Crop Science. 7: 327–329.

87. HALISKY, P. M., and C. R. FUNK. 1966. Environmental factors affecting growth and sporulation of *Helminthosporium vagans* and its pathogenicity to *Poa pratensis*. Phytopathology. 56: 1294–1296.

88. HALPERIN, M. 1931. A taxonomic study of *Poa bulbosa* L. University of California Publications in Botany. 16(6): 171–183.

89. ———.1933. The taxonomy and morphology of bulbous bluegrass, *Poa bulbosa vivipara*. Journal of the American Society of Agronomy. 25: 408–413.

90. HANDOLL, C. 1967. Grass variety trials. Journal of the Sports Turf Research Institute. 43: 12–22.

91. HANSON, A. A. 1965. Grass varieties in the United States. USDA Agricultural Handbook No. 170. pp. 1–102.

92. HANSON, A. A., and H. L. CARNAHAN. 1956. Breeding perennial forage grasses. USDA Technical Bulletin No. 1145. pp. 1–116.

93. HARDISON, J. R. 1945. Observations on grass diseases in Kentucky. Plant Disease Reporter. 29(3): 76–85.

94. HARPER, J. C., and H. B. MUSSER. 1950. The effect of watering and compaction on fairway turf. Pennsylvania State College 19th Annual Turf Conference. pp. 43–52.

95. HARRISON, C. M., and C. W. HODGSON. 1939. Response of certain perennial grasses to cutting treatment. Journal of the American Society of Agronomy. 31: 418–430.

96. HARTLEY, W. 1950. The global distribution of tribes of the *Gramineae* in relation to historical and environmental factors. Australian Journal of Agricultural Research. 1: 355–373.

97. HARTLEY, W., and R. J. WILLIAMS. 1956. Centres of distribution of cultivated pasture grasses and their significance for plant introduction. Proceedings of the 7th International Grassland Congress. 7: 190–201.

98. HARTWELL, B. L., and S. C. DAMON. 1917. The persistence of lawn and other grasses as influenced especially by the effect of manures on the degree of soil acidity. Rhode Island Agricultural Experiment Station Bulletin 170. pp. 1–24.

99. HARTWIG, H. B. 1938. Relationships between some soil measurements and the incidence of the two common *Poas*. Journal of the American Society of Agronomy. 30: 847–861.

100. HAWK, V. B., and W. D. SHRADER. 1964. Soil factors affecting crownvetch adaptation. Journal of Soil and Water Conservation. 19: 187–190.

101. HEALY, M. J., M. P. BRITTON, and J. D. BUTLER. 1965. Stripe smut damage on 'Pennlu' creeping bentgrass. Plant Disease Reporter. 49(8): 710.

102. HENSON, P. R. 1963. Crownvetch, a soil conserving legume and a potential pasture and hay plant. Crops Research (ARS) 34–53. pp. 1–9.

103. HILLMAN, F. H. 1921. South German Mixed bent seed described. Bulletin of USGA Green Section. 1(3): 37–39.

104. HITCHCOCK, A. S. 1905. North American species of *Agrostis*. USDA Bureau of Plant Industry Bulletin No. 80. pp. 1–68.

105. ———. 1950. Manual of the grasses of the United States. USDA Miscellaneous Publication 200. 2nd ed. revised by A. Chase. pp. 1–1051.

106. HOLT, E. C., and R. R. DAVIS. 1948. Fairways with bent grasses may be next. Park Maintenance. 1(5): 8–10.

107. HOLT, E. C., and R. L. DAVIS. 1948. Differential responses of Arlington and Norbeck bent grasses to kinds and rates of fertilizers. Journal of the American Society of Agronomy. 40: 282–284.

108. HOLT, E. C., and K. T. PAYNE. 1952. Variation in spreading rate and growth characteristics of creeping bentgrass seedlings. Agronomy Journal. 44: 88–90.

109. HOVIN, A. W. 1957. Variations in annual bluegrass. Golf Course Reporter. 25(7): 18–19.

110. JACKLIN, A. W. 1961. An appraisal of strains, varieties and kinds of Kentucky bluegrass. Proceedings of the 15th Annual Northwest Turf Conference. pp. 89–91.

111. JACKLIN, D. W. 1967. Introducing 0217 brand, Fylking Kentucky bluegrass. Weeds, Trees, and Turf. 6(6): 27–29.

112. JACKSON, N. 1959. Evaluation of some grass varieties. Journal of the Sports Turf Research Institute. 10(35): 13–28.

113. ———. 1960. Further notes on the evaluation of some grass varieties. Journal of the Sports Turf Research Institute. 10(36): 156–160.

114. ———. 1962. Further notes on the evaluation of some grass varieties. Journal of the Sports Turf Research Institute. 10(38): 394–400.

115. ———. 1963. Further notes on the evaluation of some grass varieties. Journal of the Sports Turf Research Institute. 40: 67–75.

116. JENKIN, T. J. 1934. Interspecific and intergeneric hybrids in herbage grasses. Initial crosses. Journal of Genetics. 28: 205–264.

117. ———. 1955. Interspecific and intergeneric hybrids in herbage grasses. IX. *Festuca arundinacea* with some other *Festuca* species. Journal of Genetics. 53: 81–93.

118. JONES, K. 1953. The cytology of some British species of *Agrostis* and their hybrids. British Agricultural Bulletin 5. p. 316.

119. ———. 1956. Species differentiation in *Agrostis*. I. Cytological relationships in *Agrostis canina* L. Journal of Genetics. 54: 370–376.

120. ———. 1956. Species differentiation in *Agrostis*. II. The significance of chromosome pairing in the tetraploid hybrids of *Agrostis canina* subsp. *montana* Hartm., *A. tenuis* Sibth. and *A. Stolonifera* L. Journal of Genetics. 54: 377–393.

121. JUHREN, M., W. NOBLE, and F. W. WENT. 1957. The standardization of *Poa annua* as an indicator of smog concentrations. I. Effects of temperature, photoperiod, and light intensity during growth of test-plants. Plant Physiology. 32: 576–586.

122. JUSKA, F. V. 1963. Shade tolerance of bentgrasses. Golf Course Reporter. 31(2): 28–34.

123. JUSKA, F. V., and A. A. HANSON. 1959. Evaluation of cool season turfgrasses. Park Maintenance. 12(9): 18–20.

124. ———. 1959. Evaluation of cool season turfgrasses alone and in mixtures. Agronomy Journal. 51: 597–600.

125. ———. 1961. The nitrogen variable in testing Kentucky bluegrass varieties for turf. Agronomy Journal. 53: 409–410.

126. ———. 1963. The management of Kentucky bluegrass on extensive turfgrass areas. Park Maintenance. 16(9): 22–32.

127. ———. 1966. Nutritional requirements of *Poa annua* L. 1966 Agronomy Abstracts. p. 35.

128. JUSKA, F. V., A. A. HANSON, and C. J. ERICKSON. 1965. Effects of phosphorus and other treat-

ments on the development of red fescue, Merion and common Kentucky bluegrass. Agronomy Journal. 57: 75–78.

129. JUSKA, F. V., J. TYSON, and C. M. HARRISON. 1955. The competitive relationship of Merion bluegrass as influenced by various mixtures, cutting heights, and levels of nitrogen. Agronomy Journal. 47: 513–518.

130. ———. 1956. Field studies on the establishment of Merion bluegrass in various seed mixtures. Quarterly Bulletin of the Michigan Agricultural Experiment Station. 38(4): 678–690.

131. KENNEDY, P. B. 1929. Proliferation in *Poa bulbosa*. Journal of the American Society of Agronomy. 21: 80–91.

132. KIRK, L. E. 1939. A dry-land turf grass. Turf Culture. 1: 105–110.

133. KJELLQVIST, E. 1961. Studies in *Festuca rubra* L. 1. Influence of environment. Botaniska Notiser. 114: 403–408.

134. KLECKA, A. 1937. Vliv seslapavani na asociaci travnatych porustu. Sborn. Csl. Akad. Zemed. 12: 715–24.

135. KOSHY, T. K. 1968. Evolutionary origin of *Poa annua* L. in the light of karyotypic studies. Canadian Journal of Genetics and Cytology. 10: 112–118.

136. ———. 1969. Breeding systems in annual bluegrass, *Poa annua* L. Crop Science. 9: 40–43.

137. KRAMER, H. H. 1947. Morphologic and agronomic variation in *Poa pratensis* L. in relation to chromosome numbers. Journal of the American Society of Agronomy. 39: 181–191.

138. KREITLOW, K. W., and W. W. MYERS. 1947. Resistance to crown rust in *Festuca elatior* and *F. elatior* var. *arundinancea*. Phytopathology. 37: 39–63.

139. KREITLOW, K. W., H. SHERWIN, and C. L. LEFEBVRE. 1950. Susceptibility of tall fescue and meadow fescue to *Helminthosporium* infection. Plant Disease Reporter. 34(6): 189–190.

140. LABEAU, J. B., and J. E. MOFFATT. 1967. A suitable grass for western greens. The Greenmaster. 4: 1–8.

141. LAMSON-SCRIBNER, F. 1898. Sand-binding grasses. Yearbook of USDA. pp. 405–420.

142. LANGVAD, B. 1968. *Poa pratensis*, one of the main turfgrass species of Sweden. Weibulls Grastips. 11: 358–370.

143. ———. 1968. Variety trials with fine leafed red fescue. Weibulls Grastips. 11: 371–376.

144. LANTZ, H. L. 1955. Cool season grasses—bluegrass, fescue, bent. Golf Course Reporter. 23(2): 32–37.

145. LAPP, W. S. 1943. A study of factors affecting the growth of lawn grasses. Pennsylvania Academy of Science. 17: 117–148.

146. LAUDE, H. M. 1953. The nature of summer dormancy in perennial grasses. Botanical Gazette. 114: 284–292.

147. LAW, A. G. 1966. Performance of bluegrass varieties clipped at two heights. Weeds, Trees and Turf. 5(5): 26–28.

148. LEWIS, I. G. 1933. A greenskeeper's guide to the grasses. 2. The genus *Festuca*. Journal of the Board of Greenkeeping Research. 3(9): 94–98.

149. ———. 1934. A greenkeeper's guide to the grasses. 3. The genus *Agrostis* (cont.). Journal of the Board of Greenkeeping Research. 3(11): 200–206.

150. LOBENSTEIN, C. W. 1962. Observing bluegrasses. Proceedings of the 1962 Midwest Regional Turf Conference. pp. 66–69.

151. ———. 1963. How bluegrass spreads. Proceedings of the Midwest Regional Turf Conference. pp. 9–14.

152. LOPEX, R. R., A. G. MATCHES, and J. D. BALDRIDGE. 1967. Vegetative development and organic reserves of tall fescue under conditions of accumulated growth. Crop Science. 7: 409–412.

153. LUCANUS, R., K. J. MITCHELL, G. G. PRITCHARD, and D. M. Calder. 1960. Factors influencing survival of strains of ryegrass during the summer. New Zealand Journal of Agricultural Research. 3: 185–193.

154. LYNCH, R. I. 1903. *Poa annua* not a self-fertilizer. The Gardeners' Chronicle. 33: 380.

155. MACVICAR, R. M. 1958. Note on Norlea perennial ryegrass. Canadian Journal of Plant Science. 38: 505.

156. M'ALPINE, A. N. 1898. Production of new types of forage plants—clovers and grasses. Transactions of the Highland and Agricultural Society of Scotland. 5: 135–165.

157. MAUDE, P. F. 1940. Chromosome numbers in some British plants. New Phytologist. 39: 17–32.

158. MAY, J. W. 1966. New fine-leafed grass varieties. 12th Rocky Mountain Regional Turfgrass Conference. pp. 55–61.

159. McKEE, G. W. 1962. Registration of varieties of other legumes. Penngift crownvetch (Reg. No. 2). Crop Science. 2: 356.

160. McKEE, G. W., and A. R. LANGILLE. 1967. Effect of soil pH, drainage, and fertility on growth, survival, and element content of crownvetch, *Coronilla varia* L. Agronomy Journal. 59: 533–536.

161. McLAUGHLIN, W. T., and R. L. BROWN. 1942. Controlling coastal sand dunes in the Pacific northwest. USDA Circular No. 660. pp. 1–46.

162. McLEAN, A. 1964. Turfgrass varieties for the interior of British Columbia. Proceedings of the 18th Annual Northwest Turfgrass Conference. pp. 19–20.

163. MILLER, R. W. 1966. The effect of certain management practices on the botanical composition and winter injury to turf containing a mixture of Kentucky bluegrass (*Poa pratensis*, L.) and tall fescue (*Festuca arundinacea*, Schreb.). Illinois Turfgrass Conference Proceedings. 7: 39–46.

164. MITCHELL, K. J. 1954. Growth of pasture species. I. Short rotation and perennial ryegrass. New Zealand Journal of Science and Technology. Sec. A. 36(3): 191–206.

165. MONTEITH, J. 1929. Demonstration turf garden reports. Bulletin of USGA Green Section. 9(12): 210–221.

166. ⸻. 1931. Experimental results at Miami Beach, Florida. Bulletin of USGA Green Section. 11(10): 190–194.

167. MONTEITH, J., and K. WELTON. 1931. Demonstration turf garden reports. Bulletin of USGA Green Section. 11(6): 122–133.

168. ⸻.1931. Demonstration turf garden results: A three-year summary. Bulletin of USGA Green Section. 11(12): 230–246.

169. ⸻. 1932. Demonstration turf garden reports. Bulletin of USGA Green Section. 12(6): 218–223.

170. ⸻. 1932. Putting tests upon bent grasses. Bulletin of USGA Green Section. 12(6): 224–227.

171. MORRISH, R. H., and C. M. HARRISON. 1948. The establishment and comparative wear resistance of various grasses and grass-legume mixtures to vehicular traffic. Journal of the American Society of Agronomy. 40: 168–179.

172. MOSER, L. E., S. R. ANDERSON, and R. W. MILLER. 1968. Rhizome and tiller development of Kentucky bluegrass (*Poa pratensis* L.) as influenced by photoperiod, cold treatment, and variety. Agronomy Journal. 60: 632–635.

173. MOWER, R. G., and J. F. CORNMAN. 1962. Pre-emergence crabgrass control. Northeastern Weed Control Conference Proceedings. 16: 489–492.

174. MUNTZING, A. 1933. Apomictic and sexual seed formation in *Poa*. Hereditas. 17: 131–154.

175. ⸻. 1940. Further studies on apomixis and sexuality in *Poa*. Hereditas. 26: 115–190.

176. MUSE, R. R., and H. B. COUCH. 1965. Influence of environment on diseases of turfgrasses. IV. Effect of nutrition and soil moisture on *Corticium* red thread of creeping red fescue. Phytopathology. 55: 507–510.

177. MUSSER, H. B. 1948. Effects of soil acidity and available phosphorus on population changes in mixed Kentucky bluegrass-bent turf. Journal of American Society of Agronomy. 40: 614–620.

178. Musser, H. B., and J. M. Duich. 1959. The extent of aberrants produced by Merion Kentucky bluegrass, *Poa pratensis* L., as determined by first and second generation progeny tests. Agronomy Journal. 51: 421–424.

179. Musser, H. B., W. L. Hottenstein, and J. P. Stanford. 1954. Penngift crown vetch for slope control on Pennsylvania highways. Pennsylvania Agricultural Experiment Station Bulletin No. 576. pp. 1–21.

180. Myers, W. M., and H. D. Hill. 1947. Distribution and nature of polyploidy in *Festuca elatior* L. Bulletin of the Torrey Botanical Club. 74: 99–111.

181. Nielsen, E. L., and L. M. Humphrey. 1937. Grass studies. I. Chromosome numbers in certain members of the tribes *Festuceae, Hordeae, Avenae, Agrostideae, Chlorideae, Phalarideae*, and *Tripsaceae*. American Journal of Botany. 24: 276–279.

182. Nissen, O. 1950. Chromosome numbers, morphology, and fertility in *Poa pratensis* L. from southeastern Norway. Agronomy Journal. 42: 136–144.

183. Nittler, L. W. 1966. Seedling characteristics useful in identifying Kentucky bluegrass. Farm Research. 32(1): 14–15.

184. North, H. F. A., and T. E. Odland. 1934. Putting green grasses and their management. Rhode Island Agricultural Experiment Station Bulletin No. 245. pp. 1–44.

185. ———.1935. The relative seed yields in different species and varieties of bent grass. Journal of the American Society of Agronomy. 27: 374–383.

186. North, H. F. A., T. E. Odland, and J. A. DeFrance. 1938. Lawn grasses and their management. Rhode Island Agricultural Experiment Station Bulletin 264. pp. 1–36.

187. O'Donnell, J. 1966. Nutrient removal and rooting pattern of Merion Kentucky bluegrass. 1966 Wisconsin Turfgrass Conference Proceedings. pp. 29–31.

188. Parks, O. C., and P. R. Henderlong. 1967. Germination and seedling growth rate of ten common turfgrasses. Proceedings of the West Virginia Academy of Science. 39: 132–140.

189. Patel, A. S., and J. P. Cooper. 1961. The influence of seasonal changes in light energy on leaf and tiller development in ryegrass, timothy, and meadow fescue. Journal of the British Grassland Society. 16: 299–308.

190. Patterson, J. K. 1958. Comparative performance of the P. N. W. bluegrass selections. Proceedings of the 12th Annual Northwest Turf Conference. pp. 39–41.

191. ———.1959. A new bluegrass variety in the making. Golf Course Reporter. 27(8): 48–49.

192. Philipson, W. R. 1937. A revision of the British species of the genus *Agrostis* Linn. Journal of the Linnean Society of London. 51: 73–151.

193. Piper, C. V. 1906. North American species of *Festuca*. Contributions from US National Herbarium. 10(1): 1–48.

194. ———.1918. The agricultural species of bent grass. Part I. Rhode Island bent and related grasses. USDA Bulletin No. 692. pp. 1–14.

195. Porter, H. L. 1958. Rhizomes in tall fescue. Agronomy Journal. 50: 493–494.

196. Reid, M. E. 1932. The effects of soil reaction upon the growth of several types of bent grasses. Bulletin of USGA Green Section. 12(5): 196–212.

197. ———. 1933. Effects of shade on the growth of velvet bent and Metropolitan creeping bent. Bulletin of USGA Green Section. 13: 131–135.

198. Renney, A. J. 1964. Preventing *Poa annua* infestations. Proceedings of the 18th Annual Northwest Turfgrass Conference. pp. 3–5.

199. Richardson, E. C., E. G. Diseker, and B. H. Hendrickson. 1963. Crownvetch for highway bank stabilization in the Piedmont uplands of Georgia. Agronomy Journal. 55: 213–215.

200. Roberts, E. C., and F. E. Markland. 1966. Nitrogen induced changes in bluegrass-red fescue turf populations. 1966 Agronomy Abstracts. p. 36.

201. Ruffner, J. D., and J. G. Hall. 1963. Crownvetch in West Virginia. West Virginia Agricultural Experiment Station Bulletin 487. pp. 1–19.

202. Schery, R. W. 1965. This remarkable Kentucky bluegrass. Annals of the Missouri Botanical Garden. 52(3): 444–451.

203. SCHMIDT, R. E., and R. E. BLASER. 1961. Cool season grasses for winter turf on Bermuda putting greens. USGA Journal and Turf Management. 14(5): 25–30.

204. SCHOTH, H. A., and M. HALPERIN. 1932. The distribution and adaptation of *Poa bulbosa* in the United States and in foreign countries. Journal of the American Society of Agronomy. 24: 786–793.

205. SCHWENDIMAN, J. L., A. L. HAFENRICHTER, and A. G. LAW. 1964. Registration of Durar hard fescue. Crop Science. 4: 114.

206. SKOGLEY, C. R. 1961. The effect of Zytron on seedling turf grasses. Northeastern Weed Control Conference Proceedings. 15: 258–263.

207. SKOGLEY, C. R., and J. A. JAGSCHITZ. 1964. The effect of various preemergence crabgrass herbicides on turfgrass seed and seedlings. Northeastern Weed Control Conference Proceedings. 18: 523–529.

208. SKOGLEY, C. R., and F. B. LEDEBOER. 1968. Evaluation of several Kentucky bluegrass and red fescue strains maintained as lawn turf under three levels of fertility. Agronomy Journal. 60: 47–49.

209. SPRAGUE, H. B. 1933. Root development of perennial grasses and its relation to soil conditions. Soil Science. 36: 189–209.

210. ———. 1934. Utilization of nutrients by colonial bent (*Agrostis tenuis*) and Kentucky bluegrass (*Poa pratensis*). New Jersey Agricultural Experiment Station Bulletin 570. pp. 1–16.

211. ———. 1939. Velvet bent grass for putting greens and other fine turf. New Jersey Agricultural Experiment Station Circular 393. pp. 1–4.

212. SPRAGUE, H. B., and G. W. BURTON. 1937. Annual bluegrass (*Poa annua* L.) and its requirements for growth. New Jersey Agricultural Experiment Station Bulletin 630. pp. 1–24.

213. SPRAGUE, H. B., and E. E. EVAUL. 1930. Experiments with turf grasses in New Jersey. New Jersey Agricultural Experiment Station Bulletin 497. pp. 1–55.

214. STEINIGER, E. R. 1968. The story of Cohansey. USGA Green Section Record. 6(3): 3–7.

215. STUCKEY, I. H. 1941. Seasonal growth of grass roots. American Journal of Botany. 28: 486–491.

216. STUCKEY, I. H., and W. G. BANFIELD. 1946. The morphological varieties and the occurrence of aneuploids in some species of *Agrostis* in Rhode Island. American Journal of Botany. 33: 185–190.

217. SULLIVAN, E. F. 1962. Effects of soil reaction, clipping height, and nitrogen fertilization on the productivity of Kentucky bluegrass sod transplants in pot culture. Agronomy Journal. 54: 261–263.

218. SUMNER, D. C., R. M. HAGAN, and D. S. MIKKELSEN. 1955. Merion bluegrass seed production. USGA Journal and Turf Management. 8(1): 27–32.

219. TERRILL, E. E. 1967. Meadow fescue: *Festuca elatior* L. or *F. pratensis* Hudson? Brittonia. 19: 129–132.

220. VAN DERSAL, W. R. 1936. The ecology of a lawn. Ecology. 17(3): 515–527.

221. VAATNOU, H. 1967. Responses of five genotypes of *Agrostis* L. to variations in environment. Ph. D. Thesis. Oregon State University. pp. 1–149.

222. VINALL, H. N., and H. L. WESTOVER. 1928. Bulbous bluegrass, *Poa bulbosa*. Journal of the Americn Society of Agronomy. 20: 394–399.

223. WATKINS, J. M., G. W. CONREY, and M. W. Evans. 1940. The distribution of Canada bluegrass and Kentucky bluegrass as related to some ecological factors. Journal of the American Society of Agronomy. 32: 726–728.

224. WATSON, J. R. 1949. Compaction and irrigation studies on established fairway turf. Pennsylvania State College 18th Annual Turf Conference. pp. 88–90.

225. WEAVER, J. E., and F. W. ALBERTSON. 1943. Resurvey of grasses, forbs and underground plant parts at the end of the great drought. Ecological Monographs. 13: 63–117.

226. WELTON, F. A., and J. C. CARROLL. 1940. Lawn experiments. Ohio Agricultural Experiment Station Bulletin 613. pp. 1–43.

227. WESTOVER, H. L., and O. B. FITTS. 1927. *Poa bulbosa.* Bulletin of USGA Green Section. 7(4): 78–82.

228. WILSON, C. G., and F. V. GRAU. 1950. Merion (B-27) bluegrass. USGA Journal and Turf Management. 2(7): 27–29.

229. WILSON, D. B. 1962. Effects of light intensity and clipping on herbage yields. Canadian Journal of Plant Science. 42: 270–275.

230. WILSON, J. D. 1927. The measurement and interpretation of the water-supplying power of the soil with special reference to lawn grasses and some other plants. Plant Physiology. 2: 385–440.

231. WIT, F. 1959. Hybrids of ryegrass and meadow fescue and their value for grass breeding. Euphytica. 8: 1–12.

232. WOOD, G. M., and J. A. BURKE. 1961. Effect of cutting height on turf density of Merion, Park, Delta, Newport and Common Kentucky bluegrass. Crop Science. 1: 317–318.

233. YOUNGNER, V. B. 1957. Fine-leaved fescues as turfgrasses in southern California. Southern California Turfgrass Culture. 7(1): 7–8.

234. ———. 1958. Tall fescue-Pensacola bahiagrass combination. Southern California Turfgrass Culture. 8(3): 24.

235. ———. 1959. Ecological studies on *Poa annua* in turfgrasses. Journal of the British Grassland Society. 14: 233–237.

236. ———. 1962. Wear resistance of cool season turfgrasses. Effects of previous mowing practices. Agronomy Journal. 54: 198–199.

237. YOUNGNER, V. B., O. R. LUNT, and F. NUDGE. 1967. Salinity tolerance of seven varieties of creeping bentgrass, *Agrostis palustris* Huds. Agronomy Journal. 59: 335–336.

238. YOUNGNER, V. B., and F. J. NUDGE. 1968. Growth and carbohydrate storage of three *Poa pratensis* L. strains as influenced by temperature. Crop Science. 8: 455–457.

chapter 4

Warm Season Turfgrasses

Introduction

Warm season turfgrasses are those species having a temperature optimum of 80 to 95°F. They are widely distributed throughout the warm humid, warm subhumid, and warm semiarid climates and are being utilized to varying degrees in the transitional zones. The winter temperature is a very important influence on the distribution and use of warm season turfgrasses (39). Approximately fourteen warm season species are utilized for turfgrass purposes throughout the world.

Unlike the cool season turfgrass species that have a common European origin, the warm season species have such diverse centers of origin as Africa, South America, and Asia (40). The origin of the *Cynodon* species is centered around the Indian Ocean ranging from eastern Africa to the East Indies. East Africa is the center of origin for most turf-type bermudagrasses and kikuyugrass. The center of origin for bahiagrass is in subtropical, eastern South America. The *Zoysia* species and centipedegrass originated in southeastern Asia, while St. Augustinegrass and carpetgrass are native to the West Indies. The semiarid warm season turfgrasses such as buffalograss, sideoats grama, and blue grama originated in the plains regions of North America.

A general comparison between such cool season turfgrasses as the bluegrasses, fescues, and ryegrasses and such warm season turfgrasses as bermudagrass and zoysiagrass reveals the following general differences. The low growing, warm season turfgrasses are substantially more tolerant of close mowing; are deeper rooted; and are more drought, heat, and wear tolerant as a group than the cool season turfgrasses. However, the former are less low temperature hardy and will discolor at low temperatures. Most of the cool season species are seeded, while most of the warm season species are established vegetatively. Exceptions to these characteristics occur.

A summary of general comparisons among six commonly used warm season turfgrasses is found in Table 4-1. Discussions in this chapter that involve plant descriptions, adaptation, use, and culture must be fairly general because of the variability within species. Comparisons within a cultivar section are made among cultivars of that species. The relative scale of resistance, tolerance, and susceptibility used is the same as in the previous chapter: superior, excellent, good, intermediate (moderate or medium), fair, poor, very poor.

The Bermudagrasses (*Cynodon* L. C. Rich)

Bermudagrass is one of the most important and widely adapted of the warm season turfgrasses. It is also known by the common name of couchgrass in certain countries. Most of the turf-type bermudagrasses originated in eastern Africa. The bermudagrasses have been introduced and are now widely distributed throughout most of the warm humid, tropical, and subtropical regions of the world (49, 61). There are four main turf-type bermudagrasses: (a) *C. dactylon* (L.) Pers., (b) *C. transvaalensis* Burtt-Davy, (c) *Cynodon* x *magennisii* Hurcombe, and (d) *C. incompletus* var. *hirautus* (Stent) de Wet et Harlan.

Plant Description

Cynodon dactylon (L.) Pers. **Vernation** folded; **sheaths** compressed, loose, sparsely pubescent with fascicled hairs at throat, split with overlapping, hyaline margins; **ligule** a fringe of white hairs, 1–3 mm long; **collar** continuous, narrow to medium broad, glabrous, sparingly ciliate; **auricles** absent; **blades** mostly flat, 1.5–3 mm wide, stiff, sparsely pubescent above, usually glabrous below, margins scabrous, tapering from base to an acute apex; **stems** compressed, erect or ascending from a prostrate base, with extensively creeping, strong, flat stolons and/or scaly, stout rhizomes that branch profusely and root at the nodes; **inflorescence** four or five digitate spikes, spikelets sessile and closely appressed in two rows on a narrow, somewhat triangular rachis.

The improved turf-type bermudagrasses form a very vigorous, aggressive turf of high shoot density. The leaf width ranges from the medium texture of common bermudagrass to the very fine texture of African bermudagrass (52). Certain bermudagrass cultivars may have multiple-leaved nodes (24). The turfgrass color ranges from a light green to a very dark green. The growth habit is quite prostrate with an intergradation of stolons and rhizomes that forms a very tight sod. The root system is fibrous, extensive, and quite deep (7, 11, 22, 57). Propagation is primarily vegetative by means of sprigs, plugs, or sod. Common bermudagrass is the only turf-type bermudagrass established from seed. Bermudagrass has the most

Table 4-1

A SUMMARY OF GENERAL COMPARISONS AMONG SIX COMMONLY USED WARM SEASON TURFGRASSES (EXCEPTIONS TO THESE NORMS OCCUR)

Characteristics	Warm season turfgrass					
	Bermudagrass	*Zoysiagrass*	*St. Augustine-grass*	*Centipedegrass*	*Carpetgrass*	*Bahiagrass*
Plant description:						
leaf texture	Fine	Medium to fine	Very coarse	Coarse	Very coarse	Very coarse
shoot density	High	Medium to high	Medium	Medium to high	Medium to high	Low
growth habit (lateral shoots)	Rhizomes and stolons	Rhizomes and stolons	Thick, flat stolons	Short stolons	Stolons	Short rhizomes and stolons
seed head development	Short, few to numerous	Short and minimal	Thick, short, and minimal	Low and inconspicuous	Tall and numerous	Tall, wiry, and numerous
Adaptation:						
heat hardiness	Excellent	Excellent	Excellent	Excellent	Excellent	Excellent
low temperature hardiness	Poor	Poor to intermediate	Extremely poor	Very poor	Very poor	Poor
drought resistance	Excellent	Excellent	Fair	Poor	Poor	Excellent
shade tolerance	Very poor	Good	Excellent	Intermediate to fair	Intermediate to fair	Good
soil adaptation	Wide range	Wide range	Wide range	Acidic, infertile	Wet, acidic	Wide range
salt tolerance	Good	Good	Excellent	Poor	Poor	Poor
wear tolerance	Very good	Excellent	Intermediate	Poor	Intermediate to poor	Good
Establishment:						
method	Vegetative	Vegetative	Vegetative	Seed or vegetative	Seed or vegetative	Seed or vegetative
rate	Excellent	Poor	Good to intermediate	Poor	Intermediate	Intermediate to poor
recuperative potential	Excellent	Excellent, but slow	Good	Poor	Intermediate	Poor

Table 4-1 (Cont.)

Culture:						
intensity	High to medium	Medium	Medium	Low	Low	Very low
cutting height (in.)	0.25–1.0	0.5–1.0	1.5–2.5	1–2	1–2	1.5–2.5
preferred mower type	Reel	Reel	Reel or rotary	Reel or rotary	Rotary	Rotary
nitrogen requirement [N (lb)/1000 sq ft/growing month]	0.8–1.8	0.5–1.0	0.5–1.0	0.1–0.3	0.2–0.4	0.1–0.4
thatching tendency	High	Medium to high	Very high	Medium to high	Low	Very low
Pests:						
diseases	Brown patch, *Helminthosporium*, dollar spot, and spring dead spot	Brown patch, dollar spot, rust, and *Helminthosporium*	Brown patch, dollar spot, SADV, and gray leaf spot	Brown patch	Brown patch	Brown patch and dollar spot
insects	Sod webworms, armyworms, mole crickets, bermudagrass mite, and frit fly	Hunting billbugs, armyworms, mole cricket, and sod webworms	Chinch bugs	Ground pearls		Mole cricket
other	Nematodes	Nematodes		Nematodes		

135

rapid growth rate of the warm season turfgrasses and also possesses a very rapid establishment rate (35). The recuperative potential is excellent and wear tolerance very good.

Adaptation. Bermudagrass is a long-lived perennial that is adapted throughout the warm humid and warm semiarid regions of the world. The heat and drought hardiness are excellent but the low temperature tolerance is poor (11). Bermudagrass discolors with the advent of cool fall temperatures and remains in a state of dormancy throughout the winter. Total loss of the pigment system in the leaves and stems results in a light-tan to whitish-tan appearance. Discoloration usually occurs at soil temperatures below 50°F (81) and persists until the soil temperature rises above this level in the spring. The bermudagrass cultivars that have improved low temperature hardiness generally discolor earlier in the fall than the less hardy cultivars. Where bermudagrass has been introduced into the cooler portions of the transitional climatic zone, the turf is seriously thinned by low temperature kill on the average of once every 4 or 5 years (1). The shade tolerance is also poor (10, 73).

Bermudagrass is best adapted to moderately well-drained, fertile soils of relatively fine texture but tolerates a wide range of soil types. Bermudagrass growth is usually better on fine textured than coarse textured soils because of the higher fertility level and soil moisture retention associated with the fine textured soils. Bermudagrass tolerates a wide soil pH range of 5.5 to 7.5. It is quite tolerant of flooding for extended periods of time but growth is minimal under waterlogged soil conditions. The salt tolerance is quite good.

Use. The improved bermudagrasses form a very dense, uniform turf of high quality when grown under proper climatic and cultural conditions. It is utilized in the warm humid and warm semiarid regions on lawns, parks, cemeteries, institutional grounds, fairways, greens, tees, roughs, roadsides, airfields, athletic fields, and other comparable general-purpose lawn areas. Bermudagrass turfs provide a very good playing surface for athletic fields, tees, fairways, polo fields, and playfields because of the excellent wear tolerance and recuperative potential (Fig. 4-1). Bermudagrass turfs are subject to thinning on football fields as a result of concentrated traffic and wear during the late fall period when dormancy occurs. Although dormant bermudagrass has no immediate recuperative capability, it can reestablish an adequate playing surface after the initiation of growth in late spring. The improved cultivars that are utilized on greens have a fine leaf texture but usually do not provide as good a playing surface as the bentgrasses. Common bermudagrass is sometimes grown in a polystand with tall fescue for use on general grounds and playfields.

Winter discoloration can be alleviated by applying a turfgrass colorant to dormant bermudagrass turfs. Dormant bermudagrass can also be overseeded with cool season turfgrass species in order to provide a green turf of acceptable playing quality during the winter period.

Bermudagrass can become a weed under certain turfgrass situations because of its vigorous, creeping growth habit (59). For example, when bermudagrass is used on the aprons around bentgrass greens, it readily encroaches onto the green,

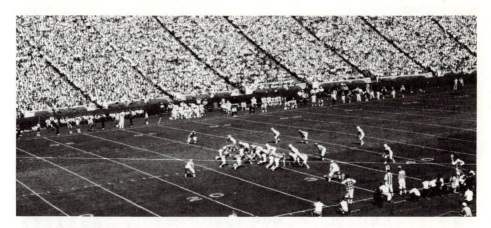

Figure 4-1. A Tiflawn bermudagrass football field in Athens, Georgia. (Photo courtesy of G. W. Burton, USDA, Georgia Coastal Plain Experiment Station, Tifton, Georgia.)

resulting in an inferior putting surface. It can also become a serious weed in flower beds, shrubbery, and driveways.

Culture. Bermudagrass requires a medium to high intensity of culture. Tolerance to close mowing is good because of a prostrate growth habit. A cutting height of 0.5 to 1.0 in. is best for bermudagrass, used as a general-purpose turf, with 0.75 in. preferred. Fairly frequent mowing is necessary in order to maintain good quality and to avoid scalping. Certain improved cultivars such as Tifgreen, Tifdwarf, Everglades, and Bayshore tolerate daily mowing at 0.25 in. (46). Cutting heights above 1.5 in. result in a more upright, stemmy growth habit that is subject to increased thatching and scalping. Since bermudagrass is quite responsive to fertilization and irrigation, a high intensity of culture is necessary in order to achieve optimum turfgrass quality. The nitrogen fertility requirement is 0.8 to 1.8 lb per 1000 sq ft per growing month (66). Bermudagrass is quite prone to thatching because of the vigorous growth rate. Periodic topdressing and/or vertical mowing are necessary to avoid excessive thatch accumulation and the associated scalping. Vertical mowing for thatch control also improves low temperature color retention.

The common disease problems of bermudagrass include *Helminthosporium* spp., brown patch, dollar spot, *Fusarium* patch, rust, *Pythium* blight, and spring dead spot. Common pests include sod webworms, armyworms, mole crickets, bermudagrass mites, frit flies, rhodesgrass scale, bermudagrass scale, and nematodes (15, 46). Bermudagrass is intolerant of the triazine herbicides (4).

Cultivars. *Cynodon dactylon* (L.) Pers. is a cross-pollinated tetraploid that is frequently referred to as common bermudagrass, couchgrass, and quickgrass. Chromosome numbers of 30, 36, and 40 have been reported (8, 37, 38, 52, 56). It was one of the first bermudagrass species widely utilized for turfgrass purposes. Common bermudagrass turfs are relatively stiff and of low shoot density with excessive seed head formation that detracts from the overall appearance.

Three species that have been introduced more recently for turfgrass use are

Table 4-2

CHARACTERISTICS OF NINETEEN VEGETATIVELY PROPAGATED, TURF-TYPE BERMUDAGRASS CULTIVARS

Cultivar	Released By	Date	Turfgrass characteristics	Adaptation	Other comments
Bayshore* (Gene Tift) (FB-3) (33)	(Florida AES development†), U.S.	—	Light green color; fine texture; tends to be stemmy; high shoot density; rapid vertical growth rate; prone to thatching	Poor low temperature hardiness and color retention; used on greens	Extensive seed head formation; moderate resistance to leaf spot but susceptible to *Pythium* blight; requires a high intensity of culture
Everglades* (FB-4) (77)	Florida AES, U.S.	1962	Medium dark green color; fine texture; low growth habit; rapid vertical growth rate	Medium poor low temperature hardiness; intermediate drought tolerance; good low temperature color retention; used on greens	Extensive seed head formation; moderate resistance to leaf spot
FB-137 (No-Mow) (3, 5, 45, 46, 47, 63)	(Florida AES development†), U.S.	—	Dark blue-green color; intermediate texture and shoot density; low growth habit; slow vertical growth rate	Improved low temperature color retention and shade tolerance; used on lawns	Extensive seed head formation; susceptible to *Helminthosporium* spp, sod webworm, armyworm, bermudagrass mite, and nematodes
Midway*	Kansas AES, U.S.	1965	Intermediate texture and shoot density; minimum thatching tendency	Superior low temperature hardiness; used on lawns	Minimal seed head formation; good resistance to *Helminthosporium* spp. and bermudagrass mite; susceptible to hunting billbug
Ormond‡ (FB-45) (15, 16, 46, 56)	Florida AES, U.S.	1962	Dark blue-green color; intermediate texture and shoot density; prostrate growth habit; tends to thatch	Excellent low temperature color retention; poor low temperature hardiness; intermediate wear tolerance; widely used on fairways; also used on lawns and sports fields	Minimal seed head formation; susceptible to dollar spot and bermudagrass mite

138

Table 4-2 (Cont.)

Pee Dee*	(South Carolina AES†), U.S.	1967	Dark green color; medium fine texture; medium low growth habit	Poor low temperature color retention, turns purple; good low temperature hardiness	Minimal seed head formation; requires a high intensity of culture
Royal Cape‡ (56)	California AES and CRD-ARS, U.S.	1960	Dark green color; fine texture; medium high shoot density	Good salt and wear tolerance; good spring green-up rate and low temperature color retention; used on lawns in hot, arid climates	Few seed heads formed; resistant to the bermudagrass mite and *Helminthosporium* spp. in arid climates
Santa Ana (RC-145) (85)	California AES, U.S.	1966	Dark blue-green color; medium leaf texture; medium high shoot density; vigorous with good recuperative potential	Excellent smog, wear, and salt tolerance; excellent low temperature color retention; poor low temperature hardiness; used on lawns, tees, fairways, and sports areas	Extensive seed head formation; good resistance to bermudagrass mite
Sunturf§ (48, 56)	Alabama, Arkansas, Oklahoma, South Carolina AES, U.S.	1956	Dark green color, very fine texture; very high shoot density; low growth habit; vigorous growth rate; spreading by stolons	Poor low temperature color retention, turns distinctly purple; medium good low temperature hardiness; excellent drought resistance, salt tolerance, wear tolerance, and recuperative potential; used on greens, lawns, and sports fields	Nominal seed head formation; susceptible to rust and dollar spot
Texturf 1F‡ (T35A) (56, 62)	Texas AES, U.S.	1957	Light green color; fine texture; medium high shoot density; vigorous growth rate; prone to thatching	Intermediate drought tolerance and low temperature hardiness; good spring green-up rate; intermediate wear tolerance; used on lawns and sports fields	Minimal seed head formation; susceptible to leaf diseases; not as tolerant to close mowing

Table 4-2 (Cont.)

| Cultivar | Released | | Turfgrass characteristics | Adaptation | Other comments |
	By	Date			
Texturf 10‡ (T47) (65)	Texas AES, U.S.	1957	Dark green color; medium fine texture; high shoot density; low growth habit; moderate rate of growth and establishment	Good low temperature color retention and spring green-up rate; used on sports fields and recreational areas	Minimal seed head formation; quite sensitive to chlorinated hydrocarbons
Tifdwarf* (12, 46, 47, 63)	Georgia AES and CRD-ARS, U.S.	1965	Dark green color; fine texture; high shoot density; low growth habit; slow growth rate	Poor low temperature color retention, turns purple; excellent low temperature hardiness; improved shade tolerance; superior tolerance to close mowing; used on greens	Minimal seed head formation; very susceptible to smog injury and sod webworm; requires a medium high intensity of culture
Tiffine* (T-127) (41, 56, 67)	Georgia AES and CRD-ARS, U.S.	1953	Medium light green color; fine texture; intermediate shoot density; low growth habit; vigorous growth rate; prone to thatching	Good rate of spring green-up; intermediate low temperature hardiness and wear tolerance; used on lawns	Excessive seed head formation
Tifgreen* (T-328) (16, 43, 46, 56, 68, 69)	Georgia AES and CRD-ARS, U.S.	1956	Dark green color; very fine texture; soft leaf blade; very high shoot density; low growth habit	Poor low temperature color retention; good low temperature hardiness and spring green-up rate; excellent drought and wear tolerance; good recuperative potential; prone to smog and 2,4-D injury; widely used on greens	Minimal seed head formation at high nitrogen levels; good resistance to *Helminthosporium* spp. and bermudagrass mite; susceptible to armyworm, sod webworm, and scale insects; requires a high intensity of culture

140

Table 4-2 (Cont.)

Tiflawn‡ (T-57) (46, 56)	Georgia AES and CRD-ARS, U.S.	1952	Medium dark green color; medium fine texture and shoot density; moderately low growing; vigorous growth rate and establishment	Intermediate low temperature color retention and hardiness; excellent drought and wear tolerance and recuperative potential; widely used on sports fields, recreational areas, and lawns	Susceptible to bermudagrass mite; somewhat resistant to nematode injury
Tifway* (T-419) (15, 16, 56)	Georgia AES and CRD-ARS, U.S.	1960	Dark green color; medium fine texture; stiff leaf blades; high shoot density; medium low growth habit; vigorous growth rate; very prone to thatching	Good low temperature color retention and hardiness; good spring green-up rate; susceptible to smog injury; used on lawns, fairways, and tees	Minimal seed head formation; tolerant of bermudagrass mite, sod webworm, and mole crickets
Tufcote‡ (16)	Maryland AES, SCS, and CRD-ARS, U.S.	1962	Medium dark green color; medium texture and shoot density; stiff leaf blades; low growing	Good low temperature color retention and hardiness; good wear and salt tolerance; used on sports fields	Minimal seed head formation; susceptible to bermudagrass mite
U-3‡ (20, 27, 46, 56, 78)	USGA Green Section and CRD-ARS, U.S.	1947	Dark grayish-green color; medium fine texture; medium high shoot density; good rate of growth and establishment	Excellent low temperature hardiness, drought tolerance, and wear tolerance; used on lawns, fairways, tees, and sports fields	Susceptible to spring dead spot
Uganda" (15, 46, 56)	(Natural selection†)	—	Light green color; very fine texture; soft leaf blades; very high shoot density; medium low growth habit; tends to be puffy	Poor low temperature color retention, turns reddish-purple; good low temperature hardiness; used on greens; requires close mowing	Moderately resistant to bermudagrass mite; very susceptible to dollar spot; requires a high intensity of culture

* *C. dactylon × C. transvaalensis.*
† Not officially released.
‡ *C. dactylon.*
§ *Cynodon × magennisii*
" *C. transvaalensis.*

(a) *Cynodon transvaalensis* (Burtt-Davy), a diploid with somatic chromosome numbers of 18 and 20 (38) and a common name of African bermudagrass; (b) *Cynodon* x *magennisii* (Hurcombe), a natural triploid having somatic chromosome numbers of 27 and 30 (38, 51, 52) and a common name, Magennis bermudagrass; and (c) *Cynodon incompletus* var. *hirsutus* (Stent) de Wet et Harlan, an aneuploid with a somatic chromosome number of 18 and the common name Bradley bermudagrass (38). The latter three species are finer in texture, darker green, and more wear tolerant but generally less low temperature hardy than *C. dactylon* (80).

C. transvaalensis has the finest texture and highest shoot density of the *Cynodon* species. It has erect leaves with a bright, light green color that turns a distinct reddish-purple at low temperatures (79). Horizontal spread occurs by short stolons and small, fleshy rhizomes (38). The shoot growth rate is comparatively slow with puffiness frequently being a problem.

Cynodon x *magennisii* is a natural triploid hybrid between *C. transvaalensis* and and *C. dactylon* (38, 52, 61). It spreads by very fine stolons with short internodes and short, shallow rhizomes (38). Characteristics include a dark green color, slow shoot growth rate, fine leaf texture, low growth habit, and very specific cultural requirements (56).

C. incompletus var. *hirsutus* spreads rapidly by fine stolons (79). There are no rhizomes (38). It forms a medium fine textured, gray-green, shallow rooted turf that is quite susceptible to insect and nematode injury (56).

Interspecific hybridization between the bermudagrass species has resulted in the development of a number of improved turf-type bermudagrass cultivars (Table 4-2). Most of the improved cultivars are vegetatively propagated to assure trueness to type. Considerable variability exists in leaf texture, color, internode length, leaf sheath length, shoot growth rate, shoot density, disease resistance, winter hardiness, and drought resistance. These variations have provided the basis for a vegetative identification key of the bermudagrass cultivars (75).

Common bermudagrass was one of the original bermudagrass selections introduced for turfgrass use and was the first to be widely utilized. The chromosome number is 36. It is the only turf-type bermudagrass propagated by seed. The establishment rate is fairly rapid. Common bermudagrass is relatively coarse textured, medium green in color, and has an intermediate shoot growth rate and density (Fig. 4-2). The shade adaptation is very poor while the wear tolerance is quite good. It is very susceptible to *Pythium* blight and bermudagrass mites (15, 33). Common bermudagrass has been one of the most widely used warm season turfgrasses but its use is now limited primarily to situations of (a) a low cultural intensity or (b) polystands with cool season turfgrass species.

The Zoysiagrasses (*Zoysia* Willd.)

The zoysiagrasses are grown primarily in the warm humid and transitional regions. Three species of *Zoysia* are utilized for turfgrass purposes: *Z. japonica*, *Z. matrella*, and *Z. tenufolia*. All three are native to tropical, eastern Asia. The

Figure 4-2. A comparison of the shoot density and leaf texture of common (left) and Tifway (right) bermudagrass turfs (Photo courtesy of G. W. Burton, USDA, Georgia Coastal Plain Experiment Station, Tifton, Georgia).

zoysiagrasses have been introduced and are now distributed throughout the warm humid and transitional climates of the world.

Plant Description

Zoysia japonica Steud. *Vernation* rolled; *sheaths* round to somewhat compressed, glabrous with fascicled hairs at the throat, split with overlapping, hyaline margins; *ligule* a fringe of hairs, 0.2 mm long; *collar* broad, continuous, pubescent at least on margins; *auricles* absent; *blades* flat, 2–4 mm wide, stiff, sparsely pubescent above near the base, glabrous below, margins glabrous; *stems* round, erect to ascending from a decumbent base, strongly stoloniferous as well as somewhat rhizomatous; *inflorescence* short, terminal, and spike-like, spikelets laterally compressed, appressed flatwise against the slender rachis.

Zoysia matrella (L.) Merr. *Vernation* rolled; *sheaths* round or somewhat compressed, glabrous with fascicled hairs at the throat, split with overlapping margins; *ligule* a fringe of hairs, 0.2 mm long; *collar* broad, continuous, pubescent at least at base, margins pubescent; *auricles* absent; *blades* 2–3 mm wide, stiff, pubescent above especially near the base, glabrous below, acute apex, margins glabrous; *stems* round, erect to ascending from a decumbent base, strongly stoloniferous as well as somewhat rhizomatous, rooting at nodes; *inflorescence* short, terminal, and spike-like, spikelets laterally compressed, appressed flatwise against the slender rachis.

Zoysiagrass forms a uniform, dense, low growing, high quality turf that has a slow rate of growth. The leaf texture and color are variable among the *Zoysia* species. The stems and leaves are very tough and stiff. This results in considerable mowing difficulties. All three *Zoysia* species have a prostrate growth habit with mascarenegrass being more diminutive than the others. Zoysiagrass spreads by an

Figure 4-3. A thick stolon of Meyer zoysiagrass with a distinctly upright type of shoot growth.

intergradation of thick stolons and rhizomes that form a very tight, vigorous, tough, prostrate growing turf. The dense, tight turf formed by the zoysiagrasses results in superior resistance to weed invasion. The stems and leaves developing from crowns, stolons, and rhizomes have a distinct, upright growth habit (Fig. 4-3). The root system of Japanese lawngrass is quite fibrous and moderately deep, while the roots of mascarenegrass are relatively shallow.

All improved zoysiagrass cultivars are propagated vegetatively by sprigs, plugs, or sod in order to assure trueness to type. A limited quantity of Japanese lawngrass seed is produced but must be hulled in order to avoid very poor germination (31). Seed head formation is not a serious problem in zoysiagrass turfs. The establishment and recuperative rate is very poor because of a notoriously slow growth rate, especially the lateral shoots (30, 35, 50). A basic chromosome number of 20 has been reported for the turf-type *Zoysia* species (29).

Adaptation. Zoysiagrass is a perennial, sod forming species that is widely adapted throughout the warm humid, warm semiarid, and transitional regions of the world. Japanese lawngrass is the most low temperature hardy of the warm season turfgrass species. Manilagrass is less low temperature hardy than Japanese lawngrass but more hardy than mascarenegrass (21). Manilagrass also ranks between the other two *Zoysia* species in the minimum growth temperature which is 60°F (21, 82). The low temperature color retention is better than for most warm season turfgrass species. Zoysiagrass discolors with the advent of 50 to 55°F temperatures and remains in a state of dormancy throughout the winter period (30, 82). It is devoid of a green pigment system during dormancy and has a distinct brownish-white color. Winter color of dormant zoysiagrass turfs can be provided by the application of a turfgrass colorant or by the winter overseeding of cool season turfgrass species.

The drought and heat hardiness of zoysiagrass is excellent. Although it is relatively winter hardy, it does not thrive in regions where the summers are short or cool. The shade tolerance is quite good when grown in the warm humid regions (50). Manilagrass is more shade tolerant than Meyer or Emerald (10). The tough, stiff nature of the leaves and stems results in zoysiagrass being the most wear tolerant of the commonly used turfgrasses when actively growing. However, wear tolerance is quite poor during the winter dormancy period. The growth rate of the zoysiagrasses is very slow. Mascarenegrass is particularly slow, manilagrass intermediate, and Japanese lawngrass the most rapid of the three (30). The recuperative rate of zoysiagrass is quite slow in comparison to that of bermudagrass.

Zoysiagrass tolerates a wide range of soil types but grows best on well-drained, relatively fine textured, fertile soils having a pH of 6 to 7 (55). It is not tolerant of poorly drained, waterlogged soil conditions. Zoysiagrass has good salt tolerance.

Use. Zoysiagrass forms a very dense, uniform turf of high quality when grown under the proper climatic and cultural conditions. It is widely used for lawns in the warm humid and transitional regions. If the slow establishment and recuperative rate is not objectionable, zoysiagrass can also be used on playgrounds, athletic fields, tees, fairways, airfields, and other intensively utilized areas. The slow growing zoysiagrasses are sometimes used as buffer strips between bentgrass greens and bermudagrass fairways or around sand traps to restrict the encroachment of bermudagrass. In Japan, Japanese lawngrass is used on the fairways and mascarenegrass on the greens.

Culture. Zoysiagrass requires a medium intensity of culture. Cutting heights of 0.5 to 1.0 in. are common for lawns. Close mowing is tolerated because of the low, prostrate growth habit. Frequent mowing at 0.75 in. or less is preferred to prevent excessive thatch accumulation and a puffy, irregular surface. Mowing is quite difficult because of the tough, wiry nature of zoysiagrass leaves. A sharp, well-adjusted reel-type mower is required to achieve acceptable mowing quality. Zoysiagrass is responsive to fertilization and irrigation, particularly when grown on coarse textured soils or in semiarid climates. The nitrogen fertility requirement ranges from 0.5 to 1.0 lb per 1000 sq ft per growing month (19, 66). Although zoysiagrass does not have a rapid growth rate, the stiff, upright growth habit can result in a serious thatching problem, particularly if maintained under intense culture and a high cutting height. Excessive thatch accumulation can be avoided by periodic vertical mowing that also reduces potential scalping problems.

Zoysiagrass is relatively free of major disease problems compared to most of the commonly used turfgrasses. Rust, *Helminthosporium* leaf spot, brown patch, and dollar spot can be serious problems under certain conditions. Nematodes and hunting billbugs have caused considerable turfgrass injury. Armyworms, sod webworms, and mole crickets can also be a problem, but zoysiagrass is generally more resistant to attack by these insects than most warm season species (19). Zoysiagrass is tolerant of most turfgrass herbicides including simazine and atrazine (4, 6).

Cultivars. *Zoysia japonica* Steud. is known by the common name of Japanese lawngrass or Korean lawngrass. It is coarser textured, lower in shoot density, and superior in low temperature hardiness to the other two turf-type *Zoysia* species (32) (Table 4-3).

Zoysia matrella (L.) Merr. is known by the common name of manilagrass. It ranks intermediate in leaf texture, shoot density, low temperature hardiness, and low temperature color retention among the three *Zoysia* species (19, 21, 30).

Zoysia tenuifolia Willd. ex Trin., common name mascarenegrass or Korean velvetgrass, is the least low temperature hardy but the finest textured and most dense of the three turf-type *Zoysia* species. The growth rate is quite slow and the root system shallow. Its use for turfgrass purposes has been quite limited except

Table 4-3

COMPARATIVE CHARACTERISTICS OF THE THREE TURF-TYPE *Zoysia* SPECIES

Characteristic	(Z. japonica)	(Z. matrella)	(Z. tenuifolia)
Color	Dark green	Light green	Very light green
Leaf width (mm)	3	1.5	1
Shoot density	Medium	Good	Excellent
Growth habit	Prostrate	Moderately prostrate	Erect
Growth rate	Slow	Very slow	Extremely slow
Puffiness	Minimal	Medium	Very high
Low temperature hardiness	Medium	Poor	Very poor
Heat hardiness	Excellent	Excellent	Excellent
Drought resistance	Good	Good	Good
Shade tolerance	Good	Good	Excellent
Wear tolerance	Excellent	Excellent	Excellent
Nutritional requirement	Medium low	Medium	Medium
Propagation	Seed or vegetative	Vegetative	Vegetative

in subtropical regions where it is usually maintained as an unmowed ground cover (30). Mascarenegrass tends to be very puffy but is quite wear tolerant.

A limited number of improved zoysiagrass cultivars have been developed for turfgrass use in recent years (Table 4-4). Interspecific hybridization between Japanese lawngrass and mascarenegrass resulted in the development of Emerald, an improved turf-type *Zoysia* cultivar. Most of the improved cultivars are vegetatively established. Considerable variability exists in leaf texture, low temperature hardiness, color, and rate of growth (Fig. 4-4).

St. Augustinegrass [*Stenotaphrum secundatum* (Walt.) Kuntze]

St. Augustinegrass is a versatile, sod forming, warm season turfgrass that is native to the West Indies. It has been widely distributed and is most frequently found in Mexico, Africa, Australia, the southern portions of the United States, and the Mediterranean area (61). It naturalizes most readily in seashore environments.

Plant Description

Vernation folded; *sheaths* compressed, keeled, loose, slightly ciliate toward the apex and along the margins; *ligule* an inconspicuous fringe of hairs about 0.3 mm long; *collar* continuous, extending through a petioled area, broad, glabrous; *auricles* absent; *blades* usually flat, petioled, 4–10 mm wide, glabrous, flexuous, bluntly acute apex; *stems* compressed and branching with extremely long, stout, creeping stolons having swollen nodes and short internodes, leaves clustered at each node; *inflorescence* a short flowering culm bearing terminal and auxillary racemes.

Table 4-4

CHARACTERISTICS OF FOUR VEGETATIVELY PROPAGATED ZOYSIAGRASS CULTIVARS

Cultivar	Released By	Date	Turfgrass characteristics	Adaptation	Other comments
Emerald* (21, 32, 46)	Georgia AES, CRD-ARS, and USGA Green Section, U.S.	1955	Medium dark green color; fine texture; high shoot density; low growth habit; tends to be puffy	Poor low temperature hardiness; intermediate shade tolerance	Susceptible to dollar spot; moderately slow rate of growth, spread, and establishment
FC 13521†	Alabama AES, U.S.	mid-1930's	Dark green color; fine texture; high shoot density	Poor low temperature hardiness and drought tolerance; good shade tolerance; used on lawns	Tends to form seed heads; very slow rate of growth, spread, and establishment
Meyer‡ (Z-52) (21, 46, 47)	CRD-ARS and USGA Green Section, U.S.	1951	Medium dark green color; intermediate texture and shoot density; reduced leaf stiffness; fairly vigorous growth rate	Good low temperature hardiness; superior wear and drought tolerance; intermediate in spring green-up rate and shade tolerance	Intermediate rate of growth and establishment; susceptible to hunting billbugs and nematodes
Midwest‡	Indiana AES, U.S.	1963	Dark green color; medium coarse texture; long stolon internode length; medium low shoot density; more open growth habit	Good low temperature color retention and spring green-up rate	Moderately rapid rate of horizontal spread and establishment

* F₁ hybrid of Z. japonica × Z. tenuifolia.
† Z. matrella
‡ Z. japonica

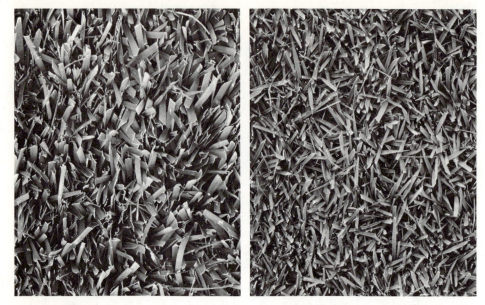

Figure 4-4. A comparison of leaf texture of Meyer (left) and Emerald (right) zoysiagrass turfs (Photo courtesy of G. C. Horn, University of Florida, Gainesville, Florida).

St. Augustinegrass forms an attractive blue-green, low growing turf of medium density and very coarse leaf texture. A unique characteristic of St. Augustinegrass is the distinct constriction and half twist at the base of the leaf blade where it joins the sheath. Also, the stems and stolons are quite flat. St. Augustinegrass is a very aggressive species that spreads by means of long, thick stolons. The root system is fibrous and of medium depth (57).

Propagation is primarily by vegetative means such as sprigs, plugs, or sod. Seed head formation is minimal in comparison to most warm season turfgrass species. Also, the few short, thick seed heads that form tend to produce relatively few viable seeds. Somatic chromosome numbers of 18, 27, and 36 have been reported (58). The aggressive spreading nature of St. Augustinegrass results in a medium good establishment rate (35). Although the wear tolerance is inferior to that of bermudagrass or zoysiagrass, St. Augustinegrass has good recuperative potential.

Adaptation. St. Augustinegrass is a perennial warm season turfgrass adapted primarily to the warmer portions of the warm humid regions of the world. It is the least low temperature hardy of the commonly used warm season turfgrasses. St. Augustinegrass discolors under low temperature conditions and remains in a brown, dormant condition throughout the winter period. It is inferior to bermudagrass and zoysiagrass in fall color retention and spring green-up rate. St. Augustinegrass remains green all year in the warmer portions of the warm humid climates.

The drought resistance ranks fair but inferior to that of bermudagrass, zoysia-grass, and bahiagrass. St. Augustine-grass is the outstanding warm season turfgrass in terms of shade tolerance (10).

St. Augustinegrass tolerates a relatively wide range of soil types, but is particularly well adapted to moist, organic soils. It tolerates soils with a pH of 6.5 to 7.5. Best growth is achieved on moist, well-drained, fertile, sandy loam soils having a pH of approximately 6.5. The salt tolerance of St. Augustine-grass is quite good.

Figure 4-5. A St. Augustinegrass turf growing in a shaded environment. (Photo courtesy of G. C. Nutter, Jacksonville Beach, Florida).

Use. St. Augustinegrass is utilized primarily in the warmer portions of the warm humid regions for lawns and similar turfgrass areas where a fine leaf texture is not required. It is widely utilized under shaded conditions (Fig. 4-5). St. Augustinegrass is also one of the main warm season turfgrass species grown for commercial sod production. It is normally not utilized on playgrounds or athletic fields.

Culture. St. Augustinegrass requires a medium to medium low intensity of culture. A cutting height of 1.5 to 2.5 in. is preferred for lawns (3). Mowing at 1 in. or less results in a thinned turf that is prone to weed invasion. The rapid vertical shoot growth rate necessitates frequent mowing in order to maintain proper turfgrass quality. The leaves of St. Augustinegrass mow rather easily. St. Augustinegrass is quite responsive to fertilization and irrigation, particularly on sandy soils and in arid climates. The nitrogen fertility requirement ranges from 0.5 to 1 lb per 1000 sq ft per growing month (66). St. Augustinegrass frequently exhibits an iron deficiency in the form of yellow, chlorotic leaf tissue that can be readily corrected through a light application of iron sulfate or chelated iron. Thatching is a severe problem, particularly at higher fertility levels (66). An annual vertical mowing is recommended for the maintenance of a high quality St. Augustinegrass turf.

Brown patch, gray leaf spot, and dollar spot can be particularly serious diseases of St. Augustinegrass (88). St. Augustinegrass decline virus (SADV) is a virus disease that is becoming serious in certain regions. Chinch bug injury is limited almost exclusively to St. Augustinegrass in the warm humid climates. It is the primary problem in growing St. Augustinegrass turfs in Florida (25, 88). Chinch bug injury is so rapid and complete that recovery seldom occurs. St. Augustinegrass is not tolerant of organic arsenicals and the phenoxy-type herbicides such as 2,4-D but is tolerant of simazine and atrazine (4, 6).

Cultivars. No improved turf-type cultivars of St. Augustinegrass were available for a number of years. The St. Augustinegrass used was a relatively coarse textured

Table 4-5

CHARACTERISTICS OF TWO VEGETATIVELY PROPAGATED, TURF-TYPE ST. AUGUSTINEGRASS CULTIVARS

Cultivar	Released		Turfgrass characteristics	Adaptation	Other comments
	By	Date			
Bitter Blue	Developed by private interests in Florida, U.S.	—	Distinct blue-green color; intermediate† texture, shoot density, internode length, and growth habit	Improved low temperature color retention; good shade tolerance; poor wear tolerance; used on lawns	Selected from the lower east coast of Florida
Floratine (64)	Florida AES, U.S.	1957	Blue-green color; finer texture; improved shoot density; shorter stolon internode length; stolons more branched; lower growth habit	Improved low temperature color retention; less low temperature hardiness; used on lawns	Selection made in 1950; improved tolerance to close mowing

† Ranks intermediate between Floratine and common types.

selection referred to as common St. Augustinegrass. It has a light green color, low shoot density, and an erect growth habit with long internodes. Roseland, a pasture-type cultivar, has a similar leaf texture, shoot density, and growth habit with even longer internodes. Several decidedly improved turf-type cultivars have become available (Table 4-5).

Centipedegrass [*Eremochloa ophiuroides* (Munro.) Hack.]

Centipedegrass is a native of southern China that has been called Chinese lawn-grass. It was introduced into the United States in 1916. This warm season species is not as widely used as bermudagrass, zoysiagrass, and St. Augustinegrass.

Plant Description

Vernation folded; *sheaths* very compressed, fused-keel, conspicuously hyaline, glabrous with grayish, fascicled hairs at the throat, margins overlapping; *ligule* a short membrane with short cilia, the cilia longer than the purplish membrane, total of about 0.5 mm; *collar* continuous, broad, constricted by the fused keel, pubescent, ciliate, tufted at lower edge; *auricles* absent; *blades* flattened, 3–5 mm wide, blunt, keeled, ciliate with margins papillose toward the base, glabrous underside; *stems* erect to ascending, compressed with thick, leafy, short-noded, opposite-branched stolons; *inflorescence* consists of spike-like racemes, glabrous, subcylindric, terminal, and axillary on slender peduncles.

Centipedegrass is a medium coarse textured, slow growing species that spreads by means of short, thick, leafy stolons having fairly short internodes (Fig. 4-6). It forms a relatively dense mat of prostrate, low growing stems and leaves. Centi-pedegrass ranks intermediate between bermudagrass and St. Augustinegrass in leaf width, shoot density, and stem size. It has a lighter green color than bermudagrass and St. Augustinegrass with shorter leaves than those of carpet-grass. The fibrous root system is generally more restricted than for most warm season turfgrasses (57).

Centipedegrass can be propagated vegetatively or by seed. Propagation by sprigs, plugs, or sod is the most common method because of the extremely slow rate of establishment from seed. The rate of establishment and recuperation is quite slow compared to most warm season turfgrass species with the exception of zoysiagrass (35, 41). Seed

Figure 4-6. Close-up of a typical centipedegrass turf mowed at 1.2 inches.

heads are formed but are low and inconspicuous in comparison to bermudagrass and bahiagrass.

Adaptation. The adaptation of centipedegrass is comparable to St. Augustinegrass with the former extending somewhat more into the cooler portions of the warm humid regions. Centipedegrass has very poor low temperature hardiness ranking between bermudagrass and St. Augustinegrass. It is well adapted to the coastal plains regions. The low temperature color retention is inferior to that of St. Augustinegrass. Discoloration commonly occurs during low temperature periods. Centipedegrass is also inferior to bermudagrass and St. Augustinegrass in drought tolerance, which is partially attributed to the limited root system. However, the recuperative rate from moisture stress is good. It is intermediate between St. Augustinegrass and bermudagrass in shade tolerance (10). Centipedegrass ranks relatively low in terms of wear tolerance.

Centipedegrass is adapted to a relatively wide range of soils with the exception of coarse textured sands. It is particularly well adapted to moderately acid soils of low fertility and fine texture. A soil pH of 4.5 to 5.5 is preferred. Tolerance to submersion, saline, and alkaline conditions is very poor (46).

Use. Centipedegrass is adapted for use on lawns and similar turfgrass areas where traffic is minimal and a relatively low intensity of culture is desired. It forms a turf of acceptable quality with a minimum of care (Fig. 4-7). Centipedegrass is not utilized on athletic fields, playgrounds, airfield runways, and similar intensely used areas because of the slow growth rate and poor wear tolerance.

Figure 4-7. A centipedegrass lawn in Gainesville, Florida. (Photo courtesy of G. C. Nutter, Jacksonville Beach, Florida).

Culture. Centipedegrass requires a low intensity of culture. A cutting height of between 1 and 2 in. is preferred for lawns. Mowing at less than 1 in. results in reduced shoot density. Centipedegrass requires less frequent mowing than most warm season turfgrass species because of the slow vertical shoot growth rate. Centipedegrass thrives on moderately acidic soils of relatively low fertility. It may actually be eliminated under intense fertilization or where the soil pH is excessively high. The nitrogen fertility requirement ranges from 0.1 to 0.3 lb per 1000 sq ft per growing month (66). Centipedegrass exhibits iron deficiency symptoms more readily than St. Augustinegrass (2). A yellow, chlorotic appearance is common, particularly after an application of nitrogen. A light application of iron is necessary to correct the iron deficiency. Centipedegrass thatches excessively if maintained at a high intensity of culture.

Centipedegrass is relatively free of insect and disease problems compared to most warm season turfgrasses. Brown patch, dollar spot, and nematodes can cause extensive injury. The insect known as ground pearls may become a persistent, serious pest under certain conditions (46). Centipedegrass is not tolerant of organic arsenicals but is tolerant of simazine, atrazine, and phenoxy-type herbicides (4, 6).

Cultivars. The development of improved turf-type cultivars of centipedegrass has received little attention, primarily because of its limited use. The common strains are sometimes distinguished by the stem colors of red, yellow, or green (45). Red stemmed strains are generally more low temperature hardy and the most widely used (46).

Oklawn is a vegetative selection of centipedegrass released by the Oklahoma AES in 1965. It forms a bluish-green, medium textured turf. Oklawn has excellent drought and heat tolerance, improved low temperature hardiness, and is adapted to partial shading (47). Best growth occurs on moderately acidic soils of medium fertility. The vertical shoot growth rate is slow.

Carpetgrass (*Axonopus* Beauv.)

Carpetgrass is a native of the West Indies and south Central America. Two species of *Axonopus* have been used to a limited extent for turfgrass purposes. *Axonopus affinis* Chase is used somewhat more widely and is known as common carpetgrass. *Axonopus compressus* (Swartz.) Beauv. is known by the common name of tropical carpetgrass. It is inferior in low temperature tolerance to common carpetgrass. Tropical carpetgrass is somewhat better adapted to droughty soil conditions, while common performs best on moist, fine textured soil conditions. The turfgrass quality is similar for both common and tropical carpetgrass when grown under lawn conditions in their regions of adaptation. Sourgrass (*Paspalum conjugatum* Bergrus.) has a growth habit similar to carpetgrass and is sometimes incorrectly called carpetgrass.

Plant Description

Axonopus compressus (Swartz.) Beauv. *Vernation* folded; *sheaths* compressed, flattened, and glabrous; *ligule* a short, pubescent fringe of hairs fused at the base, about 1 mm long; *collar* continuous, narrow, glabrous to sparingly pubescent; *auricles* absent; *blades* 4–8 mm wide, glabrous or sparsely ciliate at base, margins ciliate toward the apex, acute apex; *stems* compressed, stout with elongated, branching stolons having short internodes, nodes densely pubescent and leafy; *inflorescence* two to five racemes, usually three, on a filiform seed stalk, spikelets sparsely pilose.

Carpetgrass forms a very coarse textured, fairly dense, low growing turf with a distinct light green color. It is somewhat similar in appearance to St. Augustinegrass but is lighter green in color. Distinctive characteristics include compressed, two-edged creeping stems and blunt leaf tips. Carpetgrass spreads readily by means of low growing stolons that root readily at the nodes to form a dense sod. The roots are quite shallow compared to most warm season turfgrasses (7, 11).

Carpetgrass can be propagated by seed or stolons and has a moderate rate of establishment. Seeding is usually preferred since it is easier and less expensive. An objectionable characteristic of carpetgrass is the production of numerous, tall seed heads throughout the summer period.

Adaptation. Carpetgrass is a perennial warm season species adapted primarily to the warmer portions of the warm humid regions. The low temperature hardiness is very poor. The drought tolerance is relatively poor compared to most warm season turfgrasses (11). The shade tolerance is inferior to St. Augustinegrass and zoysiagrass and similar to centipedegrass. It has poor wear tolerance. Carpetgrass is adapted to wet, acidic, sandy or sandy loam soils of low fertility. Best growth is made at a pH of 4.5 to 5.5. It prefers moist soils but does not grow well under waterlogged soil conditions. The salt tolerance is poor.

Use. Carpetgrass is utilized primarily on lawns and similar turfgrass areas having minimal traffic. The adaptation to infertile, acidic soils and the ease of establishment from seed results in carpetgrass being well adapted for use in erosion control on steep slopes and roadsides.

Culture. Carpetgrass requires a low intensity of culture. The preferred cutting height for lawns is 1 to 2 in. The frequency of mowing is less than that required for bermudagrass and zoysiagrass because of the slow vertical shoot growth rate. A rotary mower is required to remove the unsightly seed heads that form throughout the summer period. The nitrogen fertility requirement ranges from 0.2 to 0.4 lb per 1000 sq ft per growing month. Excessive fertilization of carpetgrass should be avoided. Thatch problems are minimal. Brown patch is the most common disease problem of carpetgrass. The seed rot and damping-off organisms can be serious problems during establishment from seed. There are no turf-type cultivars of carpetgrass available.

Bahiagrass (*Paspalum notatum* Flugge.)

Bahiagrass is a native of subtropical, eastern South America. Its use for turfgrass purposes is rather limited and specific. Bahiagrass is the outstanding warm season turfgrass species under a low intensity of culture.

Plant Description

Vernation rolled or folded; *sheaths* compressed, sometimes pubescent, keeled, overlapping; *ligule* membranous, 1 mm long, truncate, entire; *collar* broad; *auricles* absent; *blades* flat to folded, 4–8 mm wide, crowded at base, margins ciliate toward base; *stems* erect to ascending with short, stout, flattened rhizomes and stolons; *inflorescence* consisting of two to three suberect racemes, spikelets solitary.

Bahiagrass forms a very coarse textured, fairly open, erect growing, tough turf (Fig. 4-8). The leaf blade has a distinct hairy margin and is one of the widest of the turfgrass species. It spreads slowly by means of short, flat, heavy stolons and rhizomes. Bahiagrass tends to be somewhat tufted in growth habit because of the short rhizome and stolon development. The root system is coarse, fairly deep, and very extensive (7, 11, 22, 57).

Propagation is primarily by seed as bahiagrass is a prolific seed producer. A very objectionable characteristic is the production of numerous, tall, rapidly growing seed heads. Bahiagrass is largely apomictic and cross-pollinated with

Figure 4-8. Close-up of a typical, unirrigated bahiagrass turf mowed at 2 inches.

chromosome numbers of 20 and 40 having been reported (9, 37). Seed germination is slow but can be greatly enhanced by scarification with acid or heat treatment (44). Establishment from seed is rather easy and inexpensive, but moderately slow.

Adaptation. Bahiagrass is a perennial, warm season species adapted to the warmer portions of the warm humid regions. The low temperature hardiness and color retention is slightly better than that of St. Augustinegrass, centipedegrass, and carpetgrass. It has good shade and wear tolerance and excellent drought resistance (10, 11). The recuperative rate from drought is excellent.

Bahiagrass is adapted to a wide range of soil types from droughty sands to poorly drained, fine textured soils. It is particularly well adapted to droughty, coarse textured, infertile sands that frequently occur along coastal regions. Slightly acidic soils with a pH of 6.5 to 7.5 are preferred. Bahiagrass is relatively intolerant of saline conditions. The tolerance to submersion is quite good.

Use. Bahiagrass forms a relatively low quality turf that is satisfactory for use on low quality, nonuse, turfgrass areas. It is particularly well suited for use on roadsides, airfields, and similar extensive, low quality turfgrass areas where minimum maintenance costs are more important than turfgrass quality (Fig. 4-9).

Culture. Bahiagrass requires a low intensity of culture. A cutting height of 1.5 to 2.5 in. is preferred for lawns (3, 74). Frequent mowing with a rotary mower is

Figure 4-9. A Pensacola bahiagrass lawn maintained at a low intensity of culture in southern Florida.

necessary to remove the numerous, tall, wiry seed stalks that are formed (54). The mower should be very sharp in order to cut the tough, fibrous vascular bundles present in bahiagrass leaves. The nitrogen fertility requirement ranges from 0.1 to 0.4 lb per 1000 sq ft per growing month (28). A fall nitrogen fertilization is particularly important. Thatch is seldom a problem.

Bahiagrass has relatively few disease and insect problems. Dollar spot and brown patch cause occasional injury. The mole cricket is the most common insect problem. Certain common turfgrass herbicides such as the organic arsenicals, simazine, and atrazine cause substantial injury to bahiagrass (4).

Cultivars. A number of cultivars of bahiagrass have been developed but only a few are adapted for turfgrass use (Table 4-6). Further improvement in the bahiagrass cultivars is needed in terms of reduced seed head formation, establishment rate, shoot density, and leaf texture.

Kikuyugrass (*Pennisetum clandestinum* Hochst. ex Chiov.)

Kikuyugrass is a native of East Africa that is now utilized for turfgrass purposes in certain warm humid, highland tropical regions of the world such as Mexico (17, 60). In the United States it is frequently a serious weed in quality turfs (17, 83, 86). The leaves are pubescent, folded in the bud shoot, and somewhat flattened near the apex. Kikuyugrass is a low growing, aggressive perennial that spreads by leafy, very thick, creeping rhizomes and stolons. It forms a dense, tough sod under close mowing. The drought tolerance is good and the heat tolerance excellent. It is inferior in low temperature hardiness to most of the warm season turfgrasses. Thus, its use is limited primarily to the tropical climatic regions of the world. The shade tolerance is moderate. Where adapted, it forms a yellowish-green, medium textured, aggressive, low growing turf that is difficult to mow and prone to thatching. It is best adapted to medium textured, fertile, imperfectly drained soils (26). Propagation is primarily vegetative but it can also be disseminated by seed. The wear tolerance is good. Kikuyugrass is quite susceptible to leaf spot, which limits its use in humid, lowland climates.

Semiarid Warm Season Turfgrass Species

Certain warm season turfgrass species are of importance primarily under warm, semiarid, unirrigated conditions. These turfgrass species are declining in impor-

Table 4-6

CHARACTERISTICS OF FOUR TURF-TYPE BAHIAGRASS CULTIVARS

Cultivar	Released By	Date	Turfgrass characteristics	Adaptation	Other comments
Argentine (46)	Florida AES and CRD-ARS, U.S.	1949	Intermediate texture; soft, hairy, succulent, leaves; improved shoot density; more prostrate growth habit; rapid vertical shoot growth rate	Intermediate low temperature color retention and hardiness; best turf-type available.	Abundant seed head formation; easier to mow; moderately resistant to dollar spot
Paraguay (46)	Not officially released	—	Medium coarse texture; grayish, hairy, fairly short, tough leaves; medium slow vertical shoot growth rate	Poor low temperature hardiness	Intermediate seed head formation; slow seed germination and establishment; susceptible to dollar spot
Pensacola (13, 47)	Florida AES, U.S.	1944	Medium fine texture; less pubescence; leafy, upright growth habit; coarse rhizomes; agressive	Intermediate drought tolerance; medium good low temperature hardiness and color retention; poor tolerance to alkaline soils; widely used on roadsides	Very abundant seed head formation; susceptible to dollar spot; fairly rapid germination and establishment and has a yellowing tendency
Wilmington	SCS, U.S.	1943*	Finest texture of the bahiagrass cultivars; semi-erect; improved shoot density; slower vertical shoot growth rate	Good low temperature hardiness and color retention	Medium low seed head formation; susceptible to dollar spot

* Year of distribution.

157

tance as irrigation of turfs becomes more widely practiced. A need still exists, however, for low quality, nonuse turfs to be used on roadsides, airfields, lawns, and similar areas that require minimum maintenance costs. The three most commonly used semiarid, warm season turfgrasses, buffalograss, blue grama, and sideoats grama, are all in the *Chlorideae* tribe.

Buffalograss [*Buchloe dactyloides* (Nutt.) Engelm.]

Buffalograss is a warm season species native to the subhumid and semiarid regions of the North American great plains. It is one of the dominant species of this shortgrass prairie region and frequently occurs in native polystands with blue grama and sideoats grama. The sod houses of the early settlers were made mostly from sod of this species. Buffalograss is the most widely used of the semiarid warm season turfgrasses.

Plant Description

Vernation rolled; *sheaths* round, open, glabrous; *ligule* a fringe of hairs, 0.5–1.0 mm long; *collar* broad, continuous, glabrous; *auricles* absent; *blades* flat, 1–3 mm wide, flexuous, sparsely pilose, gray-green, margins scabrous; *stems* erect, round with well-developed stolons; *inflorescence* involves male and female flowers usually on a separate plant, the pistillate spikelets are in clusters of four or five that occur in sessile heads, partly hidden among the leaves, the staminate inflorescences have two or three short spikes on slender, erect culms, elevated above the leaves.

Figure 4-10. Close-up of a typical, unirrigated buffalograss turf in central Oklahoma which is mowed at 1 inch. (Photo courtesy of W. W. Huffine, Oklahoma State University, Stillwater, Oklahoma).

Buffalograss forms a fine textured, low growing, soft, grayish-green turf of fairly good shoot density (Fig. 4-10). The grayish-green color is due to the fine hairs that cover the leaves. Curling of the leaf blades is a distinctive feature. Buffalograss spreads vegetatively by numerous stolons that branch profusely and form a tight sod. The root system is rather shallow (18). Propagation may be vegetatively or by seed. Vegetative propagation has been the main method of establishment, primarily because of a seed shortage (71). Buffalograss is dioecious with the male and female flowers occurring on separate plants. Female plants are preferred for vegetative propagation since the male plants bear flower stalks that are two to three times longer. The seed borne on pistillate plants is enclosed in a hard burr containing one or two seeds. The burrs have a very low germination percentage that can be improved by chilling and dehulling.

Adaptation. Buffalograss is a perennial, warm season species adapted to the transitional, warm semiarid, and warm subhumid regions. It has excellent hardiness to high temperatures and tolerates fairly low temperatures in comparison to most warm season turfgrasses. The spring green-up rate and low temperature color retention are intermediate to fair. The excellent drought resistance of buffalograss is one of its most outstanding characteristics (72). It is best adapted to regions having 12 to 25 in. of rainfall per year. Buffalograss can utilize shallow soil moisture resulting from light, infrequent rains to support rapid, horizontal stolon growth (18). It has the ability to go dormant with the advent of severe drought and to initiate new growth after prolonged periods of moisture stress. The shade tolerance is quite poor (27). Buffalograss is adapted to a wide range of soil types but grows best on finer textured soils. It is fairly tolerant of alkaline soils. The submersion tolerance is excellent (65).

Use and culture. Buffalograss is well adapted for use on unirrigated lawns, parks, cemeteries, athletic fields, roadsides, and airfields in the warm and transitional zones of the subhumid and semiarid regions (23). It requires a low intensity of culture. A cutting height of 0.5 to 1.2 in. is preferred for lawn-type situations (27). The mowing interval is generally infrequent due to a slow vertical shoot growth rate. A dense stand of buffalograss requires minimum watering, mowing, and weed control. The nitrogen fertility requirement ranges from 0.1 to 0.4 lb per 1000 sq ft per growing month (27). Thatch is seldom a problem. The establishment rate is intermediate to fair but can be enhanced by irrigation. No turf-type cultivars of buffalograss have been developed.

Blue Grama [*Bouteloua gracilis* (H.B.K.) Lag. ex Steud.]

Blue grama is a warm season species native to the great plain regions of North America. Its use for turfgrass purposes is quite limited and specific.

Plant Description

Vernation rolled; *sheaths* round, glabrous, split hyaline margins; *ligule* a fringe of hairs, 0.1–0.5 mm long, truncate; *collar* medium broad, continuous, pubescent margins; *auricles* absent; *blades* flat or loosely involute, 1–2 mm wide, soft, pubescent, and scabrous above, acuminate apex, margins toothed; *stems* round, erect with stout, short, scaly rhizomes; *inflorescence* usually has two, sometimes one to three, spikes on an erect culm, spike 2.5–5 mm long, falcate-spreading at maturity, the rachis does not project beyond the spikelets.

Blue grama is a low growing, densely tufted species with distinct grayish-green basal leaves. A curling tendency of the leaves is evident. The short, stout rhizomes result in a relatively high shoot density. The root system is fibrous, dense, and extensive but rather shallow. Propagation is primarily by seed. Chromosome numbers of 28, 35, 42, 61, and 77 have been reported (34). Establishment can be achieved quite rapidly and easily.

Adaptation. Blue grama is a perennial, warm season species adapted throughout the transitional, warm subhumid, warm semiarid, and warmer portions of the cool subhumid regions. It has good hardiness to heat stress and excellent drought resistance (72, 76). Although considered a warm season species, blue grama has good tolerance to low temperatures and is utilized well into the cool semiarid and subhumid regions. The spring green-up rate is slow. Blue grama is adapted to a wide range of soil types but grows best on finer textured, upland soils. The tolerance to alkaline soil conditions is quite good.

Use and culture. The use of blue grama is limited primarily to roadsides and similar unirrigated, nonuse turfgrass areas where the intensity of culture is low and the level of turfgrass quality is not important. Blue grama requires a comparatively low intensity of culture. Relatively infrequent mowing at a height of 2 to 3 in. is preferred. The nitrogen fertility requirement is 0.1 to 0.3 lb per 1000 sq ft per growing month. No turf-type cultivars have been developed.

Sideoats Grama [*Bouteloua curtipendula* (Michx.) Torr.]

Sideoats grama is a perennial, warm season species native to the semiarid Great Plains of North America. It is erect growing with short, scaly rhizomes. The bunch-type growth results in a relatively weak sod compared to blue grama. It is also coarser textured and taller. The leaves are usually quite hairy. The root system is moderately deep, fibrous, and multibranching with a wide lateral spread. Sideoats grama can be propagated vegetatively or by seed, with seeding generally preferred. Chromosome numbers of 28, 35, 40, 42, 45, 56, 70, and 98 have been reported (34). The rhizomes are not particularly aggressive, which results in a relatively slow rate of horizontal spread.

Sideoats grama is more widely distributed than blue grama. It is best adapted to the subhumid central Great Plains of the United States. The drought tolerance is inferior to that of blue grama. It is well adapted to a wide range of soil types if moisture is available. The tolerance to alkaline soil conditions is not as good as for blue grama. Sideoats grama requires a low intensity of culture comparable to blue grama. In general, blue grama is preferred over sideoats grama for low quality, nonuse turfgrass purposes, where the primary objective is soil stabilization (23). No turf-type cultivars have been developed but a number of improved cultivars developed for forage use are occasionally used for turfgrass purposes.

Dichondra (*Dichondra* Forst.)

Although dichondra is mowed and maintained similarly to most turfgrasses, it is a dicotyledon belonging to the *Convolvulaceae* family (87). Dichondra is native to the southern coastal plains regions of North America. *Dichondra micrantha* Urb. is the primary species used on lawns. It is considered a weed in certain warm humid regions and is widely propagated for use under mowed lawn conditions in other areas (84).

Plant Description

Dichondra can be readily recognized by the pale green, kidney-shaped leaves borne on creeping stems that form a low, dense mat or cover. The flowers are quite small and pale green in color. Dichondra is highly self pollinated. Turfgrass quality is usually poor during the flowering period. The growth rate is fairly aggressive. Dichondra can be established vegetatively or from seed which should be mechanically scarified due to an impervious seed coat. The seeding rate is one lb per 1000 sq ft. Spring is preferred.

Adaptation. Dichondra is a perennial that is adapted primarily to the warmer portions of the warm humid regions. The low temperature tolerance is poor, while the shade tolerance is good. It is moderately drought resistant. Dichondra is adapted to fine textured, slightly acidic, moist soils of relatively low fertility. The tolerance to compacted, wet soil conditions and salinity is poor. Dichondra tolerates only a limited amount of traffic.

Culture. Dichondra requires a medium to high intensity of culture. Monostands of dichondra are difficult to maintain. It forms a dense, uniform cover on lawns if grown within its range of adaptation and properly maintained (Fig. 4-11). The cutting height is normally 0.5 to 1.0 in. The nitrogen fertility requirement is 0.5 to 1.0 lb per 1000 sq ft per growing month with lower rates used in midsummer. Dichondra generally requires supplemental irrigation for best growth. The growth habit results in it becoming a weed in certain closely mowed turfgrasses. Under these conditions it is very difficult to control selectively. Diseases such as *Alternaria* leaf spot can be a serious problem on dichondra, especially in humid climates. Other problem pests include cutworms, flea beetles, lucerne moth, vegetable weevil, mites, nematodes, slugs, and snails (53). No specific, named cultivars of dichondra have been officially released in the turfgrass seed trade.

Figure 4-11. Close-up of dichondra in southern California which is mowed regularly at 0.3 inch. (Photo courtesy of V. B. Youngner, University of California at Riverside, Riverside, California).

References

1. ANONYMOUS. 1960. Winter kill problems with bermudagrass. Mid-Continent Turfletter, USGA Green Section. No. 3. pp. 1–2.

2. Bryon, O. C. 1933. Yellowing of centipede grass and its control. Florida Agricultural Experiment Station Press Bulletin 450. pp. 1–2.

3. Burt, E. O. 1963. Summary of south Florida turf research. Proceedings of the University of Florida Turf-Grass Management Conference. 11: 164–166.

4. ———. 1964. Tolerance of warmseason turfgrasses to preemergence herbicides: Preliminary report. Proceedings of the Soil and Crop Science Society of Florida. 24: 137–141.

5. ———. 1964. Summary of south Florida research. Proceedings of the University of Florida Turf-Grass Management Conference. 12: 167–169.

6. ———. 1965. Warmseason turfgrasses and preemergence herbicides. Golf Course Reporter. 33(3): 54–56.

7. Burton, G. W. 1943. A comparison of the first year's root production of seven southern grasses established from seed. Journal of the American Society of Agronomy. 35: 192–196.

8. ———. 1947. Breeding bermuda grass for the southeastern United States. Journal of the American Society of Agronomy. 39: 551–569.

9. ———. 1948. The method of reproduction in common bahia grass, *Paspalum notatum*. Journal of the American Society of Agronomy. 40: 443–452.

10. Burton, G. W., and E. E. Deal. 1962. Shade studies on southern grasses. Golf Course Reporter. 30: 26–27.

11. Burton, G. W., E. H. DeVane, and R. L. Carter. 1954. Root penetration, distribution and activity in southern grasses measured by yields, drought symptoms and P^{23} uptake. Agronomy Journal. 46: 229–233.

12. Burton, G. W., and J. E. Elsner. 1965. Tifdwarf—A new bermudagrass for golf greens. USGA Green Section Record. 2: 8–9.

13. Burton, G. W., G. M. Prine, and J. E. Jackson. 1957. Studies of drought tolerance and water use of several southern grasses. Agronomy Journal. 49: 498–503.

14. Burton, G. W., and B. P. Robinson. 1952. Tifton 57 bermuda grass for lawns, athletic fields, and parks. Georgia Coastal Plain Experiment Station Mimeograph Paper No. 78. pp. 1–3.

15. Butler, G. D., and A. A. Baltensperger. 1963. The bermudagrass eriophyid mite. California Turfgrass Culture. 13: 9–11.

16. Butler, G. D., and W. R. Kneebone. 1965. Variations in response of bermudagrass varieties to bermudagrass mite infestations with and without chemical control. Report of Arizona Turfgrass Research. Arizona Agricultural Experiment Station Report 230. pp. 7–10.

17. Cebrian, A. R. 1966. Kikuyugrass can be disastrous to southern courses. The Golf Superintendent. 34: 26–54.

18. Chamrod, A. D., and T. W. Box. 1965. Drought-associated mortality of range grasses in south Texas. Ecology. 46: 780–785.

19. Childers, N. F. 1947. Manila grass for lawns. Federal Experiment Station of Puerto Rico Circular No. 26. pp. 1–16.

20. Dale, J. L., and C. Diaz. 1963. A new disease of bermudagrass lawns and turf. Arkansas Farm Research. 12: 6.

21. Daniel, W. H. 1955. Zoysias for midwest lawns. Midwest Regional Turf Foundation Conference Proceedings. pp. 34–35.

22. Doss, B. D., D. A. Ashely, and O. L. Bennett. 1960. Effect of soil moisture regime on root distribution of warm season forage species. Agronomy Journal. 52: 569–572.

23. Dudeck, A. E., and J. O. Young. 1968. Establishment and use of turf and other ground covers. 1967 Annual Report of the Nebraska Highway Research Project 1. Progress Report 62. pp. 1–25.

24. Duell, R. W. 1961. Bermudagrass has multiple-leaved nodes. Crop Science. 1: 230–231.

25. Eden, W. G. 1960. Controlling chinch bugs on St. Augustine grass lawns. Auburn University Agricultural Experiment Station Progress Report Series No. 79. pp. 1–3.

26. EDWARDS, D. C. 1937. Three ecotypes of *Pennisetum clandestinum* Hochst. (Kikuyu grass). Empire Journal of Experimental Agriculture. 5: 371–376.

27. ELDER, W. C. 1954. Turf grasses, their development and maintenance in Oklahoma. Oklahoma Agricultural Experiment Station Bulletin No. B-425. pp. 1–32.

28. ENGIBOUS, J. C., W. J. FRIEDMAN, Jr., and M. B. GILLIS. 1958. Yield and quality of pangola-grass and bahiagrass as affected by rate and frequency of fertilization. Soil Science Society of America Proceedings. 22: 423–425.

29. FORBES, I. 1952. Chromosome numbers and hybrids in *Zoysia*. Agronomy Journal. 44: 194–199.

30. FORBES, I., and M. H. FERGUSON. 1947. Observations on the zoysia grasses. The Greenkeepers' Reporter. 15: 7–9.

31. ———. 1948. Effects of strain differences, seed treatment, and planting depth on seed germination of *Zoysia* spp. Journal of the American Society of Agronomy. 40: 725–732.

32. FORBES, I., B. P. ROBINSON, and J. M. LATHAM. 1955. Emerald zoysia—An improved hybrid lawn grass for the south. USGA Journal and Turf Management. 7: 23–25.

33. FREEMAN, T. E. 1958. Progress in disease problems. Proceedings of the University of Florida Turf Management Conference. 6: 109–112.

34. FULTS, J. L. 1942. Somatic chromosome complements in *Bouteloua*. American Journal of Botany. 29: 45–55.

35. GARY, J. E. 1967. The vegetative establishment of four major turfgrasses and the response of stolonized 'Meyer' zoysiagrass (*Zoysia japonica* var. Meyer) to mowing height, nitrogen fertilization, and light intensity. M.S. Thesis. Mississippi State University. pp. 1–50.

36. HANSON, A. A. 1965. Grass varieties in the United States. USDA Handbook No. 170. pp. 1–102.

37. HANSON, A. A., and H. L. CARNAHAN. 1956. Breeding perennial forage grasses. USDA Technical Bulletin No. 1145. pp. 1–116.

38. HARLAN, J. R., J. M. J. DE WET, W. W. HUFFINE, and J. R. DEAKIN. 1970. A guide to the species of *Cynodon* (*Gramineae*). Oklahoma Agricultural Experiment Station Bulletin B-673. pp. 1–37.

39. HARTLEY, W. 1950. The global distribution of tribes of the *Gramineae* in relation to historical and environmental factors. Australian Journal of Agricultural Research. 1: 355–373.

40. HARTLEY, W., and R. J. WILLIAMS. 1956. Centres of distribution of cultivated pasture grasses and their significance for plant introduction. Proceedings of the 7th International Grassland Congress. pp. 190–201.

41. HEIN, M. A. 1953. Registration of varieties and strains of bermuda grass, II. [*Cynodon dactylon* (L.) Pers.]. Tiffine bermuda grass (Reg. No. 3). Agronomy Journal. 45: 572–573.

42. ———. 1953. Registration of varieties and strains of bermuda grass, II. [*Cynodon dactylon* (L.) Pers.]. Tiflawn bermuda grass (Reg. No. 4). Agronomy Journal. 45: 573.

43. ———. 1961. Registration of varieties and strains of bermudagrass, III. Tifgreen bermuda-grass [*Cynodon dactylon* (L.) Pers.]. (Reg. No. 5). Agronomy Journal. 53: 276.

44. HODGSON, H. J. 1949. Effect of heat and acid scarification on germination of seed of bahia grass, *Paspalum notatum*, Flugge. Agronomy Journal. 41: 531–533.

45. HORN, G. C. 1960. Evaluation and improvement of turf grasses for Florida. Proceedings of the University of Florida Turf-Grass Management Conference. 8: 156–161.

46. ———. 1967. Turfgrass variety comparisons. Proceedings of the Florida Turf-Grass Management Conference. 15: 91–99.

47. ———. 1968. Research in turf at the University of Florida in Gainesville. Proceedings of the Florida Turf-Grass Management Conference. 16: 96–99.

48. HUFFINE, W. W. 1957. Sunturf bermuda, a new grass for Oklahoma lawns. Oklahoma Agricultural Experiment Station Bulletin B-494. pp. 1–7.

49. ———. 1966. Bermudagrass around the globe. Turf-Grass Times. 1: 18–24.

50. HUME, E. P., and R. H. FREYRE. 1950. Propagation trials with manilagrass, *Zoysia matrella*, in Puerto Rico. Proceedings of the American Society for Horticultural Science. 55: 517–518.

51. HURCOMBE, R. 1946. Chromosome studies in *Cynodon*. South African Journal of Science. 42: 144–146.

52. ———. 1948. A cytological and morphological study of cultivated *Cynodon* species. Experiments with *Cynodon dactylon* and other species. South African Turf Research Station. pp. 36–47.

53. JEFFERSON, R. N., and A. S. DEAL. 1964. Dichondra pests in southern California. California Turfgrass Culture. 14: 17–20.

54. JOHNSON, J. T. 1967. Managing bahiagrass for turf. M. S. Thesis. University of Georgia. pp. 1–61.

55. JUSKA, F. V. 1959. The response of Meyer zoysia to lime and fertilizer treatments. Agronomy Journal. 51: 81–83.

56. JUSKA, F. V., and A. A. HANSON. 1964. Evaluation of bermudagrass varieties for general purpose turf. USDA Agricultural Handbook No. 270. pp. 1–54.

57. LAIRD, A. S. 1930. A study of the root systems of some important sod-forming grasses. Florida Agricultural Experiment Station Bulletin 211. pp. 1–27.

58. LONG, J. A., and E. C. BASHAW. 1961. Microsporogenesis and chromosome numbers in St. Augustinegrass. Crop Science. 1: 41–43.

59. MADISON, J. H. 1962. The effect of management practices on invasion of lawn turf by bermudagrass (*Cynodon dactylon* L.). Proceedings of the American Society for Horticultural Science. 80: 559–564.

60. MANTELL, A., and G. STANHILL. 1966. Comparison of methods for evaluating the response of lawngrass to irrigation and nitrogen treatments. Agronomy Journal. 58: 465–468.

61. MATHEWS, J. W. 1935. Lawn grasses on trial at Kirstenbasch. Journal of the Botanical Society of South Africa. 21: 11–13.

62. MCBEE, G. G. 1966. Effects of mowing heights on Texturf 1F bermudagrass turf. Texas A & M University, Department of Soil and Crop Sciences. Departmental Technical Report No. 10. pp. 1–4.

63. ———. 1967. Studies on the association of light intensity, quality and performance of selected turfgrasses. 1967 Agronomy Abstracts. p. 53.

64. NUTTER, G. C., and R. J. ALLEN. 1962. Floratine St. Augustinegrass. Florida Agricultural Experiment Station Circular S-123A. pp. 1–14.

65. PORTERFIELD, H. G. 1945. Survival of buffalo grass following submersion in playas. Ecology. 26: 98–100.

66. PRITCHETT, W. L., and G. C. HORN. 1962. Turf fertilization research. Proceedings of the University of Florida Turf-Grass Management Conference. 10: 209–215.

67. ROBINSON, B. P., and G. W. BURTON. 1953. Tiffine (Tifton 127) turf bermuda grass. Southern Turf Foundation Bulletin No. 4. pp. 1–2.

68. ROBINSON, B. P., and J. M. LATHAM. 1956. Tifgreen—A hybrid turf bermudagrass. Georgia Coastal Plain Experiment Station Mimeograph Series N. S. 23. pp. 1–5.

69. ———. 1956. Tifgreen—a hybrid turf bermudagrass. Georgia Agricultural Experiment Station Mimeograph Series N. S. 23. pp. 1–6.

70. ———. 1956. Tifgreen—an improved turf bermudagrass. USGA Journal and Turf Management. 9: 26–28.

71. Savage, D. A. 1933. Buffalo grass for fairways in the plains states. USGA Green Section Bulletin. 13: 144–149.

72. SAVAGE, D. A., and L. A. JACOBSON. 1935. The killing effect of heat and drought on buffalo grass and blue grama grass at Hays, Kansas. Journal of the American Society of Agronomy. 27: 566–582.

73. SCHMIDT, R. E., and R. E. BLASER. 1969. Effect of temperature, light, and nitrogen on growth and metabolism of 'Tifgreen' bermudagrass (*Cynodon* spp.). Crop Science. 9: 5–9.

74. SMALLEY, R. R. 1962. Mowing heights for St. Augustinegrass and Pensacola bahiagrass. Proceedings of the University of Florida Turf-Grass Management Conference. 10: 107–109.

75. THOMPSON, W. R. 1968. Vegetative identification of bermudagrass. Turf-Grass Times. 3: 5.

76. WEAVER, J. E., and F. W. ALBERTSON. 1943. Resurvey of grasses, forbs, and underground plant parts at the end of the great drought. Ecological Monographs. 13: 63–117.

77. WHITE, R. W., W. L. PRITCHETT, and G. C. NUTTER. 1957. Effect of rate and frequency of fertilization on the year-round performance of Bayshore and Everglades 1 bermudagrass greens. Proceedings of the University of Florida Turf Management Conference. 5: 55–57.

78. WYCKOFF, C. G., and V. T. STOUTEMYER. 1954. Evaluation of bermudagrass strains. Southern California Turfgrass Culture. 4: 2–4.

79. YOUNGNER, V. B. 1956. Evaluation of new bermudagrass species and strains. Southern California Turfgrass Culture. 6: 13–14.

80. ———. 1958. Bermudagrass for turf in the southwest. Southern California Turfgrass Culture. 8: 21–23.

81. ———. 1959. Growth of U-3 bermudagrass under various day and night temperatures and light intensities. Agronomy Journal. 51: 557–559.

82. ———. 1961. Growth and flowering of *Zoysia* species in response to temperatures, photo-periods, and light intensities. Crop Science. 1: 91–93.

83. ———. 1961. Observations on the ecology and morphology of *Pennisetum clandestinum*. Bulletin of the Orton Society. 16: 77–84.

84. ———. 1966. Dichondra; past, present and future. California Turfgrass Culture. 16: 21–23.

85. ———. 1966. Santa Ana, A new turf bermudagrass for California. California Turfgrass Culture. 16: 23–24.

86. YOUNGNER, V. B., and J. R. GOODIN. 1961. Control of *Pennisetum clandestinum*, Kikuyugrass. Weeds. 9: 238–242.

87. YOUNGNER, V. B. and S. E. SPAULDING. 1964. Dichondra in California. Lasca Leaves. 14(3): 51–53.

88. ZUMMO, N., and A. G. PLAKIDAS. 1958. Brown patch of St. Augustine grass. Plant Disease Reporter. 42: 1141–1147.

chapter 5

Turfgrass Communities

Introduction

An individual turfgrass plant does not live alone but in close association with many other plants that may or may not be of the same genus, species, or cultivar. A **turfgrass community** is an aggregation of individual turfgrass plants that have mutual relationships with the environment as well as among the individual plants. The ecology of communities is referred to as **synecology**, while **autecology** involves the relationship between an individual plant and its environment.

The turfgrass community can be classified into two basic types of populations: (a) the monostand and (b) the polystand. The **monostand community** is composed of turfgrass plants of only one cultivar. A **polystand community** is composed of turfgrass plants of two or more cultivars and/or species. The polystand turfgrass community is further subdivided into blends and mixtures. A **blend** is composed of two or more cultivars of a single turfgrass species. In contrast, a **mixture** is composed of two or more different turfgrass species. The ultimate in a high quality turfgrass community is one that contains only desirable turfgrasses. Sometimes the turfgrass community contains undesirable species called weeds, which cause a reduction in the turfgrass quality.

Competition

Each turfgrass plant within a turfgrass community affects and is affected by the adjacent plants within the community. Individual plants have a certain minimum requirement for light, water, nutrients, carbon dioxide, and oxygen, which must be met in order to survive. **Competition** between two or more plants within the turf-grass community occurs when the supply of light, water, nutrients, etc., is less than that required for growth of each individual plant. The intensity of competition depends on the severity of the deficit. Competition occurs between biotypes of similar genetic constitution or between different cultivars, species, or genera. Competition may involve the mechanical crowding out of a weaker cultivar or species by a stronger cultivar or species that is more vigorous in growth rate, particularly the rate of vertical leaf extension. The germination rate and seedling vigor are major factors influencing the competitive ability of a cultivar during turfgrass establishment. The quantity of carbohydrate reserves in the seed varies greatly among species and affects the seedling vigor.

When competition occurs, the plant that gains an initial advantage subsequently increases its competitive advantage in a cumulative way. For example, a plant that initiates a root system more rapidly, extensively, and/or deeply is able to obtain a larger amount of a limited water or nutrient supply to the detriment of an adjacent plant having a weaker root system. As a result of the increased supply of water and/or nutrients, the more vigorously rooted plant is capable of supporting greater shoot growth and number than the weaker rooted plant. The increased leaf production also results in a positive competitive advantage in light utilization. This, in turn, enables the more vigorous plant to supply its root system with larger quantities of carbohydrates for continued root growth. Thus, the initial competitive advantage increases in an additive way.

Monostand competition. Competition among plants of a single cultivar is very intense because each plant in a monostand community has the same type of growth habit, root system, and requirements for light, water, and nutrients. Each environment has a maximum potential capacity for supporting a population of turfgrass plants. The **stand**, or number of plants contained within the population, is adjusted through competition to a level where deficiencies in light, water, and nutrients are corrected. For example, most seeds will germinate and initiate growth when planted at an excessively high rate, but the seedlings will tend to remain in a relatively immature, stunted state (Fig. 5-1). Eventually, the weaker plants die and the stronger plants survive to form a monostand where each individual plant is able to compete effectively for the available light, water, and nutrients.

Polystand competition. When a mixture of turfgrasses is planted, the polystand community may be composed of individual plants possessing substantial differences in genetic constitution, growth habit, root system, and requirements for light, water, and nutrients. For example,

Figure 5-1. Seven week old seedlings of Kentucky bluegrass from a stand seeded at 1 (left) versus 3 (right) pounds per 1,000 square feet.

certain turfgrasses such as Italian ryegrass may have a very rapid vertical shoot growth rate (11), while others may have a very deep root system. The survival ability of each cultivar or species within the polystand depends on its relative competitive ability and aggressiveness (16). The botanical composition of the turfgrass community in this competitive situation eventually adjusts to a level where the individual plants within the polystand are using the available environmental resources at or near capacity. Cultivars or species that are not able to compete successfully within a given polystand community become extinct.

The relative competitive ability of the turfgrass cultivars or species within a polystand community is not necessarily the same over an extended period of time as during the initial establishment period (10). During the establishment of polystands from seed, Italian ryegrass and most perennial ryegrasses compete severely with Kentucky bluegrass, while red fescue ranks intermediate (1, 12, 17, 28, 29, 44). Italian ryegrass is more competitive than perennial ryegrass during the establishment period (18). The degree of competition between two turfgrass cultivars or species can be controlled by increasing or decreasing the percentage composition of one component in the original seed mixture (17, 28).

Turfgrasses grown under trees and shrubs must compete for light, water, and nutrients. The intensity of competition depends, on the rooting depth and canopy density of the tree or shrub involved (51). A detailed discussion of turfgrass shade ecology is presented in Chapter 6.

Factors Controlling Competition

The factors that are effective in bringing about changes in the composition of a turfgrass community are extremely complex. The relative competitive ability of individual cultivars and the ultimate composition and stand of a turfgrass community is determined by the (a) turfgrass environment, including the climatic and soil phases (4); (b) cultural system, including the height of cut, fertility level, and irrigation practices (5, 6, 7, 12, 13, 14, 18, 25, 28, 29, 31, 32, 49, 53); (c) influence of turfgrass pests, including the relative degree of resistance of various cultivars, the severity of pest attacks, and whether control measures are practiced (4, 15, 27); (d) intensity and type of use the turf receives (30); and (e) genetic constitution of each individual cultivar, including the type of growth habit and root system (1, 12, 21, 27, 28, 32).

Much of the remainder of this text is devoted to a detailed discussion of the influence of the various climatic, soil, cultural, and biotic factors. The relative importance of these factors in influencing the composition of a turfgrass community varies greatly. Factors influencing the competitive ability of an individual turfgrass plant in a given environment include the (a) vertical shoot growth rate; (b) form; (c) leaf area; (d) leaf orientation relative to the incident sunlight; (e) growth habit, including bunch, stoloniferous, and rhizomatous types; (f) rooting depth, extent, and extension rate; (g) nutrient uptake capability; and (h) crown-internode height. The relative importance of each of these factors varies, depending on the turfgrass environment and the intensity of culture.

Succession. Usually a turfgrass mixture does not remain stable after it has been planted. The weaker or less competitive species or cultivars die out, while the more competitive types within a given environment and cultural habitat survive and may actually increase in the percentage composition of the polystand. Through competition a sequence of plant communities occurs that replace one another. This sequence is referred to as **succession**.

Following the initial establishment period, a succession of communities occurs until a more stable polystand develops (29, 38, 47). The botanical composition of the semistable turfgrass community that emerges is strongly influenced by the climate, soil, and cultural conditions of the specific habitat. For example, on loam soils in the cool humid regions, red fescue tends to be dominant in shaded environments, while Kentucky bluegrass is frequently dominant on sites receiving full sunlight (Fig. 5-2). The rate at which succession occurs following initial planting until a semistable community develops varies from less than 1 to more than 4 years (17, 21, 48).

Because of the competitive relationship within the turfgrass community, a certain degree of succession may continue after the semistable state of the polystand occurs. The older, more mature plants become weak and eventually die. Their space within the community is then occupied by younger, more vigorous individuals, possibly of a different genetic type. A stable turfgrass community is able to maintain a relatively constant number of individual plants unless adversely affected or injured by environmental stress, pests, or damage by man.

The composition of the turfgrass community is generally in a constant state of change but the rate of change varies greatly depending on the age of the community as well as the environmental and cultural conditions to which it is exposed. Given specific soil, climatic, and cultural conditions, the general composition of a turfgrass community at equilibrium can be predicted. Thus, a stable turfgrass community of the desired composition can be achieved and maintained through proper adjustment of the environment and intensity of culture.

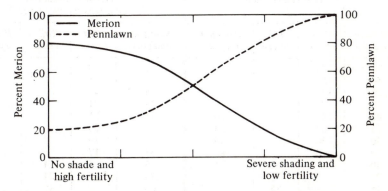

Figure 5-2. Idealized diagram of the influence of shade and nitrogen fertility level on the relative percentages of Merion Kentucky bluegrass and Pennlawn red fescue in a turfgrass community.

Turfgrass Mixtures and Blends

Mixtures or blends are more frequently used for establishing turfs in cool humid than in warm humid climates. Each component included in a seed mixture or blend should have one or more unique attributes that contribute to improved performance of the polystand. Simple mixtures or blends of two or three components are usually preferred.

Mixtures. The use of a mixture is often advantageous because of the increased range in genetic diversity and adaptive potential that is achieved (4, 27). For example, in a given lawn situation there may be areas that are shaded and others that receive full sunlight. In addition, certain portions of the lawn may have a droughty, coarse textured sand and others a very fine textured, imperfectly drained clay. A mixture of Kentucky bluegrass and red fescue contains species that can readily adapt and become dominant under each of these environmental conditions. The red fescue usually becomes dominant in shaded areas and on acidic, infertile, droughty soils, whereas Kentucky bluegrass dominates in full sunlight and on imperfectly drained, moist, fertile soils (13, 15, 25, 28, 49). Red fescue may also become dominant when the Kentucky bluegrass is seriously injured and thinned by disease or vice versa.

The competitive ability of a specific species in a mixture is best evaluated on a seed number basis rather than on a weight basis. Seed mixtures are usually prepared and sold on a weight basis. This can be misleading to the layman due to the large variation in seed size and weight among the turfgrass species (Table 5-1).

Table 5-1

A COMPARISON OF TWELVE SELECTED SEED MIXTURES ON AN ACTUAL SEED NUMBER VERSUS A WEIGHT BASIS. (THE PERCENT SEED NUMBER IS BASED ON THE FOLLOWING NUMBER OF SEEDS PER POUND: KENTUCKY BLUEGRASS, 2,200,000; RED FESCUE, 600,000; TALL FESCUE, 230,000; AND PERENNIAL RYEGRASS, 230,000.)

Percent composition on a weight basis				Percent composition on a seed number basis			
Kentucky bluegrass	Red fescue	Tall fescue	Perennial ryegrass	Kentucky bluegrass	Red fescue	Tall fescue	Perennial ryegrass
75	25			92	8		
50	50			79	21		
25	75			55	45		
10	90			29	71		
40	40		20	75	21		4
33	33		33	73	20		7
30	30		40	71	19		10
20	20		60	63	17		20
40		60		86		14	
25		75		76		24	
15		85		63		37	
5		95		33		67	

Blends. A blend of cultivars is frequently advantageous. The use of a blend can be quite advantageous in habitats where the environment is variable and a number of different disease and/or insect problems exist (24). This procedure is valid where no one cultivar is resistant to all the major diseases within that habitat but where a group of cultivars, each of which is resistant to one or more of the major diseases, can be used. If one cultivar is available that has superior resistance to all the significant pest problems and superior tolerance to all the major environmental stresses within a given habitat, then there is no advantage to using a blend (19, 20). However, with the turfgrass cultivars available and the wide range of environmental stresses and disease problems commonly found, the blending of turfgrass cultivars has been a valuable and effective practice. The most widely used blend in the cool humid regions is one composed of Kentucky bluegrass cultivars. The use of red fescue blends is becoming a more common practice. Blends are also used in one or more of the species components of seed mixtures.

Compatability Within a Turfgrass Community

The specific species or cultivars selected for use in a mixture or blend should have a similar (a) leaf texture, (b) growth habit, (c) color, (d) shoot density, and (e) vertical growth rate in order to achieve an acceptable level of turfgrass quality. For example, redtop is not particularly acceptable for use in mixtures with Kentucky bluegrass and red fescue since it has a wider leaf texture, a bunchy growth habit, and a gray-green color that disrupt the appearance of a Kentucky bluegrass-red fescue polystand (13, 14). Similarly, the coarse leaf texture and light green color of Italian ryegrass are objectionable for use in mixtures with Kentucky bluegrass and red fescue.

Medium dense growth types are preferred for the best compatibility of two or more cultivars or species in a turfgrass community. A dense cultivar may dominate the turfgrass community to the detriment of the turf. For example, creeping bentgrass is not used in mixtures with Kentucky bluegrass and red fescue. The dense, creeping habit of growth results in the creeping bentgrass forming unsightly clumps or patches that eventually predominate to an undesirable degree (12, 13, 14, 27, 38). A noncreeping bentgrass can be utilized effectively in mixtures with Kentucky bluegrass and red fescue.

There are certain other cool season turfgrasses, such as redtop, perennial ryegrass, and Italian ryegrass, which may also be included in turfgrass seed mixtures. In the past, these species have been considered temporary species to be used primarily during the establishment period for rapid germination, establishment, and vertical growth in order to protect against soil erosion (50). However, a scattering of redtop plants can persist in a Kentucky bluegrass-red fescue turf for over 10 years. The grayish clumps of redtop disrupt the turfgrass uniformity. Thus, redtop

Figure 5-3. Appearance of a 50-50 Kentucky bluegrass-red fescue (above) versus a 33-33-33 Kentucky bluegrass-red fescue-perennial ryegrass (below) mixture in May following seeding the previous August. Note thinning due to winter injury of the ryegrass and lack of Kentucky bluegrass and red fescue in the 33-33-33 mixture which is caused by the severe ryegrass competition during establishment.

should not be included in quality turfgrass mixtures.

The ryegrasses are frequently used as a secondary component in turfgrass mixtures under conditions where there is a high erosion probability or during periods of the year when environmental conditions are unfavorable for establishment (18). If the seeding is made during an optimum period such as late summer, the temporary species should be excluded since the rapid germination and vertical growth rate result in excessive competition to the desirable, permanent Kentucky bluegrass and red fescue species (14, 37) (Fig. 5-3). This classical viewpoint regarding the perennial ryegrasses is changing with the development of more diminutive, slow growing cultivars that do not compete excessively. Such cultivars as Manhattan show more promise for use in mixtures with Kentucky bluegrass, particularly for the warmer portions of the warm humid and transitional regions.

Turfgrass Communities of the Cool Climates

At equilibrium, there is a specific turfgrass community that will commonly occur under certain climatic, soil, and cultural conditions. A series of stable polystand turfgrass communities common to the cool humid regions are listed and discussed in the following section. The first turfgrass species listed is usually dominant and the second is usually present as a subdominant component. Any of the five typical communities noted may contain secondary species in addition to the dominant and subdominant components. The more common secondary cool season turfgrass species that may occur in Kentucky bluegrass-red fescue communities are perennial ryegrass, tall fescue, bentgrass, redtop, rough bluegrass, or annual bluegrass.

Kentucky bluegrass-red fescue. This community is found widely throughout the cool humid regions of the world on medium to fine textured, moist, neutral to slightly acidic, fertile soils and in environments having a minimum of shade (21, 27, 48). A medium to high intensity of culture is commonly associated with this community that usually includes a cutting height of 1.2 in. or higher, medium to high levels of fertilization, and irrigation (13, 15, 18, 25, 28, 40, 49) (Fig. 5-4).

Red fescue-Kentucky bluegrass. This community is most common in shaded environments and/or on acidic, coarse textured, droughty soils where the intensity of culture is quite low (2, 21, 27). Typically there would be a minimal fertilization program, no irrigation, and a cutting height of 1.2 in. or higher (13, 15, 25, 28, 40, 49). Red fescue may also become dominant where the Kentucky bluegrass is severely thinned by disease.

Rough bluegrass-red fescue. Rough bluegrass is dominant in moist, shaded environments that are intensely irrigated and/or where there is an imperfectly drained, fine texture soil condition and a cutting height of 1 in. or higher. In

Figure 5-4. Close-up of a Kentucky bluegrass-red fescue turfgrass community (60-40) cut at 1.5 inches.

addition, traffic must be minimal in order for the rough bluegrass to persist.

Bentgrass-annual bluegrass. Communities characterized by this composition receive a high intensity of culture, including a cutting height of 0.6 in. or less, a high fertility level, and frequent irrigation (13, 14, 25, 49). Under certain cultural conditions the order of dominance of these two species is frequently reversed with the annual bluegrass becoming dominant.

Tall fescue-Kentucky bluegrass. This polystand community is most commonly found in the humid transitional zone between the warm humid and cool humid regions where periodic heat and drought stress occur. A high intensity of traffic may also be associated with this type of community. Since tall fescue is primarily a bunch-type species, it should be utilized as the major component in mixtures with Kentucky bluegrass. The tall fescue should compose 85 to 90% of the seed mixture on a weight basis in order to achieve sufficient uniformity of stand (12). In the cooler portions of the cool humid regions where low temperature kill occurs, tall fescue frequently dies out and the Kentucky bluegrass becomes dominant, especially under close mowing and high nitrogen levels (27, 36). A few tall fescue plants usually persist as weeds in scattered unsightly clumps (12).

Turfgrass Communities of the Warm Climates

Communities containing mixtures of warm season turfgrass species are seldom used during the normal growing season of these species. Included in this group are bermudagrass, zoysiagrass, St. Augustinegrass, centipedegrass, and bahiagrass.

When one of these species occurs within the population of another, it is frequently considered a weed and behaves in this manner. For example, bermudagrass is quite patchy and disrupts the uniformity of a St. Augustinegrass turf.

Mixtures of warm season and cool season species. Mixtures of warm season and cool season turfgrass species such as (a) bentgrass-bermudagrass, (b) Kentucky bluegrass-bermudagrass, (c) Kentucky bluegrass-zoysiagrass, and (d) tall fescue-Pensacola bahiagrass have been tried from time to time but their use has not been widely adopted (2, 3, 12, 26, 33, 45, 52). Generally, the warm season species compete excessively during the hot summer period to the extent that the cool season species thins out and is lost, especially in warmer climates. In the cooler portions of the warm humid and transitional regions the warm season species is prone to periodic thinning caused by low temperature injury. As a result the cool season species tend to predominate.

The more open, less dense bermudagrass cultivars have been successfully used in combination with Kentucky bluegrass on sports turfs in the warm, semiarid climates. In this case an annual late fall overseeding of Kentucky bluegrass is usually necessary followed by an annual spring overseeding of bermudagrass. A mixture of cool season and warm season turfgrass species that has been utilized under a medium to low intensity of culture in warm climates is composed of tall fescue and an open, lawn-type bermudagrass cultivar. The tall fescue component contributes a more favorable winter color. This type of turfgrass community is used on institutional grounds and general park areas.

Winter overseeding. Cool season species are frequently overseeded into warm season species in the fall in order to provide a green turf during the winter period when the warm season species are normally brown and dormant (Fig. 5-5). With the advent of high temperatures the following spring, the cool season turfgrass species gradually thin out and the warm season species initiate growth and eventually predominate. This practice is referred to as **winter overseeding**. To achieve effective

Figure 5-5. A bermudagrass green with the right half overseeded and the left half not overseeded for winter play. (Photo courtesy of Milwaukee Sewerage Commission, Milwaukee, Wisconsin).

winter overseeding, the mixture of cool season turfgrass species selected should provide a minimum fall transition time from a green, warm season turf to a green, actively growing cool season turf and a similar minimal transition time back to the green, warm season turf in the spring.

For many years ryegrass was the commonly used cool season turfgrass species for winter overseeding. However, research in the 1960's has shown that overseeding mixtures containing two to four species such as bentgrass, red fescue, ryegrass, rough bluegrass, or Kentucky bluegrass are more desirable in terms of (a) a minimum transition period and (b) overall performance and turfgrass quality during the winter dormancy period (8, 9, 22, 23, 34, 35, 39, 41, 42, 43). Ryegrass contributes good wear tolerance and improves the late fall transition. Red fescue also improves the fall transition and the stiff leaves protect the soft tissues of other species such as rough bluegrass. Rough bluegrass provides improved spring and fall transition and good winter color. Bentgrass contributes improved spring transition. Fine textured Kentucky bluegrass cultivars provide good winter color.

A mixture is preferable for winter overseeding since no one species is available that possesses all the desirable characteristics (35). An overseeding mixture containing bentgrass, red fescue, and rough bluegrass has been quite effective on greens in the cooler portions of the warm humid regions, while Kentucky bluegrass is frequently added in the warmer portions of the warm humid regions (9, 34, 35). Kentucky bluegrass, ryegrass, and tall fescue have been used for winter overseeding into the more open bermudagrass cultivars on lawns and sports turfs. Other species that have been tried and found inferior for winter overseeding include bulbous bluegrass and redtop (43). Annual bluegrass, which behaves as a winter annual in the warm humid regions, may volunteer and provide a certain degree of green winter vegetation in closely mowed, intensely cultured warm season turfs such as bermudagrass (45, 46, 53). Differentials in adaptability to winter overseeding have been observed among the bermudagrass cultivars.

References

1. AHLGREN, H. L., and O. S. AAMODT. 1939. Harmful root interactions as a possible explanation for effects noted between various species of grasses and legumes. Journal of the American Society of Agronomy. 31: 982–985.

2. BEAL, W. J. 1893. Mixtures of grasses for lawns. Proceedings of the 14th Annual Meeting of the Society for the Promotion of Agricultural Science. 14: 28–33.

3. ———. 1898. Lawn-grass mixtures as purchased in the markets compared with a few of the best. Proceedings of the 19th Annual Meeting of the Society for the Promotion of Agricultural Science. 19: 59–63.

4. BEARD, J. B. 1965. Factors in the adaptation of turfgrasses to shade. Agronomy Journal. 57: 457–459.

5. BLACKMAN, G. E. 1932. An ecological study of closely cut turf treated with ammonium and ferrous sulfates. Annals of Applied Biology. 19: 204–220.

6. ———. 1932. A comparison between the effects of ammonium sulphate and other forms of nitrogen on the botanical composition of closely cut turf. Annals of Applied Biology. 19: 443–461.

7. ———. 1934. The ecological and physiological action of ammonium salts on the clover content of turf. Annals of Botany. 48: 975–1001.

8. BROWN, M. A. 1962. A comparison of the performance of fifteen winter grasses overseeded on Tifgreen bermudagrass under putting green conditions. M.S. Thesis. University of Florida. pp. 1–53.

9. BROWN, M., and G. C. HORN. 1961. Winter greens, A symposium. Part 2. Overseeding in the southeast. Golf Course Reporter. 29: 16–26.

10. CHIPPINDALE, H. G. 1932. The operation of interspecific competition in causing delayed growth of grasses. Annals of Applied Biology. 19: 221–242.

11. DAVIES, J. G. 1935. Seed rates and interspecific competition in seed mixtures. Report of the Australian and New Zealand Association for the Advancement of Science. 22: 316.

12. DAVIS, R. R. 1958. The effect of other species and mowing height on persistance of lawn grasses. Agronomy Journal. 50: 671–673.

13. ———. 1961. Turfgrass mixtures—influence of mowing height and nitrogen. Proceedings of the 1961 Midwest Regional Turf Conference. pp. 27–29.

14. ———. 1961. Turfgrass mixtures—influence of mowing height and nitrogen. Illinois Turfgrass Conference Proceedings. pp. 42–45.

15. ———. 1967. Population changes in Kentucky bluegrass-red fescue mixtures. 1967 Agronomy Abstracts. p. 51.

16. DONALD, C. M. 1958. The interaction of competition for light and for nutrients. Australian Journal of Agricultural Research. 9: 421–435.

17. ERDMANN, M. H., and C. M. HARRISON. 1947. The influence of domestic ryegrass and redtop upon the growth of Kentucky bluegrass and chewing's fescue in lawn and turf mixtures. Journal of the American Society of Agronomy. 39: 682–689.

18. FUNK, C. R., and R. E. ENGEL. 1951. Effect of cutting height and fertilizer on species composition and turf quality ratings of various turfgrass mixtures. Rutgers University Short Course on Turfgrass Management. pp. 47–56.

19. FUNK, C. R., R. E. ENGEL, and P. M. HALISKY. 1968. Ecological factors affecting turf performance and disease reaction of varietal blends of Kentucky bluegrass. 1968 Agronomy Abstracts. p. 63.

20. FUNK, C. R., R. E. ENGEL, P. M. HALISKY, and W. K. Dickson. 1969. Evaluation of Kentucky bluegrass blends for turf performance, disease reaction and varietal composition. Rutgers University Turfgrass Conference Proceedings. pp. 19–30.

21. GARNER, E. S., and S. C. DAMON. 1929. The persistence of certain lawn grasses as affected by fertilization and competition. Rhode Island Agricultural Experiment Station Bulletin 217. pp. 1–22.

22. GILL, W. J. 1964. An evaluation of overseeding procedures for southern lawns. M. S. Thesis. Mississippi State University. pp. 1–60.

23. GILL, W. J., W. R. THOMPSON, and C. Y. WARD. 1967. Species and methods for overseeding bermudagrass greens. The Golf Superintendent. 35: 10–17.

24. HANSON, A. A., R. J. GARBER, and W. M. MYERS. 1952. Yields of individual and combined apomictic strains of Kentucky bluegrass (Poa pratensis L.). Agronomy Journal. 44: 125–128.

25. HARPER, J. C., and H. B. MUSSER. 1950. The effects of watering and compaction on fairway turf. Pennsylvania State College 19th Annual Turf Conference. pp. 43–52.

26. HART, S. W., and J. A. DeFRANCE. 1955. Behavior of Zoysia japonica Meyer in cool-season turf. USGA Journal and Turf Management. 8: 25–28.

27. JUSKA, F. V., and A. A. HANSON. 1959. Evaluation of cool-season turfgrasses alone or in mixtures. Agronomy Journal. 51: 597–600.

28. JUSKA, F. V., J. Tyson, and C. M. HARRISON. 1955. The competitive relationship of Merion bluegrass as influenced by various mixtures, cutting height, and levels of nitrogen. Agronomy Journal. 47: 513–518.

29. ———.1956. Field studies on the establishment of Merion bluegrass in various seed mixtures. Quarterly Bulletin of the Michigan Agricultural Experiment Station. 38: 678–690.

30. KLECKA, A. 1937. Influence of treading on associations of grass growth. Sborn. Csl. Akad. Zemed. 12: 715–725.

31. MADISON, J. H. 1962. Turfgrass ecology. Effects of mowing, irrigation, and nitrogen treatments of *Agrostis palustris* Huds., 'Seaside' and *Agrostis tenuis* Sibth., 'Highland' on population, yield, rooting and cover. Agronomy Journal. 54: 407–412.

32. ———. 1962. The effect of management practices on invasion of lawn turf by bermudagrass (*Cynodon dactylon* L.). Proceedings of the American Society for Horticultural Science. 80: 559–564.

33. MAHDI, Z. 1956. The bermudagrass-bentgrass combination for an all-year putting or lawn bowling green. Southern California Turfgrass Culture. 6: 8.

34. MEYERS, H. G., and G. C. HORN. 1967. Selection of grasses for overseeding. Proceedings of the Florida Turf-Grass Management Conference. 15: 47–52.

35. ———. 1968. Selection of cool season grass mixtures for overseeding golf course greens. Proceedings of the Florida Turf-Grass Management Conference. 16: 33–37.

36. MILLER, R. W. 1966. The effect of certain management practices on the botanical composition and winter injury to turf containing a mixture of Kentucky bluegrass (*Poa pratensis*, L.) and tall fescue (*Festuca arundinacea*, Schreb.). 7th Illinois Turfgrass Conference Proceedings. pp. 39–46.

37. MORRISH, R. H., and C. M. HARRISON. 1948. The establishment and comparative wear resistance of various grasses and grass-legume mixtures to vehicular traffic. Journal of the American Society of Agronomy. 40: 168–179.

38. MUSSER, H. B. 1948. Effects of soil acidity and available phosphorus on population changes in mixed Kentucky bluegrass-bent turf. Journal of the American Society of Agronomy. 40: 614–620.

39. NOER, O. J., C. G. Wilson, and J. M. LATHAM. 1961. Winter grass overseeding on bermudagrass putting greens. Proceedings of the 15th Annual Southeastern Turfgrass Conference. pp. 33–41.

40. ROBERTS, E. C., and F. E. MARKLAND. 1966. Nitrogen induced changes in bluegrass-red fescue turf populations. 1966 Agronomy Abstracts. p. 36.

41. SCHMIDT, R. E. 1962. Overseeding winter greens in Virginia. Golf Course Reporter. 30: 44–47.

42. SCHMIDT, R. E., and R. E. BLASER. 1961. Cool season grasses for winter turf on bermudagrass putting greens. USGA Journal and Turf Management. 14: 25–29.

43. SCHMIDT, R. E., and J. F. SHOULDERS. 1964. Overseeding bermudagrass with cool season grasses for winter turf. 1964 Agronomy Abstracts. p. 102.

44. STAPLEDON, M. A., W. DAVIES, and A. R. BEDDOWS. 1927. Seeds mixture problems; soil germination, seedling and plant establishment with particular reference to the effects of environmental and agronomic factors. I. Garden trials. Welsh Plant Breeding Station Bulletin Series H. pp. 5–38.

45. STOUTEMYER, V. T. 1953. Grass combinations for turfs. California Agriculture. 7: 9–10.

46. ———. 1954. Annual bluegrass as a cool season grass for bermudagrass mixtures. Southern California Turf Culture. 4: 1–2.

47. TABOR, P. 1962. Permanent plant cover for road cuts and similar conditions by secondary succession. Agronomy Journal. 54: 179.

48. VAN DERSAL, W. R. 1936. The ecology of a lawn. Ecology. 17: 515–527.

49. WATSON, J. R. 1949. Compaction and irrigation studies on established fairway turf. Pennsylvania State College 18th Annual Turf Conference. pp. 88–90.

50. WELTON, F. A., and J. C. CARROLL. 1940. Lawn experiments. Ohio Agricultural Experiment Station Bulletin 613. pp. 1–43.

51. WHITCOMB, C. E. 1968. Grass and tree root relationships. Proceedings of the Florida Turf-Grass Management Conference. 16: 46–50.

52. YOUNGNER, V. B. 1958. Tall fescue-Pensacola bahiagrass combination. Southern California Turfgrass Culture. 8: 24.

53. ———. 1959. Ecological studies on *Poa annua* in turfgrasses. Journal of the British Grassland Society. 14: 233–237.

part II

THE TURFGRASS ENVIRONMENT

The turfgrass plant lives in a dynamic environment. **Environment** is the aggregate of all surrounding conditions influencing a grass plant or turfgrass community. The many atmospheric, soil, and biotic factors that combine to influence plant responses are known as the controls of turfgrass quality.

The condition of the atmosphere at a specific time and place is called **weather**. Elements utilized in describing weather include temperature, precipitation, humidity, air pressure, wind, and cloudiness. **Climate** is the composite state of the atmosphere for a particular region over a period of many years and encompasses the weather variations. Climates vary with latitude, topography, and proximity to large bodies of water.

The **microenvironment** involves the conditions and influences in the immediate vicinity of the turfgrass plant, including the interchanges of energy, gases, and water. It encompasses the zone from the surface of the turf to the depth of root penetration into the soil and can vary considerably in turfs located only a few yards apart.

Most standard weather bureau temperature measurements in the United States are taken 5 ft above the ground. The environment up to 5 ft above the soil is of most significance in human activities and is called the **macroenvironment**. It is representative of the overall climate and has considerable influence on the turfgrass microenvironment. However, temperatures 5 ft aboveground are quite different from those in the immediate vicinity of the turf. For example, the temperature at the surface of a turf may exceed 120°F. when the air temperature at 5 ft is only 95°F.

The complex nature of the turfgrass microenvironment is not fully understood. The environmental factors influencing turfgrass growth and development can be classified into three major groupings: (a) climatic, (b) edaphic, and (c) biotic. Grouped within climate are the light, temperature, water, and air factors; the edaphic grouping encompasses the soil aspects; and the biotic grouping includes the cultural practices superimposed on grasses by man. In order to ensure an orderly discussion, each major environmental factor is considered individually. However, it should be recognized that turfgrasses do not respond in relation to only one environmental factor but to the combined effect of many factors. For example, the influence of light on turfgrass growth and development varies with the temperature. Also, the interrelationships among various factors are not constant but ever changing.

Turfgrass culture involves manipulation of the plant environment. Thus, it is imperative that turfmen have a thorough understanding of atmospheric, soil, and biotic factors affecting the turfgrass environment. Turfgrasses are maintained in a relatively unnatural environment due to (a) close, frequent mowing, (b) high fertilization rates, (c) high plant density, and (d) intense human traffic. The complex, ever-changing character of turfgrass culture offers a stimulating challenge to the professional turfman.

chapter 6

Light

Introduction

The sun is the energy source for the support of plant life on earth. Solar radiation influences energy distribution and movement in the atmosphere and is the most important meteorological control affecting weather patterns. It also influences cell differentiation and plant distribution as well as being the energy source for growth. Radiant energy from the sun is converted to chemical energy in green plants by the photochemical process of photosynthesis. Mowed turfs are capable of absorbing and converting to chemical energy only 1 to 2% of the total incident radiant energy. A major portion of the incident radiant energy is absorbed and reradiated at longer wavelengths with the release of heat that can significantly affect turfgrass temperatures.

The growth and developmental responses of turfgrass to the quality, duration, and intensity of light are discussed in this chapter. In considering the effects of light, it is to be assumed that the water, nutrient, carbon dioxide, and temperature levels are not limiting. However, high temperatures are associated with intense incident radiation under many conditions.

Light Absorption

Maximum light absorption by turfgrass leaves is vital since a major portion of the photosynthetically active leaf area is removed in mowing. Incident radiation can be absorbed, reflected, or transmitted. Light transmission through turfgrass

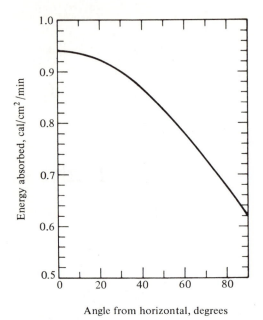

Angle from horizontal, degrees

Figure 6-1. The effect of leaf orientation on light energy absorption. The incident solar radiation is 1.20 cal $cm^{-2}min^{-1}$ and is perpendicular to the horizontal (After Gates—70).

leaves varies from 15 to 30 % (10). The relative degree of light absorption and reflection is affected by the orientation of the leaf surface to incident radiation. On a diurnal basis, light penetration into the turfgrass canopy is highest at or near midday (26). The amount of light absorption decreases as the angle between the incident radiation and the leaf surface decreases (Fig. 6-1).

The stand geometry of most dense turfs involves primarily an upright leaf orientation that is not particularly favorable for light absorption. Cultural practices are designed to enhance this type of growth habit since dense, vertical leaf growth is desired for high turfgrass quality.

Only the uppermost leaves in a turfgrass community are fully exposed to the incident solar radiation. Long turfgrass leaves with a semi-vertical orientation are subjected to considerable light intensity variations along their length. Light absorption occurs on the lower side of a leaf as well as the upper surface (111). A significant amount of reflected light is absorbed by the lower leaf surface in turfs having vertical leaf orientation. Other factors affecting light absorption include (a) leaf surface geometry including pubescence; (b) leaf thickness; (c) nature and distribution of pigments in the leaf; (d) location, number, and orientation of chloroplasts within the leaf; and (e) degree of shading due to interleaf orientation.

The albedo of actively growing turfs generally ranges from 22 to 26 (85, 108). Glossy leaf surfaces result in increased reflection. Also, young leaves reflect a higher percentage of the incident radiation than old leaves. An albedo as high as 37 has been observed when dew is present on the turfgrass leaves or just after an irrigation (85). A green, succulent turf has a lower albedo than a dry, dormant turf (10). The albedo for dark cultivated soils is 7 to 10 and from 50 to 75 for snow (85). Clean, newly fallen snow has higher values (10). The amount of reflection is generally higher at low heights of the sun near sunrise and sunset.

Light Quality

Plant responses to light quality are common to many species. The violet, blue, and ultraviolet (UV) regions are the most important in influencing anthocyanin synthesis and phototropic responses. The violet and blue wavelengths produce a short, sturdy growth habit, while yellow and red enhance shoot elongation and spindly growth (98). The 630 to 780 mμ region is important in promoting or inhibiting flowering, stem elongation, seed germination, leaf enlargement, rhizome development, and numerous other photomorphogenic plant responses (76, 77). A photoreversible pigment called **phytochrome** is involved.

The specific effects of light quality on turfgrass growth and development have not been investigated in detail. Turfgrass quality is better when grown under the blue and green wavelengths than the red (98). The first leaf of a grass seedling remains folded in darkness. The unfolding of grass leaves is promoted by red light of 660 mμ and inhibited by 710 mμ (151). Mesocotyl elongation of tall fescue and sheep fescue is inhibited about 1000 times more by red light (624 mμ) than by blue light (436 mμ) (153). Light quality is also a factor in the sporulation of certain turfgrass diseases. For example, low intensity UV wavelengths of 265 to 325 mμ are important in stimulating sporulation of *Typhula* blight (124).

Some diurnal and seasonal variability in light quality does occur. The proportion of orange and red wavelengths increases at dawn and dusk while the violet and blue wavelengths decrease. The angle of incidence of the sun's rays to the surface of the earth is less at dawn and dusk. The light must pass through a thicker atmospheric layer during these periods resulting in the screening of shorter wavelengths. The longer, infrared wavelengths are of special concern to the turfgrass environment due to their heating properties. The longer the wavelength, the greater the heating effect. A high moisture content in the atmosphere results in increased absorption of the red and near infrared wavelengths.

The seasonal and diurnal variations in light quality do not cause any major changes in turfgrass growth and development when grown in full sunlight. One reason is that most plant responses have a degree of sensitivity to all wavelengths in the visible spectrum. Also, the wavelength variations are small in relation to the total daily radiation. Turfgrass responses to light quality are more often observed in shaded environments where selective reflection, absorption, and transmission occur. This aspect of light quality is discussed in the shade ecology section. The light quality regime within a solarium or domed stadium may also be drastically altered depending on the reflection and transmission properties of the dome. The influence of light quality on turfs growing in full sunlight is not as significant as the intensity or duration.

Light Duration

Light duration is particularly significant in affecting geographic plant distribution. Native grasses adapt and proliferate through natural selection in those regions having specific day lengths that occur in seasons most favorable for growth, flowering, and reproduction (115). The day length factor for reproduction is more important in annual, bunch-type grasses than in perennial, vegetatively propagated turfgrasses. Light duration also influences plant growth and development.

Figure 6-2. Effect of 8 (left) and 16 (right) hour day lengths on the flowering of annual bluegrass.

Flowering response. Day length effects are most frequently associated with the reproductive phase of plant development. Many species exhibit a sensitivity to day length that determines whether the plant continues vegetative growth or produces flowers (Fig. 6-2). Vegetative growth is desired in normal turfgrass culture. Seed head development disrupts turfgrass uniformity, increases the mowing difficulty, and impairs the smoothness of greens. The most desirable, seeded turfgrass cultivars do not form seed heads under turfgrass conditions but can produce high yields when grown specifically for seed. This is most easily accomplished if the day length under which flowering occurs differs from the day lengths under which the cultivar is to be grown. The selection and use of non-flowering cultivars is particularly desirable if they are to be established vegetatively.

Flower induction of many cool season, perennial turfgrasses occurs during the short, cool days of autumn. The general light and temperature requirements for maximum floral induction of eighteen turfgrasses are presented in Table 6-1. Short-

Table 6-1

ENVIRONMENTAL REQUISITES FOR MAXIMUM FLORAL INDUCTION OF EIGHTEEN TURFGRASSES

Turfgrass	Scientific name	Requirement for maximum floral induction Photoperiod	Temp*	References
Annual bluegrass	*Poa annua*	Short†	Cool	82, 146
Bermudagrass	*Cynodon dactylon*	Long†	Warm	166
Bulbous bluegrass	*Poa bulbosa*	Long	Cool	125, 163
Canada bluegrass	*Poa compressa*	Long	Cool	60, 140
Colonial bentgrass	*Agrostis tenuis*	Short	Cool	45, 46
Creeping bentgrass	*Agrostis palustris*	Short	Cool	45, 46
Fairway wheatgrass	*Agropyron cristatum*	Short	Cool	67
Kentucky bluegrass	*Poa pratensis*	Short	Cool	45, 59, 67, 120, 125, 126
Manilagrass	*Zoysia matrella*	Short	Warm	34, 63, 164
Meadow fescue	*Festuca elatior*	Short	Cool	46
Perennial ryegrass	*Lolium perenne*	Short†	Cool	41, 42, 43, 44, 45, 46, 57, 58, 67
Red fescue	*Festuca rubra*	Short	Cool	46, 67
Redtop	*Agrostis alba*	Short	Cool	45, 46, 67
Rough bluegrass	*Poa trivialis*	Short	Cool	46
Sheep fescue	*Festuca ovina*	Short	Cool	46, 160
Tall fescue	*Festuca arundinacea*	Short	Cool	67, 145
Velvet bentgrass	*Agrostis canina*	Short	Cool	45, 46
Zoysiagrass	*Zoysia japonica*	Short	Warm	63, 164

* "Cool" temperatures are from 40 to 50°F and "warm" from 75 to 85°F.
† Species are capable of flowering over a wide range of photoperiods.

day plants require day lengths of 12 hr or less, while long-day plants require more than 12 hr. Some cool season, perennial turfgrasses remain vegetative due to a short day requirement (67). The day length requirement for floral induction is not absolute in some species that are called day-neutral plants. For example, maximum flowering of annual bluegrass is induced by short days but flowering can occur over a wide range of photoperiods (82, 146). Flowering of bermudagrass also occurs over a wide range of day lengths (166). The day length requirement for maximum flowering may also vary within species (42, 44, 115).

Temperature interacts with the photoperiod in inducing flowering. Cool temperatures can greatly alter or even replace the photoperiod requirement of certain turfgrasses (44, 78, 126). Temperatures of 32 to 40°F can substitute for the short day photoperiod requirement of perennial ryegrass (44) and tall fescue (145). In contrast, the *Agrostis* species have a short day induction requirement but show no response to cold treatment alone (45). Certain turfgrasses, such as Kentucky bluegrass and red fescue, have a juvenile stage during which they will not respond to floral induction conditions (45).

The optimum photoperiod for floral induction is not necessarily the most favorable for floral development. Long day lengths favor floral development of many perennial cool season turfgrasses (41, 74, 163). Floral development of Kentucky bluegrass occurs under a range of temperatures and photoperiods, but maximum floral development occurs under long day lengths and moderately cool temperatures (120). Floral development of the warm season *Zoysia* species is favored by short day lengths (34, 63, 163).

Developmental responses. The growth habit of turfgrasses is affected by day length (Table 6-2). Some typical responses of turfgrasses grown under short day lengths are

Table 6-2

THE EFFECTS OF 8 AND 16 HR DAY LENGTHS ON THE GROWTH CHARACTERISTICS
OF MERION KENTUCKY BLUEGRASS (EXPRESSED AS THE AVERAGE OF 20 PLANTS
PER LIGHT TREATMENT)

Plant characteristic	Photoperiod, hr	
	8	16
Leaf length, mm	86	220
Leaf width, mm	2.8	3.2
Number of shoots per plant	21	16
Total dry weight of shoots per plant, g	3.6	5.2
Number of rhizomes per plant	1.3	1.1
Total dry weight of roots and rhizomes per plant, g	1.7	1.9
Leaf growth habit, degrees to horizontal	47	75
Succulence, % water	70	65

1. Increased shoot density (59, 60, 120, 152).
2. Increased tillering (87, 107, 114, 115, 128, 144, 149).
3. Increased leaf appearance rate (107, 128, 144).
4. Reduced leaf and shoot length (54, 60, 107, 114, 120, 128, 134, 139, 142, 144, 152, 164).

5. Reduced internode length (60, 114, 164).
6. A prostrate growth habit with individual plants frequently growing in a rosette shape (5, 11, 59, 60, 120, 142, 152).

Figure 6-3. The growth habit of Kentucky bluegrass grown under 16 (left) and 8 (right) hours of daily illumination.

The six morphologic responses to short day lengths were observed on short-day turfgrasses. Long-day turfgrasses may not show a similar response (20). Many of the morphogenic responses to day length are probably controlled by a reversible photoreaction involving phytochrome or a similar photoreversible pigment. The prostrate growth of Kentucky bluegrass shoots and leaves under short days (Fig. 6-3) is similar to the growth habit observed during the autumn season (11). There is a striking increase in the wall thickness of fibers in the ribs of bentgrass leaves grown under long day lengths (142). This probably contributes to the increased rigidity and upright growth of turfgrass leaves grown under long day lengths. The increased leaf length under long day lengths is the result of greater cell elongation rather than an increase in cell number (142).

The night illumination of recreational and sports turfs for football, softball, baseball, soccer, and golf produces an artificially long photoperiod. The illumination level is low, ranging from (a) 5 to 10 lumens on golf courses, (b) 20 to 50 lumens on most baseball and football fields, and (c) as high as 150 to 200 lumens on major league fields (38, 39). Turfgrass responses to night lighting are primarily developmental in nature rather than growth orientated.

Growth responses. A longer day length results in more radiant energy available for absorption and utilization in photosynthesis. Turfgrass plants grown under long days have a higher carbohydrate level (152) and generally exhibit greater shoot growth than when grown under short days at the same temperature (20, 66, 82, 92, 107, 114, 115, 118, 120, 121, 139, 140, 152, 164). Physiologically there is a shift to lower soluble nitrogen and amide contents under long day lengths.

Data concerning the effect of day length on root and lateral shoot growth are variable and conflicting. Greater root, stolon, and rhizome growth (110, 123) generally occurs at longer day lengths of 13 to 16 hr with some variability occurring due to temperature and species (Table 6-2). Zoysiagrass root and rhizome production is highest at a 14-hr day length and decreases at longer and shorter day lengths (164). Shoot growth is more responsive to long day lengths than root growth. This causes a decrease in the root-shoot ratio under long day lengths (92, 120). Stolon growth of creeping bentgrass (5) and zoysiagrass (164) is enhanced by day lengths in excess of 13 hr. Thus, vegetative establishment of stoloniferous turfgrasses is enhanced by planting when photoperiods are longer. Day length is also a factor controlling the summer dormancy of certain grasses. Long days and high temperatures induce summer dormancy in *Poa bulbosa, P. scabrello,* and *P. secundo* (91).

Light Intensity

The light intensity varies greatly depending on the (a) season of year, (b) latitude, (c) time of day, (d) degree of atmospheric screening, and (e) topography. Light intensities are highest during the summer season and decrease with increasing latitude. The diurnal variation in light intensity is characterized by a broad curve with a minimum level at sunrise and sunset, increasing to a maximum at midday (Fig. 6-4). Light intensities on a clear day in temperate regions are commonly around 1.5 cal per sq cm per min (10,000 lumens). Maximum light intensities of 1.8 to 2.0 cal per sq cm per min have been reported.

Humid areas with extensive cloud or fog cover and urban areas with smoke, dust, or smog problems are subject to extensive atmospheric screening of incoming solar radiation. Smoke can screen out as much as 90% of the incoming radiation and a cloud cover up to 96% (49). This degree of screening can significantly reduce the photosynthetic rate.

Topography can cause localized variations in light intensity because it affects the angle at which radiation strikes the earth. Turfgrasses growing on locations positioned normal to the incident radiation receive the highest intensities. The angle of incidence is large and the light intensity lowest on slopes facing poleward or away from the incoming radiation, while the highest intensities occur on slopes facing the equator. Mountainous regions are subject to higher light intensities since the intensity increases proportionally with altitude.

A vertical light intensity gradient exists within the turfgrass canopy that is affected by the cutting height and shoot density. The highest light intensity occurs

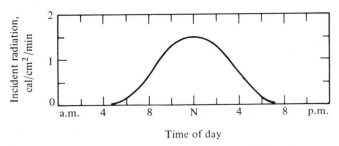

Figure 6-4. The diurnal variation in incident radiation for a clear day in July at East Lansing, Michigan.

at the top of the turfgrass stand. A large decrease in light intensity occurs over a short vertical distance downward through the turfgrass canopy (10). A minimal amount of light actually reaches the soil surface in dense turfs. The percentage penetration of light into the turfgrass canopy is greater during periods of water stress when the upper leaves are rolled or folded.

Growth responses. Photosynthesis exceeds respiration by a large margin under nonstress growing conditions. Carbohydrate reserves are utilized for respiration in darkness and at light intensities below the compensation point. The carbohydrate reserves of a turfgrass plant could be exhausted during extended periods with light intensities below the compensation point. A decline in plant vigor, recuperative potential, and turfgrass quality would result. A compensation point of 0.03 cal per sq cm per min has been reported for individual bermudagrass leaves (4). The compensation point for most turfgrasses is between 2 and 5% of full sunlight. The older, lower leaves in a dense turf are under varying degrees of partial shade. Leaves continuously exposed to light intensities below the compensation point are likely to die or must obtain carbohydrates by translocation from other plant parts.

Light intensities above the compensation point are required for growth to occur without utilization of the carbohydrate reserves. Net assimilation of individual bermudagrass leaves increases proportionally with light intensity up to 0.45 cal per sq cm per min (Fig. 6-5). The leaf was light saturated at this point since no increase in net assimilation occurred at higher intensities (4). The maximum photosynthetic rate of individual turfgrass leaves occurs at light intensities about one-third of full sunlight.

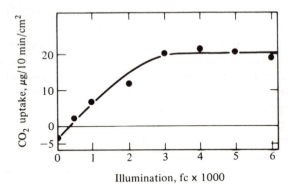

Figure 6-5. Effect of light intensity on the carbon dioxide fixed per unit area of an isolated bermudgrass leaf (After Alexander and McCloud—4).

Most leaves within the turfgrass community are not oriented perpendicular to the incoming radiation. As a result, very high light intensities are required to reach light saturation. Light saturation of bermudagrass turfs cut daily at 1 and 2 in. does not occur at intensities less than 1.0 cal per sq cm per min (Fig. 6-6). Light saturation occurred at 0.75 cal per sq cm per min when cut daily at 8 in. Light saturation at the 8-in. cutting height resulted primarily from less interleaf interference to the incident radiation. There was no difference in the leaf area index at

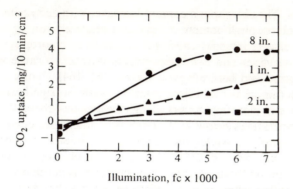

Figure 6-6. Influence of light intensity on the carbon dioxide uptake of bermudagrass turfs cut daily at 1, 2, and 8 in. (After Alexander and McCloud—4).

the 2 and 8 in. cutting heights. The 2-in. cut had vertically oriented leaves, while the 1-in. cut had many short, overlapping leaf stubs (4). Thus, a large increase in the incident light intensity is required for even a small increase in absorption by individual leaves. Light saturation of close cut turfs occurs infrequently, if at all, because of the leaf orientation and shading.

The light intensity impinging on a leaf is not necessarily constant. Large, rapid variations in light intensity may occur due to the passing of clouds or the movement of tree leaves caused by wind. The photosynthetic process responds quite rapidly to such changes in light intensity.

Increased net assimilation at higher light intensities (131) results in greater leaf, stem, rhizome, stolon, and root growth (2, 3, 21, 22, 23, 24, 31, 68, 103, 105, 106, 123, 130, 131, 152, 159). A higher root-shoot ratio indicates that the root system is more responsive to increased light intensities than the shoots (105, 106, 123). The shoots utilize the limited carbohydrate reserves at the expense of the root system at low light intensities (24). A positive factor affecting the carbohydrate reserves is the reduced respiration rate at low light intensities (130, 131).

Physiological responses. Plants grown at low light intensities exhibit distinct physiological responses (49). In this discussion a light intensity of less than 0.4 cal per sq cm per min is considered low. Typical physiological changes observed in plants grown under low light intensities are

1. Higher chlorophyll content.
2. Lower respiration rate.
3. Lower compensation point.
4. Lower carbohydrate reserve.
5. Lower carbohydrate-to-nitrogen ratio.
6. Reduced transpiration rate.
7. Higher tissue moisture content.
8. Lower osmotic pressure.

Light is necessary for the production of chlorophyll. A relatively low light intensity is sufficient to stimulate chlorophyll synthesis. High light intensities increase the rate of chlorophyll breakdown and cause a decrease in the chlorophyll content of leaves. The chlorophyll content increases with decreasing light intensity. The maximum chlorophyll content occurs at a relatively low light intensity with any further decrease in intensity causing a reduction in chlorophyll content.

High light intensities and low temperatures interact to cause winter discoloration of zoysiagrass and bermudagrass leaves (161, 164). Plants exposed to 34°F and a 0.15 cal per sq cm per min light intensity show no significant discoloration. However, severe discoloration occurs at light intensities above 0.75 cal per sq cm per min and temperatures below 45°F. High light intensities cause degradation of the existing chlorophyll, while low temperatures impair chlorophyll synthesis. The result is typical winter discoloration since the chlorophyll degradation rate exceeds the rate of synthesis. A similar mechanism is probably operative in the discoloration of other warm season turfgrasses such as St. Augustinegrass, centipedegrass, and carpetgrass. Discoloration of Meyer zoysiagrass occurs at slightly higher temperatures and lower light intensities than for U-3 bermudagrass, Emerald zoysiagrass, and manilagrass (161, 164).

Light is also responsible for stimulating the guard cells of stomata to open. High light intensities are usually associated with high transpiration rates. Other plant water relations affected by low light intensities include increased succulence (31, 83, 84) and a lower cellular osmotic pressure.

The lower carbohydrate-to-nitrogen ratio at reduced light intensities is attributed primarily to a decreased carbohydrate level (24). The total nitrogen level (64, 131, 132), especially the nitrate component (3), increases at low light intensities, while the protein content declines (64). High light intensities result in increased carbohydrate reserves, assuming other factors are not limiting (64, 130, 131). The respiration rate tends to decrease at lower light intensities (130).

Developmental responses. Heat and moisture stress are frequently associated with high light intensities. It is difficult to determine whether the morphological responses are the direct result of a low light intensity or are due to a combination of light, heat, and moisture effects. Numerous morphological changes occur at reduced light intensities (Table 6-3). Typically observed are

1. Thinner leaves with less weight per unit area (61, 101, 103, 105, 106).
2. Reduced leaf width (61, 65, 101, 103, 105, 106, 123).
3. Increased leaf length and plant height (2, 21, 61, 65, 69, 101, 105, 106, 110, 123, 152).
4. Reduced shoot density (30, 69, 83, 84, 93, 94, 99, 123, 152).
5. Longer internodes (99, 123, 152).
6. Reduced tillering (2, 61, 101, 103, 105, 106, 118, 138, 149).
7. Reduced stem diameter (99).
8. Reduced appearance rate of successive leaves on the stem (101, 103, 105, 106, 118).
9. More upright growth habit (68, 69, 90, 99, 117).

The leaf structure is modified by the light intensity under which it is grown. Specific effects of a high light intensity include (a) thicker cell walls, (b) more fully developed supporting tissues and vascular system, and (c) a thicker cutin layer.

Table 6-3

THE EFFECTS OF A 60% REDUCTION IN LIGHT INTENSITY ON THE GROWTH
CHARACTERISTICS OF MERION KENTUCKY BLUEGRASS (EXPRESSED AS THE
AVERAGE OF TWENTY INDIVIDUAL PLANTS PER LIGHT TREATMENT)

| | Light environment | |
Plant characteristic	Full sunlight	Shade
Leaf length, mm	135	254
Leaf width, mm	3.1	2.8
Shoot number per plant	11	7
Root number per plant	73	53
Rhizome number per plant	4	3
Rhizome length, mm	33	40
Leaf growth habit, degrees to horizontal	65	77
Succulence, % water	77	80

The thinner, more delicate structure of leaves grown at low light intensities is advantageous since the light absorption capability is increased. However, the delicate, poorly developed structure and increased succulence also result in reduced wear, disease, heat, drought, and cold tolerance.

Increased shoot elongation, poorly differentiated tissues, widely spaced internodes, and a reduced leaf number occur in darkness or at minimal light intensities. A low light intensity reduces the tiller number but the average weight per tiller increases due to the lack of fourth- and third-order tillers (105). The inhibition of tiller, shoot, and rhizome development at low light intensities is related to the total quantity of light energy available for photosynthesis (103). The longer leaves, internodes, and overall plant height are due to an increase in cell elongation that commonly occurs at low light intensities. The reduced leaf appearance rate and subsequent slowing of plant maturation is not of great concern in turfgrass culture.

Low light intensities favor vegetative development, while flowering and seed production are enhanced by high light intensities (127, 129). The emergence and upright growth of rhizomes and stolons are also stimulated by low light intensities (68, 69, 90, 99, 117). High light intensities induce horizontal growth of bermudagrass stems. It is hypothesized that light controls the direction of stem growth by increasing or decreasing the relative strength of the positive geotropic response (117). The light stimulus for controlling the positive geotropic response can be transmitted to stolons growing in the dark. The more upright growth of shaded turfgrasses results in a greater percentage of the plant being removed during mowing. The reduced leaf area results in a weaker, more open turf having a slow establishment rate (68). Turfgrass development and form can show distinct changes within 4 to 7 days after exposure to reduced light intensities (90, 98).

Leaf orientation and movement are due to the influence of light on auxin synthesis and degradation. Leaves tend to orient perpendicular to the incident light. The auxin level is greater on the shaded side when a stem or leaf is exposed to light from only one side. The net result is stimulation of cell elongation on the shaded side that causes the stem or leaf to turn toward the light. Phototropic responses are limited under modern turfgrass culture.

Diseases. Turfgrass disease development is usually minimal at high light intensities. This is partially attributed to the inhibition of spore germination. The indirect affect of low relative humidities and high temperatures associated with high light intensities may also be involved. Low light intensities result in succulent, poorly developed leaf tissues that are more readily infected by turfgrass fungi. Mycelial development of *Typhula* blight is enhanced by a low light intensity (48, 122). The sporulation of certain turfgrass fungi such as *Fusarium* patch also requires light of a relatively low intensity (48, 137). Higher light intensities stimulate perithecia formation of *Ophiobolus* patch (72) and certain disease coloration symptoms such as the reddish color of red thread and the copper color of *Fusarium* patch (72).

Light exclusion. Turfs can be extensively damaged by total light exclusion when equipment, hoses, or other materials are left on a turf for an extended period. An accumulation of clippings or an excessively thick layer of topdressing can also cause injury by light exclusion. Sometimes light is excluded from turfgrass leaves by certain fungal organisms such as slime mold. The accumulation of plasmodium on the leaf surface can become large enough to exclude light.

The turf turns yellowish and eventually whitish when light is excluded due to the loss of chlorophyll. The leaves and stems also become etiolated. The turf is capable of recovering quickly if the object or cover is removed soon enough. The aboveground tissues are killed if light is excluded for a sufficient length of time. Rhizomatous turfgrasses whose shoots are killed by light exclusion are capable of recovery from the underground nodes, while bunch-type turfgrasses lack this recuperative capability.

The duration of light exclusion that results in death varies with the temperature and physiological state of the turfgrass plant. Succulent tissues are injured more quickly by light exclusion, especially at higher temperatures. A cover over the turf may also enhance disease development.

Injury due to light exclusion can be avoided by not permitting objects to remain in one location on a turf for an extended period of time. Also, accumulations of clippings should be removed and excessively thick topdressings avoided. It is a common practice to place opaque tarps on athletic fields prior to scheduled games. The possibility of turfgrass injury due to light exclusion increases when a tarp is left on the field for an extended period of time. Light exclusion injury also occurs to turfs on campgrounds where tents and ground cloths are left in place for an extended period of time.

Seed germination. Seeds of certain turfgrass species have a light requirement for germination, particularly after imbibition of water has occurred. Kentucky bluegrass (8, 15, 36, 77, 113), Canada bluegrass (7, 109, 113), rough bluegrass (35), wood bluegrass (96, 113), annual bluegrass (53, 112), and hair fescue (86) require light for optimum seed germination. Light exposure is beneficial to the germination of chewings fescue, red fescue, tall fescue, meadow fescue, timothy, and bermudagrass (109) when temperatures are not conducive for complete germination (35, 36, 86).

The light requirement is of concern in turfgrass seeding practices. Seed germination of certain species may be impaired if planted too deeply (36). Light intensity decreases rapidly with soil depth. Light can penetrate to a depth of 6 to 8 mm in sandy soils. Light penetration is quite limited in fine textured soils. Wetted seeds that are exposed to light and then dried retain the capability to germinate. The lack of light at deeper planting depths is of no concern under these conditions.

The light requirement for germination of some turfgrass species is lost or reduced after 6 months storage (15, 35). Subjecting seeds to temperature changes, oxygen, acids, or nitrates can sometimes replace the light requirement for germination. Immature turfgrass seeds are generally more light responsive than ripe seeds (15, 86).

Crabgrass (148) and goosegrass (147) are common turfgrass weeds requiring light for seed germination (Fig. 6-7). Germination and growth of these annual weedy grasses can be seriously impaired in healthy dense turfs where light penetration to the soil is impaired. The absence of crabgrass under a tree canopy illustrates the light intensity requirement.

A light intensity of 0.01 cal per sq cm per min is usually sufficient to stimulate germination of turfgrass seeds having a light requirement. Light quality

Figure 6-7. The comparative germination rates of crabgrass under light intensities of 0.75 (left) and 0.08 cal per sq cm per min (right).

also affects seed germination of certain species. Red light (670 mμ) stimulates seed germination, while near-far-red light (720 mμ) is inhibitory (77, 97). It is evident that a reversible photochemical reaction is involved. A germination mechanism of this type probably exists in most light-dependent turfgrass seeds but has been demonstrated only in Kentucky bluegrass (77).

Light Relations In Turfgrass Communities

Light competition is quite critical during turfgrass establishment. The rate of germination and vertical leaf extension are key factors affecting light competition during establishment (73). Also of importance is the inclination of the first leaf initials produced. A general comparison of the competitive ability for light of turfgrass seedlings from least to the most is bentgrass < Kentucky bluegrass < redtop < red fescue < rough bluegrass < meadow fescue < tall fescue < perennial ryegrass < Italian ryegrass (141). The order of species competition is positively correlated with the germination rate, seedling vigor, and vertical leaf extension rate. These three factors cause increased shading and attrition of the less vigorous species (139). The superior seedling vigor of certain ryegrass cultivars results in excessive light competition with subsequent failure of weaker species such as Kentucky bluegrass (51). Red fescue planted in polystands with Kentucky bluegrass is not as severe a competitor for light as ryegrass.

The intensity of culture following establishment tends to mask the effects of light competition among turfgrass species. However, a degree of light competition always exists, even within a monostand. The growth habit and ability to compete for light are important factors in the succession of turfgrass communities. Certain prostrate growing perennials such as bentgrass and bermudagrass have a more favorable leaf orientation for light absorption. The ability of bentgrass and bermudagrass to invade other turfgrass species is due, in part, to a greater capability to compete for light. Many common weeds have a prostrate growth habit and leaf orientation that favor competitive light absorption within a turfgrass community. The very low light intensity at the soil surface of dense turfs may be inadequate for the growth of weed seedlings. A high leaf area index results in greater competition for the available light and is important in weed suppression within a turfgrass community.

Excessively high seeding rates result in weak, spindly growth of individual plants within the turfgrass community. Severe competition for light between individual plants is a key factor, causing the weak immature growth habit. On the other hand, suboptimal seeding rates result in a high light intensity at the soil surface that encourages the germination and development of weedy species.

Turfgrass Shade Ecology

Shading involves the partial or complete interception of direct solar radiation. The shade ecology of turfgrasses is discussed separately since shading involves the alteration of environmental factors other than light intensity. Turfgrasses are grown in association with trees in most landscapes. It is estimated that 20 to 25% of the existing turfs must be maintained under some degree of shade from trees, shrubs, or buildings. Considerable difficulty can be experienced in the culture of turfs in enclosed sports stadiums having a drastic alteration of the light regime (12).
Shade microenvironment. Shading adversely alters the microenvironment in which turfgrasses must grow. The most obvious change is the reduction in light intensity. However, a number of other important environmental factors must also be considered in turfgrass shade ecology (Table 6-4). Included are

1. Alteration of light quality.
2. Moderation of extremes in diurnal and seasonal temperatures.
3. Restricted wind movement.
4. Increased relative humidity.
5. Increased carbon dioxide level.
6. Tree root competition for water and nutrients.

The degree of light intensity reduction is affected by numerous factors. The crown density of a species and the spacing of trees are important factors influencing the amount of light reaching the turf. Trees characterized by a dense crown are maple (*Acer*), linden (*Tilia*), and oak (*Quercus*) (38). Locust (*Gleditsia*) and poplar (*Populus*) have a relatively low crown density. Small holes in the tree canopy result

Table 6-4

A Comparison of the Microenvironment of an Unirrigated Red Fescue
Turf in Full Sunlight and Intense Shade at East Lansing, Michigan,
on a Mid-August day

Environmental factor	Sunlight	Shade*
Light intensity at 12 noon, (cal per sq cm per min)	1.6	0.15
Total incident radiation (g–cal/sq cm)	659	38
Average air temperature at 1 in., °F	74	62
Average soil temperature at 6 in., °F	72	60
Maximum air temperature at 1 in., °F	96	73
Minimum air temperature at 1 in., °F	52	53
Wind velocity at 2 in. (12 noon) (mph)	4	0
Percent relative humidity at 2 in., 12 noon	45	68
Atmospheric carbon dioxide (ppm) at 1 in., 12 noon	276	305
Duration of dew, hr	0	12.5

* An enclosed stand of mature sugar maples (*Acer saccharum* Mersh.).

in sun flecks. The crown density affects the frequency and distribution of sun flecks. The occurrence of sun flecks is most frequent during a 4- to 5-hour period near midday. Sun flecks can be a significant source of light for turfs growing under a tree canopy (56).

There are fewer turfgrass cultural problems under individual trees if the lower limbs are trimmed to a height of 8 to 10 ft. The changing angle of incident sunlight enables most of the turf under a tree to be exposed to direct solar radiation at some time during the day. Turfs adjacent to the north side of buildings are also subjected to a substantial reduction in light intensity due to the diurnal southerly rotation of the incident sunlight.

The light quality under a dense stand of trees can be altered since the canopy acts as a spectrally selective filter. The quality alteration is most evident under deciduous trees, whereas conifers are more like neutral filters (40, 89, 150). The major factors affecting the light composition under a tree canopy are the composition of the incident radiation and the density of the tree crowns. Other factors include the spectral properties of the tree leaves, tree height, and height of the sun above the horizon.

Two distinct conditions must be considered in relation to changes in light quality under trees. A "green shade" predominates when reflection and transmission are the primary sources of the shade light. The green deciduous leaf canopy is characterized by maximum reflection and transmission in the green, yellow, and dark red regions of the visible spectrum and absorption of the orange, red, and blue (16, 89, 136). Thus, light under a tree canopy is depleted of the photosynthetically active wavelengths. The proportion of near-infrared wavelengths, longer than 700 mμ, is also much higher under trees than in full sunlight (9, 52, 55). A "green shade" is more common under very dense deciduous canopies no matter whether there is full sunlight or cloudy conditions (1, 133, 150). Under less dense canopies and on sunny days, the red wavelengths are reduced much more than the blue, giving a "blue shade" (150).

A tree canopy screens out a significant portion of the incident radiation, resulting in reduced temperatures in and adjacent to the turf (71) (Table 6-5). Also, the canopy restricts the nocturnal cooling process by inhibiting the loss of heat as outgoing, long wave radiation. The net effect is a moderation of extremes in air and soil temperature. Water transpired from the tree and turfgrass leaves increases the relative humidity under the tree canopy (50). Generally, the highest relative humidity levels occur during the night and gradually decline throughout the day.

Table 6-5

THE VERTICAL TEMPERATURE AND RELATIVE HUMIDITY CONDITIONS IN A DENSE FIR STAND
[AFTER GEIGER (71)]

	Temperature, °F		Relative humidity, %		
Height of measurement, ft	Daily average	Daily fluctuation	Daily average	Daily fluctuation	Average for overcast days
32.8 (above the canopy)	72	29.5	63	58	76
16.4 (in treetop area)	71	34.9	63	62	80
9.8 (in crown area)	70	34.2	70	62	84
0.6 (at the forest floor)	65	25.2	79	45	90

Dense tree and shrub plantings that encircle a turfgrass area can seriously restrict wind movement (62) (Fig. 6-8). A lack of air movement results in temperature and relative humidity stratification with the highest levels occurring adjacent to the turf. The mixing action of wind is generally desirable and can be enhanced by avoiding dense encircling shrubs or wind barriers.

The intensity and frequency of dews are less under a tree canopy. When dew formation does occur, it remains for a longer duration under shade. The reduction in wind movement and light intensity decrease the rate at which the dew evaporates. Large drops of water drip through the tree canopy during intense dews releasing

Figure 6-8. A turf growing under a tree canopy and surrounded by a dense screen of young trees and shrubs.

a virtual shower of dew on the turf. The intensity and frequency of dews is less under trees, but it is still a negative microenvironmental factor favoring disease development in shaded turfs.

The carbon dioxide content of the air under a woodland canopy is generally higher than outside the canopy (156). Concentrations exceeding twice the normal atmospheric level have been measured (55). The rate of photosynthesis is a function of the light intensity and carbon dioxide concentration. A higher carbon dioxide level would increase the photosynthetic rate of shaded turfs. The alterations in atmospheric gas concentrations under a tree canopy may increase the net assimilation rate of turfgrasses and act as a positive factor in shade adaptation. However, the importance of such a response is minimal. The mixing action of wind can reduce the atmospheric carbon dioxide content to more normal levels (156) and negates any positive response to the higher carbon dioxide levels.

Soil moisture stress in the turfgrass root zone is more severe in full sunlight than under shade during extended drought. The reduced evapotranspiration rate and higher atmospheric relative humidity are responsible for the increased efficiency of soil water utilization. Turfs on the northeast side of trees are generally the coolest and have the highest soil moisture level (158).

Roots of trees and shrubs compete with turfgrass roots for soil moisture and nutrients and can nullify the higher soil moisture levels that sometimes occur under trees (155). The degree of competition varies with the species. Trees having a fibrous, shallow root system include the willow, silver maple, Norway maple, sweet gum, cottonwood, and Australian pine. The tree canopy also intercepts and collects a significant portion of the rainfall. The degree of interception varies with the tree species and age.

Turfgrass responses to shade. The reduced light intensities under shaded conditions limit the carbohydrate reserves and the growth of roots, shoots, rhizomes, and stolons. Turfs shaded in the forenoon and then suddenly exposed to full sunlight wilt more quickly than turfs growing in full sunlight throughout the day (123). The reduced root system, thin cuticle, and poorly developed vascular system of turfgrasses growing in the shade result in an increased wilting tendency.

The morphological and physiological changes resulting from a shade environment cause an overall deterioration in plant vigor and hardiness. Shaded turfgrasses have a more delicate structure and are more succulent (49, 83, 84). The net result is reduced tolerance to heat, cold, drought, and wear stress as well as increased susceptibility to diseases and insects (12, 93, 155). The rhizomes and stolons of shaded turfgrasses tend to grow upright (68, 90, 99, 117). The establishment rate under shaded conditions is seriously inhibited due to the reduced rhizome number, more upright growth habit, decreased tillering, and limited carbohydrate reserves.

Turfgrass shade adaptation. Turfgrass adaptation to shade does not involve a single factor but a complex microenvironmental regime. The low light intensities weaken the turf due to reduced carbohydrate reserves and cause a decrease in the shoot density. A complete loss of turf occurs if the light reduction is severe enough.

However, the light intensity is sufficiently above the compensation point in many shade situations to permit maintenance of an acceptable turf. Several hours of full sunlight plus the diffuse light from leaf reflection and transmission are usually adequate for turfgrass growth under properly maintained individual trees. When trees are grouped, the resulting dense canopy is more likely to limit turfgrass quality and to increase the problems of turfgrass culture.

The loss of turf under tree canopies is usually more complex than just a light deficiency. The more favorable microclimate for disease development and the lack of disease-resistant cultivars are key factors limiting the shade adaptation of certain cool season turfgrasses (16). Pathogen activity in the shade is enhanced by a longer dew period, higher atmospheric relative humidity, reduced wind movement, altered light quality, low evapotranspiration rate, and more succulent plant tissue. Adequate disease resistance is an important factor in the development of shade adapted cool season turfgrasses. Similar disease problems have been observed with certain warm season turfgrass species grown under shade. However, the disease problem is not as critical as with the cool season turfgrasses.

The ability to overcome disease problems contributes significantly to shade adaptation (16, 73). Other factors that may be involved in shade adaptation are (a) a lower compensation point, respiration rate, or carbohydrate requirement (27); (b) lower nutrient and water requirements; (c) greater competitive ability for light and nutrients; and (d) a lower light requirement. The relative importance of these four factors has not been determined. The survival of seedlings at low light intensities is greater in species such as red fescue that have inherently slow rates of dry matter accumulation (72). Also, the growth of red fescue is not influenced as much by reduced light intensities as for some other turfgrass species (157). Plants grown at low light intensities have a lower compensation point and become light saturated at considerably lower intensities than when grown in full sunlight (27). Evidently the photosynthetic apparatus of certain shade tolerant turfgrasses can adjust more readily to a low light intensity so that the available light is utilized more efficiently.

In certain climates, shading affords a more favorable microenvironment in which turfgrass species may persist. For example, Kentucky bluegrass grows and persists under shade in hot humid climates but not in full sunlight. This is contrasted to the cooler climates where zoysiagrass grows in full sunlight but not in shaded sites. The lower temperatures of shaded sites can favor or impair the adaptation of turfgrass species depending on the temperature optimum of the species and the dominant climate of the region. A delay in fall discoloration of warm season turfgrasses also occurs due to the moderation of low temperatures and reduced light intensity under the tree canopy.

Turfgrass shade tolerance. Red fescue continues to be the preferred species for shaded environments in cool climates (16) (Table 6-6). The bentgrasses are satisfactory where a preventive fungicide program is followed and the soil is moist (83). Rough bluegrass is favored by a wet, shaded environment. Tall fescue and ryegrass are adapted to shade in the warmer portions of a cool humid climate where winterkill is not a problem (16, 84, 135).

Table 6-6

THE RELATIVE SHADE ADAPTATION OF THE COMMON COOL SEASON AND WARM SEASON
PERENNIAL TURFGRASSES IN THEIR RESPECTIVE REGIONS OF ADAPTATION

	Relative adaptation to severe shade			
Adaptation	*Excellent*	*Good*	*Medium*	*Poor*
Cool season turfgrasses	Red fescue Velvet bentgrass	Rough bluegrass Creeping bentgrass Tall fescue	Colonial bentgrass Redtop Perennial ryegrass Meadow fescue	Kentucky bluegrass
Warm season turfgrasses	St. Augustine- grass Manilagrass	Zoysiagrass	Centipede- grass Carpetgrass	Buffalograss Bermudagrass

The shade adaptation of St. Augustinegrass is excellent in warm climates (28, 30, 110), while the *Zoysia* species show good shade adaptation (68, 80, 143, 165). The bermudagrasses, as a group, have not shown adequate shade adaptation (132). Certain improved bermudagrass cultivars, such as FB-137 and Tifdwarf, show promise for use in shaded environments (98, 99). A short internode length is correlated with shade adaptation of bermudagrasses (99).

It is best to plant shade adapted ground covers other than turfgrasses where the light reduction is extreme. English ivy (*Hedera helix* L.), myrtle (*Vinca minor* L.), and pachysandra (*Pachysandra terminalis* Michx.) are commonly used in many areas.

Modifying the shade microenvironment. The underlying principles of turfgrass shade culture involve modification of the microenvironment to improve conditions for turfgrass growth and development. Pruning of tree limbs below 10 ft improves the turfgrass quality under individual trees (Fig. 6-9). Light intensity can also be improved through the selective pruning of limbs in the crown of the tree. Additional reductions in light intensity can be avoided by the immediate removal of excess clippings and fallen tree leaves. If the landscape plan requires a dense, low growing shrub or tree that is not to be pruned, one should not attempt to maintain a turf under the canopy.

Wind movement can be improved by judicious pruning and the elimination of thick underbrush and shrub plantings that form dense encircling screens. Maximum air drainage is achieved if selective clearing of underbrush is done in relation to the prevailing winds. The beauty of a landscape is sometimes enhanced by new vistas through selective shrub or tree removal. Improved temperature and relative humidity regimes are the benefits of unimpaired air movement. Disease development is also reduced due to the lower relative humidity and more favorable condi-

Figure 6-9. Shade environment with lower tree limbs pruned and screens of trees or shrubs eliminated to improve air movement and light intensity.

tions for drying. Pruning of shallow tree roots is an effective practice in alleviating competition for nutrients and water (Fig. 6-10). Water competition is especially critical in dryer climates and in situations where trees are growing adjacent to an irrigated turf such as a green.

Figure 6-10. Effect of tree root competition with left side showing the turfgrass response from pruning of tree roots versus right side showing the restricted growth from tree root competition.

Turfgrass shade culture. Cultural practices can be modified to improve the growth and shoot density of shaded turfs. Key principles in turfgrass shade culture include (a) raising the cutting height, (b) avoiding excessive nitrogen fertilization, (c) deep infrequent irrigation, (d) judicious traffic control, and (e) use of fungicides when needed. A higher height of cut increases the leaf area index, thus providing a greater capability to absorb light and synthesize carbohydrates. A cutting height of 2 to 2.5 in. is beneficial for shaded lawn turfs (29, 30).

Carbohydrate depletion is moderated by using the minimum nitrogen fertilization level that meets the requirements of the species. Excessive nitrogen fertilization produces a succulent tissue that is more prone to injury from disease and wear. Surface fertilization of trees is not desirable where the associated turfgrass species has a low fertility requirement. Red fescue is a shade adapted species that does not tolerate excessive nitrogen fertilization. The loss of red fescue turfs can be minimized by deep root feeding of trees.

Proper irrigation ensures adequate moisture for turfgrass growth. Deep watering to a minimum soil depth of 6 in. is preferred. A critical aspect of irrigation is avoiding the enhancement of pathogen activity caused by excessive or improper timing of water applications. The water application rate should not exceed the infiltration rate of the soil. Irrigations should be timed so that the water is present on the turfgrass leaves for a minimum period of time. The result is a less favorable microenvironment for infection by fungi. When a disease occurs under shade conditions, control can be achieved by use of the appropriate fungicide.

Shaded turfs should be protected as much as possible through the redirection or control of traffic because of the greater susceptibility to wear injury and reduced ability to recover from traffic damage. Early autumn establishment of cool season turfgrasses is definitely preferred under deciduous trees. Autumn seedings provide the longest period of direct sunlight, extending from tree leaf abscision in midfall until the initiation of new leaves in midspring. The immediate removal of fallen tree leaves is especially important during turfgrass establishment.

References

1. Akulova, E. A., V. S. Khazanov, Y. L. Tsel'niker, and D. N. Shishov. 1964. Light transmission through a forest canopy depending on the incident radiation and density of the tree crowns. Fiziologiya Rasteniii. 11(5): 818–823.

2. Alberda, T. 1957. The effects of cutting, light intensity and night temperatures on growth and soluble carbohydrate content of *Lolium perenne* L. Plant and Soil. 8: 199–230.

3. ———. 1965. The influence of temperature, light intensity, and nitrate concentration on dry matter production and chemical composition of *Lolium perenne* L. Netherlands Journal of Agricultural Science. 13(4): 355–360.

4. Alexander, C. W., and D. E. McCloud. 1962. CO_2 uptake (net photosynthesis) as influenced by light intensity of isolated bermudagrass leaves contrasted to that of swards under various clipping regimes. Crop Science. 2(2): 132–135.

5. Allard, H. A., and M. W. Evans. 1941. Growth and flowering of some tame and wild grasses in response to different photoperiods. Journal of Agricultural Research. 62(4): 193–228.

6. Allard, H. A., and W. W. Garner. 1940. Further observations on the response of various species of plants to length of day. USDA Technical Bulletin. No. 727. pp. 1–64.

7. Anderson, A. M. 1947. Some factors influencing the germination of seed of *Poa compressa* L. Proceedings of the 37th Annual Meeting of the Association of Official Seed Analysts. 37: 134–143.

8. ———. 1957. The effect of certain fungi and gibberellin on the germination of Merion Kentucky bluegrass seed. Proceedings of the Association of Official Seed Analysts. 47: 145–153.

9. Anderson, M. C. 1964. Studies of the woodland light climate. II. Seasonal variation in the light climate. Journal of Ecology. 52(3): 643–663.

10. Angstrom, A. 1925. The albedo of various surfaces of ground. Geografiska Annater. 7: 323–342.

11. ANONYMOUS. 1940. Day length affects bluegrass. Progress of Agricultural Research in Ohio, 1938–1939. Bulletin 617. pp. 13–14.

12. ANONYMOUS. 1967. Grass thrives under Plexiglas sheets. Florist and Nursery Exchange. 147: 16–37.

13. ATKINS, W. R. G. 1932. The measurement of daylight in relation to plant growth. Empire Forestry Journal. 11: 42–52.

14. ATKINS, W. R. G., and H. H. POOLE. 1931. Photo-electric measurements of illumination in relation to plant distribution. Part 4. Changes in the colour composition of daylight in the open and in shaded situations. Scientific Proceedings of the Royal Dublin Society. 20(4): 13–48.

15. BASS, L. N. 1951. Effect of light intensity and other factors on germination of seeds of Kentucky bluegrass (*Poa pratensis*). Proceedings of the 11th Annual Meeting of the Association of Official Seed Analysts. 41: 83–86.

16. BEARD, J. B. 1965. Factors in the adaptation of turfgrasses to shade. Agronomy Journal. 57(5): 457–459.

17. ———. 1967. Shade grasses and maintenance. 38th International Turfgrass Conference Proceedings. pp. 31–39.

18. BEARD, J. B., and W. H. DANIEL. 1966. Relationship of creeping bentgrass (*Agrostis palustris* Huds.) root growth to environmental factors in the field. Agronomy Journal. 58: 337–339.

19. ———. 1967. Variations in the total, nonprotein, and amide nitrogen fractions of *Agrostis palustris* Huds. leaves in relation to certain environmental factors. Crop Science. 7(2): 111–115.

20. BENEDICT, H. M. 1940. Effect of day length and temperature on the flowering and growth of four species of grasses. Journal of Agricultural Research. 61(9): 661–671.

21. ———. 1941. Growth of some range grasses in reduced light intensities at Cheyenne, Wyoming. Botanical Gazette. 102(3): 582–589.

22. BLACKMAN, G. E., and J. N. BLACK. 1959. Physiological and ecological studies in the analysis of plant environment. XI. A further assessment of the influence of shading on the growth of different species in the vegetative phase. Annals of Botany N. S. 23(89): 51–63.

23. BLACKMAN, G. E., and W. G. TEMPLEMAN. 1938. The interaction of light intensity and nitrogen supply in the growth and metabolism of grasses and clover (*Trifolium repens*). II. The influence of light intensity and nitrogen supply on the leaf production of frequently defoliated plants. Annals of Botany, N. S. II(7): 765–791.

24. ———. 1940. The interaction of light intensity and nitrogen supply in the growth and metabolism of grasses and clover (*Trifolium repens*). IV. The relation of light intensity and nitrogen supply to the protein metabolism of the leaves of grasses. Annals of Botany N. S. IV(15): 533–587.

25. BORTHWICK, H. A., and S. B. HENDRICKS. 1960. Photoperiodism in plants. Science. 132(3435): 1223–1228.

26. BROUGHAM, R. W. 1958. Interception of light by the foliage of pure and mixed stands of pasture plants. Australian Journal of Agricultural Research. 9: 39–52.

27. BURNSIDE, C. A., and R. H. BOHNING. 1957. The effect of prolonged shading on the light saturation curves of apparent photosynthesis in sun plants. Plant Physiology. 32: 61–63.

28. BURTON, G. W. 1956. Lawn grasses for the south. Plants and Gardens. 12(2): 156–161.

29. ———. 1962. The shade problem and what you can do about it. Proceedings of the 16th Annual Southeastern Turfgrass Conference. pp. 31–33.

30. BURTON, G. W., and E. E. DEAL. 1962. Shade studies on southern grasses. Golf Course Reporter. 30(8): 26–27.

31. BURTON, G. W., J. E. JACKSON, and F. E. KNOX. 1959. The influence of light reduction upon the production, persistence and chemical composition of coastal bermudagrass, *Cynodon dactylon*. Agronomy Journal. 51(9): 537–542.

32. BUTLER, J. D. 1962. Some effects of different levels of shade on several grasses and weeds. Illinois Turfgrass Conference Proceedings. pp. 22–25.

33. CABLER, J. F. 1963. Chemical growth substances as substitutes for high light intensities on 'Tifgreen' bermudagrass. Proceedings of the Florida State Horticultural Society. 76: 470–474.

34. CHILDERS, N. F., and D. G. WHITE. 1947. Manila grass for lawns. Puerto Rico Agricultural Experiment Station (USDA). Circular No. 26. pp. 1–16.

35. CHIPPINDALE, H. G. 1932. The operation of interspecific competition in causing delayed growth of grasses. Annals of Applied Biology. 19: 221–242.

36. ———. 1949. Environment and germination in grass seeds. Journal of the British Grassland Society. 4: 57–61.

37. CLARKE, G. L. 1965. Elements of ecology. John Wiley & Sons, Inc., New York. pp. 1–560.

38. CLERKE, P. N. 1965. Engineering aspects of night lighting a golf course. Turf Clippings (Conference Proceedings) of the Stockbridge School. 1(10): A57–A64.

39. COOK, W. L. 1963. A light on golf. Parks and Recreation. 46(9): 312–313.

40. COOMBE, D. E. 1957. The spectral composition of shade light in woodlands. Journal of Ecology. 45: 823–830.

41. COOPER, J. P. 1950. Day-length and leaf formation in the ryegrasses. Journal of the British Grassland Society. 5: 105–112.

42. ———. 1951. Studies on growth and development in *Lolium*. II. Pattern of bud development of the shoot apex and its ecological significance. Journal of Ecology. 39: 228–270.

43. ———. 1959. Selection and population structure in *Lolium*. I. The initial populations. Heredity. 13: 317–340.

44. ———. 1960. Short-day and low-temperature induction in *Lolium*. Annals of Botany N. S. 24(94): 232–246.

45. COOPER, J. P., and D. M. CALDER. 1962. Flowering responses of herbage grasses. Report of the Welsh Plant Breeding Station for 1961. pp. 20–22.

46. ———. 1964. The inductive requirements for flowering of some temperate grasses. Journal of the British Grassland Society. 19: 6–14.

47. COUCH, H. B. 1962. Diseases of turfgrasses. Reinhold Publishing Corp., New York. pp. 1–289.

48. DAHL, A. S. 1934. Snowmold of turf grasses as caused by *Fusarium nivale*. Phytopathology. 24: 197–214.

49. DAUBENMIRE, R. F. 1959. Plants and environment. John Wiley & Sons, Inc., New York. pp. 1–422.

50. DENMEAD, O. T. 1964. Evaporative sources and apparent diffusivities in a forest canopy. Journal of Applied Meteorology. 3: 383–389.

51. DONALD, C. M. 1958. The interaction of competition for light and for nutrients. Australian Journal of Agricultural Research. 9: 421–435.

52. EGLE, K. 1937. Zur Kenntnis des Lichtfeldes der Pflange und der Blattfarbstoffe. Planta. 26: 546–583.

53. ENGEL, R. E. 1967. Temperatures required for the germination of annual bluegrass and colonial bentgrass. Golf Superintendent. 35: 20–23.

54. ETTER, A. G. 1951. How Kentucky bluegrass grows. Annals of the Missouri Botanical Garden. 38: 293–375.

55. EVANS, G. C. 1939. Ecological studies on the rain forest of southern Nigeria. II. The atmospheric environmental conditions. Journal of Ecology. 27: 436–482.

56. ———. 1956. An area survey method of investigating the distribution of light intensity in woodlands, with particular reference to sunflecks. Journal of Ecology. 44: 391–427.

57. EVANS, L. T. 1960. The influence of environmental conditions on inflorescence development of some long-day grasses. New Phytologist. 59: 163–174.

58. ———. 1960. The influence of temperature on flowering in species of *Lolium* and in *Poa pratensis*. Journal of Agricultural Science. 54: 410–416.

59. EVANS, M. W. 1949. Vegetative growth, development, and reproduction in Kentucky bluegrass. Ohio Agricultural Experiment Station Research Bulletin No. 681. pp. 4–39.

60. EVANS, M. W., and J. M. WATKINS. 1939. The growth of Kentucky bluegrass and of Canada bluegrass in late spring and in autumn as affected by the length of day. Journal of the American Society of Agronomy. 31: 767–774.

61. EVANS, P. S. 1964. A comparison of some aspects of the anatomy and morphology of Italian ryegrass (*Lolium multiflorum* Lam.) and perennial ryegrass (*L. perenne* L.). New Zealand Journal of Botany. 2: 120–130.

62. FONS, W. L. 1940. Influence of forest cover on wind velocity. Journal of Forestry. 38: 481–486.

63. FORBES, I., JR. 1952. Chromosome numbers and hybrids in *Zoysia*. Agronomy Journal. 44(4): 194–199.

64. FORD, D. R. 1967. The influence of temperature and light intensity on the photosynthetic rate and chemical composition of *Festuca arundinacea* and *Dactylis glomerata*. M.S. Thesis. Purdue University. pp. 1–59.

65. FORDE, B. J. 1966. Effect of various environments on the anatomy and growth of perennial ryegrass and cocksfoot. New Zealand Journal of Botany. 4: 455–468.

66. FRAZIER, S. L. 1960. Turfgrass seedling development under measured environment and management conditions. M.S. Thesis. Purdue University. pp. 1–61.

67. GARDNER, F. P., and W. E. LOOMIS. 1953. Floral induction and development in orchard grass. Plant Physiology. 28(2): 201–217.

68. GARY, J. E. 1967. The vegetative establishment of four major turfgrasses and the response of stolonized 'Meyer' zoysiagrass (*Zoysia japonica* var. Meyer) to mowing height, nitrogen fertilization, and light intensity. M.S. Thesis. Mississippi State University. pp. 1–50.

69. GASKIN, T. 1964. Growing turfgrass in shade. Park Maintenance. 17(10): 90–94.

70. GATES, D. M. 1965. Energy, plants, and ecology. Ecology. 46(1): 1–13.

71. GEIGER, R. 1965. The climate near the ground. Harvard University Press, Cambridge, Mass. pp. 1–611.

72. GOULD, C. J. 1963. How climate affects our turfgrass diseases. Proceedings of the 17th Annual Northwest Turfgrass Conference. pp. 29–43.

73. GRIME, J. P., and D. W. JEFFREY. 1965. Seedling establishment in vertical gradients of sunlight. Journal of Ecology. 53: 621–642.

74. HANSON, A. A., and V. G. SPRAGUE. 1953. Heading of perennial grasses under greenhouse conditions. Agronomy Journal. 45: 248–251.

75. HARRISON, C. M. 1934. Responses of Kentucky bluegrass to variations in temperature, light, cutting, and fertilizing. Plant Physiology. 9: 83–106.

76. HENDRICKS, S. B. 1958. Photoperiodism. Agronomy Journal. 50: 724–729.

77. HENDRICKS, S. B., V. K. TOOLE, and H. A. BORTHWICK. 1968. Opposing actions of light in seed germination of *Poa pratensis* and *Amaranthus arenicola*. Plant Physiology. 43: 2023–2028.

78. HIESEY, W. M. 1953. Growth and development of species and hybrids of *Poa* under controlled temperatures. American Journal of Botany. 40(4): 205–221.

79. HOOVER, W. H. 1937. The dependence of carbon dioxide assimilation in a higher plant on wave length of radiation. Smithsonian Miscellaneous Collections. 95(21): 1–13.

80. HUFFINE, W. W. 1963. Grasses. Turfgrass Research, Oklahoma State University. pp. 2–6.

81. JUHREN, M., W. M. HIESEY, and F. W. WENT. 1953. Germination and early growth of grasses in controlled conditions. Ecology. 34: 288–300.

82. JUHREN, M., W. NOBLE, and F. W. WENT. 1957. The standardization of *Poa annua* as an indicator of smog concentrations. I. Effects of temperature, photoperiod, and light intensity during growth of the test-plants. Plant Physiology. 32: 576–586.

83. JUSKA, F. V. 1963. Shade tolerance of bentgrasses. Golf Course Reporter. 31(2): 28–34.

84. JUSKA, F. V., A. A. HANSON, and A. W. Hovin. 1969. Kentucky 31 tall fescue—a shade tolerant turfgrass. Weeds, Trees and Turf. 8(1): 34–35.

85. KALITIN, N. N. 1930. The measurements of the albedo of a snow cover. Monthly Weather Review. 58(2): 59–61.

86. KEARNS, V., and E. H. TOOLE. 1939. Temperature and other factors affecting the germination of fescue seed. USDA Technical Bulletin No. 638. pp. 1–35.

87. KNIGHT, W. E., and H. W. BENNETT. 1953. Preliminary report of the effect of photoperiod and temperature on the flowering and growth of several southern grasses. Agronomy Journal. 45(6): 268–269.

88. KNIGHT, W. E. 1955. The influence of photoperiod and temperature on growth, flowering, and seed production of Dallisgrass, *Paspalum dilatatum* Poir. Agronomy Journal. 47: 555–559.

89. KNUCHEL, H. 1924. Spektrophotometrische untersuchungen in walde. Mitteil. Schweiz. Centralanst. Forst. Versuch. 11: 1–94.

90. LANGHAM, D. G. 1941. The effect of light on growth habit of plants. American Journal of Botany. 28: 951–956.

91. LAUDE, H. M. 1953. The nature of summer dormancy in perennial grasses. Botanical Gazette. 114: 284–296.

92. LOVVORN, R. L. 1945. The effect of defoliation, soil fertility, temperature, and length of day on the growth of some perennial grasses. Journal of the American Society of Agronomy. 37: 570–582.

93. LUCANUS, R., K. J. Mitchell, G. G. Pritchard, and D. M. Calder. 1960. Factors influencing survivial of strains of ryegrass during the summer. New Zealand Journal of Agricultural Research. 3: 185–193.

94. MADISON, J. H. 1962. Turfgrass ecology. Effects of mowing, irrigation, and nitrogen treatments of *Agrostis palustris* Huds., "Seaside" and *Agrostis tenuis* Sibth., "Highland" on population, yield, rooting, and cover. Agronomy Journal. 54: 407–412.

95. MAEKUBO, N., and Y. CHIBE. 1968. Effects on the light for turfgrasses. Turf Research Bulletin. Kansai Golf Union Green Section Research Center. 12(1): 37–46.

96. MAIER, W. 1932. Untersuchungen zur froze der lichturrkung auf die deinmung einiger Poa-arten. Jahrb. Wiss. Bot. 77: 321–392.

97. MAYER, A. M., and A. POLJAKOFF-MAYBER. 1963. The germination of seeds. Macmillan Company, New York. pp. 1–236.

98. McBEE, G. G. 1969. Association of certain variations in light quality with the performance of selected turfgrasses. Crop Science. 9: 14–17.

99. McBEE, G. G., and E. C. HOLT. 1966. Shade tolerance studies on bermudagrass and other turfgrasses. Agronomy Journal. 58: 523–525.

100. McGINNIES, W. J. 1966. Effects of shade on the survival of crested wheatgrass seedlings. Crop Science. 6: 482–484.

101. MITCHELL, K. J. 1953. Influence of light and temperature on the growth of ryegrass (*Lolium* spp.). I. Pattern of vegetative development. Physiologia Plantarum. 6: 21–46.

102. ———. 1953. Influence of light and temperature on the growth of ryegrass (*Lolium* spp.). II. The control of lateral bud development. Physiologia Plantarum. 6: 425–443.

103. ———. 1954. Growth of pasture species. I. Short rotation and perennial ryegrass. New Zealand Journal of Science and Technology. Sec. A. 36(3): 191–206.

104. ———. 1954. Influence of light and temperature on growth of ryegrass (*Lolium* spp.). III. Pattern and rate of tissue formation. Physiologia Plantarum. 7: 51–65.

105. ———. 1955. Growth of pasture species. II. Perennial ryegrass (*Lolium perenne*), cocksfoot (*Dactylis glomerata*) and paspalum (*Paspalum dilatatum*). New Zealand Journal of Science and Technology. Sec. A. 37(1): 8–26.

106. MITCHELL, K. J., and S. T. J. COLES. 1955. Effects of defoliation and shading on short-rotation ryegrass. New Zealand Journal of Science and Technology. Sec. A. 36(6): 586–604.

107. MITCHELL, K. J., and R. LUCANUS. 1962. Growth of pasture species under controlled environment. III. Growth at various levels of constant temperature with 8 and 16 hours of uniform light per day. New Zealand Journal of Agricultural Research. 5(1): 135–144.

108. MONTEITH, J. L. 1959. The reflection of short-wave radiation by vegetation. Quarterly Journal of the Royal Meteorological Society. 85: 386–392.

109. MORINAGA, T. 1926. Effect of alternating temperatures upon the germination of seeds. American Journal of Botany. 13: 141–166.

110. MOSER, L. E., S. R. ANDERSON, and R. W. MILLER. 1968. Rhizome and tiller development of Kentucky bluegrass, *Poa pratensis* L., as influenced by photoperiod, cold treatment, and variety. Agronomy Journal. 60: 632–635.

111. MOSS, D. N. 1964. Optimum lighting of leaves. Crop Science. 4: 131–135.

112. NEIDLINGER, T. 1965. *Poa annua* L.: Susceptibility to several herbicides and temperature requirements for germination. M.S. Thesis. Oregon State University. pp. 1–82.

113. NELSON, A. 1927. The germination of *Poa* spp. Annals of Applied Biology. 14(2): 157–174.

114. OLMSTED, C. E. 1943. Growth and development in range grasses. III. Photoperiodic responses in the genus *Bouteloua*. Botanical Gazette. 105(2): 165–181.

115. ———. 1944. Growth and development in range grasses. IV. Photoperiodic responses in twelve geographic strains of sideoats grama. Botanical Gazette. 106(1): 46–74.

116. ORMROD, D. P. 1964. Light responses of turfgrasses. Proceedings of the 18th Annual Northwest Turfgrass Conference. pp. 50–51.

117. PALMER, J. H. 1956. The nature of the growth response to sunlight shown by certain stoloniferous and prostrate tropical plants. New Phytologist. 55: 346–355.

118. PATEL, A. S., and J. P. COOPER. 1961. The influence of seasonal changes in light energy and tiller development in ryegrass, timothy, and meadow fescue. Journal of the British Grassland Society. 16: 299–308.

119. PETERSON, M. L. 1946. Physiological response of Kentucky bluegrass to different management treatments. Ph. D. Thesis. Iowa State College. pp. 1–87.

120. PETERSON, M. L., and W. E. LOOMIS. 1949. Effects of photoperiod and temperature on growth and flowering of Kentucky bluegrass. Plant Physiology. 24: 31–43.

121. POHJAKALLIO, O. 1951. On the effect of the intensity of light and length of day on the energy economy of certain cultivated plants. Acta Agriculturae Scandinavica. 1: 153–175.

122. POTATOSOVA, E. G. 1960. Conditions of germination of the sclerotia of the fungi of the genus *Typhula*. Zashch. Rast. Moskva. 4: 40.

123. REID, M. E. 1933. Effects of shade on the growth of velvet bent and metropolitan creeping bent. Bulletin of the USGA Green Section. 13: 131–135.

124. REMSBERG, R. E. 1940. Studies in the genus *Typhula*. Mycologia. 32: 52–96.

125. ROBERTS, R. H., and B. E. STRUCKMEYER. 1938. The effects of temperature and other environmental factors upon the photoperiodic responses of some of the higher plants. Journal of Agricultural Research. 56(9): 633–677.

126. ———. 1939. Further studies of the effects of temperature and other environmental factors upon the photoperiodic responses of plants. Journal of Agricultural Research. 59(9): 699–709.

127. RYLE, G. J. A. 1961. Effects of light intensity on reproduction in S.48 timothy (*Phleum pratense* L.). Nature. 191: 196–197.

128. ———. 1963. Effect of the environment on leaf growth and tillering in herbage grasses. Experiments in Progress at the Grassland Research Institute. 17: 31–33.

129. ———. 1966. Physiological aspects of seed yield in grasses. In: The Growth of Cereals and Grasses, ed. F. L. MILTHORPE and J. D. IVINS. Butterworths, London. pp. 106–120.

130. SCHMIDT, R. E. 1965. Some physiological responses to two grasses as influenced by temperature, light and nitrogen fertilization. Ph. D. Thesis. Virginia Polytechnic Institute. pp. 1–116.

131. SCHMIDT, R. E., and R. E. BLASER. 1967. Effect of temperature, light, and nitrogen on growth and metabolism of 'Cohansey' bentgrass (Agrostis palustris Huds.). Crop Science. 7: 447–451.

132. ———. 1969. Effect of temperature, light, and nitrogen on growth and metabolism of 'Tifgreen' bermudagrass (Cynodon spp.). Crop Science. 9: 5–9.

133. SEYBOLD, A. 1936. Uber den Lichtfaktor photophysiologischer prozesse. Jb. Wiss. Bot. 82: 741–795.

134. SHEPHERD, H. R. 1938. Studies in breaking the rest period of grass plants by treatments with potassium thiocyanate and in stimulating growth with artificial light. Transactions of the Kansas Academy of Science. 41: 139–149.

135. SHIM, S. Y., and S. S. LEE. 1964. Shade tolerance of forest land conserving grasses. Research Report of the Office of Rural Development. Suwon, Korea. 7(1): 72–78.

136. SHULL, C. A. 1929. A spectrophotometric study of reflection of light from leaf surfaces. Botanical Gazette. 87(5): 583–607.

137. SMITH, J. D. 1963. Fungi and turf diseases. Journal of the Sports Turf Research Institute. 8(29): 230–252.

138. SOPER, K., and K. J. MITCHELL. 1956. The developmental anatomy of perennial ryegrass (Lolium perenne L.). New Zealand Journal of Science and Technology. Sec. A. 37: 484–504.

139. SPRAGUE, V. G. 1943. The effects of temperature and day length on seedling emergence and early growth of several pasture species. Soil Science Society of America Proceedings. 8: 287–294.

140. ———. 1948. The relation of supplementary light and soil fertility to heading in the greenhouse of several perennial forage grasses. Journal of the American Society of Agronomy. 40: 144–154.

141. STAPLEDON, R. G., W. DAVIES, and A. R. BEDDOWS. 1927. Seeds mixture problems: soil germination, seedling and plant establishment with particular reference to the effects of environmental and agronomic factors. I. Garden trials. Welsh Plant Breeding Station Bulletin Series H. pp. 5–38.

142. STUCKEY, I. H. 1942. Some effects of photoperiod on leaf growth. American Journal of Botany. 29(1): 92–97.

143. STURKIE, D. G., and H. S. FISHER. 1942. The planting and maintenance of lawns. Alabama Agricultural Experiment Station Circular 85. pp. 1–20.

144. TEMPLETON, W. C., JR., G. O. MOTT, and R. J. BULA. 1961. Some effects of temperature and light on growth and flowering of tall fescue, Festuca arundinacea Schreb. I. Vegetative development. Crop Science. 1: 216–219.

145. ———. 1961. Some effects of temperature and light on growth and flowering of tall fescue, Festuca arundinacea Schreb. II. Floral development. Crop Science. 1(4): 283–286.

146. TINCKER, M. A. H. 1925. The effect of length of day upon the growth and reproduction of some economic plants. Annals of Botany. 39: 721–753.

147. TOOLE, E. H., and V. K. TOOLE. 1940. Germination of seed of goosegrass, Eleusine indica. Journal of the American Society of Agronomy. 32(4): 320–321.

148. ———. 1941. Progress of germination of seed of Digitaria as influenced by germination temperature and other factors. Journal of Agricultural Research. 63(2): 65–90.

149. VAARTNOU, H. 1967. Response of five genotypes of Agrostis L. to variations in environment. Ph. D. Thesis. Oregon State University. pp. 1–149.

150. VEZINA, P. E., and D. W. K. BOULTER. 1966. The spectral composition of near ultraviolet and visible radiation beneath forest canopies. Canadian Journal of Botany. 44: 1267–1284.

151. VIRGIN, H. I. 1962. Light-induced unfolding of the grass leaf. Physiologia Plantarum. 15: 380–389.

152. WATKINS, J. M. 1940. The growth habits and chemical composition of bromegrass, *Bromus inermis* Leyss., as affected by different environmental conditions. Journal of American Society of Agronomy. 32: 527–538.

153. WEINTRAUB, R. L., and L. PRICE. 1947. Developmental physiology of grass seedlings. II. Inhibition of mesocotyl elongation in various grasses by red and by violet light. Smithsonian Miscellaneous Collections. 106(21): 1–15.

154. WELTON, F. A. 1933. Crabgrass controlled by shading. 51st Annual Report of the Ohio Agricultural Experiment Station. Bulletin 516. pp. 30–31.

155. WHITCOMB, C. E. 1968. Grass and tree root relationships. Proceedings of the Florida Turfgrass Management Conference. 16: 46–50.

156. WIANT, H. V. 1964. The concentration of carbon dioxide at some forest micro-sites. Journal of Forestry. 62(11): 817–819.

157. WILSON, D. B. 1962. Effects of light intensity and clipping on herbage yields. Canadian Journal of Plant Science. 42: 270–275.

158. WILSON, J. D. 1927. The measurement and interpretation of the water-supplying power of the soil with special reference to lawn grasses and some other plants. Plant Physiology. 2: 385–440.

159. WOOD, G. M., and H. J. HURDZAN. 1967. Evaluating turfgrasses for shade tolerance. 1967 Agronomy Abstracts. p. 55.

160. WYCHERLEY, P. R. 1954. Vegetative proliferation of floral spikelets in British grasses. Annals of Botany N. S. 18(69): 119–127.

161. YOUNGNER, V. B. 1959. Growth of U-3 bermudagrass under various day and night temperatures and light intensities. Agronomy Journal. 51(9): 557–559.

162. ———. 1960. Temperature and light in growth of turf. Golfdom. 34(4): 70.

163. ———. 1960. Environmental control of initiation of the inflorescence, reproductive structures, and proliferations in *Poa bulbosa*. American Journal of Botany. 47: 753–757.

164. ———. 1961. Growth and flowering of *Zoysia* species in response to temperatures, photoperiods, and light intensities. Crop Science. 1(2): 91–93.

165. YOUNGNER, V. B., and M. H. KIMBALL. 1962. Zoysiagrass for lawns. California Turfgrass Culture. 12(3): 23–24.

166. YOUNGNER, V. B., and S. E. SPAULDING. 1963. Influence of several environmental factors on flowering of bermudagrass. 1963 Agronomy Abstracts. p. 120.

chapter 7

Temperature

Introduction

The germination, growth, and development of turfgrasses are restricted to a specific temperature range. This range can vary considerably among turfgrass species, cultivars, and individual plants. Turfgrass pests, including diseases, insects, nematodes and weeds, also have a temperature response range. As temperatures are increased or decreased from the optimum range, growth is proportionally reduced until it ceases. Death occurs due to destruction of the protoplasm if the turfgrass plant is subjected to a further increase or decrease in temperature. There is relatively little biological activity above 125°F or below 32°F. Turfgrass growth and development is usually confined to a temperature range of 40 to 105°F. Temperature regimes within this range are a major factor influencing the adaptation of turfgrasses (231). This temperature effect is most obvious in the latitudinal distribution of turfgrass species.

The ultimate temperature of a turf or any specific plant part is governed by the relative amounts of heat gained and lost. One of the major components of the turfgrass heat load is the absorbed solar radiation. Another component is the thermal radiation received from the atmosphere and the earth's surface, including buildings and trees. There is also a certain amount of diffuse skylight that is absorbed, particularly in the shade. The absorbed energy components contribute to the total radiant heat load of a turf. The temperature of the turf would soon reach a lethal level if a substantial portion of the heat load was not dissipated immediately. This situation is avoided in healthy, growing turfs by evapotranspiration, reradiation, conduction, and convection. The energy not dissipated by these transfer processes remains to affect the specific heat balance and temperature of the turf.

Temperature Variations

Much of the variation in temperature with (a) latitude, (b) altitude, (c) topography, (d) season of the year, and (e) time of day is the result of fluctuations in the amount of heat received from the sun. The average annual temperature decreases and the seasonal variation increases with increasing latitude and elevation. There are also short duration temperature fluctuations caused by the passing of clouds and the movement of sun flecks across turfs growing under trees.

The direction of slope exposure in regions of variable topography causes temperature variations even in locations of similar elevation (73). Effective heating decreases proportionally as the angle of incidence increases from the perpendicular. In the northern hemisphere, slopes facing north have the lowest temperatures because the amount of insolation received per unit area is less than on slopes facing south or even a flat location (2, 39, 40, 73, 149). Diurnally, the maximum temperature at a 1-in. soil depth on slopes of equal elevation shifts from the east in the morning to the south at noon and to the slopes facing west in late afternoon (73). Slopes facing west normally have the highest average air and soil temperatures. The effects of slope exposure on turfs are most readily observed on steep roadside cuts.

Temporal variations in temperature. The diurnal temperature variation is correlated with the diurnal variation in insolation but with a lag (20, 73, 76, 199). The temperature increases rapidly at sunrise and continues to increase until about 2:00 p.m. (Fig. 7-1). During this period, there is a net gain in energy because the amount of incoming solar radiation is greater than the amount of outgoing reradiation. Subsequently, there is a steady decrease in temperature with the minimum occurring around 6:00 to 7:00 a.m. or just prior to sunrise. Outgoing reradiation far

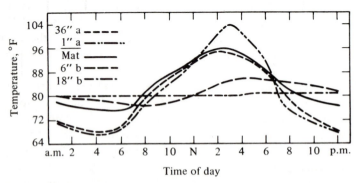

Figure 7-1. The diurnal variation in temperature at five vertical heights above (a) and below (b) a Toronto creeping bentgrass turf cut at 0.25 in. Clear day in July at East Lansing, Michigan.

surpasses incoming solar radiation during this cooling period. A diurnal temperature variation of as much as 60°F can occur in stolons (174). The typical diurnal temperature variation can be altered by cloudiness or air masses of a much different temperature. Temperature extremes are greater during the summer season than in the winter (199).

Vertical temperature variations. Pronounced vertical temperature gradients occur within the turfgrass shoot and root microenvironment (19, 20, 66, 73, 174, 182, 198, 199, 213, 214). Maximum diurnal temperatures usually occur just below the surface of dense, close cut turfs because this is the region of radiation absorption that results in heat accumulation (Fig. 7-2). Temperatures usually decrease with distance from the surface of the turf during clear days. The decrease is due to a decline in effective heating by reradiation and conduction as well as increased air turbulence. The air and soil adjacent to the turf are warmed by conduction. The heat contained in the adjacent air layer is then transferred by convection to layers farther above the turfgrass surface. There is a lag in the heating of more distant air layers.

Heat energy is also lost by long wave reradiation from the turfgrass surface. Between sunrise and midafternoon the loss is more than compensated for by the incoming solar radiation. At night and in late afternoon a major portion of the heat is lost through long wave reradiation. The surface of the turf is cooled and the temperature of the adjacent air is lowered by conduction. An inversion results with the adjacent air layers being cooler than the upper, more distant layers.

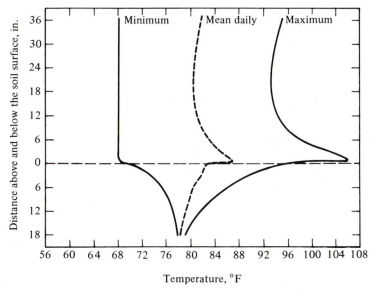

Figure 7-2. The vertical gradients in maximum, minimum, and mean daily temperatures above and below a Toronto creeping bentgrass turf cut at 0.25 in. Sunny day in mid-August at East Lansing, Michigan.

Soil warming occurs by conduction that restricts the depth of heating and causes a substantial lag in temperatures (40, 73, 119). The root zone soil temperature lag increases with depth. The vertical soil temperature gradient reverses depending on the season of the year (40, 73, 194). A downward gradient of decreasing temperature occurs in the summer, while the subsoil is warmer than the soil surface during the winter due to a more rapid loss of heat from the surface layers.

The standard meteorological data commonly reported involve temperature measurements made at an above ground height of 5 ft. Turfmen should recognize that temperatures of the turfgrass microenvironment may be much higher on sunny days and significantly lower at night than the standard weather reports (19). Temperature observations of the turfgrass microenvironment are best made by placing a shaded temperature sensing element at the turf-soil interface. The average daily soil temperature at the 6-in. depth is the best indicator of turfgrass root growth (20).

The greatest extremes in seasonal temperatures occur just below the surface of dense, close cut turfs (20, 73, 199) (Fig. 7-3). The vertical temperature gradient above a turf is frequently less during the winter season than in the summer (20, 73, 174, 199).

Vertical temperature gradients are much less under windy or cloudy conditions (73, 199). Wind activity minimizes the vertical temperature gradient through a turbulent mixing process. A cloud cover reduces vertical air temperature differences by 30 to 60% of that observed on a clear day (174). The turfgrass microenvironment is cooler under cloudy or foggy conditions than on clear days because the water present in the atmosphere absorbs and reflects a major portion of the incoming solar radiation before it reaches the turf. The presence of a cloud cover at night also impairs energy loss in the form of outgoing long wave reradiation. This slows the nocturnal cooling process and moderates the minimum temperature. Rain generally depresses turfgrass temperatures with a decrease in the order of 1 to 2°F frequently occurring (174). The moisture and cooler temperatures following a period of rain frequently stimulate turfgrass growth and improved color. A low atmospheric relative humidity enhances radiation exchange and increases temperature extremes (174).

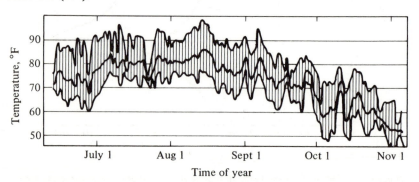

Figure 7-3. The seasonal variation in the maximum, minimum, and average daily soil temperatures at 0.5 in. below a creeping bentgrass turf cut at 0.25 in. West Lafayette, Indiana (20).

Temperature variations within a turf. Temperature extremes in and above a bare soil are generally greater than when a turfgrass cover is present (40, 45). The turf absorbs a substantial portion of the incident radiation and has an insulating effect that reduces soil temperature extremes. The degree of temperature reduction is proportional to the height of cut (45, 119, 149, 171, 214). For example, soil temperature extremes are greater under a bentgrass turf cut at 0.25 in. than at 1.5 in. Putting greens become frozen earlier in the winter than adjacent higher cut aprons (6).

The vertical temperature distribution within a grass canopy can be illustrated from observations made within a 24 in. high roadside-type turf (Fig. 7-4). The air layer having the highest temperature at midday is about 10 to 12 in. above the soil (213, 214). This is the middle zone where the density of the grass stand is increasing substantially and is immediately above the region where the intensity of the incident radiation declines sharply. Above and below this middle air layer are zones of cooler air. The shoot density of the turf affects the exact location of the middle warm air layer. The prominence of a middle warm air layer decreases proportionally with the cutting height. The temperature extremes are elevated almost to the surface of turfs cut at 0.3 in.

Other factors influencing temperature variations. The temperatures of turfs growing under a tree canopy are substantially lower than temperatures of unshaded turfs during the day and are somewhat higher at night (66, 73). Turfs in dense shade can be as much as 10 to 20°F cooler than adjacent turfs that are exposed to full sunlight (66). Also, the vertical temperature gradient is less pronounced under dense shade (70). The cooler daytime temperatures of shaded turfs are due primarily to absorption and reflection of the incoming radiation by the tree canopy before it reaches the turf. Another factor is the higher atmospheric relative humidity

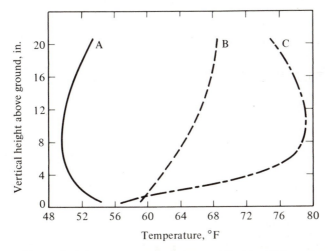

Figure 7-4. The vertical air temperature variations at (A) night, (B) late afternoon, and (C) midday, within a grass stand 24 in. high. (After Waterhouse—214).

under a tree canopy that increases the quantity of heat required to raise the air temperature. Nocturnal heat loss by long wave reradiation from the turf is also inhibited by the tree canopy. Thus, the turfgrass temperatures under a tree canopy are higher at night than adjacent unshaded turfs.

Color is also a factor influencing the temperature of turfs. Color, in this context, refers to the entire spectrum rather than just the visible impression. Color affects the proportion of incident radiation that is reflected versus absorbed and influences heat accumulation at the turfgrass surface. White or light colored surfaces remain comparatively cool due to a low degree of absorption. Species with shiny leaf surfaces such as ryegrass have a high degree of reflection (66). Dull and dark colored surfaces are warmed more readily due to a comparatively high absorption level. The temperature of a dark surface may be as much as 10 to 15°F warmer than a light colored surface when exposed to full sunlight (66). Dark surfaces have a greater reradiation capability than light colored surfaces, which results in a more rapid nocturnal cooling.

The color factor in heat accumulation is frequently utilized in the dissipation of ice and snow covers. A black material such as lampblack or a dark colored fertilizer can be applied to the ice or snow. The black particles absorb the incoming radiation, accumulate heat, and transfer the accumulated heat by conduction to the adjacent ice or snow wherein melting occurs.

The rate of temperature change is slower in wet than in dry soil because the specific heat of water is much greater than that of mineral soil. Temperature extremes are also substantially greater when a soil is dry than when wet (66). Well-drained dry soils thaw much earlier in the spring than poorly drained wet soils. Wet soils such as poorly drained, lowland mucks are generally cooler during the summer than adjacent dry soils. This favors sod production of cool season turfgrasses on muck soils having a high water table.

Snow is a poor heat conductor and serves as an insulating cover (40). A snow cover increases the mean minimum temperatures and decreases the mean maximum temperatures (174). The insulating effect is important in protecting turfs against lethal low temperatures. Loose snow is superior to compacted snow in its insulating effect (Table 7-1). The presence of a snow cover also delays thawing of the turf and underlying soil.

Table 7–1

EFFECT OF A SNOW COVER ON THE SOIL TEMPERATURE AT A 3-IN. DEPTH ON JANUARY 30, EAST LANSING, MICHIGAN [AFTER BOUYOUCOS (40)]

Types of soil cover	Temperature, °F	
	Daily mean	Minimum
Bare soil	16.3	7.5
Bare soil covered with compacted snow	18.4	15.6
Bare soil covered with uncompacted snow	27.9	27.0
Vegetation with a layer of uncompacted snow	32.6	32.2

Mulches and natural organic layers such as thatch function in moderating soil temperature extremes and temperature fluctuations. The moderating effect is attributed to (a) the materials being poor conductors and (b) protection against direct contact with cold or hot air currents. The insulating effect of mulches can be important in the midsummer establishment of cool season turfgrasses. Light colored mulches are more effective in protecting sensitive, young seedlings against lethal high temperatures than dark colored mulches.

Urban areas have a higher temperature than nearby areas of low population density. The dust, smoke, and gases discharged into the atmosphere by the concentration of human activities reduce the amount of radiation reaching the surface of the earth and inhibit heat loss by reradiation. Thus, urban climates are characterized by higher temperatures in which turfgrasses must grow. The higher temperatures also decrease the relative humidity and stimulate thawing of ice and snow. Higher temperatures in urban areas are also associated with the heat given off from buildings. The release of heat from buildings and roads is illustrated in Fig. 7-5. The air temperature increases with nearness to the building or concrete apron (127). The warming influence extends out over the adjacent turf for a considerable distance.

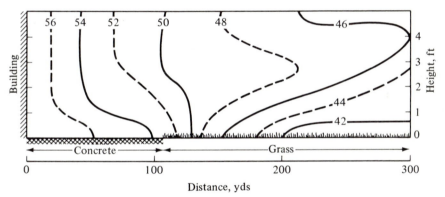

Figure 7-5. The vertical variation in temperature above a turf growing adjacent to the concrete apron of an airfield (After Knochenhauer—127).

Turfgrass Response to Temperature

Turfgrass Temperatures

The temperature of a turfgrass plant or its parts is determined by the total environment. Temperatures of the underground portions of a turfgrass plant are usually identical with the adjacent soil temperatures. Aboveground plant parts tend to follow the surrounding air temperature but with a degree of variability caused by environmental factors that affect energy exchange. The environmental factors include (a) light intensity, (b) thermal radiation, (c) air temperature, (d) atmospheric water vapor content, and (e) wind. Turfgrass plants receive energy by radiation, by convection if the plants are cooler than the air, and by condensation if the plant temperature is less than the dew point. Turfgrass plants lose energy by reradiation, by convection if the plants are warmer than the air, and by transpiration. The plant is cooled if more energy is given up than is received; it becomes warmer if more energy is received than is given off.

Radiation is the major source of energy for warming of turfgrasses. The radiation may originate from a solar or thermal source. Thermal radiation is always present while solar radiation is a factor only during the daylight hours. The energy content and temperature of a turfgrass plant increase as the absorption of incident radiation is increased. The plant temperature is primarily dependent on the relative rates of radiation absorption and energy loss by reradiation when transpirational cooling is impaired and convectional cooling is minimal. The aboveground plant temperatures can be much higher than the adjacent air temperature during periods of high light intensity due to a lack of transpirational and convectional cooling.

Air temperature influences the energy content of a turfgrass plant by the convection of energy to and away from the plant surface. Free convection occurs across a thin atmospheric zone called the **boundary layer**. The boundary layer surrounds the leaf surface and may be as much as 1 cm thick in still air. Gas molecules in contact with the leaf surface gain energy by conduction and are warmed when the air is cooler than the leaf. The increase in energy of the gas molecules causes the air to expand and rise, with cooler air moving to the leaf surface (Fig. 7-6). The direction of energy transfer is reversed at night when the leaf is cooler than the air. The rate of energy transfer across the boundary layer is controlled by the temperature gradient between the leaf and the atmosphere as well as by the thickness of the boundary layer.

Energy exchange by free convection and by forced convection as wind depends on the size, shape, and orientation of the plant surface as well as the rate of air flow (99). The heating and cooling rate of thick leaves is slower than for thin leaves (11). Thus, the efficiency of energy transfer by convection is greater for small plant parts such as the fine leaves of red fescue. Such fine textured turfgrasses are more strongly affected by the air temperature than the broader, thicker leaves and stems of tall fescue or St. Augustinegrass.

Transpiration is an efficient means of outward heat transfer from a turf. The

Figure 7-6. A schlieren photograph illustrating the cooling of a sunlit Kentucky bluegrass turf by rising convection currents which appear as dark shadows in the photograph. (Photo courtesy of D. M. Gates, Missouri Botanical Gardens, St. Louis, Missouri).

evaporation of water results in cooling of the leaf surface. Wind movement, a low atmospheric water vapor content, and adequate available soil moisture stimulate transpirational cooling (11). Turfgrass temperatures may be 2 to 12°F lower than the adjacent air temperature under these conditions (76). Should transpiration be impaired by a plant water deficit, the plant temperature can rise substantially above the adjacent air temperature when exposed to sunlight (11).

Each part of a turfgrass plant can be at a different temperature because of differences in the energy transfer rate between the various plant parts and the surrounding environment. The temperatures of leaves can fluctuate quite rapidly in response to changes in light intensity and wind (11). Temperatures as high as 118°F have been observed within a close cut creeping bentgrass turf (20). An individual leaf within a Kentucky bluegrass turf was found to have a temperature of 124°F, when the adjacent air temperature was 40°F warmer (99). Temperatures of 125°F have been observed in the vicinity of exposed grass stolons (119). Higher tissue temperatures usually occur in the thicker leaves, stems, stolons, and rhizomes.

Temperature Optima

Turfgrass species from warmer native habitats generally have higher temperature optima. The temperature optimum for maximum growth is not synonymous with the temperature optimum for turfgrass quality. The maximum shoot growth rate usually occurs at a higher temperature than the optimum for turfgrass quality. Temperatures are within the optimum range for a limited period of time because of diurnal variability. Turfgrass species vary in the degree of diurnal temperature variability that can be tolerated in terms of maintaining optimum growth or quality (107). Kentucky bluegrass can sustain fairly high levels of shoot growth over a comparatively wide temperature range (44, 107).

Shoot growth. The meristematic tissues of low growing turfgrass species from which the shoots are initiated are usually located near or within the soil. Thus, the surface soil temperatures are particularly important in influencing shoot development. The optimum temperature range for sustained shoot growth varies among the turfgrass species. Cool season turfgrasses having an optimum range of 60 to 75°F include Kentucky bluegrass (13, 20, 45, 65, 105, 107, 117, 141, 166, 197), Canada bluegrass (44, 107, 117), creeping bentgrass (75, 116), colonial bentgrass (157, 159, 197, 202), meadow fescue (197), timothy (13, 197), perennial ryegrass (4, 153, 155, 156, 157, 159, 173, 203, 218), Italian ryegrass (157, 175), wheatgrass (32), and annual bluegrass (106, 118). On the other hand, the optimum temperature range for warm season turfgrasses, such as zoysiagrass (230), manilagrass (230), bermudagrass (44, 141), and carpetgrass (141) is 80 to 95°F. Turfgrasses growing in the optimum temperature range have increased nutrient and water requirements and also require more frequent mowing.

Shoot growth decreases as temperatures are decreased or increased from the optimum range (13, 44, 45, 65, 75, 105, 107, 141, 155, 156, 157, 159, 175, 197, 202, 203, 218, 227, 230). The effect is particularly striking at higher temperatures. Shoot growth is reduced, shoot density declines, and the leaves become spindly and dark green (20, 65, 203). Shoot growth of Kentucky and Canada bluegrass occurs at temperatures as low as 40°F (44). A relationship appears to exist between the growth habit and the time when shoot growth is initiated in the spring. Certain bunch-type turfgrasses, such as Italian ryegrass, initiate shoot growth earlier in the spring than those possessing a creeping growth habit, including creeping bentgrass and Kentucky bluegrass. How valid this relationship is across a wide range of turfgrass species is yet to be determined.

Tillering, succulence, leaf number, leaf width, leaf length, leaf area, and the new leaf appearance rate have an optimum temperature range similar to the optimum for shoot growth and also decrease as the temperature is increased or decreased from the optimum range (Fig. 7-7) (13, 20, 53, 153, 154, 155, 156, 157, 158, 159). Optimum tillering appears to be favored by temperatures slightly lower than the optimum for shoot growth (4, 75, 156, 178). Tillering of perennial ryegrass is enhanced by night temperatures of 37°F when day temperatures are 68°F, while a high night temperature of 82°F inhibits tillering (3). Similarly, tillering of tall fescue is favored by a daily exposure to 45°F compared to continuous temperatures of 72°F (207). Tillers of annual bluegrass grown at optimum temperatures have a horizontal growth habit, while tillers grown under supraoptimum temperatures have a more upright growth habit (106). The horizontal growth habit of tillers is most evident at suboptimal temperatures.

Studies with zoysiagrass show greater shoot growth, leaf blade length, internode length, and stolon length at 80°F than at 70°F (230). The temperature minimum for shoot growth of Emerald zoysiagrass and manilagrass is 55°F, while the minimum for Meyer zoysiagrass is near 60°F. Shoot growth of U-3 bermudagrass ceases at mean daily soil temperatures of approximately 50°F or lower (44, 227, 228). Certain improved bermudagrass cultivars such as Ormond and Tifgreen have a temperature minimum somewhat lower than 50°F. Bermudagrass can maintain moderate shoot growth and color at night temperatures as low as 34°F provided the daytime temperatures are 70°F or higher (227).

The optimum temperature ranges refer to shoot growth over an extended period of time. Shoot growth is actually more rapid at higher temperatures when measured for only a short period of 2 to 3 weeks (3, 106, 114, 116). For example, when perennial ryegrass is exposed to a day temperature of 68°F and night temperatures of 37, 50, 68, and 82°F, the initial shoot growth rate is highest at 82°F and declines with decreasing night temperature. Subsequently, the shoot growth rate at 82°F declines in relation to the rate at lower temperatures. The result is a higher level of sustained shoot growth at the lower temperatures.

Root growth. Turfgrass root growth can be maintained at relatively high air temperatures as long as the soil temperature remains favorable (44). Moderate soil temperatures persist at relatively high daytime air temperatures as long as the nocturnal air temperatures are low. Temperatures above 90°F in the surface inch of soil can cause a restriction in Kentucky bluegrass root and rhizome growth similar to that observed when the entire root system is subjected to 90°F. The mean

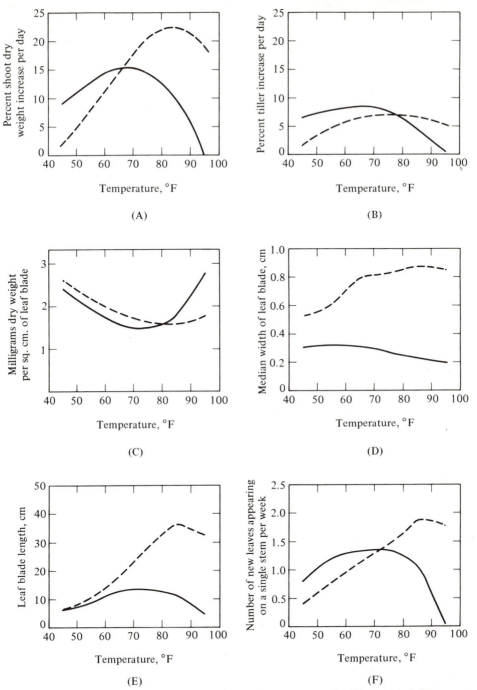

Figure 7-7. The influence of six constant temperatures on the (A) percent daily increase in shoot dry weight, (B) percent daily increase in number of tillers, (C) dry weight per cm² of leaf blade, (D) median leaf blade width, (E) length of leaf blade, and (F) rate of new leaf appearance of the cool season turfgrass, *Agrostis tenuis* (solid line), and the warm season turfgrass, *Paspalum dilatatum* (dashed line) (After Mitchell—157).

daily soil temperature at the 6-in. depth is highly correlated with root growth of creeping bentgrass and is the best predictor of seasonal variations in rooting (28).

The maximum and minimum temperatures for root growth of cool season turfgrasses are lower than for shoot growth (44, 166, 197). Roots of many cool season turfgrasses continue to grow in the autumn as long as the soil is not frozen (201). Cell division can occur in the root tips of cool season turfgrasses at temperatures just above 32°F (201). Rooting vigor at low temperatures varies among turfgrass species. For example, Kentucky bluegrass is capable of root growth at midwinter soil temperatures near 32°F (103), whereas the root growth of creeping bentgrass is minimal (20).

The optimum temperature range for sustained root growth varies among turfgrass species. Two broad groups are evident with the lower optimum temperature range represented by the cool season turfgrasses and the higher optimum encompassing the warm season species. Maximum sustained root growth for most cool season turfgrasses occurs between 50 to 65°F. Species included in this range are Kentucky bluegrass (44, 45, 65, 105, 141, 166, 197, 202), Canada bluegrass (44), creeping bentgrass (27, 28, 191), colonial bentgrass (197, 202), meadow fescue (197), timothy (197, 202), perennial ryegrass (155), annual bluegrass (106), and wheatgrass (32). Variability in temperature optima exists within this group. For example, root growth is greater at 50°F for Canada bluegrass and at 60°F for Kentucky bluegrass (44). Optimum rooting of creeping bentgrass occurs at a slightly higher temperature than for Kentucky bluegrass (195). The optimum temperature range for root growth of warm season turfgrasses is also somewhat lower than the optimum for shoot growth (228). From the limited information available, maximum sustained root growth of most warm season turfgrasses ranges from 75 to 85°F (141, 230).

Total root growth decreases as soil temperatures are either decreased or increased from the optimum range (20, 28, 44, 105, 141, 191, 197, 202). Temperatures at or slightly below the optimum range for root growth result in a higher, more favorable root-shoot ratio than temperatures above the optimum (4, 141, 156, 166, 197).

The optimum temperature ranges referred to are for sustained root growth over an extended period of time. Root elongation is more rapid at higher temperatures on a short term basis of 1 to 2 weeks (20, 28, 186). For example, the initial rate of root development and extension from a bentgrass sod is highest at 80°F, less at 70°F, and the least at 60 and 90°F (20, 27). At 80°F, however, the roots mature and cease growth more rapidly than at the lower temperatures. Total root production increases as temperature is decreased from 90 to 60°F after a period of 4 to 6 weeks (Fig. 7-8). Yet, the root elongation rate is similar at soil temperatures of 60, 70, and 80°F. Evidently the root maturation rate is accelerated as the soil temperature increases (20, 27, 65, 202).

A detailed examination of turfgrass root systems grown at various soil temperatures reveals striking differences. Roots growing in the optimum range are thick, white, and multibranched (20, 27, 65, 106, 202). Roots formed at sub-optimum temperatures are thick, white, less branched, shorter, and slower growing. Turf-

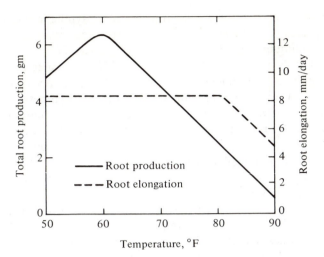

Figure 7-8. The influence of soil temperature on the total root production and rate of root elongation of a Toronto creeping bentgrass turf 45 days after transplanting.

grass roots tend to mature more rapidly and become brown, spindly, and inactive as temperatures are increased above the optimum.

Seasonal observations of bentgrass root growth indicate that most root initiation occurs in early spring at optimum soil temperatures (20, 28, 45, 195, 201). The midsummer loss of the root system of cool season turfgrasses is highly correlated with supraoptimal soil temperatures (20, 28, 45, 195, 201). Little root initiation occurs at high soil temperatures except following a 2- to 3-day period of cooler soil temperatures (28). This suggests that near optimum soil temperatures are important in promoting root initiation or are a prerequisite for root initiation.

Rhizome growth. Optimum temperatures for rhizome development of turfgrasses are similar to the optimum for root growth (44, 45, 65, 105, 230). Rhizome development includes the number, length, and total dry matter production (44). Soil temperatures above 80°F stimulate the emergence of the growing point of Kentucky bluegrass rhizomes above the soil (44, 105). The highest percentage of rhizomes remains below the soil at soil temperatures of 60 to 70°F. Exposure to temperatures of 32 to 35°F results in increased tillering and reduced rhizome development of Kentucky bluegrass (164).

Physiological responses. The rates of a majority of the physiological reactions and processes occurring in turfgrasses are temperature dependent. There is an optimum temperature range for the rate of most physiological reactions and processes. The activity of the physiological process declines as the temperature is increased or decreased from the optimum (4, 175, 189).

The temperature responses of photosynthesis and respiration are of particular interest in relation to their affect on net assimilation. Creeping bentgrass leaves, grown at 68/50, 86/68, and 104/86°F and tested at the higher light period temperature, had photosynthetic and respiratory activities that ranked as 104 > 86 > 68°F

(75). An adaptive mechanism is evident in the photosynthetic rate when grown at high temperatures (75). Apparent photosynthesis under relatively low light intensities has a maximum (Fig. 7-9) rate at approximately 77°F for Seaside creeping bentgrass and 95°F for common bermudagrass (152). The rates would probably be higher if an optimum light intensity had been used (89). It appears from the limited data available that the optimum temperatures for photosynthesis, respiration, and apparent photosynthesis are significantly higher than the optimum temperature ranges for shoot and root growth. Also, different temperature optima for apparent photosynthesis exist among turfgrass species with the warm season turfgrasses having a higher optimum than the cool season species (112, 152). The optimum temperature for these processes shifts depending on the conditions under which the turfgrass plant develops (75).

The literature concerning temperature affects on carbohydrate reserves of turfgrasses is variable and conflicting due to inconsistencies in carbohydrate extraction techniques plus the failure to keep other factors such as light intensity at nonlimiting levels. Carbohydrate reserves are relatively low at the temperature optimum for shoot growth (115, 232). Carbohydrate reserves tend to accumulate as temperatures are increased or decreased from the optimum for shoot growth (3, 4, 5, 20, 75, 115, 116, 173, 192, 233). The carbohydrate reserve can decrease at supraoptimal temperatures following a severe defoliation (203) as well as when light or nutrients are limiting.

The amide level in ryegrass is highest at 67°F and declines as temperature is increased or decreased (173). The decline in the amide level is most evident in the glutamine fraction (21, 29, 200). Asparagine accumulates at extremely high temperatures, when the rate of protein degradation is rapid (200). High temperatures also cause an increase in the total soluble nitrogen and free ammonia levels (21, 29, 190, 200). Among the free amino acids, substantial increases in glycine, alanine, valine, isoleucine, serine, threonine, and lysine occur at supraoptimal temperatures (200).

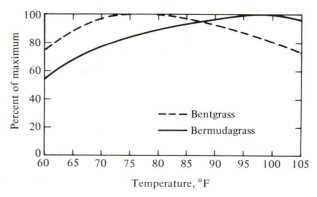

Figure 7-9. Influence of temperature on the rate of apparent photosynthesis of bermudagrass and bentgrass relative to the maximum rate found for each under a light intensity of 2000 fc (After Miller—152).

Temperature also influences turfgrass color. The maximum chlorophyll content of creeping bentgrass leaves occurs between 75 to 85°F and declines as the temperature is increased or decreased (75). The visual color of bentgrass leaves appears darker green at 95°F than at 85°F but the chlorophyll content is actually lower. Turfs also appear darker green at suboptimal soil temperatures that restrict shoot growth.

The temperature influence on the chlorophyll content of warm season turfgrasses is quite striking. Winter discoloration occurs at mean daily temperatures below 55 to 60°F when the light intensity is above 0.6 cal per sq cm per min (230). Bermudagrass can retain a green color at 34°F as long as the light intensity is less than 0.15 cal per sq cm per min.

Flowering. Floral initiation of certain turfgrass species does not occur even when grown under a favorable photoperiod without prior exposure to a low temperature vernalization. The most favorable temperature range for vernalization is 40 to 50°F. The vernalization stimulus is localized in the shoot apex (96). Thus, even embryos of imbibed seeds of certain turfgrass species such as perennial ryegrass can be vernalized by cold treatment (38, 52, 94, 177). The rate of vernalization in perennial ryegrass increases as the temperature is raised from 40 to 50°F (83). The longer the exposure to cold treatment, the more positive the vernalization process (48). Once fully vernalized, the shoot apex is capable of retaining this condition until exposed to day lengths that stimulate floral initiation (51, 224). If the shoot apex is only partially vernalized, however, it can be devernalized by subsequent exposure to temperatures of 80°F or higher (140). The vernalization response of turfgrasses is impaired when certain other environmental factors are limiting, such as an internal water deficit or an oxygen deficiency.

Turfgrass species vary considerably in the temperature requirement for floral induction (Table 7-2). Vernalization responses have been observed only within the subfamily *Festuciodeae*. Most perennial cool season turfgrasses show some degree of response to vernalization. Kentucky bluegrass, perennial ryegrass, and sheep fescue generally require vernalization for floral initiation. Many of the other perennial cool season turfgrasses exhibit an increase in floral initiation when exposed to a cold treatment. The annual turfgrasses, such as annual bluegrass and Italian ryegrass, show a minimal response to cold treatment, while the warm season turfgrasses have no vernalization requirement. Actually, temperatures below 55°F inhibit flowering of most perennial warm season turfgrasses (125, 126). Flower initiation, development, and seed set of zoysiagrass are favored by temperatures of 80°F (230).

Temperature and photoperiod frequently interact in inducing flowering. Certain temperatures can reduce the duration of the photoperiod required for floral initiation or may even act as a substitute. For example, vernalization can replace the short day photoperiod requirement in tall fescue (208), while a short photoperiod can be substituted for the cold treatment in perennial ryegrass (83). Following floral initiation, the floral development rate is accelerated as the temperature is increased to slightly above the optimum temperature for shoot growth (51, 56, 83, 178, 229).

Table 7–2

THE VERNALIZATION REQUIREMENT FOR MAXIMUM FLORAL INDUCTION
OF TWENTY-ONE TURFGRASSES

Vernalization (40–50°F) requirement for floral induction			
Determinant	*Favorable*	*Minimal*	*Inhibitory*
Kentucky bluegrass (96, 107, 140, 178, 183, 184)	Bulbous bluegrass (184, 229)	Annual bluegrass (224)	Bahiagrass (126)
	Canada bluegrass (107)	Italian ryegrass (50, 53, 54, 83)	Bermudagrass (126)
Perennial ryegrass (50, 52, 53, 54, 55, 56, 83, 224)	Colonial bentgrass (55, 56)		Carpetgrass (126)
	Creeping bent-grass (55, 56)		
Sheep fescue (225)			Manilagrass (88, 230)
	Meadow fescue (56)		Zoysiagrass (88, 230)
	Red fescue (56)		
	Redtop (55, 56)		
	Rough bluegrass (56)		
	Smooth brome (56)		
	Tall fescue (208)		
	Velvet bentgrass (55, 56)		

Seeds. The primary influence of temperature on seeds is twofold. First, temperature is a factor in maintaining the viability of turfgrass seeds during extended periods of storage. Second, temperature is one of the major factors affecting the rapidity and vigor of seed germination.

Temperature can be a factor influencing the viability of turfgrass seeds when stored for extended periods of time. High temperatures cause a loss in viability of dormant seeds because of the more rapid depletion of carbohydrate reserves in the grass seed by respiration. The loss of viability in chewings fescue seed increases as the temperature is raised from 32 to 86°F (92, 95, 111, 121). Dry seeds have a greater tolerance to high temperature than moist seeds (121).

Turfgrass seed germination is influenced by the temperature at or just beneath the soil surface. Different turfgrass species have different temperature ranges within which seed germination occurs, assuming other factors are not limiting (43, 104, 163, 196, 210). Within the germination range there is an optimal temperature (Table 7-3). Seed germination is delayed at temperatures above or below the optimum.

The optimum temperature for seed germination varies with the (a) seed age, (b) seed source, (c) cultivar, (d) seed lot, and (e) duration of the germination period (104). Fescue seed germinates over a wider temperature range as it ages (120).

Table 7–3

THE OPTIMUM TEMPERATURES FOR SEED GERMINATION OF TWENTY-NINE
TURFGRASSES AS ESTABLISHED BY THE ASSOCIATION OF OFFICIAL
SEED ANALYSTS (8)

Turfgrass species	Optimum temperatures for seed germination*	Prechilling of new seed suggested, 38–50°F
Bahiagrass	86–95	
Bentgrass:		
colonial	59–86	✕
creeping	59–86	✕
velvet	68–86	
Bermudagrass	68–95	
Bluegrass:		
annual	68–86	
bulbous	50	✕
Canada	59–86	
Kentucky	59–86	✕
rough	68–86	✕
Buffalograss	68–95	✕
Carpetgrass	68–95	
Fescue:		
chewings	69–77	✕
hair	50–77	
meadow	68–86	
red	59–77	✕
sheep	59–77	
tall	68–86	✕
Grama:		
blue	68–86	
sideoats	59–86	
Manilagrass	95–68	
Orchardgrass	68–86	✕
Redtop	68–86	
Ryegrass:		
Italian	68–86	✕
perennial	68–86	✕
Smooth brome	68–86	✕
Timothy	68–86	✕
Wheatgrass:		
western	59–86	
Zoysiagrass	95–68	

* Temperatures separated by a dash indicate an alternation of temperature; the first numeral is for approximately 16 hr and the second for approximately 8 hr.

The optimum germination temperature of certain species may not involve a specific temperature but a rhythmic alternation of temperatures. Examples of the latter include Kentucky bluegrass (91, 104, 167), creeping bentgrass (91), Canada bluegrass (163, 167), redtop (214), bermudagrass (104, 163), and bahiagrass (59). Germination of freshly harvested seed of certain species is stimulated by 8 weeks prechilling at 38°F (210, 219). The maximum and minimum temperatures for seed germination are poorly defined because of the extreme slowness of germination, especially for the minimums.

Disease activity. The occurrence and severity of disease development on turfgrasses is influenced by temperature. The seasonal occurrence of certain diseases indicates a strong relationship with temperature. Each fungal pathogen has an optimum temperature range for the mycelia spread and infection (Table 7-4). Variability from the cardinal temperatures given in Table 7-4 may occur. For example, cool temperature strains of *Rhizoctoni solani* (77, 143) and *Sclerotinia homoeocarpa* (35) have been reported. The fungicide program should be adjusted in relation to the existing temperature conditions in order to ensure effective control of the diseases most likely to occur with the greatest severity.

Table 7–4

THE OPTIMUM, MAXIMUM, AND MINIMUM TEMPERATURES FOR THE GROWTH OF EIGHT
TURFGRASS FUNGAL PATHOGENS

Fungal pathogen	Minimum	Optimum	Maximum	References
Typhula itoana	20–35	45–60	70–75	110, 181
Fusarium nivale	32	65–70	90	33, 62, 78, 110
Corticium fuciforme	35–40	65–70	85–90	34, 78, 80, 110
Sclerotinia homoeocarpa	35–40	70–80	90–95	35, 77, 78, 110
Ustilago striiformis	40–45	70–80	90–95	68, 110, 128
Rhizoctonia solani	40–45	75–90	95–100	60, 61, 72, 77, 78, 110, 160, 161
Pythium ultimum	40	80–85	100	150, 162
Pythium aphanidermatum	50	90–95	110–115	77, 78, 92, 110, 150

(Temperature range for fungal growth, °F)

The optimum temperature for fungal growth is not necessarily the optimum for injurious infections of turfgrasses. For example, temperatures of 32 to 40°F result in very severe injury to bentgrass from *Fusarium nivale*, while fungal growth of the organism is most active at 65 to 70°F (62). Frequently, the degree of disease injury, especially from facultative parasites, is more severe at temperatures above or below the optimum for growth of the turfgrass.

Thermoperiodism

The continuous, diurnal cycle of light and darkness is the basis for numerous rhythmic environmental and biological variations. Environmentally there is a diurnal temperature maximum usually near midday and a nocturnal period of low temperatures. Some plants have become acclimated to this rhythmic diurnal temperature cycle to the extent that a diurnal period of low temperature is required for normal growth and development. The response of plants to rhythmic fluctuations in temperature is called **thermoperiodism**.

Certain turfgrass species are reported to exhibit a thermoperiodic response (107). Shoot growth of Kentucky bluegrass and bermudagrass is increased when grown under either a 60/80 or 50/90°F diurnal temperature fluctuation compared to the growth at a constant temperature of 70°F (44). The thermoperiodic growth response has a physiological basis in that photosynthesis and growth are distinct processes having different cardinal temperatures. The former is operative during the day and the latter is active primarily at night when temperature conditions are more favorable.

High Temperature Stress

Turfs are exposed to high temperature stress during midsummer periods when the intensity of use is also frequently the highest. Midsummer injury of turfgrasses is usually the combined result of heat, wear, desiccation, and disease. High temperature stress is the most difficult of the four to prevent or control. Heat stress also weakens the turf to the extent that it is more easily killed or injured by one of the other common midsummer stresses.

Heat and drought stress often occur in association under unirrigated conditions. For this reason it is extremely difficult to distinguish between the two stresses. Heat stress problems are more common among the cool season than the warm season turfgrasses. The high temperature limit is a major factor restricting the use of certain desirable cool season turfgrasses in the warm humid regions of the world. Plant stresses caused by supraoptimal temperatures are of two basic types: (a) indirect growth stoppage and (b) direct kill.

Indirect High Temperature Stress

Turfgrass growth is impaired at supraoptimal temperatures that are not immediately fatal. Death due to indirect causes usually occurs only after an extended period of exposure. As temperatures increase above the optimum range for a particular turfgrass species, one of the first stress effects observed is an increase in the rate of root maturation (20, 27, 65, 202) followed by death of the root system (20, 28, 45, 195, 201). Roots under high temperature stress appear brown, spindly, and weak (20, 27, 65, 86, 202). The initiation of new roots from the meristematic tissues of the crown is also blocked (20, 28).

The next high temperature stress effect is a decline in shoot growth. Associated with this is a reduction in leaf length, width, and area; rate of new leaf appearance; and succulence (13, 20, 53, 75, 153, 155, 156, 157). The leaves become dark green to blue-green (20, 75). A decrease in turfgrass density is associated with the cessation of shoot growth at supraoptimal temperatures (75).

The decrease in shoot growth is caused by high temperature blockage of one or more critical physiological processes. High temperatures probably cause (a) destruction of certain heat-sensitive enzymes or (b) an imbalance between certain metabolic processes due to differences in the Q_{10}. The imbalance usually involves acceleration in the utilization or breakdown of an essential metabolite.

The most commonly proposed imbalance is where the temperature optimum for photosynthesis is considerably lower than the optimum for respiration. The result is a decline in the apparent photosynthetic rate as temperature increases.

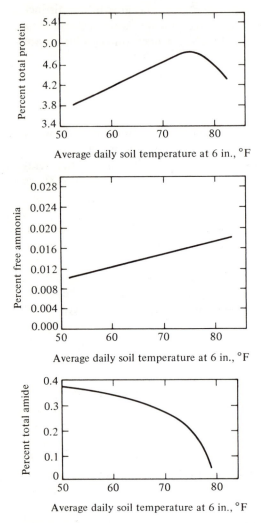

Figure 7-10. The influence of the average daily soil temperature at 6 in. on the percent total protein, percent free ammonia, and percent total amide.

If the temperature is increased high enough, the available carbohydrate reserves may eventually be exhausted causing growth stoppage. This hypothesis of indirect high temperature stress may occur in cool season turfgrass species under certain conditions. However, further detailed studies are needed to substantiate this hypothesis as a common occurrence in normal turfgrass culture. The available data are confounded with various factors that have resulted in misinterpretations.

Studies with certain cool season turfgrasses indicate that high temperature growth reduction is not caused by a photosynthesis-respiration imbalance resulting in carbohydrate exhaustion. For example, the temperature optima for root and shoot growth of Toronto creeping bentgrass are significantly lower than the optimum for net photosynthesis (27, 28, 75, 116, 152, 191). Root and shoot growth reductions occur at temperatures that are optimum for apparent photosynthesis. Also, the reserve carbohydrate level is not exhausted at temperatures causing reduced shoot growth (75). For example, carbohydrate accumulation occurs in Toronto creeping bentgrass at temperatures above the optimum for shoot growth (75, 115, 116). High temperature growth stoppage is associated with (a) a decline in the protein level, (b) an increase in free ammonia, and (c) a severe reduction in the amide level, particularly glutamine (Fig. 7-10).

High temperatures also influence other physiological processes of significance such as stomatal closure. The result is impaired transpirational cooling and a substantial increase in the leaf temperature to potentially lethal levels. Temperatures also affect the degree of activity and selectivity of herbicides. For example, the absorption and degree of kill from 2,4-D is increased at high temperatures (122, 145).

Direct High Temperature Injury

Direct heat kill occurs if turfgrass tissues are subjected to a sufficiently high temperature. This is a relatively infrequent occurrence among turfgrasses but may be more common than previously thought. High temperature kill of heat-sensitive, cool season turfgrasses, such as annual bluegrass, is most likely to occur during midday when transpirational cooling has been impaired by an internal plant water deficit or stomatal closure.

Soil temperatures are frequently more important in direct high temperature kill than air temperatures (44, 47). The temperature of the surface soil layer tends to influence the turfgrass root system to the same degree as if the entire root–soil complex were exposed to that temperature. Thus, turfgrasses can tolerate relatively high air temperatures during the day as long as the night temperatures are sufficiently cool so that an excessive heat buildup does not occur in the soil.

Low density turfs are subject to higher surface soil temperatures than dense turfs. Lethal temperatures may occur at the soil surface of low density turfs, particularly new seedings (49). Injury to the stem tissues at the soil surface eventually results in death of the whole plant (39). This type of injury is most likely to occur on dry, dark colored soils such as mucks or where dark colored mulches are used (74). The closer the heat exposure is to the time of emergence, the greater the heat kill (132, 197). The heat stress causes a reduction in percent seedling emergence as well as delayed emergence of the surviving plants. Perennial ryegrass seedlings rank as the most heat tolerant followed in order by tall fescue, meadow fescue, timothy, colonial bentgrass, and Kentucky bluegrass.

The accepted hypothesis for the mechanism of direct heat injury involves the denaturation of protoplasmic proteins in living cells. The order of cellular changes observed during direct high temperature kill is (a) protoplasmic granulation, (b) protoplasmic coagulation against the cell wall, (c) cell wall breakdown, and finally (d) total cell wall collapse (86) (Fig. 7-11).

Gross symptoms of heat injury appear as browning and decay of the affected tissue. Detailed observations of annual bluegrass provide the following information regarding heat injury. The first injury symptoms appear at the junction of the leaf sheath and leaf blade of the second and third youngest leaves (Fig. 7-12). The sheath of the affected leaf usually elongates to its normal length but the blade segment withers and turns brown from the base toward the tip. Further heat exposure results in injury to the fourth youngest leaf with the older leaves showing damage with additional heat exposure.

Heat hardiness. The mechanism of heat hardiness is believed to involve (a) increased heat stability of the protoplasmic proteins, (b) increased resynthesis of the protoplasmic proteins, or (c) a combination of the two. Heat hardiness varies with the (a) environmental conditions under which the plant is grown, (b) intensity of culture, (c) age, and (d) type of tissue.

Heat hardiness in grasses is reduced substantially when grown under shaded conditions compared to full sunlight (142). This response may be attributed to

Figure 7-11. Cross section of the third newest leaf of annual bluegrass just below the junction of the leaf sheath and leaf blade. Note the clearness of the untreated plant (left) compared to the granular appearance of the protoplasm of the mesophyll cells which were exposed to 100°F for 10.5 hours at a relative humidity of 100 percent (right) (86).

Figure 7-12. Cross section of an annual bluegrass plant at the junction of the leaf sheath and leaf blade of the second youngest leaf. Note the total collapse of the mesophyll of the second youngest leaf and the partial collapse of the mesophyll of the third youngest leaf following exposure to 108°F for 50 minutes at 100 percent relative humidity (86).

the greater succulence of grasses grown in the shade. Turfgrass species adapted to shaded environments are generally less heat hardy than those adapted to hot, sunny conditions. Prior exposure to moderately high temperatures does not increase heat hardiness. Exposure to low temperatures or moisture stress decreases the tissue hydration level resulting in increased heat hardiness (119). Thus, turfgrasses grown on wet soils or under irrigated conditions are less heat hardy than when grown under water stress.

The nutritional level affects heat

hardiness. Heat hardiness is reduced at excessive nitrogen levels (10, 47, 142, 176). The proper balance of phosphorus and potassium in relation to the nitrogen level is equally important in achieving maximum heat hardiness (10, 176). The heat tolerance of turfgrasses is also decreased when grown under acidic soil conditions. This is primarily an indirect influence affecting nutrient availability. Another cultural factor of concern in heat hardiness of turfgrasses is the mowing height. Heat hardiness is reduced at short cutting heights (119).

Heat hardiness is significantly affected by age of the tissue. The lower portion of the crown, the youngest leaf, and the apical meristem are more heat tolerant than the older tissues (86). Heat hardiness also varies depending on the specific organ involved. For example, semidormant buds on rhizomes and stolons are more heat tolerant than actively differentiating and elongating tissues (105). High temperature kill of Kentucky bluegrass rhizomes is first evident at the distal end with the lateral buds at the proximal end being the most heat hardy (44, 105). Mature, dry seeds are even more heat hardy. The lower the moisture content of the turfgrass seed, the higher the heat hardiness (111). Considerable loss of germination can occur if freshly harvested turfgrass seeds are permitted to heat up. For example, prolonged exposure to temperatures above 120°F is lethal to Kentucky bluegrass seed, while a short exposure to 140°F temperatures results in loss of viability (18, 97). No appreciable loss of germination occurred when red fescue seed was exposed to a temperature of 200°F for 30 min. (111, 169).

Burning. The burning of dormant turfs is sometimes used as a means of eliminating dead foliage that would otherwise contribute to thatch accumulation. This practice is most frequently used with the warm season species such as zoysiagrass and bermudagrass. Soil temperatures at any one location will exceed 100°F for only 30 to 90 sec if the burning is rapid (172). Significant temperature increases usually occur only in the upper 10 mm of soil with a maximum of 140°F reported. The moisture content and quantity of material to be burned determines the duration and degree of heat that occurs. Brief high temperature exposures during burning are of no significance in terms of permanent injury to dormant turfgrasses. Early spring burning should be avoided, especially on cool season species (101).

Annual burning is commonly practiced on grass seed production fields to control diseases and insects that would otherwise cause a substantial reduction in seed yield. Seed fields of most cool season turfgrass species respond favorably to burning if done just after seed harvest (165). Seed production of such warm season turfgrasses as bermudagrass and bahiagrass is increased by burning during the winter dormancy period (46).

Heat hardiness of turfgrasses. Heat hardiness varies greatly among turfgrass species and cultivars (47, 119, 132, 142, 197). A relative ranking of heat hardiness of turfgrasses that are in a hardened state is presented in Table 7-5.

Direct high temperature kill of cool season turfgrasses generally occurs at tissue temperatures of 100 to 130°F. For example, annual bluegrass is killed by a 2-hr exposure to a tissue temperature of 108°F and a 100% humidity (86). The actual lethal temperature varies with the duration of exposure and hardiness level of the tissue. The longer the high temperature exposure period, the lower the lethal

Table 7-5

THE RELATIVE HEAT HARDINESS OF NINETEEN
TURFGRASSES

Heat hardiness ranking	Turfgrass species
Excellent	Zoysiagrass
	Bermudagrass
	Buffalograss
	Carpetgrass
	Centipedegrass
	St. Augustinegrass
Good	Tall fescue
	Meadow fescue
Medium	Colonial bentgrass
	Creeping bentgrass
	Kentucky bluegrass
Fair	Canada bluegrass
	Chewings fescue
	Red fescue
	Annual bluegrass
	Perennial ryegrass
	Redtop
Poor	Italian ryegrass
	Rough bluegrass

temperature (47, 86, 119, 197). Also, the degree of injury increases in proportion to the relative humidity at any one temperature level. The plants are beneficially cooled by the increased transpiration rate at the lower relative humidities.

Summer dormancy. Turfgrasses are frequently exposed to an extended period of heat and moisture stress during midsummer that may cause complete cessation of growth. The leaves usually turn brown and die. When more favorable growing conditions occur, the perennial grasses are capable of initiating new growth from the crown meristem as well as from buds on the rhizomes and stolons. This cessation of growth and death of the leaf tissue of perennial turfgrasses due to heat and moisture stress is termed **summer dormancy**. Associated with summer dormancy is a distinct decline in the monosaccharide content of the turfgrass leaves (102).

Summer dormancy is controlled mainly by the availability of water in most perennial cool season turfgrasses, such as Kentucky bluegrass, red fescue, tall fescue, perennial ryegrass, Canada bluegrass, and creeping bentgrass. Certain grass species enter summer dormancy when exposed to high temperatures, however, even though the soil moisture is not limiting (131). For example, high temperatures and long day lengths induce summer dormancy in *Poa scabrella* even in the presence of adequate moisture. Dormancy of this species is broken by a relatively short exposure to moderate temperatures and water.

High temperature protection. No practical techniques are available for completely eliminating high temperature stress of turfgrasses. There are cultural practices available, however, that minimize the chance of heat injury. A primary consideration is the use of cultural practices that ensure adequate transpirational cooling of the turf (58). Basically, this means avoiding internal plant water deficits by maintaining an adequate (a) available soil moisture level and (b) water absorption capability. Leaf temperatures can remain quite cool at high air temperatures as long as the transpiration process is operative.

The use of more heat tolerant turfgrass species and cultivars is important in regions where heat stress problems are anticipated. Proper air movement over the turf is also a factor in preventing stratification of the highest temperatures and relative humidities adjacent to the turf (76). The positioning of trees and shrubs should not impair air movement across the turf. The cutting height of closely mowed bentgrass greens is sometimes raised during periods of high temperature stress. This practice provides more insulation against extremes in high soil temperature by increasing the depth of the turfgrass canopy.

From the standpoint of the day-to-day cultural practices, syringing can be utilized to moderate the potential midday temperature maximums (28, 76). For example, the application of 0.25 in. of water to a creeping bentgrass turf at 12 noon reduces the maximum turfgrass temperature by 4°F and the soil temperature at 2 in. by 3°F (Fig. 7-13). A light application of water not only causes a depression in the prevailing turfgrass temperature but also greatly moderates the midday heat accumulation potential. This latter effect is especially significant because the resultant midday temperature depression from the potential maximum can be in excess of 10°F (71). The temperature effects observed are related primarily to transpirational cooling since the temperature of the water droplets usually comes into equilibrium with the air temperature before contacting the turf (188). Cool, intense rains or frequent irrigations of 0.75 in. or more result in a certain degree of soil cooling.

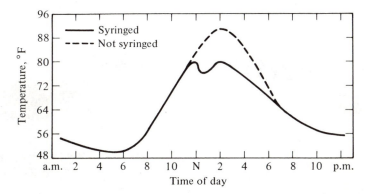

Figure 7-13. The effect of syringing at 11:00 A.M. on the temperature of a Toronto creeping bentgrass turf cut at 0.25 in.

Low Temperature Stress

Winterkill is a nonspecific term commonly used to represent any injury occurring to a turf during the winter period. It is best to avoid the term winterkill whenever possible and to cite the specific cause or causes. The specific cause of winter injury must be distinguished since the control measures vary according to the cause. Direct low temperature stress, winter desiccation, and low temperature fungi are the three major causes of injury to turfgrasses during the winter season (25). Only the first is considered in this chapter; the latter two are discussed in Chapters 8 and 17, respectively.

Direct low temperature kill is a problem common to both cool season and warm season turfgrasses (Fig. 7-14). Low temperature stress is the major factor affecting the northern adaptation limits of the warm season species. Direct low temperature kill occurs most commonly during periods of alternating freezing and thawing and is aggravated by increased crown hydration levels caused by standing water.

The turfgrass plant becomes semidormant at temperatures below the minimum for growth. Respiration continues at a substantially reduced rate and photosynthesis may sometimes be operative in the cool season turfgrasses. As the heat loss continues, tissue temperatures occur at which the protoplasm is fatally injured. Direct low temperature injury involves ice crystal formation of either an intracellular or extracellular nature.

Intracellular freezing is usually a nonequilibrium process, which results in the explosive growth of ice crystals in tissues having a high hydration level. The large ice crystals cause a mechanical disruption of the protoplasmic structure and eventual death of the tissue. Nonequilibrium intracellular ice formation is rapid and usually results in death of the particular tissues involved.

Equilibrium freezing processes involve extracellular ice formation. There is a redistribution of water during equilibrium freezing. Water is drawn from within the cell to the extracellular regions because of the lower vapor pressure of the ice

Figure 7-14. Direct low temperature kill of a bentgrass-annual bluegrass fairway and tee. (Photo courtesy of E. Johanningsmeier, Detroit, Michigan).

compared to the liquid. Frost desiccation of the protoplasm occurs if this equilibrium process continues for a sufficient length of time. Accompanying frost desiccation is a contraction of the cell and the protoplasm. The protoplasm becomes
brittle under extreme dehydration and is subjected to tensions that may ultimately
result in mechanical damage to the protoplasm. Injury may also occur during the
thawing period when the cell walls expand more rapidly than the protoplasts, thus
creating tension. Equilibrium freezing injury usually occurs at quite low temperatures compared to nonequilibrium freezing.

Low Temperature Hardiness

Most cool season perennial turfgrasses have the ability to achieve a certain
level of low temperature hardiness. In this discussion hardiness is associated with
freezing stresses determined by the redistribution of water in contrast to chill
stresses that involve metabolic imbalances. Turfgrasses grown in warm temperatures are considered tender because kill occurs immediately upon freezing, with
ice growing into the protoplasts causing mechanical destruction. Many turfgrasses
are capable of becoming hardy through internal metabolic changes when grown
under low temperature conditions. The
main feature that distinguishes a hardy
from a tender tissue is the ease with
which ice penetrates into the protoplasts.
Many types of physiological and morphological responses are correlated with
the degree of low temperature hardiness
including cell size, carbohydrate level,
types of proteins, and hydration level.

Undifferentiated tissues in the dense
crown are associated with increased low
temperature hardiness. Tissues consisting of small, closely packed, thin-walled
cells have a higher percentage of the
total water content contained within the
protoplasm. The upper portion of the
annual bluegrass crown is more low
temperature hardy than the lower portion (30). Thus, the lower meristematic
area responsible for initiating roots is
injured more readily than the upper
portion (Fig. 7-15).

Certain variations in morphological
development occur during the hardening
process. The plants become darker green,
smaller in size, reduced in leaf area,

Figure 7-15. Longitudinal cross section of the
crown of an annual bluegrass plant showing the
differential direct low temperature kill of the
lower portion of the crown and the freedom
from damage of the upper portion (After Beard
and Olien—30).

less succulent, and more prostrate in growth habit (170). Various plant tissues differ in the level of hardiness that can be achieved. Tissues composed of closely packed, thin-walled cells generally have a higher hardiness level. The cell size is further reduced during the hardening process. Dehydrated seeds are extremely low temperature hardy, but these seeds are readily killed by freezing when in a hydrated state. Young turfgrass seedlings are quite prone to low temperature kill until the four-leaf stage is reached (12, 14, 220). Seedings should be made early enough in the autumn for seedlings to mature and achieve an acceptable degree of low temperature hardiness for winter survival.

The leaves and roots are generally more sensitive to low temperature injury than the stems. Young leaves are more low temperature hardy than old leaves or the leaf apex (14). The low temperature kill of turfgrass leaves is not critical. The turf can readily recover during the optimum growing conditions of early spring as long as the meristematic regions of the crown are not injured. Serious, permanent loss of turfs can occur when the meristematic tissues are injured (30, 212).

Actively growing plants generally have a minimum level of low temperature hardiness. Shoot growth of the turfgrasses slows and eventually ceases as the autumn soil temperatures decrease below 45°F. Carbohydrate accumulation occurs during this period of minimal shoot growth. Enzymes convert the insoluble carbohydrates to soluble sugars that accumulate in the vacuole and cause an increase in the osmotic potential (48, 108). Changes in the protoplasmic proteins result in an increased capability to bind water (47, 67). The net result is a very significant reduction in the water content of the protoplasm that enables the tissues to achieve a maximum level of low temperature hardiness (26, 48). A 3 to 4 week period of soil and air temperatures slightly above freezing is required for cool season turfgrasses to achieve maximum low temperature hardiness.

The degree of low temperature hardiness varies through the winter period (26). Maximum hardiness generally occurs during early winter followed by a slight decrease in hardiness during February. A drastic reduction in low temperature hardiness is evident in late winter (Fig. 7-16). The crown tissue hydration level is inversely correlated with the seasonal variation in low temperature hardiness. A majority of the direct low temperature kill to turfs occurs during late winter and early spring when the hardiness level is lowest. The thawing of snow during this period frequently results in standing water and increased crown hydration levels. This condition further accentuates the proneness to low temperature kill. Also, prematurely warm temperatures can initiate early spring growth. Serious injury can occur if this warm period is followed by a rapid decrease in temperatures to below 20°F. The warm season turfgrasses, such as bermudagrass and zoysiagrass, are particularly prone to low temperature kill during the late winter-early spring period. Low temperature kill may also occur in certain regions during a midwinter thaw. Injury is likely to occur if the thaw period and associated standing water is immediately followed by a sudden decrease in temperature to below 20°F.

Environmental influences. The degree of low temperature hardiness achieved can be modified by the environment. Environmentally, one of the first prerequisites for

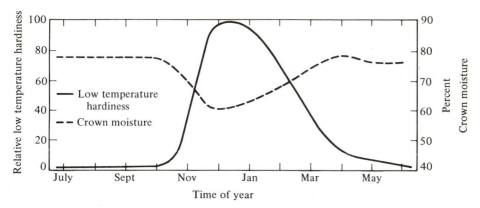

Figure 7-16. Variations in the low temperature hardiness and crown tissue hydration level of Toronto creeping bentgrass during a typical winter season at East Lansing, Michigan.

inducing hardiness is a low average daily air and soil temperature. Temperatures between 34 and 40°F are optimum for the hardening process (48). The second critical factor is light intensity. Hardiness cannot be achieved in the absence of light. This suggests that light is required during hardening for utilization in the photosynthetic process. Further evidence for the role of photosynthesis is that plants grown in carbon dioxide-free air are not capable of achieving normal low temperature hardiness.

Cultural influences. Any cultural practice that stimulates growth also reduces the low temperature hardiness level. Actively growing plants lack hardiness because the growth processes reduce the level of carbohydrate accumulation and increase the hydration of the protoplasm. The latter effect is particularly significant. Cultural practices that decrease the low temperature hardiness level include (a) a failure to provide surface and subsurface drainage, (b) nitrogen applied at excessive rates or in late fall, (c) inadequate potassium application rates, (d) excessive late fall irrigation, (e) lack of thatch control, and (f) a close cutting height.

Low temperature hardiness of turfgrasses. The low temperature hardiness of turfgrass species and cultivars varies greatly (12, 14, 26, 36, 47, 108, 151, 185, 212, 226). The relative low temperature hardiness of twenty turfgrasses is presented in Table 7-6. Differences in low temperature hardiness are also evident among cultivars (26, 226). For example, Nugget and Merion Kentucky bluegrass are quite low temperature hardy. They go off color and cease growth early in the fall and thus are able to achieve a maximum level of hardiness. Newport is much less low temperature hardy. It continues to grow late into the fall and has a much shorter period of time to achieve an acceptable level of hardiness.

The absolute temperature at which a particular cultivar is killed varies depending on the (a) hardiness level of the plant, (b) freezing rate, (c) thawing rate, (d) number of times frozen, (e) length of time frozen, and (f) post-thawing treatment.

Table 7-6

THE RELATIVE LOW TEMPERATURE HARDINESS
OF TWENTY AUTUMN HARDENED TURFGRASSES

Low temperature hardiness ranking	Turfgrass species
Excellent	Rough bluegrass
	Creeping bentgrass
Good	Timothy
	Kentucky bluegrass
	Canada bluegrass
	Colonial bentgrass
	Redtop
Medium	Annual bluegrass
	Red fescue
	Tall fescue
	Meadow fescue
	Zoysiagrass
Poor	Perennial ryegrass
	Manilagrass
	Bermudagrass
Very poor	Italian ryegrass
	Bahiagrass
	Centipedegrass
	Carpetgrass
	St. Augustinegrass

The more rapid the freezing or thawing, the higher the temperature at which kill is observed. Also, the greater the number of times frozen or the longer the tissue is frozen, the higher the temperature at which kill occurs (212). The post-thawing treatment has an important effect on the ability of the plant to recover from partial injury. A tissue can be partially injured or certain portions of the tissue killed during nonequilibrium freezing processes. The chance of recovering from partial injury is greatly reduced if (a) immediately exposed to high temperatures that stimulate growth and transpiration or (b) the plant is forced into growth by an excessive, early spring nitrogen fertilization.

The soil temperature is much more critical than the air temperature in direct low temperature kill (47, 85, 179). Turfs can survive comparatively low air temperatures as long as the soil temperature remains above the lethal point. The vital crown tissues are protected by the surrounding and underlying warmer soil temperatures.

Direct low temperature kill symptoms can be subdivided into two characteristic types. One involves immediate kill of the entire turfgrass plant. In this situation, the turf is completely dead at the time of spring thaw. This type of low temperature kill is most common in depressional areas and other poorly drained, compacted sites where water concentrates.

The second type of low temperature kill involves partial injury to the lower portion of the turfgrass crown that is responsible for initiating roots (30, 212). The freezing injury is usually most evident in the vicinity of the xylem vessels.

The leaves, stems, and upper portion of the crown may appear healthy at the time of spring thaw. However, the advent of warmer temperatures promotes growth and transpiration of the leaf tissue as well as degeneration of the injured root and lower crown tissues. No water absorption capability exists if the lower portion of the crown and root system has been severely injured. An internal water deficit develops and the aboveground plant tissues die due to atmospheric desiccation. Desiccation is a secondary cause of death, however, with the primary cause being the differential, low temperature injury to the lower portion of the crown.

Prevention of Low Temperature Kill

A number of cultural practices can be utilized to minimize direct low temperature kill of turfgrasses. A basic consideration is the selection of low temperature hardy species. There are also differentials in low temperature hardiness between cultivars that should be utilized when possible.

Cultural factors. Cultural practices should ensure that the turf is healthy, disease-free, and well rooted as the winter season approaches. The key principle in minimizing direct low temperature kill involves maintaining a low crown hydration level.

Compacted or fine textured soils should be avoided or corrected in order to minimize conditions where water can accumulate. On intensely used turfs where low temperature injury can be critical, it is important to utilize coarse textured, well-drained soils that have favorable infiltration and percolation rates. A properly designed and installed tile drainage system may be required to remove the excess water.

Adequate surface drainage and open catch basins are particularly important since internal drainage is frequently lacking during extended winter periods when the frozen soil is impervious to water. Surface contours should be designed to ensure maximum possible runoff of water from rains as well as from snow and ice thaws. Water accumulation may also occur behind dams of ice or snow. Drainage channels should be cut through the dams in such situations to provide an outlet for the melted ice and snow.

Late fall cultivation is sometimes practiced on imperfectly drained soils in order to enhance drainage of water away from the turfgrass crowns. This practice has proved beneficial under certain situations. However, severe turfgrass injury due to winter desiccation may occur in the vicinity of the coring holes during open winters of minimal snow cover.

The nutritional level is another factor influencing low temperature hardiness. There is a minimum level of nitrogen nutrition needed to ensure adequate plant vigor and metabolic activity so that the normal hardening process is not limited. Any increases above this nitrogen level cause a reduction in hardiness (1, 7, 14, 31, 47, 48, 57, 133, 151, 212). Nitrogen stimulates growth causing increased hydration of the tissues and a decrease in the carbohydrate level (48). The result is decreased low temperature hardiness. The inhibitory effects of high nitrogen nutrition on low temperature hardiness can result from (a) excessive nitrogen application(s) or (b) a late nitrogen application made during the hardening period (48).

Adequate levels of potassium and phosphorus are important in ensuring maximum low temperature hardiness (1, 31). The balance among nutrients is most critical (1, 31). It is not just a low nitrogen level or a high potassium level that is critical in maximum low temperature hardiness but the interrelationship between the two nutrients (Fig. 7-17). A nitrogen-to-potassium ratio in the order of 2 to 1 or 3 to 1 results in maximum hardiness for Kentucky bluegrass (31).

The autumn irrigation practices can be a factor affecting low temperature hardiness. Irrigation rates should be adjusted to maintain adequate levels of soil moisture in order to minimize the chance of winter soil drought. However, excessive irrigation rates that stimulate leaf growth and increase the hydration level should be avoided (7). Waterlogged soil conditions should also be avoided if at all possible.

Cultural practices should be adjusted to minimize thatch accumulation and thus reduce the proneness to low temperature kill. The presence of thatch tends to elevate the crown meristematic area above the soil and subjects the crown to the greater atmospheric temperature extremes.

Figure 7-17. The influence of 4 pounds (left) and 8 pounds (right) of nitrogen per 1,000 square feet per year and five potassium levels on the low temperature hardiness of Kentucky bluegrass exposed to a soil temperature of 0°F.

The mowing practices can be a contributing factor in minimizing direct low temperature kill (Fig. 7-18). Shorter cutting heights result in increased low temperature injury of certain species (6, 7, 31, 151, 212). Warm season turfgrasses, such as bermudagrass and zoysiagrass, are particularly prone to low temperature kill at lower cutting heights. Where winter use is minimal, the cutting height of bermudagrass is sometimes elevated to 1.5 in. to minimize low temperature injury. Two mechanisms may contribute to a reduction in low temperature kill at higher cutting heights. One involves the increased leaf area available for carbohydrate synthesis. This results in a greater capability to accumulate carbohydrates, which is important in low temperature hardiness. The second is the greater quantity of plant material, which increases the insulating effect against extremely low temperatures.

Certain growth regulators affect winter shoot growth and low temperature hardiness. During winters when there are prolonged warm periods followed by repeated freezing, an application of maleic hydrazide has effectively reduced low temperature injury to certain warm season turfgrasses including bermudagrass and bahiagrass (187).

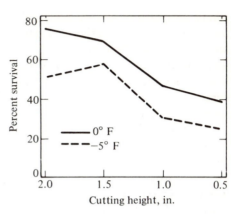

Figure 7-18. The effect of cutting height on the low temperature survival of Kentucky bluegrass at soil temperatures of 0° and −5°F.

Traffic problems. Traffic on turfs during periods of low temperature stress can result in increased injury. Two specific types of winter traffic damage commonly occur. One results from human or vehicular traffic over the turf during periods when the leaf and stem tissues are frozen. The injury mechanism involves disruption of the brittle protoplasm caused by the pressure of the traffic on the rigid frozen tissues. Leaf injury does not result in permanent damage to the turf since new leaves are readily initiated from the turfgrass crowns. This damage can be avoided by (a) withholding or diverting traffic from turfgrass areas during periods when the leaf tissue is frozen or (b) applying a light application of water in the early morning to thaw the frozen turfgrass tissues. The latter practice is most effective when the soil is not frozen and the air temperature is above freezing.

The second type of damage involves traffic on turfs that are covered with a wet slush (23). A rapid temperature decrease to below 20°F following traffic over slush covered turfs can cause serious injury to the leaves, crowns, rhizomes, and stolons of certain turfgrass species (Fig. 7-19). Damage is frequently extensive enough to negate the recuperative potential. The exact mechanism of kill has not been clarified. Traffic should be withheld whenever a wet slushy condition exists on turfgrass areas. This practice is especially important when temperatures below 20°F are anticipated.

Low temperature protection. A snow cover has excellent insulating properties for protecting turfs against low temperature extremes because of an extremely low thermal conductivity (64, 84, 85, 226). The snow cover actually serves a threefold function by (a) protecting against extremes in low temperature, (b) moderating the frequency of freezing and thawing, and (c) protecting against

Figure 7-19. Total kill of Kentucky bluegrass leaves, crowns, and rhizomes resulting from 4 pounds per square inch of foot traffic distributed uniformly throughout the plot area when an inch of wet slush was present and after which a decrease in temperature to below 20°F occurred. (After Beard—23).

winter desiccation. A snow fence or brush can be utilized to enhance snow accumulation on critical turfgrass areas such as greens and tees. Periodic inspections and adjustment of the brush are necessary during the winter period to ensure a uniform snow cover. The snow cover should be 6 in. or more in depth for adequate low temperature protection. The insulating effect increases in proportion to the depth of the snow cover (226). The insulating effect of ice or of old, compacted snow is less than that of light, new fallen snow (226).

Organic mulches and synthetic winter protection covers, such as the viscose-rayon fiber cover, processed wood fiber cover, plastic screens, and polyethylene also serve a threefold function (139, 215, 216, 217). The viscose-rayon fiber and processed wood fiber covers provide superior low temperature insulation compared to the other synthetic covers. These two types of covers also provide early spring green-up of established turfs without an excessive amount of shoot growth that is produced under a polyethylene cover. Straw and similar organic mulches have been used on sports turfs as an antifrost measure (6). Straw mulching of bermudagrass greens is effective in protecting against low temperature kill (7, 79). The use of atmospheric methods of temperature control in low temperature protection of turfs, such as smudging, artificial atmospheric heating, or forced convection by wind machines to counteract temperature stratification, are not particularly effective or practical for turfgrass areas.

Soil warming. Artificial heating from below the surface of intensively cultured turf-grass areas, especially sports turfs, has proved economically and agronomically feasible (6, 15, 16, 17, 63, 81, 82, 113, 130, 134, 146, 148). The benefits of turfgrass soil warming include (a) protection against frost and frozen soil, (b) controlled snow removal, (c) good winter color, (d) enhancement of root growth, and (e) promotion of shoot growth. Soil warming has been particularly valuable on outdoor sports turfs used for football, soccer, baseball, horse racing, and golf due to the improved playing conditions provided during the normally dormant late fall and winter periods when soils are frozen.

A major benefit of soil warming is protection against the freezing of turfs and soils. Frost protection results in superior footing for outdoor sports activities and a drier playing surface. Rapid, internal drainage of water through the unfrozen soil is achieved during periods of precipitation or snow thaw (64, 81, 82). The immediate removal of excess water minimizes soil compaction, ensures better footing for sports activities, and prevents the accumulation of water on the surface of the playing field that results in muddy, undesirable playing conditions.

The removal of moderate snow accumulations through controlled thawing can also be accomplished by soil warming (63, 82, 113, 134). Mechanical snow removal is required when heavy, rapid accumulations of snow occur (64). Snow thawing by soil warming is advantageous since it occurs at a controlled, relatively slow rate that permits the drainage of water from the turf as rapidly as it is formed (Fig. 7-20).

A green, winter color can be maintained during a normally dormant period through soil warming. Adequate color of cool season turfgrasses has been achieved

Figure 7-20. The effect of soil warming in melting a snow accumulation on a Kentucky bluegrass turf (Photo courtesy of W. H. Daniel, Purdue University, West Lafayette, Indiana).

even under intermittent periods of snow cover (64). Soil warming is also effective in maintaining the green color of warm season turfgrasses if mowed at 0.5 in. or less (15, 146, 148). In addition, soil warming protects against direct low temperature kill of certain low temperature-sensitive turfgrass species.

A significant amount of root growth of cool season turfgrasses occurs during the winter period if the root zone temperature is maintained above 35°F. Thus the winter desiccation problem is reduced since the active root system is capable of absorbing water at all times. Higher levels of soil warming have actually resulted in leaf growth under relatively severe midwinter conditions. Mowing has been required on Kentucky bluegrass turfs maintained at soil temperatures above 50°F during midwinter (16). The costs for this level of soil warming become a limiting factor.

The primary objectives of soil warming are to protect against frost and to maintain a green color. These can be achieved at soil temperatures in the range of 35 to 40°F for cool season turfgrasses. Soil temperatures must be maintained in the range of 60 to 65°F to retain the color of warm season turfgrasses (15, 146, 148). An additional benefit of soil warming is the capability to germinate seed or to stimulate sod rooting earlier in the spring (6, 63, 64, 81, 130). Worn, thinned football fields can be renovated and brought up to an acceptable level of turfgrass quality prior to the spring baseball season through soil warming.

The exact design of the heating installation depends on the size, climate, proposed use, turfgrass species, and the available power. Various methods of subsurface soil warming have been investigated including (a) electric cables, (b) heated air ducts, and (c) hot water heating through pipes. Electric heating cables have proved most practical and effective under a wide variety of conditions. Nylon insulated, polyvinyl chloride cable is preferred due to the superior tolerance to deterioration in the soil environment (17, 64, 146). The recommended cable spacing in the soil is approximately 6 to 12 in., while the most practical depth is approximately 6 to 8 in. (6, 17, 63, 64, 81, 146, 148). A 4-in. depth provides effective, rapid surface warming with the additional advantage that the quantity of soil to be heated is smaller. However, a 6-in. minimum depth is necessary to prevent damage to the cables from mechanical operations such as turfgrass cultivation. An electrical load of 5 to 10 watts per sq ft has proved most satisfactory (6, 17, 81, 82, 113, 148). The heating requirement is greater with thin, short cut turfs than with dense, high cut turfs due to the reduced insulating effect (6, 82). The heating cables can be installed into existing turfgrass areas from the surface using specialized tractor mounted equipment. Soil warming is jointly controlled by air and soil thermostats. Shaded air thermostats located 5 ft above the turf provide advanced warning of rapid decreases in temperature, thus avoiding a lag in soil warming (146). Time clocks can be used to permit heating only during nocturnal periods when electrical costs are lower. Continuous, thermostatically controlled soil warming is preferred over intermittent warming during very cold weather or just prior to use of the turf (81).

Soil warming influences the incidence of disease on turfs during the winter period. Serious injury to turfs from disease can occur if temperatures are maintained in a range that favors disease activity (15, 134). On the other hand, temperatures held at 32°F resulted in no injury to bentgrass from a low temperature Basidiomycete or *Fusarium* patch (134).

A tarp or polyethylene cover used in conjunction with soil warming reduces the heat loss, decreases the cost of soil warming, and protects against winter desiccation (6, 63, 64, 113). However, the use of covers (a) increases labor, (b) complicates temperature control due to the possibility of a lethal heat buildup under the cover during periods of high light intensity, and (c) increases the disease problems.

Winter Dormancy

Figure 7-21. A brown, dormant bermudagrass turf (left) adjacent to a green Kentucky bluegrass turf (right). January 20 in Manhattan, Kansas (Photo courtesy of R. Keen, Kansas State University, Manhattan, Kansas).

Winter dormancy involves the cessation of growth and death of the leaf tissue of certain perennial turfgrasses due to a combination of low soil temperatures and a high light intensity. The affected turfgrass species are not capable of growth during this dormant period. The turf appears brown because there is no chlorophyll in the leaf tissue (Fig. 7-21). Most perennial, warm season turfgrasses enter winter dormancy when exposed to extended periods of low temperature. Typical examples are bermudagrass and zoysiagrass. In the spring these dormant warm season species break dormancy and initiate new root and shoot growth from the nodes of stolons, rhizomes, and crown meristematic areas when soil temperatures rise above 50°F.

Most cool season turfgrasses do not have a well-defined winter dormancy period comparable to the warm season turfgrasses. The former are capable of leaf growth at any time during the winter period when temperature, moisture, and light are favorable. Most of them retain substantial levels of chlorophyll in the leaf tissue if not injured by winter desiccation. These tissues can initiate photosynthesis during the winter period whenever sufficiently high atmospheric temperatures and light intensities occur.

Winter dormancy can be broken and spring green-up stimulated by the application of (a) nitrogen, (b) gibberellic acid, or (c) a combination of the two (109, 137, 138, 221, 222). Bluegrass and bermudagrass are the most responsive to low temperature growth stimulation caused by gibberellic acid. Bentgrass, red fescue, and tall fescue are moderately responsive; ryegrass is only slightly responsive; while zoysiagrass shows no response (221). The application of a soluble nitrogen fertilizer has proved effective where early spring green-up is desired. The chance of direct low temperature kill is increased, however, if the turf is subjected to temperatures of 20°F or less following this practice.

Clear polyethylene covers placed over the turf have also been utilized to break winter dormancy. A substantial buildup of heat occurs under the clear cover due to the greenhouse effect. The cover may have to be removed occassionally in order to avoid injury from lethal high temperatures.

Colorants. Turfgrass colorants are sometimes applied to winter dormant, warm season turfgrass species that lack chlorophyll. Colorants have been used as an alternate to winter overseeding. The result is an acceptable green color with no injury to the turfgrasses if properly applied (168, 209, 223). The colorants are of two basic types: (a) dyes and (b) pigments. Many of the dyes have a relatively short effective life because they are prone to fading in the sun and washing. Properly formulated emulsifiable pigments usually persist throughout the winter period until leaf growth is initiated and the painted leaves removed by mowing (168, 209, 223).

The turf should be completely dormant at the time the colorant is applied. The dormant turf should be mowed, clippings removed, and the turf thoroughly raked prior to the application. The application should be made when air temperatures are above 40°F and free water is absent from the leaf tissue. The shade of green obtained is governed by the number of applications. Colored dormant turfs should be protected from winter desiccation injury by watering as needed. Dormant warm season turfs that have received a colorant green-up earlier in the spring (168, 223). The earlier spring growth is attributed to increased light absorption and the resultant warmer temperatures.

Ice Covers

Winter injury of turfs is frequently associated with an ice cover. The ice cover may form (a) as a result of water accumulating in depressional areas, (b) during a sleet storm when a uniform cover of ice builds up over an extensive area, or

(c) from a snow cover during periods of thawing and freezing. The ice layer formed during a sleet storm may accumulate to a thickness of 2 in. or more.

Direct ice cover injury is due to the ice functioning as a barrier to the exchange of gases between the turfgrass tissue and the atmosphere. Two mechanisms are hypothesized for direct ice cover kill. One hypothesis is that kill results from oxygen suffocation caused by an exhaustion of the oxygen supply required for the respiration process. A second hypothesis involves the accumulation of toxic gases such as carbon dioxide and cyanide that have evolved from the respiration or oxidation of living tissues, thatch, soil organic matter, soil organisms, or certain fungi.

The accumulation of toxic gases is the more likely cause of direct ice cover injury to turfs (93). No conclusive evidence is available to support the oxygen suffocation theory. Certain low temperature Basidiomycetes can produce lethal levels of cyanide gas in the host tissues under an ice cover (135, 136). Kill of bentgrass by toxic accumulations of cyanide gas is an acknowledged cause of winter injury in the western Canadian provinces.

Detailed studies of ice cover effects as they typically occur in the United States have not shown them to be a major cause of winter injury (22, 23). The amount of injury caused by the direct effect of ice covers is minimal compared to winter desiccation, low temperature fungi activity, and direct low temperature kill (Fig. 7-22). There are numerous examples where the turf is flooded and covered with ice for

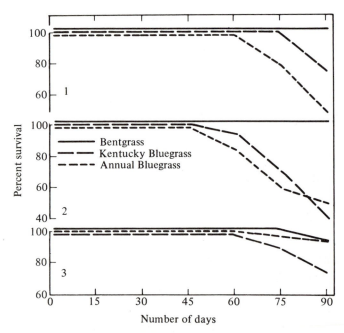

Figure 7-22. Percent survival of hardened Toronto creeping bentgrass, Kenblue Kentucky bluegrass, and annual bluegrass after being: 1. Flooded, then frozen and held at 25°F for intervals up to 90 days. 2. Frozen, then layered with ice and held at 25°F for intervals up to 90 days. 3. Frozen, then layered with ice over snow and held at 25°F for intervals up to 90 days. (After Beard—22).

skating rinks. Kentucky bluegrass or creeping bentgrass turfs are seldom injured by the ice cover if provisions are made for rapid drainage of the area during thaws.

Most perennial, cool season turfgrasses can tolerate continuous ice coverage for over 60 days without significant injury (22, 23). Winter conditions where an ice cover is in place for more than 60 days are quite rare in the United States (Fig. 7-23). Toronto creeping bentgrass has survived for 150 days within a solid ice block. Creeping bentgrass is quite resistant to direct ice cover injury. In contrast, annual bluegrass can be significantly injured when an ice cover remains in place longer than 75 days. Kentucky bluegrass is intermediate in ice cover tolerance between creeping bentgrass and annual bluegrass. Differences in tolerance to extended periods of ice coverage are also evident among cultivars. The seeded creeping bentgrasses, especially Seaside, are less tolerant of ice coverage than the vegetatively established creeping bentgrasses such as Cohansey and Toronto (24). Astoria colonial bentgrass is much less tolerant of extended ice coverage than the creeping bentgrasses.

The type of ice affects the degree to which the ice cover impairs gaseous diffusion. Ice of high density, sometimes termed "clear ice," impairs gaseous diffusion much more than low density ice, called "white or milk ice." The degree of potential injury is significantly reduced when a snow layer is present between the ice cover and the turf (22, 23).

Although turfgrass injury caused by the direct effects of ice coverage is minimal, much of the low temperature kill is associated with an ice cover. These effects are indirect in nature. For example, water may not be removed from the turf area during a thaw due to a lack of surface or internal soil drainage. The turfgrass meristematic tissues could stand in water for an extended period of time. A substantial increase in the hydration level of the submerged crowns would occur (22). The potential for direct low temperature kill.is increased substantially in this situation should a rapid decrease in temperature to below 20°F occur.

Figure 7-23. Survival of Kentucky bluegrass (left) and Toronto creeping bentgrass (right) after 51 days of direct coverage with an ice sheet of 2 inches in depth. April 20, 1963, East Lansing, Michigan (After Beard—23).

Ice and snow removal. The removal of ice and snow covers is beneficial under certain conditions. The ice cover should be removed from annual bluegrass turfs after 50 to 60 days of ice coverage. Most creeping bentgrasses can tolerate over 100 days of ice coverage. When direct damage from the ice cover is anticipated due to an extended period of coverage, the ice can be mechanically removed. Ice and snow removal should also be done just prior to an anticipated thaw if a potential crown hydration problem exists. All the ice and snow should be removed except the last 0.5 to 1 in. A major portion of the water is mechanically removed through this practice with minimum damage to the turf. The remaining ice or snow also provides protection against injury from winter desiccation and low temperature extremes (226).

Another method of ice removal involves enhancing the thawing process by (a) punching holes through the ice at 1- to 3-ft intervals; (b) an application of a black organic material such as soot, lampblack, or a fertilizer; or (c) a combination of the two. The black particles lying on the ice or snow absorb the incoming solar radiation. The resulting heat buildup causes more rapid melting of the ice adjacent to the particles. Snow thawing is also expedited by nontoxic materials that reduce the surface tension (129). A thick film of water tends to form around the snow particles during thawing. The resulting competition for energy needed in the evaporative process reduces the melting rate. Surface tension reducing materials prevent the formation of thick water films and thus stimulate water movement down through the snow. The black energy absorbing materials and the surface tension reducers are quite effective in snow melting when used in combination (129). Adequate surface and internal soil drainage is a necessity for the melting technique to be utilized successfully.

Frost Heaving

The upward lifting of plants from their normal position that exposes the root and crown tissues to atmospheric desiccation is termed **heaving**. Young turfgrass seedlings can be lifted completely above the soil surface (123, 124). It may be difficult for the young seedling to become rooted again if this occurs, because of inadequate soil contact. Thus, the exposed root and associated shoot die from atmospheric desiccation.

Specific prerequisites are necessary for frost heaving to occur. Surface soil temperatures must be below 32°F. The capillary pore size in the soil must be sufficiently large to permit a reasonable rate of water movement to the basal site of freezing. Finally, free water must be available in the soil for movement by molecular cohesion to the site of freezing. Frost heaving is most common (a) in fine textured soils, (b) in soils having a high water content, (c) during periods when the temperature fluctuates above and below freezing, and (d) when no snow cover exists (124).

The frost heaving mechanism involves the growth of ice formed from water that is drawn to the bottom of the ice crystal and not merely the expansion of water during freezing (41, 147, 204, 205, 206). A thin layer of ice develops at the soil surface during gradual temperature drops below freezing. Vertically oriented ice lenses or columns are formed as water freezes at the bottom of this layer. They continue to grow in length as water moves through the soil to the site of freezing. Additional layers of ice build up from below on successive nights of freezing. A frozen mass of segregated water lifts the first layer of ice higher and higher. A greater quantity of water is concentrated at the soil surface when the heaved ice thaws than was present before freezing. Thus more water is available for the next heaving cycle. Heaving continues as long as water is available and the heat released during freezing of the water escapes upward. Alternate freezing and thawing are not required for heaving to occur. Turfgrass stems that are frozen fast to the upper ice crust are gradually lifted out of the ground.

Injury caused by tissue breakage may also occur during frost heaving (123, 124). The breakage may involve leaf, stem, or root tissues with stem breakage frequently causing total plant kill (123). The position of injury affects the survival ability of the plant. The plant may produce adventitious roots and survive if breakage is below the shoot apex. The plant usually dies if the break is above the shoot apex.

Heaving injury is of no significance on established turfs, as far as permanent injury is concerned (69). It can result in extensive disruption of the soil surface, which may require rolling in early spring. Late fall seedings are quite prone to heaving damage, especially on newly seeded sod production fields established on organic soils (41, 42, 193). Young seedlings having a root system of 1 in. or less in depth can be lifted completely out of the ground by frost heaving (Fig. 7-24). Turfgrasses should be seeded sufficiently early in the autumn so that individual

Figure 7-24. The frost heaving kill of Kentucky bluegrass seedlings planted in late fall on an organic soil at East Lansing, Michigan. Viewed on May 15. (Photo courtesy of P. E. Rieke, Michigan State University, East Lansing, Michigan).

plants can become well rooted prior to the winter heaving period (12, 14, 220). A 1-in. planting depth results in more seedling injury than a 0.5-in. depth where injury is caused primarily by leaf or stem tissue breakage (123). However, root breakage and heaving of turfgrasses is greater at a shallow planting depth.

The degree of heaving depends on the (a) turfgrass species, (b) stage of plant growth, (c) soil water content, (d) soil type, (e) frequency of freezing and thawing, and (f) amount of surface insulation. Cultural practices that increase the number, size, and strength of the turfgrass roots result in increased tolerance to frost heaving. Proper drainage is important in minimizing frost heaving since water accumulations near the soil surface favor heaving (205). The soil type affects the size of soil pores and the rate of water movement.

Heaving damage can be reduced by applying a mulch cover to young turfgrass seedlings planted in late fall (37). Soil heaving is limited by the rate of heat loss from the surface. A mulch cover moderates soil temperature fluctuations above and below freezing and reduces the rate of heat loss from the soil surface. Practices that enhance snow accumulation also provide protection against frost heaving (226). When frost heaving of turfgrass seedlings occurs, rolling just after thawing of the soil surface tends to press the roots back into contact with the soil and thus reduces the chance of desiccation injury.

References

1. ADAMS, W. E., and M. TWERSKY. 1959. Effect of soil fertility on winter killing of Coastal bermudagrass. Agronomy Journal. 52: 325–326.

2. AIKMAN, J. M. 1940. The effect of aspect of slope on climatic factors. Iowa State College Journal of Science. 15: 161–167.

3. ALBERDA, T. 1957. The effects of cutting, light intensity and night temperature on growth and soluble carbohydrate content of Lolium perenne L. Plant and Soil. 8: 199–230.

4. ———. 1965. The influence of temperature, light intensity and nitrate concentration on dry matter production and chemical composition of Lolium perenne L. Netherlands Journal of Agricultural Science. 13(4): 335–360.

5. ———. 1965. The problems of relating greenhouse and controlled environmental work to sward conditions. Journal of the British Grassland Society. 20: 41–48.

6. ANONYMOUS. 1951. Electrical soil warming as an anti-frost measure for sports turf. Journal of the Sports Turf Research Institute. 8(27): 25–44.

7. ANONYMOUS. 1952. Factors affecting winter survival of bermuda grass. Golf Course Reporter. 20(4): 7.

8. ANONYMOUS. 1960. Rules for testing seeds. Proceedings of the Association of Official Seed Analysts. 49(2): 1–71.

9. ANONYMOUS. 1963. Rotavators to remove ice. Parks, Golf Courses and Sports Grounds. 28(8): 588.

10. ANONYMOUS. 1968. Manage fertilizer for more heat tolerant bent. Weeds, Trees, and Turf. 7: 35–38.

11. ANSARI, A. Q., and W. E. LOOMIS. 1959. Leaf temperatures. American Journal of Botany. 46: 713–717.

12. ARAKERI, H. R., and A. R. SCHMID. 1949. Cold resistance of various legumes and grasses in early stages of growth. Agronomy Journal. 41: 182–185.

13. BAKER, B. S., and G. A. JUNG. 1968. Effect of environmental conditions on the growth of four perennial grasses. I. Response to controlled temperature. Agronomy Journal. 60: 155–158.

14. BAKER, H. K., and G. L. DAVID. 1963. Winter damage to grass. Agriculture. 70(8): 380–382.

15. BALTENSPERGER, A. A. 1962. Reduce dormancy of bermudagrass by soil heating. 1962 Report of Turfgrass Research, Arizona Agricultural Experiment Station Report 212. pp. 18–22.

16. BARRETT, J. R., and W. H. DANIEL. 1965. Electrically warmed soils for sports turfs—second prograss report. Midwest Turf Research Report. 33: 1–6.

17. ———. 1966. Turf heating with electric cable. Agricultural Engineering. 47(10): 526–529.

18. BASS, L. N. 1953. Relationship of temperature, time and moisture content to the viability of seeds of Kentucky bluegrass. Proceedings of the Iowa Academy of Science. 60: 86–88.

19. BAUM, W. A. 1949. On the relation between mean temperature and height in the layer of air near the ground. Ecology. 30: 104–107.

20. BEARD, J. B. 1959. The growth and development of *Agrostis palustris* roots as influenced by certain environmental factors. M. S. Thesis. Purdue University. pp. 1–75.

21. ———. 1961. The independent and multiple contribution of certain environmental factors on the seasonal variation in amide nitrogen fractions of grasses. Ph.D. Thesis. Purdue University. pp. 1–116.

22. ———. 1964. Effects of ice, snow and water covers on Kentucky bluegrass, annual bluegrass, and creeping bentgrass. Crop Science. 4: 638–640.

23. ———. 1965. Effects of ice covers in the field on two perennial grasses. Crop Science. 5: 139–140.

24. ———. 1965. Bentgrass (*Agrostis* spp.) varietal tolerance to ice cover injury. Agronomy Journal. 57: 513.

25. ———. 1966. Winter injury. Golf Superintendent. 34(1): 24–30.

26. ———. 1966. Direct low temperature injury of nineteen turfgrasses. Quarterly Bulletin of the Michigan Agricultural Experiment Station. 48(3): 377–383.

27. BEARD, J. B., and W. H. DANIEL. 1965. Effect of temperature and cutting on the growth of creeping bentgrass (*Agrostis palustris* Huds.) roots. Agronomy Journal. 57: 249–250.

28. ———. 1966. Relationship of creeping bentgrass (*Agrostis palustris* Huds.) root growth to environmental factors in the field. Agronomy Journal. 58: 337–339.

29. ———. 1967. Variations in the total, nonprotein, and amide nitrogen fractions of *Agrostis palustris* Huds. leaves in relation to certain environmental factors. Crop Science. 7: 111–115.

30. BEARD, J. B., and C. R. OLIEN. 1963. Low temperature injury in the lower portion of *Poa annua* L. crowns. Crop Science. 3: 362–363.

31. BEARD, J. B., and P. E. RIEKE. 1966. The influence of nitrogen, potassium and cutting height on the low temperature survival of grasses. 1966 Agronomy Abstracts. p. 34.

32. BENEDICT, H. M. 1940. Effect of day length and temperature on the flowering and growth of four species of grasses. Journal of Agricultural Research. 61: 661–671.

33. BENNETT, F. T. 1933. *Fusarium* species on British cereals. Annals of Applied Biology. 20: 272–290.

34. ———. 1935. *Corticium* disease of turf. Journal of the Board of Greenskeeping Research. 4(12): 32–39.

35. ———. 1937. Dollarspot disease of turf and its causal organism, *Sclerotinia homoeocarpa* N. SP. Annals of Applied Biology. 24: 236–257.

36. BERGGREN, F. 1952. MRTF turf researcher freeze-tests zoysia. Midwest Turf News and Research. 6(3): 3–4.

37. BISWELL, H. H., A. M. SCHULTZ, D. W. HEDRICK, and J. I. MALLORY. 1953. Frost heaving of grass and brush seedlings on burned chamise brushlands in California. Journal of Range Management. 6: 172–180.

38. BOMMER, D. 1961. Samen-vernalisation perennierender Graserarten. Zeitschrift Für Planzeuzuchtung. 46: 105–111.

39. BOSSHART, R. P. 1967. The effects of soil moisture, mulch, slope-facing, and surface temperature on grass seedlings. M.S. Thesis. Virginia Polytechnic Institute. pp. 1–68.

40. BOUYOUCOS, G. J. 1916. Soil temperature. Michigan Agricultural Experiment Station Technical Bulletin No. 26. pp. 1–133.

41. BOUYOUCOS, G. J., and M. M. McCOOL. 1928. The correct explanation for the heaving of soils, plants and pavements. Journal of the American Society of Agronomy. 20: 480–491.

42. BRINK, V. C., J. R. MACKAY, S. FREYMAN, and D. G. PEARCE. 1967. Needle ice and seedling establishment in southwestern British Columbia. Canadian Journal of Plant Science. 47: 135–139.

43. BROWN, E. 1902. Germination of Kentucky blue grass. Proceedings of the 15th Annual Convention of the Association of American Agricultural Colleges and Experiment Stations. USDA Bulletin No. 115. pp. 105–110.

44. BROWN, M. E. 1939. Some effects of temperature on the growth and chemical composition of certain pasture grasses. Missouri Agricultural Experiment Station Research Bulletin 299. pp. 1–76.

45. ———. 1943. Seasonal variations in the growth and chemical composition of Kentucky bluegrass. Missouri Agricultural Experiment Station Research Bulletin 360. pp. 1–56.

46. BURTON, G. W. 1944. Seed production of several southern grasses as influenced by burning and fertilization. Journal of the American Society of Agronomy. 36: 523–529.

47. CARROLL, J. C. 1943. Effects of drought, temperature, and nitrogen on turf grasses. Plant Physiology. 18: 19–36.

48. CARROLL, J. C., and F. A. WELTON. 1939. Effect of heavy and late applications of nitrogenous fertilizer on the cold resistance of Kentucky bluegrass. Plant Physiology. 14: 297–308.

49. CHAMPNESS, S. S. 1950. Effect of microclimate on the establishment of timothy grass. Nature. 165: 325.

50. COOPER, J. P. 1951. Studies on growth and development in *Lolium*. II. Pattern of bud development of the shoot apex and its ecological significance. Journal of Ecology. 39: 228–270.

51. ———. 1952. Studies on growth and development of *Lolium*. III. Influence of season and latitude on ear emergence. Journal of Ecology. 40: 352–379.

52. ———. 1956. Developmental analysis of populations in the cereals and herbage grasses. I. Methods and techniques. Journal of Agricultural Science. 47: 262–279.

53. ———. 1957. Developmental analysis of populations in the cereals and herbage grasses. II. Response to low-temperature vernalization. Journal of Agricultural Science. 49: 361–383.

54. ———. 1960. Short-day and low-temperature induction in *Lolium*. Annals of Botany N.S. 24: 232–246.

55. COOPER, J. P., and D. M. CALDER. 1962. Flowering response of herbage grasses. Report of the Welsh Plant Breeding Station for 1961. pp. 20–22.

56. ———. 1964. The inductive requirement for flowering of some temperate grasses. Journal of the British Grassland Society. 19: 6–14.

57. CORDUKES, W. E., J. WILNER, and V. T. ROTHWELL. 1966. The evaluation of cold and drought stress of turfgrasses by electrolytic and ninhydrin methods. Canadian Journal of Plant Science. 46: 337–342.

58. CRAIG, R. B. 1966. Air-conditioning fairways. 1966 Midwest Regional Turf Conference Proceedings. pp. 33–34.

59. CULLIMAN, B. 1941. Germinating seeds of southern grasses. Proceedings of the Association of Official Seed Analysts. pp. 74–76.

60. DAHL, A. S. 1933. Effect of temperature on brown patch of turf. Phytopathology. 23: 8.

61. ———. 1933. Effect of temperature and moisture on occurrence of brownpatch. Bulletin of USGA Green Section. 13(3): 53–61.

62. ———. 1934. Snowmold of turf grasses as caused by *Fusarium nivale*. Phytopathology. 24: 197–214.

63. DANIEL, W. H., and J. R. BARRETT. 1966. Electricity warms soils for sport turf. Weeds, Trees, and Turf. 5(2): 14–16.

64. DANIEL, W. H., J. R. BARRETT, and L. H. COOMBS. 1964. Electrically warmed soils for sports turfs—a progress report. Midwest Turf News and Research. 28: 1–6.

65. DARROW, R. A. 1939. Effects of soil temperature, pH, and nitrogen nutrition on the development of *Poa pratensis*. Botanical Gazette. 101: 109–127.

66. DAUBENMIRE, R. F. 1943. Temperature gradients near the soil surface with reference to techniques of measurement in forest ecology. Journal of Forestry. 41: 601–603.

67. DAVIS, D. L., and W. B. GILBERT. 1967. Changes in the soluble protein fractions during cold acclimation of bermudagrass. 1967 Agronomy Abstracts. p. 50.

68. DAVIS, W. H. 1924. Spore germination of *Ustilago striaeformis*. Phytopathology. 14: 251–267.

69. DECKER, A. M., and T. S. RONNINGEN. 1957. Heaving in forage stands and in bare ground. Agronomy Journal. 49: 412–415.

70. DENMEAD, O. T. 1964. Evaporation sources and apparent diffusivities in a forest canopy. Journal of Applied Meteorology. 3: 383–389.

71. DE VRIES, D. A., and J. W. BIRCH. 1961. The modification of climate near the ground by irrigation for pastures on the Riverine plain. Australian Journal of Agricultural Research. 12: 260–272.

72. DICKINSON, L. S. 1930. The effect of air temperature on the pathogenicity of *Rhizoctonia solani* parasitizing grasses on putting-green turf. Phytopathology. 20(8): 597–608.

73. DREIBELBIS, F. R. 1950. A summary of data on soil and air temperatures at the North Appalachian experimental watershed, Coshocton, Ohio. Proceedings of the Soil Science Society of America. 15: 394–399.

74. DUDECK, A. E., N. P. SWANSON, and A. R. DEDRICK. 1966. Protecting steep construction slopes against water erosion. II. Effect of selected mulches on seedling stand, soil temperature and soil moisture relations. 1966 Agronomy Abstracts. p. 38.

75. DUFF, D. T. 1967. Some effects of supraoptimal temperatures upon creeping bentgrass (*Agrostis palustris* Huds.). Ph.D. Thesis. Michigan State University. pp. 1–61.

76. DUFF, D. T., and J. B. BEARD. 1966. Effects of air movement and syringing on the microclimate of bentgrass turf. Agronomy Journal. 58: 495–497.

77. ENDO, R. M. 1961. Turfgrass diseases in southern California. Plant Disease Reporter. 45(11): 869–873.

78. ———. 1963. Influence of temperature on rate of growth of five fungus pathogens of turfgrass and on rate of disease spread. Phytopathology. 53: 857–861.

79. ENTRIKEN, H. L. 1925. Bermuda grass experiences at Enid, Oklahoma. Bulletin of the USGA Green Section. 5(11): 245.

80. ERWIN, L. E. 1941. Pathogenicity and control of *Corticium fuciforme* (Berk.) Wakef. Rhode Island Agricultural Experiment Station Bulletin 278. pp. 1–34.

81. ESCRITT, J. R. 1954. Electrical soil warming as an anti-frost measure for sports turf—a further report. Journal of the Sports Turf Research Institute. 8(30): 354–364.

82. ———. 1959. Electrical soil warming as an anti-frost measure for sports turf—a further report. Journal of the Sports Turf Research Institute. 10(35): 29–41.

83. EVANS, L. T. 1960. The influence of temperature on flowering in species of *Lolium* and in *Poa pratensis*. Journal of Agricultural Science. 54: 410–416.

84. FERGUSON, A. C. 1966. Winter injury north of the 49th. Golf Superintendent. 34(8): 38–39.

85. ———. 1967. Some observations on the winter injury in turfgrass experiments at the University of Manitoba. Summary of the 18th Annual RCGA National Turfgrass Conference. pp. 7–12.

86. FISCHER, J. A. 1967. An evaluation of high temperature effects on annual bluegrass (*Poa annua* L.). M.S. Thesis. Michigan State University. pp. 1–42.

87. FITTS, O. B. 1925. A preliminary study of root growth of five grasses under turf conditions. Bulletin of USGA Green Section. 5: 58–62.

88. FORBES, I. 1952. Chromosome numbers and hybrids in *Zoysia*. Agronomy Journal. 44: 194–199.

89. FORD, D. R. 1967. The influence of temperature and light intensity on the photosynthetic rate and chemical composition of *Festuca arundinacea* and *Dactylis glomerata*. M.S. Thesis. Purdue University. pp. 1–59.

90. FOY, N. R. 1934. Deterioration problems in New Zealand chewings fescue. New Zealand Journal of Agriculture. 49: 10–24.

91. FRAZIER, S. L. 1960. Turfgrass seedling development under measured environment and management conditions. M.S. Thesis. Purdue University. pp. 1–61.

92. FREEMAN, T. E. 1960. Effects of temperature on cottony blight of ryegrass. Phytopathology. 50: 575.

93. FREYMAN, S. 1967. The nature of ice-sheet injury to forage plants. Ph.D. Thesis. University of British Columbia. pp. 1–100.

94. FRISCHKNECHT, N. C. 1959. Effects of presowing vernalization on survival and development of several grasses. Journal of Range Management. 12: 280–286.

95. GANE, R. 1948. The effect of temperature, humidity and atmosphere on the viability of chewings's fescue grass seed in storage. Journal of Agricultural Science. 38: 90–92.

96. GARDNER, F. P., and W. E. LOOMIS. 1953. Floral induction and development in orchard grass. Plant Physiology. 28: 201–217.

97. GARMAN, H., and E. C. VAUGHN. 1916. The curing of blue-grass seeds as affecting their viability. Kentucky Agricultural Experiment Station Bulletin No. 198. pp. 27–40.

98. GASKIN, T. A. 1966. Testing for drought and heat resistance in Kentucky bluegrass. Agronomy Journal. 58: 461–462.

99. GATES, D. M. 1965. Heat transfer in plants. Scientific American. 213: 76–84.

100. GEIGER, R. 1950. The climate near the ground. Harvard University Press, Cambridge, Mass. pp. 1–611.

101. GRABER, L. F. 1926. Injury from burning off old grass on established bluegrass pastures. Journal of the American Society of Agronomy. 18: 815–819.

102. GREEN, D., and J. B. BEARD. 1969. Seasonal relationships between nitrogen nutrition and soluble carbohydrates in the leaves of *Agrostis palustris* Huds. and *Poa pratensis* L. Agronomy Journal. 61(1): 107–110.

103. HANSON, A. A., and F. V. JUSKA. 1961. Winter root activity in Kentucky bluegrass (*Poa pratensis* L.). Agronomy Journal. 53: 372–374.

104. HARRINGTON, G. T. 1923. Use of alternating temperatures in the germination of seeds. Journal of Agricultural Research. 23(5): 295–332.

105. HARRISON, C. M. 1934. Responses of Kentucky bluegrass to variations in temperature, light, cutting, and fertilizing. Plant Physiology. 9: 83–106.

106. HAWES, D. T. 1965. Studies of the growth of *Poa annua* L. as affected by soil temperature, and observations of soil temperature under putting green turf. M.S. Thesis. Cornell University. pp. 1–79.

107. HIESEY, W. M. 1953. Growth and development of species and hybrids of *Poa* under controlled temperatures. American Journal of Botany. 4: 205–221.

108. HODGSON, H. J. 1964. Performance of turfgrass varieties in the subarctic. 1964 Agronomy Abstracts. p. 101.

109. HOSHIZAKI, T. 1953. Nitrogen uptake and temperature response of U-3 bermuda grass. M.S. Thesis. University of California at Los Angeles. pp. 1–70.

110. HOWARD, F. L., J. B. ROWELL, and H. L. KEIL. 1951. Fungus diseases of turf grasses. Rhode Island Agricultural Experiment Station Bulletin 308. pp. 1–56.

111. HYDE, E. O. C. 1935. Chewings fescue seed. The influence of temperature and moisture content upon the rate of loss of its germinating capacity. New Zealand Journal of Agriculture. 38(1): 40–42.

112. IYAMA, J., Y. MARATA, and T. HOMMA. 1964. Studies on the photosynthesis of forage crops. III. Influence of the different temperature levels on diurnal changes in the photosynthesis of forage crops under constant conditions. Proceedings of the Crop Science Society of Japan. 33: 25–28.

113. JANSON, L., and B. LANGVAD. 1968. Forlangning av vegetationsperioden for turfgrass genom artificiell tillforsel av varme i rotzonen. Weibulls Grastips. 10: 318–354.

114. JEWISS, O. R., and M. J. ROBSON. 1962. Studies on the growth of ecotypes of tall fescue. Experiments in progress. Grassland Research Institute. 14: 23–24.

115. JORDAN, E. E. 1958. Carbohydrate production and balance in turf. 1958 Midwest Regional Turf Conference. pp. 26–27.

116. ———. 1959. The effect of environmental factors on the carbohydrate and nutrient levels of creeping bentgrass (*Agrostis palustris*). M.S. Thesis. Purdue University. pp. 1–63.

117. JUHREN, M., W. H. Hiesey, and F. W. Went. 1953. Germination and early growth of grasses in controlled conditions. Ecology. 34: 288–300.

118. JUHREN, M., W. NOBLE, and F. W. WENT. 1957. The standardization of *Poa annua* as an indicator of smog concentrations. I. Effects of temperature, photoperiod, and light intensity during growth of the test-plants. Plant Physiology. 32: 576–586.

119. JULANDER, O. 1945. Drought resistance in range and pasture grasses. Plant Physiology. 20: 573–599.

120. KEARNS, V., and E. H. TOOLE. 1939. Temperature and other factors affecting the germination of fescue seed. USDA Technical Bulletin No. 638. pp. 1–36.

121. ———. 1939. Relation of temperature and moisture content to longevity of chewings fescue seed. USDA Technical Bulletin No. 670. pp. 1–27.

122. KELLY, S. 1949. The effect of temperature on the susceptibility of plants to 2,4-D. Plant Physiology. 24: 534–536.

123. KINBACHER, E. J. 1956. Resistance of seedlings to frost heaving injury. Agronomy Journal. 48: 166–170.

124. KINBACHER, E. J., and H. M. LAUDE. 1955. Frost heaving of seedlings in the laboratory. Agronomy Journal. 47: 415–418.

125. KNIGHT, W. E. 1955. The influence of photoperiod and temperature on growth, flowering, and seed production of dallisgrass, *Paspalum dilatatum* Poir. Agronomy Journal. 47: 555–559.

126. KNIGHT, W. E., and H. W. BENNETT. 1953. Preliminary report of the effect of photoperiod and temperature on the flowering and growth of several southern grasses. Agronomy Journal. 45: 268–269.

127. KNOCHENHAUER, O. W. 1934. Inwieweit sind die Temperatur- und Feuchtigkeitsmessungen unserer Flughafen repräsentativ? Erf. Ber. dt D. Flugwetterd. 9. Folge Nr. 2: 15–17.

128. KREITLOW, K. W. 1943. *Ustilago striaeformis*. II. Temperature as a factor influencing development of smutted plants of *Poa pratensis* L. and germination of fresh chlamydospores. Phytopathology. 33: 1055–1063.

129. LANGVAD, B. 1964. Berol snow-kill. Weibulls Grastips. 7: 204.

130. ———. 1966. Jagersro galoppbana. Weibulls Grastips. Maj. pp. 289–295.

131. LAUDE, H. M. 1953. The nature of summer dormancy in perennial grasses. Botanical Gazette. 114: 284–292.

132. LAUDE, H. M., J. E. SHRUM, and W. E. BIEHLER. 1952. The effect of high soil temperatures on the seedling emergence of perennial grasses. Agronomy Journal. 44: 110–112.

133. LAWRENCE, T. 1963. The influence of fertilizer on the winter survival of intermediate wheat-grass following a long period of drought. Journal of the British Grassland Society. 18: 292–294.

134. LEBEAU, J. B. 1964. Control of snow mold by regulating winter soil temperature. Phytopathology. 54: 693–696.

135. ———. 1966. Pathology of winter injured grasses and legumes in western Canada. Crop Science. 6: 23–25.

136. LEBEAU, J. B., and M. W. CORMACK. 1956. A simple method for identifying snow mold damage on turf grasses. Phytopathology. 46: 298.

137. LEBEN, C., and L. V. BARTON. 1957. Effects of gibberellic acid on growth of Kentucky bluegrass. Science. 125: 494–495.

138. LEBEN, C., E. F. ALDER, and A. CHICKUK. 1959. Influence of gibberellic acid on the growth of Kentucky bluegrass. Agronomy Journal. 51: 116–117.

139. LEDEBOER, F. B., and C. R. SKOGLEY. 1967. Plastic screens for winter protection. Golf Superintendent. 35: 22–23.

140. LINDSEY, K. E., and M. L. PETERSON. 1962. High temperature suppression of flowering in *Poa pratensis* L. Crop Science. 2: 71–74.

141. LOVVORN, R. L. 1945. The effect of defoliation, soil fertility, temperature, and length of day on the growth of some perennial grasses. Journal of the American Society of Agronomy. 37: 570–582.

142. LUCANUS, R., K. J. MITCHELL, G. G. PRITCHARD, and D. M. CALDER. 1960. Factors influencing survival of strains of ryegrass during the summer. New Zealand Journal of Agricultural Research. 3: 185–193.

143. MADISON, J. H., L. J. PETERSEN, and T. K. HODGES. 1960. Pink snowmold on bentgrass as affected by irrigation and fertilizer. Agronomy Journal. 52: 591–592.

144. MARCHBANKS, W. W. S. 1953. Limits of viability of roughpea and dallisgrass seed. Embryology of *Lathyrus hirsutus* L. Ph.D. Thesis. Mississippi State College. pp. 1–109.

145. MARTH, P. C., and F. F. DAVIS. 1945. Relation of temperature to the selective herbicidal effects of 2,4-dichlorophenoxyacetic acid. Botanical Gazette. 106: 463–472.

146. McBEE, G. C., W. E. McCUNE, and K. R. BEERWINKLE. 1968. Effect of soil heating on winter growth and appearance of bermudagrass and St. Augustinegrass. Agronomy Journal. 60(2): 228–231.

147. McCOOL, M. M., and G. J. BOUYOUCOS. 1929. Causes and effects of soil heaving. Michigan Agricultural Experiment Station Special Bulletin No. 192. pp. 1–11.

148. McCUNE, W. E., K. R. BEERWINKLE, and G. G. McBEE. 1965. The effect of soil heating on winter growth and appearance of warmseason turfgrasses. Texas Agricultural Experiment Station. PR-2360: 1–17.

149. McKEE, W. H., R. E. Blaser, C. K. CURRY, and R. B. COOPER. 1964. Effect of microclimate on adaptation of species along Virginia highway banks. 1964 Agronomy Abstracts. p. 102.

150. MIDDLETON, J. T. 1943. The taxonomy, host range and geographic distribution of the genus *Pythium*. Memoirs of the Torrey Botanical Club. 20(1): 1–171.

151. MILLER, R. W. 1966. The effect of certain management practices on the botanical composition and winter injury of turf containing a mixture of Kentucky bluegrass (*Poa pratensis* L.) and tall fescue (*Festuca arundinacea* Schreb.). 7th Illinois Turfgrass Conference Proceedings. 7: 39–46.

152. MILLER, V. J. 1960. Temperature effects on the rate of apparent photosynthesis of Seaside bentgrass and bermudagrass. Proceedings of the American Society for Horticultural Science. 75: 700–703.

153. MITCHELL, K. J. 1953. Influence of light and temperature on the growth of ryegrass (*Lolium* spp.). I. Pattern of vegetative development. Physiologia Plantarum. 6: 21–46.

154. ———. 1954. Growth of pasture species. I. Short rotation and perennial ryegrass. New Zealand Journal of Science and Technology. Sec. A. 36(3): 191–206.

155. ———. 1954. Influence of light and temperature on growth of ryegrass (*Lolium* spp.). III. Pattern and rate of tissue formation. Physiologia Plantarum. 7: 51–65.

156. ———. 1955. Growth of pasture species. II. Perennial ryegrass (*Lolium perenne*), cocksfoot (*Dactylis glomerata*) and paspalum (*Paspalum dilatatum*). New Zealand Journal of Science and Technology. 37: 8–26.

157. ———. 1956. Growth of pasture species under controlled environment. I. Growth at various levels of constant temperature. New Zealand Journal of Science and Technology. Sec. A. 38: 203–216.

158. MITCHELL, K. J., and R. LUCANUS. 1960. Growth of pasture species in controlled environment. II. Growth at low temperatures. New Zealand Journal of Agricultural Research. 3: 647–655.

159. ———. 1962. Growth of pasture species under controlled environment. III. Growth at various levels of constant temperature with 8 and 16 hours of uniform light per day. New Zealand Journal of Agricultural Research. 5(1): 135–144.

160. MONTEITH, J. 1926. The brown-patch disease of turf: Its nature and control. Bulletin of USGA Green Section. 6(6): 127–142.

161. MONTEITH, J., and A. S. DAHL. 1928. A comparison of some strains of *Rhizoctonia solani* in culture. Journal of Agricultural Research. 36: 897–903.

162. MOORE, L. D., H. B. COUCH, and J. R. BLOOM. 1963. Influence of environment on diseases of turfgrasses. III. Effect of nutrition, pH, soil temperature, air temperature, and soil moisture on *Pythium* blight of Highland bentgrass. Phytopathology. 53: 53–57.

163. MORINAGA, T. 1926. Effect of alternating temperatures upon the germination of seeds. American Journal of Botany. 13: 141–158.

164. MOSER, L. E., S. R. ANDERSON, and R. W. MILLER. 1968. Rhizome and tiller development of Kentucky bluegrass (*Poa pratensis* L.) as influenced by photoperiod, cold treatment, and variety. Agronomy Journal. 60: 632–635.

165. MUSSER, H. B. 1947. The effect of burning and various fertilizer treatments on seed production of red fescue, *Festuca rubra* L. Journal of the American Society of Agronomy. 39: 335–340.

166. NAYLOR, A. W. 1939. Effects of temperature, calcium and arsenous acid on seedlings of *Poa pratensis*. Botanical Gazette. 101: 366–379.

167. NELSON, A. 1927. The germination of *Poa* spp. Annals of Applied Biology. 14(2): 157–174.

168. NEWTON, J. P., J. P. CRAIGMILES, S. V. STACY, and J. M. ELROD. 1961. Winter lawn colorants. Georgia Agricultural Research. 3(1): 12.

169. NICHOLAS, J. E., and H. B. MUSSER. 1941. Seed drier uses infrared electric lamps. Agricultural Engineering. 22 (12): 421–426.

170. NITTLER, L. W., and T. J. KENNY. 1967. Response of seedlings of *Festuca rubra* varieties to environmental conditions. Crop Science. 7: 463–465.

171. NORMAN, M. J. T., A. W. KEMP, and J. E. TAYLOR. 1957. Winter temperature in long and short grass. Meteorological Magazine. 86: 148–152.

172. NORTON, B. E., and J. W. McGARITY. 1965. The effect of burning of native pasture on soil temperature in northern New South Wales. Journal of the British Grassland Society. 20: 101–105.

173. NOWAKOWSKI, T. Z., R. K. CUNNINGHAM, and K. F. NIELSEN. 1965. Nitrogen fractions and soluble carbohydrates in Italian ryegrass. I. Effects of soil temperature, form and level of nitrogen. Journal of the Science of Food and Agriculture. 16: 124–134.

174. ORGELL, W. H. 1952. Temperature relations of the microclimate. M. S. Thesis. Pennsylvania State College. pp. 1–57.

175. PARKS, W. L., and W. B. FISHER. 1958. Influence of soil temperature and nitrogen on ryegrass growth and chemical composition. Soil Science Society of America Proceedings. 22: 257–259.

176. PELLETT, H. M., and E. C. ROBERTS. 1963. Effects of mineral nutrition on high temperature induced growth retardation of Kentucky bluegrass. Agronomy Journal. 55: 473–476.

177. Peterson, M. L., J. P. Cooper, and L. F. Bendixen. 1961. Thermal and photoperiodic induction of flowering in Darnel (*Lolium temulentum*). Crop Science. 1: 17–20.

178. Peterson, M. L., and W. E. Loomis. 1949. Effects of photoperiod and temperature on growth and flowering of Kentucky bluegrass. Plant Physiology. 24: 31–43.

179. Pfeifer, R. P., and J. P. Kline. 1960. A major cause of winterkill of winter oats. Agronomy Journal. 52: 621–623.

180. Power, J. F., D. L. Gruhes, G. A. Reichman, and W. O. Willis. 1964. Soil temperature effects on phosphorus availability. Agronomy Journal. 56: 545–548.

181. Remsberg, R. E. 1940. Studies of the genus *Typhula*. Mycologia. 32: 52–96.

182. Rider, N. E., and G. D. Robinson. 1951. A study of the transfer of heat and water vapour above a surface of short grass. Quarterly Journal of the Royal Meteorological Society. 77(33): 374–401.

183. Roberts, R. H., and B. E. Struckmeyer. 1938. The effects of temperature and other environmental factors upon the photoperiodic responses of some of the higher plants. Journal of Agricultural Research. 56(9): 633–677.

184. ———. 1939. Further studies of the effects of temperature and other environmental factors upon the photoperiodic responses of plants. Journal of Agricultural Research. 59: 699–709.

185. Rogler, G. A. 1943. Response of geographical strains of grasses to low temperatures. Journal of the American Society of Agronomy. 35(7): 547–559.

186. Rosenquist, D. W., and D. H. Gates. 1961. Response of four grasses at different stages of growth to various temperature regimes. Journal of Range Management. 14: 198–202.

187. Ruelke, O. C. 1961. The role of growth inhibitors in reducing winter injury in Florida's pastures. Soil and Crop Science Society of Florida Proceedings. 21: 136–139.

188. Sale, P. J. M. 1965. Changes in water and soil temperature during overhead irrigation. Weather. 20: 242–245.

189. Savage, D. A., and L. A. Jacobson. 1935. The killing effect of heat and drought on buffalo grass and blue grama grass at Hays, Kansas. Journal of the American Society of Agronomy. 27: 566–582.

190. Schmidt, R. E. 1965. Some physiological responses of two grasses as influenced by temperature, light, and nitrogen fertilization. Ph.D. Thesis. Virginia Polytechnic Institute. pp. 1–116.

191. Schmidt, R. E., and R. E. Blaser. 1967. Effect of temperature, light, and nitrogen on growth and metabolism of 'Cohansey' bentgrass (*Agrostis palustris* Huds.). Crop Science. 7: 447–451.

192. ———. 1969. Effect of temperature, light, and nitrogen on growth and metabolism of 'Tifgreen' bermudagrass (*Cynodon* spp.). Crop Science. 9: 5–9.

193. Simpson, M. J. A., and L. B. Moore. 1955. Seedling studies in fescue—tussock grassland. 1. Some effects of shading, cultivation, and frost. New Zealand Journal of Science and Technology. Sec. A. 37: 93–99.

194. Smith, A. 1932. Seasonal subsoil temperature variations. Journal of Agricultural Research. 44: 421–428.

195. Sprague, H. B. 1933. Root development of perennial grasses and its relation to soil conditions. Soil Science. 36: 189–209.

196. Sprague, V. G. 1940. Germination of freshly harvested seeds of several *Poa* species and *Dactylis glomerata*. Journal of the American Society of Agronomy. 32: 715–721.

197. ———. 1943. The effects of temperature and day length on seedling emergence and early growth of several pasture species. Soil Science Society of America Proceedings. 8: 287–294.

198. Sprague, V. G., A. V. Havens, A. M. Decker, and K. E. Varney. 1955. Air temperatures in the microclimate at four latitudes in the northeastern United States. Agronomy Journal. 47: 42–44.

199. Sprague, V. G., H. Neuberger, W. H. Orgell, and A. V. Dodd. 1954. Air temperature distribution in the microclimatic layer. Agronomy Journal. 46: 105–108.

200. Stoin, H. R. 1966. The effects of high temperature on the soluble nitrogen fraction of Kentucky bluegrass (*Poa pratensis* L.). M.S. Thesis. Michigan State University. pp. 1–28.

201. Stuckey, I. H. 1941. Seasonal growth of grass roots. American Journal of Botany. 28: 486–491.

202. ———. 1942. Influence of soil temperature on the development of colonial bent grass. Plant Physiology. 17: 116–122.

203. Sullivan, J. T., and V. G. Sprague. 1949. The effect of temperature on the growth and composition of the stubble and roots of perennial ryegrass. Plant Physiology. 24: 706–719.

204. Taber, S. 1929. Frost heaving. Journal of Geology. 37: 428–461.

205. ———. 1930. Freezing and thawing of soils as factors in the destruction of road pavements. Public Roads. 11: 113–132.

206. ———. 1930. The mechanics of frost heaving. Journal of Geology. 38: 303–317.

207. Templeton, W. C., G. O. Mott, and R. J. Bula. 1961. Some effects of temperature and light on growth and flowering of tall fescue, *Festuca arundinacea* Schreb. I. Vegetative development. Crop Science. 1: 216–219.

208. ———. 1961. Some effects of temperature and light on growth and flowering of tall fescue, *Festuca arundinacea* Schreb. II. Floral development. Crop Science. 1: 283–286.

209. Thompson, W. R., and C. M. Johnson. 1961. Grass colorants pass durability, appearance tests. Mississippi Farm Research. 24(11): 2.

210. Toole, E. H., and V. K. Toole. 1941. Progress of germination of seed of *Digitaria* as influenced by germination temperature and other factors. Journal of Agricultural Research. 63(2): 65–90.

211. Vaartnov, H. 1967. Responses of five genotypes of *Agrostis* L. to variations in environment. Ph.D. Thesis. Oregon State University. pp. 1–149.

212. Vorst, J. J. 1966. The effects of certain management practices on winter injury to turf containing a mixture of Kentucky bluegrass (*Poa pratensis* L.) and tall fescue (*Festuca arundinacea* Schreb.). M.S. Thesis. Ohio State University. pp. 1–39.

213. Waterhouse, F. L. 1950. Humidity and temperature in grass microclimates with reference to insolation. Nature. 166(4214): 232–233.

214. ———. 1955. Microclimatological profiles in grass cover in relation to biological problems. Quarterly Journal of the Royal Meteorological Society. 81: 63–71.

215. Watson, J. R. 1968. Blankets to protect golf greens against winter injury. 1968 Agronomy Abstracts. p. 61.

216. Watson, J. R., H. Kroll, and L. Wicklund. 1960. Protecting golf greens against winterkill. Golf Course Reporter. 28(7): 10–16.

217. Watson, J. R., and L. Wicklund. 1962. Plastic covers protect greens from winter damage. Golf Course Reporter. 30(9): 30–38.

218. Weihing, R. M. 1963. Growth of ryegrass as influenced by temperature and solar radiation. Agronomy Journal. 55: 519–521.

219. Weisner, M., and L. A. Kanipe. 1951. Delayed germination of *Lolium multiflorum*—common ryegrass. Proceedings of the Association of Official Seed Analysts. 41: 86–88.

220. White, W. J., and W. H. Horner. 1943. The winter survival of grass and legume plants in fall sown plots. Scientific Agriculture. 23: 399–408.

221. Wittwer, S. H., and M. J. Bukovac. 1957. Gibberellin and higher plants. V. Promotion of growth in grass at low temperatures. Quarterly Bulletin of the Michigan Agricultural Experiment Station. 39: 682–686.

222. Wittwer, S. H., M. J. Bukovac, and B. H. Gregsby. 1957. Gibberellin and higher plants. VI. Effects on the composition of Kentucky bluegrass (*Poa pratensis*) grown under field conditions in early spring. Quarterly Bulletin of the Michigan Agricultural Experiment Station. 40: 203–206.

223. WISE, L. N., and C. M. JOHNSON. 1959. "Painted" lawns given winter test. Mississippi Farm Research. 22(11): 3.

224. WYCHERLEY, P. R. 1952. Temperature and photoperiod in relation to flowering in three perennial grass species. Mededelingen Van De Landbouwhogeschool Te Wageningen. 52(2): 75–92.

225. ———. 1954. Vegetative proliferation of floral spikelets in British grasses. Annals of Botany N. S. 18: 119–127.

226. YLIMAKI, A. 1962. The effect of snow cover on temperature conditions in the soil and over-wintering of field crops. Annales Agriculturae Fenniae. 1: 192–216.

227. YOUNGNER, V. B. 1959. Growth of U-3 bermudagrass under various day and night temperatures and light intensities. Agronomy Journal. 51: 557–559.

228. ———. 1960. Temperature, light, and growth of turfgrass. Proceedings of the 31st Annual Golf Course Superintendent Association of America Conference. 28: 37–39.

229. ———. 1960. Environmental control of initiation of the inflorescence, reproductive structures, and proliferations in *Poa bulbosa*. American Journal of Botany. 47: 753–757.

230. ———. 1961. Growth and flowering of *Zoysia* species in response to temperatures, photoperiods, and light intensities. Crop Science. 1: 91–93.

231. YOUNGNER, V. B., J. H. MADISON, M. H. KIMBALL, and W. B. DAVIS. 1962. Climatic zones for turfgrasses in California. California Agriculture. 16(7): 2–4.

232. YOUNGNER, V. B., and F. J. NUDGE. 1968. Growth and carbohydrate storage of three *Poa pratensis* L. strains as influenced by temperature. Crop Science. 8: 455–457.

233. ZANONI, L. J., L. F. MICHELSON, W. G. COLBY, and M. DRAKE. 1969. Factors affecting carbohydrate reserves of cool season turfgrasses. Agronomy Journal. 61: 195–198.

chapter 8

Water

The Role of Water in Turfgrasses

Water is one of the most unique, mobile, and abundant compounds on earth. It is a vital constituent of all living plants no matter how simple or complex. The water content of actively growing turfgrasses is generally from 75 to 85% by weight (14). The water content varies with the turfgrass cultivar, type of plant tissue, weather, location, intensity of culture, time of day, and time of year. Young tissues are higher in water content than mature tissues because of a lower dry matter content, thin cell walls, and a highly hydrated protoplasm. The roots are lowest in water content; the leaves, intermediate; and the stems, highest.

Water is a vital constituent of protoplasm. Death occurs if the water content of the protoplasm decreases below a critical level. The lethal water content varies with the physiological condition of the turfgrass plant. A 10% reduction in water content from 80 to 70% may be sufficient to cause death. Turfgrass species and cultivars vary in tolerance to water loss but none can survive extreme dehydration.

Water, along with carbon dioxide and energy, is required for the photosynthetic process. Water serves as the solvent or catalyst for many metabolic processes occurring in living cells. The ionization properties of water also permit it to be involved directly in many hydrolytic reactions within the plant. For example, water is utilized in the hydrolysis of reserve carbohydrates such as the conversion of starch into monosaccharides. It is also an end product of respiration.

Water functions as the transport medium or solvent by which nutrients, organic compounds, and gases enter and move through the turfgrass plant. The cell vacuoles

261

contain an aqueous solution that serves as a storage pool for excess materials. The high tensile strength, surface tension, and density enable water to withstand the suction force that pulls it to the upper portion of plants.

The high specific heat of water moderates the rate of temperature change of the protoplasm. This characteristic protects the turf from sudden changes in temperature. The high heat of vaporization of water is important during periods of high temperature stress. The large energy requirement for evapotranspiration functions in cooling the turfgrass plant and protects it against heat injury. Water also functions in maintaining the turgidity of cells and the opening of stomata through which water vapor and gaseous exchange occurs. Turgid cells result in improved wear tolerance of turfgrasses. Water is required for the germination of turfgrass seeds. It is also required by the beneficial bacteria and fungi involved in thatch decomposition.

Atmospheric Moisture

Water vapor is a relatively small, invisible constituent of the earth's atmosphere, but it is a vital one since it is the indirect source of water for plants. The amount present in the air affects the (a) precipitation potential, (b) evapotranspiration rate, and (c) rate of heat loss from the earth. The water vapor content of the atmosphere varies greatly, ranging from near zero up to 5%.

Locational, Seasonal and Diurnal Variations

Water vapor fluxes in and near the turfgrass canopy are governed by radiation, temperature, humidity, and wind speed. The water vapor content of the atmosphere varies with latitude and season of the year. The absolute humidity is generally highest near the equator and decreases toward the poles. The absolute humidity varies in direct relation to the temperature in most locations. Thus, the average water vapor content is highest during the summer and lowest in the winter. Within a season, it is generally greater on cloudy or rainy days than on clear days.

Diurnal variations in the atmospheric water vapor content are observed (171). Typically, the vapor pressure deficit (vpd) above a Kentucky bluegrass turf cut at 1.5 in. has a maximum near midday and a minimum in early morning (Fig. 8-1). When expressed in terms of relative humidity, the maximum occurs just before sunrise and the minimum between 12:00 noon and 3:00 p.m. The relative humidity above a turf seldom reaches 100% except during a prolonged rainy or foggy period.

Vertical Variations

Vertical variations in water vapor content are of particular concern in the turfgrass microenvironment (171). The moisture content of the air is highest adjacent to irrigated turfs (Fig. 8-2) and decreases with height (154, 171). The

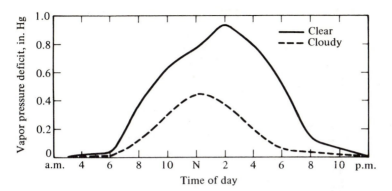

Figure 8-1. The diurnal variation in vapor pressure deficit at 1.5 in. above a turf on an average clear and an average cloudy day (After V. G. Sprague—171).

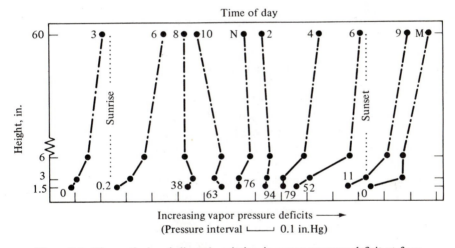

Figure 8-2. The vertical and diurnal variation in vapor pressure deficit at four heights above a Kentucky bluegrass turf cut at 1.5 in. (After V. G. Sprague—171).

water content in and adjacent to the turf may be depressed when the soil moisture is limiting (182, 183). The greatest vertical variation is evident in the initial 6-in. zone above the turf. The maximum water vapor variation occurs just above the turf. The least variation with vertical height generally occurs at noon and the greatest at sunset. The vertical variation is less on cloudy days.

Other Effects

The highest water vapor content occurs within the turfgrass canopy. A higher shoot density results in a proportionally higher atmospheric water vapor content. There is a net outward movement of water vapor from the turf during daylight hours when the grass is actively transpiring. The upward transport of water vapor occurs primarily through the mass exchange process of eddy diffusion. The vertical

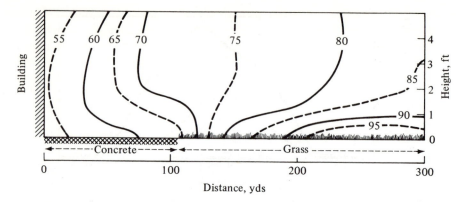

Figure 8-3. The vertical atmospheric water vapor distribution above a turf growing adjacent to a concrete apron of an airfield (After Knochenhauer—99).

and horizontal distribution of water vapor above a turf growing adjacent to a concrete strip is of interest (Fig. 8-3). Over a turf the air is more moist while over the concrete it is drier. The dry, warm air builds up over the concrete and flows out over the turf. Turfs growing adjacent to similar hard surfaces are more prone to drought stress because of this effect.

The water vapor content in the turfgrass canopy is increased when the atmospheric water vapor content is increased or eddy diffusion is restricted due to a lack of air movement. Shaded or enclosed turfgrass areas where air movement is restricted are characterized by water vapor stratification with the highest levels occurring adjacent to the turf (44).

Waterlogged soils due to poor drainage, intense rainfall, or excessive irrigation tend to elevate the water vapor content within and near the turf. The water vapor content above an irrigated turf frequently is 5 to 10% higher than a comparable unirrigated turf (45). The high temperature tolerance of turfgrasses decreases as the water vapor content of the surrounding air increases.

Disease Responses

A high atmospheric water vapor content favors the penetration and infection of certain turfgrass pathogens. Most fungi require a high relative humidity for mycelial growth and the production of spores. The activity of brown patch (34, 39), *Fusarium* patch (77, 169), powdery mildew (34), slime mold (34), *Pythium* blight (34), copper spot (34), dollar spot (34), red thread (34, 77), and *Typhula* blight (33) is stimulated by a high atmospheric water vapor content.

Condensation of Atmospheric Water Vapor

Condensation is dependent on the relative humidity of the air and the degree of cooling that occurs. The capacity of the air to hold water is reduced as it is cooled. If sufficiently cooled, the relative humidity of the air reaches 100% even though the absolute humidity has not changed. Hygroscopic particles in the atmosphere serve as microscopic nuclei around which water droplets form by condensation. Salt spray from the ocean and smoke are particularly effective as condensation nuclei. The minute water droplets formed around the nuclei may occur as visible atmospheric moisture such as clouds or fog. Dew is another form of condensation which can be an important consideration in turfgrass culture.

Clouds. Clouds consist of a visible assemblage of minute water droplets formed primarily by ascending air currents. Adiabatic cooling is the major process through which condensation of water vapor occurs and clouds are formed. Clouds are usually associated with altitudes well above the surface of the earth but may occur at all elevations from the earth's surface up to 40,000 ft. The degree of cloud cover varies with the time of year, latitude, location, and time of day.

The presence of a cloud cover can greatly restrict the quantity and quality of incoming solar radiation, as well as the outgoing, long wave radiation from the earth. Extended periods of cloud coverage can restrict carbohydrate assimilation of turfgrasses and may even result in a depletion of carbohydrate reserves. Clouds also prevent the reradiation of heat away from the earth, thus inhibiting the nocturnal cooling of the turf. Frost formation is less when nocturnal reradiation is impaired by a cloud cover. A reduction of morning frost can be significant, particularly on greens.

Fog. Fog consists of visible water droplets similar to clouds, but it occurs adjacent to the earth. Fog formation involves the cooling of moisture-laden air masses that are adjacent to the ground. Fog most commonly occurs when warm, moist air moves over cold land or water. Coastal areas are quite subject to fogs. Inland fogs are usually associated with lowland areas and valleys.

The frequent occurrence of fog can restrict incoming solar radiation causing negative growth effects on turfgrasses, similar to a cloud cover. Cooler temperatures are also associated with fog. Fog may serve as a source of water for plant growth under certain conditions. Since fog occurs adjacent to the turf, the frequency and severity of turfgrass diseases are increased due to the high water vapor content and the extended duration of leaf wetness. Fog is dissipated primarily by reevaporation of the moisture into water vapor when exposed to solar radiation.

Dew. Dew is moisture formed by the condensation of water vapor on the cool surface of turfgrass leaves (126). Dew formation requires a high atmospheric vapor pressure with a decreasing gradient extending to the condensing surface. This is primarily a nocturnal occurrence that is dependent on radiation cooling of exposed surfaces. Conditions favoring dew formation are (a) a clear sky, (b) moderate wind movement, and (c) a moist atmosphere adjacent to the turf. A cloudless sky permits the maximum possible rate of nocturnal reradiation from the turf, thus permitting rapid cooling of the surface (111). Maximum dew formation occurs at wind speeds of 10 to 12 ft per sec that aid in water vapor transfer to the air layer adjacent to the turf (176). Wind velocities in excess of 15 ft per sec impair dew formation.

Condensation occurs (a) as dew when the temperature of the adjacent air layer is cooled to below the dew point which is still above 32°F and (b) as white frost when the dew point is below 32°F. The amount of dew formation is controlled by

(a) the moisture content of the air, (b) air temperature, (c) turfgrass species, (d) intensity of culture, (e) position of the turfgrass leaves, and (f) microturbulence of the air adjacent to the turf. Dew deposition is greater on the upper exposed side of leaves than the lower side. In addition to condensation from the atmosphere, distillation of water vapor from the warm soil onto the cooler turfgrass leaves in the form of dew has been observed on calm nights (5, 52, 56, 126).

As much as 0.3 to 0.4 mm of dew may be formed in a single night under favorable conditions (7). A dew accumulation of 6 to 10 in. per year has been measured on a bluegrass sod growing in a cool humid climate (82, 83). Topography affects the occurrence of dew. The cooler, north slopes and lowland areas are subject to more frequent dews of a longer duration (111). Also, dew formation occurs less frequently under trees but when formed it persists for a longer time (111).

Two beneficial effects have been attributed to dew. Dewfall can serve as a source of water for the support of growth of certain species (53, 175, 176). It is important to differentiate between dew originating as dewfall from the atmosphere or as distillation from the soil. Dewfall decreases soil moisture extraction by turfgrass roots, while distillation increases soil moisture depletion. A portion of the dewfall deposited on turfgrass leaves may be absorbed and used directly in rehydration. Thus, less dew is frequently observed on young than on old leaves. It is yet to be proved that dew is a significant source of water for turfgrass growth, except possibly in semiarid regions or during severe soil droughts.

The second beneficial effect of dew involves a delay in the onset of both transpiration and the rise in temperature during the early morning hours. The net result is to (a) conserve plant and soil moisture (82) and (b) serve as a cooling agent. The presence of dew on turfgrass leaves also functions as a medium through which certain foliar applied pesticides are absorbed into the leaf.

The benefits of dew are frequently overshadowed by the detrimental effects involving the enhancement of disease development (54, 127, 161, 203). Many pathogenic fungi require free water or leaf wetness for spore germination and penetration into the turfgrass tissue. The presence of dew is especially favorable for brown patch infections on bentgrass (91). The occurrence and severity of diseases are influenced by the intensity, duration, and frequency of dews. The duration of the dew is more important than the amount that occurs on turfs. The duration of the dew depends on (a) the degree of exposure to sunlight, (b) temperature, (c) air turbulence, and (d) atmospheric water vapor content. Conditions exist where dew formation is initiated several hours before sunset and continues until after sunrise (7). Turfgrass leaves may be covered with dew for as much as 15 hr per day (52, 111). Dew frequently persists through the late morning hours under shaded conditions. Structures that impair wind movement also favor dew formation and subsequent disease development.

It is a common practice to drag, brush, pole, or syringe closely mowed, intensively cultured turfgrass areas such as greens in order to reduce the disease problem caused by dew (161). The incidence and severity of brown patch on bentgrass greens is substantially reduced by early morning watering (64). This practice facilitates

Figure 8-4. The early morning dispersal of dew from a bentgrass green by dragging a hose over the turf.

the drying process by knocking the water droplets from the leaf and by distributing the water more evenly over the leaf surface (Fig. 8-4). Dew dispersal practices should be accomplished soon after sunrise for maximum effectiveness (161). The use of a wetting agent also reduces the incidence of dew.

Precipitation

Precipitation is the major source of water for turfgrass growth in many locations. It originates from concentrations of condensed water vapor that exist in the atmosphere as clouds. Rapid condensation occurs when an air mass containing clouds is sufficiently cooled below the dew point. The minute water droplets increase in size until they are heavy enough to respond to the pull of gravity. The droplets then fall to earth in the form of snow, sleet, hail, or rain depending on the air temperature. Not all clouds produce rain since the size of water droplets is governed by the cooling rate, which depends on the velocity of the ascending air mass.

Snow. Snow is a solid form of precipitation that occurs at temperatures below freezing. A snow cover has a beneficial function in protecting or insulating the turf against winter desiccation and extremely low temperatures. Snow is readily transported by the wind, which results in more unequal distribution of moisture compared to rain. A minimal snow cover on exposed, high locations may cause a water deficit while accumulations in low areas and protected sites may result in a surplus of moisture. Turfs covered with snow are more prone to winter injury by low temperature parasitic fungi because of the favorable microenvironment under the snow.

Sleet. Sleet is frozen or partially frozen water that freezes when it strikes the turf. This form of precipitation can result in the accumulation of thick ice layers over a turf. The particles and resulting ice layer appear as clear ice. There is no immediate turfgrass injury resulting from the ice cover.

Hail. Hail is the largest and heaviest form of precipitation. It occurs as pellets or small, usually rounded, masses of ice and is a periodic occurrence of the warm season. Hail can cause extensive temporary injury to the turfgrass leaves, and less frequently even to the crown tissue. Under favorable growing conditions however, most perennial turfgrasses are capable of recovering completely within a relatively short time.

Rain. Rain is precipitation in the form of water droplets. This form of precipitation is of primary concern in turfgrass culture (82). A major portion of the rain eventually runs off through streams into the lakes and oceans. The remainder either evaporates into the atmosphere or infiltrates into the soil. A portion of the water entering the soil is absorbed by plants and returned to the atmosphere through transpiration.

Turfs are adversely affected by deficiencies or excesses of rain. The frequency, intensity, duration, and distribution of rain are all important factors influencing the quantity of water available for turfgrass growth and the amount of irrigation that is required. The effectiveness of precipitation depends on the degree of water infiltration into the soil. Low intensity rains of long duration are preferred for maximum infiltration into the soil (194). High intensity rains result in excessive water loss by surface runoff. Low intensity, short duration rains are much less effective in supplementing the soil moisture level, especially if the evaporation rate is high (194). Short duration rains during midday high temperatures are beneficial in turfgrass cooling.

The seasonal distribution of rain influences the effectiveness of precipitation (2). Two locations may receive the same total annual rainfall but the distribution determines how useful the water is in turfgrass growth (Fig. 8-5). Some regions have a relatively uniform monthly rainfall distribution, while others have a major-

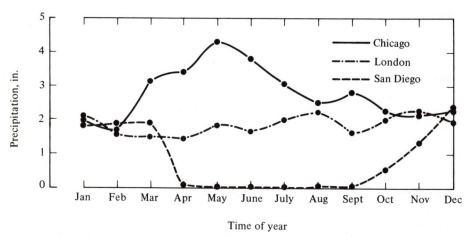

Figure 8-5. Seasonal distribution of rainfall at San Diego and Chicago, U.S.A., and London, England.

ity of the precipitation concentrated in a 2- to 3-month period that is followed by an extended dry season (23). The rainfall distribution during the growing season is particularly important in determining the optimum time for seeding turfgrasses. Seedings are usually avoided during a midsummer drought period.

Rainfall distribution varies widely with location, ranging from nearly zero to over 400 in. per year. Locations subject to landward breezes from large bodies of water frequently receive large amounts of precipitation. Areas on the windward side of mountains are also regions of high rainfall. The leeward side of a mountain is generally a low rainfall region.

Waterlogged soils and high atmospheric relative humidities associated with wet weather favor such turfgrass pathogens as *Pythium* blight, *Ophiobolus* patch, copper spot, and the leaf spot phase of the *Helminthosporium* spp. (27, 34, 125). An increased incidence of brown patch is frequently associated with rainy weather (39, 40, 41), as is the development of mushrooms from fairy rings (34). The dissemination of certain disease causing organisms such as *Pythium* blight and copper spot is enhanced by the splashing action of raindrops (34, 125).

The adjustment of turfgrass cultural practices to the weather conditions is not simple because weather forecasts are not 100% accurate. An elementary understanding of meteorology will assist the professional turfman in making decisions concerning cultural practices such as irrigation. The time and rate of irrigation should be adjusted in relation to the effective precipitation received or anticipated. Knowledge of the causes and characteristics of pressure systems, air masses, and frontal systems is of great value in anticipating the occurrence and amount of precipitation.

Turfgrass Water Relations

Water Absorption

There is a continuous transfer of water within and among the soil, the turfgrass plant, and the atmosphere. The transfer of water may occur in the liquid or vapor state. Water is absorbed in far greater quantities than any other substance required for growth. However, only 1 to 3% of the absorbed water is actually utilized in plant metabolic processes. As much as 600 lb of water may be required to produce 1 lb of dry matter (48, 82, 83, 89). Most of the absorbed water is lost by transpiration. The internal plant water balance is influenced by the intricately interrelated effects of water absorption, internal translocation, and transpiration.

Root absorption. Turfgrass plants absorb water primarily from the soil through the root system. Water diffuses from the soil into the root along a potential gradient. Most water absorption occurs in the root hair zone located a short distance behind the root apex. Root hairs increase the absorption surface of roots manifold. The effectiveness of water absorption by a root system is a function

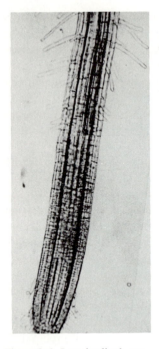

Figure 8-6. Longitudinal section of the primary root of a Seaside creeping bentgrass seedling showing the root hairs. (Photo courtesy of R. M. Endo, University of California at Riverside, Riverside, California).

of the number of root tips (Fig. 8-6). The root hair zone is continually replaced as the root elongates. Root hairs live for only a few weeks and are then destroyed by the suberization process that occurs as the root matures. The number of root hairs decreases as the soil moisture level increases. The length of the root hair zone varies with the turfgrass species, the environmental conditions existing during root development, and the age of the root.

The quantity of water absorbed depends on the (a) depth of the root system, (b) root number, (c) amount of available soil water in the root zone, (d) root extension rate, (e) transpiration rate, and (f) soil temperature. Turfgrasses typically have a large number of roots but the depth is limited by mowing. The rooting depth generally increases in proportion to the cutting height resulting from the associated increase in carbohydrate synthesis. Root activity and water absorption are restricted by (a) excessive nitrogen fertilization, (b) overwatering, or (c) acidic, compacted soils. Shallow rooting also occurs when nutrient distribution is limited to the upper portion of the soil profile.

Water is first absorbed at a shallow depth beneath the turf and then at successively deeper depths as the water supply is depleted (51). Thus, continued root extension is important in contacting new supplies of available soil water. An active root hair zone is also maintained through continued root growth and requires a continuing supply of carbohydrates (94). Optimum soil permeability, temperature, nutrition, pH, and aeration favor maximum root hair activity and water absorption.

Some water absorption and translocation occurs through a dead root system (100). The absorption rate is much less, however, possibly due to plugging of the xylem caused by substances originating from the dead roots. The water absorption capability of dead roots during periods of rapid transpiration is frequently not adequate, which results in an internal water stress.

The amount of available water in the root zone is governed by the (a) precipitation pattern, (b) irrigation practices, (c) depth of the water table, (d) water retaining and transfer properties of the soil, and (e) salt concentration of the soil solution.

A low soil temperature restricts water absorption by reducing the plant water potential, slowing root extension, lowering root membrane permeability due to increased protoplasmic viscosity, and reducing the rate of water movement to the roots due to the increased viscosity of the water (178). Reduced water absorption at low temperatures is more common in the warm season than in the cool

season turfgrasses. Wilting of bermudagrass occurs more readily at soil temperatures below 70°F (20).

The presence of parasitic diseases, insects, and nematodes that attack the turfgrass roots can seriously impair water absorption. Certain pesticides are also injurious to the root system, thus restricting water absorption and decreasing drought
tolerance (61, 62).

Foliar absorption. Water can be absorbed by the leaves and stems of turfgrasses
(88). It may enter leaves in the liquid or vapor state. The path of water movement
is usually through the epidermal cells rather than the stomata. The water absorption rate through leaves depends on the thickness and permeability of the cuticle
as well as the water potential of the cells. Water deposited on the leaves from dew,
fog, precipitation, or irrigation provides a favorable water potential gradient for
absorption. The degree of foliar water absorption varies with the turfgrass species
and is more rapid in young than in old leaves. The quantity of water absorbed
through the turfgrass shoots under normal growing conditions is a minor portion
of the total water absorbed. Under droughty conditions, however, the water
absorbed by leaves can be of vital significance. During periods of plant water stress,
the small quantity of water absorbed through the leaves after a light syringing or
a morning dew may be of major significance in maintaining a favorable plant water
balance and preventing serious desiccation injury.

Internal Water Transport

The xylem is the principal water conducting tissue in plants and is continuous
from the root hair zone of the roots, through the stem, and to the mesophyll cells
of leaves. Actually, the xylem is a continuous system of overlapping vessels along
which water diffuses through cross walls that may or may not be perforated. The
xylem elements of many turfgrass species are small in number and diameter, which
increases the resistance to diffusion (57, 200).

The upward translocation of water in plants is explained primarily by the trans-
piration-cohesion-tension theory. A hydrostatic gradient develops between the
evaporation zone of the leaves and the water absorption region of the roots with
the water flowing along gradients of decreasing water potential. The evaporation
of water from the leaf surface causes a decrease in the water potential of the
mesophyll cells. This causes water to diffuse toward the mesophyll cells from
the xylem and creates a tensile strain throughout the length of the xylem due
to the strong cohesion of water molecules. This negative pressure or tension pulls
the water upward through the xylem tissue and causes water to diffuse into the
lower regions of the xylem from the adjacent root cells. As a result of the low
water potential of the cells adjacent to the xylem in the root, water enters the root
epidermal cells or root hairs and diffuses passively through the cells of the cortex,
the single layer of endodermal cells, and the pericyclic cells into the central vascular
zone of the root. There are certain conditions where water diffuses upward due to
a positive pressure from the roots even when transpiration is negligible. The amount

of water translocated upward by root pressure is small compared to the amount resulting from the cohesion-tension mechanism.

As water is transported from the root epidermal cells to the mesophyll cells of the stomata, it may be diverted from the main transpirational stream to numerous types of specialized tissues depending on the water potential gradient that is established. The water molecules at these sites may function in a diversity of metabolic and physical processes including cell enlargement. A diurnal variation in the water translocation rate is usually characterized by a maximum in early afternoon that diminishes to a minimum during the night. The water absorption rate tends to lag behind the transpiration rate even when the soil moisture is adequate. The lag is caused by resistance to water movement through the plant, which occurs primarily in the living cells of the roots (200).

Transpiration

Most transpirational water loss occurs through the leaves although some may occur through any plant part exposed to the atmosphere. The amount of water contained in turfgrasses at any one time is only a small portion of the quantity absorbed and transpired.

Cuticular transpiration. Water is lost primarily by evaporation from the epidermal cells of the leaf when the stomata are closed. The external surface of most leaf epidermal cells is covered with a wax-like layer called the **cuticle**. The cuticle is composed primarily of esters of long chain fatty acids that are relatively impermeable to water. Wax-like hydrocarbons impregnate the cuticular layer as the plant ages. Dehydration decreases the water permeability of the cuticle.

The amount of cuticular transpiration is controlled to a large degree by the thickness of the cuticle layer. The lower epidermis of the turfgrass leaf is more heavily cutinized than the upper epidermis. The thickness of the cuticle also varies with the turfgrass species and the environmental conditions existing during leaf development. A low light intensity, low atmospheric humidity, or high calcium content tend to increase the thickness of the cuticle (105). The cuticle thickness of turfgrass leaves developed under frequent irrigation is reduced considerably (121).

Stomatal transpiration. The stomata are important structures facilitating the gaseous exchange of carbon dioxide and oxygen so vital to photosynthesis. The same stomatal features that enhance efficient gaseous exchange also result in extensive water loss by transpiration. Although composing only 2 to 3% of the turfgrass leaf area, the stomata are responsible for as much as 90% of the total water lost to the atmosphere by transpiration.

Stomatal transpiration involves the evaporation of water from the wetted surfaces of the mesophyll cells into the intercellular spaces. The water vapor then diffuses along a vapor pressure gradient through the intercellular spaces into the stomatal cavity and eventually to the external atmosphere.

The outward diffusion rate of water vapor depends on the (a) internal leaf diffusion resistance, (b) external boundary layer diffusion resistance surrounding the leaf, and (c) vapor pressure gradient. The internal diffusion resistance to water vapor is approximately 0.4 sec per cm for the turfgrasses, while the external diffusion resistance is approximately 1.2 sec per cm (109). The diffusion resistance values vary among turfgrass species with the more drought tolerant species having higher values. The stomatal transpiration rate is a function of the vapor pressure gradient which increases with (a) a decrease in the atmospheric vapor pressure adjacent to the leaf, (b) an increase in wind speed adjacent to the leaf, (c) a high leaf moisture content, and (d) an increase in temperature.

The stomatal density and distribution varies with (a) turfgrass species, (b) leaf position and surface, and (c) environment under which the stomata developed. Water stress and high light intensities increase the number of functional stomata differentiated from epidermal cells (121). The stomata density of most turfgrass species ranges from 1000 to 6000 per sq cm on the lower leaf surface and 4000 to 10,000 on the upper surface. Stomata of turfgrass leaves are usually arranged in longitudinal rows interspersed among other epidermal cells. The spacing of stomata on a leaf results in a more rapid outward diffusion rate per unit of evaporating area than would occur from a free water surface. Stomata are found on stem tissues but with a substantially reduced density compared to the leaves.

The stomata of turfgrasses are distinctly elongated and relatively small (22) (Fig. 8-7). Stomata possess the unique ability to open and close through the action of the physiologically regulated guard cells. Light stimulates a carbon dioxide control system that causes the stomata to open while darkness causes closure. Other factors influencing the stomatal aperture include temperature, internal plant water stress, pH, and certain chemicals. Partial or complete closure of stomata may occur during midday due to a plant water deficit or extremely high temperatures.

Significance of transpiration. The primary benefit of transpiration is the cooling effect resulting from the evaporative process. This cooling is particularly significant for the cool season turfgrasses. Transpirational cooling may determine whether the plant lives or dies when the air temperature approaches the lethal level. Transpirational cooling at nonlethal, supraoptimal temperatures may also be important in influencing the rates of vital metabolic processes. Transpiration also expedites ion transport through the plant but is not a necessary prerequisite. High transpiration rates are correlated with a reduced incidence of certain turfgrass diseases such as brown patch (39).

Perhaps the most significant aspect of transpiration is the potentially detrimental effects. Environmental and cultural conditions under which the water loss rate through transpiration exceeds the absorption rate by the root system can result in wilt and ultimately desiccation of the turf.

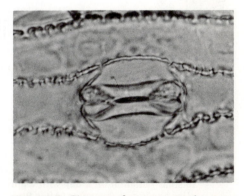

Figure 8-7. Close-up of a stoma on the upper surface of a zoysiagrass leaf blade.

Evapotranspiration

The loss of water from the soil by evaporation and by transpiration from the plants growing thereon is termed **evapotranspiration**. Transpiration accounts for most of the water lost by evapotranspiration from dense, uniform turfgrass areas. As much as 80 to 85% of the soil moisture depletion can be attributed to evapotranspiration (82).

The transpiration rate of a turf varies seasonally, diurnally, and even from hour to hour. Maximum evapotranspiration rates usually occur during the summer period with a substantial decline occurring in the winter (56, 82, 83, 89, 145, 153). The diurnal evapotranspiration cycle corresponds to the daily solar radiation curve (Fig. 8-8). The evapotranspiration rate rises steadily after daybreak, increasing to a maximum at or slightly after midday, and then diminishing to a minimum level at darkness (56, 145).

The evapotranspiration rate is influenced by the (a) light duration, (b) temperature, (c) atmospheric vapor pressure, (d) wind, (e) water absorption rate, and (f) soil moisture tension. Solar radiation is the main source of energy for the evaporation of water. The amount of energy available for evapotranspiration depends on the relative partitioning of incoming solar energy among evapotranspiration, reflection, and heat transfer by conduction. The internal leaf vapor pressure increases as the leaf temperature rises due to increased solar radiation. There is a corresponding increase in the transpiration rate due to the increased vapor pressure gradient. Extremely high temperatures may cause temporary closure of the stomata, which restricts transpiration. The evapotranspiration rate of turfgrasses growing in the shade is reduced because of the decreased incident radiation.

The influence of atmospheric water vapor content and wind speed on the evapotranspiration rate can be substantial. The evapotranspiration rates from a perennial ryegrass turf are compared in Fig. 8-9 for 2 days having similar solar

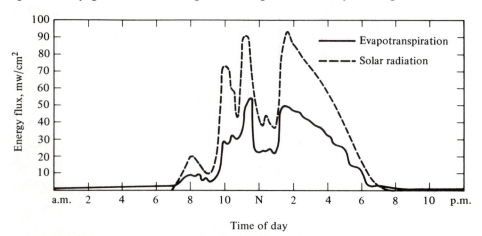

Figure 8-8. The evapotranspiration rate from a perennial ryegrass turf in relation to solar radiation on a day with highly variable cloudiness up to 2: 10 P.M. and clear conditions thereafter (After Pruitt—145).

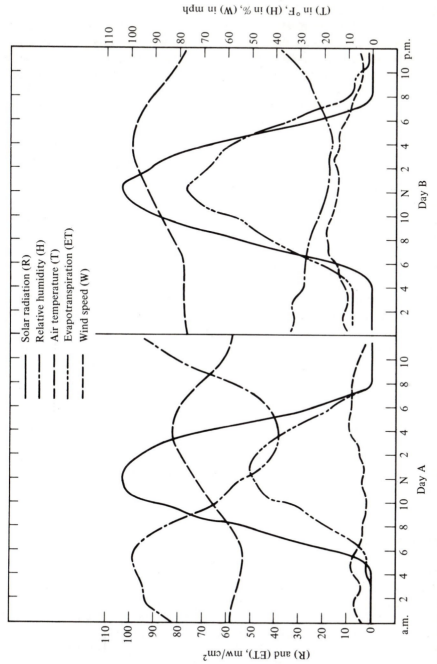

Figure 8-9. The comparative evapotranspiration rates from a perennial ryegrass turf on two clear days. Day A was cool and humid with a gentle breeze while day B was hot, dry, and windy. (After Pruitt—145).

275

radiation patterns. Day A is comparatively cool, calm, and humid with a low total evapotranspiration rate of 0.23 in., while day B is characterized by hot, dry, strong winds and a total evapotranspiration rate of 0.46 in. The water vapor diffusion rate from a leaf is influenced by the difference between the vapor pressure in the intercellular spaces and the vapor pressure of the external atmosphere. A decrease in the atmospheric water vapor content results in a more rapid transpiration rate due to a larger vapor pressure gradient (145). Transpired water vapor tends to accumulate in the boundary layer adjacent to the leaf in still air. The mixing action of wind reduces the thickness of the water vapor boundry layer, which lowers the external vapor pressure and increases transpiration (145).

The water absorption rate is a function of the (a) soil water availability, (b) soil temperature, and (c) extent and effectiveness of the root system. The available soil moisture level depends on the retaining and transmitting properties of the soil. A lack of available soil moisture lowers the internal plant vapor pressure and causes a reduction in the vapor pressure gradient with the external atmosphere (55). The resulting plant water deficit causes a reduction in the transpiration rate (83).

The transpiration rate varies among turfgrass species. For example, redtop has a very rapid transpiration rate; Kentucky bluegrass is intermediate; and the rates of timothy, Canada bluegrass, and chewings fescue are low (140, 191). Species variations in the transpiration rate are due to inherent differences in (a) rooting depth; (b) total number of roots; (c) root-shoot ratio; (d) total leaf surface area; (e) cuticle thickness; (f) osmotic pressure of the cells in the leaf; (g) leaf morphology; (h) leaf orientation; (i) internal leaf structure; (j) structure, spacing, size, and location of stomata; and (k) the leaf rolling or folding capability.

The pesticides and fertilizers used in turfgrass culture can have a significant effect on the transpiration rate. Certain herbicides, soluble salts, and copper-containing fungicides increase the transpiration rate and reduce drought tolerance (158, 196, 197). In contrast, other herbicides cause a reduction in the transpiration rate (166). Little is known about the influence of insecticides on turfgrass transpiration rates. The transpiration rate of turfs exposed to smog is reduced due to closure of the stomata.

Water Use Rate

The **water use rate** is defined as the total amount of water required for turfgrass growth plus the quantity lost by transpiration and evaporation from soil and plant surfaces. The evapotranspiration rate of a turf is much greater than the evaporation rate from bare soil (9, 191, 194). This is due to transpiration through the extensive leaf surface area plus a root system capable of removing water from a larger portion of the soil profile. The water use rate of most turfgrasses is 0.1 to 0.3 in. per day (50, 56, 80, 81, 82, 89, 113, 146, 179, 204). Water use rates in excess of 0.45 in. occur occasionally. A number of factors influence the water use rate of turfs including the (a) evapotranspiration rate, (b) length of the growing season, (c) growth rate, (d) turfgrass species or cultivar, (e) intensity of culture, (f) intensity of traffic, (g) soil type, (h) rainfall, and (i) available soil moisture (50).

The total seasonal water use rate of a turf is a function of the length of the growing season. The longer the growing season, the greater the annual water use rate of a turf. The water use rate also varies with the season of the year (50). Seasonal conditions favoring rapid shoot growth and transpiration cause an increase in the water use rate. Peak water use rates generally occur in early to midsummer in most regions and decline to relatively low levels during the winter (3, 56, 82, 83, 89, 153, 204).

Turfgrass species and cultivars vary in the amount of water used (51, 113). The water use rate of a chewings fescue turf is much less than Kentucky bluegrass, while the rate for Washington creeping bentgrass is slightly less than for Kentucky bluegrass (190). The water use rate is not necessarily related to the drought tolerance of a turfgrass species.

Cultural practices influence the water use rate of turfs. For example, the water use rate increases as the cutting height is raised (50, 51, 114, 124, 159, 172, 189, 191). A reduction in the leaf area causes a decrease in the total transpiration rate per plant but the water loss rate per unit of leaf area actually increases. Cutting with a dull, improperly adjusted mower mutilates the leaf tissue and increases the water use rate of the turf. Nitrogen fertilization increases the total water use of turfgrasses (102, 115). When expressed in terms of the water used per unit of growth produced, however, increasing the fertility level results in more efficient water use (24, 102, 172). The water use rate declines as the moisture content of the soil decreases (81, 83, 106, 146). Turfs that are irrigated less frequently have a lower water use rate. Water use rates also decrease as the depth to the water table is increased (204). Certain turfgrass diseases such as rust disrupt the leaf epidermis causing an increase in the water use rate. Turfs subjected to intense traffic also have an increased water use rate.

The turf must be irrigated to prevent a plant water deficit whenever the water use rate exceeds the effective precipitation for a period of time (146, 153). A knowledge of the water use rate of turfs is necessary in designing and utilizing irrigation systems. The irrigation system should be designed in relation to the water use rates and effective precipitation anticipated under the environmental conditions occurring in each locality. The difference between total monthly precipitation and the total monthly water use rate of turfs has been determined for a number of locations in North America (3).

Water Deficits

The water content of a turfgrass plant is determined by the balance between water absorption and transpiration. The plant water balance is favorable or positive as long as water absorption exceeds transpiration. Whenever transpiration exceeds water absorption, the water balance is negative causing an internal water deficit or stress. The negative water balance may be transient or permanent depending on the available soil moisture level. Water deficits most commonly occur during the months when many recreational turfs receive maximum usage.

Plant responses. The growth rate and turgidity of a turfgrass plant declines under moisture stress (24, 27, 51, 76, 106, 112, 190, 198). The effects of water deficits range from death to less severe morphological and physiological modifications of the turfgrass plant (Fig. 8-10). A plant water deficit causes such morphological modifications as

1. Increased rooting depth (16, 27, 49, 134, 188).
2. Increased root-shoot ratio (188).
3. Decreased tillering (134).
4. Decreased leaf number (134).
5. Reduced shoot elongation (134).
6. Decreased size and total area of leaves (78, 188).
7. Thinner leaves.
8. Smaller cells in the leaves (78).
9. Thicker cuticle (121).
10. Thicker cell walls.
11. Smaller intercellular spaces (78, 121).
12. Smaller xylem cells (78).

Figure 8-10. Merion Kentucky bluegrass plants grown under optimum (left) and minimum (right) soil moisture levels.

Many phases of growth are affected during plant water stress. The influence of moisture stress on cell enlargement is quite pronounced since water is a vital constituent required for maintaining turgidity during the enlargement process. A water deficit restricts leaf and stem elongation. The influence of moisture stress on cell division is minimal causing only a slight reduction in cell number. The cells are much smaller in size, however, causing a decrease in leaf size, especially total leaf area.

Moderate soil moisture stress increases root growth (27). A severe water deficit affects the roots much less than the shoots, which increases the root-shoot ratio. A plant water deficit also increases the cell maturation rate and enhances leaf senescence. The leaf number per plant is reduced under water stress because death of the older, lower leaves occurs earlier than when grown under optimum moisture levels. Evidently, the apical meristem and younger leaves have a higher diffusion pressure deficit and are preferentially supplied with water during periods of severe soil drought, sometimes at the expense of older leaves (135).

The influence of water deficits on physiological processes is not as obvious as the morphological modifications but the effects are just as important. Physiological changes resulting from a plant water deficit include

1. Decreased succulence (97, 174).
2. Higher osmotic pressure (131, 174).
3. Decreased photosynthetic rate (164).
4. Increased soluble carbohydrate content (95).
5. Decreased protein content (10, 97, 142, 143).
6. Increased bound water (131, 192).

Turfgrass plants grown under a continual water deficit have a higher osmotic pressure and a lower tissue water content. A water deficit causes a general reduction in physiological activity. The effect of a water stress in decreasing the photosynthetic rate is quite striking. An internal water stress causes stomatal closure and increases the mesophyll resistance to the inward diffusion of carbon dioxide required for photosynthesis (164). Dehydration decreases respiration in seeds and certain mature tissues but stimulates respiration in actively growing tissues.

Associated with the metabolic changes are increases in the soluble carbohydrate, free amino acid, amide, and bound water content (97). A water deficit enhances the hydrolysis of starch causing an increase in the water-soluble carbohydrates. The decreased protein content of bermudagrass and ryegrass when under moisture stress is attributed to increased proteolysis and an inhibition of protein synthesis (10, 97).

Diurnal and seasonal variations. A diurnal variation in the water content of grass plants is observed (174). The osmotic pressure of leaves is usually lowest during the night and increases substantially during the daylight hours (Fig. 8-11). The increased osmotic pressure during the daylight hours is caused by (a) an increase in the concentration of soluble carbohydrates and other organic compounds resulting from photosynthesis and (b) a decrease in the water content caused by the transpiration rate exceeding the absorption rate. The leaf water content varies inversely with the osmotic pressure, with the highest turgidity usually occurring at night and the minimum at midday. Wilting of turfs is most likely to occur during the latter period.

The diurnal variation in osmotic pressure is usually greater when the available soil moisture level is low. The diurnal variation is also greater in the turfgrass shoots than in underground tissues. Young tissues have a lower osmotic pressure and higher water content than old tissues (174).

Seasonal variations in the water content are common in turfgrasses growing in regions subject to periodic suboptimal temperatures or soil drought. A substantial decrease in the tissue water content occurs during winter dormancy (14). This is related to a higher osmotic pressure caused by an increase in the soluble carbohydrate level that occurs during the cold hardening process. The decrease in tissue hydration levels is of vital importance in low temperature survival of turfgrasses.

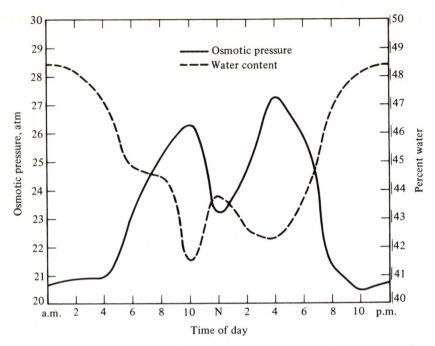

Figure 8-11. The diurnal variation in the osmotic pressure and water content of *Andropogon scoparius* leaves. (After Stoddart—174).

Increased osmotic pressures and plant water deficits are also common in regions subject to extended seasonal soil drought. The tissue water content is usually higher during rapid growth periods such as early spring. Succulent tissues are more prone to injury if high temperature stress occurs.

Disease responses. Disease development in certain turfgrasses is influenced by plant water deficits. For example, the *Pythium* blight susceptibility of Highland colonial bentgrass increases under soil moisture stress (35, 128). Similarly, soil moisture stress increases the dollar spot susceptibility of Kentucky bluegrass (34, 35, 36, 91). The crown and root rot phases of many *Helminthosporium* diseases are favored by water stress (34). In contrast, soil moisture stress reduces the leaf spot phase of *Helminthosporium sativum* on Kentucky bluegrass (27, 34). Also, Rainier red fescue is more tolerant of red thread when grown under soil moisture stress (35, 130).

Wilt

The visible drooping, rolling, or folding of turfgrass leaves resulting from a loss of turgidity is termed **wilt** (Fig. 8-12). It is the first visible symptom of water stress with a gray to blue-green or slate coloration frequently evident on the leaves (87, 195). Wilt occurs when the rate of water loss by transpiration exceeds the absorption rate of the roots. A restriction in water translocation caused by increased resistance of the water conducting tissues to water flow may occasionally cause wilt. Wilt may occur as small localized spots or may extend over a large turfgrass area. It can occur at any time during the growing season but is most common during midsummer.

"Foot printing" of turfs is a visual symptom of an internal plant water deficit indicating that wilt is imminent. This technique involves walking across a turf and observing the rate at which the

Figure 8-12. A comparison of wilted (left) and turgid (right) Kentucky bluegrass plants.

leaves return to their original upright position. Slow or partial recovery indicates that (a) leaf turgidity is lacking and (b) visible wilt symptoms will appear shortly.

Wilt that occurs even when the available soil moisture level is adequate is termed **wet wilt**. It occurs when the transpirational water loss exceeds absorption by the roots even though the available soil moisture level is adequate. Wet wilt is usually caused by a limited root system or a poor water absorption capability. Wilt of turfs may occur while growing in waterlogged soil. The lack of oxygen for respiration causes death or serious injury to the vital root hair zone of the turfgrass roots. The resulting inhibition of water absorption produces a plant water deficit (107). Such a mechanism is probably involved in the "wet wilt" of greens that occurs during or after an extended period of high temperatures, relative humidities, and precipitation.

Environmental and cultural influences. Factors that stimulate transpiration or restrict water absorption increase the wilting tendency of turfs. Conditions favoring transpiration and potential wilting are (a) high temperatures, (b) wind movement, (c) solar radiation, and (d) a low external vapor pressure (87). Wilt of turf is usually a more frequent problem on ridges, knolls, and high spots that are exposed to the drying action of winds and where water infiltration is reduced.

Water absorption may be impaired by a lack of available soil moisture or a limited, nonfunctioning root system. Shallow rooted turfs are more prone to wilt. Common causes of poor rooting include (a) lack of aeration, (b) compaction, (c) waterlogged soil, (d) excessive nitrogen fertilization, (e) severe leaf defoliation, or (f) a high soluble salt level in the soil. Cultural practices that stimulate shoot growth at the expense of the root system tend to reduce the wilt tolerance of turfgrasses. Turfs grown at a low nitrogen level have greater wilt tolerance than when grown at a high level (26). Turfgrasses hardened by growing under reduced soil moisture levels are also more wilt tolerant (26). For example, the wilt susceptibility of annual bluegrass increases when grown under frequent irrigation and high fertilization rates (121). Thatched turfs are more likely to wilt due to the shallow root system. Root feeding insects and nematodes can also damage the root system and reduce the wilt tolerance.

Wilting tendency of turfgrasses. Little is known about the relative wilting tendency of various turfgrasses. The thin turfgrass leaves result in an increased transpiration rate and greater wilt susceptibility. The soil water supplying power at which wilt of turfgrasses occurs has been investigated (195). Only the water supplying power of the soil is measured in this approach and not the actual tissue moisture content. However, some indication of the relative wilting tendency of eleven cool season turfgrasses is provided (Table 8-1). Among the warm season turfgrasses, bermudagrass has the least wilting tendency, Pensacola bahiagrass ranks intermediate, while carpetgrass has a high tendency to wilt (23). Other turfgrasses having a low wilting tendency are zoysiagrass and tall fescue (78). The greater wilting tendency of certain turfgrass species may be due to a low osmotic pressure of the cells plus a shallow root system.

Table 8-1

THE WATER SUPPLYING POWER OF THE SOIL AT WHICH
WILTING OF ELEVEN COOL SEASON TURFGRASSES
OCCURS [AFTER WILSON AND LIVINGSTON (195)]

Wilting tendency	*Turfgrass species*
Very low	Sheep fescue
	Red fescue
Low	Meadow fescue
	Redtop
	Chewings fescue
Medium	Canada bluegrass
	Kentucky bluegrass
	Creeping bentgrass
	Perennial ryegrass
High	Velvet bentgrass
	Timothy

Turfgrass responses. Temporary wilting can result in numerous detrimental effects that are not visible. Temporary water stress inhibits many vital metabolic processes in the plant. It causes stomatal closure that seriously restricts photosynthesis. Transpirational cooling of the leaf is also restricted by stomatal closure. The resulting rapid rise in leaf temperatures may cause the temporary impairment of numerous metabolic processes or even death. This is most likely to occur during periods of high solar radiation.

Turgid leaves resist the pressure of foot and vehicular traffic. Traffic over turfgrasses that are wilted frequently results in injury to the aboveground tissues (Fig. 8-13). Vehicular traffic is of most concern since the concentrated weight of the wheels frequently causes death of wilted tissues. Vehicular traffic should be withheld from wilted turfs until the condition is corrected.

The most serious aspect is that wilt is the first stage leading to desiccation. Wilt and even desiccation can occur within a matter of hours on close cut bent-

Figure 8-13. Injury of a bentgrass turf by tire traffic of golf cars when the turf was in a wilted condition. (Photo courtesy of J. R. Watson, Toro Manufacturing Corp., Minneapolis, Minnesota.)

grass greens having a limited root system. The rapidity of wilt occurrence may not allow turfgrass plants time to adapt to the moisture stress through physiologic or xeroplastic changes.

Wilt Prevention and Control

Wilt prevention is a difficult phase of turfgrass culture due to the complex interaction of plant, soil, and atmospheric factors controlling water absorption and transpiration. The first consideration is to ensure an adequate supply of available soil moisture. The frequency and rate of water application should be adjusted to the evapotranspiration rate. Water absorption is enhanced by a deep, extensive, and actively growing root system through (a) good soil aeration; (b) a high cutting height and the proper mowing frequency; (c) adequate nitrogen, phosphorus, and potassium levels that are balanced but avoid an excessively high nitrogen level; and (d) thatch control. Excessively high soil solution concentrations or substances toxic to the roots should be avoided or corrected. Certain preemergence herbicides are injurious to turfgrass roots (61, 62) and should not be used where wilt is a continuing problem.

Corrective measures should be taken when the first visible symptoms of wilt occur. The most commonly used corrective measure is syringing. It reduces the transpiration rate of the turf by lowering the temperature and increasing the atmospheric water vapor content surrounding the leaves. Midday syringing maintains the turgor of leaves and prevents stomatal closure. The cooling effect of syringing also reduces the respiration rate. The proper timing of syringings is critical in wilt prevention. The turfman must be constantly alert for wilt, especially during periods of high wilt probability, and be prepared to syringe immediately. Syringing may be necessary more than once per day under severe atmospheric drought stress.

Antitranspirants. Methods of controlling transpiration at the leaf-air interface are receiving much attention since wilt may occur when the soil moisture is adequate. The antitranspirant methods may involve (a) chemically induced stomatal closure, (b) covering the mesophyll surface with a thin, monomolecular layer, or (c) covering the leaf surface, including the stomata, with a thick film that is relatively impermeable to water vapor. Methods of controlling transpiration at the leaf-air interface are acceptable only if they (a) do not impair gaseous exchange to the extent that growth is restricted and (b) do not seriously impair transpirational cooling. A certain degree of growth inhibition of turfs can be tolerated if the antitranspirant is effective in wilt reduction. Similarly, transpirational cooling should not be reduced to the point that leaf temperatures can rise readily to lethal levels.

Chemically Induced Closure of Stomata

Investigations with nongrass species have shown that transpiration can be reduced to a significantly greater degree than photosynthesis or growth by the use of certain chemicals (163, 164, 166, 167, 173, 181, 205, 206, 207). The effective chemicals include phenylmercuric acetate, certain metabolic inhibitors such as the α-hydroxysulfonates, and 8-hydroxyquinoline sulfate. Phenylmercuric acetate is one of the more effective, low cost chemicals in inducing stomatal closure (167). It also has a low mobility that is usually limited to the leaf epidermis, thus minimizing any toxic effects to the vital metabolic systems within the leaf (166, 167).

Chemical closure of stomata to prevent wilt is a promising technique. However, detailed studies are lacking on turfgrass species. Additional information is needed regarding the (a) most effective chemical for stomatal closure of specific turfgrass species, (b) optimum application rate, (c) presence of undesirable side effects on plant metabolism and growth, (d) degree to which gas interchange is restricted, (e) duration of the effect, and (f) effect on leaf temperature heat balance (177, 191). Certain stomatal closing antitranspirants have remained effective for as long as 14 days (206). A residual effect of this duration probably can not be achieved with turfs that are mowed frequently. The techniques used for applying the antitranspirant are important since the degree of transpiration control depends on whether only the upper side or both sides of the leaf are sprayed (180). An additional, potential benefit from chemical closure of stomata is a reduction in the incidence of certain pathogens that enter the host plant primarily through open stomata (152).

Monomolecular Layers

Certain highly polar alcohol molecules can form a monomolecular layer over the surface of water reservoirs that is impermeable to water vapor. The layer reduces the evaporation rate. These materials are basically long chain fatty alcohols such as hexadecanol and octadecanol. Such materials have been suggested for use in suppressing transpiration of plants by forming a similar water impermeable film. However, results with the monomolecular layer type of antitranspirant have been erratic in suppressing transpiration (6, 67, 68, 133, 136, 141, 155, 166). Also, the photosynthetic reduction is substantially more than the reduction in transpiration.

Thick Films

Transpiration control has also been attempted with the relatively thick films such as a wax emulsion or plastic base of polyethylene, vinyl acrylate, or a polyvinyl chloride complex (193). The entire leaf, including the stomatal pore, is covered with a film that is impermeable to water vapor. Films of this type have been used in transplanting woody plant species (1, 29, 116, 122, 123). Proper coverage of the leaf surface is very critical in order for the thick films to be effective antitranspirants. Detailed studies regarding the use of such materials are lacking for turfs. The polyethylene films have the disadvantage that the permeability to carbon dioxide is restricted to a greater extent than water vapor (202). Due to the necessity of mowing, the use of thick films is not a promising method of controlling transpiration in actively growing turfs. Frequent leaf removal results in the production of new, untreated tissue within a short time. This means that the antitranspirant film must be applied quite frequently at considerable cost. The thick wax or plastic films have also been proposed for use as a winter antidesiccant for dormant turfs. The limited studies using these films for winter desiccation prevention of turfs have not been promising.

Drought

Plant growth is slowed; the tissues wilt; and death eventually occurs if a water deficit continues for a sufficient length of time (110, 194, 198). A prolonged water stress that limits or prevents turfgrass growth is termed **drought**. **Atmospheric drought** results from a plant water deficit causing wilt and desiccation of a turf due to the transpiration rate exceeding the absorption rate even though the available soil water level is adequate. Atmospheric drought causes wilt and sometimes death if not corrected. Plant water deficits caused by atmospheric drought are usually less severe and of shorter duration than those caused by soil drought.

A plant water deficit causing wilt and desiccation of the turf due to a lack of available soil water is called **soil drought**. Extended periods without precipitation, a high evapotranspiration rate, and the lack of an irrigation capability contribute to soil drought (190). The extent and frequency of soil drought on unirrigated turfs is greater in the more arid regions. The only protection against extended periods of soil drought is the availability and effective use of stored water supplies from lakes, reservoirs, rivers, or underground sources.

The severity of soil drought depends on the (a) duration without rain, (b) evaporative power of the air, and (c) soil type. Turfs on the upper portions of slopes are subject to severe drought (194). The reduced soil moisture on these sites is the result of higher evapotranspiration rates and reduced infiltration rates.

Drought frequently occurs during the midsummer period when high temperatures are common. Summer dormancy of turfs is usually caused by the combined effect of heat and water stress. However, the primary cause of summer dormancy in turfgrasses is usually a lack of water. High temperature and water deficit effects are difficult to distinguish between because they commonly occur together. The effects of heat and moisture stress are additive, causing increased injury when occurring in combination (112).

Mechanism of drought injury. The accepted working hypothesis for the drought injury mechanism is the mechanical explanation of Iljin (93). Death is attributed to the mechanical injury of cells resulting from the drying and remoistening processes rather than to the lack of water itself. The outward diffusion of water causes the vacuole to shrink and the protoplasm to be pulled inward. The cell wall resists collapse if it is relatively rigid. The protoplasm is subjected to stress in this situation caused by the inward pull of the shrinking vacuole and the outward pull resulting from adherence to the rigid cell wall. The resulting mechanical disruption of the dehydrated protoplasm causes death of the plant tissue. Similar stresses may occur during the remoistening process when the cell wall distends more rapidly than the protoplasm.

Drought Resistance

The ability of a plant to survive an unfavorable external moisture stress is termed **drought resistance**. Turfgrasses can survive drought by (a) escape, (b) dormancy, (c) an increased water absorption capability, (d) xeromorphic features, or (e) a physiological capability to endure dehydration. The escape and dormancy mechanisms are specialized means for turfgrasses to avoid soil drought. The increased water absorption capability and the xeromorphic features act by delaying the onset of dehydration.

Escape. Some annual turfgrasses escape soil drought by means of a short life cycle that is completed during the rainy season. Annual bluegrass has this escape capability to germinate, establish, mature, and produce seed within a short time when the soil moisture level is favorable. Certain weeds also evade drought by passing through stress periods in the form of seeds.

Dormancy. Turfgrasses can survive drought periods in a dormant state (Fig. 8-14). Shoot growth ceases and death of the aboveground leaves occurs under severe soil drought (103, 194). Buds in the crowns, stolons, and rhizomes of dormant grasses survive the drought and initiate new growth when favorable soil moisture conditions develop (8, 135). Buds are extremely drought hardy because of the small cells that are devoid of vacuoles. Kentucky bluegrass can break dormancy rapidly with recovery frequently evident within 3 to 5 days after a substantial rain (107, 194). Typical examples of turfgrasses that have a dormancy capability for drought survival are Kentucky bluegrass and bermudagrass.

Figure 8-14. Turgrass response to an extended period of drought. A green, irrigated fairway turf (upper right) surrounded by a brown, dormant turf.

The brown, dead leaves of a dormant turf serve as a mulch to reduce water loss by evapotranspiration, thereby providing additional protection for the dormant plant tissues (194). Dormant turfs caused by soil drought are also protected against high temperature stress.

Water absorption capability. Certain turfgrasses have a greater potential for drought survival because of characteristics that result in a greater water absorption capability. The xeromorphic factors influencing water absorption include the (a) rooting depth, (b) root number, (c) degree of branching, (d) extent of the root hair zone, and (e) root growth activity (30, 165). A deep, extensively rooted turf is not necessarily drought hardy. It only has the capability of absorbing water from a lower soil horizon and a greater volume of soil, thus prolonging the length of time before the turf is subject to wilt. A deep, extensive root system provides a greater capability to survive drought only if water is present at the lower soil depths.

Turfgrasses vary genetically in their rooting characteristics. Many have a relatively shallow but fibrous, dense root system that can exhaust the available soil moisture supply in the root zone within 1 to 4 weeks (21). For example, the root system of bentgrass may be only 2 to 4 in. deep in midsummer when maintained under putting green conditions. Among the cool season turfgrasses, tall fescue is relatively deep rooted, while Kentucky bluegrass is intermediate. The drought resistant hard fescue is quite deep rooted and has a high root-shoot ratio (170). The root system of certain warm season turfgrasses may extend as deep as 6 to 8 ft (80). Bermudagrass has a deep, extensive root system, while bahiagrass ranks intermediate, and the drought susceptible carpetgrass ranks low in rooting depth (23, 49).

Xeromorphic features. Some turfgrasses possess inherited structural modifications that reduce the water loss by transpiration when a water deficit occurs. Typical examples are bermudagrass, zoysiagrass, and tall fescue (78). Xeromorphic features that reduce transpiration include (a) decreased leaf surface area; (b) altered size, spacing, number, and location of stomata; (c) increased cuticle thickness; (d) presence of surface hairs; (e) less intercellular space; (f) diminutive conducting tissues; and (g) rolling or folding of the leaves.

Detailed knowledge of the xeromorphic characteristics of turfgrasses is quite limited. The narrow leaves of red fescue contribute to a low leaf area index, a reduced transpiration rate, a lower demand for the available soil moisture, and a greater potential for drought survival (190). Drought resistant grasses frequently have a higher root-shoot ratio (30). Certain species may have a reduced leaf area but show no reduction in transpiration due to an increase in the size and number of stomata. The indentation of stomata in leaves is another xeromorphic feature that can restrict transpiration. Some turfgrass species also have stomata that close much earlier in the development of a plant water deficit. This is one mechanism of increasing drought resistance since transpiration is greatly reduced by stomatal closure.

In order to reduce cuticular transpiration, the cuticle must not only be thick but should be dense and composed of materials that are impermeable to water. The pubescence on certain turfgrasses increases the thickness of the water vapor boundary layer on the leaf. This lengthens the diffusion pathway and increases resistance to water vapor diffusion, which reduces the transpiration rate. The leaf hair length of certain zoysiagrass and bermudagrass species ranges from 1 to 2 mm (92). The epidermal hairs on leaves are of two types. One type is dead and nonfunctioning, while the second type is alive, contains protoplasm, and is an area of active transpiration (160). Turfgrasses with living epidermal hairs may have a transpiration rate from the leaf hair that negates the beneficial effect of an increased boundary layer thickness. The transpiration reduction attributed to leaf hairs is not as significant as the other xeromorphic features discussed.

The capability of leaf rolling or folding during water stress reduces the amount of leaf area exposed to the atmosphere and decreases the water loss rate (28, 108, 135). The structural mechanism of leaf folding and rolling is attributed to the bulliform or motor cells. They are (a) triangular shaped with the greater depth being opposite the leaf surface, (b) considerably larger than normal epidermal cells, and (c) occur in rows that extend the length of the upper leaf. Bulliform cells have a more rapid rate of transpirational water loss than adjacent epidermal cells because of a much thinner cell wall and cuticle.

Collapse of the bulliform cells due to a water deficit causes the turfgrass leaf to fold or roll depending on how the cells are distributed. Some turfgrass species have two parallel groups of bulliform cells, with one group located on each side and immediately adjacent to the midrib, which facilitates folding. Other species have numerous groups of bulliform cells distributed in parallel rows over the leaf, which facilitates a rolling action. Kentucky bluegrass, rough bluegrass, and annual bluegrass are examples of a folding type of mechanism, while bentgrass, zoysiagrass, tall fescue, and redtop have a rolling mechanism (28, 118) (Fig. 8-15). An additional xerophytic feature of certain grasses is the arrangement of stomata along the sides of the bulliform channels so that the stomata are not exposed when the blade rolls or folds.

Figure 8-15. Diagram of transverse sections of Kentucky bluegrass (left), creeping bentgrass (center), and red fescue (right) leaves when under moisture stress (above) adequate moisture (below).

Drought hardiness. The fifth means of achieving drought resistance is through **drought hardiness** which is the ability of a plant to survive desiccation. The drought hardiness mechanism can be interpreted in relation to Iljin's theory of mechanical injury. Drought hardy tissues usually have a small cell size with a small vacuole that reduces the amount of contraction and associated protoplasmic stress that can occur. Thus, the protoplasm of small cells is subjected to less mechanical stress during desiccation. The cell shape is also important since long narrow cells are subjected to a minimal contraction. The hydration properties of the protoplasm influence the degree of stress that can be endured. Drought hardy species generally have a high cell sap concentration and high osmotic pressure. The high osmotic pressure increases the ability of cells to retain water, which lessens the degree of cell contraction. Drought hardened tissues also have increased protoplasmic permeability to polar substances.

The diffusive resistance of the mesophyll cells to water loss increases as the internal plant water stress increases. This lowers the transpiration rate. The diffusion resistance of the more drought resistant kikuyugrass is much greater than for ryegrass (109). This high resistance to water vapor diffusion may be another drought adaptive mechanism.

Factors in Drought Hardiness

The drought hardiness of turfgrasses varies with the stage of development. Seeds are extremely drought hardy since the protoplasm is in a resting state. Most turfgrass seeds can survive extended periods of exposure to air-dry conditions without a significant loss in viability (11, 70). Semiresting dormant buds on rhizomes and stolons are also quite drought hardy. When the resting stage of dormant organs is broken and seedling growth is initiated, the tissues still retain a considerable degree of drought tolerance when compared to mature tissues.

The growth rate and resulting physiological state of the tissue are also factors influencing drought hardiness. Slow growing tissues with a small cell size and a high carbohydrate content are more drought hardy (95). Cultural practices that avoid excessive growth stimulation increase drought hardiness. Factors that reduce drought hardiness include (a) high nitrogen fertilization rates, (b) a potassium deficiency, (c) shading, (d) intense traffic, and (e) excessive irrigation (25, 46). Excessive nitrogen fertilization stimulates rapid shoot growth, enlarges cell size, increases tissue hydration, and causes a general reduction in drought hardiness (66, 112). In contrast, a potassium deficiency should be avoided, as drought hardiness may be reduced (112). Turfs growing under shaded conditions are less drought hardy due to the reduced carbohydrate content, increased succulence, and the more delicate cell structure.

Light, frequent irrigations prevent hardening and exhaust the carbohydrate reserves of the plant (21). Turfgrasses grown under a limited soil moisture level have greater drought hardiness than when grown under an adequate moisture level (25, 194). This response is an important consideration in turfgrass irrigation practices. It is better to restrict irrigation than to intensively irrigate for a portion of a drought period and then miss several weeks. If the latter situation occurs, the turf may be thinned due to severe desiccation injury of the hydrated tissue that was not permitted to harden. The unirrigated turf is in a brown dormant state during the soil drought but produces a dense healthy turf quite rapidly when the drought is terminated. A brown dormant turf may actually be in better condition physiologically during a droughty period than an excessively or inadequately irrigated turf (21).

The relative drought hardiness of grass species varies considerably. The literature on turfgrass drought hardiness is limited and quite variable. No data are available where the relative drought hardiness of various turfgrasses is compared to known internal tissue water deficits. This approach gives the only accurate measure of drought hardiness. Several studies have simulated drought conditions by withholding water and exposing the turfgrasses to air temperatures of 95°F or higher (25, 26, 162). The results of drought studies where high temperatures are employed vary considerably from tests where water is withheld and the plants are not exposed to heat stress.

Drought resistance of turfgrasses. The relative drought resistance of the commonly used turfgrasses is presented in Table 8-2 (8, 23, 80, 97, 113, 117, 129, 162, 187).

Since many plant characteristics contribute to drought resistance, a ranking among species is difficult and must be very general in nature. Some variability in drought resistance occurs within turfgrass species (13, 73, 201). Among the Kentucky blue-grasses, Cougar has superior drought resistance, while Pennlawn is one of the more drought resistant red fescue cultivars (13). Merion has fairly good drought resistance that is partially attributed to a deeper root system (66, 80). The drought resistance of chewings fescue and Park Kentucky bluegrass seedlings is good during establishment (201). Turfgrasses ranked on a basis of drought hardiness do not necessarily have a similar drought resistance ranking.

Table 8-2

THE RELATIVE DROUGHT RESISTANCE OF
TWENTY-TWO TURFGRASSES

Drought resistance	Turfgrass species
Excellent	Buffalograss
	Bermudagrass
	Zoysiagrass
	Bahiagrass
Good	Crested wheatgrass
	Hard fescue
	Sheep fescue
	Tall fescue
	Red fescue
Medium	Kentucky bluegrass
	Redtop
	Timothy
	Canada bluegrass
Fair	Perennial ryegrass
	Meadow fescue
	St. Augustinegrass
Poor	Centipedegrass
	Carpetgrass
	Italian ryegrass
	Creeping bentgrass
	Rough bluegrass
	Velvet bentgrass

Winter Desiccation

Semidormant turfs may be seriously damaged by winter desiccation during the winter season (Fig. 8-16). It is one of the three major causes of turfgrass winterkill (12). Winter desiccation is usually most severe (a) on elevated sites, (b) in areas exposed to excessive wind movement, (c) where surface runoff of precipitation is high, and (d) when air temperatures are above 32°F (194). Winter desiccation may be the result of either soil or atmospheric drought.

Winter desiccation caused by soil drought is most common in regions of low rainfall or where the seasonal precipitation distribution is low during the winter

Figure 8-16. Winter desiccation injury on a bentgrass green.

period. Such locations lack the snow cover needed to prevent extensive loss of soil water by evapotranspiration. Wide, deep cracks sometimes develop in greens and similar turfgrass areas due to soil shrinkage caused by soil drought (132).

Turfs may also be injured by atmospheric winter desiccation where adequate soil moisture is present. Desiccation occurs when conditions favor rapid evapotranspiration but water absorption is inadequate due to (a) a restricted root system, (b) increased viscosity of water at low temperatures, (c) decreased root membrane permeability, or (d) the soil water being in a frozen, unavailable state for root absorption.

A certain degree of atmospheric desiccation occurs every winter. It is usually quite superficial and causes no permanent damage to the turf. This superficial injury to the leaf tissue is sometimes referred to as "wind burn." It commonly occurs during the winter after the snow cover thaws. The turfgrass leaves appear green and healthy immediately after the thaw. Shortly thereafter, the green leaves turn brown due to atmospheric desiccation during a period of rapid evapotranspiration. Wind burn or related superficial injuries are not critical as long as the water deficit does not injure the vital meristematic tissues of the crown. Desiccation of the crown tissue causes serious damage to the turf, which may not recover except from the nodes of surviving rhizomes and stolons.

Cool season turfgrasses are more frequently damaged by winter desiccation but warm season turfgrasses such as bermudagrass may also be injured (132). Species or cultivars that maintain or initiate physiological activity and growth at lower temperatures are more prone to winter desiccation injury than dormant turfgrass species. Turfgrasses having a noncreeping, bunch-type growth are also more prone to permanent damage by winter desiccation. Other species, such as Kentucky bluegrass, have the capability of surviving winter desiccation and initiating new growth from the nodes of lateral shoots. Differences within species are also observed. For example, Seaside is more tolerant of winter desiccation than most bentgrass cultivars.

Winter Desiccation Prevention

Winter desiccation injury can be prevented by (a) the application of water to dry soils or (b) controlling evapotranspiration from the turf. The soil moisture level should be near field capacity when the soil freezes or the first permanent snow occurs. A late autumn irrigation of dormant turfs may be required. Saturated soil conditions should be avoided in order to minimize hydration of the turfgrass crown tissue. Dormant turfs subjected to soil drought must be watered at periodic intervals during the winter to maintain an adequate soil moisture level. Hauling water to the site is costly and time consuming but effective (132). This has been a common practice on greens.

Windbreaks. Winter desiccation prevention through evaporation control can be accomplished in numerous ways. Turfs located on sites prone to desiccation injury can be protected from the drying action of strong, prevailing wind through the strategic placement of windbreaks. An additional benefit is the snow accumulation on the leeward side of windbreaks that also protects against desiccation. Effective windbreaks can be achieved by constructing protective banks, planting protective vegetative screens of trees and shrubs, proper placement of snow fences, or piling brush on the site.

Mulches, brush, and topdressing. Loss of water by evaporation can be controlled by placing straw, pine needles, or a similar organic mulch on critical turfgrass areas such as greens and tees that have a history of serious desiccation. The mulches provide a protective microenvironment against the drying action of winds. Such materials are difficult to remove in the spring and leave an unsightly appearance.

The piling of brush is preferred to organic mulches because the brush enhances snow accumulation and is more easily removed in the spring. An adequate quantity of brush is not readily available in certain areas. A late autumn topdressing is also effective in preventing winter desiccation. The topdressing should be applied at 0.3 to 0.4 cubic yards per 1000 sq ft but not matted or raked into the turf. The soil mix used in topdressing should be similar to the existing soil. A preventive fungicide should be applied prior to putting the mulch, brush, or topdressing in place since the microenvironment under these materials favors snow mold disease development.

Synthetic protective covers. The use of synthetic protective covers for desiccation control is a relatively recent innovation (Fig. 8-17). Synthetic protective covers are a means of preventing winter injury on turfs subject to (a) minimal snow

Figure 8-17. A putting green with the right three-fourths of the turf covered with clear polyethylene (above) and the relative amount of injury which occurred on the covered and uncovered sections (below). (Photo courtesy of J. R. Watson, Toro Manufacturing Corp., Minneapolis, Minnesota).

accumulation, (b) low amounts of winter precipitation, (c) severe drying winds, (d) severe low temperature stress, (e) negligible winter use, and (f) a history of serious desiccation and low temperature injury. They are most commonly used on intensively cultured, high quality turfgrass areas. An effective dual-purpose winter protection cover should (a) prevent winter desiccation injury of turfs by trapping or retaining soil moisture, (b) modify extremes in low temperature which may cause direct injury of turf, (c) reduce the loss of fungicides from turfgrass leaves and crowns, (d) permit sufficient light penetration and energy exchange to stimulate early spring green-up of the turf, and (e) provide a degree of temperature insulation so that there is no lethal heat buildup or excessive shoot growth.

Winter protection covers should be installed after the turf has ceased growth and entered a hardened state but before the soil freezes or the first permanent snow fall occurs. All clippings and other debris should be removed from the turfgrass surface prior to installing the cover. A fungicide should also be applied to prevent snow mold disease development since the microenvironment under the cover favors pathogen activity. The cover should be unrolled across the area and secured by means of 5 in. wire staples or by placing wooden lath at 20 ft spacings over the cover and along the overlapping edges. Spikes can be driven through predrilled holes in lath and into the underlying soil. The cover should be drawn tight to minimize the whipping action of winds. It is sometimes necessary to provide a restraining fence on turfgrass areas subject to human traffic or snowmobile activity. The stability of the covers is such that they can be severely damaged or destroyed.

Proper timing in the removal of winter protection covers is quite important. Removal should be accomplished when there is a relatively low probability of climatic conditions occurring that favor desiccation or direct low temperature injury and after sufficient spring green-up of the covered turf has occurred. It should also be removed if the heat accumulation under the cover approaches the lethal level. This is most commonly a problem under polyethylene covers. The timing of cover removal is also governed to some degree by the quantity of leaf growth produced under the cover. The cover should be removed early enough that the initial mowing does not scalp the turf. Mowing should be delayed for 3 days after removal of the cover in order to permit the turf to harden and to improve the mowing quality (104).

Types of covers. The most effective dual purpose winter protection covers are (a) a viscose-rayon fiber cover, (b) a plastic shade cloth or screen giving at least 90% shade, (c) a processed wood fiber cover, and (d) four- to six-mil low density polyethylene (104, 108). The first three covers provide dual purpose winter protection against both desiccation and low temperature kill, while the polyethylene cover is effective in preventing winter desiccation but lacks adequate low temperature insulation characteristics. In addition, polyethylene is more prone than the other three covers to wind displacement, lethal heat buildups, and excessive early spring shoot growth. Comparisons among the viscose-rayon fiber, plastic screen, and processed wood fiber covers show that the process wood fiber cover is somewhat inferior to the other two in desiccation protection while the plastic screen is slightly inferior in low temperature protection.

Detailed comparisons among various types of plastic screens reveal that the polypropylene and saran materials with shade potentials of 30 to 50% are suitable under moderate desiccation stress (104). Plastic screens with a shade potential of more than 90% should be used where severe desiccation is anticipated. Comparisons of various colors, types, and thicknesses of polyethylene indicate that a four- to six-mil low density polyethylene is preferred in order to withstand the winter rigors of wind, animals, and human traffic (71, 185, 186). Clear or green polyethylene covers are preferred where rapid early spring green-up and growth are desired. The opaque white polyethylene cover is preferred where controlled early spring green-up and growth are desired.

Physiological Drought

Physiological drought of turfs results from an internal plant water deficit caused by a high external salt concentration. Turfgrasses are damaged by physiological drought in two primary ways. The first involves a high salt concentration in the soil solution surrounding the turfgrass roots. The increased osmotic pressure of the soil solution causes a decrease in water availability to the roots. Thus, turf-grasses growing on saline soils have an increased wilting tendency and require more frequent irrigation in order to prevent drought. Newly emerged turfgrass seedlings are quite prone to physiological drought if a fertilizer high in water soluble salts is applied to the soil at an excessive rate. Other adverse effects of salinity are discussed in Chapter 10.

The second type of physiological drought is caused by the presence of water soluble salts on the surface of turfgrass leaves and stems. It is sometimes called "foliar burn" and can occur during the active growing season or winter dormancy. The higher osmotic pressure of salt particles in contact with the leaves and stems causes the exosmosis of water from the tissue to the salt particles. Wilt and eventually death of the tissue adjacent to the salt particle occurs as water diffuses from the tissue. Leaves are most frequently damaged, with the injured tissue exhibiting a whitish, bleached appearance. This type of physiological drought can be prevented by avoiding excessive application rates when using water soluble fertilizers. The probability of foliar fertilizer burn can also be minimized by washing the salt particles off the shoots and into the soil (Fig. 8-18). The turf should be irrigated immediately after the fertilizer application.

Damage to turfs may occur following a water soluble fertilizer application made during the winter when the soil is frozen or when no water is available to wash the fertilizer into the soil. The blockage of fertilizer movement into the soil results in a high salt concentration around the turfgrass tissues during freezing. Physiological drought injury or kill of the turf is likely to occur in this situation.

Fertilizers vary in their tendency to cause physiological drought (Table 8-3). The **salt index** measures the effect of fertilizers on the soil solution concentration and is expressed as the ratio of increase in osmotic pressure produced by a material to that produced by the same weight of sodium nitrate, which is given a salt index of 100. Better comparisons for applied purposes can be made using the **partial salt**

Figure 8-18. The effect of irrigation (left) immediately after an application of two pounds of actual nitrogen per 1,000 square feet as ammonium nitrate versus the foliar burn which occurred when not irrigated (right).

Table 8-3

THE RELATIVE FOLIAR BURNING TENDENCY AND SALT INDEX OF EIGHT FERTILIZERS COMMONLY
USED ON TURFS [AFTER RADER, WHITE, AND WHITTAKER (147)]

Foliar burn tendency	Fertilizer	Salt index	Partial salt index number
High	Potassium nitrate, 12% N + 33% K	74	5.34
	Ammonium sulfate, 20% N	69	3.25
	Ammonium nitrate, 33% N	105	2.99
Medium	Potassium chloride, 50% K	116	1.94
	Urea, 45% N	75	1.62
Low	Potassium sulfate, 42% K	46	0.85
	Organic nitrogen carriers, 5% N	4	0.70
	Superphosphate, 8% P	8	0.39

index number, which is the salt index per unit of plant nutrient applied. Where
the problem of foliar burn is of primary concern, the use of such fertilizers as the
natural organic nitrogen carriers, superphosphate, and potassium sulfate would
be preferred.

Excess Water

Excessive water is just as detrimental as a deficit. Excess water may result from (a)
poor internal soil drainage; (b) inadequate surface drainage; (c) excessive rainfall;
(d) excessive irrigation; (e) a high water table; or (f) flooding. A detailed discussion
of soil water relations can be found in Chapter 10. Injury to flooded or waterlogged turfgrasses is seldom caused by the direct effect of water but by the lack of
aeration. The diffusion of air is inhibited by saturated soil conditions causing
a rapid decrease in the oxygen content of the soil atmosphere and an increase in
the carbon dioxide concentration. The resulting oxygen deficiency in the soil
restricts root growth (188) and causes a general decline in turfgrass quality, vigor,
and rooting depth (16, 41, 49, 137). There are exceptions such as redtop, which
adapts fairly well to waterlogged soil conditions (96). Low soil oxygen levels are
generally associated with flooded or waterlogged soils. Temporary low oxygen
levels may also occur in compacted or poorly drained soils following an intensive
rain or excessive irrigation since only 1 to 2 hours are required to deplete the soil
oxygen level.

Flooding

It is a common practice to locate parks, golf courses, and recreational areas on
lowlands adjacent to rivers. Turfs on these sites are subject to periodic flooding for
varying lengths of time (Fig. 8-19). Turfs located near coastal areas are also subject
to flooding caused by hurricanes (148) and abnormally high tides. An additional
problem of coastal flooding is the salt content of the water. Flood damage can be

Figure 8-19. A lowland golf course with over sixty percent of the turfgrass area under water.

of three distinct types: (a) erosion, (b) deposition of soil, salt, and debris, and (c) direct submersion injury.

Erosion. The loss of turf and soil by erosion occurs when the turf is subjected to flood waters of a relatively high velocity. Critical turfgrass areas such as greens and tees can be protected from this type of damage by the use of dikes, diversion ditches, and waterways.

Deposition of soil, salt, and debris. The second type of damage results from the soil, salt, and debris deposited on turfs by flood waters (76). Deposition is generally more severe in slow moving water. All wood, metal, and similar debris must be immediately removed at considerable expense in order to avoid interference with the mowing operation and turfgrass injury by light exclusion.

Deposition of the less desirable soil fractions may kill the turf by burial or may disrupt future water movement into the soil. Deposits of 2 in. or more should be removed immediately or incorporated into the soil below by plowing. Removal should be done quickly so that the buried turf is not killed. Turfgrass species probably vary in their tolerance to the depth of soil deposition but no data are available on this aspect of flood damage. It is usually not possible to remove thin layers of 1 in. or less except on close cut greens. Thin depositions of silt, fine sand, or clay may not cause permanent injury to turfs. An undesirable soil layer results, however, that reduces the water infiltration rate and causes a gradual deterioration in soil structure.

The turf should be (a) irrigated to wash off the salt and soil and (b) cultivated by coring or slicing as soon as the soil is removed (148). A layer of silt or salt on the turfgrass leaves and stems can cause serious injury unless washed off as soon as possible. The probability of injury is greater at high temperatures.

Direct injury from submersion. The third type of flood damage involves direct injury to turfs from submersion under water. Some aspects of submersion injury are similar to those observed on waterlogged soils. Injury from submersion is usually not due to any one factor but to a combination of interacting forces (101).

Complete submersion results in soil oxygen depletion within a matter of hours. The lowered oxygen concentration inhibits root respiration. The root hairs die and are not replaced, causing a marked reduction in water and nutrient uptake. The yellowing of submerged turfgrass plants is attributed to a nitrogen deficiency caused by impaired nutrient uptake. Submersion injury may also result from the (a) buildup of certain potentially toxic materials such as ferrous and sulfide ions that are formed by reduction under anaerobic soil conditions, (b) accumulation of toxic organic compounds such as methane or carbon dioxide produced by anaerobic decomposition of soil organic matter, and (c) accumulation of toxic by-products within the plant tissue under anaerobic conditions. The relative importance of these three injury mechanisms has not been investigated for turfgrasses.

Factors influencing submersion injury. The degree of injury from submergence depends on the (a) turfgrass species, (b) duration of submergence, (c) depth of submergence, (d) physiological condition of the plant tissue, (e) temperature, and (f) light intensity.

The turfgrass species vary greatly in submersion tolerance (Table 8-4). Turf-grasses can be killed in 1 day from submersion at a high water temperature of 86°F as in the case of Pennlawn red fescue or may survive submersion durations of more than 60 days at a low water temperature of 50°F as in the case of Toronto creeping bentgrass. The range in submersion tolerance is great enough that the more tolerant turfgrass species should be planted on locations where flooding is anticipated. The physiological basis for submersion tolerance of turfgrass species is not understood. It has been suggested that submersion tolerance involves (a)

Table 8-4

THE RELATIVE SUBMERSION TOLERANCE OF TWELVE TURFGRASSES

Submersion tolerance	Turfgrass species	References
Excellent	Buffalograss	139, 144, 150, 151
	Bermudagrass	43, 69, 150, 151
	Creeping bentgrass	15, 41, 42, 148, 149
Good	Timothy	17, 42, 119, 149
	Rough bluegrass	149
Medium	Meadow fescue	17, 42, 119
	Kentucky bluegrass	15
Fair	Crested wheatgrass	17, 119
	Annual bluegrass	15, 41
	Perennial ryegrass	42, 148
Poor	Red fescue	15, 42, 119, 148, 149
	Centipedegrass	90

a lower oxygen requirement and/or (b) a greater capability to develop intercellular air spaces within the tissue that facilitates air transfer (4).

The degree of injury increases with the depth of submergence (150, 151). Grasses with the leaf tips extending above the water are able to survive much longer than when totally submerged (42). Creeping bentgrass stolons have survived an entire growing season while floating in water with the leaves extending above the surface. Submerged grasses are killed more readily under stagnant water than under running water (42). Repeated frequent flooding also causes increased injury (150, 151). Turfgrasses grown under a high nutritional level tolerate submergence better than when grown at a low level.

Grasses that are dormant or semidormant when flooded tolerate a longer duration of submersion than actively growing plants (150, 151). Dormant buffalograss has survived 19 months of continuous submergence (144). Submersion injury to actively growing turfgrasses usually occurs to tissues that are in the process of development at the time of flooding. The leaf tissues are generally the first to die. The prevailing temperatures at the time of flooding are of major importance in influencing the degree of submersion injury (15, 63, 75, 119). Injury increases as the water temperature is increased (Fig. 8-20). Thus, flooding during midsummer is likely to result in greater injury due to the higher temperatures.

Algae and anaerobic microorganisms. Associated with static flooding and water-logged soils are favorable conditions for the growth of algae and anaerobic microorganisms. Algae are green plants adapted to wet, waterlogged soil conditions. Wet, bare soils or turfs thinned to the extent that light reaches the soil tend to enhance algal development. The algae cover the area with a thick green scum that can smother emerging turfgrass shoots. When dry, the scum forms a tough black layer that may crack and flake. The quantity of growth produced by these orga-

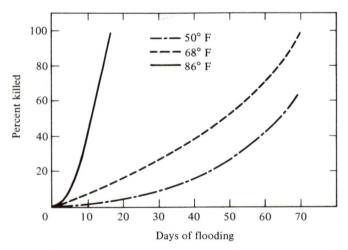

Figure 8-20. The effect of water temperature on the flooding survival of Toronto creeping bentgrass.

nisms can be so great that the surface of the soil is mechanically sealed and the soil pores clogged. This greatly reduces the infiltration and percolation rates.

Turfs injured or weakened by flooding and waterlogged soil conditions should be cultivated as deeply as possible. The removal of cores (a) improves soil aeration, (b) disrupts any surface seal caused by the dead algae and/or microorganisms, and (c) facilitates water movement downward through the soil. Additional spiking, slicing, or raking may be required to break up the scum layer more completely. The application of 2 to 3 lb of hydrated lime per 1000 sq ft serves to (a) neutralize any toxic gases or materials that may have accumulated in the soil, (b) check algal growth, and (c) act as a desiccant. Irrigation should be kept at the minimum required for turfgrass growth to discourage algal growth. Cultural practices that encourage rapid recovery and growth of the turf will discourage the algae because of shading by the grass.

Scald

Scald is the condition that exists when a turfgrass plant collapses and turns brown under standing water, high temperatures, and intense light (30). The typical scald symptom is a burned or scorched appearance on the turfgrass leaves. Injury can range from a few leaf lesions to total kill of the plant (Fig. 8-21). The exact mechanisms of scald injury to turfgrasses has not been studied in detail. The two prerequisites for scald are (a) standing water and (b) high temperatures. Scald injury is more severe at higher than at lower atmospheric relative humidities. Succulent, rapidly growing tissues are more susceptible to scald. No data are available on the relative scald tolerance of various turfgrass species and cultivars.

Scald prevention and treatment. Preventative measures can be taken to minimize the chance of scald injury to turfs. Contours should be designed to permit rapid surface drainage of excess water in order to prevent ponding. Dry wells or tile drainage assists in the rapid removal of surface water in depressions. Soils having a low infiltration rate or poor internal drainage should be avoided, if possible, or steps taken to correct this problem. The frequency and

Figure 8-21. A creeping bentgrass-annual bluegrass green with scald injury in the depressional areas. (Photo courtesy of J. R. Watson, Toro Manufacturing Corp., Minneapolis, Minnesota).

rate of irrigation should be adjusted to the soil infiltration rate in order to avoid surface ponding of water. Small, shallow pools of water should be removed immediately by means of a squeegee type device if they occur on a turfgrass area during high temperature periods.

A scum of algae frequently occurs where the turf has been thinned or killed by scald. The recovery of scald damaged turfs is enhanced by cultivation and possibly the application of hydrated lime to counteract the toxic gases and compounds produced under anaerobic conditions. The specific details are described under flooding, p. 301.

Leaf Exudates

Droplets of water are frequently observed at the leaf tips of turfgrasses. The water droplets may form by guttation from the hydrathodes at the ends of uncut leaves or may be the result of direct wound exudation from the freshly cut ends of turfgrass leaves (Fig. 8-22). Leaf exudation usually occurs during the night or early morning hours although it has been observed to occur in full sunlight on turfgrasses (37). The exudate droplets may persist as long as 14 hr (157). Sometimes the exudation is erroneously called dew. The two can be differentiated by the location of the water droplets. Exudates usually occur at the leaf tip, while dew forms on all leaf surfaces exposed to the cooled atmosphere.

Exudation is enhanced by conditions that stimulate rapid water absorption from the roots and at the same time restrict transpiration. These conditions increase the "root pressure," causing a higher turgor pressure at the leaf tip. Wound exudation occurs readily from freshly cut leaf blades. Exudation of turfs is stimulated by (a) frequent irrigation, (b) high nitrogen fertilization, and (c) close frequent mowing (86). Warm days followed by cool nights favor leaf exudation.

The leaf exudate contains mineral salts, sugars, amino acids, amides, and other organic compounds. Tip burn of leaves from a high solute concentration may result when the water droplet evaporates or is reabsorbed by the leaf (38). The presence of these solutes in the exudate is quite significant in turfgrass culture. The exudation fluid is an ideal medium for the enhancement of fungal and bacterial growth as well as for the infection of turfgrass leaf tissues (84, 85, 86, 199). For example, *Helminthosporium sorokinianum* spores exhibit an increased percentage and rate of germination as well as accelerated germ tube growth when cultured in the guttation fluid of Seaside creeping bentgrass or Merion Kentucky bluegrass (58, 59, 85, 86). The presence of an exudate on bentgrass leaf tips increases the incidence of brown patch (157). In this case, the exudation fluid stimulates the growth and spread of *Rhizoctonia*

Figure 8-22. Exudation droplets on the leaf tips of annual bluegrass leaves.

fungal threads (60). A bridging of fungal threads from one exudate droplet to another can be observed.

The degree of pathogen infection varies with the solute concentration and composition of the exudation fluid (59). Exudates produced by turfgrasses growing under a very low light intensity or a low nitrogen fertility level have a reduced tendency to enhance disease infection (59). Perennial ryegrass exudes considerable quantities of glutamine from the cut ends of leaves when grown under a high ammonium fertilization level (38, 79). Leaf spot infections of creeping bentgrass are stimulated by glutamine (84, 86). Glutathione, another organic constituent of wound exudates, tends to inactivate certain fungicides such as the phenyl mercury compounds (157).

Exudate removal. The exudation droplet may (a) roll, blow off, or be manually removed; (b) evaporate and leave a deposit; or (c) be reabsorbed into the leaf. The break up and removal of exudates is a desirable practice, especially when conditions are favorable for disease development (60, 157). A 60% reduction in brown patch was obtained by daily poling in early morning (47). A determination should be made as to whether the water droplets present in the morning are leaf exudates or dew. If an exudate is present, the preferred method of removal is by syringing since the organic compounds are washed from the leaf surface. Compared to syringing, dragging, brushing, or poling are not as effective in reducing disease development because the fungal mycelium and organic materials are distributed over the entire leaf surface (157). Turfs treated with phenyl mercury acetate are relatively free of leaf exudation (121). Proper air movement, judicious nitrogen fertilization, and minimal irrigation are important factors in preventing leaf exudation problems.

References

1. ALLEN, R. M. 1955. Foliage treatments improve survival of longleaf pine plantings. Journal of Forestry. 53: 724–727.

2. ANONYMOUS. 1931. Moisture requirements of grass, with figures on rainfall for 1925–1930, inclusive. Bulletin of USGA Green Section. 11(8): 154–162.

3. ANONYMOUS. 1966. Rainfall-evapotranspiration data for the United States and Canada. Toro Manufacturing Corporation, Minneapolis, Minn. pp. 1–63.

4. ARIKADO, H. 1964. Studies on the development of the ventilating system in relation to the tolerance against excess-moisture injury in various crops. Nihon Sakumotsu Gakkai Proceedings. 32: 353–357.

5. ASLYNG, H. C. 1965. Rain, snow and dew measurements. Acta Agriculturae Scandinavica. 15: 275–283.

6. AUBERTIN, G. M., and G. W. GORSLINE. 1964. Effect of fatty alcohol on evaporation and transpiration. Agronomy Journal. 56: 50–52.

7. BAIER, W. 1966. Studies on dew formation under semi-arid conditions. Agricultural Meteorology. 3: 103–112.

8. BAILEY, L. F. 1940. Some water relations of three western grasses. II. Drought resistance. III. Root developments. American Journal of Botany. 27: 129–135.

9. BALDWIN, H. I. 1928. A comparison of the available moisture in sod and open soil by the soil-point method. Bulletin of the Torrey Botanical Club. 55: 251–255.

10. BARNETT, N. M., and A. W. NAYLOR. 1966. Amino acid and protein metabolism in bermuda grass during water stress. Plant Physiology. 41(7): 1222–1230.

11. BASS, L. N. 1953. Relationships of temperature, time and moisture content to the viability of seeds of Kentucky bluegrass. Proceedings of the Iowa Academy of Science. 60: 86–88.

12. BEARD, J. B. 1966. Winter injury. Golf Superintendent. 34(1): 24–30.

13. ———. 1966. Selected turfgrass variety evaluations. Winterkill and drought tolerance. Michigan Turfgrass Report. 3(1): 7–9.

14. ———. 1966. Direct low temperature injury of nineteen turfgrasses. Quarterly Bulletin of the Michigan Agricultural Experiment Station. 48(3): 377–383.

15. BEARD, J. B., and D. P. MARTIN. 1968. Submersion tolerance of creeping bentgrass, Kentucky bluegrass, annual bluegrass and red fescue as affected by water temperature. 1968 Agronomy Abstracts. p. 62.

16. BENNETT, O. L., and B. D. DOSS. 1960. Effect of soil moisture level on root distribution of cool-season forage species. Agronomy Journal. 52: 204–207.

17. BOLTON, J. L., and R. E. MCKENZIE. 1946. The effect of early spring flooding on certain forage crops. Scientific Agriculture. 26(3): 99–105.

18. BRIGGS, L. J., and H. L. SHANTZ. 1912. The wilting coefficient for different plants and its indirect determination. USDA Bureau of Plant Industry Bulletin No. 230. pp. 1–83.

19. ———. 1914. Relative water requirement of plants. Journal of Agricultural Research. 3(1): 1–64.

20. BROWN, E. M. 1939. Some effects of temperature on the growth and chemical composition of certain pasture grasses. Missouri Agricultural Experiment Station Research Bulletin 299. pp. 1–56.

21. ———. 1943. Seasonal variations in the growth and chemical composition of Kentucky bluegrass. Missouri Agricultural Experiment Station Research Bulletin 360. pp. 1–56.

22. BROWN, W. V., and S. C. JOHNSON. 1962. The fine structure of grass guard cells. American Journal of Botany. 49: 110–115.

23. BURTON, G. W., E. H. DEVANE, and R. L. CARTER. 1954. Root penetration, distribution, and activity in southern grasses measured by yields, drought symptoms and P^{32} uptake. Agronomy Journal. 46: 229–233.

24. BURTON, G. W., G. M. PRINE, and J. E. JACKSON. 1957. Studies of drought tolerance and water use of several southern grasses. Agronomy Journal. 49: 498–503.

25. CARROLL, J. C. 1943. Effects of drought, temperature, and nitrogen on turf grasses. Plant Physiology. 18: 19–36.

26. ———. 1943. Atmospheric drought tests of some pasture and turf grasses. Journal of the American Society of Agronomy. 35: 77–79.

27. CHEESMAN, J. H., E. C. ROBERTS, and L. H. TIFFANY. 1965. Effects of nitrogen level and osmotic pressure of the nutrient solution on incidence of *Puccinia graminis* and *Helminthosporium sativum* infection in Merion Kentucky bluegrass. Agronomy Journal. 57: 599–602.

28. CLOUSTON, D. 1962. Identification of grasses in non-flowering condition. Published by the Sports Turf Research Institute, Bingley, Yorkshire, England. pp. 1–31.

29. COMAR, C. L., and C. G. BARR. 1944. Evaluation of foliage injury and water loss in connection with use of wax and oil emulsions. Plant Physiology. 19: 90–104.

30. COMMITTEE on Crop Terminology. 1962. Summary of terms. Crop Science. 2(1): 85–87.

31. COOK, C. W. 1943. A study of the roots of *Bromus inermis* in relation to drought resistance. Ecology. 24: 169–182.

32. CORDUKES, W. E., J. WILNER, and V. T. ROTHWELL. 1966. The evaluation of cold and drought stress of turfgrasses by electrolytic and ninhydrin methods. Canadian Journal of Plant Science. 46: 337–342.

33. CORMACK, M. W., and J. B. LEBEAU. 1959. Snow mold infection of alfalfa, grasses, and winter wheat by several fungi under artificial conditions. Canadian Journal of Botany. 37: 685–693.

34. COUCH, H. B. 1962. Diseases of turfgrasses. Reinhold Publishing Corp., New York. pp. 1–289.

35. ———. 1966. Relationship between soil moisture, nutrition and severity of turfgrass diseases. Journal of the Sports Turf Research Institute. 11(42): 54–64.

36. COUCH, H. B., and J. R. BLOOM. 1960. Influence of environment on diseases of turfgrasses. II. Effect of nutrition, pH, and soil moisture on *Sclerotinia* dollar spot. Phytopathology. 50: 761–763.

37. CURTIS, L. C. 1943. Deleterious effects of guttated fluids on foliage. American Journal of Botany. 30: 778–781.

38. ———. 1944. The exudation of glutamine from lawn grass. Plant Physiology. 19(1): 1–5.

39. DAHL, A. S. 1933. Effect of temperature and moisture on occurrence of brownpatch. Bulletin of USGA Green Section. 13(3): 53–61.

40. ———. 1933. Effect of watering putting greens on occurrence of brownpatch. Bulletin of USGA Green Section. 13(3): 62–66.

41. ———. 1934. The relation between rainfall and injuries to turf—season 1933. Greenkeepers' Reporter. 2(2): 1–4.

42. DAVIS, A. G., and B. F. MARTIN. 1949. Observations on the effect of artificial flooding on certain herbage plants. Journal of the British Grassland Society. 4: 63–64.

43. DE GRUCHY, J. H. B. 1952. Water fluctuation as a factor in the life of the higher plants of a 3300 acre lake in the Permian Red Beds of Central Oklahoma. Ph.D. Thesis. Oklahoma State University. pp. 1–117.

44. DENMEAD, O. T. 1964. Evaporation sources and apparent diffusivities in a forest canopy. Journal of Applied Ecology. 3: 383–389.

45. DE VRIES, D. A., and J. W. BIRCH. 1961. The modification of climate near the ground by irrigation for pastures on the Riverine plain. Australian Journal of Agricultural Research. 12: 260–272.

46. DEXTER, S. T. 1937. The drought resistance of quack grass under various degrees of fertilization with nitrogen. Journal of the American Society of Agronomy. 29: 568–576.

47. DICKINSON, L. S. 1930. The effect of air temperature on the pathogenicity of *Rhizoctonia solani* parasitizing grasses on putting-green turf. Phytopathology. 20(8): 597–608.

48. DILLMAN, A. C. 1931. The water requirement of certain crop plants and weeds in the northern Great Plains. Journal of Agricultural Research. 42(4): 187–238.

49. DOSS, B. D., D. A. ASHLEY, and O. L. BENNETT. 1960. Effect of soil moisture regime on root distribution of warm season forage species. Agronomy Journal. 52: 569–572.

50. DOSS, B. D., O. L. BENNETT, and D. A. ASHLEY. 1964. Moisture use by forage species as related to pan evaporation and net radiation. Soil Science. 98: 322–327.

51. DOSS, B. D., O. L. BENNETT, D. A. ASHLEY, and H. A. WEAVER. 1962. Soil moisture regime effect on yield and evapotranspiration from warm season perennial forage species. Agronomy Journal. 54: 239–242.

52. DRUMMOND, A. J. 1945. The persistence of dew. Quarterly Journal of the Royal Meteorological Society. 71: 415–417.

53. DUVDEVANI, S. 1964. Dew in Israel and its effect on plants. Soil Science. 98: 14–21.

54. DUVDEVANI, S., J. REICHERT, and J. PALTI. 1946. The development of downy and powdery mildew of cucumber as related to dew and other environmental factors. Palestine Journal of Botany, Rehovot Series. 5(2): 127–151.

55. EAGLEMAN, J. R., and W. L. DECKER. 1965. The role of soil moisture in evapotranspiration. Agronomy Journal. 57: 626–629.

56. EKERN, P. C. 1966. Evapotranspiration by bermudagrass sod, *Cynodon dactylon* L. Pers., in Hawaii. Agronomy Journal. 58: 387–390.

57. EMERSON, W. W. 1954. Water conduction by severed grass roots. Journal of Agricultural Science. 45: 241–245.

58. ENDO, R. M., and J. J. OERTLI. 1964. Stimulation of fungal infection of bentgrass. Nature. 201(4916): 313.

59. ENDO, R. M., and R. H. AMACHER. 1964. Influence of guttation fluid on infection structures of *Helminthosporium sorokinianum.* Phytopathology. 54(11): 1327–1334.

60. ENDO, R. M. 1967. The role of guttation fluid in fungal disease development. California Turfgrass Culture. 17(2): 12–13.

61. ENGEL, R. E., and L. M. CALLAHAN. 1964. Kentucky bluegrass response to pre-emergence herbicide residues in a turfgrass soil. 1964 Agronomy Abstracts. p. 105.

62. ———. 1967. Merion Kentucky bluegrass response to soil residue of preemergence herbicides. Weeds. 15(2): 128–130.

63. FINN, B. J., S. J. BOURGET, K. F. NIELSEN, and B. K. DOW. 1961. Effects of different soil moisture tensions on grass and legume species. Canadian Journal of Soil Science. 41(1): 16–23.

64. FITTS, O. B. 1924. Early morning watering as an aid to brownpatch control. Bulletin of the Green Section of USGA. 4(7): 159.

65. FREE, G. R. 1936. A comparison of soil moisture under continuous corn and bluegrass sod. Journal of the American Society of Agronomy. 28: 359–363.

66. FUNK, C. R., and R. E. ENGEL. 1963. Effect of cutting height and fertilizer on species composition and turf quality ratings of various turfgrass mixtures. Rutgers University Short Course on Turfgrass Management. pp. 47–56.

67. GALE, J. 1961. Studies on plant antitranspirants. Physiologia Plantarum. 14: 777–786.

68. GALE, J., A. POLJAKOFF-MAYBER, I. NIR, and I. KAHANE. 1964. Preliminary trials of the application of antitranspirants under field conditions to vines and bananas. Australian Journal of Agricultural Research. 15: 929–936.

69. GAMBLE, M. D. 1958. Inundation tolerance of bermudagrass. 1958 Agronomy Abstracts. p. 37.

70. GANE, R. 1948. The effect of the temperature, humidity and atmosphere on the viability of chewing's fescue grass seed in storage. Journal of Agricultural Science. 38: 90–92.

71. GARD, H. 1962. Quebec tests of plastic covers used on greens. Golf Course Reporter. 30(9): 40–43.

72. GARDNER, J. L. 1942. Studies in tillering. Ecology. 23: 162–174.

73. GASKIN, T. A. 1966. Testing for drought and heat resistance in Kentucky bluegrass. Agronomy Journal. 58: 461–462.

74. GIBEAULT, V. A., and C. R. SKOGLEY. 1967. Effects of DMPA (Zytron) on colonial bentgrass, Kentucky bluegrass, and red fescue root growth. Crop Science. 7: 327–329.

75. GILBERT, W. B., and D. S. CHAMBLEE. 1965. Effect of submersion in water on tall fescue, orchardgrass, and ladino clover. Agronomy Journal. 57: 502–504.

76. GLOVER, W. H. 1937. Flood waters at Shawnee C. C. Greenkeepers' Reporter. 5(2): 12–40.

77. GOULD, C. J. 1963. How climate affects our turfgrass diseases. Proceedings of the 17th Annual Northwest Turfgrass Conference. pp. 29–43.

78. GRANT, C. L. 1938. Plant structure as influenced by soil moisture. Proceedings of the Indiana Academy of Science. 48: 67–70.

79. GREENHILL, A. W., and A. C. CHIBNALL. 1934. The exudation of glutamine from perennial rye-grass. Biochemical Journal. 28(2): 1422–1427.

80. HAGAN, R. M. 1954. Water requirements of turfgrasses. Proceedings of the 8th Annual Texas Turfgrass Conference. pp. 9–20.

81. ———. 1955. Watering lawns and turf and otherwise caring for them. Water, the Yearbook of Agriculture for 1955, USDA. pp. 462–477.

82. Harrold, L. L., and F. R. Dreibelbis. 1951. Agricultural hydrology as evaluated by monolith lysimeters. USDA Technical Bulletin No. 1050. pp. 1–149.

83. ———. 1955. Evaluation of agricultural hydrology by monolith lysimeters. USDA Technical Bulletin No. 1179. pp. 1–166.

84. Healy, M. J. 1965. Factors influencing disease development on putting green turf. 6th Illinois Turfgrass Conference Proceedings. pp. 1–7.

85. ———. 1967. Factors affecting the pathogenicity of selected fungi isolated from putting green turf. Ph.D. Thesis. University of Illinois. pp. 1–59.

86. Healy, M. J., and M. P. Britton. 1967. The importance of guttation fluid on turf diseases. 8th Illinois Turfgrass Conference Proceedings. 8: 40–46.

87. Henrici, M. 1926. Wilting and osmotic phenomena of grasses and other plants under arid conditions. 11th Report of the Department of Agriculture, Union of South Africa. 11: 618–668.

88. ———. 1926. Transpiration of grasses and other plants under arid conditions. 11th Report of the Department of Agriculture, Union of South Africa. 11: 669–702.

89. Horn, G. C. 1965. Excess water and turfgrass management. I. Florida Turf-grass Association Bulletin. 12(2): 1–3.

90. ———. 1966. Excess water and turfgrass management. II. Florida Turf-grass Association Bulletin. 13(1): 3–8.

91. Howard, F. L., J. B. Rowell, and H. L. Keil. 1951. Fungus diseases of turf grasses. Rhode Island Agricultural Experiment Station Bulletin 308. pp. 1–56.

92. Hurcombe, R. 1948. A cytological and morphological study of cultivated *Cynodon* species. Experiments with *Cynodon dactylon* and Other Species at the South African Turf Research Station. pp. 36–47.

93. Iljin, W. S. 1931. Austrocknungeresistenz des Farnes Notochlaena Marantae R. Br. Protoplasma. 13: 322–330.

94. Jackson, W. T. 1962. Effect of sugars on rate of elongation of root hairs of *Agrostis alba* L. Physiologia Plantarum. 15: 675–682.

95. Julander, O. 1945. Drought resistance in range and pasture grasses. Plant Physiology. 20: 573–599.

96. Kauter, A. 1933. Beiträge zur Kenntnis der Wurzelwachstums der Gräser. Ber. Schweiz. Bot. Ges. 42:37–109.

97. Kemble, A. R., and H. T. Macpherson. 1954. Liberation of amino acids in perennial rye grass during wilting. Biochemical Journal. 58: 46–49.

98. Klomp, G. J. 1939. A comparison of the drought resistance of selected native and naturalized grasses. M.S. Thesis. Iowa State College. pp. 1–73.

99. Knochenhauer, O. W. 1934. Inwieweit sind die Temperatur-und Feuchtigkeitsmassungen unserer Flughafen repräsentativ? Erf. Ber. d. Dt. Flugwetterd. 9. Folge Nr. 2: 15–17.

100. Kramer, P. J. 1933. The intake of water through dead root systems and its relation to the problem of absorption by transpiring plants. American Journal of Botany. 20: 481–492.

101. ———. 1951. Causes of injury to plants resulting from flooding of the soil. Plant Physiology. 26: 722–736.

102. Krogman, K. K. 1967. Evapotranspiration by irrigated grass as related to fertilizer. Canadian Journal of Plant Science. 47: 281–287.

103. Laude, H. M. 1953. The nature of summer dormancy in perennial grasses. Botanical Gazette. 114: 284–292.

104. Ledeboer, F. B., and C. R. Skogley. 1967. Plastic screens for winter protection. Golf Superintendent. 35(8): 22–33.

105. Lee, B., and J. H. Priestley. 1924. The plant cuticle. Annals of Botany. 38: 525–545.

106. LE ROUX, M., and H. WEINMANN. 1948. Studies on the water relations of *Cynodon dactylon*. Experiments with *Cynodon dactylon* and Other Species at the South African Turf Research Station. pp. 52–55.

107. LETEY, J., O. R. LUNT, L. H. STOLZY, and T. E. SZUSZKIEWICZ. 1961. Plant growth, water use, and nutritional response to rhizosphere differentials of oxygen concentration. Soil Science Society of America Proceedings. 25: 183–186.

108. LEWTON-BRAIN, L. 1903. On the anatomy of the leaves of British grasses. Transactions of the Linnean Society of London, Botany. 6: 315–359.

109. LINACRE, E. T. 1966. Resistance impeding the diffusion of water vapor from leaves and crops. Ph.D. Thesis. University of London. pp. 1–81.

110. LIVINGSTON, B. E., and I. OHGA. 1926. The summer march of soil moisture conditions as determined by porous porcelain soil points. Ecology. 7: 427–439.

111. LLOYD, M. G. 1961. The contribution of dew to the summer water budget of northern Idaho. Bulletin of the American Meteorological Society. 42: 572–580.

112. LUCANUS, R., K. J. MITCHELL, G. G. PRITCHARD, and D. M. CALDER. 1960. Factors influencing survival of strains of ryegrass during the summer. New Zealand Journal of Agricultural Research. 3: 185–193.

113. MADISON, J. H. 1959. Drouth—1959. Proceedings of Northern California Turfgrass Institute for 1959. pp. 12–19.

114. MADISON, J. H., and R. M. HAGAN. 1962. Extraction of soil moisture by Merion bluegrass (*Poa pratensis* L. 'Merion') turf, as affected by irrigation frequency, mowing height, and other cultural operations. Agronomy Journal. 54: 157–160.

115. MANTELL, A. 1966. Effect of irrigation frequency and nitrogen fertilization on growth and water use of a kikuyugrass lawn (*Pennisetum clandestinum* Hochst.). Agronomy Journal. 58: 559–561.

116. MARSHALL, H., and T. E. MAKI. 1946. Transpiration of pine seedlings as influenced by foliage coatings. Plant Physiology. 21: 95–101.

117. McALPINE, A. N. 1890. How to know grasses by the leaves. Darien Press, Edinburgh, Scotland. pp. 1–92.

118. McALISTER, D. F. 1944. Determination of soil drought resistance in grass seedlings. Journal of the American Society of Agronomy. 36: 324–336.

119. McKENZIE, R. E. 1951. The ability of forage plants to survive early spring flooding. Scientific Agriculture. 31: 358–367.

120. McKENZIE, R. E., L. J. ANDERSON, and D. H. HEINRICHS. 1949. The effect of flooding on emergence of forage crop seeds. Scientific Agriculture. 29: 237–240.

121. MEUSEL, H. W. 1964. What makes grass wilt? Golf Course Reporter. 32(3): 24–38.

122. MILLER, E. J., V. R. GARDNER, H. G. PETERING, C. L. COMAR, and A. L. NEAL. 1950. Studies on the development, preparation, properties, and applications of wax emulsions for coating nursery stock and other plant materials. Michigan Agricultural Experiment Station Technical Bulletin 218. pp. 1–78.

123. MILLER, E. J., J. A. NEILSON, and S. L. BANDEMER. 1937. Wax emulsions for spraying nursery stock and other plant materials. Michigan Agricultural Experiment Station Special Bulletin No. 282. pp. 1–39.

124. MITCHELL, K. J., and J. R. KERR. 1966. Differences in rate of use of soil moisture by stands of perennial ryegrass and white clover. Agronomy Journal. 58: 5–8.

125. MONTEITH, J. 1933. A *Pythium* disease of turf. Phytopathology. 23(1): 23–24.

126. MONTEITH, J. L. 1957. Dew. Quarterly Journal of the Royal Meteorological Society. 83: 322–341.

127. MONTEITH, J., and A. S. DAHL. 1932. Turf diseases and their control. Bulletin of USGA Green Section. 12(4): 86–187.

128. MOORE, L. D., H. B. COUCH, and J. R. BLOOM. 1963. Influence of environment on diseases of turfgrasses. III. Effect of nutrition, pH, soil temperature, air temperature, and soil moisture on *Pythium* blight of Highland bentgrass. Phytopathology. 53: 53–57.

129. MUELLER, I. M., and J. E. WEAVER. 1942. Relative drought resistance of seedlings of dominant prairie grasses. Ecology. 23: 387–398.

130. MUSE, R. R., and H. B. COUCH. 1965. Influence of environment on diseases of turfgrasses. IV. Effect of nutrition and soil moisture on *Corticium* red thread of creeping red fescue. Phytopathology. 55: 507–510.

131. NEWTON, R., and W. M. MARTIN. 1930. Physico-chemical studies on the nature of drought resistance in crop plants. Canadian Journal of Research. 3: 336–427.

132. NOER, O. J. 1962. Spring and winter: desiccation problems. Golf Course Reporter. 30(3): 60–64.

133. OERTLI, J. J. 1963. Effects of fatty alcohols and acids on transpiration of plants. Agronomy Journal. 55: 137–138.

134. OLMSTED, C. E. 1941. Growth and development in range grasses. I. Early development of *Bouteloua curtipendula* in relation to water supply. Botanical Gazette. 102: 499–519.

135. ———. 1942. Growth and development in range grasses. II. Early development of *Bouteloua curtipendula* as affected by drought periods. Botanical Gazette. 103: 531–542.

136. OLSEN, S. R., F. S. WATANABE, W. D. KEMPER, and F. E. CLARK. 1962. Effect of hexadecanol and octadecanol on efficiency of water use and growth of corn. Agronomy Journal. 54: 544–545.

137. OSVALD, H. 1919. Untersuchungen über die Einwirkung des Grundwasserstands auf die Bewurzelung von Wiesenpflanzen auf Moorboden. Frühlings Landw. Z. 68: 321–386.

138. PALTRIDGE, T. B., and H. K. C. MAIR. 1936. Studies of selected pasture grasses. The measurement of the xerophytism of any species. Commonwealth of Australia Council for Scientific and Industrial Research. Bulletin No. 102. pp. 1–38.

139. PARKER, J. M., and C. J. WHITFIELD. 1941. Ecological relationships in playa lakes in the southern great plains. Journal of the American Society of Agronomy. 33: 125–129.

140. PARTRIDGE, N. L. 1941. Comparative water usage and depth of rooting of some species of grass. Proceedings of the American Society for Horticultural Science. 39: 426–432.

141. PETERS, D. B., and W. J. ROBERTS. 1963. Use of octa-hexadecanol as a transpiration suppressant. Agronomy Journal. 55: 79.

142. PETRIE, A. H. K., and J. G. WOOD. 1938. Studies on the nitrogen metabolism of plants. I. The relation between the content of protein, amino-acids, and water in the leaves. Annals of Botany N. S. 2: 33–60.

143. ———. 1938. Studies on the nitrogen metabolism of plants. III. On the effect of water content on the relationship between proteins and amino-acids. Annals of Botany N. S. 2: 887–898.

144. PORTERFIELD, H. G. 1945. Survival of buffalo grass following submersion in playas. Ecology. 26: 98–100.

145. PRUITT, W. O. 1964. Evapotranspiration—a guide to irrigation. California Turfgrass Culture. 14(4): 27–32.

146. QUACKENBUSH, T. H., and J. T. PHELAN. 1965. Irrigation water requirements of lawns. Journal of the Irrigation and Drainage Division, ASCE. 91(IR2): 11–19.

147. RADER, L. F., L. M. WHITE, and C. W. WHITTAKER. 1943. The salt index—a measure of the effect of fertilizers on the concentration of the soil solution. Soil Science. 55: 201–218.

148. RADKO, A. M. 1956. Hurricane damage in the northeast. USGA Journal and Turf Management. 9(3): 13–16.

149. REYNTENS, H. 1949. Onderzoek betreffende de weerstand tegen overstroming van verschillende gras—in klaversoortan en waarden der grasflora in Dender—, Schelde—en Durme-

valleri. Mededenlingen van de Landbouurhogeschool en de Opzoekingsstations van de Staat te Gent. Deel XIV, Nr. 3.

150. RHOADES, E. D. 1964. Inundation tolerance of grasses in flooded areas. Transactions of the ASAE. 7(2): 164–169.

151. ———. 1967. Grass survival in flooded pool areas. Journal of Soil and Water Conservation. 22(1): 19–21.

152. RICH, S. 1963. The role of stomata in plant disease. Connecticut Agriculture Experiment Station Bulletin 664. pp. 102–114.

153. RICHARDS, S. J., and L. V. WEEKS. 1963. Evapotranspiration for turf measured with automatic irrigation equipment. California Agriculture. 17(7): 12–13.

154. RIDER, N. E., and G. D. ROBINSON. 1951. A study of the transfer of heat and water vapor above a surface of short grass. Quarterly Journal of the Royal Meteorlogical Society. 77(33): 375–401.

155. ROBERTS, E. C., and D. P. LAGE. 1965. Effects of an evaporation retardant, a surfactant, and an osmotic agent on foliar and root development of Kentucky bluegrass. Agronomy Journal. 57(1): 71–74.

156. ROSEVEARE, G. M. 1947. Reseeding flooded lands. Journal of the British Grassland Society. 2: 226–229.

157. ROWELL, J. B. 1951. Observations on the pathogenicity of *Rhizoctonia solani* on bentgrass. Plant Disease Reporter. 35(5): 240–242.

158. RUNNELS, H. A., and J. D. WILSON. 1934. The influence of certain spray materials, herbicides and other compounds on the desiccation of plant tissue. Ohio Agricultural Experiment Station Bimonthly Bulletin. 19(166): 104–109.

159. SAVAGE, D. A., and L. A. JACOBSON. 1935. The killing effect of heat and drought on buffalo grass and blue grama grass at Hays, Kansas. Journal of the American Society of Agronomy. 27: 566–582.

160. SAYRE, J. D. 1919. The relation of hairy leaf coverings to the resistance of leaves to transpiration. Ohio Journal of Science. 20(1): 55–75.

161. SCHARDT, A. 1925. Brown-patch control resulting from early-morning work on greens. Bulletin of the Green Section of USGA. 5(11): 254–255.

162. SCHULTZ, H. K., and H. K. HAYES. 1938. Artificial drought tests of some hay and pasture grasses and legumes in sod and seedling stages of growth. Journal of the American Society of Agronomy. 30: 676–682.

163. SHIMSHI, D. 1963. Effect of chemical closure of stomata on transpiration in varied soil and atmospheric environments. Plant Physiology. 38: 709–712.

164. ———. 1963. Effect of soil moisture and phenylmercuric acetate upon stomatal aperture, transpiration, and photosynthesis. Plant Physiology. 38: 713–721.

165. SILKER, T. H. 1941. Effect of clipping upon the forage production, root development, establishment, and subsequent drought resistance of western and crested wheatgrass seedlings. M.S. Thesis. Iowa State College. pp. 1–69.

166. SLATYER, R. O., and J. F. BIERHUIZEN. 1964. The effect of several foliar sprays on transpiration and water use efficiency of cotton plants. Agricultural Meteorology. 1: 42–53.

167. ———. 1964. The influence of several transpiration suppressants on transpiration, photosynthesis, and water-use efficiency of cotton leaves. Australian Journal of Biological Sciences. 17: 131–146.

168. SMITH, D., and K. P. BUCHOLTZ. 1964. Modification of plant transpiration rate with chemicals. Plant Physiology. 39(4): 572–578.

169. SMITH, J. D. 1953. Fungi and turf diseases. Journal of the Sports Turf Research Institute. 8(29): 230–252.

170. SPRAGUE, H. B. 1933. Root development of perennial grasses and its relation to soil conditions. Soil Science. 36: 189–209.

171. SPRAGUE, V. G. 1955. Distribution of atmospheric moisture in the microclimate above a grass sod. Agronomy Journal. 47: 551–555.

172. SPRAGUE, V. G., and L. F. GRABER. 1938. The utilization of water by alfalfa (*Medicago sativa*) and by bluegrass (*Poa pratensis*) in relation to managerial treatments. Journal of the American Society of Agronomy. 30: 986–997.

173. STODDARD, E. M., and P. M. MILLER. 1962. Chemical control of water loss in growing plants. Science. 137: 224–225.

174. STODDART, L. A. 1935. Osmotic pressure and water content of prairie plants. Plant Physiology. 10: 661–680.

175. STONE, E. C. 1957. Dew as an ecological factor. I. A review of the literature. Ecology. 38: 407–413.

176. ———. 1963. The ecological importance of dew. Quarterly Review of Biology. 38: 328–341.

177. TANNER, C. B. 1963. Plant temperatures. Agronomy Journal. 55: 210–211.

178. TEW, R. K., S. A. TAYLOR, and G. L. ASHCROFT. 1963. Influence of soil temperature on transpiration under various environmental conditions. Agronomy Journal. 55: 558–560.

179. VAN BAVEL, C. H. M., and D. G. HARRIS. 1962. Evapotranspiration rate from bermudagrass and corn at Raleigh, North Carolina. Agronomy Journal. 54: 319–322.

180. WAGGONER, P. E. 1965. Relative effectiveness of change in upper and lower stomatal openings. Crop Science. 5(4): 291–294.

181. WAGGONER, P. E., J. L. MONTEITH, and G. SZEICZ. 1964. Decreasing transpiration of field plants by chemical closure of stomata. Nature. 201: 97–98.

182. WATERHOUSE, F. L. 1950. Humidity and temperature in grass microclimates with reference to insolation. Nature. 166(4214): 232–233.

183. ———. 1955. Microclimatological profiles in grass cover in relation to biological problems. Quarterly Journal of the Royal Meteorological Society. 81: 63–71.

184. WATSON, J. R. 1968. Blankets to protect golf greens against winter injury. 1968 Agronomy Abstracts. p. 61.

185. WATSON, J. R., and L. WICKLUND. 1962. Plastic covers protect greens from winter damage. Golf Course Reporter. 30(9): 30–38.

186. WATSON, J. R., H. KROLL, and L. WICKLUND. 1960. Protecting golf greens against winterkill. Golf Course Reporter. 28(7): 10–16.

187. WEAVER, J. E., and F. W. ALBERTSON. 1943. Resurvey of grasses, forbs, and underground plant parts at the end of the great drought. Ecological Monographs. 13: 63–117.

188. WEAVER, J. E., and W. J. HIMMEL. 1930. Relation of increased water content and decreased aeration to root development in hydrophytes. Plant Physiology. 5: 69–92.

189. WEAVER, R. J. 1941. Water usage of certain native grasses in prairie and pasture. Ecology. 22: 175–192.

190. WELTON, F. A., and J. D. WILSON. 1931. Water-supplying power of the soil under different species of grass and with different rates of water application. Plant Physiology. 6: 485–493.

191. ———. 1938. Comparative rates of water loss from soil, turf, and water surfaces. Bimonthly Bulletin of the Ohio Agricultural Experiment Station. 23(190): 13–16.

192. WHITMAN, W. C. 1941. Seasonal changes in bound water content of some prairie grasses. Botanical Gazette. 103: 38–63.

193. WILLIAMSON, R. E. 1963. The effect of a transpiration-suppressant on tobacco leaf temperature. Soil Science Society of America Proceedings. 27: 106.

194. WILSON, J. D. 1927. The measurement and interpretation of the water-supplying power of the soil with special reference to lawn grasses and some other plants. Plant Physiology. 2: 384–440.

195. WILSON, J. D., and B. E. LIVINGSTON. 1932. Wilting and withering of grasses in greenhouse cultures as related to water-supplying power of the soil. Plant Physiology. 7: 1–34.

196. WILSON, J. D., and H. A. RUNNELS. 1931. Bordeaux mixture as a factor increasing drought injury. Phytopathology. 21: 729–738.

197. ———. 1934. Transpirational response of various plants to bordeaux mixture. Bimonthly Bulletin of the Ohio Agricultural Experiment Station. 19(166): 198–202.

198. WILSON, J. D., and F. A. WELTON. 1935. The use of an evaporation index in watering lawns. Bimonthly Bulletin of the Ohio Agricultural Experiment Station No. 174. 20(174): 112–119.

199. WILSON, J. K. 1923. The nature and reaction of water from hydrathodes. Cornell University Experiment Station Memoir 65. pp. 1–11.

200. WIND, G. P. 1955. Flow of water through plant roots. Netherlands Journal of Agricultural Science. 3: 259–264.

201. WOOD, G. M., and H. E. BUCKLAND. 1966. Survival of turfgrass seedlings subjected to induced drouth stress. Agronomy Journal. 58: 19–23.

202. WOOLLEY, J. T. 1967. Relative permeabilities of plastic films to water and carbon dioxide. Plant Physiology. 42(5): 641–643.

203. YARWOOD, C. E. 1939. Relation of moisture to infection with some downy mildews and rusts. Phytopathology. 29: 933–945.

204. YOUNG, A. A., and H. F. BLANEY. 1942. Use of water by native vegetation. California Department of Public Works, Division of Water Resources Bulletin No. 50. pp. 1–160.

205. ZELITCH, I. 1961. Biochemical control of stomatal opening in leaves. Proceedings of the National Academy of Science. 47(9): 1423–1433.

206. ZELITCH, I., and P. E. WAGGONER. 1962. Effect of chemical control of stomata on transpiration and photosynthesis. Proceedings of the National Academy of Sciences. 48(7): 1101–1108.

207. ———. 1962. Effect of chemical control of stomata on transpiration of intact plants. Proceedings of the National Academy of Sciences. 48(8): 1297–1299.

chapter 9

Air

Introduction

The aerial environment surrounding a turf encompasses the atmosphere and its constituents. It is a turbulent stratum where the climate and weather patterns of the earth are generated. The gaseous phase of the atmosphere is a mechanical mixture containing approximately 78% nitrogen (N), 21% oxygen (O), 0.93% argon (Ar), and 0.03% carbon dioxide (CO_2) by volume. Of these four gases, only CO_2 shows any degree of variation that might be significant in plant growth. Water vapor, dust, smoke, industrial gases, smog, and microorganisms also occur in the atmosphere. The concentration varies greatly with location and season.

Plant Carbon Dioxide and Oxygen Relations

The levels of CO_2 and O_2 in the atmosphere are surprisingly constant from year to year in view of the dynamic gaseous exchange of living organisms. The O_2 evolved from the photosynthetic process of plants balances the O_2 utilized in respiration, combustion, and certain geologic processes. The CO_2 balance in the atmosphere is maintained primarily by the oceans, which have a great buffering potential due to the large capacity for dissolving CO_2.

Local variations in the CO_2 content are observed with higher levels occurring in the vicinity of urban areas having extensive industrial activity. Microvariations in CO_2 are common. Higher concentrations are found in wooded areas and adjacent to decaying organic matter such as logs and stumps (45). The soil is also a source of CO_2 that is evolved from the respiration of soil microorganisms. Micro-differences in CO_2 can be minimized by wind turbulence and eddy currents (10, 23, 30, 45).

A diurnal CO_2 fluctuation exists with high levels occurring during the night, a maximum in early morning, and a steady decline with the advent of full sunlight (10, 22, 31, 43) (Fig. 9-1). At midday, CO_2 levels 10 to 12% below normal have been found in green canopies having a very active photosynthetic rate (10, 30, 31, 43). No detailed information is available on the CO_2 regime within a turfgrass community.

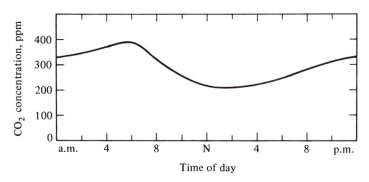

Figure 9-1. The average diurnal variation in the atmospheric carbon dioxide content within a Kentucky bluegrass turf for a selected period in May.

The CO_2 assimilated by turfs is obtained primarily by downward turbulent transfer from the atmosphere and by upward diffusion from the soil. The exchange rate of gases between the turfgrass leaf and the atmosphere is influenced by atmospheric and internal plant resistances. The resistances to diffusion include the (a) presence or absence of air movement, (b) stomatal size and distribution, (c) presence of a cuticle layer, (d) extent of intercellular space, and (e) mesophyll cells.

Photosynthesis can be increased by raising the CO_2 concentration above the normal atmospheric level of approximately 300 ppm (21, 31, 41). The response to enriched CO_2 concentrations is greatest at high light intensities (21, 31) (Fig. 9-2). An inadequate CO_2 level might cause a temporary restriction in photosynthesis on days characterized by a high light intensity and minimal turbulent transfer of CO_2. It is doubtful whether CO_2 deficiencies are a critical factor in maintaining quality turfs. Further study is needed to clarify this aspect of the turfgrass microenvironment.

Enrichment techniques for CO_2 are not, as yet, feasible on turfs. The most practical way of preventing depressions in the CO_2 level is to encourage vigorous air movement since net assimilation increases with wind speed (12, 46). Turfs growing in the shade where wind movement is restricted are likely to have an altered

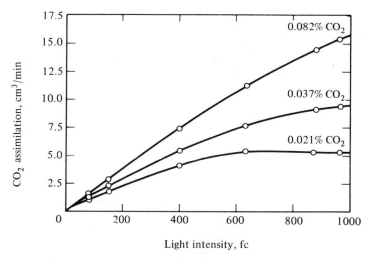

Figure 9-2. The interrelationships of light intensity and carbon dioxide concentration in affecting carbon dioxide assimilation of *Triticum aestivum* leaves (After Hoover, et al.—21).

gaseous microenvironment. The CO_2 concentration under a woodland tree canopy is frequently above normal (15, 45) (Fig. 9-3).

The atmospheric O_2 level is adequate for the plant parts growing aboveground. An O_2 deficiency of aboveground plant tissues has been theorized during extended periods of coverage with dense ice. The available information indicates that turf kill from O_2 deficiencies under an ice cover rarely occurs (4, 5, 6, 18). Oxygen deficiencies are most likely to occur within the soil environment.

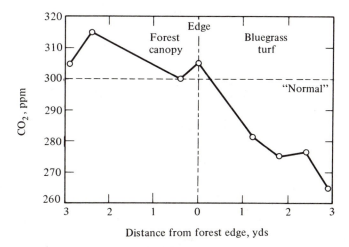

Figure 9-3. Distribution of carbon dioxide at the forest edge measured 3 in. above the soil. (After Wiant—45).

Wind

The importance of wind in affecting turfgrass growth varies with location. Winds influence turfs directly by (a) cooling; (b) increasing transpiration; (c) enhancing CO_2 exchange; (d) abrasive action; and (e) displacement and transport of soil, sand, salt, snow, spores, pollen, seeds, and propagules. Wind also has an important indirect influence on the control of climate through the redistribution of heat and water vapor.

Wind is essentially air in motion with both velocity and directional components. It consists of a succession of gusts and lulls rather than a uniform velocity. Wind is usually caused by differences in the density or pressure of the atmosphere. Atmospheric pressure and wind are not as important as temperature and precipitation in directly influencing turfgrass growth; but both exert a major indirect influence in controlling temperature and precipitation.

A diurnal variation in wind velocity is frequently observed in temperate climates. The maximum velocity generally occurs around noon with minimums at daybreak and dusk (Fig. 9-4). This is significant in herbicide use where the drift should be minimized to protect adjacent, sensitive species. The proper timing of irrigation in relation to the diurnal wind variation is also important particularly in locations characterized by strong prevailing winds. The scheduling of irrigation at times when there is a maximum probability of low wind velocity ensures a more effective water distribution pattern.

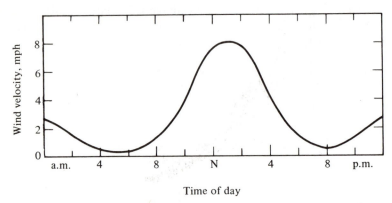

Time of day

Figure 9-4. The average diurnal variation in wind velocity at 6 in. above a Kentucky bluegrass turf for a selected period in mid-July.

Vertical variations in wind velocity also exist. The horizontal wind velocity is zero at the top of a close cut turf and increases with height above the surface (34) (Fig. 9-5). The stronger the wind, the more rapid the rate of increase at vertical heights near the turf. On infrequently mowed roadside turfs of 2 to 24 in. in height, the wind penetrates over halfway into the canopy, but with rapidly decreasing velocity and degree of fluctuation as depth in the canopy increases (44).

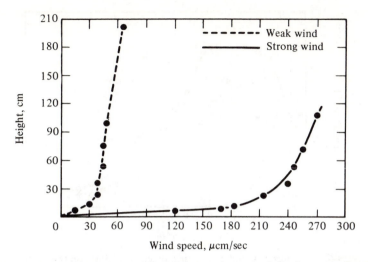

Figure 9-5. Typical vertical profiles of weak and strong winds above a turf cut at 0.8 in. (After Rider and Robinson—34).

On a micro basis, wind velocity is influenced by the vegetation. The type, form, and location of the vegetation affect wind velocity. The drag or resistance due to surface roughness that a close cut turf exerts on wind is small and aerodynamically may be called smooth. The drag coefficient determined for a turf (cut at 0.4 in) is 5.54×10^{-3} compared to 1.60×10^{-2} for grass cut 4 in. high and 2.11×10^{-3} for ice or smooth snow (36). There is no increase in the drag coefficient of mowed turfs with increasing wind velocity.

Effects of wind on turfgrasses. Many plants exhibit a deformed growth habit resulting from continuous exposure to strong prevailing winds. Low growing turfgrasses are generally not subject to such physical effects of wind. The mixing action of wind is important in affecting the CO_2, temperature, and water vapor microenvironment immediately above the turf. The turbulent transfer of air increases proportionally with wind velocity. Turbulent transfer generated by wind tends to bring the CO_2 concentration adjacent to the leaves more nearly in equilibrium with the normal atmospheric level of 300 ppm (10, 23, 30, 45). The influence of turbulence on the CO_2 supply near the leaf surface is demonstrated by the increased net assimulation that occurs with increasing turbulence (12, 46).

Temperature stratification occurs in the absence of wind movement. Stratification accentuates temperature extremes immediately adjacent to the turf. The mixing action of wind is effective in cooling turfs during periods of heat stress (13, 44). For example, the turf is cooled 13°F by a 4-mph wind on a moderately warm day with an 86°F maximum ambient air temperature (Fig. 9-6). Significant cooling can occur to a soil depth of 2 in. Cooling increases with wind velocities up to 5 mph (9). Gusts of wind can cause rapid fluctuations in temperature (13). The cooling results from both reduced temperature stratification and an increased transpiration rate.

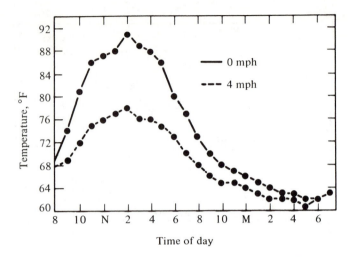

Figure 9-6. The effect of 0 and 4 mph air movement regimes on the temperature of a Toronto creeping bentgrass turf over a 24-hr period. (After Duff and Beard—13).

The observation that a wilted, nontranspiring turf heats up rapidly suggests a significant cooling effect resulting from transpiration. The transpirational cooling effect stimulated by air movement is an important factor in maintaining a healthy turf. Excessive midday heating of turfs can be modified through good air circulation.

The mixing action of wind is equally important in the redistribution of stratified water vapor. Turbulent transfer reduces the water vapor content immediately adjacent to the turf to a level less favorable for pathogen activity. A reduction in the water vapor boundary layer adjacent to the leaf is most significant for turfs surrounded by trees and underbrush or in wet, hollow depressions.

The transpiration rate is increased by air movement (8, 27). The water vapor accumulation adjacent to the leaf is removed by air movement causing a decrease in the external diffusion resistance to water vapor loss (25). An increase in the diffusion gradient results in increased transpiration. The water requirement per unit of dry matter produced increases proportionally with the wind velocity (17).

The effect of wind in increasing the transpiration rate is negative when desiccation occurs. Desiccation of turfs is most common on exposed slopes and in regions having strong prevailing winds with a low water vapor content. Turfs can be protected from the drying action of winds by using windbreaks. The proper placement of trees and shrubs can serve as a protective windbreak. Other practices include the sloping of greens away from the prevailing winds and the construction of protective banks on the windward side (26). Greens and other critical turfgrass areas should be constructed with gentle contours since steep slopes are more prone to damage by the desiccating action of winds.

Wind is also a factor in the transfer and deposition of snow. Elevated, exposed areas do not receive as much winter precipitation as hollow or protected areas due to snow removal from these areas by wind. Thus, areas of maximum exposure are characterized by dry soils and severe winter desiccation problems. Snow fences, brush, or synthetic protective covers can be used to protect turfs from the drying action of wind as well as to provide more uniform snow distribution.

Other physical influences of wind. The loss of turfgrass seedings by wind erosion is a problem in certain regions. Seedings made on light sandy or organic soils can be completely lost during a brief period of high winds. The seeds and even newly germinated seedlings can be transported from the planting site. Practices that minimize this problem include (a) maintaining a moist seedbed through irrigation, (b) ridging the seedbed perpendicular to the prevailing winds, and (c) using windbreaks. The orientation of wind barriers should be in relation to the direction of maximum wind erosion force.

Serious wind erosion of soil may occur during extended periods of drought, particularly in arid regions. A dense turf cover is one of the best means of preventing wind erosion. The deposition of windblown soil or sand in excess of 1 in. can cause extensive injury to turfs. In addition, windblown silt can be deposited as a distinct layer that disrupts vertical air and water movement. Annual accumulations of more than 1 in. can occur in arid regions (Fig. 9-7).

The abrasive action of windblown sand and soil particles can result in serious injury to turfgrasses. The extent of injury depends on the (a) wind velocity; (b) duration of exposure; (c) air temperature and humidity; (d) gustiness of the wind; and (e) size, shape, and density of the abrasive particles. The abrasive force is strongest in the first inch above the soil. Turfgrass plants are frequently sheared off at the soil surface by the abrasive action. Beachgrass is quite resistant to sandblasting because of the tough leaf sheath that protects the stem.

Salt spray is a problem along seacoasts. The action of ocean waves dashing against the rocks and shoreline produces a salt spray that is carried inland by winds. The highest salt concentrations occur near the ocean and diminish inland. Sensitive turfgrass species growing adjacent to the ocean are prone to foliar injury from the wind-borne salt spray. The intensity of salt spray is greatest during severe storms. Wind blown salt spray problems are reduced by the use of windbreaks and salt tolerant turfgrass species. Periodic washing of salt films from the leaves is also practiced.

Figure 9-7. A one inch layer of wind blown silt deposited on a green at Manhattan, Kansas. (Photo courtesy of R. Keen, Kansas State University, Manhattan, Kansas).

Wind dissemination of weed seeds and propagules is an avenue for the continual introduction of weeds into quality turfs. Light seeds or seeds with wing-like structures are adapted to wind transport. The parachute-like pappus of the dandelion is a typical example. Wind is also important in the dissemination of spores of such turfgrass pathogens as rust, powdery mildew, slime mold, and stripe smut.

Atmospheric Pollution

Atmospheric pollutants are of increasing concern in turfgrass culture. Continuing industrialization and urbanization are major factors contributing to the pollution problem. Pollutants may occur as (a) gases, (b) minute solid particles, or (c) aerosols in the form of small water droplets containing high concentrations of acids or salts. The injurious effects of air pollution on plants may result in direct toxicity or may be indirect through the screening of incoming solar radiation. In addition, turfgrasses injured or weakened by atmospheric pollutants are usually more readily damaged by diseases, insects, and nematodes.

Natural pollution has existed for many years. Dust, sea salt, and smoke from forest fires and volcanic eruptions are the main inorganic forms of natural pollution. Inorganic particles can remain in the atmosphere for extended periods of time and may cause a serious reduction in light intensity. Natural organic pollutants such as spores and pollen are also found in the atmosphere.

Artificial pollution originating from human activities is most serious in the vicinity of large metropolitan areas. A major portion of the turfgrass acreage is also concentrated in these urban, industrial areas. Smoke, dust, and fly ash are produced from the incomplete combustion of coal and petroleum. Such particulate matter (0.01 to 1.0 μ) can screen out a significant portion of the sun's radiation (11). Artificial pollutants are the most common cause of direct toxic injury to turfs. The most important types include sulfur dioxide gas, fluoride-containing gases, and toxic gases associated with smog.

Sulfur dioxide. Sulfur dioxide (SO_2) has been recognized as a major artificial pollutant for a long time (11, 37). The primary sources of SO_2 are from the burning of coal and oil high in sulfur content and from the smelting of sulfide ores. Concentrations of SO_2 below the toxic level may be beneficial to turfs growing in sulfur deficient soils (40).

Visual symptoms of SO_2 injury are classified as acute or chronic (1, 37, 38, 39). Acute injury is the result of a short duration exposure to a relatively high SO_2 concentration. Symptoms involve collapse of the marginal and interveinal tissues resulting in a dull, water soaked appearance. Subsequently, drying and bleaching to an ivory color occurs in most species. Necrosis appears initially at the terminal portion of the leaf blade. Chronic injury is caused by prolonged exposure to SO_2 concentrations too low to cause acute injury. Typical symptoms are a gradual yellowing or bleaching of the leaves in the interveinal area. Most of the pigment system is eventually destroyed but the tissue does not collapse.

The destruction of chlorophyll by SO_2 results in the inhibition of photosynthesis and growth restriction. The injury is not systemic in nature. A high light intensity, temperatures above 40°F, and a high relative humidity stimulate physiological activity that results in greater SO_2 toxicity. Creeping bentgrass and red fescue are quite sensitive to acute injury by SO_2 while annual bluegrass, Kentucky bluegrass, and perennial ryegrass rank intermediate (1, 3, 39). Bermudagrass, zoysiagrass, and redtop are relatively insensitive to SO_2 exposure.

Fluorides. Toxic fluoride-containing gases such as hydrogen fluoride (HF) and silicon tetrafluoride (SiF_4) are a problem in the vicinity of certain types of industry. Fluorine injury to plants occurs at extremely low concentrations in the parts per billion range. Visual injury symptoms include a gray-green, water soaked appearance at the tips of turfgrass leaves (1, 37, 38, 39). The lesions then turn light tan to reddish-brown and gradually extend down the blade in a fairly uniform front. Succulent, young leaves are most sensitive to fluoride injury. Fluoride injury occurs primarily by absorption from the atmosphere rather than uptake from the soil (20). Annual bluegrass, perennial ryegrass, meadow fescue, and red fescue are quite sensitive to HF (3). Redtop is fairly insensitive to HF while Kentucky bluegrass ranks intermediate.

Other gaseous pollutants include ammonia (NH_3), chlorine (Cl_2), ethylene (C_2H_4), hydrogen cyanide (HCN), hydrogen sulfide (H_2S), mercury vapors (Hg), and the oxides of nitrogen (NO, NO_2, etc.). Any of these may be a problem in local areas adjacent to the source. The Air Pollution Handbook provides specific details for these less common atmospheric pollutants (39). The toxicity of gases to plants is in the following decreasing order: $HF > HCl > SO_2 > NH_3 > H_2S$ (35). The degree of toxic gas pollution from industry is related to the discharge rate and height of the smokestack as well as the velocity and direction of the prevailing winds (16, 38).

Smog. A new and more complex type of air pollution has come to the forefront. **Smog** is defined as a combination of smoke, chemical fumes, and fog, but in common usage it includes many types of atmospheric pollution. Fog is not an essential constituent of smog. Smog is a unique blending of solid and liquid aerosols in a fine particulate state with numerous inorganic and organic gases. Polymerization of oxidants to aerosols causes a reduction in light intensity.

The phytotoxic components of smog are not necessarily the direct products of industrial and urban activities. The more important toxic constituents are formed by slow photochemical reactions occurring in the atmosphere between the unsaturated hydrocarbons and small concentrations of certain nitrogen oxides (38, 39). The resultant toxic oxidants include ozone, ozonides, peroxynitrates, and peroxyacids. Peroxyacetyl nitrate (PAN) and ozone are the most common phytotoxic constituents of smog. The three essential prerequisites for a serious smog problem are (a) sources of nitrogen oxides and unsaturated hydrocarbons; (b) a low wind velocity and a large scale, semipermanent, nondiurnal temperature inversion that prevents movement of pollutants from the area; and (c) solar radiation for the photochemical oxidation reactions (32).

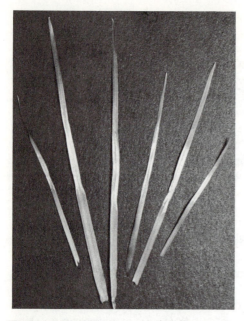

Figure 9-8. Typical transverse banding symptoms of annual bluegrass leaves caused by smog injury. (Photo courtesy of the Los Angeles County Department of Arboreta and Botanic Gardens, Arcadia, California).

Injury from the oxidizing constituents of smog involves a series of visual symptoms (7, 33, 37, 38). Initially, the lower surface of the leaf appears oily. A metallic sheen or glazed appearance develops in 1 to 3 days. Both the interveinal tissue and the veins are affected. The silvering or bronzing of leaves is the result of blisters or air pockets on the lower epidermis (28). Subsequently, the chlorotic markings extend all the way through the leaf. Cells adjacent to the substomatal chambers are damaged first. Cellular injury is characterized by disintegration of the chloroplasts, cell plasmolysis, and finally dehydration.

Acute ozone toxicity symptoms on turfgrasses usually appear as a bleaching and necrosis of the terminal portion of the leaf blade. An exception is red fescue which has minute, dark brown stipples. Annual bluegrass is quite sensitive to smog (7, 28, 29) and has been used as a bioassay indicator plant (24, 33). Injury symptoms from both ozone and PAN appear as a tan spotting or transverse banding of the leaf (7) (Fig. 9-8). The banding is a glossy, purple-brown necrosis in ryegrass (33). Banding results from injury to cells of a specific physiological age (33). The most susceptible region of the leaf is the zone where the cells have just completed maximum expansion and the stomata have become functional (7). Senescent and very young leaves are usually not sensitive. Sensitivity is influenced by the (a) stomatal distribution and activity, (b) balance of intercellular air space, and (c) cell age (7). It is not correlated with the plant size or growth rate (24). Sensitivity is highly correlated with the area over which the stomata are wholly or partially open and is highest under optimum growth conditions. Smog damage is increased by higher concentrations and extended periods of exposure (35).

The oxidizing constituents of smog cause a reduced photosynthetic rate, lower chlorophyll content, increased respiration rate, and increased cell permeability (38, 42). It may inhibit growth without any visible injury to the plant. Smog sensitivity varies among and within the turfgrass species. On a relative basis, annual bluegrass and creeping bentgrass are quite sensitive to ozone injury; Kentucky bluegrass, red fescue, Italian ryegrass, and perennial ryegrass rank intermediate; while zoysiagrass and bermudagrass are fairly resistant (39). Santa Ana bermudagrass was developed for superior smog resistance (47). Highland colonial bentgrass and Kingstown velvet bentgrass are less sensitive to ozone injury than Cohansey,

Seaside, and Penncross creeping bentgrasses. Smog will probably be an increasing problem affecting turfgrass culture in urban areas.

References

1. ADAMS, D. F. 1956. The effects of air pollution on plant life. AMA Archives of Industrial Health. 14: 229–245.

2. ANSARI, A. Q., and W. E. LOOMIS. 1959. Leaf temperatures. American Journal of Botany. 46(10): 713–717.

3. ANTIPOV, V. G. 1959. Gas resistance of lawn grasses. Botanicheskii Zhurnal. 44: 990–992.

4. BEARD, J. B. 1964. Effects of ice, snow and water covers on Kentucky bluegrass, annual bluegrass and creeping bentgrass. Crop Science. 4: 638–640.

5. ———. 1965. Effects of ice covers in the field on two perennial grasses. Crop Science. 5(2): 139–140.

6. ———. 1965. Bentgrass (*Agrostis* spp.) varietal tolerance to ice cover injury. Agronomy Journal. 57: 513.

7. BOBROV, R. A. 1955. The leaf structure of *Poa annua* with observations on its smog sensitivity in Los Angeles County. American Journal of Botany. 42(5): 467–474.

8. BRIGGS, L. J., and H. L. SHANTZ. 1916. Hourly transpiration rate on clear days as determined by cyclic environmental factors. Journal of Agricultural Research. 5(14): 583–650.

9. BROWN, H. T., and W. E. WILSON. 1905. On the thermal emissivity of a green leaf in still and moving air. Proceedings of the Royal Society of London, Series B. 76(507): 122–137.

10. CHAPMAN, H. W., L. S. GLEASON, and W. E. LOOMIS. 1954. The carbon dioxide content of field air. Plant Physiology. 29(6): 500–503.

11. CROWTHER, C., and A. G. RUSTON. 1911. The nature, distribution, and effects upon vegetation of atmospheric impurities in and near an industrial town. Journal of Agricultural Science. 4(1): 25–55.

12. DECKER, J. P. 1947. The effect of air supply on apparent photosynthesis. Plant Physiology. 22: 561–571.

13. DUFF, D. T., and J. B. BEARD. 1966. Effects of air movement and syringing on the microclimate of bentgrass turf. Agronomy Journal. 58: 495–497.

14. EHLERS, J. H. 1915. The temperature of leaves of *Pinus* in winter. American Journal of Botany. 2(1): 32–70.

15. EVANS, G. C. 1939. Ecological studies on the rain forest of southern Nigeria. II. The atmospheric environmental conditions. Journal of Ecology. 27: 436–482.

16. FAITH, W. L. 1959. Air pollution control. John Wiley & Sons, Inc., New York. pp. 1–259.

17. FINNELL, H. H. 1928. Effect of wind on plant growth. Journal of the American Society of Agronomy. 20(11): 1206–1210.

18. FREYMAN, S., and V. C. BRINK. 1966. Ice-sheet injury to turf. Proceedings of the 20th Annual Northwest Turfgrass Conference. pp. 64–68.

19. HAAGEN-SMIT, A. J., E. F. DARLEY, M. ZAITLIN, H. HULL, and W. NOBLE. 1952. Investigation on injury to plants from air pollution in the Los Angeles area. Plant Physiology. 27(1): 18–34.

20. HANSEN, E. D., H. H. WIEBE, and W. THORNE. 1958. Air pollution with relation to agronomic crops. VII. Fluoride uptake from soils. Agronomy Journal. 50: 565–568.

21. HOOVER, W. H., E. S. JOHNSTON, and F. S. BRACKETT. 1933. Carbon dioxide assimilation in a higher plant. Smithsonian Miscellaneous Collections. 87(16): 1–19.

22. HUBER, B. 1952. Über die vertikale Reechweite vegetationsbedingter Tagesschwankungen im CO_2—Gehalt der Atmosphäre. Forstv. C. 71: 372–380.

23. ———. 1952. Der Einfluss der Vegetation auf die Schwankungen des CO_2—Gehalts der Atmosphäre. Arch. F. Met. (B). 4: 154–167.

24. JUHREN, M., W. NOBLE, and F. W. WENT. 1957. The standardization of *Poa annua* as an indicator of smog concentrations. I. Effects of temperature, photoperiod, and light intensity during growth of the test-plants. Plant Physiology. 32(6): 576–586.

25. LINACRE, E. T. 1966. Resistances impeding the diffusion of water vapor from leaves and crops. Ph.D. Thesis. University of London. pp. 1–81.

26. LOCKWOOD, H. 1956. Altitude is turf problem on Wyoming course. Park Maintenance. 9(8): 14–16.

27. MARTIN, E. V., and F. E. CLEMENTS. 1935. Studies of the effect of artificial wind on growth and transpiration in *Helianthus annus*. Plant Physiology. 10(4): 613–636.

28. MIDDLETON, J. T., J. B. KENDRICK, and H. W. SCHWALM. 1950. Injury to herbaceous plants by smog or air pollution. Plant Disease Reporter. 34(9): 245–252.

29. ———. 1950. Smog in the south coastal area. California Agriculture. 4(11): 7–10.

30. MONTEITH, J. L., G. SZEICZ, and K. YABUKI. 1964. Crop photosynthesis and the flux of carbon dioxide below the canopy. Journal of Applied Ecology. 1: 321–337.

31. MOSS, D. N., R. B. MUSGRAVE, and E. R. LEMON. 1961. Photosynthesis under field conditions. III. Some effects of light, carbon dioxide, temperature, and soil moisture on photosynthesis, respiration, and transpiration of corn. Crop Science. 1(2): 83–87.

32. NEIBURGER, M. 1957. Weather modification and smog. Science. 126(3275): 637–645.

33. NOBLE, W. M. 1955. Pattern of damage produced on vegetation by smog. Journal of Agricultural and Food Chemistry. 3(4): 330–332.

34. RIDER, N. E., and G. D. ROBINSON. 1951. A study of the transfer of heat and water vapour above a surface of short grass. Quarterly Journal of the Royal Meteorological Society. 77(33): 375–401.

35. SCHMIDT, F. H. 1963. Atmospheric pollution. In: Physics of Plant Environment, edited by W. R. Van Wijk. North-Holland Publishing Company, Amsterdam. pp. 1–382.

36. SHEPPARD, P. A. 1947. The aerodynamic drag of the earth's surface and the value of von Karman's constant in the lower atmosphere. Royal Society of London, Proceedings. 188: 208–222.

37. THOMAS, M. D. 1951. Gas damage to plants. Annual Review of Plant Physiology. 2: 293–322.

38. ———. 1958. Air pollution with relation to agronomic crops. I. General status of research on the effects of air pollution on plants. Agronomy Journal. 50: 545–550.

39. THOMAS, M. D., and R. H. HENDRICKS. 1956. Effect of air pollution on plants. In: Air Pollution Handbook, edited by P. L. Magill, F. R. Holden, and C. Ackley. McGraw-Hill Book Company, New York. pp. 9–1 to 9–44.

40. THOMAS, M. D., R. H. HENDRICKS, T. R. COLLIER, and G. R. HILL. 1943. The utilization of sulphate and sulphur dioxide for the sulphur nutrition of alfalfa. Plant Physiology. 18: 345–371.

41. THOMAS, M. D., and G. R. HILL. 1949. Photosynthesis under field conditions. In: Photosynthesis in Plants, ed. by J. Franck and W. E. Loomis. Iowa State College Press, Ames, Iowa. pp. 19–52.

42. TODD, G. W., J. T. MIDDLETON, and R. F. BREWER. 1956. Effects of air pollutants. California Agriculture. 10(7): 7–14.

43. VERDUIN, J., and W. E. LOOMIS. 1944. Absorption of carbon dioxide by maize. Plant Physiology. 19: 278–293.

44. WATERHOUSE, F. L. 1955. Microclimatological profiles in grass cover in relation to biological problems. Quarterly Journal of the Royal Meteorological Society. 81: 63–71.

45. WIANT, H. V. 1964. The concentration of carbon dioxide at some forest micro-sites. Journal of Forestry. 62(11): 817–819.

46. WILSON, J. W., and R. M. WADSWORTH. 1958. The effect of wind speed on assimilation rate— a re-assessment. Annals of Botany N. S. 22: 285–290.

47. YOUNGNER, V. B. 1966. Santa Ana, a new turf bermudagrass for California. California Turfgrass Culture. 16(3): 23–24.

chapter 10

Soil

Introduction

The soil is a complex medium of organic and inorganic materials that functions as a major source of water and nutrients for turfgrass growth as well as for anchorage of the turfgrass roots. Turfgrass soils should provide a firm, resilient surface that resists compaction when subjected to intense traffic (59). Knowledge of the physical, chemical, and biological characteristics and functions of the soil is a necessary prerequisite in developing a basic understanding of the principles of turfgrass culture.

Soils vary greatly depending on the parent material, climate, topography, and vegetation under which they were formed. Many turfgrass soils do not possess the typical profile developed through many years of weathering, leaching, and microorganism activity. Soils around homes, buildings, and other construction sites have usually been disturbed during the developmental phases. Soils on greens, tees, and athletic fields are frequently modified to minimize compaction. Thus, the classical soil profile may not apply to turfgrass situations.

The characteristics that directly or indirectly affect turfgrass growth are referred to as the **edaphic** factors. Soil physical properties such as texture, structure, and porosity are quite important in influencing the infiltration, retention, and movement of water as well as soil aeration. The most significant chemical properties of the soil are soil reaction, fertility, and salt effects. The four major components of the soil that influence the edaphic factors are (a) mineral, (b) organic, (c) water, and (d) air. The relative proportion of each component found in the soil varies greatly from location to location.

The Mineral Component

The mineral fraction constitutes the major component of the soil. The mineral fraction has specific physical and chemical properties that vary depending on the parent material and conditions under which the soil was formed. Such physical properties as bulk density, water retention, hydraulic conductivity, and porosity are strongly influenced by the soil texture and structure.

Physical Properties

Soils are classified into various inorganic particle size groups called soil separates. The physical characteristics of seven soil separates are shown in Table 10-1. The surface area of a given amount of soil increases significantly as the particle size decreases. This effect strongly influences the physical and chemical reactions occurring in the soil. Knowledge of the soil texture gives an indication of the aeration, drainage, water retention, and inherent fertility (8). The three major soil separates of significance in turfgrass culture are sand, silt, and clay.

Table 10-1

THE SIZE AND SURFACE AREA OF SEVEN SOIL SEPARATES [TABLE COURTESY OF MILLAR, TURK AND FOTH (75)]

Soil separate	Diameter, mm*	Number of particles per gram	Surface area in 1 g, sq cm
Very coarse sand	2.00–1.00	90	11
Coarse sand	1.00–0.50	720	23
Medium sand	0.50–0.25	5700	45
Fine sand	0.25–0.10	46,000	91
Very fine sand	0.10–0.05	722,000	227
Silt	0.05–0.002	5,776,000	454
Clay	<0.002	90,260,853,000	8,000,000

* United States Department of Agriculture System.

Sand. Quartz is the predominant mineral in the sand fraction of most soils. Sand particles have a small surface area per unit weight, low water retention, and minimal chemical activity compared to silt and clay (24, 51, 92). The sand fraction functions as a structural framework or matrix for aeration and water movement. Because of these two characteristics sand is widely used in soil modification to minimize compaction problems under turfs (47).

Silt. The silt size group is intermediate between sand and clay. It has a relatively limited surface area and minimal chemical activity. Water retention is quite high. Silt is generally undesirable because of proneness to compaction under intense traffic as well as poor soil water movement and aeration.

Clay. The clays have a very large surface area and are quite active chemically, particularly the cation exchange capacity. The large surface area also results in good water retention, but much of the water is not available to the turfgrass plant. The swelling, plasticity, and cohesion of clays are greater than for silt and sand. Soils with a high clay content are not particularly desirable for turfgrass areas exposed to intense traffic because they are very prone to compaction (47). The aeration, infiltration, and percolation of weakly structured clays is limited, especially when compacted (57). The clay minerals are composed primarily of aluminum, silica, and oxygen that are grouped in certain distinct crystal lattices. For example, montmorillonite clay has an expanding crystal lattice structure with two silican tetrahedral layers enclosing an aluminum octahedral layer while kaolinite has a 1:1 aluminosilicate crystal lattice structure. The chemical and physical properties of a given type of clay are determined to a great extent by the arrangement of the crystal lattices. The montmorillonite clays exhibit extensive swelling properties compared to the kaolinites. Swelling exposes a much larger surface area and increases ion absorption.

Texture. Soil texture is used to describe the soil particle sizes. The relative proportions of the various soil separates in a soil determines the **soil texture class**. The graph shown in Fig. 10-1 can be used to describe the soil texture class of a given soil provided the sand, silt, and clay percentages are known. The content of sand,

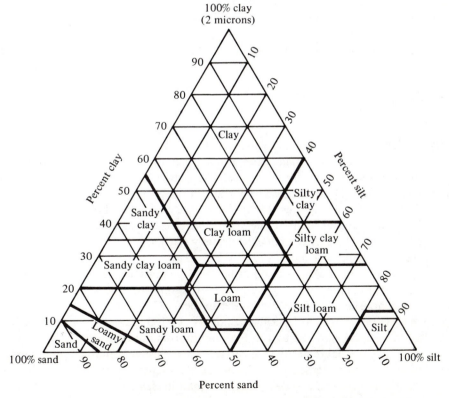

Figure 10-1. The soil texture triangle (Courtesy of Millar, Turk, and Foth—27).

silt, and clay in a given soil sample can be determined by mechanical analysis. The loamy sands, coarse sandy loams, and loams are the preferred soil texture classes for turfgrass culture. The soil texture can vary vertically between horizons within a soil profile or horizontally from location to location.

Structure. The arrangement of soil particles into larger aggregates is referred to as **soil structure**. Clays, and to a certain extent silts, are capable of being combined into larger units having different physical properties from the corresponding mass of individual particles. The soil structure types are described as granular, crumb, platy, blocky, columnar, prismatic, and subangular blocky. The granular and crumb types of structure are preferred for turfgrass growth because of the improved aeration and water movement compared to the structureless clays.

The primary clay particles are drawn together by cohesive forces to form the larger aggregates. The materials that provide aggregate adhesion are colloidal in nature and include organic matter, colloidal oxides of iron and aluminum, and certain clay minerals. The development of soil structure is enhanced by (a) turfgrass roots, (b) microbial activity, (c) wetting and drying, and (d) freezing and thawing. Destruction of soil structure results from intense traffic.

Soil structure contributes significantly to more favorable soil conditions for turfgrass growth and development on unmodified soils having a low intensity of traffic. Such properties as good water retention and cation exchange capacity of the clays are combined with improved infiltration, percolation, and aeration. Soil modification with a high percentage of sand or a similar coarse, stable aggregate is frequently necessary to assure adequate aeration and water movement where continuous, intense traffic is anticipated.

Chemical Properties

The mineral component is composed primarily of silica, aluminum, and iron oxides which do not contribute to the nutritional needs of turfs. A small amount of calcium, potassium, and magnesium is contained in the mineral fraction but not in a readily available form for plant use. The nutrient absorption capacity of the mineral component is much more important in turfgrass nutrition than the actual nutrient content of the minerals. The colloidal size clay minerals retain nutrients by the process of cation exchange. The sum total of exchangeable cations absorbed by a soil is called the **cation exchange capacity** (CEC).

Fertile soils generally have a high CEC. Cations applied to coarse textured soils are readily removed by leaching because of the low CEC. Soils high in colloids such as clay or organic matter have a high CEC compared to sands that have a low colloid content. There are great differences in the CEC even within the clay mineral group. The CEC, expressed in milliequivalents per 100 grams of soil, ranges from 3 to 15 for kaolinite clay, 80 to 120 for montmorillonite clay, and 150 to 500 for organic matter. The large number of cation exchange sites contained on the extensive surface area of clay minerals is the major factor contributing to the high CEC. The B-horizon of the soil profile is usually highest in clay content and therefore has

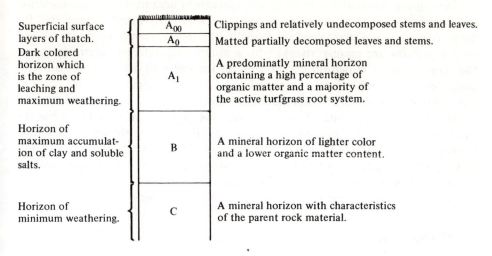

Superficial surface layers of thatch.	A_{00}	Clippings and relatively undecomposed stems and leaves.
	A_0	Matted partially decomposed leaves and stems.
Dark colored horizon which is the zone of leaching and maximum weathering.	A_1	A predominatly mineral horizon containing a high percentage of organic matter and a majority of the active turfgrass root system.
Horizon of maximum accumulation of clay and soluble salts.	B	A mineral horizon of lighter color and a lower organic matter content.
Horizon of minimum weathering.	C	A mineral horizon with characteristics of the parent rock material.

Figure 10-2. Diagram of a typical soil profile and the characteristics found under an irrigated creeping bentgrass turf.

the higher cation exchange capacity (Fig. 10-2). Calcium, magnesium, and potassium are supplied to turfgrass roots primarily through cation exchange.

The Organic Component

The organic component of the soil can be subdivided into the living and nonliving fractions. Both the organic matter content and microorganism population of a soil contribute significantly to turfgrass growth and development even though they constitute a relatively small portion of the soil complex.

Organic Matter

The soil organic matter fraction is a major factor in turfgrass growth and development because it contributes to improved soil structure, aeration, water retention, water movement, and nutrient availability (51, 101, 102). The extensive, fibrous root system of turfgrasses contributes to the soil organic matter content through continual decomposition of the older roots (22, 105, 119). The organic matter content in the A-horizon of the soil profile is maintained or even increased in this way. Decomposition of the plant organic matter contributes nutrients for turfgrass growth as well as colloidal forms that serve as adhesive agents in soil aggregation.

The organic matter content of mineral soils normally is in the range of 1 to 8% by weight. Soils containing over 20% organic matter in the upper 1 ft of the profile are called **organic soils**. There are different classifications within the organic soil category. **Peat soils** are largely undecomposed or only slightly decomposed. **Muck soils** are well decomposed and in a finely divided state.

The soil organic matter content depends on rate of organic matter accumulation and decomposition. Soil microorganisms are the primary agents responsible for the decomposition of plant residues. Initially, the sugars, starch, amino acids, and certain proteins are readily attacked by a number of different microorganisms. The more resistant structural components of the cell wall are decomposed relatively slowly. The more inert compounds such as lignin impart the dark color to soils containing a significant organic matter content.

The decomposition rate of organic materials depends on how favorable the microenvironment is for microorganism activity. Root, stem, rhizome, and stolon decomposition is enhanced by (a) temperatures of 85 to 105°F, (b) moist conditions, (c) good soil aeration, (d) a favorable carbon-nitrogen ratio, (e) a pH near neutral, and (f) freedom from toxic compounds. A carbon-nitrogen ratio of 25: 1 to 30: 1 is usually most favorable for the decomposing microorganisms.

Soil Organisms

The soil contains numerous plants and animals ranging from microscopic bacteria to large soil animals such as earthworms. Included within the soil microorganism group are bacteria, fungi, actinomycetes, algae, protozoa, and nematodes. The soil organisms function in a wide range of activities, which can be beneficial or detrimental to turfgrass growth and development. Beneficial activities include (a) thatch and organic matter decomposition, (b) nitrogen fixation, (c) transformation of essential elements from one form to another, (d) soil aggregation, and (e) improved soil aeration and drainage. Bacteria are the smallest and most numerous microorganisms. They make an important contribution to organic matter decomposition, nitrogen fixation, and the transformation of sulfur and nitrogen. The fungi and actinomycetes contribute beneficially to organic matter decomposition. The soil macrofauna group includes earthworms that function in the incorporation of organic matter into the soil as well as directly improving aeration and water movement by means of their soil channels. Soil fungi, nematodes, or insects that feed on roots and lateral shoots to the detriment of turfgrass growth may have to be controlled. The effects of turfgrass pesticides on the soil microorganism population have not been investigated in detail.

The Water Component

Most of the water utilized in turfgrass growth is absorbed from the soil through the root system. The soil can function in the storage of water for future plant use. Thus, a knowledge of soil moisture relations is vital in turfgrass culture. Natural precipitation, irrigation, and flooding are the primary sources of soil moisture.

Soil water retention. Soil water exists in the pore space between the solid particles of minerals and organic matter. The tenacity with which the water is held in the soil by absorptive and capillary forces is called the **soil suction**. The soil is saturated when all the pore spaces are filled with water and the soil suction is zero. This is usually a temporary condition since a portion of the water drains from the larger pores quite rapidly in response to the forces of gravity. Saturation of the soil is most likely to occur after an intense rain or excessive irrigation. The turfgrass root system can be seriously injured due to a lack of oxygen and/or the accumulation of toxic gases if the soil remains saturated for an extended period of time.

The air porosity increases as the water is removed from the large pores by gravitational forces. The rate of downward water movement decreases as the soil suction increases. Water is held in the soil pores by the forces of cohesion between water molecules as well as the adhesive forces between the water molecules and soil particles. The water is spread as a film over the surfaces of all soil particles. The film thickness varies with the amount of water present and the total surface area of the soil particles. A continuing decrease in the soil water content occurs under actively growing turfs primarily because of the upward loss by evapotranspiration. A decreasing water film thickness on the soil particles causes an increase in the soil suction and a greater energy requirement for the removal of water and its absorption by the turfgrass roots. The point is reached where the turfgrass plant wilts due to the inability of the roots to absorb the soil water because it is held too tightly by the soil particles.

Soil characteristics associated with high water retention include (a) fine soil texture, (b) good structure, and (c) high organic matter content. One of the primary principles of turfgrass irrigation is to maintain the soil water content at a water suction level that is low enough to readily satisfy the water absorption requirements of the turfgrass plant without creating a significant internal plant water stress.

Soil water movement. The downward entry of water into the soil is termed **infiltration**. Factors affecting the infiltration rate include the surface condition, texture, structure, organic matter content, and soil water percolation rate. The infiltration velocity is normally quite rapid when water initially contacts the soil, barring the presence of a hydrophobic soil or thatch layer. It gradually declines after a period of time to a level controlled by the rate of soil water movement. The infiltration rate is usually lower in fine textured soils, particularly if compacted or low in organic matter (Fig. 10-3). Sands have a much higher infiltration rate than the clays. When the precipitation rate exceeds the infiltration velocity of the soil, the excess water is lost by surface runoff or becomes ponded in depressional areas.

The direction of water movement through the soil is determined by the relative magnitudes of the gravitational forces and the soil suction. The rate of water movement is also determined by these two forces along with the water conductivity coefficient. This coefficient is called the hydraulic conductivity in saturated soils where the flow is mainly dependent on the size of the pores available for water movement. The resistance to saturated and unsaturated flow increases as the pore size decreases. The coefficient is referred to as the unsaturated conductivity in unsaturated soils where a significant portion of the pores are not filled with water. Unsaturated water flow is strongly influenced by the thickness and continuity of

Figure 10-3. The effect of three soil textures on the water infiltration rate when the water is applied at the same rate and time. (Photo courtesy of W. H. Gardner, Washington State University, Pullman, Washington).

water films. The resistance to unsaturated flow increases as the water film thickness becomes greater. The water film continuity between soil particles is greater in fine textured soils than in coarse textured soils. Sands have a comparatively low number of contact points between soil particles compared to clays. Sandy soils have a rapid rate of saturated flow and a slow rate of unsaturated water flow compared to the finer textured soils such as clays.

The downward movement of water through the soil is termed **soil water percolation**. It mainly involves the saturated or near saturated water flow to the ground water table or drain tile. The rate of soil water percolation is affected by the (a) size, number, and continuity of pores; (b) hydration of the pores; and (c) resistance of entrapped air. Percolation is quite slow in clay soils since the pores are small, whereas coarse textured sands have fairly rapid percolation and good internal drainage properties. Clays and water-soluble nutrients tend to be leached downward into the B-horizon of the soil profile during percolation.

An elevated soil layer of a distinctly different texture can seriously impede water movement. An impervious clay layer impairs both the upward and downward movement of water due to a lack of pore space. A coarse-textured sand or gravel layer also disrupts water movement because of the poor continuity of the water films between the fine-textured soil and the underlying coarse-textured soil particles. The water content in the fine textured soil zone just above the coarse textured soil layer must approach saturation before downward water movement occurs. Thus, water can be held above the coarse layer, giving a perched water table effect (Fig. 10-4). This principle has been utilized in the USGA Green Section method of complete soil modification.

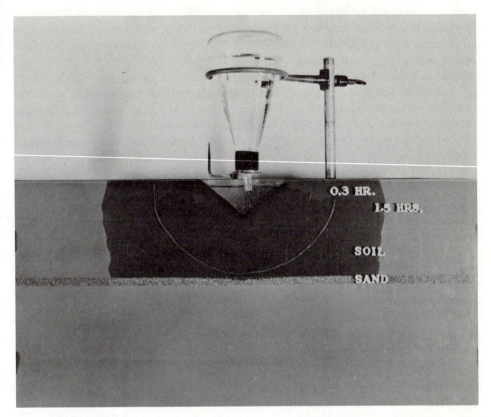

Figure 10-4. The pattern of water movement in which a fine textured soil overlays a coarse sand. Note how the water accumulates above the sand. (Photo courtesy of W. H. Gardner, Washington State University, Pullman, Washington).

Water movement in the soil occurs primarily in an upward or lateral direction when the downward movement caused by gravitational forces becomes minimal. The upward movement of water by the forces of soil suction and water vapor diffusion along differential gradients can be an important source of water for the turfgrass plant. The direction and rate of unsaturated water flow depends on the moisture gradient from wet to dry zones in the soil and is affected by the soil texture and structure.

Drainage

Drainage involves the removal of excess water from the root zone. Rapid drainage of excess surface water is important in ensuring the maximum use potential for the turfgrass area. The general effects of waterlogged soils include (a) a shallow root system, (b) reduced turfgrass vigor and quality, (c) poor soil aeration, (d) increased disease activity, and (e) increased compaction proneness under intense traffic. In addition, poorly drained, wet soils are slower to warm up in the spring. Drainage functions in washing soluble salts out of the root zone and prevents salt accumulation to potentially toxic levels in arid regions. Thus, ensuring adequate drainage is an important phase of turfgrass culture.

Surface drainage. Effective surface drainage is necessary during periods of intense precipitation. Surface drainage should be designed and constructed when the turfgrass area is being established. The cost of developing adequate surface drainage is usually less than for subsurface drainage and it also contributes to more effective subsurface drainage. It is important to avoid depressional areas where water accumulations might occur. The installation of slit trenches, dry wells, or surface drains connected to drain tile assists in the removal of surface water from depressional areas that cannot be avoided. Proper surface grading of the area to be established will minimize depressions. Smooth, wide, turfed surface waterways should be constructed where periodic concentrations of flowing surface water occur. Rapid surface drainage is particularly important on athletic fields that are subjected to intense traffic under all types of weather and soil moisture conditions. Improved playing conditions and reduced soil compaction proneness are achieved through proper surface drainage. Most sports turfs such as football, soccer, and baseball should be constructed with a crown arrangement having a 1 to 2 % slope to the sidelines or to appropriately located surface catch basins. Lawn areas should have a 1 % slope from the building to the street or surface drain.

Open ditches or channels carry large volumes of water and serve as surface outlets for drain tile systems. The open ditches are used on sod production fields as a water reservoir for subsurface irrigation and water control as well as for drainage. Ditch banks should be kept free from weedy perennial grasses that could infest adjacent turfgrass areas. The ditch can be lined with gravel, cobble, or masonry where soil erosion is anticipated due to high water velocities.

Subsurface drainage. Consideration should be given to the installation of artificial subsurface drainage where the natural internal soil water drainage is not adequate. Subsurface drainage is usually achieved by (a) the proper soil texture and structure for rapid percolation and/or (b) a drain tile system for more rapid removal of excess water from the root zone.

Tile drainage is necessary when the water cannot percolate to deeper soil layers where it is removed by natural subsurface lateral flow. The presence of a water table within 6 ft of the soil surface indicates the need for improved subsurface drainage. A properly designed and installed drain tile system assists in rapid removal of excess soil water from the turfgrass root zone. It should be installed before the irrigation system. Design of the tile system should be based on the (a) potential water discharge loads, (b) vertical soil textural characteristics, (c) water table depth, (d) water removal rate desired, and (e) topography of the area. The drain tile system normally consists of small, parallel lateral lines connected to larger main lines. A tile drainage system should not be installed until an adequate outlet is available for removal of the water. The size of the main tile lines is determined by the amount of surface area to be drained by the connecting laterals. The lateral drain tile are normally 4 in. in diameter or less. The laterals are normally arranged in a herring-bone or parallel gridiron arrangement on level sports turf areas. On more rolling topography such as golf courses, drain tile are usually placed in low areas as determined by the contours and along the base of slopes for the interception of seepage water.

The depth and spacing of drain tile depends on the (a) source of water to be removed, (b) soil texture and structure, (c) presence of impervious layers, and (d) rooting depth of the turf. Tile are normally placed at a depth of 3 ft or more on extensive turfgrass areas. Drain tile are placed as deep as 6 ft where the leaching of salts is necessary to prevent salinization. The water movement rate to the drain tile depends on the permeability of the soil. The system cannot function properly if an impervious soil layer exists between the drain tile and the soil surface. The spacing between tile lines ranges from 15 to 60 ft. The spacing should be reduced if the tile are placed at shallow depths or the lateral water movement is poor due to a fine soil texture. Placing a layer of coarse sand or gravel over the drain tile prior to backfilling enhances percolation to the tile. This porous layer should extend to the top of an impermeable subsoil. Subsoiling prior to turfgrass establishment will temporarily break up an impervious subsoil or hardpan layer in the upper 1 to 2 ft.

Clay and concrete are the standard types of drain tile used. Perforated plastic tubing has become available and proved effective for subsurface drainage of turfgrass areas. Tile lines are generally preferred to open ditches because of the reduced maintenance requirement and more complete drainage achieved, particularly on finer textured soils.

Slit trenches and dry wells are two types of subsurface vertical well drains utilized on turfgrass areas (44). The slit trench technique involves a trench 2 to 3 in. wide and 2 to 4 ft deep. A vertical column of gravel or crushed stone is then placed in the trench and filled to the soil surface, which is left open (Fig. 10-5).

Figure 10-5. A herringbone slit trench installation in a depressional area on a fairway (Photo courtesy of L. Record, USGA Green Section, Chicago, Illinois).

The turf grows rapidly over the surface of the trench leaving little evidence of its presence. This type of drainage has been effective (a) on poorly drained, fine texture soils where the infiltration rate is quite slow, (b) where an impermeable clay layer exists over a permeable soil, and (c) in depressional areas. A slit trench provides an open vertical column in which excess soil water can move downward quite rapidly. The placement of slit trenches over tile lines is desirable to facilitate lateral drainage.

A dry well consists of a fairly deep hole, 1 to 2 ft in diameter and 3 to 5 ft deep. Large stones are placed in the hole and the immediate surface covered with a layer of coarse textured soil over which the sod is placed. The large spaces between the stones serve as a storage reservoir for excess surface water that cannot be removed from depressional areas by other means.

The Soil Air Component

Turfgrass roots and many soil organisms require oxygen for respiration and the maintenance of life processes. Oxygen is also required for the absorption of certain essential elements by turfgrass roots (64). Thus, the air component of soils is an important factor in turfgrass growth. The soil oxygen level is quite variable ranging from 0 to 21%, while the soil carbon dioxide level varies from 0.03 to 21% or higher. The soil water vapor content is usually in the range of 80 to almost 100%.

Roots and microorganisms not only absorb oxygen but release carbon dioxide during respiration. An accumulation of carbon dioxide or other toxic gases could result in toxicity to the organisms growing in the soil. Thus, it is important to maintain a balance of oxygen and other potentially toxic gases in the root zone through the exchange of gases with the aboveground atmosphere. The process by which air in the soil is replaced by air from the atmosphere is called **soil aeration**. Poor soil aeration results in a much lower oxygen content and higher carbon dioxide content than the atmosphere above the soil. Waterlogged or compacted soil conditions limit soil aeration causing shallow rooting and reduced turfgrass vigor (113).

Soil aeration occurs primarily by diffusion through the large pores of the soil. The diffusion rate is directly proportional to **air porosity** which is the space that is actually filled with air at any given time. Both the volume and continuity of pores affect the soil aeration rate. A certain amount of oxygen, dissolved in rain and irrigation water, is also brought into the soil. Air is also drawn downward from the aboveground atmosphere as the excess soil water percolates through the soil. Thus, periodic rains or irrigation can cause a pumping action that enhances soil aeration.

Soil pore space. Roots, some microorganisms, air, and water are contained between the mineral particles and organic matter of the soil matrix. The total space not occupied by soil particles in a bulk volume of soil is termed **pore space**. The pore space can vary from 35 to 70% of the total soil volume. The greater the pore space of a given soil, the lower the soil bulk density. Compacted or poorly drained soils are lower in pore space (Fig. 10-6).

The pore size distribution in the soil varies greatly depending on the texture and structure (8). Coarse textured soils contain a higher percentage of large pores. Water and air move quite readily through the larger pores. The small pores are usually filled with water. The proportion of smaller pores in the pore space determines the water retention properties of the soil. The smaller the pore, the greater the force with which water is held because of increased adhesion.

Sandy soils have a very low percentage of small pores which results in a low water retention level. In contrast, fine textured clay soils have a larger total pore space than sands with a major portion of the pore space composed of small pores having a high water retention level. The rate of air and water movement in clay soils is usually much slower, however, because the percentage of large pores is quite small, particularly if the soil is compacted (61, 63, 115).

The relative amounts of large and small pore space are more important than the total pore space (18). The amount of large pores is more critical than the amount of small pores (57). A total pore space of approximately 40 to 50% is most favorable for turfgrass growth on soils not subjected to intense traffic (29, 91). The optimum total pore space is reduced to 35 to 40% where resistance to compaction is a major concern due to intense traffic. Best results have been obtained with 14 to 20% large pores and 18 to 24% small pores (29, 57, 91).

Soil oxygen. The soil oxygen concentration and oxygen diffusion rate (ODR) are normally the highest near the soil surface and decrease with increasing soil depth (64, 115). The oxygen diffusion rate through the soil is more significant in terms of oxygen utilization by turfgrass roots than the soil oxygen concentration. In order for oxygen to reach the root, it not only diffuses through the large pores of the soil but also through the water layer surrounding the root.

The oxygen requirement varies with (a) the physiological condition of the roots and (b) the turfgrass species. The oxygen requirement of roots usually increases proportionally with temperature. Most turfgrasses have good tolerance to a low ODR compared to other plants. Species differences are observed (31, 48, 62, 115, 124) (Table 10-2). Vegetative growth of Newport Kentucky bluegrass is not

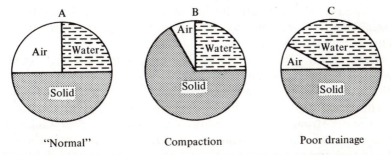

Figure 10-6. The effect of compaction (B) and poor drainage (C) on the volume of air space in the soil. (After Letey—61).

Table 10-2

THE SOIL OXYGEN DIFFUSION RATE BELOW WHICH SERIOUS IMPAIRMENT OF ROOT GROWTH
OCCURS IN FOUR TURFGRASS CULTIVARS

Oxygen diffusion rate, $gm\ cm^{-2}\ min^{-1}$	Turfgrass cultivar	References
3×10^{-8}	Penncross creeping bentgrass	115
5 to 9×10^{-8}	Merion Kentucky bluegrass	115
15×10^{-8}	Common bermudagrass	62
20×10^{-8}	Newport Kentucky bluegrass	63, 64
$> 20 \times 10^{-8}$	Most crop species	61

affected by a soil oxygen content as low as 2%. A 1% oxygen concentration restricts root growth of Kentucky bluegrass but not creeping bentgrass (64, 115). Kentucky bluegrass roots grown in soils having poor aeration are thicker and less branched than roots grown under normal aeration (115). Since turfgrass root growth can occur at a very low soil oxygen level and ODR, it is likely that soil oxygen is not as serious a problem in turfgrass root growth as has been claimed in the past. A low soil oxygen level and/or ODR caused by a waterlogged soil condition is probably a more common cause of poor root growth.

Soil Reaction

Soil reaction is the degree of acidity or alkalinity of a soil expressed as a pH value. A soil pH of 6 is ten times as acidic as a soil pH of 7 since pH values are expressed as logarithms. The pH of the A-horizon in fertile soils normally ranges from 3.5 to 8. There are exceptions to this in arid regions where a high sodium content results in pH's of 8.5 to 10. Soils tend to become acidic in regions where there is sufficient precipitation to leach the soluble basic salts such as calcium and magnesium downward out of the A-horizon of the soil profile. The A-horizon also has a fairly high organic matter content and a concentration of soil microorganisms that produce organic acids. As a result, soil acidity is usually greatest at the surface and decreases down through the soil profile (18). The inability to thoroughly cultivate soils under established turfs further accentuates soil pH stratification. The use of acidifying fertilizers such as ammonium sulfate also causes an increase in soil acidity (19, 25, 39, 43, 77, 100).

Effects of soil acidity. The soil reaction has many direct and indirect effects on turfgrass growth such as (a) nutrient availability, (b) solubility of toxic elements, (c) rooting, and (d) microorganism activity (3). The soil reaction has considerable influence on the nutrient availability in soils. For example, phosphorus availability is reduced under strongly acidic conditions because the higher soluble iron and aluminum concentrations form relatively insoluble complexes with the phosphate ion (88). Phosphorus availability is highest at a soil pH of 6 to 7. Strongly acid soils are also depleted of calcium and magnesium ions normally associated with the clay colloid complex (95). Iron, manganese, copper, and zinc availability is reduced at a soil pH above 7. The most rapid release of nitrogen from natural organic and ureaformaldehyde fertilizers occurs at a neutral soil reaction. Overall nutrient availability is the highest at a soil pH of 6.5.

The higher aluminum solubility is a direct concern since it can be toxic to plant growth (88). Strongly acid soil conditions cause a brown, restricted root system and reduced turfgrass vigor. The toxic effects on roots are more likely the result of a higher soluble aluminum concentration than an increased hydrogen ion concentration (38, 95).

A neutral soil pH favors the activity of beneficial soil microorganisms involved in (a) transformations of certain essential elements, (b) organic matter and thatch decomposition, and (c) nitrogen fixation. Increased thatch accumulation is associated with strongly acid soil conditions (25, 26). The structure of soils having a neutral soil reaction is usually superior to that under acidic or alkaline conditions (25). Improved soil water retention and aeration results. Earthworm activity is also restricted under strongly acid soil conditions.

Effects on turfgrass species. Many turfgrasses are adapted to moderately acid soil conditions. Satisfactory growth can usually be made in the pH range of 5.5 to 7. Turfs growing in soils with a pH below 6 may require higher rates of fertilization to maintain an adequate level of available nutrients. The detrimental effects observed under more acidic conditions include a general decline in overall turfgrass shoot growth, vigor, and competitive ability (3, 4, 25, 53, 54, 66, 77, 95, 117). The effects of soil acidity are more evident on root growth than shoot growth (53, 88). The root system becomes short, brown, and spindly. The turf usually becomes darker green in color and thatching increases. Decreased tolerance to environmental stress and recuperative potential also occur. The reduction in drought tolerance is quite noticeable.

Variations in tolerance to soil reaction are observed among the turfgrasses (17, 21, 43, 52, 54, 77, 83, 88, 90, 95, 100, 108, 112, 118) (Table 10-3). Legumes such as alfalfa, white clover, and crownvetch are best adapted to neutral soil pH's (43, 74, 90, 95, 112). The composition of a turfgrass community can be significantly influenced by the soil reaction (4, 43, 83, 112). Kentucky bluegrass tends to be dominant at soil pH's near neutral, while bentgrass and red fescue are more dominant at pH's below 6.0.

Many turfgrass weeds are suppressed when grown under strongly acidic soil conditions (4, 19, 37, 38, 43, 112, 118). Some individuals advocate the maintenance of acidic soil conditions as a cultural method for controlling turfgrass weeds. Although the population of certain weeds declines under acidic soil conditions, the vigor, competitive ability, wear tolerance, and drought tolerance of the desirable turfgrass species are also reduced (4, 99). It is usually best to maintain the soil pH in the most favorable range for turfgrass growth and to utilize other cultural and chemical methods for controlling weeds (100, 117).

Table 10-3

OPTIMUM SOIL pH RANGE FOR THE GROWTH OF TWENTY ONE TURFGRASSES

Optimum soil pH range	Turfgrass species
7.5–6.5	St. Augustinegrass
	Bahiagrass
7.0–6.0	Kentucky bluegrass
	Zoysiagrass
	Timothy
	Rough bluegrass
	Perennial ryegrass
	Italian ryegrass
	Bermudagrass
6.5–5.5	Annual bluegrass
	Tall fescue
	Canada bluegrass
	Creeping bentgrass
	Colonial bentgrass
	Red fescue
	Chewings fescue
6.0–5.0	Velvet bentgrass
	Carpetgrass
	Redtop
5.5–4.5	Sheep fescue
	Centipedegrass

Correcting Soil Acidity

Liming is the practice of correcting soil acidity through the application of lime. The primary objectives of liming are to (a) neutralize acids, (b) correct calcium deficiencies in the soil colloidal complex, and (c) precipitate soluble compounds of iron and aluminum that are toxic to turfgrasses. Neutralizing soil acids results in a more favorable environment for microorganism activity, increased availability of certain nutrients such as phosphorus and molybdenum, and improved soil structure. Lime is composed of compounds of calcium, magnesium, or both.

Liming Materials

The most common forms of lime are the oxides, hydroxides, or carbonates of calcium and magnesium (Table 10-4). The oxide and hydroxide forms change rapidly to the carbonate or bicarbonate forms once applied to the soil. Carbonates are insoluble in water but are subject to leaching from the soil when converted to bicarbonates.

Carbonates of calcium and magnesium. Calcium carbonate is the most widely used material for liming. It occurs naturally in limestone. Limestone rock that is mined, ground, and processed for use in liming is called agricultural limestone. The rate at which limestone corrects acidity increases proportionally with the fineness of the material because of low water solubility. Calcitic limestones are predominately

Table 10-4

THE RELATIVE NEUTRALIZING VALUE OF SIX COMMON FORMS OF LIME [AFTER MILLAR, TURK, AND FOTH (75)]

Liming material	Chemical formula	Relative neutralizing value, %
Magnesium oxide	MgO	250
Calcium oxide	CaO	178
Magnesium hydroxide	$Mg(OH)_2$	172
Calcium hydroxide	$Ca(OH)_2$	135
Magnesium carbonate	$MgCO_3$	119
Calcium carbonate	$CaCO_3$	100

calcium carbonate with a minimum of magnesium. Dolomitic limestones contain a significant percentage of magnesium. Magnesium causes the limestone to be much harder and more resistant to acids. Thus, dolomitic limestone affects the soil reaction at a much slower rate than calcitic limestone. Both have a longer residual action than the hydroxide and oxide forms.

Other liming materials that are used occasionally include marl, chalk, shells, coral, and marble. All five materials contain varying percentages of lime in the carbonate form and should be finely ground to achieve effectiveness as rapidly as possible. Marl contains primarily calcium carbonate although some magnesium carbonate may occur. Common impurities in marl are organic matter and soil, particularly clay.

Hydroxides of calcium and magnesium. The hydroxides of calcium and magnesium rank second to the carbonate forms as commonly used liming materials. The hydroxide forms are available commercially as slaked, caustic, and hydrated lime. The neutralizing value of the hydroxides ranks intermediate between the oxide and carbonate forms of lime. The rate of action is much more rapid than for the carbonate forms. The fineness of the material is not an important factor in the rate of action since they are quite water soluble. However, fineness does assist in achieving more uniform distribution in the soil. The hydroxides are caustic, powdery, and objectionable to handle.

Oxides of calcium and magnesium. The oxide forms of lime include burned lime, quicklime, caustic lime, and unslaked lime. The calcium and magnesium oxides have a much higher neutralizing value than calcium carbonate (Table 10-4). The oxide forms have a much more rapid rate of action than the carbonate forms. The fineness of oxide forms permits more thorough distribution in the soil but has no effect on the rate of action since they are highly water soluble. The oxides are highly caustic, powdery, and objectionable to handle.

Liming Procedures

Liming procedures involve selection of the liming material, application rate, time of application, and application method. Specific considerations must be made for each step to ensure effective liming of turfs.

Selecting the liming material. Factors to consider in selecting the form of lime include (a) fineness, (b) rate of action, (c) cost, (d) presence of magnesium, (e) ease of handling, and (f) caustic nature. Finely ground limestone is more readily available than coarser materials but does not have as long a residual response. Hydrated lime is somewhat more soluble than the carbonate forms. The caustic effects to plants from the oxide and hydroxide forms are of particular concern on sandy soils. A carbonate form is preferred for sandy soils.

Application rate. Certain visual symptoms such as the presence of red sorrel indicate an acidic soil condition (117). However, the amount of lime needed is best determined by a chemical soil test that measures the pH and percent base saturation. The amount of lime required varies depending on the (a) degree of soil acidity, (b) soil buffering capacity, (c) percent base saturation of the soil, (d) turfgrass species, and (e) fineness of the limestone material. A general guideline for calcium carbonate applications is presented in Table 10-5. Soils with the same pH may have different lime requirements depending on the cation exchange capacity and buffering capacity of the soil. The lime requirement is usually lower for soils having a low organic matter or clay content.

Table 10-5

GENERAL GUIDELINES FOR THE AMOUNT OF CALCIUM CARBONATE, LB PER 1000 SQ FT, REQUIRED TO CORRECT VARYING DEGREES OF SOIL ACIDITY

Soil pH	Fescues and bentgrasses		Bluegrasses, bermudagrasses, and ryegrasses	
	Sands and sandy loams	Loams and clays	Sands and sandy loams	Loams and clays
6.3–7.0	0	0	0	0
5.8–6.2	0	0	25	35
5.3–5.7	25	35	50	75
4.8–5.2	50	75	75	100
4.0–4.7	75	100	100	150

Only enough lime should be applied to correct the acidic condition. Alkaline soil conditions caused by an excessive lime application are just as detrimental as an acidic condition in limiting the availability of certain essential nutrients. Shoot growth can be retarded by high rates of lime applied to established turfs growing on strongly acid soil. Under these conditions it is best to apply no more than 25 to 50 lb per 1000 sq ft of finely ground limestone per year until the desired soil pH is achieved. Coarser limestone materials can be applied at higher rates. No more than 25 lb per 1000 sq ft should be applied per year if the hydroxide or oxide forms are used. Dolomitic materials are preferred on soils having a magnesium deficiency, but they should be avoided if the soil magnesium level is high.

Time of application. Lime only corrects acidity in the immediate vicinity of the soil particles since the vertical movement is limited. Whenever possible the lime should be applied at a time when it can be thoroughly mixed with the soil. Thus, it is particularly important to correct an acidic soil condition prior to turfgrass establishment. The incorporation of lime into the soil should be accomplished as long before planting as possible because of the slow reaction rate.

The liming frequency varies considerably with the degree of soil leaching. The quantity of calcium and magnesium removed by leaching depends on the intensity of precipitation and irrigation, the soil texture and structure, and the soil base exchange properties. Other factors affecting the liming frequency are the (a) amount of calcium and magnesium removed in clippings, (b) quantity of calcium and magnesium added in irrigation water, (c) amount of calcium released by weathering of soil minerals, and (d) acidifying effect of the fertilizers used. The irrigation water from wells in certain regions contains sufficient calcium and magnesium to maintain a desirable pH for turfgrass growth. There is a continuing increase in soil acidity in areas subject to intense leaching. Continued use of acidic fertilizers such as ammonium sulfate can also cause a substantial decrease in the soil pH (76, 77, 104). Periodic liming may be required at intervals of 3 to 5 years under these conditions. A soil test taken every 2 to 3 years can determine the need for liming.

From the convenience standpoint, lime is best applied to established turfs during the late fall, winter, or early spring periods. It is preferable to apply lime in the late fall or early winter in cooler climates. Alternate freezing and thawing during the winter months enhance the movement of lime into the soil. Since the oxide and hydroxide forms are caustic, they are best applied in late fall or early winter in order to minimize turfgrass injury. Applications made just prior to an anticipated rain are ideal in terms of washing the lime off the turfgrass shoots and into the soil. Irrigation can also be effectively used for this purpose.

Hydrated lime should not be applied within 10 to 14 days of an ammonia-type nitrogen fertilizer application. These two materials react causing the release of ammonia gas that is quite toxic to turfgrasses. Liming should also be avoided after or prior to the application of a soluble phosphate fertilizer or an inorganic arsenate herbicide. This minimizes fixation of the soluble arsenic and phosphate into unavailable forms.

Application techniques. The lime should be applied uniformly and mixed thoroughly into the soil if the application is made prior to turfgrass establishment. The standard gravity-type lime spreaders are effective in accomplishing uniform distribution. Bulk application by trucks with a specially designed distribution system may also be used on soils that can support these heavy vehicles without causing an objectionable degree of rutting or compaction. Heavy equipment can be used on frozen soils with minimum compaction. Cultivation of established turfs at the time of application encourages movement of the lime into the soil.

Alternate uses. Light, frequent applications of lime can be used to increase the pH of the thatch. This practice enhances microorganism activity and results in increased thatch decomposition. Hydrated lime is sometimes applied to discourage algae growth and to counteract toxic accumulations of organic compounds produced by anaerobic microorganisms under waterlogged soil conditions. This technique can be utilized even at neutral soil reactions if these problems exist. A 2 to 5 lb calcium hydroxide application per 1000 sq ft is normally used. It can be applied dry by mixing with sand or sprayed on with water. Repeat applications may be necessary if the problems persist.

Correcting Moderately Alkaline Soil Conditions

Moderately alkaline pH's of 7.5 to 8.4 may develop in soils that are waterlogged, rarely leached, irrigated with water containing calcium and magnesium, or limed excessively. A deficiency of iron, manganese, copper, zinc, or boron is likely to occur under moderately alkaline conditions. A decline in turfgrass vigor usually occurs on alkaline soils.

Unfavorable alkaline pH's of 7.5 to 8.4 can be corrected by applying an acidifying material. This can be accomplished by the use of (a) elemental sulfur, (b) certain acidifying fertilizers such as ammonium sulfate and iron sulfate, or (c) aluminum sulfate (4, 104). Aluminum sulfate can be quite toxic to turfgrasses if not properly used. Moderately alkaline soil reactions are most easily corrected by thoroughly mixing elemental sulfur into the soil prior to turfgrass establishment. An alkaline soil problem can be readily corrected by applying elemental sulfur according to the guidelines presented in Table 10-6. Elemental sulfur is best applied to established turfs by mixing with sand or topdressing material. The rate of a single application should not exceed 5 lb per 1000 sq ft. Do not apply sulfur during midsummer stress periods.

Table 10-6

AMOUNT OF ELEMENTAL SULFUR (95%), LB PER 1000 SQ FT, REQUIRED TO LOWER THE pH OF SOILS AT THREE LEVELS OF ALKALINITY TO ABOUT 6.5 [AFTER COLLINGS (14)]

Original soil pH	Sandy soil	Clay soil
7.5	10–15	20–25
8.0	25–35	35–50
8.5	35–50	40–50

Acidifying fertilizers can serve a dual function in the turfgrass fertilization program if an increase in the soil pH is anticipated due to the calcium and magnesium contained in the irrigation water. Correction of slightly alkaline soil conditions by the use of elemental sulfur or aluminum sulfate is usually not necessary if a preventive program such as this is followed. Ammonium sulfate, ammonium chloride, and ammonium phosphate are quite effective in reducing the pH of slightly alkaline soils (4, 77, 104). The use of ammonium chloride also involves the addition of chloride ions that may be objectionable.

Saline and Sodic Soils

Saline and sodic soils contain sufficient soluble salts, exchangeable sodium, or both to impair normal turfgrass growth and development. Severe saline or sodic conditions commonly occur in arid or semiarid climates. Such soils necessitate special cultural practices to maintain an acceptable level of turfgrass quality. Occasional salinity problems occur along coastal regions and in humid climates where the presence of saline ground water is a factor. The three types of salt-affected soils are (a) saline, (b) sodic, and (c) saline-sodic.

Saline soils. Saline soils have a conductivity of the saturation extract greater than 4 mmhos per cm at 25°C and an exchangeable-sodium percentage less than 15. The pH of these soils is usually 8.4 or less. Saline soils contain soluble salt concentrations sufficiently high to impair turfgrass growth or quality. The chief anions present are chloride, sulfate, and sometimes nitrate; the primary cations present are calcium and magnesium. Boron may prove to be the toxic salt in certain situations. The relatively low sodium level of saline soils results in good flocculation and water permeability.

A white salt crust is frequently observed at the surface of saline soils. **Salinization** is the process of soluble salt accumulation in soils. The source of accumulated salts can be from drainage water, ground water associated with a high water table, irrigation water, or the weathering of existing soil parent material. The primary factors affecting the salt accumulation rate under various conditions are the (a) evaporation rate, (b) rate of water movement to the surface, (c) salt content of the ground water, (d) amount of irrigation water applied, (e) salt content of the irrigation water, (f) impermeability of the soil to water, (g) quantity of surface water draining into basins with no outlet, and (h) salt content of the drainage water.

Salt accumulation is most common in arid and semiarid climates where drainage has been restricted due to a lack of surface drainage and/or poor soil permeability to water. The salts contributing to salinization are usually carried into drainage basins possessing no outlet by surface drainage water. Impairment of downward leaching, plus the evaporation of accumulated water, concentrates the salts at the soil surface.

Salt can be brought to the soil surface by a net upward water movement from ground water that is high in soluble salts and is near the soil surface. Continued evaporation of water from the surface and the movement of salt to the surface result in an increasing concentration of soluble salts.

A soil salinity problem can also develop as a result of irrigation. Irrigation waters may contain from 0.1 to 5 tons of salt per acre-ft of water. Thus, a substantial amount of soluble salt can be added to a soil in a relatively short time.

A temporary salinity condition may also occur in soils growing adjacent to roadsides and walks where salt is used for winter ice removal. Turfgrass injury can occur during the late winter or early spring period due to a high salt concentration (110). This salinity problem is usually temporary, however, because the salt level is readily dissipated by the leaching action of spring and early summer rains.

Sodic soils. This group includes soils having an exchangeable sodium percentage greater than 15 and/or a pH of 8.5 or higher. Sodic soils contain sufficient sodium to interfere with turfgrass growth or quality. The effects of sodium become dominant in sodic soils due to the absence of calcium and magnesium. This occurs when sodium composes more than half of the soluble cations in the soil solution and becomes the predominant absorbed cation on the soil colloids. The dominance of sodium impairs flocculation and causes deterioration of soil structure (107). Sodic soils are characterized by a relatively coarse textured, thin surface soil zone. Beneath is a dense layer of clay accumulated from the horizon above, which has an extremely low water permeability.

Saline-sodic soils. The saline-sodic soils have a conductivity of the saturation extract greater than 4 mmhos per cm at 25°C and an exchangeable-sodium percentage greater than 15. The soil pH seldom exceeds 8.5 after salinization has occurred. The properties tend to shift toward those of a sodic soil if the excess soluble salts are leached downward. Saline-sodic soils are usually somewhat more alkaline than saline soils but the presence of substantial quantities of calcium and magnesium prevents the sodium from adversely affecting flocculation and soil structure.

Effects of Saline and Sodic Soils on Turfgrasses

Saline effects. Soluble salts produce harmful effects on turfs by increasing the salt content of the soil solution and the exchangeable sodium percentage on the soil colloids. The salt concentration above which turfgrass growth is impaired depends on the (a) turfgrass species, (b) soil texture, (c) salt distribution in the soil profile, and (d) composition of the salts present. A high salt content in the soil solution impairs the absorption of water and essential nutrients by the turfgrass root system. The turfgrass plant is more prone to wilt and desiccation when water absorption by turfgrass roots is restricted by the high osmotic pressure of the soil solution. Impaired seed germination and poor stands frequently occur on saline soils (23, 33, 73, 76).

The first visual symptoms of salinity effects on turfgrasses are wilting and a blue-green appearance followed by an irregular stunting of growth (73, 124). Tip burn may occur. A distinct thinning of shoots eventually occurs. Final diagnosis of a salinity problem should involve a soil test as well as visual symptoms.

The salinity tolerance of turfgrass species varies considerably (13, 33, 73, 76, 107, 110, 122) (Table 10-7). The less tolerant turfgrass species tend to accumulate chloride ions more rapidly in the leaf tissue than the more tolerant species (73). Differences in salinity tolerance also occur within a given species (122, 123). For example, Seaside, Arlington, Pennlu, and Old Orchard creeping bentgrasses are the most tolerant; Congressional and Cohansey, intermediate; and Penncross, the least salt tolerant (124). Seaside also has the best recuperative potential from the effects of salinity.

Sodic effects. Indirect effects of sodic soils on turfgrass growth include those related to soil structure. The high sodium content causes deflocculation of the soil colloids and a general deterioration in soil structure. A severe reduction in soil

Table 10-7

THE RELATIVE SALINITY TOLERANCE OF ELEVEN COMMONLY USED TURFGRASSES

Relative salinity tolerance	Electrical* conductivity, $ECe \times 10^{-3}$	Turfgrass
Good	8–16	Bermudagrass
		Zoysiagrass
		Creeping bentgrass
		St. Augustinegrass
Medium	4–8	Tall fescue
		Perennial ryegrass
Poor	< 4	Meadow fescue
		Red fescue
		Kentucky bluegrass
		Colonial bentgrass
		Centipedegrass

* Electrical conductivity of the saturation extract in millimhos per centimeter at 25°C (77°F).

aeration and water infiltration results. Compaction proneness is increased substantially. All of these indirect effects on the physical properties of the soil are detrimental to turfgrass growth and development. The absorption of iron, manganese, and phosphorus is also adversely affected in sodic soils.

The first visual symptoms of a sodic soil problem in turfgrasses is a reduction in shoot growth (72). Turfgrass species exhibit different tolerances to sodic conditions. The wheatgrasses have excellent tolerance to high sodium levels compared to most turfgrasses (86). Seaside creeping bentgrass has a higher tolerance to sodium than tall fescue, Kentucky bluegrass, and bermudagrass (72).

Boron toxicity. Specific ions can also occur in sufficiently high concentrations in saline soils to cause direct toxic effects. For example, boron toxicity may occur in saline soils in arid and semiarid regions. Injury symptoms first appear as a necrosis at the leaf tip where the highest boron concentration occurs (85). Frequent mowing removes the high boron content in the leaf tip and is thought to increase the ability of turfgrasses to tolerate high soil boron concentrations (85). The boron tolerance of turfgrasses is comparatively good. Turfgrass species that are rapid accumulators of boron are also the first to show injury. They rank in the following order in terms of the rate of boron accumulation: bermudagrass > Japanese lawngrass > Kentucky bluegrass > tall fescue > perennial ryegrass > creeping bentgrass (85). Calcium, magnesium, chloride, and bicarbonate ions also cause a more specific toxicity in saline soils.

Correcting Soil Salinity and Sodic Problems

Reclamation of saline soils is possible if the internal soil drainage is satisfactory. It involves the removal of excess salts from the root zone by the application of sufficient water to leach the salts from the root zone into lower soil depths. The irrigation water used should have a low salt content and the amount of water used in leaching should be controlled. Leaching is usually accomplished by flood irrigation with a specified depth of water. The amount of irrigation water applied should exceed the rate of water loss by surface evaporation. Frequent, light irrigations do not leach the salts from the turfgrass root zone and may actually increase the salt accumulation rate. The natural precipitation is usually adequate to leach the soluble salts downward in humid regions.

The common procedure for correcting a sodic soil problem is to amend with materials that replace the exchangeable sodium on soil colloids. Sulfur and gypsum are frequently used because of low cost but are relatively slow acting. A more rapid response can be obtained from sulfuric acid or calcium chloride but at a higher cost.

When reclaiming saline-sodic soils, leach most of the soluble salts from the root zone first and then apply the chemical amendment. Leaching the salts should not be practiced prior to amending if a substantial decrease in soil permeability may occur. The amendments can be broadcast and incorporated into the soil or applied through irrigation water. The soil is irrigated after the application to move the amendment downward and to leach the soluble sodium salts from the root zone. Leaching should be delayed for a period of time after amending with sulfur and the incorporation should be thorough to encourage rapid sulfur oxidation.

Soil Modification

It is possible to maintain a high quality turf on most soil textures ranging in extremes from sands to clays (11). The difficulty of maintenance is much greater and the proneness to turfgrass loss from environmental stresses is higher on certain soil types such as clay. Fertilization, irrigation, cultivation, and disease control practices must be varied substantially depending on the particular soil texture. Turfs grown on soils high in clay or silt content require a much more delicate manipulation of cultural practices compared to soils having a high sand content.

Soil modification involves the incorporation of texture and/or structural improving materials, which may or may not involve the existing soil, for the purpose of improving the soil physical and chemical conditions of the turfgrass root zone. Soil modification is most commonly practiced on soils (a) having a high clay or sand content or (b) subjected to intense traffic. Coarse textured materials are incorporated to improve soil aeration, percolation, and infiltration and to reduce the compaction tendency. In contrast, finer textured materials are incorporated to enhance water and nutrient retention.

Characteristics desired in a modified turfgrass root zone include (a) minimum compaction tendency, (b) good soil water infiltration and percolation rates, (c) adequate aeration for deep rooting, (d) freedom from toxic chemicals, (e) an active microorganism population, (f) a certain degree of resiliency, (g) high cation exchange capacity, and (h) adequate water retention (49, 50). There are few soil types that meet all these criteria. The primary objective is to minimize soil compaction by the incorporation of coarse textured materials on turfs subjected to intense traffic. Nutrient and water retention are sacrificed to achieve a minimal compaction tendency and adequate soil aeration, percolation, and infiltration. Improved water retention can be achieved on sandy soils subjected to minimal traffic by amending with fine textured materials.

Materials Utilized in Modification

The objectives of soil modification must be ascertained for each situation. Factors to be considered in selecting the appropriate amendments for soil modification include the (a) effect on soil texture, (b) effect on soil structure, (c) effect on soil chemical properties, (d) long term stability, (e) availability, (f) amount required, and (g) cost. The preferred root zone soils for turfgrass culture are the loamy sands, sandy loams, and loams (32). These soils contain a reasonably good balance of the desired properties.

Organic materials. Organic soil amendments function in enhancing soil structure and aeration as well as contributing to improved nutrient and water retention (24, 46, 51, 62, 65, 96, 102). Soil resiliency can also be increased with organic amendments. There are a number of organic materials available for use in soil modification. Materials that decay over an extended period of time are preferred. Organic materials having a higher proportion of slowly decomposable constituents such as lignin have a slower decomposition rate.

The common types of peat available for soil modification are (a) peat humus, (b) reed-sedge peat, (c) hypnum peat, and (d) sphagnum moss peat (Table 10-8). The best all-around organic material for use in soil modification has been decomposed peat (65, 97). The peat selected for use can be fibrous but should be relatively fine in order to facilitate uniform mixing. The organic matter content should be above 90% with a high moisture absorption capacity. A reasonably dry material is preferred to minimize handling costs. Peats are subject to compaction when wet.

Other potential organic amendments include manure, unprocessed sewage sludge, spent mushroom soil, sawdust, hulls, barks, leaves, various animal and vegetable by-products, and green manures. These materials can be in a relatively undecomposed state unless composted (101, 102). A very large quantity of undecomposed organic material must be added in order to add a satisfactory quantity of residual humus. Thorough mixing of the undecomposed materials during soil modification is frequently difficult. These materials utilize nitrogen from the soil for decomposition. This competition for the available nitrogen may create a turfgrass nitrogen deficiency unless unusually high nitrogen rates are applied. A positive factor is the release of some essential plant nutrients during decay (102, 111).

Decomposed organic materials are preferred for soil modification. Organic materials that can be effectively utilized if adequately decomposed include lignified wood, shredded bark, and sawdust (9, 62, 80, 81, 82, 111). Undecomposed sawdust from certain species of trees contains compounds that are toxic to turfgrasses (116). Sawdust from ash and red oak trees is the most toxic. The toxicity dissipates after

Table 10-8

CHARACTERISTICS OF FIVE COMMON TYPES OF PEAT AVAILABLE FOR TURFGRASS SOIL MODIFICATION [AFTER LUCAS, ET AL. (68)]

Type of peat	State of decomposition	pH	Water absorbing capacity, %	Volume weight, lb/cu ft	Nitrogen, %	Desirability for soil modification
Peat humus	Advanced	5.0–7.5	150–500	20–40	2.0–3.5	Excellent
Reed-sedge peat, high pH	Intermediate	5.1–7.5	400–1200	10–18	2.0–3.5	Good
Reed-sedge peat, low pH	Intermediate	4.0–5.0	500–1200	10–15	1.5–3.0	Good
Hypnum moss peat	Least	5.0–7.0	1200–1800	5.0–10	2.0–3.5	Good
Sphagnum moss peat	Least	3.0–4.0	1500–3000	4.5–7.0	0.6–1.4	Poor

6 to 7 months of weathering. Burned organic materials should be avoided. Spent mushroom soil is an effective source of organic matter and soil for the modification of sandy soils (101, 102).

Sand. Great variations exist among the sands in terms of the actual (a) shape, (b) particle size, and (c) relative distribution of particle sizes (5, 24, 94). Not all sands are desirable for soil modification. The chief functions of sand are associated with the larger particle size that enhances aeration, infiltration, and percolation as well as reducing the compaction tendency (24, 47, 69, 109).

Medium to coarse sands having a diameter of 0.25 to 1.2 mm are preferred for use in soil modification where the primary objective is to reduce the compaction tendency and increase the rooting depth (47, 55, 56, 57, 69, 114) (Fig. 10-7). Better moisture retention is achieved with a major portion of the sand particles in the 0.25 to 0.5 mm range (47, 55, 69). This is particularly desirable in arid climates. The sand selected should have a graded, uniform texture with fines removed. Sands containing predominantly particle sizes less than 0.2 mm are usually not satisfactory when mixed with silt or clay because they tend to compact. The pH of the sand should be checked since it can vary greatly. Appropriate correction of the soil reaction can be made prior to establishment if necessary.

Calcined clay. Calcined clay is a granular clay mineral that is fired to a very high temperature, crushed, and screened to a specific size fraction. The addition of calcined clay to fine textured soils causes a decrease in bulk density and an increase in pore space, water retention, percolation, and infiltration (16, 41, 46, 62, 78, 79, 80, 81, 82, 96, 97). Although calcined clay has good water retention, much of it is not available to the turfgrass plant (114). Calcined clays have a relatively large particle size that mini-

Figure 10-7. Deep rooting of bentgrass cut at 0.25 inch when grown on a green modified with sand to minimize compaction. (Photo courtesy of the USGA Green Section).

mizes compaction (46, 62, 78, 80, 82, 96). The cation exchange capacity is low but greater than for sands.

Calcined clay has been utilized for partial modification of turfgrass root zones particularly on sports turfs and greens. It provides temporary improvement of problem greens containing fine textured, compacted soil. Root and rhizome growth is significantly increased on a short term basis (58). The long term stability has not been fully determined. There are wide variations in the physical properties of the calcined clays available for soil modification (16, 79). Types with a long life expectancy in terms of mechanical strength and resistance to breakdown are preferred.

Processed mica. Processed micas such as vermiculite have been available for use in soil modification. Vermiculite is a porous, laminated mineral produced by exposing washed, graded crude ore to very high temperatures. The high temperature causes expansion or lamination of the ore. Initially, it has a large particle size, is quite porous, and has good water retention (27, 36, 40). The cation exchange capacity is greater than for the sands. The long term stability to soil weathering is limited and there is a lack of mechanical strength (40). The plate-like particles produced during the deterioration of vermiculite tend to orient perpendicular to the compaction pressure. A layering effect results causing impaired water movement. This also occurs when vermiculite is used as a topdressing. A lack of large pore space and water permeability becomes evident in soil-vermiculite mixes after a period of time (97). The use of processed micas such as vermiculite for modification of fine textured soils has been minimal. It has been effective in improving the water retention, small pore space, and cation exchange capacity of sandy soils when amended at 10% by volume (45, 46, 96).

Other materials used in soil modification. Soil modification materials that have been used with varying degrees of success include (a) expanded shale, (b) colloidal phosphate, (c) perlite, (d) waste ash, and (e) blast furnace slag. These materials function in soil modification by diluting the soil rather than affecting soil structure.

Graded expanded shale is available for soil modification in certain areas (91). Soils modified with expanded shale have greater pore space than corresponding soils modified with sand (92).

Colloidal phosphate is used for modifying soils having a high sand content. It increases the small pore space and cation exchange capacity (45, 46, 96). There is also a decrease in the hydraulic conductivity and large pore space of soil-colloidal phosphate mixes (96, 97). Colloidal phosphate improves the wettability of hydrophobic sandy soils. It should be mixed with hydrophobic sands at a rate of 5 to 8% by volume (45, 96).

Expanded perlite is mined from lava flows, crushed, screened, and heated in furnaces where the natural perlite expands into porous particles. Perlite possesses a relatively large particle size that improves the water retention, aeration, porosity, and infiltration rate when mixed with soil (24, 40, 114). Its use for turfgrass soil modification has been very limited due to poor mechanical strength and limited life expectancy. Perlite contributes very little to soil improvement that cannot be accomplished with coarse sand.

Waste ash or fly ash from the burning of coal fuels can be used for soil modification if topsoil is incorporated into the upper portion of the root zone (7, 15). The ash must be free of potentially toxic materials or a high salt content.

Blast furnace slag is a by-product of the steel manufacturing industry. It is quite porous with good moisture absorption, pore space, and water retention (114). The long term stability is not well documented. The soil reaction may have to be adjusted if slag is used in soil modification.

Soil Modification Procedures

Soil improvement may involve partial or complete modification. **Partial modification** involves the incorporation of foreign material into the existing soil. In contrast, **complete modification** involves the preparation of a root zone mix that may or may not contain soil from the existing site. Effective modification must improve the soil physical and chemical characteristics of the root zone to a depth of 8 to 12 in. (71, 79).

Nonprofessionals frequently attempt to improve a soil by light applications of sand or peat over the surface of the turf. This approach is generally ineffective and may prove detrimental if a distinct layer is formed that can seriously impair soil water movement. At least 2 and preferably 4 in. of material must be incorporated into the upper 6 to 8 in. of the existing soil in order to achieve significant soil improvement. Light applications of 0.1 to 0.5 in. thick are seldom effective and generally a waste of time.

Many different procedures can be effectively utilized in soil modification. Eight distinct approaches are covered in detail. The selection of the appropriate modification procedure depends on the (a) previous experience regarding the effectiveness of a modification procedure in that area, (b) cost, (c) anticipated intensity of traffic, and (d) relative importance of the characteristics desired.

Partial modification of sandy soils. Partial modification of droughty soils containing predominantly sand is desirable in order to improve water and nutrient retention. The addition and incorporation of a loam soil high in organic matter is preferred (Fig. 10-8). A clay content of 5 to 8% is usually sufficient to provide the desired degree of nutrient and water retention. Organic matter, usually in the form of decomposed peat, can be incorporated into the soil mixture, particularly if the loam soil utilized is relatively low in organic matter content (51, 89, 101, 102). No sub-

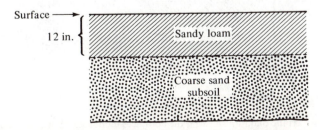

Figure 10-8. Profile diagram showing the partial modification of a sandy soil.

surface drain tile are installed since the percolation of most sands is quite good, assuming the sand profile is deep with adequate lateral drainage.

Partial modification of clay soils. The primary objective in modifying clay soils is to improve soil aeration and water movement. Partial modification of clays is an alternate method when adequate monies are not available for complete modification. It involves mixing coarse sand and organic matter into the upper 10 to 14 in. of the root zone. Subsurface drainage should be provided by means of drain tile, preferably with a vertical column of sand or gravel that extends from the tile upward to the sandy loam root zone (Fig. 10-9).

Complete modification of clay soils. Complete modification of a clay soil involves the preparation of a subgrade on the clay that ensures lateral drainage of water toward tile spaced at intervals in somewhat lower microtrenches. Gravel is then placed around the tile in the microtrench and a 10- to 14-in. layer of coarse sand placed above this (Fig. 10-10). The root zone layer is composed of a sandy loam soil 10 to 14 in. deep. Organic matter is usually incorporated into the surface root zone.

California Method. This method was developed and tested at the University of California at Los Angeles. It was one of the earliest procedures developed from actual research rather than trial and error (69, 70, 71). The subgrade is contoured for rapid lateral drainage to drain tile spaced at 10- to 15-ft intervals (Fig. 10-11). Above the tile is 20 to 24 in. of a prepared soil root zone mixture composed of 85

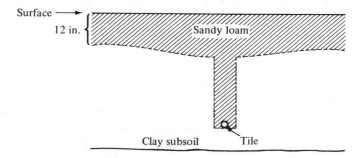

Figure 10-9. Profile diagram showing the partial modification of a clay soil.

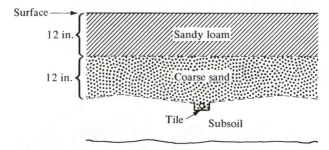

Figure 10-10. Profile diagram showing the complete modification of a clay soil.

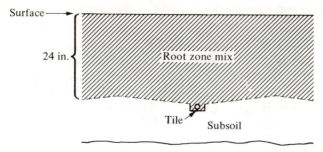

Figure 10-11. Profile diagram showing the California Method of complete soil modification.

to 90% sand, 5 to 7.5% well aggregated clay, and 5 to 7.5% peat. The sand used in the California Method has a very narrow size range. Specifications for the sand fraction include (a) 50% or more smaller than 0.4-mm diameter, (b) 25% or more smaller than 0.25 mm, and (c) less than 10% smaller than 0.1 mm (71). Organic matter is incorporated into the upper 4 to 5 in. of the root zone mixture.

Weigrass Method. The Weigrass Method was developed in Sweden for use on football and soccer fields (60). The clay soil subgrade is prepared with 2-in. diameter plastic drain tile placed in narrow, 8 in. deep trenches spaced at 15 ft intervals. The soil between the trenches is cultivated and the subgrade shaped with a 1% slope to enhance lateral water movement. The trench is filled to the subgrade with coarse gravel, 2 to 8 mm in size.

An 8-in. layer of 0.2 to 0.4 mm diameter sand is evenly distributed over the subgrade (Fig. 10-12). Fertilizer is incorporated into the upper 4 in. of the sand and a 2-in. layer of topsoil is added to the surface. The topsoil is composed of 60% sand and 40% peat by volume and prepared by off-site mixing. Football field construction by this method is relatively low in cost and has proved effective on football grounds in the Scandinavian countries.

USGA Green Section Method. A soil modification procedure for greens involving a perched water table principle has evolved from research sponsored by the USGA

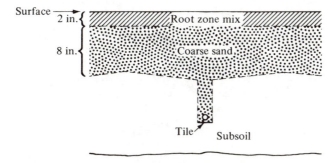

Figure 10-12. Profile diagram showing the Weigrass Method of complete soil modification.

Green Section (2, 28, 29). The basic construction involves subsurface drain tile placed in microtrenches in the subgrade. The subgrade contours should conform closely with the proposed final surface grade with rapid lateral drainage ensured. Filled areas in the subgrade must settle thoroughly. The tile lines are spaced at 10-ft intervals in a herringbone or gridiron arrangement. Washed gravel about 0.25 in. in size or crushed stone is placed over the tile and to a 4 in. depth above the subgrade. Above this is added a 1.5- to 2.0-in. layer of washed coarse sand that is 1.0 mm or larger in size. On top of this layer is placed 12 in. of the root zone soil mixture (Fig. 10-13).

The root zone soil mixture composition is determined by specific physical testing procedures outlined by the USGA Green Section (30, 47, 56). These laboratory measurements involve primarily an evaluation of the soil permeability and pore space relationships of the mixture. Other helpful information includes the texture, structure, water retention, mineral derivation, and bulk density that influence the soil permeability and pore space relationships. The USGA Green Section criteria in determining a satisfactory root zone soil mixture are as follows. After compaction, a core of the soil mixture should permit water to pass through at a rate not less than 0.5 in. nor more than 1.5 in. per hr when subjected to a hydraulic head of 0.25 in. The mixture should have a minimum total pore space of 33%, comprised of 12 to 18% large pores and 15 to 21% small pores.

The modification must be accomplished as outlined in order to achieve the perched water table effect. The root zone layer is placed over a coarser textured sand layer located above a gravel layer. The abrupt change in texture below the root zone soil mixture disrupts the water film continuity. Under these conditions gravitational drainage of water from the root zone layer occurs only after it becomes saturated with water. Water drains downward through the coarse sand and gravel layers to the drain tile when the accumulation of water becomes sufficiently large. The perched water table results in a certain quantity of water being retained in the root zone soil mixture where it can be readily absorbed by the root system. **Purr-Wick System.** The Purr-Wick System involves holding soil water above an impermeable plastic layer (87). This system was developed for use on turfgrass

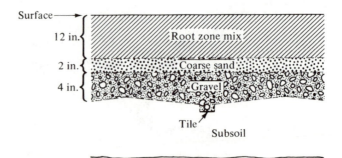

Figure 10-13. Profile diagram showing the USGA Green Section Method of complete soil modification.

areas subjected to intense traffic with the objective of maintaining adequate soil aeration and water movement with a minimal compaction tendency.

Construction involves the preparation of a subgrade composed of a series of level terraces having an elevation difference of 6 in. and an elevated lip of 4 in. at the outer edge of each terrace. Polyethylene sheets of 6 to 10 mil are then laid across the terraces starting from the highest terrace working toward the lowest terrace and over the outer retaining edges. A 2-in. diameter plastic pipe with numerous slits on alternate sides is then laid level above the plastic sheet of each terrace. A drain plug and adjustable riser for controlling the soil water level are connected to the pipe in a pit at the outer edge of each terrace. A sealant should be used where the drain pipe is inserted through the plastic. Coarse sand is applied over the slits in the pipe. Bulk washed, medium sand is placed above this to the desired root zone depth and compacted. Peat is incorporated into the upper 2 in. of soil. The Purr-Wick System was proposed in 1966 and has received only limited evaluation under field conditions. Further studies must be conducted before a full assessment of its potential effectiveness can be made.

Partial modification by vertical columns. Poor soil water movement and aeration can be partially alleviated by modification using vertical columns (10, 106). In this case, 0.5- to 1.0-in. cores are removed to a depth of 4 to 6 in. and the holes filled with a sandy loam soil. The spacing interval between vertical columns varies from 3 to 12 in. depending on the degree of modification desired and the time available. This method has resulted in improved turfgrass quality, rooting depth, infiltration, and aeration (10, 106). It is not as satisfactory as complete soil modification but is an alternate method where funds are not available for complete modification.

Root Zone Soil Mixtures

The selection of an effective root zone soil mixture has not developed to an exact science. The amount of valid experimental data available is quite limited. The amounts of soil, sand, and organic matter by volume to be utilized in preparing a root zone soil mixture vary greatly depending on the specific types available. The appearance and feel of a mixture in a dry condition can be misleading. Selection of the appropriate sand-soil-organic matter ratio is best made by a combination of (a) physical soil tests (89); (b) a knowledge of the physical and chemical characteristics of the individual sand, soil, and organic matter components to be utilized; and (c) practical experience regarding the performance of the sand-soil-organic matter mixture when combined in a given ratio.

One of the primary objectives in soil modification is to develop a root zone soil mixture capable of maintaining an adequate turfgrass cover under intense traffic. This is achieved with a coarse textured mixture that has adequate soil aeration and percolation plus a minimum compaction tendency. Water and nutrient retention is sacrificed to achieve these objectives (94, 114). It would be desirable for a root zone soil mixture to possess all of these characteristics, including adequate nutrient and water retention. This ultimate soil mixture is seldom achieved.

The amount of sand required to achieve a specified total sand content in the soil mixture varies from 40 to 80% depending on the (a) sand content of the soil component and (b) particle uniformity in the sand fraction. Selection of the exact quantity of each fraction should be based on previous practical experience concerning the performance of various sand-soil-peat mixtures in the area and laboratory analyses of the soil physical characteristics (30, 47, 56, 89, 97).

Representative samples of the sand, soil, and organic matter components can be submitted to a reputable soil laboratory for analysis. The primary physical determinations made are the hydraulic conductivity and pore space. Determinations are also made of the aggregation, texture, water retention, mineral derivation, and bulk density (89). Mixtures are then prepared using various proportions of sand, soil, and organic matter based on the results of the physical determinations. These mixtures are evaluated relative to a known optimum standard. A recommendation is given based on the results of these analyses as to the appropriate quantities of each of the components to utilize in preparing a satisfactory turfgrass root zone mixture for use under intense traffic.

The actual volume percentages of sand, soil, and organic matter needed in preparing a soil mixture vary greatly depending on the particle size distribution and characteristics of the three components. Thus, a general sand-soil-organic matter ratio recommendation cannot be made. The root zone soil mixture commonly used under intense traffic is in the loamy sand soil textural category (32). A soil mixture composed of 85 to 90% sand of 0.2 to 0.4 mm in diameter, 5 to 7.5% well-aggregated clay, and 5 to 7.5% peat has been effective on intensely used turfs in California (69). Similar results were reported from Oklahoma (36). The soil mixture should be as low as possible in silt and fine sand (36).

An organic material is usually incorporated into the soil mixture, particularly on soils having a high sand content or poor structure. The organic matter improves the water and nutrient retention, soil structure, and resiliency (102). Relatively decomposed peat is the most common organic matter source although other materials may be used if available locally and are well decomposed. The quantity of peat added depends on the organic matter content of the soil utilized. As the peat content is increased in sand mixtures, the bulk density is lowered and the amount of available moisture is increased (51). Also, the onset of severe wilting is more gradual and occurs at higher tensions. Generally 10 to 20% peat by volume is satisfactory. Higher amounts should be used on sandy soils. Adding more than 20% peat should be avoided since an excessive quantity of moisture is retained at the soil surface causing soggy, unfavorable soil conditions (36, 65).

Soil Mixing Procedures

A uniform soil mixture throughout the root zone is extremely important. The two primary mixing procedures utilized are (a) off-site mixing and (b) on-site mixing. The materials to be utilized in off-site mixing are brought to a convenient site and mixed with some type of mechanical equipment in the required proportions

(Fig. 10-14). Soil shredders or power screens are commonly used but are not entirely satisfactory since many are not designed specifically for soil mixing. To achieve any degree of uniform mixing with a shredder or rotary screen, the components to be mixed must be fed through the machine at a constant rate in the proper proportion.

A three-compartment blending bin with an adjustable feeder gate facilitates mixing (84). The soil is run through a shredder and screen prior to placement in the blending bin. The soil, sand, and organic matter are fed in the desired proportions onto a conveyer that feeds into a mixing unit. A high capacity mechanical soil mixing machine is needed for off-site turfgrass soil modification. The root zone soil mixture is transported to and spread on the actual construction site after mixing. Assuming the materials are added in the appropriate ratios and that no separation occurs during subsequent handling, a very uniform soil mixture can be achieved by off-site mixing.

On-site mixing involves the uniform distribution of the materials in layers over the surface of the actual construction site. Organic material should be placed at the bottom. The soil is usually placed over the area with the sand spread over the top. The materials are then mixed by the use of cultivating equipment. The area must be cultivated repeatedly until a uniform soil mixture is achieved with no layers or pockets of a particular soil textural size. Tillers, cultivators, disks, and harrows can be used for mixing but none are entirely satisfactory. Such devices must be passed over the area to be mixed many times in order to achieve uniform soil mixing. The soil profile should be thoroughly examined at frequent intervals and tillage continued until an acceptable degree of uniformity is attained. Disks are particularly prone to leaving pockets on turns. Also, the high speed rotary tiller results in a separation of particle sizes at the soil surface. If a rotary tiller is used, a low speed type is preferred to avoid this problem. On-site mixing reduces the amount of time required in handling large volumes of material. There is more likelihood of nonuniformity, however, unless great care is taken to ensure uniform

Figure 10-14. Off-site mixing of materials for complete modification of a green. (Photo courtesy of A. M. Radko, USGA Green Section, Highland Park, New Jersey.)

mixing. Lighter materials tend to float toward the surface. On-site mixing causes greater compaction and is inferior to off-site mixing.

Composting

Decomposed organic matter can be developed for soil modification or topdressing by composting. **Compost** encompasses organic residues, or a mixture of organic residues and soil, that have been piled, moistened, and allowed to undergo decomposition. A well decomposed compost contributes organic matter to improve the physical characteristics of the soil, particularly the structure and water retention. Compost may also be used as a mulch for improving surface soil moisture retention.

Compost contains varying amounts of plant nutrients depending on the degree of decomposition and leaching during the composting process. It contains approximately 2% nitrogen but may vary from 1.5 to 2.5% of the dry matter. The phosphorus content normally ranges from 0.5 to 1% while the potassium content varies from 1 to 2%. The phosphorus and potassium levels can be increased by adding a mixed fertilizer when making the compost pile. Plant nutrients from compost become available to turfgrasses at a relatively slow rate and should not be considered as a major source of plant nutrients in the cultural program.

The composting process. Composting is a microbiological process in which organic materials are partially decomposed by the activity of microorganisms. The water soluble substances and protein are broken down very rapidly. The celluloses and hemicelluloses are also decomposed to a considerable extent. On the other hand, lignin is quite resistant to decay and composes a major portion of the organic fraction upon completion of the composting process.

The composting process proceeds most rapidly under conditions favoring microorganism activity. These include good aeration of the compost pile and temperatures above 75°F. Heat is evolved by the microorganisms during the decomposition process. Temperatures will gradually rise over a 2 to 3 week period to as high as 140 to 170°F in a properly insulated compost pile. A decline in the temperature of the compost pile to ambient air temperature indicates that the composting process has been completed.

The process produces an organic material that is in an advanced state of decomposition and contains a higher nitrogen content. This reduces potential problems with a nitrogen shortage when the organic compost material is incorporated into the soil. The physical nature of the organic material is also altered during composting. It becomes friable, crumbly, and easy to handle, thus facilitating soil incorporation.

Composting Procedures

The composting procedure involves piling the organic materials in layers to a height of no more than 6 ft. The 6 ft maximum ensures adequate aeration to the bottom of the pile. The width and length can be adjusted to facilitate convenient handling and turning of the compost. Sources of compostable materials include leaves, straw, hay, animal manures, municipal refuse such as garbage and street refuse, organic wastes from cities, wood residues such as sawdust and wood chips, and grass clippings. Grass clippings should be permitted to wilt before incorporating into a compost pile.

Rapid decomposition is ensured by proper aeration, water content, nutrient level, and temperature within the compost pile. The organic material should be piled and moderately compressed to the extent that adequate aeration occurs but without large air spaces that result in the rapid loss of moisture and heat from the pile. The optimum moisture content of the compost pile is in the range of 50 to 70%. Water is best applied to the layers as the pile is built. Lower water contents impair microorganism activity while higher water contents result in anaerobic conditions that are unfavorable for organic matter decomposition. Favorable high temperatures are ensured by (a) constructing large compost piles or (b) insulating the outer surface of the pile.

The composting process can sometimes be accelerated by adding nutrients, particularly nitrogen, phosphorus, potassium, and calcium. As a general guideline, approximately 15 lb of actual nitrogen should be incorporated per ton of dry organic material. Calcium cyanamide and urea are preferred to ammonium sulfate since they also give a slightly basic reaction that favors microorganism activity. Liming may be required to neutralize the acidic effect of the sulfate anion if ammonium sulfate is used. Phosphorus and potassium should be added to organic materials such as sawdust or leached plant residues that are low in these nutrients. Approximately 20 lb of superphosphate and 10 lb of potassium chloride should be added per ton of organic material. An alternate procedure is the use of a mixed fertilizer having an analysis of approximately 10–6–4.

Small amounts of soil can be mixed into the compost pile but are not necessary. The soil may function in conserving potential plant nutrient loss, particularly nitrogen. A thin soil layer on the outer surface of the compost pile aids in moisture retention and temperature insulation.

The decomposition process in the compost pile is hastened by periodic turning to improve aeration. As a general rule the compost should be turned at least once every 3 weeks. More frequent mixing stimulates rapid, uniform decomposition throughout the compost pile. Small compost piles may require as long as 3 months for the process to be completed under favorable temperature and moisture conditions. The composting process may be completed within 2 to 3 weeks in larger composting operations where the pile is turned every 3 to 4 days.

References

1. ANONYMOUS. 1957. Soil. The 1957 Yearbook of Agriculture. USDA. pp. 1–784.
2. ANONYMOUS. 1960. Specifications for a method of putting green construction. USGA Journal and Turf Management. 13(5): 24–28.
3. ARNON, D. I., W. E. FRATZKE, and C. M. JOHNSON. 1942. Hydrogen ion concentration in relation to absorption of inorganic nutrients by higher plants. Plant Physiology. 17: 515–524.

4. BENGTSON, J. W., and F. F. DAVIS. 1939. Experiments with fertilizers on bent turf. Turf Culture. 1: 192–213.

5. BINGAMAN, D. 1968. Sands for root zones. Proceedings of the 1968 Midwest Regional Turf Conference. pp. 68–70.

6. BINGAMAN, D. E. 1970. Relationships between pore space and water behavior of compacted sands and three constituent sand particle dimensions. M.S. Thesis.Purdue University. pp. 1–181.

7. BREDAKIS, E., and H. PAGE. 1966. Can flyash support plant life? Massachusetts Turf Bulletin. 3(3): 3–4.

8. BROOKS, C. R. 1966. Growth of roots and tops of bermudagrass as related to aeration and drainage in stratified golf green soils. Ph.D. Thesis. Texas A & M University. pp. 1–114.

9. BURTON, G. W., E. H. DEVANE, and R. L. CARTER. 1953. Sawdust as a source of organic matter for top-dressing. Golf Course Reporter. 21(5): 5–11.

10. BYRNE, T. G., W. B. DAVIS, L. J. BOOHER, and L. F. WERENFELS. 1965. A further evaluation of the vertical mulching method of improving old greens. California Turfgrass Culture. 15(2): 9–11.

11. CALE, E. B. 1950. Grass establishment on stabilized airport soils. Florida Greenkeeping Superintendents Association Turf Management Conference Proceedings. pp. 24–26.

12. CALHOUN, C. R., and E. C. ROBERTS. 1969. "Artificial" soil mixtures for putting greens. 1969 Agronomy Abstracts. p. 52.

13. CHITTENDEN, D. B., and E. C. ROBERTS. 1969. Relationship between sodium as a percent of total soluble salts in the nutrient solution and response of two roadside grasses grown in solution culture. 1969 Agronomy Abstracts. p. 52.

14. COLLINGS, G. H. 1955. Commercial fertilizers. 5th ed. McGraw-Hill Book Company, New York. pp. 1–617.

15. COPE, F. 1964. The establishment of playing fields using power-station waste-ash. Journal of the Sports Turf Research Institute. 40: 51–66.

16. DANIEL, W. H. 1963. Prescription root zones with calcined clays. Proceedings of the 18th Annual Texas Turfgrass Conference. pp. 17–18.

17. DARROW, R. A. 1939. Effects of soil temperature, pH, and nitrogen nutrition on the development of *Poa pratensis*. Botanical Gazette. 101: 109–127.

18. DAVIS, R. R. 1952. Physical conditions of putting-green soils and other environmental factors affecting greens. USGA Journal and Turf Management. 5(1): 25–27.

19. DAWSON, R. B. 1933. Report on results of research work carried out in connection with soil acidity and the use of sulfate of ammonia as a fertilizer for putting greens and fairways. Journal of the Board of Greenkeeping Research. 3(9): 65–78.

20. DAWSON, R. B., and B. M. BOYNS. 1938. The use of sulfur in improving the physical condition of clay soils on golf courses. Journal of the Board of Greenkeeping Research. 5(18): 189–200.

21. DAWSON, R. B., and T. W. EVANS. 1931. The establishment of grasses on very acid morland with a view to turf formation. Journal of the Board of Greenkeeping Research. 2(5): 107–111.

22. DEVANE, E. H., M. STELLY, and G. W. BURTON. 1952. Effect of fertilization and management of different types of bermuda and bahia grass sods on the nitrogen and organic matter content of Tifton sandy loam. Agronomy Journal. 44: 176–179.

23. DEWEY, D. R. 1962. Germination of crested wheatgrass in salinized soil. Agronomy Journal. 54: 353–355.

24. DUICH, J. M. 1967. Soil modification research. 38th International Turfgrass Conference Proceedings. pp. 25–30.

25. EDMOND, D. B., and S. T. J. COLES. 1958. Some long-term effects of fertilizers on a mown turf of browntop and chewing's fescue. New Zealand Journal of Agricultural Research. 1: 665–674.

26. ENGEL, R. E., and R. B. ALDERFER. 1967. The effect of cultivation, topdressing, lime, nitrogen, and wetting agent on thatch development in ¼-inch bentgrass turf over a ten-year period.

1967 Report on Turfgrass Research at Rutgers University. New Jersey Agricultural Experiment Station Bulletin 818. pp. 32–45.

27. Escritt, J. R. 1955. Exfoliated vermiculite in turf production and maintenance. Journal of the Sports Turf Research Institute. 9(31): 90–93.

28. Ferguson, M. H. 1955. When you build a putting green make sure the soil mixture is a good one. USGA Journal and Turf Management. 8(6): 26–29.

29. ———. 1965. After five years: The Green Section specification for a putting green. USGA Green Section Record. 3(4): 1–7.

30. Ferguson, M. H., L. Howard, and M. E. Bloodworth. 1960. Laboratory methods for evaluation of putting green soil mixtures. USGA Journal and Turf Management. 13(5): 30–33.

31. Finn, B. J., S. J. Bourget, K. F. Nielsen, and B. K. Dow. 1961. Effects of different soil moisture tensions on grass and legume species. Canadian Journal of Soil Science. 41: 16–23.

32. Fitts, O. B. 1925. A preliminary study of the root growth of fine grasses under turf conditions. USGA Green Section Bulletin. 5: 58–62.

33. Foresberg, D. E. 1953. The response of various forage crops to saline soils. Canadian Journal of Agricultural Science. 33: 542–549.

34. Gardner, W. R., and C. F. Ehlig. 1962. Some observations on the movement of water to plant roots. Agronomy Journal. 54: 453–456.

35. Garman, W. L. 1950. Physical properties of soil as applied to turf management. Oklahoma-Texas Turf Conference. pp. 89–94.

36. ———. 1952. Permeability of various grades of sand and peat and mixtures of these with soil and vermiculite. USGA Journal and Turf Management. 5(1): 27–28.

37. Garner, E. S., and S. C. Damon. 1929. The persistence of certain lawn grasses as affected by fertilization and competition. Rhode Island Agricultural Experiment Station Bulletin 217. pp. 1–22.

38. Gilbert, B. E., and F. R. Pember. 1935. Tolerance of certain weeds and grasses to toxic aluminum. Soil Science. 39: 425–429.

39. Goss, R. L. 1967. The effects of urea on soil pH and calcium levels. Northwest Turfgrass Topics. 9(3): 1–4.

40. Hagan, R. M., and J. R. Stockton. 1952. The effect of porous soil amendments on water retention characteristics of soils. USGA Journal and Turf Management. 5(1): 29–31.

41. Hansen, M. C. 1962. Physical properties of calcined clays and their utilization for rootzones. M. S. Thesis. Purdue University. pp. 1–64.

42. Hart, S. W., and J. A. DeFrance. 1956. The effect of synthetic soil conditioners in turfgrass soils. Golf Course Reporter. 24(2): 5–8.

43. Hartwell, B. L., and S. C. Damon. 1917. The persistence of lawn and other grasses as influenced especially by the effect of manures on the degree of soil acidity. Rhode Island Agricultural Experiment Station Bulletin 170. pp. 1–24.

44. Holmes, J. L. 1968. Better drainage through slit trenches. USGA Green Section Record. 6(1): 8–10.

45. Horn, G. C. 1965. Soil and water relationships in turfgrass management. Proceedings of the University of Florida Turf-Grass Management Conference. pp. 28–31.

46. Horn, G. C., and H. G. Meyers. 1968. Soil amendment study. Florida Turfgrass Research Unit Department of Ornamental Horticulture Research Plot Layout and Progress Report I. pp. 1–1a.

47. Howard, H. L. 1959. The response of some putting green soil mixtures to compaction. M.S. Thesis. Texas A & M College. pp. 1–97.

48. Hughes, T. D., J. F. Stone, W. W. Huffine, and J. R. Gingrich. 1966. Effect of soil bulk density and soil water pressure on emergence of grass seedlings. Agronomy Journal. 58: 549–553.

49. HUMBERT, R. P., and F. V. GRAU. 1949. Soil and turf relationships. A report on some studies of the physical properties of putting-green soils as related to turf maintenance. USGA Journal. 2(2): 25–32.

50. ———. 1949. Soil and turf relationships. Part II. USGA Journal. 2(3): 28–29.

51. JUNCKER, P. H., and J. J. MADISON. 1967. Soil moisture characteristics of sand-peat mixtures. Soil Science Society of America Proceedings. 31(1): 5–8.

52. JUSKA, F. V. 1959. Response of Meyer zoysia to lime and fertilizer treatments. Agronomy Journal. 51: 81–83.

53. JUSKA, F. V., and A. A. HANSON. 1969. Nutritional requirements of *Poa annua* L. Agronomy Journal. 61: 466–468.

54. JUSKA, F. V., A. A. HANSON, and C. J. ERICKSON. 1965. Effects of phosphorus and other treatments on the development of red fescue, Merion, and common Kentucky bluegrass. Agronomy Journal. 57: 75–78.

55. KEEN, R. A. 1968. Modified rootzone research. Proceedings of the 1968 Midwest Regional Turf Conference. pp. 64–65.

56. KUNZE, R. J. 1956. The effects of compaction of different golf green soil mixtures on plant growth. M.S. Thesis. Texas A & M College. pp. 1–75.

57. KUNZE, R. J., M. H. FERGUSON, and J. B. PAGE. 1957. The effects of compaction on golf green soil mixtures. USGA Journal and Turf Management. 10(6): 24–27.

58. LAGE, D. P., and E. C. ROBERTS. 1964. Influence of nitrogen and a calcined clay soil conditioner on development of Kentucky bluegrass turf from seed and sod. Proceedings of the American Society for Horticultural Science. 84: 630–635.

59. LANGVAD, B. 1968. Ball-bouncing and ball-rolling as a function of mowing height and kind of soil have been studied at Weibullsholm. Weibulls Grastips. 10: 355–357.

60. ———. 1968. The Weigrass-method for construction of football-grounds in grass. Weibulls Grastips. 10: 377–379.

61. LETEY, J. 1961. Aeration, compaction and drainage. California Turfgrass Culture. 11(3): 17–21.

62. LETEY, J., W. C. MORGAN, S. J. RICHARDS, and N. VALORAS. 1966. Physical soil amendments, soil compaction, irrigation, and wetting agents in turfgrass management. III. Effects on oxygen diffusion rate and root growth. Agronomy Journal. 58: 531–535.

63. LETEY, J., L. H. STOLZY, O. R. LUNT, and N. VALORAS. 1964. Soil oxygen and clipping height effects on the growth of Newport bluegrass. Golf Course Reporter. 32(2): 16–26.

64. LETEY, J., L. H. STOLZY, O. R. LUNT, and V. B. YOUNGNER. 1964. Growth and nutrient uptake of Newport bluegrass as affected by soil oxygen. Plant and Soil. 20: 143–148.

65. LONGLEY, L. E. 1936. Influence on grass growth of various proportions of peat in lawn soils. Proceedings of the American Society for Horticultural Science. 34: 649–652.

66. LONGNECKER, T. C., and H. B. SPRAGUE, 1940. Rate of penetration of lime in soils under permanent grass. Soil Science. 50: 277–288.

67. LOVE, J. R. 1965. Soil physical conditions and water movement in soils. 1965 Wisconsin Turfgrass Conference Proceedings. pp. 1–5.

68. LUCAS, R. E., P. E. RIEKE, and R. S. FARNHAM. 1965. Peats for soil improvement and soil mixes. Michigan Cooperative Extension Bulletin No. 516. pp. 1–11.

69. LUNT, O. R. 1956. A method for minimizing compaction in putting greens. Southern California Turfgrass Culture. 6(3): 17–20.

70. ———. 1958. Soil types for putting greens. Southern California Turfgrass Culture. 8(2): 13–16.

71. ———. 1961. Soil mixes and turfgrass management. California Turfgrass Culture. 11(3): 23–24.

72. LUNT, O. R., C. KAEMPFFE, and V. B. YOUNGNER. 1964. Tolerance of five turfgrass species to soil alkali. Agronomy Journal. 56: 481–483.

73. LUNT, O. R., V. B. YOUNGNER, and J. J. OERTLI. 1961. Salinity tolerance of five turfgrass varieties. Agronomy Journal. 53: 247–249.

74. MCKEE, G. W., and A. R. LANGILLE. 1967. Effect of soil pH, drainage, and fertility on growth, survival, and element content of crownvetch, *Coronilla varia* L. Agronomy Journal. 59: 533–536.

75. MILLAR, C. E., L. M. TURK, and H. D. FOTH. 1965. Fundamentals of soil science, 4th ed. John Wiley & Sons, Inc., New York. pp. 1–491.

76. MILLINGTON, A. J., G. H. BURVIL, and B. B. MARSH. 1951. Salt tolerance, germination, and growth tests under controlled salinity conditions. Journal of Agriculture of Western Australia. 2nd Series. 28(1): 198–211.

77. MONTEITH, J. 1932. Soil acidity and lime for bent turf. Bulletin of USGA Green Section. 12(5): 190–195.

78. MONTGOMERY, R. H. 1961. The evaluation of calcined clay aggregates for putting green rootzones. M.S. Thesis. Purdue University. pp. 1–49.

79. MONTGOMERY, R. 1961. The evaluation of calcined clay aggregates for putting green rootzones. Proceedings of the 1961 Midwest Regional Turf Conference. pp. 88–91.

80. MORGAN, W. C., J. LETEY, S. J. RICHARDS, and N. VALORAS. 1965. Effects of soil amendments and irrigation on soil compactability, water infiltration, and growth of bermuda grass. 1965 Agronomy Abstracts. p. 46.

81. ———. 1966. Physical soil amendments, soil compaction, irrigation, and wetting agents in turfgrass management. I. Effects on compactability, water infiltration rates, evaporation, and number of irrigations. Agronomy Journal. 58: 525–535.

82. ———. 1966. The use of physical soil amendments, irrigation and wetting agent in turfgrass management. California Turfgrass Culture. 16(3): 17–21.

83. MUSSER, H. B. 1948. Effects of soil acidity and available phosphorus on population changes in mixed Kentucky bluegrass-bent turf. Journal of the American Society of Agronomy. 40: 614–620.

84. ———. 1966. Soil mixing operations. Turf-Grass Times. 1(6): 6–7.

85. OERTLI, J. J., O. R. LUNT, and V. B. YOUNGNER. 1961. Boron toxicity of several turfgrass species. Agronomy Journal. 53: 262–265.

86. PEARSON, G. A., and L. BERNSTEIN. 1958. Influence of exchangeable sodium on yield and chemical composition of plants. II. Wheat, barley, oats, rice, tall fescue, and tall wheatgrass. Soil Science. 86: 254–261.

87. RALSTON, D. S. 1969. Putting greens of sand over plastic. New York Turfgrass Association Bulletin 85. pp. 329–331.

88. REID, M. E. 1932. The effects of soil reaction upon the growth of several types of bent grasses. Bulletin of USGA Green Section. 12(5): 196–212.

89. RICHARDS, S. J., J. E. WARNEKE, A. W. MARSH, and F. K. ALJIBURY. 1964. Physical properties of soil mixes. Soil Science. 98: 126–132.

90. ROBINSON, R. R., and W. H. PIERRE. 1938. Response of permanent pastures to lime and fertilizers (1930–1936). West Virginia Agricultural Experiment Station Bulletin 289. pp. 1–47.

91. SCHMIDT, R. E. 1967. VPI's putting greens test modified soils. Weeds, Trees and Turf. 6(2): 17–18.

92. SCHMIDT, R. E., and R. E. BLASER. 1967. Soil modification for heavily used turfgrass areas. 1967 Agronomy Abstracts. p. 54.

93. SHOOP, G. J. 1967. Effects of various coarse textured material and peat on the physical properties of Hagerstown soil for turfgrass production. Ph. D. Thesis. Pennsylvania State University. pp. 1–225.

94. ———. 1967. Soil mixtures for golf course greens. Proceedings of the West Virginia Turfgrass Conference. pp. 27–33.

95. SHOOP, G. J., C. R. BROOKS, R. E. BLASER, and G. W. THOMAS. 1961. Differential responses

of grasses and legumes to liming and phosphorus fertilization. Agronomy Journal. 53: 111–115.

96. SMALLEY, R. R., and W. L. PRITCHETT. 1962. Soil amendments for putting greens. Golf Course Reporter. 30(4): 37–53.

97. SMALLEY, R. R., W. L. PRITCHETT, and L. C. HAMMOND. 1962. Effects of four amendments on soil physical properties and on yield and quality of putting greens. Agronomy Journal. 54: 393–395.

98. SMITH, J. D. 1958. The effect of lime application on the occurrence of *Fusarium* patch disease on a forced *Poa annua* turf. Journal of the Sports Turf Research Institute. 9(34): 467–470.

99. SPRAGUE, H. B. 1933. Root development of perennial grasses and its relation to soil conditions. Soil Science. 36: 189–209.

100. SPRAGUE, H. B., and G. W. BURTON. 1937. Annual bluegrass (*Poa annua* L.) and its requirements for growth. New Jersey Agricultural Experiment Station Bulletin 630. pp. 1–24.

101. SPRAGUE, H. B., and J. F. MARRERO. 1931. The effect of various sources of organic matter on the properties of soils as determined by physical measurements and plant growth. Soil Science. 32: 35–47.

102. ———. 1932. Further studies on the value of various types of organic matter for improving the physical condition of soils for plant growth. Soil Science. 34: 197–208.

103. SPURWAY, C. H. 1941. Soil reaction (pH) preferences of plants. Michigan Agricultural Experiment Station Special Bulletin 306. pp. 1–36.

104. SPURWAY, C. H., and C. E. WILDON. 1943. Controlling the reaction (pH) of greenhouse soils. Michigan Agricultural Experiment Station Quarterly Bulletin. 26: 115–121.

105. STARK, R. H., A. L. HAFENRICHTER, and W. A. MOSS. 1950. Adaptation of grasses for soil and water conservation at high altitudes. Agronomy Journal. 42: 124–127.

106. STOLZY, L. H., W. MORGAN, and J. LETEY. 1964. Renovation of golf greens by vertical mulching. 1964 Agronomy Abstracts. p. 103.

107. STOUTEMYER, V. T., and F. B. SMITH. 1936. The effects of sodium chloride on some turf plants and soils. Journal of the American Society of Agronomy. 28: 16–23.

108. SULLIVAN, E. F. 1962. Effects of soil reaction, clipping height, and nitrogen fertilization on the productivity of Kentucky bluegrass sod transplants in pot culture. Agronomy Journal. 54: 261–263.

109. SWARTZ, W. E., and L. T. KARDOS. 1963. Effects of compaction on physical properties of sand-soil-peat mixtures at various moisture contents. Agronomy Journal. 55: 7–10.

110. THOMAS, L. K., and G. A. BEAN. 1965. Winter rock salt injury to turf. Turf Bulletin. 2(13): 11–13.

111. VALORAS, N., W. C. MORGAN, and J. LETEY. 1966. Physical soil amendments, soil compaction, irrigation, and wetting agents in turfgrass management. II. Effects on top growth, salinity, and minerals in the tissue. Agronomy Journal. 58: 528–531.

112. VAN DERSAL, W. R. 1936. The ecology of a lawn. Ecology. 17: 515–527.

113. VOSE, P. B. 1962. Nutritional response and shoot/root ratio as factors in the composition and yield of genotypes of perennial ryegrass, *Lolium perenne* L. Annals of Botany. N. S. 26: 425–427.

114. WADDINGTON, D. V. 1967. Soil modification and results of further testing. Stockbridge School Turf Clippings, Conference Proceedings. pp. A-25 to A-30.

115. WADDINGTON, D. V., and J. H. BAKER. 1965. Influence of soil aeration on the growth and chemical composition of three grass species. Agronomy Journal. 57: 253–257.

116. WADDINGTON, D. V., W. C. LINCOLN, JR., and J. TROLL. 1967. Effect of sawdust on the germination and seedling growth of several turfgrasses. Agronomy Journal. 59: 137–139.

117. WELTON, F. A. 1932. Soil reaction not an efficient method by which to control some lawn weeds. 50th Annual Report of the Ohio Agricultural Experiment Station Bulletin 497. pp. 4–41.

118. WHEELER, H. J., and J. A. TILLINGHAST. 1900. Effect of lime upon grasses and weeds. Rhode Island Agricultural Experiment Station Bulletin 66. pp. 135–147.

119. WHITT, D. M. 1941. The role of bluegrass in the conservation of the soil and its fertility. Soil Science Society of America Proceedings. 6: 309–311.

120. WILLIAMSON, R. E. 1963. Root aeration and fescue growth. Plant Physiology (Proceedings). 38: xiix.

121. WILSON, J. D. 1927. The measurement and interpretation of the water-supplying power of the soil with special reference to lawn grasses and some other plants. Plant Physiology. 2: 385–440.

122. YOUNGNER, V. B. 1966. Salinity tolerance and your turfgrass species. Proceedings of the California Turf, Landscape, Tree and Nursery Conference. pp. 4–1 to 4–3.

123. YOUNGNER, V. B., and O. R. LUNT. 1964. Salinity tolerance of bermudagrass strains and varieties. 1964 Agronomy Abstracts. p. 103.

124. YOUNGNER, V. B., O. R. LUNT, and F. NUDGE. 1967. Salinity tolerance of seven varieties of creeping bentgrass, *Agrostis palustris* Huds. Agronomy Journal. 59: 335–336.

chapter 11

Traffic

Introduction

Man developed turfs for functional, recreational, and ornamental purposes. During the utilization and enjoyment of turfs, man can cause serious injury and even loss of turfs. Traffic results in four problems: (a) turfgrass wear, (b) soil compaction, (c) soil displacement, and (d) turf removal or divots. Turfgrass wear and soil compaction are the most common problems. Wear involves the direct effects on turfgrass.

Turfs can be injured, thinned, or completely killed through concentrated foot or vehicular traffic. Man can also damage turfs by the action of certain specialized types of shoes utilized in sports activities. The pressure, scuffing, and turning action of shoes equipped with spikes, lugs, or cleats can result in varying degrees of tearing and even removal of divots from the turf. The action of the golf club in striking the turf may result in the removal of divots of varying sizes. The action of golf, tennis, or cricket balls that strike the playing surface can also cause turfgrass injury, particularly if the turf is not properly repaired.

Compaction is a soil problem that indirectly affects turfgrass growth. It may involve actual soil displacement with resulting foot printing or rutting of the soil by human or wheeled traffic, respectively. Surface disruptions by soil displacement occur primarily on water saturated soils.

Turfgrass Wear

The injurious effects of concentrated traffic on a turf are collectively termed **turfgrass wear**. Direct pressure on the turf tends to crush the leaves, stems, and crowns

368

of the plant. Damage is greatly accentuated by the scuffing or tearing action frequently associated with traffic. Turfgrass leaves and stems injured by traffic are more prone to disease infections since the bruised areas are potential penetration points for pathogenic organisms. The wear tolerance of a turf varies depending on the (a) turfgrass species, (b) intensity of culture, (c) environment, and (d) intensity and type of traffic.

Turfgrass influence. Turfgrasses possess anatomical and morphological differences that affect the degree of wear tolerance. For example, the quantity and location of sclerenchyma and similar strengthening tissues as well as the lignin content are factors affecting wear tolerance. Considerable variation in wear tolerance is exhibited among the turfgrass species (7, 20, 31, 37, 38, 40) (Table 11-1). The warm season turfgrass species as a group are more wear tolerant than the cool season species. The more wear tolerant turfgrasses possess a tough, coarse stem and leaf structure as well as a high shoot density and an increased lignin content. In contrast, the less tolerant bentgrasses have a soft, succulent type of shoot growth. Variations in wear tolerance also occur among cultivars of a single species (6, 37, 38) (Table 11-2). The selection of a wear tolerant turfgrass species or cultivar is important on areas where concentrated traffic is anticipated.

Young seedlings are much less wear tolerant than mature turfgrasses. Also, a vigorous, actively growing turf has better wear tolerance and recuperative potential than a weakened or dormant turf. The recuperative potential from severe wear is equally important in the long term performance of turfs under concentrated traffic. Zoysiagrass and bermudagrass have excellent recuperative potential from injury caused by concentrated traffic (38). However, the recovery rate is slower for

Table 11-1

THE RELATIVE WEAR TOLERANCE OF SEVENTEEN TURFGRASSES WHEN GROWN IN THEIR RESPECTIVE REGIONS OF ADAPTATION

Relative wear tolerance	Turfgrass species
Excellent	Zoysiagrass
	Bermudagrass
Good	Bahiagrass
	Tall fescue
Medium	Perennial ryegrass
	Kentucky bluegrass
	Meadow fescue
	Canada bluegrass
	Red fescue
	St. Augustinegrass
Poor	Carpetgrass
	Creeping bentgrass
	Redtop
	Colonial bentgrass
	Centipedegrass
Very poor	Timothy
	Rough bluegrass

Table 11-2

THE RELATIVE WEAR TOLERANCE OF TEN
BERMUDAGRASS CULTIVARS

Relative wear tolerance	Bermudagrass cultivar
Superior	Tiflawn
	Tifway
	Tifgreen
	Santa Ana
	Sunturf
	U-3
Excellent	Ormond
	Texturf 1-F
Very good	Uganda
	Common

zoysiagrass than bermudagrass. The rhizomatous Kentucky bluegrasses have a more rapid recuperative rate than most cool season turfgrasses.

Cultural influences. Many cultural practices influence turfgrass wear tolerance. Wear tolerance generally increases as the cutting height is raised (6, 25, 38, 39). The increased tolerance at higher mowing heights is attributed to more vigorous shoot growth and a greater total quantity of vegetative tissue. Wear tolerance also increases as the thatch depth accumulates to a moderate amount (33, 37). This has been attributed to a cushioning effect.

Any cultural practice that increases the succulence and delicate nature of the turfgrass shoots causes a reduction in wear tolerance. Excessive nitrogen fertilization or irrigation stimulates rapid shoot growth and increases the succulence (25). High potassium levels are reported to increase wear tolerance because of reduced succulence and increased leaf turgidity. This effect has not been documented other than by casual observation. Nutrient deficiencies, which restrict the shoot density and verdure, result in reduced wear tolerance.

Environmental influences. The low light intensity under a tree canopy results in succulent, delicate leaf tissues that are more prone to injury from traffic. An excess or deficiency of soil moisture also results in decreased wear tolerance (25, 38). Wilted turfs are quite susceptible to damage from traffic because of the brittle nature of the protoplasm when under water stress (7).

Traffic damage on frosted turfgrass leaves results from the mechanical disruption of the protoplasm by ice crystals. Turfgrass rhizome and root injury can result from traffic over areas where only the immediate soil surface is thawed. Damage is caused by the shearing action of the soft, unfrozen surface moving across the frozen subsurface soil layer. Turfs in an inactive state of winter or summer dormancy lack wear tolerance. Winter traffic over turfs covered with a wet slush can cause complete kill of leaves, stems, crowns, and rhizomes if the temperature drops below 20°F shortly afterward (5). Winter traffic damage can also be caused by snowmobiles. Serious injury and thinning of turfs can occur if snowmobile traffic is not controlled or restricted under certain conditions. Damage to dormant turfs is most common (a) during periods of minimal snow accumulation, (b) during wet slushy periods of alternate freezing and thawing, or (c) on trails where the traffic is intense. The minimum protective depth is less for dense, compacted snow than for loose snow.

Compaction

The visible effects of traffic are readily seen in terms of turfgrass wear. A second, far less obvious effect occurs to the underlying soil that is sometimes referred to as the "hidden effect." The mechanical pressure applied by human and vehicular traffic results in varying degrees of soil compaction. **Compaction** is the pressing together of soil particles into a more dense soil mass. Compaction of soils having a high clay content orients the clay particles into flat plates that are layered together in an overlapping arrangement. This overlapping restricts air and water movement. Most compaction in turfgrass situations occurs within 2 to 3 in. of the soil surface, with the highest soil densities occurring in the upper inch (7, 9, 24, 26). Even though only a thin surface layer of soil is compacted, this can adversely affect turfgrass growth by impairing soil aeration and water movement.

Compaction is most common on turfgrass areas subjected to intense traffic. Soil compaction can also occur prior to turfgrass establishment, if heavy earth moving equipment or trucks have been driven over the area. Compaction is not necessarily caused by human or vehicular traffic. Falling raindrops or droplets of irrigation water can cause compaction, especially on bare soil. The presence of a turf provides considerable protection against this type of compaction.

Effects of Compaction

Compaction alters the soil physical properties that affect turfgrass growth and development. In addition, compacted soils lack the resiliency needed for certain sports. Compaction results in the destruction of soil structure. Soil density increases in relation to the degree of compaction (8, 36). Associated with the increased density is a decrease in large pores (1, 2, 3, 8, 21, 24, 34, 36) (Fig. 11-1). Low air porosity reduces soil aeration (30). The soil atmosphere can become saturated with carbon dioxide and other gases that are toxic to the turfgrass root system. In addition, soil water infiltration and percolation are reduced and the amount of surface runoff is increased (1, 2, 3, 4, 8, 21, 24, 27, 28, 29, 32, 34). The heat conductivity of compacted soils is enhanced, permitting greater soil temperature extremes to occur.

Soil compaction decreases the overall turfgrass quality, growth, and vigor (4, 8, 19, 28). As the degree of soil compaction increases, soil aeration is decreased inversely to the point where root growth is restricted. Death of the root system may eventually occur (11, 12, 23, 28). Severely compacted soils having an extremely low porosity can also mechanically impede root penetration through the soil. Soil densities of 1.5 g per cc or higher seriously impair root growth.

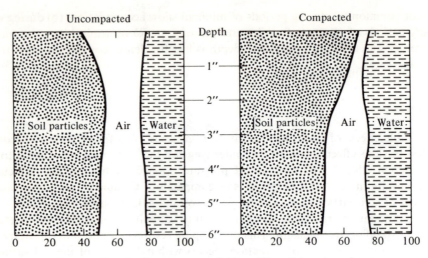

Percent of total soil volume occupied by solid particles, air, and water

Figure 11-1. The influence of compaction on the percent of the total soil volume occupied by solid soil particles, air, and water. (After Alderfer—2).

Factors Influencing the Degree of Soil Compaction

The potential for soil compaction is influenced by the (a) soil texture, (b) soil water content, (c) severity of the pressure applied, (d) frequency at which the force is applied, and (e) amount of vegetation.

Soils vary greatly in compaction tendency depending on the texture. Most soils can be compacted but considerable variation occurs in the degree of compactability. Fine textured soils such as clays are far more prone to compaction and the resulting adverse affects on turfgrasses than coarse textured soils (Fig. 11-2). Even certain fine textured sands can become compacted to the extent that turfgrass root growth is impaired. In contrast, compaction of certain coarse textured sands can result in improved water retention. Potential soil compaction problems can be reduced by soil modification with coarse textured sands or similar coarse materials such as calcined clay (28). Soil modification is most critical in the surface 2 or 3 in. Very coarse sands can be used on turfed pathways, highway shoulders, and maintenance access roadways where traffic is concentrated. Such stabilized soils have a minimum compaction tendency.

The soil water content is a very important factor influencing the degree of compaction that occurs (34). Each soil particle in moist soils is surrounded by a film of water. This film acts as a lubricant that assists in the compacting or sliding together of the plate-like clay particles into a dense, impervious state. The compaction tendency remains minimal at lower soil water contents and then begins to increase as the water content rises to the sixty-centimeter percentage (Fig. 11-2).

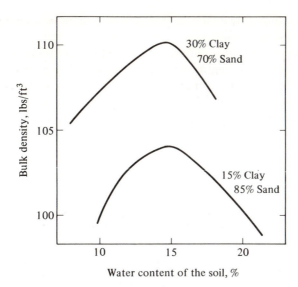

Figure 11-2. Compactability of two soil mixes at different soil water contents. (After Lunt—24).

A saturated soil is not compacted as readily as a moist soil but the soil structure is more completely destroyed (27). The soil is readily displaced with a minimum degree of compaction because the pores are filled with water under saturated conditions (27). Soil compaction occurs more frequently in the spring in most regions because of the higher surface soil water contents during this period. Since the water content has such an important effect on the compacting tendency, irrigation should be timed so that the soil surface is not excessively wet during periods when concentrated traffic is anticipated.

The greater the intensity and frequency of pressure applied by traffic, the greater the degree and depth of soil compaction that occurs (7). **Pressure** is computed as the weight of the pedestrian or vehicle divided by the surface area actually in contact with the soil. The area of soil contact should be as large as possible so that the pressure is distributed over a greater surface area thus minimizing compaction. This can be illustrated with a 200-lb man wearing football shoes versus the same individual wearing regular street shoes (36). Football shoes with seven, 0.9 cm diameter cleats have an effective surface area of 4.45 sq cm in contact with the soil surface. This is only 1.3 sq in. Thus, a 200-lb football player exerts 150 lb per sq in. of static pressure when wearing football shoes with cleats. He may exert two to three times this pressure when running. In contrast, street shoes contain approximately 32 sq in. of effective surface area. As a result, this 200-lb man would exert only 6.25 lb per sq in. when wearing street shoes. It has been estimated that there is a momentary exertion of approximately 17 lb per sq in. during the normal process of walking when the heel is placed down first. The frequency of compaction by tires can be reduced by continually alternating the mowing pattern rather than repeating the same one (16).

The amount of vegetation influences the soil compaction tendency (2, 3) and is a function of the shoot density and cutting height. A greater amount of vegetation is produced at higher cutting heights and when adequate levels of nutrition and water are provided (2, 3, 8). Higher cutting heights or thatched turfs dissipate a portion of the compaction pressure because of the cushioning effect. The physical effects of freezing and thawing assist in alleviating soil compaction in the cooler climates.

Types of Traffic

The severity of vehicular traffic effects depends on (a) the type and number of tires utilized and (b) how the equipment is operated. The number, width, pressure, and type of surface of the tires are important considerations on mechanized vehicles (6, 7). Tires with lugs reduce the surface area in contact with the turf, which concentrates the pressure on a smaller turfgrass area. Low pressure tires that distribute the weight over as large a surface area as possible are preferred on powered vehicles such as golf cars and maintenance equipment (7). Light-weight vehicles and a greater number of tires also reduces the pressure per unit area of turf. Narrow wheeled, hand pulled golf carts are widely used and result in concentrated wear that can be just as great a problem as the self propelled cars and vehicles.

Proper equipment operation also minimizes turfgrass injury. Sudden starts, quick stops, and sharp turns at excessive speeds should be avoided to prevent slipping and skidding (7). The use of vehicles during periods of high soil water content should be restricted or prohibited especially on fine textured soils that compact readily.

The type of shoe soles, cleats, or spikes also affects the severity of wear from human foot traffic. A scuffing, abrasive type of wear from a flat or ripple soled shoe results in less damage than a punching, tearing action caused by shoes with spiked soles (25, 38). Turfgrass injury is greater from a shoe having spikes than from a flat shoe sole without spikes (39). The greater injury is caused by direct spike damage to the turfgrass crown. Cleats on football shoes can remove large divots of turf during play when sharp turns are involved. The longer the cleat, the greater the tendency for divots to be removed.

There are a number of different types of soles available for use on golf shoes. Included are (a) a spiked shoe with a shoulder, (b) a ripple soled shoe, and (c) a spiked shoe without a shoulder. The sole with a spike and shoulder causes the greatest injury (15). The damage is caused primarily by the intense pressure of the shoulder that projects out from the shoe sole. Most of the individual's weight is supported on the shoulders. In contrast, a spike without a shoulder results in the weight being distributed throughout the entire area of the shoe sole. Ripple soled shoes cause the least wear injury of the three types.

Traffic Control

A certain amount of wear and compaction results whenever traffic occurs. Severe wear and compaction can be prevented through proper traffic control, soil modification, and turfgrass cultural practices. For example, concentrated traffic that severely thins or compacts the turf should be avoided whenever possible (Fig. 11-3). Design techniques can be employed to disperse the traffic. Foot and vehicular traffic can be guided and controlled through the proper selection and placement of trees, shrubs, traps, walks, roadways, and contours. The use of ornamental plants, hedges, and shrubs is preferred to fences, chains, posts, and wires because they enhance rather than detract from the natural beauty. The design should offer as large a number of alternate routes from one location to another as possible. For example, large, wide concrete areas can be used at the entrances to buildings to disperse the exit and entrance points. Severe wear frequently results when pedestrians take a short cut between two walks that join. A wide curved area of concrete or paving at the juncture of the two walks assists as does the placement of shrubs or a similar barricade.

On golf courses, the placement of traps so close to a green that traffic on the apron is concentrated in a narrow zone should be avoided. Cup setting locations on putting greens can be planned so that the exit and entrance patterns to the green are varied (13). The location of traps and mounds in front of a green should not restrict approaching traffic to a narrow entrance area.

Mechanized vehicles should use the stabilized service roads or paths when available. A deliberate effort should be made to randomly distribute the traffic over turfgrass areas where no roadways are provided. Changing the mower width on hydraulic mounted mowers or the hookup of pull-type mowers avoids constant travel in the same tracks.

Recreational areas. On turfs where such sports as soccer, football, and baseball are played, it is desirable to have a sufficiently large area that the concentration of play can be rotated. It should be large enough that by the time play is rotated back to the original site, the turf has had time to completely recover. The rotation interval is determined by the intensity of wear.

Golf cars. Motorized golf cars are a relatively recent development on golf courses. Their widespread use has created additional problems in golf course maintenance and traffic control. A few

Figure 11-3. A bare, compacted path caused by concentrated traffic due to faulty sidewalk design.

general guidelines include the following: (a) no car operation on tees, approaches, aprons, and greens; (b) traffic control on fairways and roughs should be minimal since the pattern of play tends to distribute the traffic fairly well; (c) car paths can be provided for sites where intense, continuous traffic occurs, such as between a green and the next tee; and (d) adequate paved parking areas should be provided for cars in confined areas.

Car paths are constructed only when absolutely necessary. Paths can be designed so that all traffic does not enter and leave the path at the same place (14). The starting and terminal points of the path can extend beyond the normal point of entrance and exit and be flared away from the normal route of travel (14). This design encourages a more random discharge of traffic over a wide area rather than at one concentrated point. Car paths should be of an adequate width, 6 to 8 ft, and follow a natural, convenient route that avoids sharp turns (Fig. 11-4). Steep inclines must be avoided or corrected in areas where car traffic is expected since damage from spinning wheels is common on such areas.

Traffic control is an important factor in minimizing turfgrass wear and soil compaction. Cultural practices such as soil modification are also available for reducing soil compaction. Since some compaction normally occurs, corrective cultural practices such as cultivation must be utilized. Cultivation and related practices are discussed in Chapter 15.

Alternate surfaces. A turf can only tolerate a certain amount of wear beyond which complete kill occurs. The traffic flow or pattern is so intensive and continuous in certain areas that it is impossible to maintain a turf. In these situations it is better to design and develop walks, paths, or roads rather than attempt to maintain turf, assuming that the traffic cannot be diverted or dispersed. Alternate, wear tolerant

Figure 11-4. Gentle, curved asphalt car path along a fairway.

surfaces include concrete, composition rubber, asphalt, wood chips, pine needles, tanbark, sand, fine stone, or gravel. The alternate surface should be capable of tolerating the anticipated traffic and have a minimum maintenance requirement.

Divot and Ball Mark Injury

Divots. The size of the divot taken varies with the player, turfgrass species, and cutting height. The recuperative potential from divots depends on the growth habit of the turfgrass species and the divot size. Rhizomatous turfgrasses are characterized by a relatively small divot and a more rapid recuperative potential since regrowth and recovery can occur from rhizomes under the center of the divot as well as from the sides. In contrast, bunch-type and certain stoloniferous turfgrass species are characterized by a larger divot size with most of the recovery limited to regrowth from the edges of the divot. The comparative divot size of three commonly used cool season turfgrasses cut at 0.5 in. ranks in the following order: bentgrass, the largest; annual bluegrass, intermediate; and Merion Kentucky bluegrass, the smallest. The surface stolons of creeping bentgrass are readily removed by the divot action. In comparison, Kentucky bluegrass possesses underground rhizomes that are less prone to removal when a divot is taken and are capable of initiating numerous shoots from the nodes of surviving rhizomes. Annual bluegrass may recover from the center of the divot by means of germinating seeds present in the soil.

On golf courses, divot damage is most intense on tees with some problems also occurring on fairways. The frequency at which tee markers are moved should be adjusted in relation to the intensity of play and severity of injury that occurs per unit of time. The tee must be large enough so that by the time the markers have been rotated around the tee to the original point, the turf has completely recovered from divot injury. Par-3 holes subject to iron play may require twice the usable tee space as par-4 and par-5 holes where divot occurrence is much less frequent. The placement of tee markers should also be based on the position in relation to the ball washer, benches, and flag location on the previous green (14).

Ball marks. Approach shots onto golf greens frequently result in ball marks that may injure the turf. A ball mark consists of a circular soil depression and an uplifting of the turf at one side that is subject to removal during mowing. Also, the bruised, uplifted turf is readily killed by desiccation unless immediately repaired. Improperly repaired ball marks disrupt the uniformity of the playing surface and interfere with the direction and distance of ball roll.

All ball marks must be repaired before mowing a green. Proper ball mark repair involves stretching the disturbed turf back to its original position. The disturbed turf should not be severed from the adjacent stable turf. The soil is then loosened beneath the ball mark and the turf pressed down in good contact with the loosened soil and level with the surface (Fig. 11-5).

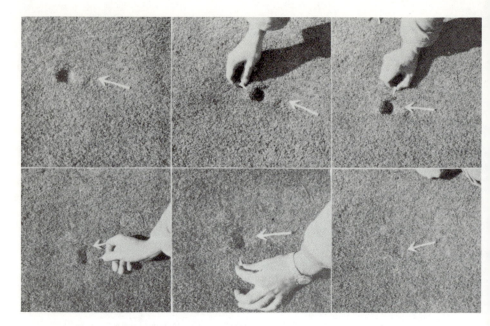

Figure 11-5. Pictorial illustration of the techniques involved in repairing a ball mark on a bentgrass putting green (Photo courtesy of the USGA Green Section— 17).

References

1. ALDERFER, R. B. 1950. The influence of soil compaction on the intake and runoff of water. The Pennsylvania State College 19th Annual Turf Conference Proceedings. pp. 52–58.

2. ———. 1951. Compaction of turf soils—some causes and effects. USGA Journal and Turf Management. 4: 25–28.

3. ———. 1951. Soil compaction—some basic causes and effects. Pennsylvania State College 20th Annual Turf Conference Proceedings. pp. 80–86.

4. ———. 1954. Effects of soil compaction on bluegrass turf. Proceedings of the 23rd Annual Penn. State Turfgrass Conference. pp. 31–32.

5. BEARD, J. B. 1965. Effects of ice covers in the field on two perennial grasses. Crop Science. 5: 139–140.

6. BURTON, G. W. 1962. Effects of traffic on turf. Proceedings of the 16th Annual Southeastern Turfgrass Conference. pp. 18–20.

7. BURTON, G. W., and C. LANSE. 1966. Golf car versus grass. Golf Superintendent. 34: 66–70.

8. CORDUKES, W. E. 1966. Soil compaction experiments with turfgrasses. Summary of the 17th Annual RCGA Sports Turfgrass Conference. pp. 9–11.

9. DAVIS, R. R. 1950. The physical condition of putting green soils and other environmental factors affecting the quality of greens. Ph.D. Thesis. Purdue University. pp. 1–104.

10. EDMOND, D. B. 1957. The influence of treading on pasture. Proceedings of the 19th Conference of the New Zealand Grassland Association. pp. 82–89.

11. ENGEL, R. E. 1951. Studies of turf cultivation and related subjects. Ph.D. Thesis. Rutgers University. pp. 1–100.

12. ———. 1951. Some preliminary results from turf cultivation. Rutgers University 19th Annual Short Course in Turf Management. pp. 17–22.

13. FERGUSON, M. H. 1959. Turf damage from foot traffic. USGA Journal. 12: 29–32.

14. ———. 1961. The need for planning for traffic control. Proceedings of the 16th Annual Texas Turfgrass Conference. pp. 16–18.

15. ———. 1963. Effects of traffic on turf. USGA Green Section Record. 1(1): 3–5.

16. FLANNAGON, T. R., and R. J. BARTLETT. 1961. Soil compaction associated with alternating green and brown stripes of turf. Agronomy Journal. 53: 404–405.

17. FULWIDER, J. R. 1968. Repair of ball marks. USGA Green Section Record. 5(6): 28.

18. GOSS, R. L., and J. ROBERTS. 1964. A compaction machine for turfgrass areas. Agronomy Journal. 56: 522.

19. HARPER, J. C., and H. B. MUSSER. 1950. The effects of watering and compaction on fairway turfs. The Pennsylvania State College 19th Annual Turf Conference Proceedings. pp. 43–52.

20. KLECKA, A. 1937. Influence of treading on associations of grass growth. Sborn. Csl. Akad. Zemed. 12: 715–724.

21. KUNZE, R. J., M. H. FERGUSON, and J. B. PAGE. 1957. The effects of compaction on golf green mixtures. USGA Journal and Turf Management. 10: 24–27.

22. LETEY, J. 1961. Aeration, compaction and drainage. California Turfgrass Culture. 11: 17–21.

23. LETEY, J., W. C. MORGAN, S. J. RICHARDS, and N. VALORAS. 1966. Physical soil amendments, soil compaction, irrigation, and wetting agents in turfgrass management. III. Effects on oxygen diffusion rate and root growth. Agronomy Journal. 58: 531–535.

24. LUNT, O. R. 1952. The compaction problem in greens. 1952 Southern California Turf Conference Proceedings. pp. 31–37.

25. ———. 1954. Relation of soils and fertilizer to wear resistance of turfgrass. Proceedings of the Northern California Turfgrass Conference. pp. 1–2.

26. ———. 1956. A method for minimizing compaction in putting greens. Southern California Turfgrass Culture. 6: 17–20.

27. MERKEL, E. J. 1952. The effect of aerification on water absorption and runoff. Pennsylvania State College 21st Annual Turf Conference. pp. 37–41.

28. MORGAN, W. C., J. LETEY, S. J. RICHARDS, and N. VALORAS. 1965. Effects of soil amendments and irrigation on soil compactability, water infiltration, and growth of Bermuda grass. 1965 Agronomy Abstracts. p. 46.

29. ———. 1966. Physical soil amendments, soil compaction, irrigation, and wetting agents in turfgrass management. I. Effects on compactability, water infiltration rates, evaporation, and number of irrigations. Agronomy Journal. 58: 525–535.

30. MORGAN, W. C., J. LETEY, and L. H. STOLZY. 1965. Turfgrass renovation by deep aerification. Agronomy Journal. 57: 494–496.

31. MORRISH, R. H., and C. M. HARRISON. 1948. The establishment and comparative wear resistance of various grasses and grass-legume mixtures to vehicular traffic. Journal of the American Society of Agronomy. 40: 168–179.

32. MUSSER, H. B. 1950. The use and misuse of water. Greenkeeper's Reporter. 18: 5–9.

33. PERRY, R. L. 1958. Standardized wear index for turfgrasses. Southern California Turfgrass Culture. 8: 30–31.

34. SWARTZ, W. E., and L. T. KARDOS. 1963. Effects of compaction on physical properties of sand-soil-peat mixtures at various moisture contents. Agronomy Journal. 55: 7–10.

35. WATSON, J. R. 1950. Irrigation and compaction on established fairway turf. Ph.D. Thesis. Pennsylvania State University. pp. 1–69.

36. ———. 1961. Some soil physical effects of traffic. Proceedings of the 16th Annual Texas Turfgrass Conference. pp. 1–9.

37. YOUNGNER, V. B. 1954. Wear resistance of turfgrasses. Proceedings of the Northern California Turfgrass Conference. pp. 1–3.

38. ———. 1961. Accelerated wear tests on turfgrasses. Agronomy Journal. 53: 217–218.

39. ———. 1962. Wear resistance of cool season turfgrasses. Effects of previous mowing practices. Agronomy Journal. 54: 198–199.

40. ———. 1969. A review of recent turfgrass research in southern California. California Turfgrass Culture. 19(1): 6–7.

part III

TURFGRASS
CULTURAL PRACTICES

Turfgrass culture is the science and practice of establishing and maintaining turf-grasses for specialized purposes. These practices can be divided into the following major areas: mowing, fertilization, irrigation, cultivation, establishment, and turf-grass pest control. Mowing and cultivation are practices unique to turfgrass culture.

The desired quality and type of turfgrass surface maintained for a particular recreational, ornamental, or functional use is achieved through proper manipulation of the turfgrass cultural principles and practices. The microenvironment of the turfgrass plant can be modified to ensure the best possible growing conditions. The composition of a turfgrass community can be maintained or altered by the types of cultural practices employed. Encroachment of turfgrass weeds is also determined to a great extent by the cultural practices. The principles of turfgrass culture are discussed in detail in the following six chapters.

Turfgrass cultural systems. The basic principles involved in developing a turfgrass cultural system can be grouped in relation to the intensity of culture. The four major groups of turfgrass cultural systems are (a) greens, (b) sports turfs, (c) lawn turfs, and (d) functional turfs (Fig. III-1). The greens category is used primarily for recreational purposes such as golf putting greens, lawn bowling courts, grass tennis courts, cricket wickets, and championship croquet courses. The sports turf cultural system grouping includes tees and fairways of golf courses, and surfaces for such sports as football, soccer, baseball, cricket, rugby, lacrosse, field hockey, and horse racing tracks. Represented in the lawn turf grouping are the grounds of homes, parks, institutions, industries, schools, cemeteries, and military installations. Roadsides, airfields, golf course roughs, ski slopes, shooting ranges, ditch

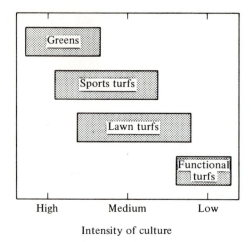

Intensity of culture

Figure III-1. A comparison of the intensity of culture among the four major groups of cultural systems.

banks, and similar minimal use, low quality turfs are included in the functional turfgrass cultural system group.

A turfgrass cultural system is developed through the integration of various cultural principles and practices with the objective of achieving a specified level of turfgrass quality and/or playability. Selection of the appropriate cultural practices and their integration requires a fundamental knowledge of the (a) turfgrasses, their adaptation, characteristics, and cultural requirements; and (b) influence of the environment on turfgrass growth and development.

Development of a turfgrass cultural system is not an exact science. Good judgment based on sound principles plus a considerable amount of practical experience must be utilized in developing an effective cultural system. Selection of a cultural system varies with the climate, soil, labor availability, type of equipment, size of budget, and capabilities of the supervisory staff. Due to the great variability in soils, climate, and turfgrasses, it is not possible to outline a turfgrass cultural system for use in all potential situations even within a given cultural intensity group or for a specific use.

The professional turfman should do some applied experimentation before incorporating any new practices, chemicals, or turfgrasses into a successful cultural system. The turfgrass research conducted by both private and public agencies eliminates those practices, chemicals, and turfgrasses that are least effective and proves those that are most promising. It is then the responsibility of the professional turfman to test those practices that the turfgrass researchers have proved most effective to determine if they can be incorporated into the established cultural system. Such experimentation is best done on a turfgrass nursery or in a small scale on a fairway or similar area rather than immediately incorporating the newer ideas into the entire cultural system. Through this experimentation, the professional turfman also has an opportunity to learn how to properly utilize the technique, chemical, or turfgrass and to determine how it can best be incorporated into the particular cultural system being used.

chapter 12

Mowing

Introduction

Mowing is the most fundamental and universal practice man utilizes in turfgrass culture. It provides a uniform surface for ornamental beautification and for many outdoor sport and recreational activities. Mowing is a defoliation process in which a portion of the turfgrass leaf is removed. Since the photosynthetic processes occur primarily in the leaf tissue, mowing strongly influences the physiological and developmental condition of the turfgrass plant. Any cutting or defoliation is detrimental to the turf. The turfgrass plant is able to survive mowing and continue new leaf development because leaf growth occurs from the base of the plant. The cut tip of a recently mowed grass blade is an ideal site for the penetration of pathogens (126). An exudation is also associated with the cut tip. Exudates contain essential elements, glutamine, carbohydrates, and other organic compounds that enhance the spore germination, penetration, and infection rates of pathogens.

Cutting Height

Cutting height is defined as the distance above the soil surface at which the turf is mowed. The **bench setting** is the height at which the cutting edge or bed knife is set above a level surface. It is the height generally referred to in the cutting height recommendations for various turfgrass species and cultivars. The **effective cutting height** is the actual height above the soil surface at which the turf is being cut. This height may vary from the bench setting because of varying thicknesses of thatch and floatation of the cutting unit. The effective cutting height is usually greater than the bench setting. Cutting height, as used in the following discussions, refers to the relative height within a range of 0.2 to 4 in. Few turfgrass species tolerate continual mowing below 0.2 in. or maintain adequate turfgrass uniformity and sod cover at cutting heights above 4 in.

The primary criteria utilized in selecting the cutting height for a given turf are the (a) physiological condition of the turf, (b) purpose for which the turf is to be utilized, and (c) growth habit of the turfgrasses involved.

Physiological, developmental, and growth effects. Mowing drastically alters the physiological, developmental, and growth responses of turfgrasses. As the cutting height is lowered, low growing turfs exhibit the following responses:

1. Decreased carbohydrate synthesis and storage (18, 35, 44, 103).
2. Increased shoot growth per unit area (3, 24, 55, 57, 67, 76, 96, 100, 101, 102, 108, 138).
3. Increased shoot density (42, 100, 101, 128).
4. Decreased leaf width (63, 106, 107, 125).
5. Increased succulence of shoot tissues (100).
6. Increased quantity of chlorophyll per unit area (100).
7. Decreased root growth rate and total root production (1, 10, 14, 16, 22, 27, 32, 44, 46, 48, 55, 58, 63, 64, 65, 66, 67, 71, 74, 75, 76, 80, 86, 100, 101, 114, 121, 123, 124, 125, 130, 138, 143).
8. Decreased rhizome growth (34, 42, 58, 63, 64, 65, 67, 74, 75, 80, 143).

These responses occur as the cutting height is reduced within the tolerance range for a given turfgrass species or cultivar. Close mowing stimulates tillering only if the stem apex is removed. Shoot density increases and the size of individual plants decreases as the cutting height is lowered but the overall quantity of shoot growth is increased. The increased shoot density at lower cutting heights results in improved turfgrass quality. Shoot growth eventually decreases if the cutting height is lower than that generally tolerated by a given species.

The decreased leaf area resulting from mowing has several negative effects on the physiological condition of the turf. **Leaf area index** (LAI) is defined as the total quantity of leaf area present per unit area of soil surface. There is a decline in LAI as the cutting height is lowered. However, the increase in shoot density as the cutting height is lowered tends to dampen the reduction in LAI. The quantity of carbohydrates synthesized by the turf decreases as the LAI declines at lower cutting heights. The shoot growth rate is increased at the same time which results in increased utilization of carbohydrates. The net result of these interacting effects is a drastic reduction in the depth, extent, and total quantity of roots produced by the turfgrass plant at lower cutting heights (Fig. 12-1). The lack of root growth at close cutting heights is commonly attributed to a carbohydrate deficiency but the potential effects of root growth regulators synthesized in the leaves may also be a factor.

The degree of root growth reduction is directly correlated with the decrease in cutting height. Cutting height has a greater effect on root growth and carbohydrate reserves than the mowing frequency. Although moderate defoliation restricts root growth, it stimulates rhizome and stolon development and sod formation. Excessively close mowing restricts rhizome and stolon development (135). Root diameter, number of root hairs, number of root protoxylem elements, and root initiation rate are also reduced by close cutting heights (14, 48, 63, 114, 125).

Figure 12-1. Effect of cutting height on the root growth of Kentucky bluegrass. (Photo courtesy of E. C. Roberts, University of Florida, Gainesville, Florida).

The reduced carbohydrate reserves, rooting, and plant vigor associated with lower cutting heights also causes decreased tolerance to environmental stresses (11, 117, 139). The decreased amount of vegetation at lower cutting heights reduces temperature insulation and protection for the vital meristematic areas of the crowns and rhizomes. The cutting height is sometimes raised during periods of heat stress, especially on putting greens. The greater drought tolerance of high cut turfs is partially attributed to a deeper root system (51). Lower cutting heights are frequently associated with increased disease development, particularly leaf spot, rust, and dollar spot (33, 38, 51, 72, 101). Turfgrass wear tolerance is also decreased at lower cutting heights (147).

The amount of regrowth following mowing is positively correlated with the cutting height (82, 136). Regrowth from normal turfgrass mowing practices consists of two phases (96, 98). The first phase, ranging from 3 to 4 days in duration, involves the elongation of existing cut leaves, the second phase involves the initiation and development of new leaves.

Influence of use requirements. The cutting height is sometimes dictated by the purpose for which the turf is maintained (120). For example, a very close cutting height of approximately 0.25 in. is required on golf greens, bowling greens, and tennis courts to provide the desired quality of surface necessary for good playability. The desired qualities include a smooth, uniform, dense, resilient playing surface free of excessive thatch. Kentucky bluegrass fairways must be mowed at a shorter height than is desirable from a physiological standpoint. The close cutting height is required for the ball to lie upon the turf in the proper position for good control of fairway shots. Shorter cutting heights are also preferred on tees and fairways because of the firmer stance provided for the golfer. In sports that involve running, such as football, soccer, and baseball, a close cutting height is desired to achieve maximum running speed. The height of cut can affect the degree of bounce, distance, and rate of ball roll in sports such as baseball, golf, and tennis (83). Usually, the cutting height used on a given sports turf is a compromise between the demands of the specific game involved and the physiological principles that influence the health and vigor of the turfgrass species.

Turfgrass species and cultivars. Turfgrass species vary greatly in tolerance to cutting height (38, 65, 98, 111, 123) (Table 12-1). The cutting height tolerance of a given species or cultivar is a function of the growth habit including both stem and leaf growth characteristics (15, 16, 94). Stem growth involves the (a) erect versus horizontal direction of growth or an intergradation between the two and (b) location of the crown apical meristem in relation to the effective cutting height. Prostrate, low growing turfgrasses such as bermudagrass and creeping bentgrass tolerate a much shorter cutting height than erect growing, noncreeping types such as tall fescue and Italian ryegrass (49, 123). Western wheatgrass and Italian ryegrass have the apical meristem elevated well above the soil surface due to the elongated stem internodes (15). Leaf growth may be erect as in the case of red fescue or more horizontal as for creeping bentgrass.

Table 12-1

PREFERRED CUTTING HEIGHTS OF NINETEEN TURFGRASSES AS DETERMINED
BY THE RESULTING TURFGRASS QUALITY AND VIGOR

Relative cutting height	Cutting height, in.	Turfgrass species
Very close	0.2–0.5	Creeping bentgrass
		Velvet bentgrass
Close	0.5–1	Colonial bentgrass
		Annual bluegrass
		Bermudagrass
		Zoysiagrass
Medium	1–2	Buffalograss
		Red fescue
		Centipedegrass
		Carpetgrass
		Kentucky bluegrass
		Perennial ryegrass
		Meadow fescue
High	1.5–3	Bahiagrass
		Tall fescue
		St. Augustinegrass
		Fairway wheatgrass
Very high	3–4	Canada bluegrass
		Smooth brome

The general range in cutting height tolerance varies with the characteristics of cultivars within a species (80). For example, certain Kentucky bluegrass cultivars such as Park, Kenblue, and Delta have a more erect growth habit as well as higher crown and apical meristem than Merion, Fylking, and Cougar (9, 32, 37, 72, 86, 123, 143). As a result, the latter three cultivars tolerate a shorter cutting height.

The turf becomes weak and less vigorous if the cutting height is lower than normally tolerated by a given species or cultivar (9). The more erect growing species are eventually eliminated through continued close, frequent mowing (34). Turfs mowed excessively close are also more readily invaded by weeds (31, 32, 38, 50, 51, 57, 61, 138). A typical example is the increased invasion of annual bluegrass into Kentucky bluegrass when the cutting height is lowered (31, 43, 51). A turf mowed at the proper height for the given species is healthy, dense, and vigorous, assuming other cultural and environmental factors are not limiting. Weeds have great difficulty invading turfs under these conditions because of the severe competition.

Cutting height can be an important factor affecting the establishment rate of stoloniferous species. Cutting heights less than 1 in. impair horizontal stolon growth of bermudagrass and zoysiagrass because the spreading, nonrooted stolons are cut off in the mowing process (52, 97). However, mowing that is not excessively close enhances the sod formation rate (81).

Mowing too high causes as many problems with certain turfgrass species as cutting too close. The prostrate growing species such as creeping bentgrass and bermudagrass must be mowed close; otherwise excessive thatch accumulation and puffiness occur (38, 102). Such thatched turfs are more prone to scalping, diseases, and injury from environmental stress. Raising or lowering the cutting height in late fall is not a desirable practice.

Scalping

The removal of an excessive quantity of green leaves at any one mowing that results in a stubbly, brown appearance is referred to as **scalping**. Scalped turfs appear stubbly and brown because most of the green leaf tissue has been removed leaving only the brown stems. Temporary thinning of the turf can occur, particularly if the scalping has been severe enough to remove some of the apical meristems. Scalping is more common on areas having (a) irregular contours, (b) excessive thatch, or (c) an infrequent mowing interval (Fig. 12-2). When turfs grow excessively long, the cutting height should be gradually lowered to the desired level by mowing at frequent intervals. Never straddle a high point or ridge during the mowing operation.

Reserve carbohydrates within the plant are utilized in shoot regrowth processes when all or a majority of the green leaves are removed by mowing (2, 122, 132). Eventually the new shoot growth

Figure 12-2. Scalping of a bermudagrass putting green which had an excessive thatch accumulation. (Photo courtesy of W. R. Thompson, Jr., American Potash Institute, State College, Mississippi).

is able to provide for the immediate carbohydrate requirements of the plant. Depletion of the carbohydrate reserves ceases at this point and accumulation eventually occurs when the rate of synthesis exceeds the assimilation rate. The depletion of carbohydrate reserves continues, however, if the mowing frequency is excessive or the degree of defoliation is too severe.

Tiller, rhizome, and root initiation stops immediately after severe defoliation (2, 27, 113). Growth of existing roots and rhizomes usually ceases (64, 113). Root growth stoppage occurs within 24 hr after defoliation and does not begin again until shoot growth recovery is well advanced (27). Removing the leaves from one-half of a grass plant results in growth stoppage of those roots associated with the defoliated half (27). Defoliation effects on root growth are minimal when no more than 40% of the leaf area is removed (27).

Leaf growth may continue but at a much lower rate following severe defoliation (4). Recovery is primarily the result of new shoot development rather than growth from the existing, partially defoliated leaves. The recovery rate depends on the (a) quantity of reserve carbohydrates contained in the tissues remaining after defoliation; (b) number of undisturbed growing points present; (c) environmental conditions, especially water, temperature, and light; and (d) availability of essential nutrients (122).

Mowing Frequency

The mowing frequency can be as important as the cutting height in the overall physiological and developmental condition of the turfgrass plant. Longer intervals between mowings improve the overall turfgrass vigor and quality, assuming no more than 40% of the leaf tissue is removed at any one mowing (96). The mowing frequency is determined by the (a) shoot growth rate, (b) environmental conditions, (c) cutting height, and (d) purpose for which the turf is to be utilized.

The growth rate of a turf is one of the most important factors influencing the mowing frequency. The frequency must be adjusted in relation to the shoot growth rate so that an excessive amount of leaf surface is not removed at any one mowing. No more than 30 to 40% of the leaf area should be removed at any one mowing (27). A severe shock to the physiological balance of the turfgrass plant occurs, if the defoliation exceeds this amount. Defoliation stimulates immediate shoot growth and tillering. The shoots have priority over the roots for the available carbohydrates under these conditions. When the carbohydrate reserve is limited and defoliation is severe, all available carbohydrates may be utilized by the shoots causing death to a major portion of the root system. Thus, shoot growth stimulation occurs at the expense of the roots. The greater the percentage of leaf removal, the longer that root growth stoppage occurs.

The vertical shoot growth rate is controlled by the environment, nutritional level, and turfgrass species involved. Irrigation, increased nutritional levels, high light intensities, and optimum growth temperatures stimulate shoot growth and increase the mowing frequency. Species variability in the rate of vertical shoot growth is substantial. For example, the rate for Alta tall fescue is three times as rapid as for bentgrass (98).

The mowing frequency generally increases as the cutting height is lowered. Greens maintained at 0.25 in. are usually mowed daily or on an every other day basis (Fig. 12-3). In contrast, Kentucky bluegrass cut at 1.5 to 2 in. requires a mowing interval of 4 to 8 days depending on the rate of shoot growth.

Mowing frequency is an important factor influencing the playability of sports turfs. Bowling and putting greens require a uniform, smooth, dense, fine textured, close cut surface. This can only be achieved through frequent mowing, usually daily. Less frequent mowing results in a coarser leaf texture and lower shoot density.

Based on shoot growth studies on the day following mowing, it has been suggested that the mowing frequency of bentgrass cut at 0.5 in. should be at 5-day intervals to achieve the best possible recuperative potential and turfgrass vigor (96, 98). Similar studies with bentgrass mowed daily at 0.25 in. indicate that a rest period at 7- to 8-day intervals is desirable in terms of improved physiological condition of the turf (99). Two consecutive days per week without mowing results in more regrowth than interspersing the 2 days of rest at intervals throughout the week (99).

Seed head formation influences the mowing frequency under certain conditions. Desirable turfgrass species such as bahiagrass, carpetgrass, St. Augustinegrass, and certain bermudagrasses produce an excessive number of seed heads that disrupt the uniformity of a turf unless mowed off regularly. Frequent mowing can also be used for cutting off the flowering parts of certain weeds before the seeds become viable (85).

Figure 12-3. Mowing a green with a three gang, hydraulic mower.

Effects of mowing frequency. The turf can be adversely affected if the interval between mowings is too short. The following turfgrass responses are generally observed as the mowing frequency is increased:

1. Increased shoot density (42, 71, 92, 99, 104, 111).
2. Decreased carbohydrate reserves (17, 68).
3. Decreased rooting (1, 3, 8, 27, 53, 56, 68, 71, 95, 100, 124).
4. Decreased rhizome development (17).
5. Decreased shoot growth (3, 8, 17, 36, 43, 53, 56, 60, 67, 76, 92, 94, 95, 96, 98, 99, 108, 111, 118, 124, 136, 138).
6. Decreased chlorophyll content (96, 100).
7. Increased succulence (100).

The stress of more frequent mowing results in an increased shoot density. However, the turf is adversely affected by excessively frequent mowing beyond that required to maintain the proper physiological condition and leaf area balance. This is demonstrated by the reduction in shoot growth, root growth, chlorophyll content, and recuperative potential of the turf. Root growth is reduced more than shoot growth by an increased mowing frequency (95). A more prostrate growth habit and an increased rate of sod formation are generally associated with an increased mowing frequency (92, 93, 104).

More frequent mowing removes shorter lengths of leaves that fall deeper into the turf rather than remaining on or near the turfgrass surface. The deeper the leaves fall into the stand, the more rapid the decomposition rate since the micro-environment is more favorable.

Turfgrass Species Influences on Mowing Effectiveness

The anatomy, structure, composition, and succulence of the shoot tissue can significantly influence the quality and difficulty of the mowing operation. For example, most ryegrass cultivars contain a tough, fibrous vascular system that is very difficult to mow cleanly even with a sharp, properly adjusted reel mower. As a result, the ryegrasses are quite inferior in terms of mowing quality. The silica composition of zoysiagrass leaves causes mowing difficulties and results in more rapid dulling of the mower due to the abrasive action. In contrast, the tissue structure, composition, and succulence of bentgrass, bermudagrass, and bluegrass permit a clean cut with minimal mowing difficulty.

A second consideration in mowing quality as influenced by the turfgrass species is the incidence of seed head formation. A rotary mower is preferred on lawns containing bahiagrass, St. Augustinegrass, and carpetgrass during periods of extensive seed head formation. It is difficult to remove tall seed heads with a reel-type mower.

A third factor is the vigor and aggressiveness of the species. Stoloniferous, warm season turfgrasses such as zoysiagrass and the improved bermudagrasses require heavier and higher powered mowing equipment. The use of lightweight, underpowered mowers on these vigorous turfgrass species results in a gradual increase in the effective mowing height and a severe thatch problem.

Clipping Removal

Leaf clippings are a by-product of the mowing operation. The return of clippings over an extended period of time tends to reduce the turfgrass quality under conditions of intensive turfgrass culture. Clippings should be removed when (a) the clippings interfere with the purpose for which the turf is maintained, (b) excessively heavy, and (c) disease development would be enhanced. Clippings are removed from golf and bowling greens since their presence would interfere with the direction and velocity of ball roll. An excessive amount of clippings may occur during periods when the growth rate has been quite rapid and environmental conditions have interrupted the normal mowing frequency. When this condition exists, the clippings must be removed in order to avoid damage to the turf by light exclusion. Random accumulations of clippings, which occur primarily when mowing wet turfs, should be dispersed by the use of whipping poles, dragging a hose over the area, recutting, or some similar means.

The return of clippings is commonly practiced on turfs grown at a low intensity of culture. This is beneficial since the clippings release nutrients to the soil for subsequent utilization by the turf. In studies at East Lansing, Michigan, where Merion Kentucky bluegrass clippings were removed, the annual nitrogen fertilization rate had to be increased by 2 lb per 1000 sq ft per year in order to maintain the color, shoot density, and turfgrass quality comparable to adjacent areas where the clippings were returned. The comparative amounts of nitrogen, phosphorus, and potassium removed in the clippings from a creeping bentgrass turf maintained under putting green conditions are shown in Table 12-2. These data illustrate that where clippings are removed the fertilization rate must be increased in order to

Table 12-2

THE TOTAL AMOUNT OF NITROGEN,
PHOSPHORUS, AND POTASSIUM REMOVED IN
THE CLIPPINGS OVER A GROWING SEASON
FROM A WASHINGTON CREEPING BENTGRASS
PUTTING GREEN IN MILWAUKEE, WISCONSIN
[AFTER NOER (112)]

Element	Total lbs per 1000 sq ft removed annually
Nitrogen	4.8
Phosphorus	0.7
Potassium	2.6

replace the quantity of nutrients being removed in the clippings (112). Clippings also contribute a certain amount of organic matter to the soil that can result in improved soil structure. The infiltration rate of water is greater where the clippings are returned compared to where they are removed (110).

Mowing Equipment

The type, size, and number of mowing units required for economical, efficient maintenance of a turfgrass area depend on (a) size of the area, (b) turfgrass species, (c) type of use for which the turf is being maintained, (d) intensity of culture, (e) landscaping, (f) topography, and (g) available labor force. The mowing unit should have good maneuverability, be adequately powered, and be easily adjusted. Mowing units range from the single 18-, 20-, or 24-in. units commonly used on home lawns to reel-type mowing units of 5, 7, or 9 gangs that are used on extensive turfgrass areas such as sod farms, parks, and golf courses (Fig. 12-4). It is important to select the type and size of mowing unit that accomplishes the mowing operation most efficiently. For example, a 30-in. mower cuts 6.1 acres in 8 hr, while a 3-gang mower cutting 76 in. wide mows the same area in 3.2 hr when operated at the same speed of 3 mph. Thus, one 3-gang unit and one operator can cut a larger turfgrass area in 1 day than two 30-in. mowers requiring two operators.

Figure 12-4. Mowing a golf course area with a seven gang, hydraulic lift, tractor mounted mower. (Photo courtesy of the Toro Manufacturing Corporation, Minneapolis, Minnesota).

The efficiency of mowing operations can be substantially improved by proper placement of trees, shrubs, traps, flower beds, and paved areas. Abrupt contours that require small hand-operated mowing units are costly to maintain. Areas that are continually being scalped should be recontoured. Contours around tees and greens should be gradual enough that multigang mowing units can be utilized to reduce mowing costs. Similarly, trees and shrubs should be spaced so that larger gang mowing units can be readily operated between and around the plantings. The use of wood chip borders or soil sterilants under fences are important practices that minimize costly trimming operations.

Mower Operation and Adjustment

Proper mower operation is quite important. Most mowing units are designed to operate within a certain range of ground speeds, usually less than 6 mph for reel types (133). Operating at excessive speeds causes the mowers to bounce and results in an uneven cut. Mowing units should be turned or the direction reversed by making a wide loop. A short spinning turn, especially with a green mower, causes injury to the turf by a bruising action. Banks and terraces should be mowed up-and-down rather than across the slope to avoid slippage and tearing of the turf.

Mowing wet grass should be avoided. By mowing when the turf is dry, it cuts more easily, does not ball up and clog the mower, gives a finer appearance, and minimizes the bunching of clippings. Mowing dry grass requires less time than for wet grass. Mowing when wet also enhances the spread of disease causing organisms.

Metallic objects, stones, and other debris should be removed from the turfgrass area prior to mowing. The bed knife and blades can be damaged by striking such objects. In addition, rotary mowers can throw solid objects for considerable distances and can cause serious injury if an individual is struck by one of these flying objects.

It is quite important to maintain the mower in proper adjustment and sharpness at all times (133). The effects of a dull, improperly adjusted mower on the growth of Kentucky bluegrass leaves are shown in Fig. 12-5. A mutilated leaf caused by an improperly adjusted, dull mower has a browned appearance that disrupts the turf-

Figure 12-5. Effect of a sharp (right) versus dull (left) reel mower on the leaf growth of Kentucky bluegrass at 5, 15, and 30 hours after cutting. (Photo courtesy of J. R. Watson, Toro Manufacturing Corp., Minneapolis, Minnesota).

grass uniformity and quality. It also bruises the leaf tip, increases moisture loss, and provides a favorable site for the penetration of pathogens into the leaf. Turfgrass vigor, quality, and leaf growth are impaired if the turf is mowed with a dull, improperly adjusted mower for an extended period of time.

Grain

The tendency for turfgrass leaves and stems to grow horizontally in one or more directions rather than vertically is referred to as **grain** (Fig. 12-6). Turfgrass leaves tend to lean or become oriented in the same direction that the mower is operated. Continual mowing in the same direction results in the formation of grain, especially on greens. Grain is minimized by alternating the mowing in two or more different directions. This is widely practiced on greens and to some extent on fairways and similar close cut turfgrass areas. Alternating the mowing direction also minimizes the wave-like ridges that develop at right angles to the direction of mowing when it is consistently done in the same direction. If the grain cannot be controlled by alternating the mowing direction, a brush or comb can be attached in front of the mower to assist in straightening or lifting the leaves and stems for more effective mowing. **Brushing** is the practice of lifting leaf or stem growth prior to mowing.

Figure 12-6. Close-up of a grainy condition on a bentgrass green cut at 0.25 inch.

Types of Mowers

Mowing can be accomplished by four basic types of cutting action: (a) reel, (b) rotary, (c) vertical, and (d) sickle bar. The quality of turf resulting from the mowing operation varies depending on the basic type of cutting action utilized. Selection of the specific mower type should be based on the size of the area, topography, and quality of turf desired.

Reel. In the reel-type cutting action the reel functions in guiding the turfgrass leaves against the bed knife (13). The bed knife is the actual cutting blade and should be maintained in proper adjustment and sharpness at all times. The cutting action of the reel against the bed knife results in a scissors-type effect in contrast to the impact action of the rotary and vertical types. Mowing quality of the reel type is determined by the "frequency of clip" that is a function of the number of blades and reel speed. The number of blades in the reel must be greater at close cutting heights and where a more uniform surface must be maintained.

Reel-type mowers require a relatively smooth soil surface for effective operation. High speed, reel gang mowers are the most efficient and economical type for mowing extensive areas. The power requirement is quite low. The reel-type mower is most effective and preferred under conditions of close cutting and where a dense turf is to be maintained. It is superior in terms of a clean, even cut, especially for intensively cultured greens. Reel mowers are also preferred for certain close cut warm season turfgrasses such as bermudagrass and zoysiagrass.

The mowing frequency is more critical with reel mowers than other types of cutting units. A reel mower cannot effectively mow grass shoots or weeds that are higher than the horizontal centerline of the reel. Thus, a 6-in. diameter reel set to cut at 1 in. cannot mow grass taller than 4 in.

Reel mowing units are coupled together in three- to eleven-gang combinations up to 26 ft in width. Wheel types available include steel, molded rubber, and pneumatic tires. Steel or molded rubber wheels with lugs give the most traction. Pneumatic tires facilitate transport over paved areas.

Rotary. The rotary-type mower cuts by the impact of a knife blade operated in a horizontal plane. It causes a certain amount of mutilation and injury to the leaf blade at the point of impact. The rotary-type cutting action is not adapted to quality mowing at a cutting height of less than 1 in. The rotary mower is a more versatile unit that can be used to cut taller grass and weeds, to mulch fallen tree leaves, and for general trimming. The power requirement is quite high. Rotaries are commonly used for mowing roadsides and similar rough areas. Scalping is more of a problem with a rotary mower where the soil surface is uneven and the mower width is wider.

There are a number of types of rotary blades available for specific uses including cleaver, parallel edge, reversible, and suction types. Maintaining sharpness is equally as important with rotary mowers as with reel-type units. Cutting with a dull rotary blade results in greater mutilation of the leaves that turn brown and detract from the turfgrass quality. The rotary blade should be balanced at all times for proper operation. It should always be checked after each sharpening operation.

Desired safety features in a rotary mower include a heavy gauge housing, slip clutch, pivot mounted blades, and protective guards at the discharge area. Ejection of stones, metal, and similar objects is a concern in terms of potential injury if they strike someone.

Multiple spindle rotary units with hinged frames are available in widths from 4 to 27.5 ft for mowing extensive areas at cutting heights ranging from 2 to 12 in. Speciality units include front, mid, side, and boom mounted types.

Vertical. The vertical-type mower may involve fixed blades or free swinging knives that operate in a vertical rather than horizontal plane. The free swinging cutting knives are most commonly used for routine mowing operations. Cutting is by impact as with the rotary mowers. The cutting height can be adjusted with gauge rollers from 0.75 to 6 in. The mowing quality varies with the spacing, number, and type of knives used. Improved mowing quality is mainly associated with the number of knives per foot of cut. Knife sharpness and rotor balance are also important factors influencing mowing quality.

Vertical mower types are best used where vegetation control is of prime concern rather than turfgrass quality. They are generally used on extensive turfgrass areas, particularly where the contours are irregular, stones and other debris are present, and appearance is not of major concern. Scalping is not as severe a problem as with rotary mowers.

Ejection of stones, metal, and other foreign objects from a vertical mower is not as great a problem as with rotary mowers, particularly if a flap or chain deflector is provided as a housing extension. The improved safety is attributed to a vertical plane of rotation, heavy gauge housing, fold-back knives, and deflector restraints.

Vertical mowers are available in centered or offset versions of both the mounted and pull type. Rear mounted units having cutting widths of 5, 6, or 7 ft are most common. Multiple trailing units are available in widths up to 19 ft.

Sickle bar. This type of mowing unit is adapted primarily to relatively coarse, bunch-type turfgrasses and higher cutting heights where only an occasional mowing is required. The cutting action is similar to that of a reel-type mower. It involves a multisectional knife with a reciprocating action across ledger plates. Sickle bar-type units mounted on tractors have been utilized primarily along roadsides where the mowing frequency ranges from one to four times per year. Side-mounted cutting units having cutting widths of 5, 6, and 7 ft are most commonly used.

Sickle bar mowers are best adapted for use on problem areas such as slopes, terraces, and ditches or areas having tall weed growth or numerous obstacles. The low power requirement and improved stability of the sickle bar-type mower are desirable on less frequently mowed, steep slopes. It is not adapted for use where high quality turfs are desired. The ability to adjust the cutting height and cutting angle while in operation contributes to the flexibility and maneuverability of these mowers. An objectionable feature is that the mower leaves the cut vegetation in a swath rather than finely chopped and uniformily distributed. Effective operation is frequently quite difficult in dense, fine textured turfs such as Kentucky bluegrass due to mower plugging. The relatively slow ground speed and very high maintenance costs result in a higher cost of mowing compared to the rotary- and vertical-type mowers. Operating adjustments are also more critical and numerous.

Growth Regulation

Mowing is a costly but important operation in maintaining quality turfs. Chemicals that inhibit or regulate turfgrass shoot growth and thus reduce the mowing frequency have attracted considerable attention since 1945. Actually, vertical leaf growth is not important under optimum growing conditions where no adversities occur. It only necessitates expensive, time-consuming mowing operations.

A chemical growth inhibitor could be useful for restricting leaf growth and thus negating the mowing requirement. This appears desirable in principle. However, it is based on the assumption that turfgrasses are capable of resisting all adversities including environmental stresses from heat, drought, and cold; turfgrass pests including insects, diseases, and nematodes; and damage from man caused by excessive traffic, divots, ball marks, and vehicles. As yet, a turfgrass cultivar capable of resisting injury from all these adversities has not been developed. Thus, turfs must possess the capability to recuperate from injury caused by adversities. The recuperative potential of a turf is negated if a growth inhibitor has been applied (39). This lack of recuperative potential is a serious disadvantage in the use of growth inhibitors on quality turfs. This will continue to be a problem until a turfgrass cultivar is developed that resists injury from the many adversities of environment, pests, and man or until an antigrowth inhibitor is developed.

One approach to ensuring adequate recuperative potential when a growth inhibitor is used is the capability of applying a chemical or cultural practice that stimulates growth should a turf be injured by adversity. Such an anti-growth regulator would counteract the growth inhibitor effects and stimulate recovery of the turf. Growth promoters such as gibberellin, naphthaleneacetic acid, and indoleacetic acid can counteract the growth retarding effect of certain growth inhibitors such as CCC but not maleic hydrazide and Phosphon (90). There is also evidence that high fertilization rates may counteract the effect of growth regulators (21). The field application of such techniques has not yet been developed.

A more reasonable approach is the use of growth retardants that decrease the shoot growth rate but do not cause total growth stoppage. A controlled rate of shoot growth would reduce the mowing frequency but maintain the recuperative potential. Selective growth retardation of turfgrasses that impairs intercalary mersitem elongation of shoots but not apical meristem activity is preferred since lateral stem, stolon, and rhizome growth would not be affected. Partial growth retardation offers such benefits as controlled growth (a) during wet seasons when mowing equipment can not be used, (b) on sites that can not be mowed with conventional equipment, and (c) on minimal use, low quality turfgrass areas where growth control is desired but a manicured surface is not required.

Types of Growth Regulators

Chemicals which have been utilized or suggested for inhibiting turfgrass growth include B-995, CCC, chlorflurenol, 2-chloroethylphosphonic acid, maleic hydrazide, and Phosphon. Only chlorflurenol and maleic hydrazide have shown any significant degree of growth retardation in turfgrasses. As yet none of these materials has been widely accepted for use in growth retardation of turfs. Chemicals having growth stimulant activity include gibberellin and kinetin. Neither of these materials have been utilized in normal turfgrass cultural programs.

B-995. The chemical name is N-dimethyl aminosuccinamic acid. Developmental effects typically associated with B-995 involve a reduction in internode elongation and internode number. Growth reduction resulting from B-995 applications is directly correlated with the application rate and is generally attributed to direct phytotoxic effects. Direct injury to the leaves results in an objectionable reduction in appearance and turfgrass quality without an acceptable degree of growth control.

CCC. The chemical name is 2-chloroethyl-trimethyl-ammonium chloride. It is a true growth regulator in that it inhibits cell division in the subapical region but not apical meristem activity. Developmental effects associated with CCC include shorter leaves, a more prostrate growth habit, and increased tillering. The effects of CCC are counteracted by indoleacetic acid. CCC activity is more positive on seedling plants than those in a mature stage of development. The effectiveness of CCC has been relatively poor on most turfgrasses except for bermudagrass. It is more active in growth inhibition than B-995 and Phosphon but also is more phytotoxic. CCC applications cause foliar toxicity of cool season turfgrasses that is positively correlated with the concentration applied.

Chlorflurenol. The chemical name is methyl 2-chloro-9-hydroxyfluorene-9-carboxylate. It is included in a group of compounds called morphactins. Chlorflurenol is absorbed by the leaves and readily translocated with a distinct systemic effect. Typical developmental effects include a darker green color, inhibition of apical bud formation, and stimulation of lateral shoot development but with a restricted growth rate.

Some shoot growth retardation of turfgrasses is achieved with chlorflurenol but phytotoxic injury to the leaves frequently occurs. This morphactin has not yet been adequately evaluated over a wide range of conditions and turfgrass species. However, preliminary information indicates it does have growth retardation activity that is slightly better than for maleic hydrazide but not as effective as is ultimately desired. The most effective growth inhibition has been achieved with a combination of chlorflurenol and maleic hydrazide. A distinct synergistic action is achieved by mixing these two materials.

Chlorflurenol has even better activity on dicotyledons than turfgrasses. This is desirable on low quality turfs containing broadleaf species. The most effective growth inhibition occurs from applications made in the spring when the turfgrasses have just initiated rapid growth and are approximately 2 to 3 in. high. A uniform application is quite important in achieving maximum potential effectiveness.

Ethrel. The chemical name is 2-chloroethyl phosphonic acid. Ethrel acts by stimulating the release of ethylene in plant tissues. It causes a temporary growth inhibition of cool season grasses that lasts for only 4 to 5 weeks at the maximum. Renewed growth occurs after this with an increase in internode elongation, leaf number, and tiller number observed. Stimulation of rhizome development has also been reported. These observations have been limited primarily to controlled experimental conditions. Additional investigations are required to determine whether the lateral shoot development effects are of any significance under normal turfgrass cultural conditions.

Maleic hydrazide. Maleic hydrazide acts by inhibiting cell division. It not only restricts stem and leaf growth but can impair flowering (7, 21, 24, 26, 28, 30, 47, 54,

59, 73, 78, 84, 115, 116, 129, 148). The use of maleic hydrazide on high quality turfs is questionable (41, 73). Although growth is restricted, the turfgrass surface is usually quite irregular (148). Maleic hydrazide has been used primarily on low quality, minimum use turfgrass areas having a low intensity of culture such as on roadsides, on steep slopes, on stream banks, and along fences and borders where mowing operations are difficult to accomplish (7, 12, 40, 41, 116, 134, 140, 144, 148, 149) (Fig. 12-7). It also impairs bermudagrass growth to the extent that the transition from warm season species to overseeded cool season species is facilitated (5, 47).

Growth inhibition and discoloration of cool season turfgrasses increase as the maleic hydrazide application rate is increased (24, 29, 40, 62, 84, 134, 148, 149). Temporary discoloration of turfgrass leaves is evident with rates as low as 2 lb per acre (39, 137). Mature turfgrasses are more tolerant to the potential phytotoxic effects (26). Root and rhizome growth can be impaired by maleic hydrazide (24). Thinning of turfs has been observed at higher rates, with repeat applications, or when mowing is practiced following an application (24, 62, 73, 119). This thinning along with the growth reduction could result in weed invasion (23, 24, 29, 41, 119).

The response to maleic hydrazide varies with the turfgrass species. Growth of certain warm season species such as manilagrass, St. Augustinegrass, centipedegrass, and carpetgrass is not reduced by maleic hydrazide to the extent observed in cool season species (28). Also, the discoloration is more severe, particularly on St. Augustinegrass and bermudagrass. Similarly, red fescue is more tolerant of maleic hydrazide than Kentucky bluegrass and colonial bentgrass (62). This differential response may alter the composition of the turfgrass community. For example, the composition of roadside turfs is altered with rhizomatous species such as Kentucky bluegrass and red fescue increasing and the dicotyledons and noncreeping grasses declining (140, 144).

Figure 12-7. Effect of various rates of maleic hydrazide (MH) on the vertical shoot growth of Kentucky bluegrass. Untreated (far left), 3#MH per acre (left), 3#MH in April plus 3#MH in May (right), 3#MH in April plus mowing in summer (far right). (Photo courtesy of F. V. Juska, USDA, Beltsville, Maryland).

The application techniques for maleic hydrazide are quite important in achieving good uniformity (12). It can be absorbed through the leaves or roots and is readily translocated (26). Considerable leaf area is required for adequate foliar absorption. A minimum of 2 and preferably 3 to 4 in. of vertical leaf growth is suggested. Approximately 36 to 48 hr is required for absorption and translocation. Mowing and irrigation should be withheld during this period.

The duration of leaf growth inhibition from maleic hydrazide varies from 6 to 14 weeks (39, 149). Two applications at a rate of 2 lb per acre are more effective than one 4-lb application in achieving growth control with a minimum of discoloration and thinning (137). Repeat applications are cumulative in effect so that the interval between successive sprayings can be lengthened (149). Spring applications are preferred to fall or midsummer treatments, especially on warm season turfgrass species (7, 23, 41, 148, 149). The timing of maleic hydrazide applications is critical in achieving maximum growth inhibition. Applying recommended rates when the grass is dormant can cause serious injury and thinning (137). Maleic hydrazide is compatible and should generally be used in combination with 2, 4-D or 2(MCPP) (29, 59, 134, 137, 140).

Phosphon. The chemical name is 2, 4-dichlorobenzyl-tributyl phosphonium chloride. Developmental effects associated with Phosphon include a restriction in vegetative shoot growth and a very distinct dark green color. Root growth is also impaired. The growth retarding effects are more evident with bermudagrass than other turfgrass species. A slight reduction in shoot growth occurs but is associated with phytotoxic injury to the leaves. The damage increases proportionally with the rate of application. Thus, the use of Phosphon on high quality turfs is very questionable due to the shoot injury (73). No growth inhibition is achieved with Phosphon on certain turfgrass species.

Gibberellin. Gibberellin is a growth regulator that stimulates turfgrass growth. Vertical leaf growth increases as the gibberellin application rate is increased up to the point where phytotoxicity occurs (25, 45, 70, 77, 88, 105, 127, 145). Associated with the increased shoot growth is a reduction in root growth and the root-shoot ratio (70). A lighter green color, reduced shoot density, and decreased turfgrass quality frequently occur when gibberellin is applied to an actively growing turf (77, 87, 105, 109, 145). Chlorosis caused by gibberellin can be corrected to a substantial degree by nitrogen fertilization (105, 127). Kentucky bluegrass and bermudagrass are most responsive to gibberellin, in terms of vertical shoot growth stimulation, with bentgrass and red fescue being intermediate (69, 77, 105, 141). The response is minimal with ryegrass and zoysiagrass.

The use of gibberellin in stimulating lateral shoot growth and vegetative establishment has been quite variable and generally not successful (69, 77, 79, 127, 145). The tillering rate is actually reduced by gibberellin applications (45, 109). Similarly, the response from gibberellin seed treatment has been variable and usually does not affect the rate of germination or establishment under optimum temperatures and light (6, 69, 70, 77, 105, 127). The rate and percent seed germination of Kentucky bluegrass and zoysiagrass are increased at suboptimal temperature and light conditions (89, 105). Gibberellin is effective in enhancing spring green-up and growth of dormant winter turfs (88, 127, 141, 142). However, midwinter applications to warm season turfgrass species are injurious (146).

Kinins. Kinetin is a growth substance that overcomes the effect of apical dominance, thus stimulating growth and development of lateral meristems. No specific, practical application for kinins has yet been developed for turfgrasses. However, there is a potential use for a kinetin-like substance in stimulating tiller, stolon, and rhizome development of turfgrasses. Such a response could significantly enhance the establishment rate, shoot density, recuperative potential, and competitive ability of turfgrasses. The rate of vegetative establishment could be substantially improved.

References

1. ANONYMOUS. 1946. Roots in airfield and roadside sod. Timely Turf Topics. 6(1): 4.

2. ALBERDA, T. 1957. The effects of cutting, light intensity and night temperature on growth and soluble carbohydrate content of *Lolium perenne* L. Plant and Soil. 8(3): 199–230.

3. ALBERT, W. B. 1927. Studies on the growth of alfalfa and some perennial grasses. Journal of the American Society of Agronomy. 19: 624–654.

4. ALEXANDER, C. W., and D. E. McCLOUD. 1962. CO_2 uptake (net photosynthesis) as influenced by light intensity of isolated bermudagrass leaves contrasted to that of swards under various clipping regimes. Crop Science. 2: 132–134.

5. ALEXANDER, P. M., and J. G. WRIGHT. 1964. Transition aided by maleic hydrazide. Golf Course Reporter. 32: 36–40.

6. ANDERSON, A. M. 1957. The effect of certain fungi and gibberellin on the germination of Merion Kentucky bluegrass seed. Proceedings of the Association of Official Seed Analysts. 47: 145–153.

7. ANDERSON, C. R. 1963. Inhibitors for economic maintenance. 22nd Short Course on Roadside Development. pp. 74–76.

8. BAUGHMAN, R. W. 1939. Effect of clipping on the development of roots and tops in various grass seedlings. M.S. Thesis. Iowa State College. pp. 1–38.

9. BEACH, G. A. 1963. Management practices in the care of lawns. Proceedings of the 10th Annual Rocky Mountain Regional Turfgrass Conference. pp. 42–45.

10. BEARD, J. B., and W. H. DANIEL. 1965. Effect of temperature and cutting on the growth of creeping bentgrass roots. Agronomy Journal. 57: 249–250.

11. BEARD, J. B., and P. E. RIEKE. 1966. The influence of nitrogen, potassium and cutting height on the low temperature survival of grasses. 1966 Agronomy Abstracts. p. 34.

12. BEASLEY, J. L. 1961. Massachusetts progress report on research with maleic hydrazide. Annual Massachusetts Turfgrass Conference Proceedings. pp. A57–A62.

13. BENSON, D. O. 1963. Mowing grass with a reel-type machine. California Turfgrass Culture. 13(2): 11–14.

14. BISWELL, H. H., and J. E. WEAVER. 1933. Effect of frequent clipping on the development of roots and tops of grasses in prairie sod. Ecology. 14: 368–390.

15. BRANSON, F. A. 1953. Two new factors affecting resistance of grasses to grazing. Journal of Range Management. 6: 165–171.

16. BREDAKIS, E. J. 1959. Interaction between height of cut and various nutrient levels on the development of turfgrass roots and tops. M.S. Thesis. University of Massachusetts. pp. 1–91.

17. BROWN, M. E. 1943. Seasonal variations in the growth and chemical composition of Kentucky bluegrass. Missouri Agricultural Experiment Station Research Bulletin No. 360. pp. 5–56.

18. BUKEY, F. S., and J. E. WEAVER. 1939. Effects of frequent clipping on the underground food reserves of certain prairie grasses. Ecology. 20: 246–252.

19. BURGESS, A. C. 1964. Height of mowing is important on greens. New Zealand Institute for Turf Culture Newsletter No. 3. pp. 14–15.

20. BUTTON, E. F. 1959. Effect of gibberellic acids on laboratory germination of creeping red fescue (*Festuca rubra*). Agronomy Journal. 51: 60–61.

21. CABLER, J. F., and G. C. HORN. 1963. Chemical growth retardants for turf-grasses. Golf Course Reporter. 31(1): 35–44.

22. CARTER, J. F., and A. G. LAW. 1948. The effect of clipping upon the vegetative development of some perennial grasses. Journal of American Society of Agronomy. 40: 1084–1091.

23. CHAMBERLIN, R. E. 1963. Inhibitors for roadside maintenance. 22nd Short Course on Roadside Development. p. 77.

24. CLAPHAM, A. J. 1967. Growth response of a mixed stand of perennial grasses treated with maleic hydrazide. M.S. Thesis. University of Rhode Island. pp. 1–81.

25. CORNS, W. G. 1958. Effects of foliage treatments with gibberellin on forage yield of alfalfa, Kentucky bluegrass and winter wheat. Canadian Journal of Plant Science. 38: 314–319.

26. CRAFTS, A. S., H. B. CURRIER, and H. R. DREVER. 1958. Some studies on the herbicidal properties of maleic hydrazide. Hilgardia. 27(22): 723–757.

27. CRIDER, F. J. 1955. Root growth stoppage resulting from defoliation of grass. U.S. Technical Bulletin No. 1102. pp. 1–23.

28. CRUZADO, H. J., and T. J. MUZIK. 1958. Effect of maleic hydrazide on some tropical lawn grasses. Weeds. 6: 329–330.

29. CURTIS, D. V., H. H. IURKA, E. W. MULLER, E. SECOR, and D. WARD. 1955. Herbicide work on New York State Highways. Proceedings of the 9th Annual Northeastern Weed Control Conference. pp. 463–470.

30. CURTIS, O. F. 1951. Maleic hydrazide for grass control in fruit plantings. Proceedings of the 5th Annual Northeastern Weed Control Conference Supplement. pp. 115–117.

31. DAVIS, R. R. 1958. The effect of other species and mowing height on persistence of lawn grasses. Agronomy Journal. 50: 671–673.

32. ———. 1961. No matter how you cut it the roots get hurt. Golfdom. 35(5): 38–40.

33. DERCHSLER, C. 1930. Leafspot and foot rot of Kentucky bluegrass caused by *Helminthosporium vagans*. Journal of Agricultural Research. 40: 447–456.

34. DEXTER, S. T. 1936. Response of quack grass to defoliation and fertilization. Plant Physiology. 11: 843–851.

35. DODD, J. D., and H. H. HOPKINS. 1958. Yield and carbohydrate content of blue grama grass as affected by clipping. Transactions of Kansas Academy of Science. 61(3): 280–287.

36. ELLETT, W. B., and L. CARRIER. 1915. The effect of frequent clipping on total yield and composition of grasses. Journal of American Society of Agronomy. 7: 85–87.

37. ENGEL, R. E. 1955. Resists diseases, close cutting—Merion bluegrass. New Jersey Agriculture. 37(2): 10–11.

38. ———. 1966. A comparison of colonial and creeping bentgrasses for $\frac{1}{2}$ and $\frac{3}{4}$ inch turf. 1966 Report on Turfgrass Research at Rutgers University. New Jersey Agricultural Experiment Station Bulletin 816. pp. 45–58.

39. ENGEL, R. E., and G. H. AHLGREN. 1950. Some effects of maleic hydrazide on turf grasses. Agronomy Journal. 42: 461–462.

40. ESCRITT, J. R. 1953. Grass growth stunting with maleic hydrazide. Journal of the Sports Turf Research Institute. 8(29): 269–273.

41. ——. 1955. Maleic hydrazide as a grass growth regulator. Journal of the Sports Turf Research Institute. 9: 76–84.

42. EVANS, M. W. 1949. Kentucky bluegrass. Ohio Agricultural Experiment Station Bulletin No. 681. pp. 4–39.

43. EVANS, T. W. 1932. The cutting and fertility factors in relation to putting green management. Journal of Board of Greenkeeping Research. 2: 196–200.

44. EVERSON, A. C. 1966. Effects of frequent clipping at different stubble heights on western wheatgrass (Agropyron smithii Rybd.). Agronomy Journal. 58(1): 33–35.

45. FEJER, S. O. 1960. Effects of gibberellic acid, indole-acetic acid, coumarin and perloline on perennial ryegrass (Lolium perenne L.). New Zealand Journal of Agricultural Research. 3: 734–743.

46. FITTS, O. B. 1925. A preliminary study of the root growth of fine grasses under turf conditions. USGA Green Section Bulletin. 5: 58–62.

47. FOLKNER, J. S. 1956. Maleic hydrazide, an aid in summer-winter turf transition. Golf Course Reporter. 24: 5–7.

48. FRAZIER, S. L. 1960. Turfgrass seedling development under measured environment and management conditions. M.S. Thesis. Purdue University. pp. 1–61.

49. FUNK, C. R., and R. E. ENGEL. 1963. Effect of cutting height and fertilizer on species composition and turf quality ratings of various turfgrass mixtures. Rutgers University Turfgrass Short Course in Turf Management. pp. 47–56.

50. ——. 1966. Influence of variety, fertility level, and cutting height on weed invasion of Kentucky bluegrass. Rutgers University Short Course. pp. 10–17.

51. FUNK, C. R., R. E. ENGEL, and P. M. HALISKY. 1966. Performance of Kentucky bluegrass varieties as influenced by fertility level and cutting height. 1966 Report on Turfgrass Research at Rutgers University. New Jersey Agricultural Experiment Station Bulletin 816. pp. 7–21.

52. GARY, J. E. 1967. The vegetative establishment of four major turfgrasses and the response of stolonized 'Meyer' zoysiagrass (Zoysia japonica var. Meyer) to mowing height, nitrogen fertilization, and light intensity. M.S. Thesis. Mississippi State University. pp. 1–50.

53. GERNERT, W. B. 1936. Native grass behavior as affected by periodic clipping. Agronomy Journal. 28: 447–456.

54. GOODIN, J. R. 1959. Chemical control of flowering of Pennisetum clandestinum, kikuyugrass. Southern California Turfgrass Culture. 9: 17.

55. GOSS, R. L., and A. G. LAW. 1967. Performance of bluegrass varieties at 2 cutting heights and 2 nitrogen levels. Agronomy Journal. 59(6): 516–518.

56. GRABER, L. F. 1931. Food reserves in relation to other factors limiting the growth of turfgrasses. Plant Physiology. 6: 43–72.

57. ——. 1933. Competitive efficiency and productivity of bluegrass (Poa pratensis L.) with partial defoliation at two levels of cutting. Journal of American Society of Agronomy. 25: 328–333.

58. GRABER, L. F., and H. W. REAM. 1931. Growth of bluegrass with various defoliations and abundant nitrogen supply. Journal of American Society of Agronomy. 23: 938–946.

59. GREENE, W. C. 1955. The progress of weed control on Connecticut State highways. Proceedings of the Northeastern Weed Control Conference. pp. 429–432.

60. HART, R. H., and G. W. BURTON. 1966. Prostrate vs. common dallisgrass under different clipping frequencies and fertility levels. Agronomy Journal. 58(5): 521–522.

61. HART, R. H., and W. S. McGUIRE. 1964. Effect of clipping on the invasion of pastures by velvetgrass, Holcus lanatus L. Agronomy Journal. 56(2): 187–188.

62. HART, S. W., and J. A. DeFRANCE. 1955. Effects of maleic hydrazide on the growth of turfgrass. Golf Course Reporter. 23(7): 5–7.

63. HARRISON, C. M. 1931. Effect of cutting and fertilizer applications on grass development. Plant Physiology. 6: 669–684.

64. ———. 1934. Responses of Kentucky bluegrass to variations in temperature, light, cutting, and fertilizing. Plant Physiology. 9: 83–106.

65. HARRISON, C. M., and C. W. HODGSON. 1939. Response of certain perennial grasses to cutting treatments. Journal of American Society of Agronomy. 31: 418–430.

66. HODGES, T. K. 1960. Factors influencing the rooting depth of *Poa pratensis*. M.S. Thesis. University of California, Davis. pp. 1–53.

67. HUGHES, H. D. 1937. Response of *Poa pratensis* L. to different harvest treatments, measured by weight and composition of forage and roots. Report of the 4th International Grassland Congress. Aberystwyth, Wales. pp. 447–452.

68. JACQUES, W. A. 1937. The effect of different rates of defoliation on the root development of certain grasses. New Zealand Journal of Science and Technology. 19(7): 441–450.

69. JUSKA, F. V. 1958. Some effects of gibberellic acid on turfgrasses. Golf Course Reporter. 26: 5–9.

70. ———. 1959. The effect of gibberellic acid on Kentucky bluegrass root production. Agronomy Journal. 51: 184–185.

71. ———. 1961. Frequency and height of cutting bluegrass. Proceedings of the 2nd Annual Missouri Lawn and Turf Conference. pp. 1–2.

72. JUSKA, F. V., and A. A. HANSON. 1963. The management of Kentucky bluegrass on extensive turfgrass areas. Park Maintenance. 16: 22–32.

73. ———. 1964. Effects of growth retardants on Kentucky bluegrass turf. Golf Course Reporter. 32: 60–64.

74. JUSKA, F. V., J. TYSON, and C. M. HARRISON. 1955. The competitive relationship of Merion bluegrass as influenced by various mixtures, cutting heights, and levels of nitrogen. Agronomy Journal. 47: 513–518.

75. KENNEDY, W. K., and M. B. RUSSELL. 1948. Relationship of top growth, root growth, and apparent specific gravity of the soil under different clipping treatments of a Kentucky bluegrass-wild white clover pasture. Journal of American Society of Agronomy. 40: 535–541.

76. KNUTTI, H. J., and M. HIDIROGLOU. 1967. The effect of cutting and nitrogen treatments on yield, protein content and certain morphological characteristics of timothy and smooth bromegrass. Journal of the British Grassland Society. 22: 35–41.

77. KOLLETT, J. R., and J. A. DEFRANCE. 1959. Some effects of gibberellic acid on various turfgrasses. Golf Course Reporter. 27(3): 30–32.

78. KOSESAN, W. H. 1956. Effect of timing of maleic hydrazide on growth inhibition of roadside grasses. Research Progress Report, Western Weed Control Conference. p. 116.

79. KRIEDEMAN, P. E. 1963. The effects of gibberellic acid on a dwarf variant of *Cynodon dactylon*. Australian Journal of Science. 25: 468.

80. KUHN, A. O., and W. B. KEMP. 1939. Effect of removing different proportions of foliage on contrasting strains of Kentucky bluegrass, *Poa pratensis* L. Journal of American Society of Agronomy. 31: 892–895.

81. LAIRD, A. S. 1930. A study of the root systems of some important sod-forming grasses. University of Florida Agricultural Experiment Station Bulletin 211. pp. 1–27.

82. LANGER, R. H. M. 1959. A study of growth in swards of timothy and meadow fescue. II. The effects of cutting treatments. Journal of Agricultural Science. 52(3): 273–281.

83. LANGVAD, B. 1968. Ball-bouncing and ball-rolling as a function of mowing height and kind of soil have been studied at Weibullsholm. Weibulls Grastips. 10: 355–357.

84. LANING, E. R., and V. H. FREED. 1952. Growth inhibition of grasses. Proceedings of the Western Weed Control Conference. p. 155.

85. LARUE, C. D. 1935. The time to cut dandelions. Science. 82: 350.

86. LAW, A. G. 1966. Performance of bluegrass varieties clipped at two heights. Weeds, Trees and Turf. 5(5): 26–28.

87. LEBEN, C., and L. V. BARTON. 1957. Effects of gibberellic acid on growth of Kentucky bluegrass. Science. 125: 494.

88. LEBEN, C., E. F. ALDER, and A. CHICKUK. 1959. Influence of gibberellic acid on the growth of Kentucky bluegrass. Agronomy Journal. 51: 116–117.

89. LeCROY, W. C. 1963. Characterizing zoysia by field and anatomical studies. Ph.D. Thesis. Purdue University. pp. 1–82.

90. LEOPOLD, A. C., and W. H. KLEIN. 1951. Maleic hydrazide as an antiauxin in plants. Science. 114: 9–10.

91. LETEY, J., L. H. STOLZY, O. R. LUNT, and N. VALORAS. 1964. Soil oxygen and clipping height. Golf Course Reporter. 32(2): 16–26.

92. LEUKEL, W. A., and J. M. COLEMAN. 1930. Growth behavior and maintenance of organic foods in bahia grass. Florida Agricultural Experiment Station Technical Bulletin 219. pp. 1–56.

93. LEUKEL, W. A., J. P. CAMP, and J. M. COLEMAN. 1934. Effect of frequent cutting and nitrate fertilization on the growth behavior and relative composition of pasture grasses. Florida Agricultural Experiment Station Bulletin 269. pp. 1–48.

94. LOVVORN, R. L. 1944. The effects of fertilization, species competition, and cutting treatments on the behavior of dallis grass, *Paspalum dilatatum* Poir., and carpet grass, *Axonopus affinis* Chase. Journal of American Society of Agronomy. 36: 590–600.

95. ———. 1945. The effect of defoliation, soil fertility, temperature, and length of day on the growth of some perennial grasses. Journal of American Society of Agronomy. 37: 570–582.

96. MADISON, J. H., Jr. 1960. The mowing of turfgrass. I. The effect of season, interval, and height of mowing on the growth of Seaside bentgrass turf. Agronomy Journal. 52: 449–452.

97. MADISON, J. H. 1962. The effect of management practices on invasion of lawn turf by bermudagrass (*Cynodon dactylon* L.). Proceedings of the American Society for Horticultural Science. 80: 559–564.

98. ———. 1962. Mowing of turfgrass. II. Responses of three species of grass. Agronomy Journal. 54: 250–252.

99. ———. 1962. Mowing of turfgrasses. III. The effect of rest on Seaside bentgrass turf mowed daily. Agronomy Journal. 54: 252–253.

100. ———. 1962. Turfgrass ecology. Effects of mowing, irrigation, and nitrogen treatments of *Agrostis palustris* Huds., "Seaside" and *Agrostis tenuis* Sibth., "Highland" on population, yield, rooting, and cover. Agronomy Journal. 54: 407–412.

101. MADISON, J. H., and R. M. HAGAN. 1962. Extraction of soil moisture by Merion bluegrass (*Poa pratensis* L. 'Merion') turf, as affected by irrigation frequency, mowing height, and other cultural operations. Agronomy Journal. 54: 157–160.

102. McBEE, G. G. 1966. Effects of mowing heights on Texturf 1F bermudagrass turf. Texas A & M University, Soil and Crop Sciences Departmental Technical Report No. 10. pp. 1–4.

103. McCARTY, E. C., and R. PRICE. 1942. Growth and carbohydrate content of important mountain forage plants in central Utah affected by clipping and grazing. USDA Technical Bulletin No. 818. pp. 1–51.

104. McKEE, W. H., R. H. BROWN, and R. E. BLASER. 1967. Effect of clipping and nitrogen fertilization on yield and stands of tall fescue. Crop Science. 7(6): 567–570.

105. McVEY, G. R., and S. H. WITTWER. 1959. Responses of certain lawn grasses to gibberellin. Quarterly Bulletin of the Michigan Agricultural Experiment Station. 42(1): 176–196.

106. MITCHELL, K. J. 1955. Growth of pasture species. II. Perennial ryegrass (*Lolium perenne*), cocksfoot (*Dactylis glomerata*), and paspalum (*Paspalum dilatatum*). New Zealand Journal of Science and Technology. Sec. A. 37(1): 8–26.

107. MITCHELL, K. J., and S. T. J. COLES. 1955. Effects of defoliation and shading on short-rotation ryegrass. New Zealand Journal of Science and Technology. Sec. A. 36(6): 586–604.

108. MORTIMER, G. B., and H. L. AHLGREN. 1936. Influence of fertilization, irrigation, and stage and height of cutting on yield and composition of Kentucky bluegrass. Journal of American Society of Agronomy. 28: 515–533.

109. MOSER, L. E. 1967. Rhizome and tiller development of Kentucky bluegrass (*Poa pratensis* L.) as influenced by photoperiod, vernalization, nitrogen level, growth regulators, and variety. Ph.D. Thesis. Ohio State University. pp. 1–143.

110. MUSSER, H. B. 1950. The use and misuse of water. Greenkeeper's Reporter. 18: 5–9.

111. NEWELL, L. C., and F. O. KEIM. 1947. Effects of mowing frequency on the yield and protein content of several grasses grown in pure stands. University of Nebraska Agricultural Experiment Station Research Bulletin No. 150. pp. 3–33.

112. NOER, O. J. 1959. Better bent greens. Turf Service Bureau of the Milwaukee Sewerage Commission Bulletin No. 2. pp. 1–33.

113. OSWALT, D. T., A. R. BERTRAND, and M. R. TEEL. 1959. Influence of nitrogen fertilization and clipping on grass roots. Soil Science Society of America Proceedings. 23: 228–230.

114. PARKER, K. W., and A. W. SAMPSON. 1930. Influence of leafage removal on anatomical structure of roots of *Stipa pulchra* and *Bromus hordeaceus*. Plant Physiology. 5: 543–552.

115. PARRIS, C. D., and E. G. RODGERS. 1954. Herbicidal eradication of unproductive centipede grass sod. Proceedings of the Southern Weed Conference. pp. 185–187.

116. PHALEN, J., and W. E. KENNEDY. 1952. Trimming equipment. Notes and Quotes Publication of National Catholic Cemetery Conference. 4: 5–6.

117. POHJAKALLIO, O., and S. ANTILA. 1955. On the effect of removal of shoots on the drought resistance of red clover and timothy. Acta Agricultural Scandainavica. 5: 239–244.

118. PRINE, G. M., and G. W. BURTON. 1956. The effect of nitrogen rate and clipping frequency upon the yield, protein content and certain morphological characteristics of coastal bermudagrass [*Cynodon dactylon* (L) Pers.]. Agronomy Journal. 48: 296–301.

119. RADKO, A. M. 1953. What about maleic hydrazide. USGA Journal and Turf Management. 6: 30–32.

120. ———. 1968. Grooming your golf course is important. USGA Green Section Record. 6(2): 1–4.

121. REID, M. E. 1933. Effect of variations in concentration of mineral nutrients upon the growth of several types of turf grasses. USGA Green Section Bulletin. 13(5): 122–131.

122. REYNOLDS, J. H., and D. SMITH. 1962. Trend of carbohydrate reserves in alfalfa, smooth bromegrass, and timothy grown under various cutting schedules. Crop Science. 2: 333–336.

123. ROBERTS, E. C. 1958. The grass plant—feeding and cutting. Golf Course Reporter. 26(3): 5–9.

124. ROBERTS, R. A., and I. V. HUNT. 1936. The effect of shoot cutting on the growth of root and shoot of perennial ryegrass (*Lolium perenne* L.) and timothy (*Phleum pratense* L.). Welsh Journal of Agriculture. 12: 158–174.

125. ROBERTSON, J. H. 1933. Effect of frequent clipping on the development of certain grass seedlings. Plant Physiology. 8: 425–447.

126. ROWELL, J. B. 1951. Observations on the pathogenicity of *Rhizoctonia solani* on bent grass. Plant Disease Reporter. 35: 240–242.

127. SARTORETTO, P. A. 1957. Gibberillic acid, amazing growth stimulant. Golf Course Reporter. 25(2): 40–44.

128. SCHERY, R. W. 1966. Remarkable Kentucky bluegrass. Weeds, Trees and Turf. 5(10): 16–17.

129. SCHOENE, D. L., and O. L. HOFFMANN. 1949. Maleic hydrazide, a unique growth regulant. Science. 109: 588–590.

130. SPRAGUE, H. B. 1933. Root development of perennial grasses and its relation to soil conditions. Soil Science. 36: 189–209.

131. SULLIVAN, E. F. 1962. Effects of soil reaction, clipping height, and nitrogen fertilization on the productivity of Kentucky bluegrass sod transplants in pot culture. Agronomy Journal. 54: 261–263.

132. SULLIVAN, J. T., and V. G. SPRAGUE. 1943. Composition of the roots and stubble of perennial ryegrass following partial defoliation. Plant Physiology. 18: 656–670.

133. THOMAS, R. J. 1966. What's the best mowing speed. Golf Superintendent. 34: 24–26.

134. TRASK, A. D. 1962. Inhibitors for economic maintenance. 21st Short Course on Roadside Development. pp. 98–101.

135. VAARTNOU, H. 1967. Responses of five genotypes of *Agrostis* L. to variations in environment. Ph.D. Thesis. Oregon State University. pp. 1–149.

136. VENGRIS, J., E. R. HILL, and D. L. FIELD. 1966. Clipping and regrowth of barnyardgrass. Crop Science. 6: 342–344.

137. WEBBER, E. R. 1954. Inhibiting grass growth with maleic hydrazide. Parks, Golf Courses and Sports Grounds. 20(1): 23–27.

138. WELTON, F. A., and J. C. CARROLL. 1940. Lawn experiments. Ohio Agricultural Experiment Station Bulletin 613. pp. 1–43.

139. WILKINS, F. S. 1935. Effect of overgrazing on Kentucky bluegrass under conditions of extreme drought. Journal of American Society of Agronomy. 27: 159–160.

140. WILLIS, A. J. 1965. Methods of weed control and the use of growth retarding substances. Parks, Golf Courses and Sports Grounds. 30(9): 733–736, 740.

141. WITTWER, S. H., and M. J. BUKOVAC. 1957. Gibberellin and higher plants: V. Promotion of growth in grass at low temperatures. Quarterly Bulletin of the Michigan Agricultural Experiment Station. 39: 682–686.

142. ———. 1958. The effects of gibberellin on economic crops. Economic Botany. 12(3): 213–255.

143. WOOD, G. M., and J. A. BURKE. 1961. Effect of cutting height on turf density of Merion, Park, Delta, Newport, and common Kentucky bluegrass. Crop Science. 1: 317–318.

144. YEMM, E. W., and A. J. WILLIS. 1962. The effects of maleic hydrazide and 2, 4-dichlorophenoxyacetic acid on roadside vegetation. Weed Research. 2(1): 24–40.

145. YOUNGNER, V. B. 1958. Gibberellic acid on zoysia grass. Southern California Turfgrass Culture. 8: 5–6.

146. ———. 1959. Effects of winter applications of gibberellic acid on bermudagrass and zoysia turf. Southern California Turfgrass Culture. 9: 7.

147. ———. 1962. Wear resistance of cool season turfgrasses. Effects of previous mowing practices. Agronomy Journal. 54: 198–199.

148. ZAK, J. M., and E. J. BREDAKIS. 1967. Use of maleic hydrazide for growth suppression of highway turf. Massachusetts Turf Bulletin No. 4. pp. 17–25.

149. ZUKEL, J. W. 1953. Temporary grass inhibition with maleic hydrazide. Agricultural Chemicals. 8: 45–143.

chapter 13

Fertilization

Introduction

The actively growing turfgrass plant contains 75 to 85% water with the remaining portion composed primarily of organic compounds frequently referred to as dry matter. These organic compounds are composed of 16 essential elements that are directly involved in plant nutrition and are necessary for the plant to complete its life cycle. The 16 commonly recognized essential elements are listed in Table 13-1.

A major portion of the plant dry matter content is composed of three elements: carbon, hydrogen, and oxygen. The plant derives carbon mainly from atmospheric carbon dioxide, while hydrogen is obtained primarily from water absorbed by the

Table 13-1

The Sixteen Essential Elements Necessary for the Nutrition of Turfgrasses

Macronutrients		Micronutrients
Obtained from carbon dioxide and water	Obtained primarily from the soil	
Carbon, C	Nitrogen, N	Iron, Fe
Hydrogen, H	Phosphorus, P	Manganese, Mn
Oxygen, O	Potassium, K	Zinc, Zn
	Calcium, Ca	Copper, Cu
	Magnesium, Mg	Molybdenum, Mo
	Sulfur, S	Boron, B
		Chlorine, Cl

408

root system. Oxygen is derived from water and atmospheric carbon dioxide. Thirteen additional essential elements occur in plant tissues in smaller amounts but they are equally important. Nitrogen, phosphorus, potassium, calcium, magnesium, and sulfur are required in relatively large amounts compared to iron, manganese, copper, zinc, molybdenum, boron, and chlorine. The latter group of micronutrients are required in concentrations of less than 2 ppm. Although the quantity required for plant growth and development is quite small, the micronutrients are equally as important as the macronutrients in overall nutrition. All 13 elements except nitrogen are absorbed by the plant from the soil and are derived from parent materials in the soil. Nitrogen is usually absorbed from the soil but is ultimately derived from atmospheric nitrogen through fixation processes.

In addition to the 16 essential elements, traces of at least 40 other elements including aluminum, arsenic, cobalt, fluorine, iodine, lead, selenium, silicon, sodium, strontium, and vanadium can be found in turfgrass tissues. These elements are usually found in much smaller quantities but can sometimes (a) interfere with the uptake of essential elements or (b) become phytotoxic.

Nutrient Uptake

The essential elements for plant growth must be present in the proper amounts and the proper proportions to one another for optimum growth and development to occur. Nutrients can be absorbed into the plant through the leaves, stems, or roots. The leaves and to a lesser degree the stems are the major surfaces for carbon dioxide absorption. The above-ground plant parts are also capable of absorbing water and mineral nutrients, primarily through the stomata. However, most water and mineral nutrient absorption occurs through the root system. Water and carbon dioxide uptake are discussed in Chapters 8 and 9. The immediate discussion is limited to nutrient uptake from the soil by roots.

The total quantity of nutrients contained in most soils is high compared to the requirements of the turfgrass plant; however, a major portion of these nutrients is tightly bound in unavailable forms. Only nutrients contained in the soil solution are available for uptake by the turfgrass roots. The soil solution is a mixture of many organic and inorganic constituents including the essential nutrients. The nutrients are originally derived from the (a) decomposition of soil organic matter, (b) weathering and breakdown of rocks and finer textured minerals in the soil, and (c) applied fertilizer.

The turfgrass root system is ideally suited for nutrient uptake because of a fibrous, extensive root system having a large surface area. The roots are surrounded by the soil solution containing nutrients and are also in intimate contact with soil particles. Nutrient ions can be absorbed into the root cells against a diffusion gradient and are accumulated within the cell. This uptake process requires energy for the transfer of ions across the impermeable outer membrane of the cell. It is not reversible and is selective in nature with certain ions being absorbed and others excluded.

More specifically, active uptake of nutrient ions entails the following series of events. Ions diffuse through the soil solution into the "outer space" of the root. Diffusion is nonselective from regions of higher to lower concentrations. This phase is reversible and passive in nature. Nutrient ions then combine with carrier molecules at binding sites located at the interior of the root "outer space". The ion is carried across the impermeable membrane and released from the carrier molecule into the inner space of the cell. Transport of the carrier molecule–ion complex past the semipermeable membrane requires energy. Thus, active ion absorption is directly related to the root respiration rate.

Each carrier molecule is specific for a given ion or group of ions. Not only is the active uptake mechanism selective but it also absorbs specific nutrients at different rates. For example, monovalent cations such as potassium are taken up more rapidly than divalent cations such as calcium and magnesium. Selectivity is demonstrated in that the root accumulates large amounts of potassium while tending to exclude other ions such as sodium. The impermeable membrane also serves as a barrier against the diffusion of ions from the higher concentration in the vacuole outward into the soil solution. Thus, the plant root is capable of retaining high concentrations of nutrient ions required for plant growth.

In contrast to active ion absorption, nonionized solutes such as water enter and leave the cells by diffusion. The solutes move from regions of higher to regions of lower concentration. The rate of movement depends on the molecule size and its relative solubility in the fatty solvents of the membrane. Thus, small water molecules enter quite rapidly, whereas larger sugar molecules that are quite low in fat solubility enter quite slowly.

Once absorption has occurred, the nutrients are translocated to the actively growing tissues of the immediate plant. Nutrient translocation to creeping lateral stems such as rhizomes and stolons is limited (90).

Factors influencing nutrient uptake. Nutrient uptake is determined by the characteristics and condition of the root system and surrounding soil. The depth and extent of the root system are important factors influencing nutrient uptake. This is particularly important with nutrients such as phosphorus that are less mobile in the soil. Phosphorus fertilization is more critical with shallow- than with deep-rooted turfgrass species.

A second factor is energy availability from root respiration. Adequate respiration rates are maintained by ensuring (a) an adequate oxygen supply, (b) satisfactory levels of respiratory substrate, and (c) optimum temperatures for root activity (167). Poor soil aeration or waterlogged soils severely restrict respiration and impair the uptake of essential nutrients (137).

Externally, the soil moisture content is important because it influences the ion diffusion rate into the "outer space" of roots. Nutrient uptake is impaired by waterlogged soil conditions or by a lack of soil moisture. It is also important to have an adequate but not excessive concentration of nutrients present in the soil solution for uptake by the root system. In addition, the concentration of other elements can affect the uptake of essential nutrients. The presence of certain ions in the external soil solution can increase or decrease root membrane permeability to essential nutrients. The hydrogen ion concentration is an important external factor influencing nutrient uptake (7, 264). The concentration of certain nutrients declines in the soil solution with increases in acidity or alkalinity from soil pH's of 6 to 7. Soil temperatures also influence the nutrient uptake rate (185). It is impaired by low soil temperatures (231).

Essential Nutrients

The essential elements have a minimal physiological effect in themselves but are active when combined in certain organic compounds or as ions in solution. Essential nutrients have a number of vital functions in plant growth and development such as (a) constituents of living tissues; (b) catalysts in certain biochemical reactions; (c) influencing the cell osmotic pressure; (d) influencing the acidity of plant tissues; and (e) affecting membrane permeability to nutrient uptake and transport.

The role of each essential nutrient in plant growth and development is quite specific but sometimes not completely understood. Each essential nutrient is equally important in that a deficiency of any one can seriously impair the overall plant growth and developmental processes.

Carbon, hydrogen, and oxygen are required for synthesis of carbohydrates, proteins, lipids, and related compounds. Since these three essential nutrients are contained in most organic compounds found in plants, they have a major function in normal plant growth and development. The remaining 13 essential nutrients are discussed in more detail since each has specific functions within the plant. These nutrients can be supplemented through soil or foliar fertilization when a deficiency occurs.

Nitrogen

Turfgrasses require nitrogen in the largest amount of any of the essential nutrients with the exception of carbon, hydrogen, and oxygen. For this reason, nitrogen is the nutrient applied in largest amounts in turfgrass fertilization programs (74, 109). It is a vital constituent of (a) the chlorophyll molecule, which is involved in photosynthesis; (b) amino acids and proteins, which compose a major portion of the protoplasm; (c) nucleic acids, which function in hereditary transfer of plant characteristics; and (d) enzymes and vitamins, which catalyze metabolic reactions within the plant. All are vitally important in turfgrass growth and development.

Turfgrasses normally contain from 3 to 6% nitrogen on a dry matter basis providing there is no soil nitrogen deficiency (9, 55, 56, 99, 121, 226, 246, 284). Cool season turfgrass species growing at soil temperatures above 70°F or at higher levels of nitrogen fertility have a higher nitrogen content. The nitrogen content is also usually higher in young, actively growing tissues such as the leaves. Nitrogen is quite mobile within the plant. The nitrogen fertilization requirement of turfgrass species and cultivars ranges from 0 to 2 lb of actual nitrogen per 1000 sq ft per growing month.

Effects of nitrogen. Nitrogen nutrition can affect a turf in a number of ways including (a) shoot growth; (b) root growth; (c) shoot density; (d) color; (e) disease proneness; (f) heat, cold, and drought hardiness; (g) recuperative potential; and (h) composition of the turfgrass community.

Nitrogen influences the growth rate of turfgrass tissues substantially (32). There is an increase in the growth rate of roots and shoots of the turfgrass plant as the level of nitrogen nutrition is increased from zero (12, 25, 27, 36, 80, 108, 124, 140, 152, 155, 163, 168, 191, 192, 194, 234, 240, 242, 243, 246, 257, 269). Higher nitrogen levels also increase the respiration rate (19). This overall plant growth response does not continue indefinitely with increased nitrogen levels. At a certain point in nitrogen growth stimulation, the quantity of carbohydrates available for protein synthesis becomes limiting. This causes a distinct suppression in root growth and carbohydrate reserves, while shoot growth continues to respond to higher nitrogen levels (Fig. 13-1) (4, 13, 18, 19, 21, 31, 51, 77, 80, 106, 107, 110, 111, 114, 115, 140, 143, 145, 152, 155, 162, 174, 186, 190, 198, 202, 221, 222, 230, 234, 235, 236, 237, 240, 251, 252, 253, 260, 265, 276, 287, 293). Shoots have priority over the roots for the available carbohydrates. Excessive stimulation of shoot growth caused by higher nitrogen levels can result in death of the turfgrass root system due to a lack of carbohydrates for maintenance of the existing roots. Rhizome growth and stolon internode length are restricted at excessively high nitrogen fertility levels (251). A decrease in cell wall thickness and an increase in cell size also occur.

It is important to maintain a controlled level of nitrogen nutrition that (a) avoids excessive leaf growth requiring more frequent mowing, (b) does not stimulate shoot growth to the extent that root growth is impaired, and (c) varies with the environmental and cultural conditions. Excessive nitrogen fertilization is usually a greater problem on intensively cultured turfs than a nitrogen deficiency. When the former occurs, the depth and extent of the root system are restricted causing decreased nutrient and water uptake. Excessively high nitrogen levels also stimulate thatch accumulation and puffiness on greens (69, 70).

The level of nitrogen nutrition is directly correlated with the color and

Figure 13-1. The effect of a high (left) versus low (right) level of nitrogen nutrition on the root growth of Kentucky bluegrass (Photo courtesy of E. C. Roberts, University of Florida, Gainesville, Florida).

shoot density of the turf (14, 60, 86, 89, 106, 125, 130, 143, 184, 186, 187, 189, 192, 199, 203, 205, 237, 249, 257, 275). Color is frequently used as an indicator in determining the timing of nitrogen fertilizations. A decrease in shoot density indicates a severe nitrogen deficiency. Higher levels of nitrogen also increase the leaf texture, while low levels result in a finer texture providing the shoot density is not adversely affected (60). The recuperative potential of turfgrasses is enhanced by moderate nitrogen levels that do not exhaust the carbohydrate reserves (175, 176). Higher nitrogen levels delay maturation and reduce seed head formation (165, 198, 234, 254, 262, 265).

The level of nitrogen nutrition can influence the severity of disease development on disease susceptible turfgrass species (188, 247). Turfs grown at higher nitrogen fertility levels are more prone to injury by (a) *Helminthosporium* leaf spot (35, 36, 86, 87, 126), (b) brown patch (1, 42, 60), (c) *Fusarium* patch (104, 105), (d) *Ophiobolus* patch (97, 103), (e) *Fusarium* blight (87), and (f) gray leaf spot (83) (Fig. 13-2). In contrast, facultative saprophytes such as dollar spot (1, 38, 39, 41, 60, 68, 70, 126, 127, 136, 192, 199, 244, 245), rust (28, 158, 244), and red thread (71, 97, 104, 209, 261) are most severe on nitrogen deficient turfs. The relationship between increased disease proneness and lower levels of nitrogen nutrition is attributed to the rate of vertical shoot growth. As a result, the rapidly growing leaves that are infected are also removed more frequently by mowing (43, 244).

Hardiness to heat, cold, and drought stress is also influenced by the level of nitrogen nutrition. The degree of tissue hydration is directly correlated with the nitrogen level (33, 34, 35, 168, 190, 191, 233). Deficiencies and particularly excesses of nitrogen result in reduced hardiness (2, 10, 33, 34, 35, 92, 114, 149, 201, 233). High nitrogen levels also increase the wilting tendency and proneness to winter desiccation.

Figure 13-2. The influence of a high (left) versus low (right) nitrogen fertility level on the incidence of *Ophiolobus* patch in *Agrostis tenuis*. (Photo courtesy of R. L. Goss, Washington State University, Puyallup, Washington).

The composition of a turfgrass community is sometimes influenced by the level of nitrogen nutrition (201). A typical example is a Kentucky bluegrass-red fescue community where Kentucky bluegrass tends to dominate at higher nitrogen levels while red fescue becomes dominant under minimal levels of nitrogen nutrition (53, 87, 155, 245, 247). Similar responses occur within blends as well as mixtures. For example, Merion Kentucky bluegrass is favored by a higher nitrogen fertility level, while Kenblue responds more favorably to a lower nitrogen level. Shifts in the species composition of a turfgrass community occur in direct relation to changes in the nitrogen fertility program, assuming other factors are not differentially limiting species growth. An excessively low nitrogen fertility program that restricts shoot growth, density, and color results in the encroachment of undesirable weeds into the turf (17, 53, 54, 60, 79, 81, 85, 86, 109, 184, 200, 203, 247, 249, 257, 283, 292). There are exceptions, such as crabgrass and annual bluegrass activity being enhanced by higher nitrogen fertility levels (85, 86, 245).

Soil relationships. Nitrogen is absorbed from the soil solution primarily in the nitrate (NO_3^-) form though turfgrasses can absorb the ammonia (NH_4^+) forms. Nitrogen uptake is quite rapid with translocation to the leaf tissue occurring within 15 hr (194). A major portion of the nitrogen utilized by plants is obtained from the soil. The source of nitrogen is from the decomposition of soil organic matter or, more commonly, from supplemental applications of fertilizer. Small quantities of oxidized nitrogen are also derived from lightning activity during rain storms.

The nitrogen content in soils varies greatly, depending on the (a) organic matter content, (b) soil texture, (c) carbon-nitrogen ratio, and (d) extent of irrigation and similar cultural practices. Coarse textured soils generally have a lower nitrogen content. The nitrogen content decreases with increasing soil depth. Temperature and moisture increase the ultimate quantity of organic matter accumulated in soils and, in turn, influence the amount of nitrogen available as the result of biological mineralization. The organic matter accumulation decreases with increasing soil temperatures and decreasing soil moisture levels. The rate of organic matter decomposition is most rapid under warm, moist conditions. Organic matter accumulates under very cold, moist conditions and the rate of nitrogen release is quite slow.

Nitrogen can be lost from the soil by leaching and volatilization unless held in organic forms other than urea. Nitrogen loss by leaching occurs primarily in the nitrate form (15). The extent of leaching depends on the amount of precipitation and irrigation, temperature, and soil texture. Excessive applications of water-soluble nitrogen fertilizers should be avoided to minimize the leaching of nitrates that can contribute to nitrate pollution of the ground water, streams, and lakes.

Certain soil bacteria are capable of reducing nitrate nitrogen and utilizing the energy obtained from the reduction. Gaseous nitrogen is released to the atmosphere during this process. Volatile loss of nitrogen occurs most commonly when there is (a) a limited oxygen supply, (b) an accumulation of nitrates, and (c) a quantity of readily decomposable organic matter. This condition would typically occur on compacted, excessively irrigated, intensely fertilized greens. Loss of nitrogen by volatilization also occurs directly from the application of certain nitrogen carriers such as urea.

Phosphorus

Phosphorus is an essential macronutrient contained in every living cell. It is involved in a number of physiological functions within the plant including (a) energy transformations within the plant in the form of adenosine triphosphate (ATP), (b) a constituent of the genetic material in the cell nucleus, and (c) carbohydrate transformations such as the conversion of starch to sugar.

Phosphorus occurs in large amounts in young tissues in the regions of cell division (138). It also occurs in large quantities in seeds as a constituent of phytin. As a plant matures, the phosphorus is transferred to the reproductive portions of the plant and eventually accumulates in the seeds. The quantity of phosphorus utilized by the turfgrass plant is considerably less than the amount of nitrogen and potassium. Turfgrasses vary in phosphorus absorption. Kentucky bluegrass ranks quite high, whereas carpetgrass, bahiagrass, and bermudagrass are relatively low (210).

Effects of phosphorus. Phosphorus affects the (a) establishment, (b) rooting, (c) maturation, and (d) reproduction of turfgrasses. It is particularly vital during the seedling stage of turfgrass growth and development (195, 196, 210). Thus, phosphorus should be placed near the seeds during planting to assure rapid establishment. It also stimulates root growth and branching (27, 153, 240, 246). Higher phosphorus levels hasten maturity, while a phosphorus deficiency delays maturity. Seed setting is also enhanced by higher phosphorus levels. A phosphorus deficiency causes a reduction in the tillering, shoot growth, and moisture content of ryegrass shoots (138). Turfgrasses respond to phosphorus applications when the soil phosphorus level is less than 5 ppm (178).

Phosphates can accumulate to very high soil levels if fertilizers containing a substantial quantity of phosphorus are used continuously (46, 47, 218). Red fescue and Kentucky bluegrass can grow on soils having phosphorus levels as high as 1500 lb of actual phosphorus per acre with no detrimental effects (153). This is well above the levels commonly reported. The lack of detrimental effects is attributed to a major portion of the phosphorus being in an unavailable form.

Soil relationships. Phosphorus is absorbed by the plant primarily as the $H_2PO_4^-$ ion. Phosphorus absorption is greatest at a soil pH of 6 to 7 and during periods of active plant growth. The phosphorus content of soils varies greatly depending on the (a) parent material from which the soil was formed, (b) type of weathering, and (c) organic matter content. Fine textured soils usually have a higher phosphorus content than sandy soils when formed under comparable climatic conditions.

The phosphorus content is usually highest in the upper portion of the soil profile. Phosphorus is quite immobile in the soil and is not leached to any extent (52, 218) since it is immobilized almost immediately when applied to soils. The rapid immobilization of applied phosphorus fertilizers in soils avoids problems with phosphate pollution of rivers and lakes. The quantity of available phosphorus in the soil solution is usually quite small. The phosphate ion combines readily with iron and aluminum to form insoluble, unavailable combinations particularly under acid soil conditions. The phosphate ion is also absorbed on the surface of certain clay minerals such as kaolinite. Cultivation of established turfs by coring prior to an application of phosphate fertilizer assists in penetration of a portion of the phosphorus into the root zone rather than just at the soil surface (113).

The interaction of phosphorus and arsenic in soils is of particular significance in turfgrass weed control (169). The quantity of arsenic absorbed by the roots decreases as the available phosphorus concentration in the soil increases (45, 248, 250). Where inorganic arsenicals are utilized for controlling annual grassy weeds, excessively high available soil phosphorus levels should be avoided in order to achieve adequate weed control at a minimal arsenic level and acceptable cost. A high available soil phosphorus level requires a higher quantity of arsenic to achieve satisfactory weed control (Fig. 13-3) (148, 150, 151).

Figure 13-3. The effect of four rates of phosphorus (A—0; B—109; C—218; and D—436 lbs of phosphorus per acre) incorporated into the soil on the degree of annual bluegrass control using 440 lbs of calcium arsenate per acre. (Photo courtesy of Juska and Hanson—148)

Potassium

Potassium is not a constituent of living cells but is essential in plant growth and developmental processes. Potassium functions in (a) carbohydrate synthesis and translocation, (b) amino acid and protein synthesis, (c) catalyzing numerous enzymatic reactions including nitrate reduction, (d) regulating transpiration, (e) controlling the uptake rate of certain nutrients, and (f) regulating the respiration rate. A potassium deficiency increases the respiration rate causing a drain on the carbohydrate reserves. The transpiration rate of potassium deficient turfs is also higher.

Turfgrasses require potassium in relatively large amounts, second only to nitrogen (121). The potassium content is quite high in young, actively growing turfgrass plants but decreases rapidly as the plant reaches maturity (139). It is absorbed and stored in turfgrass tissues in much larger quantities than needed for normal growth and development, when large amounts of available potassium are present in the soil. This excessive absorption is called "luxury consumption." Potassium is prone to leaching from leaf tissues because it is retained in the ionic form.

Effects of potassium. Turfgrasses may not exhibit a visual response to potassium in terms of shoot color, density, or growth (60, 127). However, potassium does influence turfgrass (a) rooting; (b) drought, heat, and cold hardiness; (c) disease proneness; and (d) wear tolerance. The increased root development, particularly branching, at higher potassium levels contributes to improved drought tolerance (12, 124, 153, 190, 202, 234, 240, 246). Stolon and rhizome growth is also positively correlated with the level of potassium nutrition (27, 202, 234, 240).

Potassium regulates the absorption and retention of water by plants, which influences the heat, cold, and drought hardiness of turfgrasses. The degree of tissue hydration decreases as the potassium level increases (190). A potassium deficiency causes a distinct reduction in cold and heat hardiness (2, 10, 92, 233, 268). High potassium levels also cause increased leaf turgor pressure that reduces the wilting tendency.

Higher potassium levels reduce the incidence of *Helminthosporium* spp. (76, 126), brown patch (42, 60), *Ophiobolus* patch (102, 103, 104), *Fusarium* patch (102, 104), red thread (104), and dollar spot (192). Increased proneness to disease resulting from a potassium deficiency is associated with (a) an excessive accumulation of nitrogen and carbohydrates, which provides a favorable media for pathogen activity; (b) a thin, delicate cell wall structure, which is easily crushed during mowing operations and provides ideal penetration sites; (c) changes in the reaction and composition of the cell sap, which enhances pathogen activity; and (d) reduced plant vigor.

Turfgrass wear tolerance is also reported to increase proportionally with the potassium level. This has not been well documented. The thicker cell walls, increased plant vigor, higher cellulose content, and increased turgor pressure associated with higher potassium levels may contribute to improved wear tolerance. The wear tolerance around cups on putting greens is frequently increased by an application of potassium.

Soil relationships. Potassium is commonly absorbed as the K^+ ion. The potassium content of soils varies considerably but is generally larger than the amount of phosphorus or nitrogen normally found. Although the total potassium content of the soil may be high, only a small portion is normally available for uptake by the turfgrass plant. Potassium can be lost from the soil by leaching. Leaching is generally greater in sandy than in fine textured soils when intensely fertilized. The use of nitrogen carriers containing ammonia also increases potassium loss by leaching since the NH^+ ion readily replaces K^+ on the exchange sites of the soil colloids.

A balance tends to exist between the exchangeable and nonexchangeable potassium in the soil. When potassium is absorbed by the turfgrass root, the potassium in the soil solution decreases causing the exchangeable potassium to move into the soil solution. This results in an increased conversion rate of nonexchangeable potassium to exchangeable forms. Similarly, there is a reverse reaction when soluble potassium fertilizers are applied to the soil that results in an increase in the nonexchangeable potassium fraction. By this equilibrium mechanism a portion of the potassium is held in the soil in a form that is not readily lost by leaching.

Calcium

Calcium is a macronutrient found in relatively large quantities in turfgrass tissues, ranking third after nitrogen and potassium (121, 225). Calcium functions as a (a) vital constituent of cell walls, (b) specific nutrient requirement for meristem growth by cell division, and (c) neutralizing factor for potentially toxic substances within the cell. Calcium occurs primarily in the leaves and stems of turfgrass plants rather than in the seeds. It is a vital constituent of calcium pectate commonly found in the middle lamellae of cells. Calcium is permanently fixed in the cell walls, which results in high calcium levels in leaf tissues. This fixation also causes calcium to be relatively immobile within the tissues. It tends to accumulate in older leaves and other plant parts.

The influence of calcium on meristematic activity is indicated by the suppression of mitotic division when a calcium deficiency occurs. Nonlimiting calcium levels enhance root growth, particularly root hair development (124, 240). Calcium ions also exert a strong influence on the absorption of other ions by the turfgrass plant. Specifically, the uptake of potassium and magnesium is modified by the concentration of calcium ions. A calcium deficiency results in increased proneness to red thread (209) and *Pythium* blight (204).

Soil relationships. Calcium is more widely recognized for its effects on soil chemical and physical properties that influence turfgrass growth than as an essential macronutrient. Calcium is normally absorbed by the plant as Ca^{++}. The quantity present in soils varies greatly depending on the soil texture, degree of leaching, and parent material from which the soil developed. The calcium content is usually lower in older, coarse textured, intensely leached soils. The leaching of calcium from the surface zone causes the development of an acidic A-horizon in the soil profile and an accumulation of calcium at lower depths. Calcium has a strong influence on soil acidity and structure. Higher calcium levels cause increased flocculation and improved soil structure, which enhance soil aeration, water retention, and infiltration. The calcium level also influences the cation exchange capacity of clay minerals formed in certain soils.

Magnesium

Magnesium is a macronutrient that functions in several vital physiological processes within the plant such as (a) being a constituent of chlorophyll, (b) the translocation of phosphorus, and (c) cofactors of many plant enzyme systems. Magnesium is essential for the maintenance of green color and growth in turfgrasses because it is a vital constituent of the chlorophyll molecule (198). Unlike most cations essential for plant growth, magnesium is an actual constituent of living cells. Enzyme systems in which it serves as a cofactor include certain dehydrogenases, carboxylases, and transphosphorylases. Since magnesium is involved in phosphorus translocation within plants, it affects phosphorus utilization.

Magnesium ions are not required in large amounts because they are reutilized for growth processes. It is relatively mobile in the plant with redistribution occurring from older to younger tissues. Magnesium levels are normally highest in the leaves although accumulations may also occur in the growing tips of stems and roots as well as in seeds. A high magnesium concentration can be toxic to the turfgrass plant. This condition is most likely to occur on sodic soils.

Soil relationships. Magnesium is normally absorbed by the plant as the Mg^{++} ion. It does not occur in soils in as large a quantity as calcium and does not exert as great an influence on the soil physical and chemical properties. The magnesium content is usually higher in fine textured than coarse textured soils. The leaching tendency of magnesium results in a higher content in the subsoil and B-horizon than at the soil surface. Magnesium ions are absorbed by soil colloids but do not enhance aggregation as observed with calcium. High concentrations may cause a decrease in the degree of flocculation under certain conditions.

Sulfur

The role of sulfur is primarily as a constituent of certain essential amino acids such as cystine, cysteine, and methionine, which are required for protein synthesis. It is also a constituent of certain other important organic compounds in plants such as thiamine and biotin. A sulfur deficiency disrupts protein synthesis and impairs growth. Sulfur occurs in plants primarily in the form of proteins, sulfates, and certain volatile compounds. It is fairly evenly distributed throughout the plant. The quantity of sulfur removed in turfgrass clippings is similar to that for phosphorus (217, 225). A sulfur deficiency in Kentucky bluegrass increases powdery mildew disease development.

Soil relationships. Sulfur is taken up by turfgrass roots primarily in the SO^{--} ion, although some is absorbed by the leaves as gaseous sulfur dioxide. A considerable portion of the sulfur in soils is contained in organic matter and, thus, is concentrated in the surface horizons of the soil profile. The amount occurring in the sulfate form is quite small because of its high solubility. The sulfur content of soils is usually lower under conditions that accentuate organic matter decomposition and where leaching is more severe.

Micronutrients

The micronutrients iron, manganese, zinc, copper, molybdenum, boron, and chlorine are just as important as the macronutrients but are required by the plant in much smaller amounts. Most soils have adequate supplies of these micronutrients since the plant requirements are relatively small (20, 262). However, certain organic soils and intensely leached, sandy soils may be deficient in one or more of the micronutrients (8). Micronutrient deficiencies of turfgrasses are more likely to occur on soils modified to a high sand content and on marginal soils commonly being used for growing turf. Intense irrigation and compacted soils also increase the proneness to micronutrient deficiencies. Certain micronutrients can become toxic if present in the soil at higher concentrations (59, 224). Manganese, zinc, copper, and boron are most likely to produce toxic effects on turfgrasses at higher concentrations.

Iron. Iron is the micronutrient that is most commonly deficient in turf. An iron deficiency is usually a result of insolubility rather than an absence of the element in the soil. Deficiencies of iron are most common in soils that are (a) alkaline; (b) high in phosphate, manganese, zinc, or arsenic (268); (c) high in organic matter content; (d) waterlogged; or (e) excessively thatched. Plants can absorb iron as either the ferric (Fe^{+++}) or ferrous (Fe^{++}) ions. Iron is physiologically active only in the ferrous state.

The functions of iron include (a) chlorophyll synthesis and (b) a constituent of certain enzymes in the respiratory system. Iron is not a constituent of chlorophyll but is required for chlorophyll synthesis. Thus, turfgrass color is influenced by the level of iron available to the plant. Discoloration occurs when there is a deficiency (59). Centipedegrass is particularly prone to iron deficiencies (30, 224) (Fig. 13-4). Iron is a constituent of certain respiratory enzymes such as catalase, peroxidase, and cytochrome oxidase. The quantity of iron found in plant tissues is relatively small and is quite immobile within the plant. It can be inactivated in plant tissues by an excess of manganese, which acts as an oxidizing agent.

Figure 13-4. An iron deficiency symptom on leaves of centipedegrass in southern Florida.

Manganese. Manganese is required in very small concentrations by turfgrasses. Manganese availability in the soil is controlled to a great degree by the solubility. Acidic soil conditions or an anaerobic soil environment results in increased manganese availability (197). Thus, the available manganese content is normally highest in acidic, imperfectly drained, poorly aerated, and/or compacted soils. Alkaline conditions or intense leaching frequently results in a manganese deficiency (8, 197). A manganese deficiency usually occurs because it is in an unavailable form rather than being absent from the soil. It is usually absorbed by plants in the divalent form, as Mn^{++}.

Manganese functions (a) in chlorophyll synthesis and (b) as a cofactor in a number of plant enzyme systems. It is not a constituent of the chlorophyll molecule but is required for chlorophyll synthesis. Thus, a manganese deficiency results in discoloration of the turf (59). It is also a cofactor for certain enzyme systems including the dehydrogenases and carboxylases. The primary importance of manganese in plant physiological processes is associated with oxidation-reduction reactions. It is usually most abundant in actively growing tissues such as the leaves. An excessively high concentration of iron in turfgrass tissues can induce a manganese deficiency. The utilization of iron and manganese in the plant is closely interrelated.

Zinc. This element is essential to plants but only in minute quantities. Higher concentrations are very toxic to turfgrass plants, particularly rhizomes (59). The zinc level in soils is usually highest near the soil surface because it is immediately fixated by the soil when returned in plant residues. Zinc availability is reduced under alkaline soil conditions.

The actual role of zinc as an essential nutrient is not well documented. Functions of zinc that have been reported include (a) a constituent of certain plant enzymes such as carbonic anhydrase and (b) a cofactor in the synthesis of certain plant hormones and auxins such as indoleacetic acid. A deficiency usually results in the restriction of leaf development. This is probably associated with an inability of the plant to synthesize certain growth promoting substances. The zinc content of Kentucky bluegrass seeds is quite high.

Copper. Copper is highly toxic to plants except when it occurs in very dilute soil concentrations. Deficiencies are most common in highly alkaline organic soils or intensely leached sandy soils. Copper functions (a) as a component of several enzyme systems and (b) in the synthesis of certain plant growth promoting substances. It is a constituent of certain oxidation-reduction enzymes such as ascorbic acid oxidase. The copper content is usually highest in actively growing tissues. A copper deficiency that impairs plant growth hormone synthesis can cause death of the axillary buds in the apical meristem of turfgrasses.

Molybdenum. Molybdenum is required in an extremely small amount by plants and is usually absorbed as the molybdate ion. The molybdenum level is usually higher at the soil surface than lower in the soil profile. The availability of molybdenum is greater under alkaline soil conditions because of increased solubility. Its primary function in plants is as a cofactor to the enzyme system involved in nitrate reduction. A deficiency results in nitrate accumulation and impaired protein synthesis. The molybdenum concentration is usually highest in the leaf blade and other active growing plant parts. Plant tissues tend to accumulate molybdenum as they mature.

Boron. Boron is usually lowest in acidic soils that are intensely leached. The availability of boron is reduced under alkaline soil conditions. Relatively low soil concentrations of boron can be extremely toxic to plants, especially under acidic conditions (59, 229). High soil concentrations seldom occur except in certain arid regions. The exact function of boron in plant growth and development is not well understood. It is thought to function in calcium utilization. Turfgrass color is sometimes poorer when a boron deficiency occurs (59). It is relatively immobile within the plant with the highest concentrations normally found in the leaf tips (229).

Chlorine. Chlorine was the last element to be accepted as an essential micronutrient. The quantity of chlorine required by plants is relatively small but it is one of the most abundant anions found in plants. A deficiency is rarely observed. Chlorine is present in plant tissues as soluble, inorganic chlorides. The exact function of chlorine in plant growth is not yet clear but is thought to involve the regulation of osmotic pressure and cation balance within plant cells. No specific role in plant metabolism has been attributed to chlorine.

Nutrient Deficiency Symptoms

Most essential nutrients exhibit certain characteristic symptoms that can be utilized as indicators of deficiency problems. Foliar deficiency symptoms may or may not be easily diagnosed. Sometimes discoloration or an abnormal characteristic is caused by a disease or some other unrelated external effect rather than a nutrient deficiency. A combination of visual deficiency symptoms, tissue tests, and/or soil tests should be utilized in determining that a nutrient deficiency exists (228).

The deficiency symptoms of nitrogen, phosphorus, potassium, calcium, magnesium, and sulfur have been studied in great detail for Seaside creeping bentgrass, Merion Kentucky bluegrass, Pennlawn red fescue, and bermudagrass (164, 173, 226). Much of the information from these studies is utilized as the basis for the turfgrass deficiency symptoms discussed in the following sections.

Nitrogen. A nitrogen deficiency is characterized by a stunting of shoot growth, including decreased tillering and leaf length. Initial visual symptoms appear on the older leaves as a pale green color that changes to a yellow hue as the deficiency symptoms progress toward the base of the blade (20, 51, 226, 242). The shoot density decreases substantially if the deficiency persists, with individual plants being spindly and weak. A copper color eventually develops at the leaf tip of older leaves and progresses down the leaf in a horizontal pattern. The first deficiency symptoms in bermudagrass are reduced shoot growth followed by yellowing and increased seed head formation (198). A slight purplish coloration develops as the deficiency progresses, followed by necrosis of the leaf blades (226).

Phosphorus. Visual symptoms of a phosphorus deficiency first appear as a dark green coloration of the lower, older leaves (20, 110). The plants tend to be spindly and dwarfed but not to the degree associated with a nitrogen deficiency (138, 240).

The dark green coloration changes to a dull blue-green as the phosphorus deficiency progresses, with a purple discoloration appearing along the entire margin of the blade as well as in the basal main veins (20, 51, 240). Merion Kentucky bluegrass does not exhibit the dull blue-green to purplish stage just described for Seaside creeping bentgrass and Pennlawn red fescue. Subsequently, the purple discoloration gives way to dull reddish tints appearing first near the leaf tip and then developing down the blade. Necrosis of the older leaf tips occurs in the advanced stages of a phosphorus deficiency and gradually moves to the base of the blade. In bermudagrass the deficiency initially appears as a dark green color that gradually changes to a pale green color with minimal effect on the shoot growth rate (198).

Potassium. Potassium deficiency symptoms for Seaside creeping bentgrass and Pennlawn red fescue initially appear as a drooping of the leaves, which also feel soft when touched (240). The blades tend to be horizontally inclined. An excessive degree of tillering is usually noted. Moderate yellowing develops in the interveinal areas, especially at the tips of older leaves (20, 51, 240). This is followed by rolling and withering of the leaf tips, which retain blotches of green coloring. As the deficiency progresses, the yellowing extends to the midvein, which remains green even though the leaf margins become scorched and the tips severely withered (20) (Fig. 13-5). Symptoms for Merion Kentucky bluegrass are similar except for the early loss of chlorophyll in the leaf tips and delayed withering of the tip and marginal scorching of the blade. Initial symptoms on bermudagrass are a distinct thinning of shoots followed by necrosis of the older leaf tips (198). Bermudagrass and zoysiagrass exhibit dwarfed growth and a brown color if the deficiency persists (268).

Calcium. The deficiency symptoms for Seaside creeping bentgrass, Merion Kentucky bluegrass, and Pennlawn red fescue vary depending on the age of individual plants. The first symptoms appear in younger leaves as a reddish-brown discoloration developing along the margins and gradually extending to the midvein (100). The first evidence of a calcium deficiency in older plants is the development of a reddish-brown discoloration of the interveinal leaf tissue. The color then fades to lighter shades of red, predominately a rose red, and the tips become withered. A deficiency in bermudagrass is characterized by a growth reduction followed by necrosis of the younger leaves (198). The stolons are quite stunted (226).

Magnesium. The deficiency symptoms for magnesium resemble those of calcium. However, visual symptoms usually appear first in the older, lower leaves with the initial discoloration being more of a cherry red color (240). Also, the coloring in 30 to 50% of the affected

Figure 13-5. A potassium deficiency (left) exhibited by a velvet bentgrass turf cut at 0.25 in. (Photo courtesy of C. R. Skogley, University of Rhode Island, Kingston, R. I.).

leaves is blotchy and results in a banded appearance that is not observed in calcium deficient turfgrass plants. Necrosis eventually develops. Bermudagrass exhibits a pale green color and reduced shoot growth (198). The older leaves turn completely yellow as the deficiency progresses and become necrotic (226). Deficiency symptoms appear at tissue concentrations below 0.4% magnesium (226).

Sulfur. A sulfur deficiency in Seaside creeping bentgrass and Pennlawn red fescue involves an initial paling of the older, lower leaves similar to a nitrogen deficiency. The blades take on a pale, yellow-green appearance as the deficiency progresses (240). Associated with this is the appearance of a faint scorching of the leaf tip that advances toward the base of the blade in a thin line along each margin. This scorching of the margin usually enlarges until the entire leaf blade is affected and withers. In the case of Merion Kentucky bluegrass, the veins do not become chlorotic, giving the leaf a striped appearance. Eventually, however, the midvein becomes chlorotic and the entire blade withers.

Iron. Initially, an iron deficiency appears as an interveinal yellowing of the youngest, actively growing leaves (164). Iron deficiency symptoms are quite similar to the symptoms for nitrogen except that it first appears on the younger leaves (226). The chlorosis eventually spreads to the older leaves if the deficiency persists. The plants also become spindly and weak. The blades become nearly white or ivory in color in the advanced stages. Necrosis is minimal. Chlorosis commonly appears when the iron content of the tissue is below 50 ppm (226).

Manganese. Initially, a manganese deficiency results in an interveinal yellowing of the leaves similar to an iron deficiency (164). However, small, distinct necrotic spots soon develop on the affected leaves. The veins tend to remain green and the leaves droop. Yellowing develops initially at some distance from the leaf tip, which tends to remain green for a period of time. The entire leaf discolors and eventually withers or rolls if the deficiency persists.

Zinc. The first symptom is a stunting of growth (164). The leaves are thin and tend to shrivel with the younger leaves being affected first. At the same time the leaves darken and become desiccated. In the case of bermudagrass, a white speckled appearance develops on the leaves that is actually a crystalline exudate from the stomatal openings.

Copper. A copper deficiency usually involves a bluish discoloration at the tips of the youngest, actively growing leaves (164). A continued deficiency results in death of the leaf tips and progresses toward the base.

Molybdenum. A molybdenum deficiency results in a paling or chlorosis that first appears on the older, lower leaves (164). The mottled yellowing occurs primarily in the interveinal areas with necrosis and withering eventually developing. Stunting of the plants occurs if the deficiency persists.

Boron. Visual symptoms of a boron deficiency do not appear for some time after the deficiency occurs and are first evident in the meristematic tissues of the plant. There is a discoloration of turfgrass shoots (59) and a stunting of the growing point (164). The leaves are stubby which results in a rosette appearance. Chlorotic streaks then develop in the interveinal area of the leaf. Bermudagrass exhibits a boron deficiency, but deficiency symptoms seldom appear on Kentucky bluegrass and bentgrass (164).

Chlorine. No visual deficiency symptoms are delineated for turfgrasses.

Fertilizers

Turfgrass fertilizers include all materials containing one or more essential plant nutrients that are added to the soil to increase the supply of available plant nutrients for the purpose of maintaining satisfactory turfgrass quality. Nitrogen, phosphorus, and potassium are the primary macronutrients contained in the commonly used turfgrass fertilizers since these three are required in the largest quantity.

The percentage composition of a fertilizer is referred to as the **analysis**. In the past the fertilizer analysis has been expressed as the percent N, P_2O_5, and K_2O. More recently the analysis description of phosphorus and potassium percentages are also being expressed on an elemental basis as is done with the micronutrients and nitrogen. The **grade** of a fertilizer is the minimum guaranteed analysis of plant nutrient content.

Fertilizer materials whose principal constituent is nitrogen, phosphorus, or potassium are commonly referred to as nitrogen, phosphorus, or potassium carriers, respectively. These three types of fertilizer carriers are applied for the purpose of supplementing one particular nutrient that is deficient.

Nitrogen Carriers

Nitrogen carriers can be subdivided into three groups on the basis of whether they are of an inorganic or organic source and whether they are from natural origin or synthesized utilizing nitrogen fixed from the atmosphere. The three classifications are (a) synthetic inorganic nitrogen carriers, (b) natural organic nitrogen carriers, and (c) synthetic organic nitrogen carriers.

Synthetic inorganic nitrogen carriers. This group of nitrogen carriers contains nitrogen obtained from the air and combined with other substances by chemical methods. A list of commonly used synthetic inorganic nitrogen carriers is presented in Table 13-2. Characteristics associated with the synthetic inorganic nitrogen carriers include (a) high water solubility, (b) rapid initial plant response (11, 62, 93, 165, 191, 230, 284), (c) nitrogen availability has minimum temperature dependence, (d) high foliar burn potential (141, 203), (e) limited residual response of 4 to 6 weeks (62, 65, 93, 183, 184, 208, 211, 212, 284), (f) subject to loss by leaching if in the nitrate form (146), (g) can be dissolved in water for application as a soil drench or foliar spray, (h) rapid cold temperature plant growth response (62, 65, 191, 212), and (i) low cost per unit of nitrogen. The turfgrass responses to the four synthetic inorganic nitrogen carriers discussed in this section are quite similar (292).

Nitrate-containing synthetic inorganic carriers release nitrogen to the soil solution in the form of nitrate. Nitrates are readily absorbed and utilized by turfgrasses

Table 13-2

CHARACTERISTICS OF THE PRIMARY SYNTHETIC INORGANIC NITROGEN CARRIERS UTILIZED
FOR TURFGRASS FERTILIZATION

Carrier	Chemical formula	Nitrogen content, %	Comments
Ammonium sulfate	$(NH_4)_2SO_4$	20.5	Increases soil acidity; contains 24% sulfur
Ammonium nitrate	NH_4NO_3	33	Should be conditioned to resist moisture absorption
Sodium nitrate	$NaNO_3$	16	Occurs as a natural inorganic
Calcium nitrate	$Ca(NO_3)_2$	15.5	Absorbs moisture very rapidly
Ammonium chloride	NH_4Cl	26	Leaves chloride residue

but are also easily leached from the soil. Leaching is caused by the high water solubility of nitrates combined with minimal fixation on the soil colloids.

Most soil nitrogen that is in the ammoniacal form is converted to nitrates by nitrification before being absorbed by the turfgrass plant. Ammonia is water soluble but is not leached as readily as nitrate because a substantial portion of the ammonia is absorbed and held by soil colloids. Some ammonia is lost by volatilization, particularly from dry or alkaline soils (160, 256, 272). Nitrogen losses by volatilization can be substantially reduced by watering immediately after the nitrogen carrier is applied (160). Ammonium carriers are not as prone to leaching as nitrates, particularly on sandy soils (29).

Ammonium nitrate is widely used as a nitrogen carrier for turfgrass fertilization. This ammonium salt of nitric acid contains 50% of the nitrogen in the nitrate form and 50% in the ammoniacal form. The high water solubility results in a very rapid plant response and a high foliar burn potential. It is slightly acidifying (14, 128, 130, 165) and has a substantially lower rate of gaseous ammonia loss than the other water-soluble nitrogen carriers listed (112, 272). Ammonium nitrate tends to harden and cake due to its hygroscopic properties unless properly conditioned during the manufacturing process. There is also a substantial fire and explosion hazard.

Ammonium sulfate is also widely used for turfgrass fertilization. Because of the strong acidifying properties (67, 74, 85, 109, 135, 263, 266, 283, 292), it is the preferred nitrogen carrier for soils having an excessively high pH. Continued use on soils with pH's of 7 to 8 lowers the soil reaction into a more favorable range. The use of ammonium sulfate carriers can also cause the soil to become too acidic. Periodic liming may be required to raise the soil pH into the favorable range. Thus, the pH should be checked regularly on soils where nitrogen fertilization involves the use of ammonium sulfate. The foliar burn potential of ammonium sulfate is greater than for the other commonly used synthetic inorganic nitrogen carriers. Leaching of the exchangeable ammonium ion is limited until changed to the nitrate form. The moisture absorption problem of ammonium sulfate is minimal if stored in a dry location.

Sodium nitrate and calcium nitrate provide a basic soil reaction (14, 109, 266, 283) but are seldom used as nitrogen carriers for turfgrass fertilization. Both are considered fire hazards. Sodium nitrate, also known as nitrate of soda, occurs as a natural inorganic salt deposit in Chile. It is absorbed very rapidly by the plant (194). Continued use of sodium nitrate can result in high sodium levels that cause deflocculation of soil colloids and loss of soil structure. Calcium nitrate is also known as nitrate of lime. A high affinity for water necessitates the shipment and handling of calcium nitrate in airtight containers. The foliar burn potential is very high.

Natural organic nitrogen carriers. The nitrogen in natural organic carriers is contained in complex organic compounds that are not readily soluble in water. Nitrogen release for turfgrass utilization depends on (a) microorganisms for decomposition of the organic compounds or (b) soil weathering through time. Decomposition is most rapid at temperatures favorable for microorganism activity and is quite limited at soil temperatures below 55°F. The dependency on microorganism decomposition also results in a much slower nitrogen release rate and a longer residual response. The nitrogen release rate by biological mineralization is more rapid under acidic soil conditions (26). The biological processes initially convert the nitrogen into ammonia, which is then transformed into nitrates in the soil solution.

Characteristics commonly associated with the natural organic nitrogen carriers include (a) a medium slow initial release rate (165), (b) low water solubility, (c) minimum foliar burn potential (141, 203), (d) higher cost per unit of nitrogen, (e) reduced loss by leaching (15, 146), (f) lower nitrogen analysis, and (g) longer residual period of 4 to 8 weeks (62, 93, 96, 165, 284). The low water solubility of the natural organics results in reduced leaching losses, a longer residual response, and a minimum chance of foliar burn during application. The duration of the residual response is intermediate between the water-soluble synthetic inorganics and the ureaformaldehydes (65, 93, 212, 267). However, certain natural organics can release nitrogen at a fairly rapid rate under high soil temperature conditions (232). A portion of the nitrogen contained in some natural organic carriers such as activated sewage sludge is retained in unavailable or slowly released forms. The immediate cold temperature response from natural organic carriers is slow because nitrogen release is dependent on warmer soil temperatures (29, 62, 65, 135, 206, 212, 258). This is a disadvantage where an early spring turfgrass response is desirable but is also quite beneficial during the autumn low temperature hardening process where limited shoot growth is preferred.

The natural organic nitrogen carriers can be grouped according to their origin into those that are by-products of (a) animals from meat and fish processing industries and (b) plant origin from the vegetable oil industries. Many natural organics contain significant quantities of phosphorus and/or potassium as well as nitrogen. The more commonly available natural organic nitrogen carriers are listed in Table 13-3.

Table 13-3

Approximate Nutrient Content of Eleven Natural Organic Nitrogen Carriers

Carrier	Content, % **			Comments
	N	P$_2$O$_5$	K$_2$O	
Sewage sludge, activated	4–7	4–6	0.4–0.7	Good physical characteristics
Sewage sludge, digested	1.5–3	1–2.5	T*	May not be in an acceptable physical state
Blood, dried	12–14	0.2–2	0.4–0.8	Fairly rapid nitrogen release
Tankage, animal	7–10	2–6	0.4–0.6	Refuse from slaughter houses after fat extraction
Bone meal, raw	3–5	15–27	—	May be steamed or raw
Fish meal, dried	7–12	2–6	T*	Ground remains after steaming and oil extraction
Hoof and horn meal	10–15	0.7–1	T*	Ground dried hoofs and horns
Cottonseed meal	5–9	2–3	1–2.5	Produced by grinding cottonseed press cakes
Castor pomace	4–6	2–3	0.5–2	Residue left after oil extraction
Soybean meal	6–8	1–2	1.5–3	Ground residue after oil extraction
Corn gluten meal	7	0.4	T*	By-product of starch extraction

* T—trace.

** These are average ranges for typical analyses. Much wider variations in nutrient content occur.

Activated sewage sludge is one of the more common natural organic nitrogen carriers used for turfgrass fertilization (215). Activated sewage products are made from sewage freed from grit and coarse solids by aerating in tanks after being innoculated with microorganisms. The resulting flocculated organic matter is withdrawn from the tanks, filtered with or without the aid of coagulants, dried in rotary kilns, ground, and screened. The preferred activated sewage sludges are steam sterilized by drying in hot ovens to kill weed seeds and harmful organisms. Activated sewage sludge is being produced and marketed by such cities as Milwaukee, Wisconsin, as Milorganite®, and Houston, Texas, as Hu-Actinite®. Milorganite is an effective (279) and widely used activated sewage sludge that also contributes significant amounts of micronutrients, such as manganese, zinc, copper, molybdenum, and boron, to support plant growth (191, 192, 239). An interesting effect on disease proneness is associated with the use of activated sewage sludge. The incidence of dollar spot on bentgrass and *Pythium* blight on ryegrass decreases when activated sewage sludge is used as the nitrogen carrier (38, 39, 192, 245, 282). The exact mechanism involved in this response is not understood. Activated sewage sludge has a minimal effect on the soil reaction.

Digested sewage sludge contains a much lower nutrient content and may or may not be in a physical state that is acceptable for effective application to turfs (82). Supplemental nitrogen, phosphorus, and potassium applications may be necessary to achieve the maximum response when digested sewage sludges are used (207). These materials may also contain viable weed seeds.

Animal by-products such as dried blood, animal tankage, bone meal, fish meal, and hoof and horn meal can be used as natural organic nitrogen fertilizers (278, 284). Odor can be a problem with certain of the tankages and meals, particularly those originating from fish (278). The nitrogen release rate of dried blood is quite rapid compared to most natural organic nitrogen carriers. Dried blood is slightly acidifying, while bone meal decreases soil acidity. Fish meal and animal tankage have a minimum effect on the soil reaction.

Seed meals such as cottonseed, castor pomace, and soybean meal are not widely utilized turfgrass nitrogen carriers. They have a slightly acidifying effect on the soil (283). Castor pomace is hazardous to store and may cause a skin rash or intense itching to the user. A natural organic nitrogen carrier obtained from plants that has been effectively used in turfgrass fertilization is corn gluten (96). It is a by-product of carbohydrate extraction from corn and contains 8% nitrogen. The overwintering characteristics of turfs receiving an autumn corn gluten application are quite good and similar to the response from Milorganite.

Synthetic organic nitrogen carriers. This group of nitrogen carriers can be subdivided into those that are (a) primarily water-soluble and (b) primarily water-insoluble. The water-soluble synthetic organic nitrogen carriers include urea [$CO(NH_2)_2$], which usually contains 45% nitrogen, and calcium cyanamide ($CaCN_2$), with 22% nitrogen. These two water-soluble synthetic organics have most of the characteristics typically associated with the synthetic inorganics (65). Urea and calcium cyanamide have (a) high water solubility, (b) rapid initial plant response (82, 93, 241), (c) relatively short residual response (93, 96), (d) tendency to leach (146), and (e) high foliar burn potential (93, 94, 96, 141, 291). They also have a higher rate of gaseous ammonia loss compared to most water-soluble nitrogen carriers, particularly when applied at high rates (11, 112, 160, 208, 256, 272, 273, 274).

Urea is eventually transformed to ammonia in moist soils, and thus is less prone to loss by leaching. The acidifying effect of urea is less than ammonium nitrate or ammonium sulfate (14, 100, 263, 266, 271, 283). Calcium cyanamide contains 39% calcium, which decreases soil acidity. Calcium cyanamide is transformed to urea under acidic, moist soil conditions and to dicyanodiamide under alkaline conditions. The foliar burn probability is quite high. Thus, calcium cyanamide is usually (a) applied during the winter when the turfgrasses are dormant or (b) incorporated into the soil prior to establishment (96). It tends to absorb moisture and to cake if not stored in closed containers.

Ureaformaldehydes. The primary nitrogen carrier in the water-insoluble synthetic organic nitrogen group is ureaformaldehyde (116). Ureaformaldehyde (UF), a reaction product of urea and formaldehyde, usually contains 38% nitrogen, which is largely in an insoluble, slowly available form (230, 290). UF's vary in the degree of availability of the insoluble nitrogen fraction (64). The nitrogen contained in UF can be separated into three classes based on solubility: (a) cold-water-soluble (CWSN), (b) cold-water-insoluble (CWIN), and (c) hot-water-insoluble (HWIN). The cold-water-soluble (77°F) fraction is immediately available for absorption by turfgrasses. It is desirable to have 25% of the total nitrogen content in the CWSN class. The total of the CWSN and CWIN fractions has been assumed to be available for use by the turfgrass plant during a single growing season. This may or may not occur. The percentage of CWIN minus HWIN contained in the CWIN fraction of ureaformaldehyde is called the **activity index** (AI). The AI indicates the rate at which the CWIN nitrifies in the soil to nitrogen forms that are available for plant use. A satisfactory UF nitrogen carrier should have a minimum AI of 40 and preferably higher (65, 93).

The characteristics of the UF's are quite similar to those of the natural organics. They are (a) an intermediate initial release rate, (b) low foliar burn potential (49, 141, 180, 206), (c) medium low water solubility, (d) reduced loss by leaching (117, 146), (e) high cost per unit of nitrogen, and (f) long residual response. The initial response from UF's is usually more rapid than from the natural organics but slower than the synthetic inorganics, particularly at cold early spring soil temperatures (5, 6, 11, 61, 65, 72, 93, 161, 206, 258, 288, 289). This is due to the CWIN fraction. The residual response is from 6 to 12 weeks (5, 6, 11, 48, 61, 75, 78, 91, 93, 94, 95, 96, 172, 177, 180, 206, 211, 212, 213, 267, 286, 288). Longer residual responses have been reported when a very high rate of UF is applied (289).

Nitrogen release from UF is dependent on the hydrolytic enzyme activity of soil microorganisms (84). Nitrogen liberation is a function of the soil microorganism population as affected by the soil temperature, pH, and nutrient level (159). Nitrogen release from UF is most rapid at soil temperatures of 90°F and is quite slow below 50°F. The release rate is also enhanced by a soil pH of 6.1 and the presence of phosphorus and potassium.

Effective use of the UF's is dependent on the buildup of relatively large reserves of insoluble nitrogen in the soil (170). A single application of a UF at a high rate is usually not sufficient to maintain an adequate level of turfgrass quality through-out the growing season (39, 74, 75, 91, 93, 96, 117, 133, 172). Exceptions have been reported when very high rates of application are used (65, 212, 289).

The proportion of the applied nitrogen recovered by the turfgrass plant is much lower for UF's than the other nitrogen carriers commonly used on turfs (39, 49, 56, 65, 75, 93, 154, 157, 161, 191, 286). Nitroform® and Ureaform® are two common ureaformaldehyde nitrogen carriers sold for turfgrass fertilization. Finely ground UF carriers release nitrogen more rapidly than granular forms but with the same overall release pattern.

Phosphate Carriers

Phosphate carriers are not as widely used for turfgrass fertilization as the nitro-gen carriers. There are certain conditions, however, that require supplemental fertilization with a phosphate carrier, particularly during turfgrass establishment on soils where turfs have not been grown previously. The phosphate carriers can be subdivided into (a) natural phosphates, (b) organic phosphates, (c) by-product phosphates, and (d) chemical phosphates (65). A list of the principal phosphate carriers is given in Table 13-4.

Natural phosphates. The natural phosphates may include treated or untreated types. **Superphosphate** is a treated natural phosphate carrier. It is the most important source of readily available phosphorus for turfgrass fertilization (74). The mono-

Table 13-4

CHARACTERISTICS OF NINE PHOSPHORUS CARRIERS AVAILABLE FOR TURFGRASS FERTILIZATION

Phosphate carrier	P content, %	P_2O_5 content, %	Comments
Superphosphate, ordinary	7–9.5	15–22	Decreases acidity
Treble superphosphate	16–23	37–53	Decreases acidity
Calcium metaphosphate	26–27	62–65	Composed mainly of $Ca(PO_3)_2$
Rock phosphate	9–15	21–34	Do not use on soils that are alkaline or low in organic matter
Bone meal, raw	6–11	15–27	Contains 3–5% nitrogen
Bone meal, steamed	8–15	18–34	Contains 1–4% nitrogen
Basic slag	4–8	10–18	Should be finely ground; contains 32% calcium
Monoammonium phosphate	18–22	41–52	Contains 11% nitrogen
Diammonium phosphate	20–23	46–54	Contains 18–21% nitrogen

calcium phosphate $[Ca(H_2PO_4)_2]$ contained in superphosphate is readily available for utilization by the turfgrass plant. In addition to containing approximately 20% P_2O_5, ordinary superphosphate contains substantial amounts of soluble calcium and sulfur. The calcium sulfate $(CaSO_4)$ acts as a dehydrating agent that improves the physical condition of superphosphate.

Treble superphosphate differs from ordinary superphosphate in that it contains very little calcium sulfate. This higher analysis superphosphate is just as effective as ordinary superphosphate and is widely used in compounding high analysis fertilizers.

Calcium metaphosphate is an effective source of available phosphorus for turf-grass plant use under acid soil conditions. The cost is relatively low. Usually available in a granular form.

Rock phosphate is a natural phosphate mined from large deposits, ground, and separated from major impurities. The phosphorus in rock phosphate is relatively insoluble. In order for significant quantities of phosphate to be available to the turf it must be (a) finely ground, (b) thoroughly mixed into the soil, and (c) used on acidic soils that contain a high level of decaying organic matter. The use of rock phosphate for turfgrass fertilization is not practiced to any extent.

Natural organic phosphate carriers. Bone meals are the most common natural organic phosphate carriers commercially available. Included in this group are raw bone meal, steamed bone meal, and precipitated bone meal. Steamed bone meal is more readily available than raw bone meal because it is ground finer and has the fat removed during the steaming process. Steamed bone meal is the most common type available commercially. Phosphate release is dependent on decomposition of the organic materials with which the phosphorus is combined. The bone meals are a relatively expensive source of phosphorus compared to the superphosphates and are most effective on acid soils. They tend to decrease soil acidity.

By-product phosphates. Basic slag is the most common example of a by-product phosphate and ranks second to the superphosphates as a major source of phosphorus, particularly in Europe. It is a by-product of steel manufacturing. The degree of phosphorus availability depends on the fineness with which the slag is ground. Basic slag has a relatively long residual response, decreases soil acidity, and contains some magnesium and manganese (74).

Chemical phosphates. Typical examples of chemical phosphates are ammoniated superphosphate, potassium phosphate, and potassium metaphosphate. Ammonium superphosphate is obtained by treating superphosphate with anhydrous or aqueous ammonia. The added ammonia tends to neutralize any excess acidity in the super-phosphate and increases the nitrogen content of the carrier without appreciably increasing the bulkiness. The ammoniated superphosphates are utilized quite commonly in high analysis fertilizers and have an acidifying effect on the soil (266, 283).

Potassium Carriers

Potassium carriers are utilized more than phosphorus carriers, but not nearly as much as the nitrogen carriers, in turfgrass fertilization. Included within this group are (a) potassium chloride, (b) potassium sulfate, (c) potassium nitrate, and

Table 13-5

CHARACTERISTICS OF SEVEN POTASSIUM CARRIERS AVAILABLE FOR TURFGRASS FERTILIZATION

Potassium carrier	Chemical formula	K content, %	K_2O content, %
Potassium carbonate	K_2CO_3	44–58	53–70
Potassium chloride	KCl	48–52	58–62
Potassium sulfate	K_2SO_4	40–44	48–53
Potassium nitrate	KNO_3	37	44
Potassium metaphosphate	KPO_3	33	40
Potassium magnesium sulfate	$K_2SO_4 \cdot 2MgSO_4$	18–22	22–26
Sodium–potassium nitrate	$NaNO_3 \cdot KNO_3$	12	14

(d) potassium metaphosphate (24, 37) (Table 13-5). All 7 potassium carriers are soluble in water.

Potassium chloride is commonly known as muriate of potash. It is widely used and is available at a relatively low cost. The salt index is very high and it contains 47% chlorine.

Potassium sulfate is a good potassium carrier for turfgrass fertilization (129, 134, 136). This is partially attributed to the sulfur component present in potassium sulfate. It is more expensive than potassium chloride but has a much lower salt index. It can contain no more than 2.5% chlorine.

Potassium nitrate contains a minimum of 13% nitrogen. Specific characteristics include high water solubility, basic soil reaction, and minimal hygroscopic action. Continued use can result in soil deflocculation similar to that caused by sodium nitrate. Potassium nitrate is generally inferior to potassium sulfate and potassium chloride as a potassium source (134, 136). There is a fire hazard associated with potassium nitrate.

Other potassium carriers that are infrequently used for turfgrass fertilization include (a) potassium metaphosphate which contains 24% phosphorus (55% P_2O_5), (b) potassium magnesium sulfate, which contains significant amounts of magnesium, and (c) sodium-potassium nitrate, which contains 15% nitrogen. The latter two are free from chloride. Potassium magnesium carbonate has a neutralizing effect on soil acidity.

Other Essential Nutrient Carriers

Essential nutrient carriers that are used much less frequently than the nitrogen, phosphorus, and potassium carriers just discussed include sulfur, calcium, magnesium, and the seven essential micronutrients. These ten elements are applied as individual carriers much less frequently because (a) they are required by the turfgrass plant in much smaller amounts and (b) the existing quantities of these nutrients contained in the soil are usually sufficient to meet the requirements of the turfgrass plant.

Calcium. Calcium is the exchangeable cation that is usually found in the largest amounts in soils, particularly those having a neutral soil reaction. It occurs in the soil primarily as carbonates, as complex calcium silicates, and in organic matter. The amount of calcium present in most soils is so large that a deficiency rarely occurs except in intensely leached soils. Soils having a low pH are more likely to have a calcium deficiency.

Calcium is usually applied in the form of lime as calcium oxide, calcium hydroxide, or calcium carbonate (Table 13-6). The most common liming materials are calcitic or dolomitic limestone that has been finely ground. Significant quantities of calcium are also applied with such fertilizer carriers as superphosphate, gypsum, calcium cyanamide, and calcium nitrate. Calcium-containing carriers that are sometimes applied include marl, chalk, coral, and ground shells. Many fertilizers also contain a substantial quantity of calcium.

Table 13-6

APPROXIMATE CALCIUM CONTENT OF ELEVEN LIMING MATERIALS
AND FERTILIZER CARRIERS

Calcium carrier	Approximate calcium content, %
Calcium oxide	53*
Calcitic limestone	32*
Dolomitic limestone	22*
Gypsum	19–23
Calcium cyanamide	38
Calcium metaphosphate	19
Calcium nitrate	19
Raw bone meal	22
Rock phosphate	33
Superphosphate, ordinary	18–21
Treble superphosphate	12–14

* Can vary greatly with the source of the material.

Magnesium. Magnesium is second to calcium as the most abundant exchangeable cation on the soil complex at soil reactions near neutral. Magnesium deficiencies are most commonly observed on acidic, sandy soils that are intensely leached. A high calcium content can also induce a magnesium deficiency in turfgrasses.

The most common magnesium carriers include (a) dolomitic limestone, (b) potassium magnesium sulfate, (c) magnesium sulfate, and (d) magnesite (Table 13-7). Liming with a dolomitic-type limestone eventually corrects a magnesium deficiency. The magnesium contained in limestone is not water soluble and becomes available relatively slowly in the soil. Calcitic limestones also contain smaller, varying amounts of magnesium. For a more immediate response, the application of such magnesium carriers as magnesite, magnesium sulfate, and potassium magnesium sulfate readily corrects a magnesium deficiency. A small but significant amount of magnesium also occurs in phosphate carriers such as superphosphate (0.3% Mg) and potassium sulfate (0.6% Mg) as well as in the nitrogen carrier,

Table 13-7
APPROXIMATE MAGNESIUM CONTENT OF FIVE LIMING MATERIALS
AND FERTILIZER CARRIERS

Magnesium carrier	Approximate magnesium content, %
Magnesite ($MgCO_3$)	27.2
Dolomitic limestone	11.8
Potassium magnesium sulfate	11.1
Magnesium sulfate ($MgSO_4$)	9.7
Basic slag, open hearth	3.4

calcium nitrate (1.5% Mg). Many mixed fertilizers also contain some magnesium that may or may not be indicated on the label. The quantity of magnesium contained in mixed fertilizers is small but significant in view of the relatively small amount required by the turfgrass plant.

Sulfur. Most sulfur compounds that occur in the soil are fairly water soluble and, thus, are subject to leaching. Considerable quantities of sulfur are also absorbed by the turfgrass plant and incorporated into organic compounds such as proteins. Thus, supplementing the sulfur level of the soil is necessary and occurs in many ways. A significant amount of sulfur is added to the soil through the absorption and removal of sulfur gases from the atmosphere by rainwater. The quantity of sulfur brought to earth from the air is quite variable but is usually greatest adjacent to urban areas having a high fuel consumption. A certain amount of gaseous sulfur dioxide is absorbed directly from the atmosphere by the turfgrass plant and the soil itself.

Sizable amounts of sulfur occur in a number of commonly used turfgrass fertilizers. Fertilizing with such common carriers as ammonium sulfate, potassium sulfate, ordinary superphosphate, or potassium magnesium sulfate results in the addition of considerable quantities of sulfur to the soil (Table 13-8). Mixed fertilizers of low analysis also contain substantial quantities of sulfur. Continued use of high analysis mixed fertilizers containing small quantities of sulfur may necessitate supplemental fertilization with a sulfur carrier. The use of sulfur-containing

Table 13-8
APPROXIMATE SULFUR CONTENT OF EIGHT FERTILIZER CARRIERS

Sulfur carrier	Approximate sulfur content, %
Sulfur, elemental	99
Ammonium sulfate	24
Ferrous sulfate	18.8
Gypsum	18.6
Potassium magnesium sulfate	18
Potassium sulfate	17.6
Ferrous ammonium sulfate	16
Superphosphate, ordinary	11.6

fertilizers not only supplements the sulfur level in the soil but also functions in lowering the pH of moderately alkaline soils.

Sulfur may also be applied to the soil in the elemental form that normally contains 85 to 99% sulfur. Elemental sulfur undergoes oxidation by microorganisms when applied to the soil and is transformed into sulfuric acid within 10 to 15 days. The response to elemental sulfur is most rapid if incorporated into the soil by cultivation. The foliar burn potential of elemental sulfur is high if applied to turfs. This necessitates immediate watering following application to wash the sulfur off the turfgrass leaves (58).

Gypsum ($CaSO_4 \cdot 2H_2O$) is a sulfur-containing mineral that decomposes slowly in the soil. It is not very soluble in water but is more soluble than limestone. Gypsum applications should be thoroughly mixed into the soil for best results.

Iron. Iron is the only essential micronutrient commonly applied to turfs as an individual carrier. Iron fertilization is most common on intensively cultured turfs such as bentgrass and annual bluegrass. Iron deficiencies are more common in soils having high calcium and magnesium contents. The application of an acidifying fertilizer frequently corrects an iron deficiency caused by alkaline conditions.

Carriers utilized to correct an iron deficiency include (a) ferrous sulfate (20% Fe), (b) ferrous ammonium sulfate, (c) ferrous oxalate (30% Fe), and (d) chelated iron. The first three are water-soluble iron carriers that correct the immediate iron deficiency within the plant. There is no long term effect, however, because they are (a) quite water-soluble and subject to loss by leaching or (b) converted to insoluble, unavailable forms, particularly under alkaline conditions.

Figure 13-6. The correction of an iron deficiency on creeping bentgrass within 10 hours after a foliar fertilization of iron sulfate. Note chlorosis of area which was covered with a cloth during the foliar application. (Photo courtesy of Milwaukee Sewerage Commission, Milwaukee, Wisconsin).

Iron sulfate ($FeSO_4 \cdot 7H_2O$) is frequently used to correct an iron deficiency by means of a foliar application to the turf (268) (Fig. 13-6). It is normally applied at a rate of 2 to 3 oz per 1000 sq ft. Iron sulfate should be mixed just before application since it oxidizes quite rapidly to an unavailable state. A disadvantage of iron sulfate is the brownish discoloration that results if it is accidently applied to walks or buildings adjacent to turfgrass areas.

Iron can be combined with an organic compound to form a stable carrier called chelated iron. It is not easily broken down in the soil and thus is less prone to loss by leaching. The iron on the stable organic complex is readily exchanged with cations on the root surface and absorbed by the turfgrass plant. Chelated iron carriers have a longer residual response in the soil than the water soluble carriers such as iron sulfate. Common chelating agents include (a) ethylenediamine tetraacetate (EDTA), (b) diethylenetriamine pentacetate (DTPA), and (c) ethylenediamine di-(o-hydroxyphenylacetate) (EDDHA). The last two are particularly effective on alkaline soils. Activated sewage sludge is also a source of naturally chelated iron. Milorganite® contains approximately 4.8% elemental iron.

Other essential micronutrient carriers. A summary of the commonly available essential micronutrient carriers is presented in Table 13-9. Metal chelates of manganese, zinc, and copper are available. These carriers hold the three micronutrients in the soil for a period of time in a form that is available for absorption by the turfgrass plant. The remainder of the micronutrient carriers listed are quite water-soluble and subject to leaching. They are usually applied to the leaf tissue at low rates and with a limited quantity of water in order to ensure foliar absorption and a rapid response. The application rate of these micronutrient carriers is quite important. A sufficient quantity should be applied to correct the nutrient deficiency and yet to avoid toxicity caused by an excessively high rate. Fertilizing with micronutrients should be practiced only as needed to correct a deficiency. Indiscriminate use of micronutrients in fertilizers can cause turfgrass injury due to toxicity at higher soil or tissue levels.

Table 13-9

SOME COMMON FERTILIZER CARRIERS FOR SIX OF THE ESSENTIAL MICRONUTRIENTS

Essential micronutrient	Carriers	Chemical formula	Common nutrient content
Manganese	Manganese sulfate	$MnSO_4 \cdot H_2O$	32% Mn
	Manganese chelate	MnEDTA	6–7% Mn
Zinc	Zinc sulfate, heptahydrate	$ZnSO_4 \cdot 7H_2O$	23% Zn
	Zinc oxide	ZnO	67–80% Zn
	Zinc chelate	ZnEDTA	6–7% Zn
Copper	Copper oxide	CuO	75% Cu
	Copper sulfate, pentahydrate	$CuSO_4 \cdot 5H_2O$	25% Cu
	Copper chelate	CuEDTA	6–13% Zn
Molybdenum	Sodium molybdate	$Na_2MoO_4 \cdot 2H_2O$	39% Mo
	Ammonium molybdate	$(NH_4)_2MoO_4$	49% Mo
Boron	Borax	$Na_2B_4O_7 \cdot 10 H_2O$	11% B
Chlorine	Potassium chloride	KCl	47% Cl

Mixed Fertilizers

Most turfs are fertilized with a mixed fertilizer at some time during the growing season. Mixed fertilizers are usually made by combining two or more of the separate fertilizer carriers. The preparation of a mixed fertilizer can vary from a relatively simple bulk mixing of several carriers to a much more involved process requiring specific quality control of chemicals and temperatures. The latter process is required when high analysis fertilizers are formulated and for this reason they cost more.

Mixed fertilizers can be classified into several groups based on the total percentage of nutrients in the fertilizer. A low analysis fertilizer is one containing less than 20% total plant nutrients. A high analysis fertilizer contains between 20 and 30% total plant nutrients, while concentrated fertilizers are those containing more than 30% total plant nutrients. Most mixed fertilizers utilized for turfgrass purposes would fall in the high analysis or concentrated fertilizer classifications.

The **fertilizer ratio** is the ratio existing among the percentages of the three major macronutrients, nitrogen-phosphorus-potassium, contained in a fertilizer. For example, a 1–1–1 ratio would include fertilizer analyses such as 10–10–10, 15–15–15, and 20–20–20 (Table 13-10). Ratios of 4–1–2, 4–1–1, 3–1–2, 5–1–2, and 2–1–1 are preferred for use on established turfs when evaluated with reference to the ratio of nutrients removed in the clippings (22, 99, 123, 216, 217, 225, 243, 246, 284). In addition to nitrogen, phosphorus, and potassium, most mixed fertilizers contain smaller amounts of calcium, magnesium, and sulfur that may not be indicated on the label. This is particularly true of the lower analysis fertilizers. Special turfgrass fertilizers are also prepared that contain one or more of the essential micronutrients. Iron is the most common micronutrient used and may even be included as the fourth number on the fertilizer analysis. Iron, manganese, copper, and zinc may be added as sulfates; boron, as a sodium salt.

The nitrogen fraction of mixed fertilizers may contain percentages of (a) imme-

Table 13-10

A List of Common Fertilizer Ratios and Some Typical Fertilizer
Analyses for Each Ratio

Ratio on an oxide basis	Approximate fertilizer analysis		Ratio on an elemental basis
	$N–P_2O_5–K_2O$	N–P–K	
1–1–1	10–10–10	10–4–8	2.5–1–2
	15–15–15	15–6–12	
	20–20–20	20–8–16	
0–1–1	0–20–20	0–8–16	0–1–2
	0–25–25	0–10–20	
1–2–2	5–10–10	5–4.4–8.3	1.1–1–1.9
	10–20–20	10–8.7–16.6	
1–4–4	4–16–16	4–7–13.3	1–1.7–3.3
2–1–1	16–8–8	16–3.5–6.6	4.6–1–2
	20–10–10	20–4.4–8.3	
4–2–1	20–10–5	20–4.4–4.2	4.8–1–1
4–1–1	20–5–5	20–2–4	9–1–2
	24–6–6	24–2.6–5	
2.5–1.5–1	10–6–4	10–2.6–3.3	3.8–1–1
4–1–2	16–4–8	20–1.7–6.6	9.2–1–3.8
3.3–1–2.3	10–3–7	10–1.3–5.8	7.7–1–4.5
5–1–2	25–5–10	25–2.2–8.3	11–1–4

diately available cold-water-soluble nitrogen and (b) cold-water-insoluble nitrogen that is not readily available and depends on microorganism activity at favorable temperatures for nitrogen release. These water-soluble and water-insoluble nitrogen fractions can be combined in mixed fertilizers to obtain a rapid initial response as well as a longer residual nitrogen release to the turf (60, 74, 149, 161, 206, 230, 288). The insoluble, slowly available fraction may be composed of natural organics, ureaformaldehyde, or IDBU. A water-insoluble nitrogen content of 50% has been suggested for mixtures of nitrogen carriers (157).

A longer residual release of water soluble carriers is also accomplished with a thin, semipermeable resinous coating around the fertilizer granules or prills (11, 16, 78, 118, 177, 193, 258). Nitrogen release from within the encapsulating membrane is dependent on diffusion through the coating (179, 227). Various release rates and an extended residual response can be accomplished by varying the membrane thickness and number of holes per capsule (3, 118). Coated fertilizers have a longer residual response when applied to the surface of established turfs rather than incorporated into the soil (177).

Various herbicides, insecticides, and/or fungicides are sometimes added to mixed fertilizers to achieve convenience, particularly for nonprofessional turfgrass users. By doing this, the fertilizer serves the dual purpose of supplying needed plant nutrients to the soil as well as controlling turfgrass pests. Uniform application techniques at the proper rate are particularly important when pesticides are combined with mixed fertilizers.

Fertilizer Practices

Few soils possess sufficient inherent fertility to maintain the desired turfgrass quality and recuperative potential throughout the growing season. The primary objective in applying a fertilizer is to supply plant nutrients that are deficient in the soil in order to maintain the turfgrass quality. In addition to supplying essential nutrients, fertilizers influence other soil properties including the (a) soil reaction, (b) nutrient availability, and (c) indirect effects on the structure and microorganism population. The indirect effects are usually related to an increase in the soil organic matter content. An effective fertilizer program involves a determination of the fertility requirements followed by selection of the appropriate fertilizer carriers, application rates, and times of application.

Determining the Fertility Requirements

The fertility status of a soil is constantly changing because of (a) nutrient loss by leaching, volatilization, and removal of clipping and (b) nutrient release from organic matter, minerals, and exchange sites on soil colloids. Certain nutrients are not removed but can be combined with other compounds in insoluble forms or soil colloids in forms that are not available to the turfgrass plant. The release rate of nutrients contained in soil minerals and organic materials varies throughout the season depending on the soil temperature, moisture, pH, and microorganism population.

Diagnostic aids that can be used in determining the nutrient status and fertility requirements of a turf include (a) visual symptoms, (b) tissue tests, and (c) soil tests. A combination of two or all three diagnostic aids is preferred in assessing the fertility requirements because of certain disadvantages inherent in each of the diagnostic aids.

Visual symptoms. Visual deficiency symptoms for 12 essential nutrients have already been presented in this chapter. Visual diagnosis of a nutrient deficiency can be effectively utilized on only a few of the essential nutrients. Diagnosis by visual deficiency symptoms is an important technique in nitrogen fertilization since soil tests for nitrogen are of little value. Nitrogen fertility requirements of turfgrasses are commonly diagnosed using such visual symptoms as (a) a chlorotic or yellowish-green appearance and (b) a reduced shoot growth rate (Fig. 13-7). Sometimes the deficiency symptoms are similar for several essential plant nutrients. For example, the development of a yellow-chlorotic appearance is also used in diagnosing iron and sulfur deficiencies in turfgrasses. When using visual symptoms for diagnosing a possible nutrient deficiency, it must be recognized that some of the symptoms utilized are not always caused by a nutritional deficiency. One must take into consideration the possibility of alternate causes such as diseases, nematodes, insects, soil compaction, waterlogged soil conditions, salinity, or unfavorable environmental conditions for growth caused by temperature or moisture stress.

Tissue tests. Several types of on-site tissue testing kits are available for determining nitrogen, phosphorus, and potassium deficiencies. These rapid tissue tests involve crushing turfgrass clippings on a paper and treating the expressed sap with the appropriate solutions. The resulting color or turbidity gives an indication of the

Figure 13-7. Chlorosis and thinning of a Toronto creeping bentgrass turf (center) caused by a nitrogen deficiency.

immediate nutrient status of the tissue. The field tissue test technique gives a rapid, fair approximation of the amount of free nutrients present in the tissue. It does not measure nutrients that have been converted into complex organic materials. Tissue tests have an advantage over soil tests in that they determine whether the turfgrass plant has actually absorbed and translocated the nutrient in question. When sampling plant tissues, it is important to test tissues of the same age and type. The surface of the turfgrass tissues must be free of dust or soil containing nutrients that would invalidate the test. On-site tissue tests can be used in conjunction with visual deficiency symptoms in ascertaining the existence of a nutrient deficiency or toxicity.

Laboratory tissue tests are a future possibility for detailed nutrient status determinations of turfgrasses. All sixteen essential elements can be analyzed using the mass spectrograph. The usefulness of this technique is still limited for turfgrass purposes. The actual analysis is extremely accurate. The limitations are in developing proper sampling procedures and in interpreting the results. Representative, uniform tissue sampling techniques, in terms of the age and type of tissue, are particularly important when this type of tissue analysis is utilized. Also, the presence of minute quantities of dust or soil particles on the tissue can adversely affect the analysis, particularly the essential micronutrients. Even if a representative sample is obtained and the analysis made, there is not enough information available to interpret the results properly. Considerable data need to be collected on the levels of essential nutrients in the tissues of various turfgrass species growing under deficient and adequate fertility levels as well as various environmental conditions. Detailed laboratory tissue analyses may become a useful tool in the diagnosis of turfgrass nutrient status and requirements when sufficient data become available.

Soil tests. Chemical tests are widely used in determining the available essential plant nutrient supply of the soil. A total chemical analysis is not valid because a substantial portion of the nutrients contained in the soil are not available for plant absorption. Thus, a rapid soil test that measures the amount of certain available plant nutrients is of much greater value. The actual test usually involves treating the soil for a specified period of time with an extracting solution. The type and strength of extract used may vary with the specific soil test and soil type being analyzed. The nutrients removed in the extract are then quantitatively measured by colorimetric or turbidity methods.

The essential nutrients normally measured in a standard soil test include phosphorus, potassium, calcium, and magnesium. Tests for determining the soluble iron and manganese contents of soils are available. In addition, the soil pH and cation exchange capacity are determined for use as a guide in correcting acidic or alkaline conditions. The level of soil salinity and sodium content can also be measured if a potential problem exists. The amount of available nitrogen in the soil is usually not determined by a soil test because it is subject to rapid change. The variable results from nitrate analyses have not given satisfactory information for use in nitrogen fertilization.

The effectiveness of soil tests in assessing the turfgrass fertility requirements depends on (a) obtaining a representative sample from the turfgrass area, (b) the accuracy of the testing method and its effectiveness on the particular soil being tested, and (c) the accuracy in interpreting the test results. One of the most critical aspects in soil testing is obtaining a representative sample. The standard sampling procedure involves collecting twelve to fifteen 1-in. diameter soil cores from a turfgrass area characterized by uniformity of soil type, topography, previous fertility treatment, drainage, and other related cultural practices. Separate samples should be collected for analysis where distinct differences in fertility practices or soil type occur.

The soil cores should be taken to a standard depth that may vary from 2 to 4 in. depending on the recommendations of the specific soil testing lab in which the sample is to be analyzed. It is important to collect the soil samples at the specified depth in order to obtain accurate fertility recommendations. The sampling depth should be measured from the soil surface. The thatch and aboveground turf are usually removed from the sample prior to mixing. The twelve to fifteen cores collected with a soil tube from a uniform area should be thoroughly mixed in a paper bag or plastic container that is free from contaminating residues (Fig. 13-8). A standard sample size of 1 to 2 pints of air-dry soil is then submitted for analysis.

The soil sample should be submitted to a reputable laboratory that utilizes a method adapted to the particular soil type being tested. It is wise to employ the same laboratory each time soil tests are made. The frequency of soil testing on established turfs varies from 1 to 4 years depending on the type of turf, soil type, and intensity of culture. Soil samples can be taken at any time but are best made in the spring or fall prior to fertilizing.

Proper interpretation of the soil test results is particularly important. Individuals making recommendations from soil tests should have a knowledge of the soil types involved as well as a basic understanding of turfgrass fertilization principles. Few detailed correlation studies have been conducted concerning turfgrass responses to fertilization with various levels of plant nutrients on soils with various inherent fertility levels. Most soil test recommendations for turfgrasses are based on soil test correlation studies conducted with field crops such as corn and certain pasture grasses. Utilizing these correlation studies as a base, modifications have arbitrarily been made in applying them to turfgrass conditions. It is unfortunate that adequate correlation studies have not been conducted for turfs. Hopefully this problem will be corrected in the future.

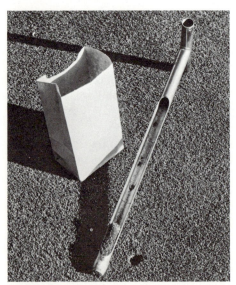

Figure 13-8. A standard soil sampler used in collecting soil cores for testing.

Certain precautions may have to be made in interpreting phosphorus soil tests. Arsenic gives the same type of reaction as phosphorus in the standard methods used in soil testing. Greens and other turfgrass areas where arsenic compounds have been applied for the control of weeds and insects have a higher phosphorus test than is actually contained in the soil. Thus, a knowledge of the previous history of the turfgrass area regarding arsenic applications is required for a proper interpretation of phosphorus soil tests. This factor is of primary concern where low or borderline soil phosphorus levels occur.

Selecting a Fertilizer

Characteristics of importance in selecting a turfgrass fertilizer include (a) spreadability, (b) dustiness, (c) bulkiness, (d) water solubility, (e) foliar burn potential, (f) initial plant response time, (g) residual response, (h) ratio, (i) efficiency, (j) effect on soil reaction, (k) cost, (l) fire hazard, and (m) storage characteristics.

Physical characteristics such as spreadability and freedom from dust are important considerations in selecting a turfgrass fertilizer, particularly for nonprofessionals. Fertilizer materials that have been granulated and screened to a specific size range can be spread more uniformly during application. Granulated materials also prevent the segregation of compounds within mixed fertilizers and reduce the tendency to cake. Excessive bulkiness can result in increased handling problems. On the other hand, a lightweight material is preferred by some nonprofessionals. Higher analysis mixed fertilizers are usually less bulky.

The water solubility of a fertilizer affects nutrient loss by leaching from the soil and whether it can be applied as a foliar spray. Another factor related to solubility is the foliar burn potential. Water-soluble fertilizers have a much higher foliar burn potential than the insoluble organic forms.

Water solubility also determines the initial plant response time. For example, water-soluble carriers give a more rapid initial plant response than natural organic carriers that depend on microorganisms for nutrient release (Fig. 13-9). On the other hand, fertilizers with the nutrients contained in insoluble, slow release organic forms have a longer residual response. That is, the nutrients are released as the organic compounds decompose over a period of time ranging from 4 to 12 weeks. Fertilizers having a longer response time usually have a higher cost per unit of nitrogen applied. The higher cost is somewhat offset by a reduced cost of application due to the decreased number of fertilizer applications required to maintain a comparable turfgrass response.

The actual ratio of the fertilizer should also be considered. It is preferable that the ratio used relate to the nutrient requirements of the particular turf, whether or not clippings are removed, and the level of available nutrients in the soil. Maintenance of the proper nutrient balance in the turfgrass plant and soil is also an important consideration in selecting the fertilizer ratio. The presence of micronutrients in the fertilizer is an additional consideration where this type of deficiency may occur.

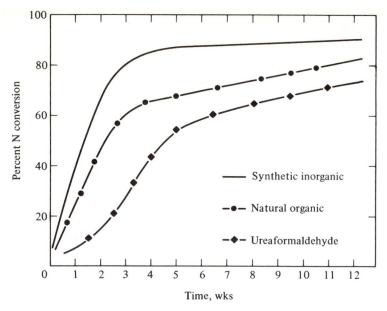

Figure 13-9. Idealized diagram of the nitrification rate for three types of nitrogen carriers.

The efficiency of a fertilizer refers to the percentage of nutrients applied in a given fertilizer that is available for absorption by the turfgrass roots. Nutrient unavailability may result from loss by volatilization, failure of nutrient containing organic complexes to decompose, or loss by leaching. Examples of these include (a) nitrogen volatilization loss from urea, (b) the lack of decomposition in a portion of the ureaformaldehyde nitrogen, and (c) the leaching proneness of the water soluble nitrate nitrogen forms.

The effects on soil reaction can be a significant factor in selecting a fertilizer. Turfs irrigated with water containing calcium and magnesium tend to increase in alkalinity through time. The pH can be maintained in the favorable range of 6 to 7 by using fertilizers with an acidifying effect. On the other hand, the soil reaction should be checked regularly to avoid excessively acid soils caused by acidifying fertilizers. This response is of most concern on highly leached soil areas that are not irrigated or are irrigated with water having a low calcium and magnesium content.

When considering fertilizer costs, one should not only determine the cost per unit of nutrient purchased but the value of other convenience factors such as reduced foliar burn potential, longer residual response, improved spreadability of granulated carriers, less mechanical wear on equipment, and freedom from dust. A carrier having a minimal foliar burn potential is preferred when the fertilizer application must be done by inexperienced or unreliable individuals. Other characteristics desired in a turfgrass fertilizer are a minimal fire hazard and good storage properties in terms of freedom from caking. Preferably the fertilizer should have a minimum tendency to absorb water from the atmosphere.

The turfgrass response from any mixed fertilizer or nutrient carrier depends on how it is used (56, 63, 64, 65, 135, 142, 149, 212, 292). A combination of several types of mixed fertilizers and/or nutrient carriers is preferred for use in a seasonal fertilizer program under many conditions (62, 132, 219). No one fertilizer possesses all the characteristics desired for turfgrass use (275). Each has specific advantages and disadvantages. Thus, it is important to select the fertilizer material or combination of fertilizers that best suits the particular conditions. Specifically, this involves soil, environmental, use, budgetary, and labor factors.

Use of Fertilizers

There are many combinations of fertilizer carriers and application frequencies that can be used to maintain an acceptable level of turfgrass quality in a given situation (56, 142, 277). No one carrier or application frequency is superior or ideal except under rather specific conditions. Even the combination of fertilizers varies as the soil and environmental conditions vary.

Factors determining the appropriate fertilizer rate, time of application, and carrier to utilize in a given situation include (a) level of available essential nutrients in the soil, (b) turfgrass requirements, (c) cost, (d) environmental conditions, (e) turfgrass quality level desired, (f) shoot growth rate desired, (g) intensity of use, (h) soil physical conditions, and (i) cultural system utilized. The relative importance of each factor varies considerably with the particular situation.

Level of available essential nutrients in the soil. The level of available phosphorus, potassium, calcium, and magnesium in the soil can be measured by a soil test and used as a guideline in determining the amount and frequency that each nutrient should be applied. Phosphorus and potassium fertilizer practices are usually governed by the available soil levels. When fertilizing with any one nutrient carrier, it is important to maintain the proper balance with other nutrients in order to ensure maximum nutrient absorption and turfgrass response (12, 77, 98, 123, 203, 242, 243). Balance is particularly important between nitrogen and potassium (10, 98, 104, 105, 126, 127, 144, 165, 182, 202, 240).

Turfgrass requirements. Variability exists in the nutrient requirements among turfgrass species (21, 23, 65, 87, 143, 147, 245, 259, 277). The variability is most significant with nitrogen. Each turfgrass species has a maximum and minimum nitrogen fertility response range to maintain an adequate level of turfgrass quality in terms of color, shoot density, and recuperative potential (Table 13-11). Certain turfgrass species may have a medium to high nitrogen requirement for optimum turfgrass quality and a narrow tolerance to lower fertility levels, while other species tolerate a wide range of fertility levels. For example, zoysiagrass responds to higher fertility but tolerates a relatively low level (27). In contrast, many red fescue cultivars have a low nitrogen fertility requirement and will actually decline in shoot density and turfgrass quality at higher nitrogen fertilization levels. Turfgrass species possessing

Table 13-11

THE RELATIVE NITROGEN FERTILITY RESPONSE RANGE
OF TWENTY-FOUR TURFGRASSES

Nitrogen fertility level	Pounds per growing month per 1000 sq ft	Turfgrass species
Very low	0.0–0.4	Blue grama
		Buffalograss
		Bahiagrass
		Centipedegrass
Low	0.2–0.6	Carpetgrass
		Chewings fescue
		Red fescue
		Canada bluegrass
		Fairway wheatgrass
Medium	0.4–1.0*	Zoysiagrass
		Meadow fescue
		Tall fescue
		Italian ryegrass
		Perennial ryegrass
		Redtop
		Rough bluegrass
		Annual bluegrass
		Colonial bentgrass
		Velvet bentgrass
		Timothy
		St. Augustinegrass
High	0.5–1.5*	Kentucky bluegrass
		Creeping bentgrass
		Bermudagrass

* The upper portion of the range represents the nitrogen levels used at higher intensities of culture.

a higher nitrogen requirement generally have a higher requirement for phosphorus and potassium as well.

Substantial variability in the nitrogen fertility requirement also occurs within turfgrass species (85, 171, 249, 259, 285). Among cultivars of Kentucky bluegrass, Merion has a nitrogen fertility requirement of 1 to 1.3 lb of nitrogen per 1000 sq ft per growing month, whereas Kenblue normally requires from 0.4 to 0.7 lb.

Cost. Unfortunately, the budget for fertilizer expenditures is sometimes a factor in determining the particular carrier utilized and the fertility level at which the turf is maintained. Higher nitrogen fertilization rates not only increase fertilizer cost but also increase the mowing and irrigation requirements. Fertilizer is one of the best investments in turfgrass culture for maintaining an acceptable level of turfgrass quality. In addition, turfs maintained at a moderate but adequate fertility level have better recuperative potential from injury and are more able to resist weed invasions. Where the available budget is minimal, the cost of a carrier may be of more concern than the total expenditure for nutrients. Fertilizer costs vary considerably depending on the particular characteristics desired. For example, fertilizers possessing a minimum foliar burn potential are generally higher in cost per unit of nitrogen but may be necessary when there is no experienced labor available to apply the materials.

Environmental conditions. Environmental conditions that result in a rapid shoot growth rate also increase the nutritional requirements to support this growth. Adequate levels of nitrogen, phosphorus, and potassium are necessary in maintaining hardiness to heat, cold, and drought stress. A minimum nutritional level should be maintained to ensure rapid turfgrass recovery from injury due to environmental stress. The timing and amount of nitrogen fertilizer used are particularly critical prior to periods of anticipated environmental stress. Nitrogen stimulates shoot growth causing an increase in the hydration level of the turfgrass tissues (33, 34, 35, 168, 190, 233). A general reduction in heat, cold, and drought hardiness results (2, 10, 33, 34, 35, 92, 114, 233). Thus, nitrogen fertilization should be withheld or applied very judiciously prior to and during such periods (34). In contrast, hardiness to heat, cold, and drought stress are enhanced by potassium (2, 10, 92, 253, 268). A nitrogen fertilizer application immediately after a drought, heat, or cold stress period facilitates rapid recovery of the turf.

Turfgrass quality level desired. The level of turfgrass quality desired governs the amount of fertilizer utilized. This is particularly important for certain recreational turfs where the quality of the surface is an important factor influencing the game. Turfs on intensively used greens demand a higher quality surface and fertility level than turfs used for parks and extensive turfgrass areas.

Growth rate desired and intensity of use. The shoot growth rate of the turfgrass plant is determined primarily by the nitrogen fertility level, assuming other factors are not limiting. Depending on the particular situation and type of use, a minimal growth rate or a relative rapid growth rate may be desired. A minimal nitrogen fertility level is preferred on ornamental areas where a dense, high quality turf exists and the primary objective is to maintain acceptable turfgrass color. In contrast, weakened turfs of low shoot density or turfs injured by environmental stress or turfgrass pests require a higher nitrogen level to stimulate shoot growth and facilitate rapid recovery of the damaged areas (166). Similarly, a rapid shoot growth rate is desired to encourage recovery on sports turfs damaged by divots of a golf club or from the turning, twisting action of cleated shoes. Thus, when selecting the fertility level, particularly nitrogen, it is important to determine whether the objective is to maintain the existing high quality turf or to grow grass in order to improve the quality or to recover from injury. An additional consideration is that the nutritional requirements are greater during establishment than after a turf is established.

Soil physical conditions. The soil physical conditions, in terms of texture and structure, determine the amount of available nutrients retained on the soil colloids and the degree of nutrient removal by leaching. Coarse textured sandy soils have minimal nutrient retention capabilities and are prone to nutrient loss by leaching. Fine textured loams and clay loams possess more desirable properties in terms of nutrient retention and degree of leaching. Soils high in clay are subject to compacted, waterlogged conditions that impair soil aeration. Nutrient absorption is restricted under these conditions and influences the fertility program. Nutrient loss by leaching in sandy soils can be partially overcome by fertilizing more frequently at lower rates and/or utilizing fertilizer carriers that retain the nutrients in an insoluble form. The nutrients are released at a controlled rate for absorption by the turfgrass root system and, thus, are less prone to loss by leaching.

Cultural system utilized. A turfgrass cultural system that involves clipping removal or intensive irrigation results in an increased fertility requirement (216, 217, 284). The rate and frequency of fertilizer application must be adjusted accordingly. For example, clipping removal from a Merion Kentucky bluegrass turf results in an increased nitrogen fertility requirement of 0.2 to 0.3 lb per 1000 sq ft per growing month.

General Considerations in Fertilizing

A minimum of two fertilizer applications per year, usually spring and fall, is generally preferred to maintain an acceptable level of turfgrass color and shoot density under most conditions (56, 154, 203, 259, 277). Supplemental nitrogen fertilizer applications may be required during the summer for turfs having a high nitrogen fertility requirement.

Fertilizer applications should be timed to favor the desirable turfgrass species rather than weeds. For example, late fall or winter fertilization of bermudagrass encourages the growth of winter active weeds such as annual bluegrass. The timing of fertilizer applications can also influence disease proneness (Fig. 13-10).

Light, frequent applications of readily available nitrogen fertilizers are sometimes preferred to heavy, infrequent applications (123, 184, 277). The latter tends to stimulate excessive shoot growth and impair root growth. Applications of readily available nitrogen exceeding one pound per 1000 sq ft will generally exhaust the plant carbohydrate reserves. In addition, the controlled level of nutrient availability is frequently not as desirable under a heavy, infrequent nitrogen fertility program. A fertilizer application containing nitrogen should be made immediately after any

Figure 13-10. Effect of the time of nitrogen application on the incidence of *Typhula* blight on Penncross creeping bentgrass.

cultural practice that temporarily disrupts turfgrass growth and quality. Typical examples are vertical mowing, coring, or grooving practices. An application of phosphorus-containing fertilizer just after coring enhances phosphorus penetration below the soil surface (113).

In cool climates the timing of the spring fertilization can vary considerably depending on the situation. Where the existing turf is of high quality and an excessive amount of mowing is to be avoided, nitrogen fertilization can be delayed until after the early flush of shoot growth has occurred. A nitrogen fertilizer application that coincides with the cool spring period of rapid shoot growth results in increased puffiness on bentgrass greens (69). One problem with late spring fertilization is that it may encourage certain annual weedy grasses (203). If the objective is to improve turfgrass quality or to recover from serious injury, an early spring nitrogen fertilization will stimulate turfgrass growth and cause a rapid improvement in the turfgrass quality. Early spring or late fall dormant fertilizations can also be utilized to break winter dormancy and stimulate earlier spring green-up (79). However, early spring nitrogen fertilization may also increase the chance of winter injury associated with low temperatures.

Summer nitrogen fertilization should be practiced judiciously, if at all, during periods of heat stress when shoot growth is minimal or ceases. This factor is a greater concern on cool than warm season turfgrass species. The same principle should be applied during periods of drought stress unless irrigation water is available and utilized effectively.

The timing of late summer and fall nitrogen fertilizer applications prior to the advent of winter dormancy may vary with the climatic region. In cooler climates where direct low temperature kill is a problem, late fall nitrogen applications that stimulate shoot growth and increase tissue hydration should be avoided to minimize direct low temperature kill (34, 114). Withholding nitrogen for 30 to 40 days prior to the advent of winter dormancy ensures maximum low temperature hardiness in the colder climates.

In contrast, low temperature hardy, cool season turfgrasses growing in regions having somewhat more moderate winter temperatures and a minimum probability of low temperature injury are adaptable to a much different fall fertility program. Late fall nitrogen fertilization under such conditions results in improved winter color retention, shoot density, and root growth plus an adequate spring and summer response (111, 236, 241). Late fall and winter nitrogen fertilization stimulates winter chlorophyll synthesis and photosynthesis at temperatures just above freezing without a comparable increase in respiration, carbohydrate reserve utilization, and shoot growth (21, 236). An adequate spring response in terms of turfgrass quality and color is also achieved without overstimulating shoot growth that causes an increased mowing frequency.

Dormant fertilization, which is practiced after the turf has ceased shoot growth, can be utilized without affecting the low temperature hardiness of turfgrasses. This practice results in a more rapid breaking of winter dormancy and stimulates early spring green-up (280). A portion of the nutrients applied by dormant fertilization can be lost by surface erosion or leaching prior to the initiation of turfgrass growth in the spring. The efficiency of the fertilizer is reduced under this situation.

Fertilizer Application

Fertilizers can be applied in a solid, dry form or in a water carrier. A fertilizer spreader is essential for uniform distribution of solid fertilizers. Factors to consider in selecting a fertilizer spreader include (a) uniformity of distribution, (b) application width, (c) range in rates and ease of adjustment, (d) hopper capacity, (e) hopper agitator, (f) ease of emptying and cleaning, (g) positive shutoff controls, and (h) tires or wheels of sufficient size and width to minimize soil compaction and rutting. Fertilizer spreaders vary considerably in application width, ease of operation, accuracy, and durability of construction. Spreaders should be thoroughly cleaned immediately after use to avoid corrosion and ensure satisfactory operation in the future.

Solid applicators. Application of solid fertilizers by broadcasting is preferred to drill or row placement in terms of maximizing nutrient absorption by established turfs (90, 214). A number of different types of broadcast fertilizer applicators are available. The two commonly used types of solid applicators are the (a) gravity- or drop-type and (b) centrifugal-type (Fig. 13-11). Solid fertilizers can also be effectively applied by airplane or helicopter on extensive turfgrass areas.

The gravity-type spreader results in a very uniform application but covers a limited area. There is a potential problem with overlapping or missed areas. Gravity feed spreaders should be equipped with a baffle board to provide uniform distribution and avoid banding or streaking of the fertilizer.

The centrifugal-type spreader covers a wider area more rapidly. Problems with overlapping or missed areas are minimal. This type of spreader tends to lack uni-

Figure 13-11. View of the two basic types of solid fertilizer applicators: gravity-type (left) and centrifugal-type (right).

formity in distribution throughout the spreader width. This problem can be overcome by a controlled degree of overlapping.

It is usually necessary to "water in" the water soluble fertilizers immediately after application. "Watering in" minimizes the probability of foliar burn by washing the fertilizer particles off the leaves. This practice also reduces the loss of nitrogen by volatilization if urea is being used as the nitrogen carrier (274). Foliar burn is greater when the applications are made to wet turf (277).

The pickup and removal of dry fertilizers during the mowing of dense, close cut greens can be a problem for a period of time after fertilizing. This problem is minimized if certain precautions are taken. First, the dry fertilizer should be applied just after mowing. The use of a brush plus irrigating encourages movement of the fertilizer into the turf. If these steps are not completely effective, mowing on the day following a dry fertilizer application can be skipped or the mowing accomplished with the catcher removed.

Liquid application. Nutrients can be applied by using a liquid carrier, which is usually water. This method of application permits flexibility in nutrient placement. Liquid fertilizer application may involve a (a) foliar feeding or (b) soil drench (255).

Foliar feeding is the application of a relatively small amount of nutrients using a limited volume of water so that the nutrients are absorbed directly by the aboveground shoot tissues. A maximum of 0.5 gal of water per 1000 sq ft is generally utilized. The concentration of nutrients that can be applied at any one time is limited in order to minimize the foliar burn potential. As a result, foliar feeding must be practiced at frequent intervals, usually 1 to 2 weeks, to avoid a nutrient deficiency. The response to foliar feeding is more rapid than soil applications. This is a relatively costly technique for routine fertilization unless it can be done in conjunction with a regular preventive program of pesticide spraying. Problems with foliar feeding can arise due to differentials in the absorption rate of different nutrients (242). For example, nitrogen is absorbed much more rapidly than phosphorus, which results in a nutritional imbalance. Foliar feeding techniques are frequently used (a) to correct a micronutrient deficiency or (b) when rapid correction of a nutrient deficiency is required. In addition, foliar feeding can be effective in correcting turfgrass nutrient deficiencies caused by compacted, waterlogged soil conditions that restrict nutrient uptake by the roots.

Soil drench or liquid fertilization involves the application of nutrients in water as a drench with a major portion of the applied nutrients washed into the soil. In this case, the nutrients are absorbed primarily through the root system rather than through the leaves. A much larger quantity of water is applied compared to foliar feeding. This technique necessitates the handling of a large volume of water and, thus, is more expensive than solid applications. Turfs should be irrigated following the soil drench fertilizer application to minimize foliar burn, which can be severe (183).

No difference in turfgrass quality is noted between liquid and solid fertilization at low rates (183). The turfgrass response to a soil drench is not as favorable as from a solid, dry application when applied at high rates (220, 223). This differential response is attributed to phytotoxicity caused by the high rate of liquid fertilizer applied.

Fertilizer injection into an irrigation system. A modified method of fertilizer application using a water carrier for foliar feeding or soil drenching involves injection into the irrigation system. There are two types of fertilizer injectors available: (a) proportioners and (b) metering pumps. An injection system is preferred. The fertilizer should not pass through the pump in order to reduce corrosion problems to the pump mechanism.

Fertilizer applications through a sprinkler irrigation system can be effectively used if the system has been designed and installed to ensure uniform water distribution throughout the turfgrass area. The rate and amount of water applied through the irrigation system should be adjusted to avoid surface runoff. Applying water at a rate in excess of the soil infiltration velocity results in inferior uniformity of nutrient application in some areas. The irrigation water must be applied during periods when wind activity does not disturb the distribution pattern. The pH of the irrigation water should be below 7 to minimize nitrogen loss as gaseous ammonia (119). Nitrogen volatilization from a sprinkler application can be reduced by using nitrate nitrogen carriers rather than ammonia types (119). Volatilization loss increases as the water temperature is increased (119).

There are three phases to liquid fertilizer injection into an irrigation system. The first phase is an irrigation interval to moisten the turf and soil. The fertilizer is then injected into the irrigation system at a predetermined rate and duration. The amount of fertilizer injected depends on the area to be covered, the type of fertilizer, and the rate of nutrient application desired. The duration of injection is determined by the amount of fertilizer to be applied, the area to be covered, and the rate at which the water is applied. The proper duration is important in avoiding foliar burn of the turf and corrosion of the irrigation system. The third phase is a period of irrigation involving only water. The objectives of this phase are to (a) wash the fertilizer from the turfgrass shoots to minimize foliar burn and (b) to rinse the fertilizer out of the irrigation system to avoid corrosion.

Establishment Fertilization

The quantity of phosphorus and potassium applied to the soil prior to turfgrass establishment by seed, sod, or stolons is best determined by a soil test. An adequate phosphorus level is particularly important in obtaining effective turfgrass establishment from seed. Since phosphorus is relatively immobile in soil, it should be mixed uniformly into the upper 4 to 6 in. of the soil root zone prior to planting. Incorporating the fertilizer just prior to seeding is preferred to incorporation 2 weeks before seeding (158). Excessive fertilizer rates or the placement of certain fertilizers immediately adjacent to the turfgrass seeds can result in phytotoxicity (125). Such an effect has been observed with urea (40, 44). A general guideline for the fertilization of soils prior to turfgrass establishment is the incorporation of 2 lb each of nitrogen, phosphoric acid, and potash per 1000 sq ft into the upper 4 to 6 in.

of the soil. A 0.5 to 1 lb per 1000 sq ft application of readily available nitrogen when the turfgrass seedlings are 1.5 to 2.5 in. tall can be very effective in enhancing the establishment rate. Additional fertilizations at 3 to 4 week intervals are usually required to achieve rapid formation of a quality sod.

References

1. ANONYMOUS. 1952. Special purpose turf. Pennsylvania Agriculture Experiment Station Bulletin 553. pp. 52–53.

2. ANONYMOUS. 1968. Manage fertilizer for more heat tolerant bent. Weeds, Trees and Turf. 7(5): 35–38.

3. AHMED, I. U., O. J. ALTOE, L. E. ENGELBERT, and R. B. COREY. 1963. Factors affecting the rate of release of fertilizer from capsules. Agronomy Journal. 55: 495–499.

4. ALBERDA, TH. 1965. The influence of temperature, light intensity and nitrate concentration on dry-matter production and chemical composition of *Lolium perenne* L. Netherlands Journal of Agricultural Science. 13(4): 335–360.

5. ARMIGER, W. H., K. G. CLARK, F. O. LUNDSTROM, and A. E. BLAIR. 1951. Urea-form: greenhouse studies with perennial ryegrass. Journal of American Society of Agronomy. 43(3): 123–127.

6. ARMIGER, W. H., I. FORBES, JR., R. E. WAGNER, and F. O. LUNDSTROM. Urea-form—A nitrogenous fertilizer of controlled availability: Experiments with turf grasses. Journal of American Society of Agronomy. 40: 342–356.

7. ARNON, D. I., W. E. FRATZKE, and C. M. JOHNSON. 1942. Hydrogen ion concentration in relation to absorption of inorganic nutrients by higher plants. Plant Physiology. 17: 515–524.

8. BAIR, R. A. 1948. Minor elements in turf production. Greenkeeper's Reporter. 16(2): 28–31.

9. BEARD, J. B. 1961. The independent and multiple contribution of certain environmental factors on the seasonal variation in amide nitrogen fractions of grasses. Ph.D. Thesis. Purdue University. pp. 1–116.

10. BEARD, J. B., and P. E. RIEKE. 1966. The influence of nitrogen, potassium and cutting height on the low temperature survival of grasses. 1966 Agronomy Abstracts. p. 35.

11. BEATON, J. D., W. A. HUBBARD, and R. C. SPEER. 1967. Coated urea, thiourea, urea-formaldehyde, hexamine, oxamide, glycoluril, and oxidized nitrogen-enriched coal as slowly available sources of nitrogen for orchardgrass. Agronomy Journal. 59: 127–133.

12. BELL, R. S., and J. A. DEFRANCE. 1944. Influence of fertilizers on the accumulation of roots by closely clipped bentgrasses and on the quality of turf. Soil Science. 58: 17–24.

13. BENEDICT, H. M., and G. B. BROWN. 1944. The growth and carbohydrate responses of *Agropyron smithii* and *Bouteloua gracilis* to changes in nitrogen supply. Plant Physiology. 19: 481–494.

14. BENGSTON, J. W., and F. F. DAVIS. 1939. Experiments with fertilizers on bent turf. Turf Culture. 1(3): 192–213.

15. BENSON, N., and R. M. BARNETTE. 1939. Leaching studies with various sources of nitrogen. Journal of American Society of Agronomy. 31: 44–54.

16. BESEMER, S. T. 1963. A comparison of resin-fertilizer with noncoated soluble fertilizers on bentgrass putting green turf. California Turfgrass Culture. 13(1): 3–5.

17. BLACKMAN, G. E. 1932. An ecological study of closely cut turf treated with ammonium and ferrous sulfates. Annals of Applied Biology. 19: 204–220.

18. BLACKMAN, G. E., and W. G. TEMPLEMAN. 1940. The interaction of light intensity and nitrogen supply in the growth and metabolism of grasses and clover (*Trifolium repens*). IV. The relation of light intensity and nitrogen supply to the protein metabolism of the leaves of grasses. Annals of Botany N. S. 4(15): 533–587.

19. BLASER, R. E., and R. E. SCHMIDT. 1964. Effect of nitrogen on organic food reserves and some physiological responses of bentgrass and bermudagrass grown under various temperatures. 1964 Agronomy Abstracts. p. 99.

20. BLASER, R. E., and W. E. STOKES. 1943. Effect of fertilizer on growth and composition of carpet and other grasses. Florida Agricultural Experiment Station Bulletin 390. pp. 1–31.

21. BLASER, R. E., R. E. SCHMIDT, and F. B. STEWART. 1967. Rate and seasons of applying nitrogen on turf quality and physiology. 1967 Agronomy Abstracts. p. 50.

22. BRANSON, R. L. 1966. Better answers with soil and plant analyses. California Turfgrass Culture. 16(1): 1–2.

23. BREDAKIS, E. J. 1959. Interaction between height of cut and various nutrient levels on the development of turfgrass roots and tops. M.S. Thesis. University of Massachusetts. pp. 1–87.

24. BREDAKIS, E. J., W. G. COLBY, and M. E. WEEKS. 1964. The utilization of potassium from different sources by turfgrass species. 1964 Agronomy Abstracts. p. 104.

25. BREDAKIS, E. J., and E. C. ROBERTS. 1959. The response of turfgrass roots to clipping and fertilization practices. 1959 Agronomy Abstracts. p. 88.

26. BREDAKIS, E. J., and J. E. STECKEL. 1960. Breakdown of organic nitrogen fertilizers used in turf production. 1960 Agronomy Abstracts. p. 69.

27. ———. 1963. Leachable nitrogen from soils incubated with turfgrass fertilizers. Agronomy Journal. 55: 145–146.

28. BRITTON, M. P. 1957. Bluegrass rust studies. Proceedings of the 1957 Midwest Regional Turf Conference. pp. 78–79.

29. BROWN, R. L., and A. L. HOFENRICHTER. 1948. Factors influencing the production and use of beachgrass and dunegrass clones for erosion control. Journal of American Society of Agronomy. 40: 677–684.

30. BRYAN, O. C. 1933. Yellowing of centipede grass and its control. Florida Agricultural Experiment Station Press Bulletin 450. pp. 1–2.

31. CALHOUN, C. R., J. T. PESEK, and E. C. ROBERTS. 1960. The effect of ureaformaldehyde nitrogen in the presence of varying phosphorus and potassium levels on the yield and nutrient content of a Kentucky bluegrass turf. 1960 Agronomy Abstracts. p. 69.

32. ———. 1964. Effect of phosphorus, potassium and ureaformaldehyde nitrogen on predictions of quality in bluegrass turf. 1964 Agronomy Abstracts. p. 105.

33. CARROLL, J. C. 1943. Effects of drought, temperature, and nitrogen on turf grasses. Plant Physiology. 18: 19–36.

34. CARROLL, J. C., and F. A. WELTON. 1939. Effect of heavy and late applications of nitrogenous fertilizer on the cold resistance of Kentucky bluegrass. Plant Physiology. 14: 297–308.

35. CHEESMAN, J. H., and E. C. ROBERTS. 1964. Effect of nitrogen level and moisture stress on *Helminthosporium* infection in Merion bluegrass. 1964 Agronomy Abstracts. p. 105.

36. CHEESMAN, J. H., E. C. ROBERTS, and L. H. TIFFANY. 1965. Effects of nitrogen level and osmotic pressure of the nutrient solution on incidence of *Puccinia graminis* and *Helminthosporium sativum* infection in Merion Kentucky bluegrass. Agronomy Journal. 57: 599–602.

37. COLBY, W. G., and E. J. BREDAKIS. 1966. The feeding power of four turf species for exchangeable and non-exchangeable potassium. 1966 Agronomy Abstracts. p. 35.

38. COOK, R. N., and R. E. ENGEL. 1959. The effect of nitrogen carriers on the incidence of disease on bentgrass turf. 1959 Agronomy Abstracts. p. 88.

39. COOK, R. N., R. E. ENGEL, and S. BACHOLDER. 1964. A study of the effect of nitrogen carriers on turfgrass disease. Plant Disease Reporter. 48(4): 254–255.

40. COOKE, I. J. 1962. Toxic effect of urea on plants. Nature. 194: 1262–1263.

41. CORNMAN, J. F. 1969. The influence of maintenance practices on thatch formation by Merion Kentucky bluegrass turf. 1969 Agronomy Abstracts. p. 53.

42. COUCH, H. B. 1958. Nitrogen applications in brown patch control. Golfdom. 31(4): 64.

43. COUCH, H. B., and J. R. BLOOM. 1960. Influence of environment on diseases of turfgrasses. II. Effect of nutrition, pH, and soil moisture on *Sclerotinia* dollar spot. Phytopathology. 50: 761–763.

44. COURT, M. N., R. C. STEPHEN, and J. S. WAID. 1962. Nitrite toxicity arising from the use of urea as a fertilizer. Nature. 194: 1263.

45. DANIEL, W. H. 1955. *Poa annua* controls with arsenic materials. Golf Course Reporter. 23(1): 5–8.

46. ———. 1957. Soil test summaries show turf needs. Proceedings of the 1957 Midwest Regional Turf Conference. pp. 82–85.

47. ———. 1961. Soil test summaries show turf needs. Midwest Turf News and Research No. 16. pp. 1–2.

48. ———. 1961. Aids in understanding nitrogen nutrition. Proceedings of the 1961 Midwest Regional Turf Conference. pp. 65–67.

49. DANIEL, W. H., and N. R. GOETZE. 1960. Heavy nitrogen fertilization of cool-season turfgrasses. 1960 Agronomy Abstracts. p. 69.

50. DARROW, R. A. 1939. Effects of soil temperature, pH, and nitrogen nutrition on the development of *Poa pratensis*. Botanical Gazette. 101: 109–127.

51. DAVIS, R. R. 1949. Nutritional studies of turf grasses. Proceedings of the 1949 Midwest Regional Turf Conference. pp. 42–44.

52. ———. 1950. The physical condition of putting-green soils and other environmental factors affecting the quality of greens. Ph.D. Thesis. Purdue University. pp. 1–104.

53. ———. 1961. Turfgrass mixtures—influence of mowing height and nitrogen. Proceedings of 1961 Midwest Regional Turf Conference. pp. 27–29.

54. ———. 1961. Turfgrass mixtures—influence of mowing height and nitrogen. Illinois Turfgrass Conference Proceedings. pp. 42–45.

55. ———. 1962. Nitrogen fertilization of turfgrasses. 1962 Agronomy Abstracts. p. 102.

56. ———. 1965. Forms of nitrogen for fertilizing Kentucky bluegrass in Ohio. Golf Course Reporter. 33(5): 42–50.

57. DAWSON, R. B. 1933. Report on results of research work carried out in connection with soil acidity and the use of sulfate of ammonia as a fertilizer for putting greens and fairways. Journal of Board of Greenkeeping Research. 3(9): 65–78.

58. DAWSON, R. B., and B. M. BOYNS. 1938. The use of sulfur in improving the physical condition of clay soils on golf courses. Journal of Board of Greenkeeping Research. 5(18): 189–200.

59. DEAL, E. E., and R. E. ENGEL. 1965. Iron, manganese, boron, and zinc: Effects on growth of Merion Kentucky bluegrass. Agronomy Journal. 57: 553–555.

60. DEFRANCE, J. A. 1938. The effect of different fertilizer ratios on colonial, creeping, and velvet bentgrass. Proceedings of the American Society for Horticultural Science. 36: 773–780.

61. ———. 1958. Ureaform fertilization on putting green turf. Golfdom. 32: 79–83.

62. DEFRANCE, J. A., and T. E. ODLAND. 1939. A comparison of nitrogen carriers for bentgrass fertilization. Proceedings of the American Society for Horticultural Science. 37: 1084–1090.

63. DUICH, J. M., and H. B. MUSSER. 1958. Response of creeping bentgrass turf under putting green management to urea-form and other nitrogenous fertilizers. 1958 Agronomy Abstracts. p. 64.

64. ———. 1959. Response of creeping bentgrass turf under putting-green management to urea-form and nitrogen fertilizers. 1959 Agronomy Abstracts. p. 89.

65. ———. 1960. Response of Kentucky bluegrass, creeping red fescue, and bentgrass to nitrogen fertilizers. Pennsylvania Agricultural Experiment Station Progress Report 214. pp. 1–20.

66. DUNN, J. H., and R. E. ENGEL. 1966. Bentgrass turf growth response to urea nitrogen impregnated on hydrocarbons and vermiculite carriers. 1966 Report on Turfgrass Research at Rutgers University. New Jersey Agricultural Experiment Station Bulletin 816. pp. 22–40.

67. EDMOND, D. B., and S. T. COLES. 1958. Some long-term effects of fertilizers on a mown turf of browntop and chewings fescue. New Zealand Journal of Agricultural Research. 1(5): 665–674.

68. ENDO, R. M. 1967. Why nitrogen fertilization controls the dollar spot disease of turfgrass. California Turfgrass Culture. 17(2): 11.

69. ENGEL, R. E. 1967. A note on the development of puffiness in ¼-inch bentgrass turf with varied nitrogen fertilization. 1967 Report on Turfgrass Research at Rutgers University. New Jersey Agricultural Experiment Station Bulletin 818. pp. 46–49.

70. ENGEL, R. E., and R. B. ALDERFER. 1967. The effect of cultivation, topdressing, lime, nitrogen, and wetting agent on thatch development in ¼-inch bentgrass turf over a ten year period. 1967 Report on Turfgrass Research at Rutgers University. New Jersey Agricultural Experiment Station Bulletin 818. pp. 32–45.

71. ERWIN, L. E. 1941. Pathogenicity and control of *Corticium fuciforme*. Rhode Island Agricultural Experiment Station Bulletin 278. pp. 1–34.

72. ESCRITT, J. R. 1961. Report on yield trials with urea formaldehyde resins—1961. Journal of Sports Turf Research Institute. 10(37): 290–296.

73. ESCRITT, J. R., and H. J. LIDGATE. 1962. An investigation of the suitability of a urea formaldehyde fertilizer product for use on turf. Journal of Sports Turf Research Institute. 10(38): 385–393.

74. ———. 1964. Report on fertilizer trials. Journal of Sports Turf Research Institute. 40: 7–42.

75. ———. 1965. Further report on the suitability of a urea formaldehyde fertilizer product for use on turf. Journal of Sports Turf Research Institute. 41: 5–24.

76. EVANS, E. M., R. D. ROUSE, and R. T. GUDAUSKAS. 1964. Low soil potassium sets up Coastal for leafspot disease. Highlight of Agricultural Research. 11: 2.

77. EVANS, T. W. 1931. The root development of New Zealand browntop, chewing's fescue and fine-leaved sheep's fescue under putting green conditions. Journal of Board of Greenkeeping Research. 2(5): 118–124.

78. FARNHAM, R. S. 1964. Nitrogen fertilization of turf grasses. Minnesota Farm and Home Science. 22(1): 11–12.

79. FERGUSON, A. C. 1967. Some observations on winter injury in turfgrass experiments at the University of Manitoba. Summary of the 18th Annual RCGA National Turfgrass Conference. pp. 7–12.

80. FITTS, O. B. 1925. A preliminary study of the root growth of fine grasses under turf conditions. USGA Green Section Bulletin. 5: 58–62.

81. FOOTE, L. E. 1966. The effects of fertilizer rates on the amount of bare ground and roadside turf components. 1966 Agronomy Abstracts. p. 38.

82. FRAPS, G. S. 1932. The composition and fertilizing value of sewerage sludge. Texas Agricultural Experiment Station Bulletin No. 445. pp. 1–23.

83. FREEMAN, T. E. 1964. Influence of nitrogen on severity of *Piricularia grisea* infection of St. Augustine grass. Phytopathology. 54: 1187–1189.

84. FULLER, W. H., and K. G. CLARK. 1947. Microbiological studies on urea-formaldehyde preparation. Soil Science Society of America Proceedings. 12: 198–202.

85. FUNK, C. R., and R. E. ENGEL. 1966. Influence of variety, fertility level, and cutting height on weed invasion of Kentucky bluegrass. Proceedings of the New Jersey Lawn and Utility Turf Three-Day Course. pp. 10–17.

86. FUNK, C. R., R. E. ENGEL, and P. M. HALISKY. 1966. Performance of Kentucky bluegrass varieties as influenced by fertility level and cutting height. 1966 Report on Turfgrass Research at Rutgers University. New Jersey Agricultural Experiment Station Bulletin 816. pp. 7–21.

87. ———. 1967. Summer survival of turfgrass species as influenced by variety, fertility level and disease incidence. 1967 Report on Turfgrass Research at Rutgers University. New Jersey Agricultural Experiment Station Bulletin 818. pp. 71–77.

88. GARNER, E. S., and S. C. DAMON. 1929. The persistence of certain lawn grasses as affected by fertilization and competition. Rhode Island Agricultural Experiment Station Bulletin 217. pp. 1–22.

89. GARY, J. E. 1967. The vegetative establishment of four major turfgrasses and the response of stolonized "Meyer" zoysiagrass (*Zoysia japonica* var. Meyer) to mowing height, nitrogen fertilization, and light intensity. M.S. Thesis. Mississippi State College. pp. 1–50.

90. GIDDENS, J., H. F. PERKINS, and L. C. WALKER. 1962. Movement of nutrients in Coastal bermudagrass. Agronomy Journal. 54: 379–382.

91. GILBERT, F. A., A. H. BOWERS, and M. D. SANDERS. 1958. Effect of nitrogen sources in complete fertilizers on bluegrass turf. Agronomy Journal. 50: 320–323.

92. GILBERT, W. B., and D. L. DAVIS. 1967. Relationship of potassium nutrition and temperature stresses on turfgrasses. 1967 Agronomy Abstracts. p. 52.

93. GOETZE, N. 1957. Nitrogen fertility research. Proceedings of the 1957 Midwest Regional Turf Conference. pp. 79–82.

94. GOETZE, N. R., and W. H. DANIEL. 1956. Response of ryegrass turf to various organic nitrogen fertilizer materials. 1956 Agronomy Abstracts. p. 89.

95. ———. 1957. Turf quality as influenced by nitrogen fertilization. 1957 Agronomy Abstracts. p. 77.

96. ———. 1957. Nitrogen fairway fertilization research. Midwest Turf News and Research. pp. 1–2.

97. GOSS, R. L. 1962. Fertility effects on disease development. Proceedings of the 16th Annual Northwest Turfgrass Conference. pp. 83–85.

98. ———. 1963. Turfgrass fertility research report. Proceedings of the 17th Annual Northwest Turfgrass Conference. pp. 50–53.

99. ———. 1965. The fate of soil applied N, P, and K at various rates on bentgrass putting green turf (*Agrostis tenuis*, Sibth). 1965 Agronomy Abstracts. p. 45.

100. ———. 1967. The effects of urea on soil pH and calcium level. Northwest Turfgrass Topics. 9(3): 1–4.

101. ———. 1967. Some effects on soil pH and calcium levels from fertilizer applications to soil producing putting green turf. Journal of Sports Turf Research Institute. 43: 23–27.

102. GOSS, R. L., and C. J. GOULD. 1967. The effect of potassium on turfgrass and its relationship to turfgrass disease. 1967 Agronomy Abstracts. p. 52.

103. ———. 1967. Some interrelationships between fertility levels and *Ophiobolus* patch disease in turfgrasses. Agronomy Journal. 59: 149–151.

104. ———. 1968. Turfgrass diseases: the relationship of potassium. USGA Green Section Record. 5(5): 10–13.

105. ———. 1968. Some inter-relationships between fertility levels and *Fusarium* patch disease of turfgrasses. Journal of Sports Turf Research Institute. 44: 19–26.

106. GOSS, R. L., and A. G. LAW. 1967. Performance of bluegrass varieties at two cutting heights and two nitrogen levels. Agronomy Journal. 59: 516–518.

107. GRABER, L. F. 1931. Food reserves in relation to other factors limiting the growth of grasses. Plant Physiology. 6: 43–72.

108. GRABER, L. F., and H. W. REAM. 1931. Growth of bluegrass with various defoliations and abundant nitrogen supply. Journal of American Society of Agronomy. 23: 938–944.

109. HALL, T. D., and D. MOSES. 1931. The fertility factor in turf production. South African Journal of Science. 28: 180–201.

110. HALLOCK, D. L., R. H. BROWN, and R. E. BLASER. 1965. Relative yield and composition of Ky. 31 fescue and Coastal bermudagrass at four nitrogen levels. Agronomy Journal. 57: 539–542.

111. HANSON, A. A., and F. V. JUSKA. 1961. Winter root activity in Kentucky bluegrass (*Poa pratensis* L.). Agronomy Journal. 53: 372–374.

112. HARDING, R. B., T. W. EMBLETON, W. W. JONES, and T. M. RYAN. 1963. Leaching and gaseous losses of nitrogen from some nontilled California soils. Agronomy Journal. 55: 515–518.

113. HARPER, J. C. 1952. The effect of aerification on phosphorus penetration. Pennsylvania State College 21st Annual Turf Conference Proceedings. pp. 32–41.

114. HARRISON, C. M. 1931. Effect of cutting and fertilizer applications on grass development. Plant Physiology. 6: 669–684.

115. ———. 1934. Responses of Kentucky bluegrass to variation in temperature, light, cutting and fertilizing. Plant Physiology. 9: 83–106.

116. HAYS, J. T. 1968. Ureaforms in turfgrass fertilization. Summary of the Nineteenth Annual RCGA National Turfgrass Conference. pp. 13–24.

117. HAYS, J. T., and W. W. HADEN. 1966. Soluble fraction of ureaforms-nitrification, leaching, and burning properties. Journal of Agricultural and Food Chemistry. 14: 339–341.

118. HEILMAN, M. D., J. R. THOMAS, and L. N. NAMKEN. 1966. Reduction of nitrogen losses under irrigation by coating fertilizer granules. Agronomy Journal. 58: 77–80.

119. HENDERSON, D. W., W. C. BIANCHI, and L. D. DONEEN. 1955. Losses of ammoniacal fertilizers from sprinkler jets. USGA Journal and Turf Management. 8(3): 28–30.

120. HILL, A. 1934. Fertilizer trials on lawns at Craibstone. Journal of Board of Greenkeeping Research. 3(10): 153–160.

121. HODGES, T. K. 1965. Nutrient absorption by plants. 6th Illinois Turfgrass Conference Proceedings. pp. 1–8.

122. HOLT, D. A., and A. R. HILST. 1969. Daily variation in carbohydrate content of selected forage crops. Agronomy Journal. 61: 239–244.

123. HOLT, E. C., and J. E. ADAMS. 1955. The response of bermudagrass turf to fertilizer treatments. 1955 Agronomy Abstracts. p. 74.

124. HOLT, E. C., and R. L. DAVIS. 1948. Differential responses of Arlington and Norbeck bentgrasses to kinds and rates of fertilizers. Journal of American Society of Agronomy. 40: 282–284.

125. HOLBEN, F. J. 1949. Effects of potash on turf establishment. Pennsylvania State College 18th Annual Turf Conference Proceedings. pp. 96–101.

126. ———. 1950. Nitrogen-potash relationships as they affect the growth and diseases of turf grasses. Pennsylvania State College 19th Annual Turf Conference Proceedings. pp. 74–93.

127. ———. 1952. Potash-nitrogen fertilization on fescue and bent turf. Pennsylvania State College 21st Annual Turf Conference Proceedings. pp. 52–56.

128. HORN, G. C. 1964. Gainsville turf research. Proceedings of the University of Florida Turf-Grass Management Conference. 7: 152–159.

129. ———. 1965. The effects of potash sources at different rates on the response of Tifway bermudagrass. Proceedings of the University of Florida Turf-Grass Management Conference. 8: 188–191.

130. ———. 1965. The relationship between three rates each of nitrogen, phosphorus and potash in factorial combination on growth of four turfgrasses. Proceedings of the University of Florida Turf-Grass Management Conference. 8: 179–187.

131. ———. 1965. The effects of soluble and insoluble nitrogen sources on Tifgreen bermudagrass putting green turf. Proceedings of the University of Florida Turf-Grass Management Conference. 8: 170–178.

132. ———. 1965. The relationship of quality of Tifgreen bermudagrass to various sources and rates of nitrogen. Proceedings of the University of Florida Turf-Grass Management Conference. 8: 170–178.

133. ———. 1967. Comparison of nitrogen sources for 'Tifgreen' bermudagrass under putting green conditions. Florida Turf-Grass Association Bulletin. 14(1): 1–3.

134. HORN, G. C., and H. G. MEYERS. 1968. Tifway (T-419) bermudagrass potassium source-rate-frequency study progress report. Florida Turfgrass Research Plot Layout and Progress Report I. pp. 15–15b.

135. ――――. 1968. Tifgreen (T-328) bermudagrass nitrogen source-rate-frequency study. Florida Turfgrass Research Plot Layout and Progress Report I. pp. 19–19b.

136. HORN, G. C., and W. L. PRITCHETT. 1968. Role of potash in turfgrass fertilization. Florida Turf. 1(2): 1–8.

137. HUMPHRIES, A. W. 1962. The growth of some perennial grasses in waterlogged soil. I. The effect of waterlogging on the availability of nitrogen and phosphorus to the plant. Australian Journal of Agricultural Research. 13: 414–425.

138. HYLTON, L. O., A. ULRICH, D. R. CORNELIUS, and K. OKHI. 1965. Phosphorus nutrition of Italian ryegrass relative to growth, moisture content, and mineral constituents. Agronomy Journal. 57: 505–508.

139. HYLTON, L. O., A. ULRICH, and D. R. CORNELIUS. 1967. Potassium and sodium interrelations in growth and mineral content of Italian ryegrass. Agronomy Journal. 59: 311–314.

140. HYLTON, L. O., D. E. WILLIAMS, A. ULRICH, and D. R. CORNELIUS. 1964. Critical nitrate levels for growth of Italian ryegrass. Crop Science. 4: 16–19.

141. JACKSON, J. E., J. M. GOOD, and G. W. BURTON. 1961. Five-year study of nitrogen sources for 328 bermuda. Golfdom. 35(9): 36–82.

142. ――――. 1963. Nitrogen sources for bermudagrass greens. Golf Course Reporter. 31(4): 38–42.

143. JAGSCHITZ, J. A., and C. R. SKOGLEY. 1965. Turfgrass response to Dacthal and nitrogen. Agronomy Journal. 57: 35–38.

144. JEFFRIES, C. D. 1958. The potash problem as related to turfgrass management. 1958 Agronomy Abstracts. p. 65.

145. JONES, D. I., H. G. GRIFFITH, and R. J. WALTERS. 1965. The effect of nitrogen fertilizers on the water-soluble carbohydrate content of grasses. Journal of Agricultural Science. 62: 323–328.

146. JUSKA, F. V. 1959. Response of Meyer zoysia to lime and fertilizer treatments. Agronomy Journal. 51: 81–83.

147. JUSKA, F. V., and A. A. HANSON. 1961. The nitrogen variable in testing Kentucky bluegrass varieties for turf. Agronomy Journal. 53: 409–410.

148. ――――. 1967. Factors affecting *Poa annua* L. control. Weeds. 15(2): 98–101.

149. ――――. 1967. Effect of nitrogen sources, rates and time of application on the performance of Kentucky bluegrass turf. Proceedings of the American Society for Horticultural Science. 90: 413–419.

150. ――――. 1967. Phosphorus and its relationship to turfgrasses and *Poa annua* L. Proceedings of the 38th Annual International Turfgrass Conference. pp. 20–24.

151. ――――. 1967. Factors affecting *Poa annua* L. control. Golf Superintendent. 35: 46–51.

152. ――――. 1969. Nutritional requirements of *Poa annua* L. Agronomy Journal. 61: 466–468.

153. JUSKA, F. V., A. A. HANSON, and C. J. ERICKSON. 1965. Effects of phosphorus and other treatments on the development of red fescue, Merion, and common Kentucky bluegrass. Agronomy Journal. 57: 75–81.

154. JUSKA, F. V., A. A. HANSON, and A. W. HOVIN. 1969. Evaluation of tall fescue, *Festuca arundinacea* Schreb., for turf in the transition zone of the United States. Agronomy Journal. 61: 625–628.

155. JUSKA, F. V., J. TYSON, and C. M. HARRISON. 1955. The competitive relationship of Merion bluegrass as influenced by various mixtures, cutting heights and levels of nitrogen. Agronomy Journal. 47: 513–518.

156. KAEMPFFE, G. C., and O. R. LUNT. 1967. Availability of various fractions of urea-formaldehyde. Journal of Agricultural and Food Chemistry. 15: 967–971.

157. KILIAN, K. C., O. J. ATTOE, and L. E. ENGELBERT. 1966. Ureaformaldehyde as a slowly available form of nitrogen for Kentucky bluegrass. Agronomy Journal. 58: 204–206.

158. KOLLETT, J. R., A. J. WISNIEWSKI, and J. A. DEFRANCE. 1958. The effect of ureaform fertilizers in seedbeds for turfgrass. Park Maintenance. 11(5): 12–20.

159. KRALOVEC, R. O. 1954. Condensation products of urea and formaldehyde as fertilizer with controlled nitrogen availability. Journal of Agricultural and Food Chemistry. 2: 92–95.

160. KRESGE, C. B., and D. P. SATCHELL. 1960. Gaseous loss of ammonia from nitrogen fertilizers applied to soils. Agronomy Journal. 52: 104–107.

161. KRESGE, C. B., and S. E. YOUNTS. 1962. Effect of nitrogen source on yield and nitrogen content of bluegrass forage. Agronomy Journal. 54: 149–152.

162. LAGE, D. P., and E. C. ROBERTS. 1964. Influence of nitrogen and a calcined clay soil conditioner on development of Kentucky bluegrass turf from seed and sod. Proceedings of the American Society for Horticultural Science. 84: 630–635.

163. LAIRD, A. S. 1930. A study of the root systems of some important sod-forming grasses. Florida Agricultural Experiment Station Bulletin 211. pp. 1–27.

164. LARSON, R. A., and J. R. LOVE. 1967. Minor element deficiency symptoms in turfgrasses. Golfdom. 41: 31–34.

165. LATHAM, J. M. 1956. Results of the turfgrass soil fertility tests. Proceedings of the 10th Southeastern Turfgrass Conference. pp. 7–16.

166. LAWRENCE, T. 1963. The influence of fertilizer on the winter survival of intermediate wheatgrass following a long period of drought. Journal of British Grassland Society. 18: 292–294.

167. LETEY, J., L. H. STOLZY, O. R. LUNT, and V. B. YOUNGNER. 1964. Growth and nutrient uptake of Newport bluegrass as affected by soil oxygen. Plant and Soil. 20: 143–148.

168. LEUKEL, W. A., J. P. CAMP, and J. M. COLEMAN. 1934. Effect of frequent cutting and nitrate fertilization on the growth behavior and relative composition of pasture grasses. Florida Agricultural Experiment Station Bulletin 269. pp. 1–48.

169. LOBENSTEIN, C. W. 1965. Arsenic and low phosphate soils. 6th Illinois Turfgrass Conference Proceedings. pp. 1–5.

170. LONG, F. L., and G. M. VOLK. 1963. Availability of nitrogen contained in certain condensation products of reaction of urea with formaldehyde. Agronomy Journal. 55: 155–158.

171. LONG, J., and G. McVEY. 1969. Response of *Poa pratensis* varieties to certain nitrogen levels. 1969 Agronomy Abstracts. p. 55.

172. LONG, J. A., and E. C. HOLT. 1957. Nitrogen management practices for putting green turf. 1957 Agronomy Abstracts. p. 77.

173. LOVE, J. R. 1962. Mineral deficiency symptoms on turfgrass. I. Major and secondary elements. Wisconsin Academy of Sciences, Arts and Letters. 51: 135–140.

174. LOVVORN, R. L. 1945. The effect of defoliation, soil fertility, temperature, and length of day on the growth of some perennial grasses. Journal of American Society of Agronomy. 37: 570–582.

175. LUNT, O. R. 1954. Relation of soils and fertilizers to wear resistance of turfgrass. Proceedings of Northern California Turfgrass Conference. pp. 1–2.

176. ———. 1959. Relation of soils and fertilizers to wear resistance of turfgrass. North California Turfgrass Institute. pp. 1–2.

177. ———. 1963. Slow release fertilizer materials. Golf Course Reporter. 31(4): 30–36.

178. LUNT, O. R., R. L. BRANSON, and S. B. CLARK. 1967. Response of five grass species to phosphorus on six soils. California Turfgrass Culture. 17(4): 25–26.

179. LUNT, O. R., and J. J. OERTLI. 1962. Controlled release of fertilizer minerals by incapsulating membranes: II. Efficiency of recovery, influence of soil moisture, mode of application and other considerations related to use. Soil Science Society of America Proceedings. 26: 584–587.

180. LUNT, O. R., and V. B. YOUNGNER. 1956. Turfgrass fertilization programs with materials of low solubility. 1956 Agronomy Abstracts. p. 90.

181. LUNT, O. R., V. B. YOUNGNER, and A. H. KHADR. 1964. Critical nutrient levels in Newport bluegrass. 1964 Agronomy Abstracts. p. 106.

182. MACLEOD, L. B., and R. B. CARSON. 1966. Influence of K on the yield and chemical compo-

sition of grasses grown in hydroponic culture with 12, 50 and 75% of the N supplied as NH₄+¹. Agronomy Journal. 58: 52–57.

183. MacLeod, N. H., E. C. Roberts, and J. E. Steckel. 1958. A comparison of turfgrass response to liquid and solid fertilizer. 1958 Agronomy Abstracts. p. 66.

184. Madison, J. H. 1960. Observations—experimental fertility plots. Proceedings Northern California Turfgrass Institute. pp. 35–37.

185. ———. 1960. The effect of the root temperature of turfgrasses on the uptake of nitrogen as urea, ammonium, and nitrate. 1960 Agronomy Abstracts. p. 71.

186. ———. 1962. Turfgrass ecology. Effects of mowing, irrigation, and nitrogen treatments of *Agrostis palustris* Huds., 'Seaside' and *Agrostis tenuis* Sibth., 'Highland' on population, yield, rooting, and cover. Agronomy Journal. 54: 407–412.

187. Madison, J. H., and A. H. Anderson. 1963. A chlorophyll index to measure turfgrass response. Agronomy Journal. 55: 461–464.

188. Madison, J. H., J. P. Lawrence, and T. K. Hodges. 1960. Pink snowmold on bentgrass as affected by irrigation and fertilizer. Agronomy Journal. 52: 591–592.

189. Mantell, A., and G. Stanhill. 1966. Comparison of methods for evaluating the response of lawngrass to irrigation and nitrogen treatment. Agronomy Journal. 58: 465–468.

190. Markland, F. E., and E. C. Roberts. 1967. Influence of varying nitrogen and potassium levels on growth and mineral composition of *Agrostis palustris* Huds. 1967 Agronomy Abstracts. p. 53.

191. ———. 1969. Influence of nitrogen fertilizers on Washington creeping bentgrass, *Agrostis palustris* Huds. I. Growth and mineral composition. Agronomy Journal. 61: 698–700.

192. Markland, F. E., E. C. Roberts, and L. R. Frederick. 1969. Influence of nitrogen fertilizers on Washington creeping bentgrass, *Agrostis palustris* Huds. II. Incidence of dollar spot, *Sclerotinia homoeocarpa*, infection. Agronomy Journal. 61: 701–705.

193. Mays, D. A., and G. L. Terman. 1969. Sulfur-coated urea and uncoated soluble nitrogen fertilizers for fescue forage. Agronomy Journal. 61: 489–492.

194. McCool, M. M., and R. L. Cook. 1930. Rate of intake, accumulation and transformation of nitrate nitrogen by small grains and Kentucky bluegrass. Journal of American Society of Agronomy. 22: 757–764.

195. McVey, G. R. 1967. Response of turfgrass seedlings to various phosphorus sources. 1967 Agronomy Abstracts. p. 53.

196. ———. 1968. How seedlings respond to phosphorus. Weeds, Trees and Turf. 7(6): 18–19.

197. McVickar, M. H. 1942. Manganese status of some important Ohio soil types and uptake of manganese by Kentucky bluegrass. Journal of American Society of Agronomy. 34: 123–128.

198. Menn, W. G., and G. G. McBee. 1967. A study of certain nutritional requirements for Tifgreen bermudagrass utilizing a hydroponic design. 1967 Agronomy Abstracts. p. 54.

199. Meyers, H. C. 1964. Effects of nitrogen levels and vertical mowing intensity and frequency on growth and chemical composition of Ormond bermudagrass. Proceedings of the 12th Annual University of Florida Turf-Grass Management Conference. pp. 77–79.

200. Miller, R. A. 1963. Bluegrass growth response to various formulations, placement and rates of plant food. Illinois Turfgrass Conference Proceedings. pp. 19–22.

201. Miller, R. W. 1966. The effect of certain management practices on the botanical composition and winter injury to turf containing a mixture of Kentucky bluegrass (*Poa pratensis*, L.) and tall fescue (*Festuca arundinacea*, Schreb.). 7th Illinois Turfgrass Conference Proceedings. pp. 39–46.

202. Monroe, C. A., G. D. Coorts, and C. R. Skogley. 1969. Effects of nitrogen-potassium levels on the growth and chemical composition of Kentucky bluegrass. Agronomy Journal. 61: 294–296.

203. Monteith, J., and J. W. Bengston. 1939. Experiments with fertilizers on bluegrass turf. Turf Culture. 1(3): 153–191.

204. MOORE, L. D., H. B. COUCH, and J. R. BLOOM. 1963. Influence of environment on diseases of turfgrasses. III. Effect of nutrition, pH, soil temperature, air temperature, and soil moisture on *Pythium* blight of Highland bentgrass. Phytopathology. 53: 53–57.

205. MOSER, L. E. 1967. Rhizome and tiller development of Kentucky bluegrass, *Poa pratensis* L., as influenced by photoperiod, vernalization, nitrogen level, growth regulators and variety. Ph.D. Thesis. Ohio State University. pp. 1–143.

206. MRUK, C. K., A. J. WISNIEWSKI, and J. A. DeFRANCE. 1957. A comparison of urea-formaldehyde material for turfgrass fertilization. Golf Course Reporter. 25(3): 5–8.

207. MULLER, J. F. 1929. The value of raw sewage sludge as fertilizer. Soil Science. 28: 423–432.

208. MUNSELL, R. I., and B. A. BROWN. 1939. The nitrogen content of grasses as influenced by kind, frequency of application, and amount of nitrogenous fertilizer. Journal of American Society of Agronomy. 31: 388–398.

209. MUSE, R. R., and H. B. COUCH. 1965. Influence of environment on diseases of turfgrasses. IV. Effect of nutrition and soil moisture on *Corticium* red thread of creeping red fescue. Phytopathology. 55: 507–510.

210. MUSSER, H. B. 1948. Effects of soil acidity and available phosphorus on population changes in mixed Kentucky bluegrass-bent turf. Journal of American Society of Agronomy. 40: 614–620.

211. ———. 1949. Tests of urea-formaldehyde formulations. Pennsylvania State College 18th Annual Turf Conference Proceedings. pp. 102–107.

212. MUSSER, H. B., and J. M. DUICH. 1958. Response of creeping bentgrass putting-green turf to urea-form compounds and other nitrogenous fertilizers. Agronomy Journal. 50: 381–384.

213. MUSSER, H. B., J. R. WATSON, JR., J. P. STANFORD, and J. C. HARPER. 1951. Urea-formaldehyde and other nitrogenous fertilizers for use on turf. Pennsylvania Agriculture Experiment Station Bulletin 542. pp. 1–14.

214. NELLER, J. R., and C. E. HUTTON. 1957. Comparison of surface and subsurface placement of superphosphate on growth and uptake of phosphorus by sodded grasses. Agronomy Journal. 49: 347–351.

215. NOER, O. J. 1926. Activated sludge: its production, composition, and value as a fertilizer. Journal of American Society of Agronomy. 18: 953–962.

216. ———. 1959. Better bent greens. Turf Service Bureau of the Milwaukee Sewerage Commission. Bulletin No. 2. pp. 1–33.

217. NOER, O. J., and J. E. HAMNER. 1956. The season's yield and chemical composition from a bermuda grass green. 1956 Agronomy Abstracts. p. 90.

218. NOER, O. J., and C. G. WILSON. 1959. Better turfgrass is through soil analysis and controlled fertility. Proceedings of the Midwest Regional Turf Conference. pp. 69–72.

219. NORTH, H. F. A., and T. E. ODLAND. 1934. Putting green grasses and their management. Rhode Island Agricultural Experiment Station Bulletin 245. pp. 1–44.

220. NOWAKOWSKI, T. Z. 1961. The effect of different nitrogenous fertilizers, applied as solids or solutions, on the yield and nitrate-N content of established grass and newly sown ryegrass. Journal of Agricultural Science. 56: 287–292.

221. ———. 1962. Effects of nitrogen fertilizers on total nitrogen, soluble nitrogen, and soluble carbohydrate content of grass. Journal of Agricultural Science. 59: 387–392.

222. NOWAKOWSKI, T. Z., R. K. CUNNINGHAM, and K. F. NIELSEN. 1965. Nitrogen fractions and soluble carbohydrates in Italian ryegrass. I. Effects of soil temperature, form, and level of nitrogen. Journal of Science of Food and Agriculture. 16: 124–134.

223. NOWAKOWSKI, T. Z., A. C. D. NEWMAN, and A. PENNY. 1967. The effect of different nitrogenous fertilizers applied as solids or solutions on dry matter yields and nitrogen uptake by grass. Part III. Journal of Agricultural Science. 68: 123–130.

224. NUTTER, G. C., W. L. PRITCHETT, and R. W. WHITE, JR. 1956. The relationship of soil reaction and minor elements on the growth and condition of four southern turfgrasses. 1956 Agronomy Abstracts. p. 91.

225. O'DONNELL, J. 1966. Nutrient removal and rooting pattern of Merion Kentucky bluegrass. 1966 Wisconsin Turfgrass Conference Proceedings. pp. 29–31.

226. OERTLI, J. J. 1963. Nutrient disorders in turfgrass. California Turfgrass Culture. 3(3): 17–19.

227. OERTLI, J. J., and O. R. LUNT. 1962. Controlled release of fertilizer minerals by incapsulating membranes: I. Factors influencing the rate of release. Soil Science Society of America Proceedings. 26: 579–583.

228. OERTLI, J. J., O. R. LUNT, and V. B. YOUNGNER. 1960. Nutrient disorders in two turfgrass species. 1960 Agronomy Abstracts. p. 71.

229. ———. 1961. Boron toxicity in several turfgrass species. Agronomy Journal. 53: 262–265.

230. PALMERTREE, H. D., E. L. KIMBROUGH, and C. Y. WARD. 1966. The evaluation of several sources and rates of nitrogen for fertilization of established roadside turf. 1966 Agronomy Abstracts. p. 39.

231. PARKS, W. L., and W. B. FISHER, JR. 1958. Influence of soil temperature and nitrogen on ryegrass growth and chemical composition. Soil Science Society of America Proceedings. 22: 257–259.

232. PATTERSON, J. K., and A. G. LAW. 1960. Nitrogen source test. Proceedings of the 14th Annual Northwest Turf Conference. pp. 17–19.

233. PELLETT, H. M., and E. C. ROBERTS. 1963. Effects of mineral nutrition on high temperature induced growth retardation of Kentucky bluegrass. Agronomy Journal. 55: 473–476.

234. PHILLIPPE, P. M. 1942. Effects of some essential elements on the growth and development of Kentucky bluegrass (*Poa pratensis*). Ph.D. Thesis. Ohio State University. pp. 1–57.

235. POWELL, A. J., R. E. BLASER, and R. E. SCHMIDT. 1967. Effect of nitrogen on winter root growth of bentgrass. Agronomy Journal. 59: 529–530.

236. ———. 1967. Physiological and color aspects of turfgrasses with fall and winter nitrogen. Agronomy Journal. 59: 303–307.

237. POWELL, A. J., and W. H. MCKEE, JR. 1965. Effect of different nitrogen sources on winter and spring response of cool season grass. 1965 Agronomy Abstracts. p. 46.

238. PRITCHETT, W. L., R. W. WHITE, JR., and G. C. NUTTER. 1956. Effects of frequencies and rates of fertilization on the soil fertility status under Bayshore and Everglades I bermudagrass putting green turf. 1966 Agronomy Abstracts. p. 91.

239. REHLING, C. J., and E. TRUOG. 1940. "Milorganite" as a source of minor nutrient elements for plants. Journal of American Society of Agronomy. 32: 894–906.

240. REID, M. E. 1933. Effect of variations in concentration of mineral nutrients upon the growth of several types of turf grasses. USGA Green Section Bulletin. 13(5): 122–131.

241. RICHARDSON, H. L. 1934. Studies on calcium cyanamide. IV. The use of calcium cyanamide and other forms of nitrogen on grassland. Journal of Agricultural Science. 24: 491–510.

242. ROBERTS, E. C. 1958. The grass plant—feeding and cutting. Golf Course Reporter. 26(3): 5–8.

243. ———. 1960. Disease control by turfgrass nutrition. Proceedings of the 31st Annual Conference of the GCSAA. p. 31–37.

244. ———. 1963. Relationships between mineral nutrition of turf-grass and disease susceptibility. Golf Course Reporter. 31(5): 52–57.

245. ———. 1965. What to expect from a nitrogen fertilizer. Summary of the 16th Annual RCGA Sports Turfgrass Conference. pp. 37–46.

246. ROBERTS, E. C., and E. J. BREDAKIS. 1960. What, why and how of turfgrass root development. Golf Course Reporter. 28: 12–24.

247. ROBERTS, E. C., and F. E. MARKLAND. 1966. Nitrogen induced changes in bluegrass-red fescue turf populations. 1966 Agronomy Abstracts. p. 36.

248. ———. 1967. Effect of phosphorus levels and two growth regulators on arsenic toxicity in *Poa annua* L. 1967 Agronomy Abstracts. p. 57.

249. ———. 1967. Fertilizing helps turf crowd out weeds. Weeds, Trees and Turf. 6(3): 12–24.

250. RUMBURG, C. B., R. E. ENGEL, and W. F. MEGGITT. 1960. Effect of phosphorus concentration on the absorption of arsenate by oats from nutrient solution. Agronomy Journal. 52: 452–453.

251. SATARI, A. M. 1967. Effects of various rates and combinations of nitrogen, phosphorus, potassium, and cutting heights on the development of rhizome, root, total available carbohydrate, and foliar composition of *Poa pratensis* L. Merion grown on Houghton muck. Ph.D. Thesis. Michigan State University. pp. 1–127.

252. SCHMIDT, R. E. 1965. The influence of night temperatures, light intensity, and nitrogen fertility on growth and physiological changes in *Agrostis* and *Cynodon* turfgrasses. 1965 Agronomy Abstracts. p. 47.

253. SCHMIDT, R. E., and R. E. BLASER. 1967. Effect of temperature, light, and nitrogen on growth and metabolism of 'Cohansey' bentgrass (*Agrostis palustris* Huds.). Crop Science. 7: 447–451.

254. SHARP, W. C. 1965. Effects of clipping and nitrogen fertilization on seed production of creeping red fescue. Agronomy Journal. 57: 252–253.

255. SILVA, J. A. 1966. Liquid vs. solid fertilizers. Weeds, Trees and Turf. 5(3): 8–16.

256. SIMPSON, D. H. M., and S. W. Melsted. 1962. Gaseous ammonia losses from urea solutions applied as a foliar spray to various grass sods. Soil Science Society of America Proceedings. 26: 186–189.

257. SINHA, J. D., L. E. FOOTE, J. A. JACKOBS, and C. N. HITTLE. 1962. The influence of rates and time of application of organic and inorganic nitrogen on the maintenance of sod on highway slopes. 1962 Agronomy Abstracts. p. 104.

258. SKOGLEY, C. R., and J. W. KING. 1968. Controlled release nitrogen fertilization of turfgrass. Agronomy Journal. 60: 61–64.

259. SKOGLEY, C. R., and F. B. LEDEBOER. 1968. Evaluation of several Kentucky bluegrass and red fescue strains maintained as lawn turf under three levels of fertility. Agronomy Journal. 60: 47–49.

260. SMITH, D., and O. R. JEWISS. 1966. Effects of temperature and nitrogen supply on the growth of timothy (*Phleum pratense* L.). Annals of Applied Biology. 58: 145–157.

261. SMITH, J. D. 1954. Fungi and turf diseases. 4. *Corticium* disease. Journal of Sports Turf Research Institute. 8(30): 365–377.

262. SPENCER, J. T., H. H. JEWETT, and E. N. FERGUS. 1949. Seed production of Kentucky bluegrass as influenced by insects, fertilizers and sod management. Kentucky Agricultural Experiment Station Bulletin 535. pp. 1–44.

263. SPRAGUE, H. B. 1933. Root development of perennial grasses and its relation to soil conditions. Soil Science. 36: 189–209.

264. ———. 1934. Utilization of nutrients by colonial bent (*Agrostis tenuis*) and Kentucky bluegrass (*Poa pratensis*). New Jersey Agricultural Experiment Station Bulletin 570. pp. 1–16.

265. SPRAGUE, H. B., and G. W. BURTON. 1937. Annual bluegrass (*Poa annua* L.), and its requirements for growth. New Jersey Agricultural Experiment Station Bulletin 630. pp. 1–24.

266. SPRAGUE, H. B., and E. E. EVAUL. 1930. Experiments with turf grasses in New Jersey. New Jersey Agricultural Experiment Station Bulletin 497. pp. 1–55.

267. STANFORD, J. P., and H. B. MUSSER. 1950. Availability of nitrogen from urea-form and other nitrogenous materials. Pennsylvania State College 19th Annual Turf Conference Proceedings. pp. 62–73.

268. STURKIE, D. G., and R. D. ROUSE. 1967. Response of zoysia and Tiflawn bermuda to P and K. 1967 Agronomy Abstracts. p. 54.

269. SULLIVAN, E. F. 1962. Effects of soil reaction, clipping height, and nitrogen fertilization on the productivity of Kentucky bluegrass sod transplants in pot culture. Agronomy Journal. 54: 261–263.

270. THOMPSON, W. R., and C. Y. WARD. 1966. Effects of potassium nutrition and clipping heights on total available carbohydrates in Tifgreen bermudagrass (*Cynodon* spp.). 1966 Agronomy Abstracts. p. 37.

271. VOLK, G. M. 1956. Efficiency of various nitrogen sources for pasture grasses in large lysimeters of Lakeland fine sand. Soil Science Society of America Proceedings. 20: 41–45.

272. ——. 1959. Volatile loss of ammonia following surface application of urea to turf or bare soils. Agronomy Journal. 51: 746–749.

273. ——. 1966. Liquid versus dry fertilizers for turf. Proceedings of the University of Florida Turf-Grass Management Conference. 14: 52–57.

274. ——. 1966. Efficiency of fertilizer urea as affected by method of application, soil moisture, and lime. Agronomy Journal. 58: 249–253.

275. VOLK, G. M., and G. C. HORN. 1965. Response of Tifgreen bermudagrass to soluble and slowly available nitrogen sources as measured by visual ratings and turf weights. Proceedings of the University of Florida Turf-Grass Management Conference. 8: 147–152.

276. VOSE, P. B. 1962. Nutritional response and shoot/root ratio as factors in the composition and yield of genotypes of perennial ryegrass, *Lolium perenne* L. Annals of Botany N. S. 26: 425–437.

277. WADDINGTON, D. V., J. M. DUICH, and E. L. MOBERG. 1969. Lawn fertilizer test. Pennsylvania Agricultural Experiment Station Progress Report 296. pp. 1–58.

278. WADDINGTON, D. V., J. TROLL, and B. HAWES. 1964. Effect of various fertilizers on turfgrass yield, color and composition. Agronomy Journal. 56: 221–222.

279. WARD, C. Y., and W. R. THOMPSON, JR. 1965. Influence of nitrogen source and frequency of application on bermudagrass (*Cynodon* spp.) golf green turf. 1965 Agronomy Abstracts. p. 47.

280. WATSON, J. R. 1956. Snowmold control. 1956 Agronomy Abstracts. p. 92.

281. WATSON, S. A., and C. W. STEWART. 1958. Improved turf from corn gluten and corn hulls. Golf Course Reporter. 26(4): 5–8.

282. WELLS, H. D. 1957. Southern turfgrass disease control. Proceedings of the 11th Annual Southeastern Turfgrass Conference. pp. 16–28.

283. WELTON, F. A. 1932. Nitrogen carriers for turf grass. Ohio Agricultural Experiment Station Bulletin 497. pp. 37–40.

284. WELTON, F. A., and J. C. CARROLL. 1940. Lawn experiments. Ohio Agricultural Experiment Station Bulletin 613. pp. 1–43.

285. WHITE, R. W., G. C. NUTTER, and W. L. PRITCHETT. 1956. The effect of frequency and rate of fertilization on the performance of Bayshore and Everglades I bermudagrass putting green turf. 1956 Agronomy Abstracts. p. 92.

286. WIDDOWSON, F. V., A. PENNY, and R. J. B. WILLIAMS. 1962. An experiment comparing urea-formaldehyde fertilizer with 'nitrochalk' for Italian ryegrass. Journal of Agricultural Science. 59: 263–268.

287. WILLARD, C. J., and G. M. MCCLURE. 1932. The quantitative development of tops and roots in bluegrass with an improved method of obtaining root yields. Journal of American Society of Agronomy. 24: 509–514.

288. WISNIEWSKI, A. J., and J. A. DEFRANCE. 1957. Results of ureaformaldehyde fertilization on lawn and fairway turf. 1957 Agronomy Abstracts. p. 79.

289. WISNIEWSKI, A. J., J. A. DEFRANCE, and J. R. KOLLETT. 1958. Results of ureaform fertilization on lawn and fairway turf. Agronomy Journal. 50: 575–576.

290. YEE, J. Y., and K. S. LOVE. 1946. Nitrification of ureaformaldehyde reaction products. Soil Science Society of America Proceedings. 11: 389–392.

291. YOUNGNER, V. B., and W. TAVENER. 1957. Calcium cyanamide fertilization of bermudagrass turf. 1957 Agronomy Abstracts. p. 79.

292. ZAHNLEY, J. W., and F. L. DULEY. 1934. The effect of nitrogenous fertilizers on the growth of lawn grasses. Journal of American Society of Agronomy. 26: 231–234.

293. ZANONI, L. J., L. F. MICHELSON, W. G. COLBY, and M. DRAKE. 1969. Factors affecting carbohydrate reserves of cool season turfgrasses. Agronomy Journal. 61: 195–198.

chapter 14

Irrigation

The total quantity and seasonal distribution of precipitation is usually not adequate to maintain a dense, green turf of acceptable quality. Irrigation practices are needed under these conditions if an adequate shoot density and color are to be maintained (38, 46) (Fig. 14-1). The microclimate of irrigated turfgrass areas is altered considerably. Both the soil and air temperatures adjacent to an irrigated turf are cooler than unirrigated areas (4, 7). The atmospheric relative humidity is also higher (7).

Irrigation is one of the most difficult aspects of turfgrass culture. Once irrigation

Sigure 14-1. Irrigation of a St. Augustinegrass lawn in southern Florida.

is initiated during a drought, it should be continued for the duration of the drought (48). Sporadic irrigation is not effective and can actually be detrimental to the turf in terms of reduced carbohydrate reserves, vigor, and drought resistance (2). Considerations in developing an irrigation program include (a) irrigation frequency, (b) quantity of water to apply, (c) water source, (d) water quality, and (e) method of irrigation.

When to Irrigate

The timing of irrigations is governed by when the turfgrass plants wilt. The preferred time to irrigate is just prior to visible wilting. This is difficult to achieve because the use pattern of the particular turfgrass area may prevent irrigation at the desired time. Turfgrass wilt occurs when an internal plant water stress develops because the water loss rate through evapotranspiration exceeds the rate of absorption through the root system. The plant, atmospheric, and soil factors influencing water absorption and evapotranspiration are discussed in detail in Chapter 8.

Irrigation must be applied prior to permanent wilting in order to avoid serious injury or permanent damage to the turf. Evidence of foot printing on a turf is an indication that wilt is imminent. The **foot printing** technique involves walking across the turfgrass area and observing the rate at which the turfgrass leaves return to their original, upright position. Turgid leaves having a positive water balance return quickly, whereas leaves possessing a negative water balance are slow to recover, leaving a distinct impression in the turf from the pressure of the footprint. An additional technique that may be utilized as a guide in timing irrigations is the use of a soil probe to check the soil moisture content and distribution in the upper 6 to 8 in. of the soil profile.

Thatched turfs present special irrigation problems. A short pre-sprinkling assists in moistening the thatch layer. As a result, the infiltration velocity is improved and the subsequent, deep irrigation can proceed in a more normal fashion. Severely thatched turfs may dictate the use of more frequent, light irrigations.

The frequency and diurnal timing of irrigation is particularly important in influencing the degree of disease development on turfgrasses. Many turfgrass pathogens require free water for spore germination and the penetration of germ tubes into the turfgrass leaf tissue. Frequent irrigation or irrigation at times that permit free water droplets to remain on the leaf for an extended period of time results in increased disease problems (21). Midday irrigation is a preferred time in terms of minimizing disease development since the evaporation rate is quite rapid. Late afternoon or evening irrigation permits free water droplets to remain on the leaf surface for the longest time. Early morning irrigation reduces the time that water droplets and leaf exudates persist on the leaves.

Irrigation during midday when evapotranspiration is usually the most rapid would be preferred from the standpoint of minimizing disease development. However, the use patterns on certain turfgrass areas, particularly golf courses and similar recreational turfs that are intensively used during midday, do not permit irrigation at this time. As a result, nocturnal irrigation may have to be practiced even though pathogen activity is enhanced. A compensating factor is a reduction in the amount of water lost by evaporation. Hence, less water is necessary when irrigating at night in comparison to the daytime. Also, poor water distribution caused by wind is minimized.

On a day-to-day basis, irrigation should be timed to avoid high soil moisture contents when intense traffic is anticipated. For example, irrigation of a football field should take place well in advance of scheduled games so that the surface soil moisture content is low enough to minimize soil compaction. Sometimes irrigation timing is controlled by water availability. This is most commonly a problem if the water is obtained from a city water system and especially when the distribution system or supply is inadequate.

Irrigation Frequency

Irrigating too frequently can be just as detrimental to turfgrass vigor and quality as not irrigating frequently enough. A greater portion of the applied water is lost by evaporation when excessively frequent irrigation is practiced. The water use rate also increases as the irrigation frequency is increased (1). The optimum irrigation frequency should maintain the soil water level in the upper 50% of the available range. As the irrigation frequency is increased from the optimum, positive plant water balance, the following turfgrass responses are observed:

1. Reduced shoot growth (1, 21).
2. Increased shoot density (21, 22).
3. Reduced chlorophyll content.
4. Reduced succulence.
5. Reduced root growth (21, 47).

The net result of an irrigation frequency in excess of that required to maintain a positive plant water balance is an overall reduction in turfgrass vigor and quality as evidenced by the decreased root and shoot growth. The weakened, less vigorous turf is more subject to weed invasion, diseases, insects, nematodes, and damage from traffic. In addition, the turfgrass plants may become lighter green; more spindly; and less tolerant to heat, cold, and drought stress. As in the case of increased mowing frequency or reduced cutting height, more frequent irrigation results in increased shoot density and verdure (20, 21, 22).

Excessively frequent irrigation cannot be defined specifically because of the great variability in soils and water use rates. Irrigating three times a week may be excessively frequent on a clay loam soil in a cool humid climate, while this irrigation frequency may be inadequate on a sandy soil in a hot arid climate. More frequent irrigation is required on droughty, coarse textured soils where the water retention is low (34, 42, 43, 45). Although more frequent irrigation is required on sandy soils, the total amount of water applied per irrigation is less.

There are two situations where frequent irrigation on a semidaily, daily, or every other day basis is necessary. The water absorption capability is quite limited in both situations. The first is on newly seeded or sodded areas. Light, midday irrigation on a daily basis is necessary to ensure effective establishment and rooting without desiccation (Fig. 14-2).

The second situation is most common on greens in midsummer where the root system is extremely shallow due to heat stress or damage by diseases, nematodes, and insects. Wet wilt and death of a bentgrass green can occur quite rapidly unless preventive syringing is practiced. In this case, the evapotranspiration rate far exceeds the water absorption capability of the restricted root system even though there is an adequate supply of soil water available.

Syringing is the application of light amounts of water for the purpose of (a) preventing wilt; (b) reducing transpiration; (c) cooling the turf; or (d) removing frost, dew, or exudations. Syringing can be accomplished with hand hoses using "rose-type" nozzles or by a brief 1 to 5 min activation of a pop-up automatic irrigation system. Syringing should only be utilized as necessary rather than as a routine, daily practice. Syringing for wilt prevention is usually done between 11:00 a.m. and 3:00 p.m. and may be necessary twice within a single day when under extreme atmospheric water stress. Syringing 2 to 3 hr prior to the diurnal temperature maximum provides the best cooling effect.

Figure 14-2. The effect of daily (left) versus every other day (right) irrigation on newly transplanted sod. Note desiccation injury of sod irrigated every other day.

Quantity of Water Applied

The amount of water applied at any one time depends on the (a) amount of water in the soil, (b) soil water retention characteristics, and (c) the water application rate relative to the soil infiltration and percolation rates. The quantity of irrigation water applied is a function of the rate and duration of application. The duration of irrigation should be sufficiently long to permit wetting of a majority of the turfgrass root zone, which is normally a minimum of 6 to 8 in. Irrigations that wet the soil to a considerable soil depth encourage deep rooting and result in a more vigorous, high quality turf. It is preferable to apply sufficient water to achieve contact between the subsurface moisture and the surface applied moisture. This contact is important in order to avoid a dry soil layer through which the turfgrass roots will not penetrate readily. A soil probe can be utilized to ascertain whether proper contact between the surface and subsoil moisture has been achieved.

The application of an excessive quantity of water should be avoided because it may result in a waterlogged condition with (a) reduced soil oxygen levels, (b) a restricted root system, (c) increased disease development, (d) increased compaction proneness, and (e) an overall reduction in turfgrass vigor and quality (46, 47). It is especially important not to overwater or saturate the soil if the internal soil drainage is poor or inadequate. An excessive water application rate is also wasteful since some water is lost by surface runoff and gravitational percolation through the soil. In addition, leaching of plant nutrients is increased.

The total quantity of water applied affects the species composition of a turfgrass community. An irrigation program that maintains a high soil moisture level encourages such species as bentgrass and rough bluegrass, whereas more judicious irrigation that permits the soil moisture to be substantially depleted encourages such species as red fescue (46, 47). Red fescue does not tolerate high soil moisture levels for an extended period of time.

Rate of water application. The water application rate should be adjusted in relation to the infiltration velocity of the soil so that a minimum amount of water is lost by surface runoff. Surface runoff of irrigation water is wasteful and costly and should be avoided if at all possible. Irrigation systems can be designed to apply water at rates ranging from 0.1 to 1.0 in. per hr. Soils vary greatly in the infiltration rate depending on the soil texture, structure, degree of compaction, and slope (10, 28). Compacted, fine textured soils of poor structure usually have a very low infiltration rate in the range of 0.05 to 0.1 in. per hr (11, 29). There are some soils that have an infiltration rate that is so slow it is difficult to fully wet the root zone in one irrigation. More frequent, light irrigations are necessary on such soils.

The water application rate is controlled by the proper spacing of heads, total operating time, and type of sprinkler heads, especially the nozzle size. It is preferable to use sprinkler heads that break up the water stream into small droplets. The smaller irrigation drops result in less soil compaction and better water distribution compared to a larger drop size but are more adversely affected by wind. Larger nozzle sizes have a higher water application rate and larger drops that strike the soil with considerable force (3).

Water Source

It is important to have an adequate, independent, high quality water source on turfgrass areas that must be irrigated regularly. The quantity of water available should be sufficient so that (a) it is not limiting the water delivery rate of the irrigation system and (b) the irrigation system can be operated at full capacity in accordance with the original design. The water supply should be located near the center of the area to be irrigated.

An independent water source is best. For example, utilizing city water is not desirable from the standpoint that the city government may place controls on the specific times when the irrigation of turfgrass areas can be practiced. As a result, the timing of irrigation is dictated by when the water is available rather than by the turfgrass needs. Generally, turfgrass irrigation systems should not be connected directly into a city water system. Governmental restrictions on water use practices will probably increase and cause more serious problems in the future.

The three common sources of irrigation water are from (a) ground water sources obtained by shallow or deep wells; (b) lakes, reservoirs, or ponds; and (c) perennial flowing streams or rivers. Another water source that may be used more commonly on turfs in the future is sewerage effluent and industrial water. This type of water is already being used successfully for turfgrass irrigation.

Wells located on the property provide an independent water source. The water yield and drawdown of the well should be ascertained. A problem sometimes associated with well water is the presence of salts and sand. These materials can cause trouble in automatic systems. The initial cost of developing a well as a water source is quite high.

Lakes, particularly if located on the property, are a dependable water source (Fig. 14-3). Larger lakes, in terms of water storage capacity, are preferred in order to have sufficient water for use during extended droughts. A determination should be made of the source and rate at which the water is supplied to the lake from springs or surface drainage. Storage ponds or reservoirs are sometimes utilized with

Figure 14-3. A storage pond with the dual purpose of functioning as a source of irrigation water and a water hazard on a golf hole.

water pumped into the pond from a well or city water supply and then pumped directly out of the pond for turfgrass irrigation. In this situation the storage pond should be large enough to provide sufficient water for at least two or preferably three irrigations. Storage ponds can be lined with plastic sheets or a clay seal to minimized seepage loss when necessary. Irrigation lakes should be kept free of algae, weeds, and other foreign debris. Filters on suction lines are necessary if the lake contains such materials.

Perennial streams and rivers are a third source of water that may or may not be acceptable depending upon their size. It should be determined whether the minimum annual flow of the stream exceeds the maximum amount of water that is required for irrigation. Prior rights for use of the river water should also be determined. The pollution problem of silt and industrial wastes in many streams and rivers is an objectionable factor for turfgrass irrigation. Filter screens are required to remove foreign material from such water sources prior to pumping into the irrigation system.

A fourth water source that may become more common is nonpotable water. This water (a) has 95% or more of the solids removed, (b) is treated to minimize the odor, and (c) has chlorine added to control bacteria. Nonpotable water usually contains significant amounts of nitrogen and phosphorus.

Water Quality

Irrigation water usually contains dissolved substances, sometimes in substantial quantities. A determination of water quality normally involves an analysis of the (a) total concentration of dissolved constituents; (b) relative proportion of sodium to other cations; (c) concentration of boron and other potentially toxic elements; and (d) bicarbonate concentration. The total concentration of soluble salts in irrigation water is usually expressed in terms of electrical conductivity (EC). An EC of 1000 mmhos per cm is equivalent to 650 ppm of dissolved salts. Typically, the dissolved salts in irrigation water are in the range of 100 to 1500 ppm (50). Water containing less than 650 ppm of soluble salts is useful under most turfgrass conditions, while levels above 2000 ppm are undesirable and can be quite injurious.

Certain dissolved materials are beneficial. For example, the nitrates, potassium, calcium, and magnesium contained in irrigation water are essential elements utilized in turfgrass growth. In addition, the calcium and magnesium exert beneficial effects on the physical characteristics of the soil. However, the detrimental effects of dissolved constituents represented by chloride, sodium, and sulfate are of great concern. Specific elements such as boron can also become toxic if the concentration in the water is high enough.

Continued irrigation with water containing salts may result in a gradual accumulation to toxic soil levels if drainage is poor or insufficient water is applied for downward leaching of the salts. This problem is even greater if conditions exist where the soluble salts cannot be leached out of the root zone. A high salt level in the soil also restricts root and shoot growth and decreases the tolerance to water stress. Irrigation water that is high in soluble salts may also cause direct injury to the turfgrass shoots through physiological drought. An accumulation of sodium is quite harmful because of the resulting deterioration in soil structure by deflocculation.

Water quality for use in irrigation systems must also involve the suspended or particulate matter. There should be freedom from sand, algae, and other foreign particles that can be quite damaging, particularly to automatic systems.

Irrigation Methods

There are three main methods of applying irrigation water to turfs: (a) overhead irrigation, (b) surface irrigation, and (c) subsurface irrigation or subirrigation. Overhead irrigation, by sprinklers, is by far the most common means of irrigating turfgrasses. Therefore, most of the discussion in this section is devoted to that type.

Overhead Irrigation

There are various methods of overhead or sprinkler irrigation. Basically, the water is applied in the form of a spray resembling that of natural rainfall. The water is delivered by means of distribution lines at either high (above 60 psi) or low (15 to 30 psi) pressure.

The advantages of sprinkler irrigation are (a) it can be effectively used on steep, irregular topography with uniform distribution and minimal soil erosion; (b) it can be used on very sandy soils with little water loss from deep percolation; (c) the water application rate is easily controlled; (d) washing and puddling of soil are minimal providing the droplet size and application rate are satisfactory; (e) it provides good economy of water use; and (f) it is adaptable to automation. The primary disadvantage of sprinkler irrigation compared to other types is a relatively high cost for the initial installation. The high pressure system is the most commonly used type. It permits the irrigation of larger areas and is preferred where the available water pressure is adequate.

Most turfgrass irrigation systems deliver the water by means of a **sprinkler head**. It is defined as the complete assembly that is connected to the distribution lines to spread water over a given area. The sprinkler head should disperse the water into fine droplets that fall uniformly on the turf as a light rain. Large drops that strike the soil at a high velocity tend to cause compaction.

The uniformity of water application depends on the (a) type of spacing, (b) type of sprinkler head, (c) wind conditions, and (d) water pressure. The uniformity of a sprinkler head can be checked by placing collection cans at set intervals throughout the area of the sprinkler pattern (25). The amount of water collected in each can after an irrigation indicates the uniformity of water application.

Three basic types of sprinkler irrigation heads are used in high pressure turf-grass irrigation: (a) rotary, (b) oscillating, and (c) fixed. Sprinkler heads are available for either full or part circle coverage.

Rotary Sprinkler Heads

Rotary sprinkler heads apply water by means of one or more long streams of spray and cover a circular area or a set portion of a circle (Fig. 14-4). The nozzles are rotated by the force of the flowing water. Rotary sprinklers may be driven by gear, cam, or impact. There are three main types of rotary sprinkler systems: (a) quick coupler rotary, (b) pop-up rotary, and (c) aboveground rotary systems.

Figure 14-4. Perimeter irrigation of a putting green with pop-up rotary sprinklers. (Photo courtesy of J. R. Watson, Toro Manufacturing Corporation, Minneapolis, Minnesota).

Quick coupler rotary system. The quick coupler system utilizes a rotary sprinkler that is placed on a coupler key for operation with quick coupler valves. The quick coupler valves are connected directly to live underground pressure lines or to aboveground portable lines by means of risers. Each head is individually controlled and operated by manually inserting the coupler key into the quick coupler valve. The labor requirement for operation of this system is greater than for the pop-up rotary system but the initial installation cost is considerably lower. In the past, the quick coupler rotary system has been the most commonly used permanent irrigation system for turfgrass areas but is now being replaced by the pop-up rotary system. *Pop-up rotary system.* The pop-up sprinkler head is one having a movable nozzle assembly that rises aboveground when activated by water pressure and drops back level with the soil surface when not in operation (Fig. 14-5). The sprinkler head is concealed below ground with a cover plate exposed at the soil surface. The pop-up rotary system is normally operated by manual or automatic controls. Groups of heads are usually activated by a single control valve. Some automatic systems are operated by remote control valves under each head where (a) there are great differ-

Figure 14-5. A pop-up rotary sprinkler head when not in operation (left) and when activated (right).

ences in soil texture and topography, or (b) heads with high application rates are used and the amount of water available limits the number of heads that can be operated at one time. The initial installation cost for a pop-up rotary system is quite high but the labor costs for operating the system are minimal if properly designed and installed.

Aboveground rotary system. Sprinkler heads that are positioned completely aboveground on a permanent riser or on a quick coupler attached to a portable pipe are called aboveground rotary heads. These types of systems are most commonly used for irrigating shrubs or turfs on sod farms. They are operated by either manual or automatic control valves.

Oscillating Sprinkler Heads

Oscillating-type sprinkler heads are most commonly used on home lawns or relatively small turfgrass areas. It has also been called a wave type sprinkler. Water is delivered through small nozzles that turn slowly back and forth in a single plane. The slow, gentle water delivery rate minimizes compaction problems. This type of head is more readily affected by wind.

Fixed Sprinkler Heads

The fixed sprinkler head has no moving parts except for the vertical lift spindle on pop-up types. It is also called the fixed spray or fixed jet sprinkler. It produces a fine spray that covers the entire area to be irrigated at once, usually in a circular arc covering an area from 15 to 50 ft in diameter. Fixed sprinkler heads (a) operate at a low pressure, (b) cover a relatively small area, and (c) are utilized primarily on narrow irregular shaped areas or small home lawns. The water application rate is usually larger than for rotary heads. The operating time is usually short in order to avoid water loss by surface runoff. This type of sprinkler head is relatively trouble free and least affected by wind.

Perforated Low Pressure Systems

Low pressure systems are used to advantage where the available water pressure is low. However, they require a greater number of settings or lateral lines to cover a given area. The perforated pipe or hose systems commonly used on home lawns are usually operated with low pressure systems. The holes in the pipe or hose are spaced so that the entire area is covered at one setting. When one area is properly watered, the pipe or hose is moved to another location. Perforated, low pressure systems are most effectively used on relatively level turfs since changes in elevation along the line alter the pressure causing uneven water distribution.

Sprinkler Irrigation System Controls

Control of the irrigation system may involve one or two basic approaches: (a) manual and/or (b) automatic. A **manual system** involves the use of men to move, place, and/or operate the sprinklers. Included are systems utilizing portable distribution lines, quick coupling sprinkler heads, hoses, or manual valve controls. This approach is commonly used with systems having quick coupling or hose attached portable sprinklers. The initial cost of manual systems is lower than an automatic system but the long term operational cost is high because of the labor required. The efficiency is also lower.

An **automatic system** requires no labor for positioning sprinklers. Operation of the sprinklers is controlled by a programmed time clock arrangement in which no manual techniques are required at the time of operation. An automatic system is commonly used with pop-up sprinkler heads. Automatic irrigation involves the use of time clocks and valve controls that can be preset to remotely control irrigation (a) on a specific day, (b) at a specific time or times within a day, and (c) for a specific duration. The automatic control system permits frequent recycling so that soils on steep slopes or soils having a low infiltration rate can be irrigated effectively. Syringing of greens can also be done in a very short time with a maximum degree of control.

Combinations of the manual and automatic systems are termed **semiautomatic systems**. Modified combinations of the quick coupling manual and a few automatic remote control valves are sometimes used to control irrigation on larger sections of turf. A loss of efficiency is always associated with any semiautomatic system in which manual components are incorporated.

Automatic controls. Remote control of pop-up and quick coupling irrigation systems is widely practiced. Clock operated, electrically timed controls are connected to a series of buried remote control valves. Each valve actuates or closes an individual sprinkler or section of sprinklers on a signal from the controller. The remote control valves can be operated by low voltage electricity or hydraulically. The design layout on extensive turfgrass areas usually includes a master control system for central programming and a group of submaster or field control units located for easy viewing of the sprinkler heads controlled by the particular unit (Fig. 14-6).

Automatic controllers can be programmed to regulate when irrigation is to occur as well as the duration of watering for each individual sprinkler or group of sprinklers. A time clock can be used to program irrigations for periods when the turfgrass area is not in use. The controls can also be designed for selected sequence sprinkling or zone irrigation, depending on the requirements of the given situation.

Fully automatic irrigation is achieved by the use of moisture-sensing elements that are integrated into the control system. The automatic irrigation system is activated when the soil moisture level declines to a predetermined, critical level as measured by the sensing element (44). The sensing element activates a time clock that controls the time of day and dura-

Figure 14-6. A submaster, automatic field control unit for irrigating selected portions of a turfgrass area. (Photo courtesy of J. W. King, Michigan State University, East Lansing, Michigan).

tion of irrigation. Characteristics to be avoided in a moisture-sensing element are a slow response time, a fragile nature, or an incomplete coverage of the moisture range.

Typical moisture-sensing devices include tensiometers and electrical resistance sensors (11, 23, 26, 27, 28, 34, 35, 36). Tensiometers are most sensitive in moist to wet soil conditions typical of intensely cultured turfs, while the electrical resistance sensors are more effective in measuring soil moisture in the moderately dry range. The placement depth of moisture-sensing devices is determined by the active root zone depth. Tensiometers are normally placed in a representative area at two depths, such as 2 and 6 in. for effective irrigation control (36). The presence of thatch or a dense, compacted soil surface interferes with the successful use of tensiometers (24).

Water Distribution Lines

The water distribution lines responsible for delivery of the water from the source to the sprinkler heads must be properly designed and installed. Factors to consider in selecting the size and type of pipe include (a) the water application requirement per unit of time, (b) comparative carrying capacity, (c) useful life expectancy, (d) cost of pipe, (e) cost of fittings, (f) cost of installation, (g) cost of pump operation, and (h) pressure at which the system is to be operated. Distribution mains and laterals may be permanent or portable.

Permanent lines. A permanent distribution system is the primary type used on most established sports and recreational turfs that need full season irrigation. Pipes in permanent installations are buried below ground with mains and laterals in sizes that are gradually reduced down from the main to the most distant sprinkler head. Both the mains and laterals are installed underground with risers located at designated intervals that connect to surface pop-up or quick coupling sprinkler heads. Risers are also located at periodic intervals that have a quick coupling head to which hoses can be attached for portable watering of turfgrass areas adjacent to the permanent main line. Permanent lines have a high initial installation cost but a low operating cost. The installation of permanent water distribution lines should be done after the subsurface drainage lines are installed at lower soil depths.

Distribution lines should carry water to the sprinkler heads (a) in the proper amount, (b) with the designated pressure, (c) with low friction loss, and (d) at a reasonable cost. Plastic, asbestos cement, cast-iron, galvanized steel, and copper pipe are utilized in permanent underground systems. Pipe having a smoother interior wall carries a larger quantity of water more efficiently. Asbestos cement, copper, and plastic have a smooth interior wall. Aluminum, steel, and copper lines or fittings are more prone to corrosion if fertilizer is applied through the irrigation system.

Cast iron and asbestos cement pipe are the most common types of pipe used for larger sized lines such as mains and submains of 4-in. diameter or larger. Cast iron pipe has good resistance to stresses but is susceptible to corrosion. It has a long life expectancy and is the only ferrous metal pipe that rust does not destroy. Cast iron pipe is available with a cement lining. Asbestos cement pipe is relatively light in weight and does not conduct electricity since it is nonmetallic. The life expectancy is very good due to excellent tolerance to corrosion and electrolysis. Care should be taken during installation because of its brittle nature. Trench bottoms should be level and excessive soil loads avoided. Contact of rocks with the pipe during backfilling operations can cause breaks.

Plastic pipe is lightweight, flexible, low in cost, and quite resistant to corrosion. Three common types of plastic pipe are (a) polyvinyl chloride (PVC), (b) acrylonitrile-butadiese-styrene (ABS), and (c) polyethylene (PE). PVC is widely used for both small and large diameter pipelines up to 4 in. because of its high pressure rating. Rubber gasket joints are used in long line installations. The flexibility of PVC pipe facilitates installation of perimeter-type irrigation systems around greens. Plastic pipe should be laid loosely in the trench because it has a high coefficient of expansion relative to temperature variation. The placement of expansion joints at intervals along the line alleviates the expansion problem. Polyethylene pipe has a discontinuous pressure rating of 80 to 100 psi that limits its use in irrigation systems having a higher operating pressure.

Galvanized steel and copper pipe are more commonly used for small diameter pipe of 3 in. or less such as laterals. The use of galvanized pipe has declined due to an unpredictable life expectancy and high installation cost. Copper pipe has a long life expectancy and is not subject to electrolysis. It is generally used in sizes of 2 in. or smaller because of the high cost. The use of copper pipe should be avoided in soils having a high salinity level or sulfur content. Electrolysis problems occur if copper and metallic pipe are mixed in the same system.

Portable lines. A portable hose attached to a single sprinkler head mounted on a stand or skid is the most common method of home lawn irrigation practiced and was used quite extensively at one time on sports turfs. The hose-end sprinklers commonly used on home lawns tend to apply water at excessively high rates and require frequent moving for efficient irrigation of turfs.

A modification of the hose distribution system involves the traveling sprinkler. A sprinkler head is mounted on a mobile wheeled device in this type of system. The unit is driven over a given turfgrass area by water powered wheels and/or a take-up wire drum assembly. The rate of travel and water application can be adjusted somewhat but is quite limited.

The hose diameter can make a substantial difference in the amount of water applied. A single 1-in. hose can carry twice as much water as two 0.75-in. hoses. Portable hose lines are usually made of rubber or plastic.

Portable irrigation lines are used on extensive turfgrass areas, particularly commercial sod production operations. The distribution lines are usually lightweight aluminum in this case (Fig. 14-7). These systems are low in initial cost but quite expensive to operate because frequent manual movement is required. In some situations a combination of portable and permanent distribution lines is used that is called a semipermanent system. In this case, the main and submain distribution lines are usually permanent and the laterals are portable.

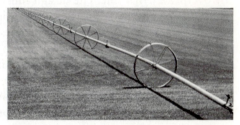

Figure 14-7. A portable irrigation system being utilized on a sod production field.

The Pump

A pump of the proper capacity must be obtained in order to achieve uniform, effective water distribution by overhead irrigation. The pump selected must deliver the required amount of water at the specified pressure necessary for efficient operation of the sprinkler heads. In accomplishing this, the delivery capabilities and water pressure that the pump develops must (a) be sufficient to lift the water from the source to the nozzles, (b) overcome the frictional losses occurring in the distribution lines, and (c) maintain the required operating pressure at the nozzle to give proper distribution and range of spray pattern designed for the particular sprinkler heads involved. The pump capacity also depends on the number of acres to be irrigated. A properly selected pump should operate at an efficiency of 80% or higher. Pumps are designed to operate efficiently under a rather limited range of discharge rates, water pressures, and revolutions per minute. Selection of the pump is influenced by the specific water source and conditions. All pumps operating from a surface water source should have screening of the proper size and capacity.

The most common kind of pump is the centrifugal pump, which may be of a horizontal or vertical type. It has a high efficiency and low initial cost but can only be used with shallow wells of no more than 15 to 20 ft or on lakes and streams. Centrifugal pumps should have a low pressure or low rate of flow cutoff in case prime is lost. Two other types of pumps frequently used are the turbine and submersible pumps. The performance characteristics are similar to the centrifugal type but they can be used in deep wells. The pump should always be operated at the recommended revolutions per minute.

The size of the power unit required for operating a pump depends on its capacity. The motor should have a nonoverloading characteristic. Electric motors are most commonly used if a satisfactory source of electrical power is available. Electric motors are dependable and have a long life. Internal combustion engines are sometimes used as the power source where electricity is not available.

Design of the Sprinkler Irrigation System

The system must be properly designed to achieve uniform distribution of water by overhead irrigation. The design must be completed before the irrigation system is purchased and installed. Basically, this involves engineering principles combined with certain turfgrass cultural principles. Turfmen generally compensate for poor water distribution caused by an improperly designed irrigation system by overwatering. This results in excess water in certain areas that causes waterlogging, increased compaction proneness, reduced turfgrass vigor and quality, and greater disease problems. An ideal system should permit irrigation of only those areas that require water. Such flexibility requires a more complex system of valve controls and a greater initial investment. The ultimate in control of an irrigation system is one with individual remote control valves located at each sprinkler head.

Considerations involved in designing an irrigation system include the (a) maximum water use rate for the turf under the specific soil and climatic condition of the area, (b) amount of water available from the source, (c) direction and intensity of prevailing winds, (d) topography of the area, (e) water infiltration characteristics of the soil, (f) number of hours available per day for irrigation, and (g) amount and frequency of irrigation anticipated. An additional factor of concern if a manual versus automatic system is being considered is labor availability. The major factor influencing the water use rate of turfs is the evapotranspiration rate (33). It, in turn, is determined by a number of interacting factors, discussed in Chapter 8. Planning the maximum water use rate is usually based on a 1 week peak need.

The wind conditions in which the irrigation system must operate are of great concern. For example, three to four times as many sprinkler heads are required to operate effectively on a given area in wind velocities of 8 to 10 mph as are required at minimal wind velocities of 0 to 3 mph. It is usually preferable to schedule irrigations during periods of minimum wind activity rather than to develop a more expensive irrigation system that operates successfully at high wind velocities. An exception would be where the system is to be used for the application of chemicals such as fertilizers. The placement of sprinkler heads in relation to the direction of prevailing winds serves as a partial compensation.

Zonal control of sprinkler head operation is particularly desirable on sloping areas where more frequent, lighter rates of water application are necessary. The irrigation system should also be designed to provide light, frequent applications of water at short intervals, using automatic controls, on fine textured, compacted soils with low infiltration and percolation rates.

The time available for irrigation of an area is controlled primarily by the time and duration of use. Wind conditions are also a factor. Many recreational turfgrass areas can only be irrigated at night. The shorter the time available for applying a certain amount of water, the larger the water delivery capability required. This affects the size of the pump and distribution lines as well as the cost of the system.

The selection of sprinkler nozzles, pipes, and pumps should be such that they are economical to install and operate as well as meeting the design requirements for the turfgrass area to be irrigated. The selection and spacing of sprinkler heads should be adjusted to ensure uniform water distribution throughout the area to be irrigated. An individual sprinkler head applies more water near the sprinkler than at the periphery of the spray pattern. Sprinkler heads are spaced with a certain percent of overlap in spray pattern in order to achieve uniform distribution. Medium pressure sprinkler heads should be spaced at 60 to 70% of their effective diameter under moderate wind conditions. A triangular spacing arrangement is preferred for uniformity of application. Consideration should also be given to the water application rate required for the specific infiltration properties of the soil and direction of the prevailing wind. Uniform water distribution by the irrigation system is particularly important where fertilizers are to be applied through the system.

Proper operating water pressure at the sprinkler heads is very important in the design of the irrigation system. Operating pressure influences the (a) diameter of the sprinkler pattern, (b) droplet size, (c) distribution pattern, and (d) wind resistance. Water droplets are large when the pressure is low in relation to the recommended operating pressure for a given sprinkler head. As a result, water distribution is abnormally great at the outer edges of the sprinkler pattern. A high water pressure in relation to the optimum operating pressure for a particular sprinkler results in fine water droplets that tend to be concentrated near the sprinkler head and are subject to wind effects. The operating pressure should be increased as the nozzle size is increased.

Installing the Irrigation System

Proper installation of the irrigation system is just as important as a good design. A 12- to 18-in. sod strip may be lifted, rolled, and stacked for relaying or a powered pipe puller used. In the former, an 8 or 12 in. wide trench is usually used in order to permit work in the trench and to effectively backfill the trench so that settling is minimal. The proper gradient in ditching permits complete drainage of the system to avoid freezing damage in colder climates. The pipes must be drained by gravity or blown empty with compressed air each fall. Manual drain valves should be placed at low points in the pipeline to ensure drainage. Care should be taken during pipe installation to avoid getting dirt into the system. Risers should be installed with a swing joint assembly to avoid damage from heavy equipment and to permit future adjustment of the attached irrigation heads or quick coupling valves to the level of the soil surface.

After laying and connecting the pipe, stone-free soil should be tamped under, around, and over the pipe to a depth of 4 to 6 in. The remainder of the backfill is handled with a backfilling machine that may involve an auger mounted backfiller or a tractor mounted blade. Settling of the disturbed soil is facilitated by moderate compaction with a mechanical tamping machine when the soil is dry. After this initial compacting, some additional soil is slightly mounded over the top in anticipation of future settling and the sod replaced.

Surface Irrigation

Flood irrigation is the primary type of surface irrigation practiced on turfgrass areas (Fig. 14-8). It is most commonly utilized on relatively level areas where a water supply is readily available at a low cost. The turfgrass area should have raised borders to contain the flood waters within the area to be irrigated. Generally, the water is introduced at a given point from an irrigation ditch and the water flows across the turfgrass area at right angles to the direction of natural slope. The slope should not exceed 1.5% in the direction of irrigation and no more than

Figure 14-8. Flood irrigation of a turf in Arizona.

0.5% across the slope. A relatively large quantity of water is required for flood irrigation with short strips of 300 ft or less being more economical to irrigate than longer ones. A large amount of water applied for a short time results in more uniform, economical water distribution. Surface irrigation has a low efficiency of water use and distribution. It is of relatively minor importance in terms of turfgrass irrigation practices throughout the world.

Subsurface Irrigation

Subsurface irrigation involves applying the water to the turfgrass plant from beneath the soil surface and without wetting the surface. Water is supplied to the turfgrass roots by capillarity. This method of irrigation minimizes soil compaction and water loss by surface evaporation and runoff. Subsurface irrigation can be accomplished by one or a combination of three methods: (a) water table control, (b) porous tile, or (c) perforated plastic paper (41). The water table control method depends on the presence of a high water table and a level soil. Clogging of the lines by slimes and a lack of uniform water distribution in certain soils can be problems when using the last two methods.

Subsurface irrigation of turfs has not been practiced to any extent except on organic soils utilized for commercial sod production. The water level is maintained relatively close to the surface by a network of tile lines, open ditches, flood gates or weirs for water control, and possibly even pumping stations for control of the subsurface ground water level. Sod production on organic soils is normally located on poorly drained sites where the subsurface water level can be easily manipulated. By a combination of flood gates and possibly a pumping station, the water level can be lowered during planting and harvesting periods and raised at other times during sod production to ensure that adequate soil moisture is available.

Surfactants

Surfactants include a wide range of surface active agents such as (a) wetting agents, (b) emulsifiers, (c) sticking agents, (d) spreaders, (e) dispersing agents, and (f) detergents (9). The type of surfactant depends on the specific use for which it is utilized. Basically, surfactants act by modifying the surface tensions existing between water and solids or other liquids. The molecules of surfactants possess hydrophilic and lipophilic regions that attract water and nonaqueous materials, respectively. Surfactants act by orienting between the water and solid surfaces in a manner that encourages more intimate contact between the two surfaces. The net result is a reduction in the interfacial tensions.

There are two major groups of surfactants: (a) nonionic and (b) ionic. The nonionic surfactants have no particle charge and, thus, possess a minimum degree of ionization or disassociation in water. These nonelectrolites are usually chemically inactive. The nonionic surfactants are most effective in hard water and at warmer temperatures. They are the common type used on turf.

The ionic surfactants ionize in water and have either a positive or negative charge. Anionic surfactants have a negative charge that is dominant. They are most effective in soft water, particularly at low temperatures. Cationic surfactants possess a positive charge that dominates. The ionic surfactants tend to be more toxic to turfgrasses and more limited in use than the nonionic types.

Wetting Agents

A **wetting agent** is a surfactant that increases the ability of a liquid such as water to moisten a solid substance such as the soil (Fig. 14-9). It acts by lowering the surface tension and thus increases the effective wetting of solid surfaces (6). The effectiveness of a wetting agent depends on how much it increases the spread of a liquid over the surface area of a solid. Different types of wetting agents vary considerably in their effects on various soils (17). This is true even within the nonionic surfactants most commonly used on turfgrass areas.

Thatch is very difficult to wet because the high surface tension repels water and causes localized dry spots to develop. A wetting agent lowers the surface tension causing water to spread more uniformily over the thatch (32). This enhances water penetration through the thatch. Less water is lost by surface runoff because of improved water penetration through the thatch and into soils. The net result is a more efficient, uniform use of the precipitation received and irrigation water applied. A similar effect occurs on difficult to wet soils and waxy leaf surfaces of plants.

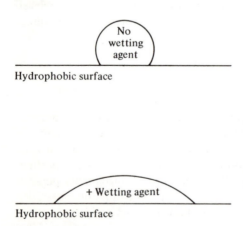

Figure 14-9. The small contact area of a water droplet (above) due to high surface tension on a hydrophobic surface versus the effect of a wetting agent (below) in increasing the surface contact or wetting which is the result of reduced interfacial tension.

The effects of wetting agents on soils are very complex and not entirely understood. They vary with the soil type. The infiltration rate into hydrophobic soils is increased by the application of a nonionic wetting agent (15, 16, 17, 19, 27, 28, 30, 32). On other soils, however, no positive effect of wetting agents in increasing infiltration has been found (15, 18, 19, 27, 32, 40, 49). Certain soils, especially clays, absorb and inactivate the wetting agent quite readily (12, 19, 49). Wetting agents may actually have an adverse effect on the structure of certain soils (32). Wetting agents can be used to improve water infiltration into hydrophobic soils but do not correct poor soil conditions caused by compaction or poor drainage. The effect of wetting agents in enhancing infiltration on hydrophobic soils and thatch is retained for two to three irrigations or rains (32).

Other effects of wetting agents include reductions in the evaporation rate, soil water retention, and dew incidence (8, 13, 14, 19, 40). Wetting agents reduce evaporative water loss by forming a diffusion barrier at the soil surface. The decreased surface tension at the soil-liquid interface reduces water movement to the soil surface. Increased thatch accumulation occurs where wetting agents are used continually because the drier thatch condition reduces decomposition by microorganisms (8). Injury to turfgrass shoots and roots may occur under certain conditions where the wetting agent is not absorbed onto clay particles (11, 31, 37, 49). Foliar injury has been most common (a) during periods of heat stress, (b) during wet weather, and (c) at higher application rates.

Wetting agents are not cure-alls for turfgrass problems. However, a wetting agent is one of the cultural practices available for use in improving water penetration and movement in hydrophobic soils, thatch, and localized dry spots (8, 11). Its use as a regular cultural practice is questionable. Further study is needed regarding the benefits and detrimental effects on soils and turfgrass plants resulting from long term, continual use of wetting agents.

References

1. ALJIBURY, F. K. 1967. Turf, water management. Proceedings of the 5th Annual Turfgrass Sprinkler Irrigation Conference. 5: 35–36.
2. BROWN, E. M. 1943. Seasonal variations in the growth and chemical composition of Kentucky bluegrass. Missouri Agricultural Experiment Station Research Bulletin 360. pp. 1–56.
3. BYRNE, J. G. 1966. A low-application sprinkler system for bowling greens. California Turfgrass Culture. 16(4): 29–31.
4. CHAMPNESS, S. S. 1950. Effect of microclimate on the establishment of timothy grass. Nature. 165(4191): 325.
5. CHRISTIANSEN, J. E. 1947. Lawn-sprinkler systems. California Agricultural Extension Service Circular 134. pp. 1–18.
6. COVEY, W. G., and M. E. BLOODWORTH. 1961. Some effects of surfactants on agricultural soils. Texas Agricultural Experiment Station Miscellaneous Publication 529. pp. 1–19.
7. DE VRIES, D. A., and J. W. BIRCH. 1961. The modification of climate near the ground by irrigation for pastures on the Riverine plain. Australian Journal of Agricultural Research. 12: 260–272.
8. ENGEL, R. E., and R. B. ALDERFER. 1967. The effect of cultivation, topdressing, lime, nitrogen, and wetting agent on thatch development in ¼-inch bentgrass turf over a ten-year period. 1967 Report on Turfgrass Research at Rutgers University. New Jersey Agricultural Experiment Station Bulletin 818. pp. 32–45.
9. ENGEL, R. E., and D. E. WOLF. 1950. Preliminary results with wetting agents in chemical control of crabgrass. Agronomy Journal. 42: 360–361.
10. GARMAN, W. L. 1949. A study of the physical properties of golf greens in Oklahoma. Oklahoma-Texas Turf Conference. pp. 44–47.
11. HEMSTREET, C. L., and F. DORMAN. 1964. Reviving old putting greens. California Turfgrass Culture. 14(3): 20–23.
12. IVARSON, K. C., and D. PRAMER. 1956. The persistence of biological effects on surface active agents in soils. Soil Science Society of America Proceedings. 20: 371–374.

13. LAW, J. P. 1963. The effect of fatty alcohols on soil moisture evaporation under controlled climatic conditions. 1963 Agronomy Abstracts. p. 63.

14. ———. 1964. The effect of fatty alcohol and a nonionic surfactant on soil moisture evaporation in a controlled environment. Soil Science Society of America Proceedings. 28: 695–699.

15. LETEY, J., R. E. PELISHEK, and J. OSBORN. 1961. Wetting agents can increase water infiltration or retard it, depending on soil conditions and water contact angle. California Agriculture. 15(10): 8–9.

16. LETEY, J., N. WELTCH, R. E. PELISHEK, and J. OSBORN. 1962. Effect of wetting agents on irrigation of water repellent soils. California Agriculture. 16(12): 12–13.

17. ———. 1963. Effect of wetting agents on irrigation of water repellent soils. California Turfgrass Culture. 13(1): 1–2.

18. LUNT, O. R., and M. R. HUBERTY. 1954. Effect of wetting agents in water on infiltration rates into soils. California Agriculture. 8(1): 12.

19. MADISON, J. H. 1966. Effect of wetting agents on water movement in the soil. 1966 Agronomy Abstracts. p. 35.

20. MADISON, J. H., and A. H. ANDERSON. 1963. A chlorophyll index to measure turfgrass response. Agronomy Journal. 55: 461–464.

21. MADISON, J. H., and R. M. HAGAN. 1962. Extraction of soil moisture by Merion bluegrass (*Poa pratensis* L. 'Merion') turf, as affected by irrigation frequency, mowing height, and other cultural operations. Agronomy Journal. 54: 157–160.

22. MANTELL, A., and G. STANHILL. 1966. Comparison of methods for evaluating the response of lawngrass to irrigation and nitrogen treatments. Agronomy Journal. 58: 465–468.

23. MARSH, A. W. 1969. Turfgrass irrigation research at the University of California. Proceedings of the 7th Annual Turfgrass Sprinkler Irrigation Conference. pp. 60–64.

24. MORGAN, W. C. 1964. Turfgrass irrigation by tensiometer-controlled system. 1964 Agronomy Abstracts. p. 102.

25. ———. 1964. Sprinkler can tests can help you. California Turfgrass Culture. 14(2): 12–13.

26. MORGAN, W. C., J. Letey, S. J. Richards, and N. Valoras. 1965. Effects of soil amendments and irrigation on soil compactability, water infiltration and growth of bermudagrass. 1965 Agronomy Abstracts. p. 46.

27. ———. 1966. Physical soil amendments, soil compaction, irrigation, and wetting agents in turfgrass management. I. Effects on compactability, water infiltration rates, evapotranspiration, and number of irrigations. Agronomy Journal. 58: 525–535.

28. ———. 1966. The use of physical soil amendments, irrigation, and wetting agents in turfgrass management. California Turfgrass Culture. 16(3): 17–21.

29. MUSSER, H. B. 1951. Water relations in turf maintenance. Proceedings of the RCGA School of Soils, Fertilization and Turf Maintenance. pp. 12–16.

30. OSBORN, J. F., R. E. PELISHEK, J. S. KRAMMES, and J. LETEY. 1964. Soil wettability as a factor in erodibility. Soil Science Society of America Proceedings. 28: 294–295.

31. PARR, J. F., and A. G. NORMAN. 1964. Effects of nonionic surfactants on root growth and cation uptake. Plant Physiology. 39: 502–507.

32. PELISHEK, R. E., J. Osborn, and J. LETEY. 1962. The effect of wetting agents on infiltration. Soil Science Society of America Proceedings. 26: 595–598.

33. PRUITT, W. O. 1964. Evapotranspiration—a guide to irrigation. California Turfgrass Culture. 14(4): 27–32.

34. RICHARDS, S. J. 1968. Water usage. 39th International Turfgrass Conference Proceedings. pp. 13–18.

35. RICHARDS, S. J., and L. V. WEEKS. 1963. Evapotranspiration for turf measured with automatic irrigation equipment. California Agriculture. 17(7): 12–13.

36. ———. 1964. Evapotranspiration for turf measured with automatic irrigation equipment. California Turfgrass Culture. 14(1): 7–8.

37. ROBERTS, E. C., and D. P. LAGE. 1965. Effects of an evaporation retardent, a surfactant, and an osmotic agent on foliar and root development of Kentucky bluegrass. Agronomy Journal. 57: 71–74.

38. ROBERTS, E. C., and F. E. MARKLAND. 1967. Fertilizing helps turf crowd out weeds. Weeds, Trees and Turf. 6(3): 12–24.

39. SCHMIDT, R. E., and R. E. BLASER. 1967. Soil modification for heavily used turfgrass areas. 1967 Agronomy Abstracts. p. 54.

40. SPURRIER, E. C., and J. A. Jacobs. 1955. Some effects of an anionic sodium sulfonate type surfactant upon plant growth. Agronomy Journal. 47: 462–465.

41. STEWART, E. H., E. O. BURT, and R. R. SMALLEY. 1965. Subirrigation of turf. Proceedings of the University of Florida Turf-Grass Management Conference. 8: 153–159.

42. TOVEY, R., J. S. SPENCER, and D. C. MUCKEL. 1969. Turfgrass evapotranspiration. Agronomy Journal. 61: 863–867.

43. TURLEY, R. H. 1964. Irrigation needs of sportsgrass turf. Proceedings of the 18th Annual Northwest Turfgrass Conference. pp. 26–32.

44. ———. 1967. Effect of frequency of irrigation on soil moisture and turf quality. Canadian Journal of Plant Science. 47: 89–97.

45. ———. 1967. Irrigation is your most costly maintenance practice. Proceedings of the 21st Northwest Turfgrass Conference. pp. 83–87.

46. WATSON, J. R. 1950. Irrigation and compaction on established fairway turf. Ph.D. Thesis. Pennsylvania State University. p. 1–69.

47. ———.1950. Irrigation and compaction on established fairway turf. USGA Journal and Turf Management. 3(4): 25–28.

48. WELTON, F. A., and J. C. CARROLL. 1940. Lawn experiments. Ohio Agricultural Experiment Station Bulletin 613. pp. 1–43.

49. WHITCOMB, C. E., and E. C. ROBERTS. 1967. Turfgrass growth with wetter water. Golf Superintendent. 35: 24–80.

50. WILCOX, L. V. 1948. Explanation and interpretation of analyses of irrigation water. USDA Circular No. 784. pp. 1–8.

chapter 15

Cultivation and Thatch

Introduction

Established turfs cannot be cultivated in the normal sense of soil tillage associated with crops. **Cultivation** as used in turfgrass culture refers to mechanical methods of selectively tilling an established turf without destroying the sod characteristics. A key difference is the selectivity of the operation. The turfgrass surface must not be disrupted or damaged to the extent that rapid recovery to its original condition prior to cultivation cannot occur. Minimal disturbance is required on greens in order to maintain playability. Aeration and aerification have been used to describe turfgrass cultivation. Unfortunately these two terms are less inclusive and are misleading since the improved soil water movement may be just as important as or even more important than the improved aeration.

Traffic over a turf eventually causes soil compaction. Turfgrass cultivation is a mechanical method of improving the exchange of air and water between the atmosphere and soil (3, 4, 11, 14, 20, 23). This exchange involves the downward movement of oxygen and water into the soil and the upward movement of carbon dioxide and other toxic gases out of the soil. Water loss by surface runoff is substantially reduced through deep cultivation (1, 9, 20, 23). The increased soil infiltration velocity and water retention resulting from cultivation also reduce the frequency and amount of irrigation needed (23). In addition, deep cultivation permits greater penetration of the more immobile fertilizer and lime materials into the soil (13, 22). The net effect is to encourage deeper, more extensive rooting (4, 7, 8, 9, 18). Increased shoot density may also result from cultivation (18).

Continued, deep cultivation gradually disrupts and destroys undesirable layers of different textures that may be present in the soil root zone. Deep cultivation also results in improved resiliency on compacted soils (4). Cultivation can stimulate thatch decomposition especially if some soil is brought to the surface where it functions as a topdressing (9, 27, 28). An additional effect is the severing of rhizomes and stolons, which stimulates new shoot and root growth from the nodes.

Cultivation should not be used as a routine cultural practice but only as needed to correct soil compaction and the associated problems such as localized dry spots. The frequency of cultivation depends on the (a) intensity and type of traffic, (b) soil water content, (c) soil texture, (d) soil structure, and (e) turfgrass species.

Cultivation may never be required on certain turfgrass areas or may be necessary on a monthly or even weekly basis. Coring is practiced as frequently as on a monthly basis in certain situations, while spiking is sometimes practiced on a weekly basis. Cultivation is best accomplished during periods of active turfgrass growth. Spring, early summer, and fall are preferred for cultivating cool season turfgrass species, while warm season species are usually cultivated during the summer. Late fall coring or grooving in which the holes are left open all winter can be desirable or disastrous. The improved water drainage resulting from the openings minimizes winter injury caused by low temperature kill of hydrated crown tissues but it also increases the proneness to desiccation injury during open, dry winters of minimal snowfall.

Cultivation is best accomplished when the soil is moist to ensure deep penetration. Cultivation of soils having a high water content should be avoided, however, in order to minimize soil compaction (20). Turfgrass cultivation is a disruptive process, especially if a core is removed (11). For this reason, it is often necessary to topdress greens and similar turfgrass areas after they have been cultivated.

In addition to improving air and water movement, turfgrass cultivation can be used for turfgrass overseeding or renovation operations. It provides good contact between the seed and soil without completely disrupting the existing, desirable turfgrass species.

Certain cultivation operations also have the negative effect of providing openings in the turf that are ideal avenues for the invasion of turfgrass weeds. For this reason, the timing of cultivation practices should be scheduled to avoid the optimum germination periods of potentially serious weeds such as annual bluegrass. Following cultivation, the need for a herbicide application to control certain serious weeds as they germinate should be considered.

A number of cultivation methods and machines have been developed over the years (19). The primary methods are (a) coring, (b) grooving, (c) slicing, (d) forking, and (e) spiking.

Coring

Coring is a form of cultivation involving the use of a hollow tine or spoon to remove soil cores that leave a hole or cavity in the sod. The core diameter may vary with common sizes being $\frac{1}{4}$, $\frac{3}{8}$, $\frac{1}{2}$, $\frac{5}{8}$, and $\frac{3}{4}$ in. The depth of penetration is 3 to 4 in. with the cores spaced on 2-, 4-, 5-, or 6-in. centers. There are two basic types of coring units (Fig. 15-1).

One type of coring unit utilizes hollow tines mounted on a shaft that operates in a vertical motion with the spacing or distance between tines determined by the forward motion of the coring machine plus the rate of vertical motion of the shaft. This type is most commonly used for the cultivation of greens.

The second type of coring unit involves semiopen spoons or tines mounted on a circular drum or reel with the cores removed by a rolling action of the reel or drum. The spoon is designed to move in an arc under the turf that results in a loosening action in addition to the removal of a core. The core spacing is a function of the number and spacing of spoons on the circular reel or drum. This latter type of coring is used more widely on extensive turfgrass areas such as athletic fields and playgrounds.

The soil cores lifted out by the coring technique may be (a) removed or (b) broken up and worked back into the turfgrass surface. The cores are usually removed on greens where the underlying soil is of an undesirable texture and structure. Removal can be accomplished by sweeping or by collecting in a box

Figure 15-1. Two types of coring units (left) and the effect of coring (right). (A) A vertical operating hollow tine type (above) and (B) a circular reel mounted spoon type (below). ("B" series photos courtesy of T. Mascaro, West Point Products Corporation, West Point, Pennsylvania).

mounted on the rear of the coring unit. Following removal of the cores, the area is topdressed and the soil matted into the holes. The cores are usually returned on extensive turfgrass areas or where the underlying soil is of the desired texture and structure. On greens this involves the use of a vertical mowing unit to break up the cores and is followed by dragging with a steel mat to work the soil into the openings. Vertical mowing is usually not used on extensive areas. The cores are partially broken up by a dragging or matting operation.

A light topdressing is usually necessary on greens 7 to 14 days after coring to fill the depressions left after the soil settles into the holes. Coring is best accomplished when the soil is moist. Penetration of the hollow tines is quite limited on dry, compacted soils. The tines of the coring unit become worn and must be replaced at regular intervals to assure the desired cultivation depth.

Grooving

Grooving is a form of cultivation involving the use of vertical rotating knives or sawteeth that cut vertical slits through the turf and into the soil. The knives may rotate in the opposite or same direction as the machine is operated. The depth of the slits or grooves can be varied down to 4 in. A fairly dry soil is preferred since a considerable quantity of soil is lifted to the surface during the grooving operation where it acts as a topdressing over the turf. Grooving is commonly used for dethatching and overseeding purposes. It is one of the most effective types of cultivation for removing an existing thatch as well as for preventive thatch control (11). The severity of grooving that can be accomplished depends on the degree and vigor of the turfgrass root system. Grooving during midsummer is frequently a problem because of the very shallow root system, which results in the sod being lifted or torn out.

Slicing

Slicing is a form of cultivation involving a deep, vertical cutting action that provides soil openings and loosening (Fig. 15-2). It is accomplished with disks or

Figure 15-2. A slicing unit in operation (left) and a close-up of the effect (right).

rigid, V-shaped knives mounted on heavy, circular wheels. The slicing unit is pulled across the turfgrass surface as the disks or knives penetrate the soil. Each row of knives on a weighted wheel should be suspended independently so that the unit operates freely over varying contours. The slicing operation penetrates to a soil depth of 3 to 4 in. but does not involve the removal of a soil core. Penetration depends on the weight of the wheels to which the knives are attached. Slicing is best accomplished on moist soils as in the case of coring.

Forking

Forking is a form of cultivation involving the use of a fork or similar solid tined device that punches small holes in the sod. This operation may involve either a powered or manual technique. The solid tines are usually mounted on a weighted drum or wheel in the case of the mechanical units. The use of forks for manual cultivation of localized dry spots is frequently practiced on greens. Forking permits deep cultivation of 6 to 8 in. It does not disrupt the turfgrass surface as severely as coring (4).

Spiking

Spiking is a form of cultivation involving the shallow perforation of a turfgrass surface by use of solid tines or blades (Fig. 15-3). The design and weight of the spiker is such that the cultivation is shallower and less effective than coring, grooving, slicing, or forking (4, 11). Both powered mechanical and manual units are available. Weighted mechanical units are definitely superior in terms of overall effectiveness. Spiking should be utilized for temporary alleviation of surface compaction problems. It can be used on greens with minimal disruption of the playing surface. Possibly just as significant as alleviating surface compaction,

Figure 15-3. A spiking machine in operation (left) and a close-up of the effect (right).

the spiking operation severs stolons and rhizomes and stimulates juvenile shoot and root growth.

Rolling

Land rolling is a method of correcting minor variations in the turfgrass surface, particularly those variations caused by winter freezing and thawing, earthworms, moles, ants, and similar insect activity. Rolling is not a method for correcting variations in surface contours caused by improper leveling prior to seeding or sodding. Rolling is also desirable for pressing turfgrass plants back into the soil after they have been heaved upward during the winter (29). This is particularly important for late autumn seedings where the exposed seedlings may die of desiccation if not pushed back into the soil immediately.

Rolling is actually a compaction operation and should only be used when absolutely necessary or where other techniques are not available for correcting the situation. If rolling is required, it should be done (a) when the soil is not excessively moist and (b) with relatively lightweight rollers in order to avoid excessive soil compaction. A water ballast-type roller is preferred because the weight can be readily adjusted for the particular soil moisture conditions by removing or adding water.

Rolling is sometimes a necessary cultural practice for certain sports such as lawn bowling, tennis, and cricket where a level, firm surface is required. A soil of relatively coarse texture is desirable where frequent rolling is practiced in order to minimize soil compaction.

Land rolling is a common practice on organic soils used in commercial sod production (Fig. 15-4). Rolling improves surface smoothness for maximum effectiveness and efficiency in sod harvesting operations. The compaction effect on organic soils is minimal compared to mineral soils.

Figure 15-4. Rolling of a Merion Kentucky bluegrass sod grown on organic soil.

Topdressing

Topdressing is the distribution of a thin layer of selected or prepared soil to a turf-grass area. Topdressing is utilized for (a) thatch control, (b) smoothing or leveling a turfgrass surface, (c) modification of the surface soil, (d) covering stolons or sprigs of vegetative plantings, and (e) winter protection of turfs. Certain nutrients and microorganisms are also provided in the topdressing that stimulate growth and improved turfgrass color (15). Topdressing is one of the most effective biological controls of thatch (27, 28). It is also the preferred method of smoothing the playing surface of greens and similar turfgrass areas that have been disrupted by ball marks, foot printing on saturated soils, and pests.

Soil modification is sometimes attempted by topdressing following coring. The topdressing is matted into the holes to provide vertical columns of soil having improved texture and structure. Some improvement in soil properties can be achieved by continuing this process over a period of time. This is a very slow process, however, and only provides a surface modification in the upper 2 to 3 in. Complete rebuilding and soil modification to the desired texture is the preferred procedure rather than surface soil modification through coring and topdressing.

Topdressing should not be used as a routine practice in the cultural system but only as needed to control thatch or to improve the smoothness of a greens surface. The frequency of topdressing varies from none to as frequently as every 3 to 4 weeks. Quality bentgrass greens have been maintained for years without topdressing, while topdressing may be a necessary cultural practice on other greens.

Topdressing rates normally range from $\frac{1}{8}$ to $\frac{3}{4}$ cu yd per 1000 sq ft, depending on the intended purpose of the topdressing (Table 15-1). The rate and frequency of topdressing is determined by the (a) thatch accumulation rate and (b) extent of irregularities in the playing surface. More frequent, heavy topdressing is required if there is a rapid rate of thatch accumulation caused by a vigorous turfgrass cultivar, infrequent mowing, or excessive nitrogen fertilization. Higher rates of topdressing are normal on new greens for the purpose of covering creeping stolons and as a smoothing operation. It is not unusual to topdress newly established greens at intervals of 2 weeks until a satisfactory turf in terms of shoot density and smoothness has been achieved.

Table 15-1

VOLUME OF SOIL REQUIRED TO TOPDRESS A 1000-SQ FT
AREA TO VARIOUS DEPTHS

Depth of topdressing, in.	Volume of soil, cu yd
0.13 ($\frac{1}{8}$)	0.40
0.25 ($\frac{1}{4}$)	0.77
0.37 ($\frac{3}{8}$)	1.14
0.50 ($\frac{1}{2}$)	1.54
0.62 ($\frac{5}{8}$)	1.91
0.75 ($\frac{3}{4}$)	2.31

The topdressing is usually matted in after each application. **Matting** involves working the topdressing into a turf using drag mats, brushes, rakes, or push boards. Powered steel drag mats have been developed more recently (Fig. 15-5).

One of the most important considerations in topdressing is the use of a soil mixture of comparable texture and composition to the underlying soil. Soil layering is one of the most serious problems in turfgrass culture because of the deleterious effects on air and water movement. The use of topdressing mixes containing textures drastically different from the underlying soil results in the development of a distinct layer that impairs air and water movement considerably and results in an overall reduction in turfgrass quality.

It is desirable for the topdressing to be mixed as much as 9 to 16 months in advance of scheduled use. The weathering and biological activity during this period produces a more homogeneous, stable soil. The topdressing soil used should be free of problem weed seeds such as annual bluegrass or else steps should be taken to kill the seeds by fumigation or sterilization. The topdressing must be dry and well shredded to be effectively applied. Thus, a dry storage area is frequently needed.

Topdressing was used less frequently during the 1940's and 50's primarily because of a limited labor supply and lack of mechanical application methods. More recently, the development of powered mechanical topdressers, mats, screens, conveyers, shredders, and mixing units has resulted in fairly complete mechanization of the topdressing operation and has significantly reduced the labor requirements. As a result, topdressing is being practiced more widely.

Figure 15-5. Dragging a green with a powered steel drag mat after coring.

Thatch

Thatch is a tightly intermingled layer of dead and living stems and roots that develops between the zone of green vegetation and the soil surface (Fig. 15-6). An allied term that is sometimes used is fiber. Physically, it is composed of sclerified vascular strands of stems and leaf sheaths plus the nodes of stems (17). Nodes of

Figure 15-6. Vertical cross section showing a 2.5 inch thick thatch taken from an 8 year old red fescue turf in Michigan.

stems and crown tissues are the most resistant to decay, while stems and roots are intermediate and leaves the least resistant. Leaf remnants occur only in the upper surface layers as a pseudo thatch (17, 26). A chemical analysis of thatch reveals a high lignin content (16, 17). Thatch generally implies an undesirable condition.

Mat is an organic layer buried and/or intermixed with soil from topdressing. It is partially decayed thatch that has become part of the soil profile. A limited mat or "cushion" of no more than 0.25 in. is desirable on greens to provide the resiliency desired for proper ball bounce. This resiliency property also provides a certain amount of protection against possible injury for active sports where participants may fall on the turfgrass surface. The wear tolerance of a turf is greater where a controlled amount of mat is present (6, 25, 30). A thatch or mat also tends to insulate the soil against temperature extremes.

Should the thatch accumulation become excessive, the disadvantages far exceed the advantages. Problems associated with an excessive accumulation include (a) increased disease and insect problems; (b) localized dry spots; (c) chlorosis; (d) proneness to scalping; (e) foot printing; and (f) decreased heat, cold, and drought hardiness.

Thatch provides an ideal microenvironment and medium for the development of disease causing organisms, particularly the facultative parasites. Thatch accumulation is associated with an increased incidence of such diseases as *Helminthosporium* leaf spot, *Typhula* blight, stripe smut, dollar spot, and brown patch. Not only does thatch enhance disease and insect activity but the effectiveness of fungicides and insecticides is also reduced. The development of certain diseases such as dollar spot and stripe smut is decreased under a vertical mowing or dethatching program.

The development of localized dry spots is frequently associated with thatch (4, 6, 11). A thatched turf becomes hydrophobic when it dries out and does not permit water to reach the underlying soil (6). Thatch can also act as a sponge that holds the water at the soil surface after an intense rain or irrigation (23). The former is a more common problem than the latter.

Thatched turfs are more prone to scalping, particularly during the heat stress periods of midsummer. Puffiness and scalping are accentuated by high nitrogen fertilization levels (10). Foot printing is common on thatched greens and is very undesirable because the trueness of ball roll is disrupted by the shoe imprint. A chlorotic condition may also be associated with thatching, particularly during midsummer periods.

Finally, a thatch condition elevates the crowns, rhizomes, and roots above the soil surface (17). Meristematic tissues of the crowns, rhizomes, and stolons that are elevated above the protective soil zone are exposed to greater temperature extremes. As a result, the proneness to heat, drought, and low temperature stress is increased. In the case of water stress, the restricted root system associated with a thatch condition also results in reduced water absorption and increased drought stress. Winter desiccation injury of turfs is particularly severe if a thatch is present.

Causes of thatch. A thatch develops when the accumulation rate of dead organic matter from the actively growing turf exceeds the rate of decomposition. Any cultural or environmental factor that stimulates excessive shoot growth or impairs the decomposition process increases the thatch accumulation rate. The major cultural factors contributing to thatch accumulation are (a) vigorous growing turfgrass cultivars, (b) acidic conditions, (c) poor aeration, (d) excessively high plant nitrogen nutritional levels, and (e) infrequent or excessively high cutting.

Many improved turfgrass cultivars such as Merion Kentucky bluegrass, Tifgreen bermudagrass, and Toronto creeping bentgrass possess a vigorous shoot growth rate that is important in tolerating intense traffic and assuring adequate recuperative potential. Associated with this increased vigor is a greater thatching tendency. This potential problem should be recognized whenever vigorous cultivars are utilized and the appropriate cultural practices followed to avoid an excessive thatch accumulation. A slow growing species can also be prone to thatching but this response is associated with the production of tissues that are more resistant to decomposition. Cultural factors such as excessive nitrogen nutrition or irrigation also stimulate shoot growth and contribute to thatch accumulation (11, 26).

The thatch decomposition rate is decreased when conditions are not favorable for microorganism activity. Acidic conditions, surface drying, presence of certain pesticides, or low oxygen levels impair microorganism activity. Soil temperatures in the 70 to 100°F range and a moist environment favor microorganism activity. The decomposition rate also varies with the types of tissue and the turfgrass species. Species containing more fibrous tissues and a high lignin content are more resistant to decomposition. Typical examples are red fescue and zoysiagrass.

Thatch control. A vertical cross section of the turf should be examined at regular intervals to ascertain the amount of thatch accumulation. Whenever the accumulation becomes excessive, such as 0.5 in. or more on a Kentucky bluegrass turf, steps must be taken to control the problem. Thatch control may involve either biological or mechanical methods.

Biological control of thatch is preferred since it is a preventive rather than a curative approach to the problem. Essentially, biological control involves the utilization of cultural practices that enhance the thatch decomposition rate and avoid excessive stimulation of turfgrass shoot growth. A controlled growth rate that maintains the shoot density, color, and recuperative potential is desired.

Figure 15-7. Applying topdressing to a bent-grass putting green with a mechanical self propelled topdressing machine (Photo courtesy of Ryan Equipment Company, St. Paul, Minnesota).

Nitrogen nutrition or irrigation practices that stimulate growth in excess of these requirements should be avoided, however, in order to minimize the thatch accumulation rate.

Cultural practices that are effective in biological thatch control are topdressing, liming, and cultivation (2, 11). Topdressing is generally more effective than cultivation or vertical mowing (11, 27, 28) (Fig. 15-7). A thick thatch should be removed by vertical mowing or grooving before the turf is topdressed. Mechanical thatch control is achieved by vertical mowing (27, 28). This is effective over a long term only if the original causes of thatch accumulation are corrected.

Factors important in ensuring rapid decomposition of accumulated organic material include optimal (a) microorganism populations, (b) pH, (c) aeration, (d) temperature, (e) moisture, and (f) carbon–nitrogen ratio. The microorganisms responsible for thatch decomposition are most active at pH's near neutral. Although the soil pH may be neutral, it is quite possible that the thatch pH is acidic due to acidification during decomposition and severe leaching that is common on frequently irrigated turfs. Acidic conditions enhance thatch accumulation (6). Frequent, light lime applications neutralize the acidic condition allowing more optimal microorganism activity and thatch decomposition (2, 11).

The accumulation of thatch is usually more rapid on poorly drained, fine textured soils than on well-aerated, coarse textured soils. Many of the microorganisms responsible for thatch decomposition are aerobic. The soil should be cultivated to improve the oxygen relationships if aeration is poor due to waterlogged conditions and compaction. The soil cores lifted out during coring function as a topdressing if broken up and matted into the thatch layer.

The critical nitrogen content for thatch decomposition is from 1.25 to 1.35%. A carbon-nitrogen ratio between 25 to 1 and 30 to 1 is required for decomposition of organic materials. The optimum carbon-nitrogen ratio varies with the turfgrass species and is lower with species having a low lignin content. The thatch is subjected to rapid leaching of nitrogen, particularly under intense irrigation. Light, frequent nitrogen fertilizations provide a more favorable carbon–nitrogen ratio and enhance the thatch decomposition rate.

The physical separation of the thatch from the soil is a major problem impairing decomposition. One of the most effective cultural practices in improving the micro-environment for thatch decomposition is topdressing (2, 11, 17, 27, 28). This practice provides good contact between the thatch and soil. The net result is a more favorable carbon–nitrogen ratio and improved pH and water relations for increased thatch decomposition. In addition, topdressing serves as a source of microorganism inoculum to further enhance thatch decomposition, assuming the topdressing has not been sterilized.

Vertical Mowing

Vertical mowing involves the use of vertically operated rigid or flexible blades or wire tines that cut or pull into the turf perpendicular to the soil surface. The powered wire tine type is sometimes referred to as a power rake. Vertical mowing is utilized in (a) mechanical thatch removal, (b) control of grain, and (c) thinning of vegetation during turfgrass renovation (Fig. 15-8). Increased shoot density has been observed after vertical mowing (26).

When biological thatch control has not been used and an excessive thatch has accumulated, it may be necessary to practice mechanical removal through vertical mowing. Biological controls are more commonly practiced on greens and similar intensively cultured turfs, while mechanical thatch control is more common on lawns or similar turfs maintained at a medium intensity of culture (27, 28). The frequency of vertical mowing for thatch removal varies with the (a) turfgrass cultivar and (b) thatch accumulation rate (21). A majority of the material removed by vertical mowing is dead rather than green vegetation if the turf is vigorous and well rooted. The severity of vertical mowing can be varied by altering the

Figure 15-8. Vertical mowing for thatch removal from a bentgrass green (Photo courtesy of Ryan Equipment Company, St. Paul, Minnesota).

spacing of the blades or tines. The material lifted to the surface must be raked or swept up and disposed of immediately after vertical mowing. The application of water and fertilizer encourages rapid recovery of the turf. A potential problem with vertical mowing may occur if rooting is poor. A considerable quantity of green vegetation may be removed or entire plants torn out. For this reason, vertical mowing should only be practiced when the rooting depth and turfgrass vigor are favorable.

The amount of thatch that can be removed in any one vertical mowing operation depends on the size of the power unit and weight of the vertical mower. The vertical mowing units rented to homeowners are relatively lightweight and have a small power unit so that they can be easily lifted and transported. As a result, such vertical mowing units may have to be operated over the area two, three, or even four times, preferably in different directions, in order to remove a significant portion of an existing thatch. In contrast, the larger vertical mowing units have sufficient weight and power to remove a considerable portion of an existing thatch in one operation. A larger vertical mowing unit is usually necessary on the more vigorous, stoloniferous warm season species such as bermudagrass and zoysiagrass.

The timing of vertical mowing can be quite important. A 30-day period of favorable growing conditions is desired after the dethatching operation is completed. For example, late summer is an ideal time in the cool humid regions since there is an extended period of cool, moist weather for rapid recovery of the turf. New shoots and roots are also being initiated at this time. Vertical mowing can also be accomplished in early spring just prior to the initiation of spring shoot growth. Dethatching at this time can also function in removing all debris and dead material that has accumulated during the dormant winter period. The rate of spring green-up is much earlier following vertical mowing. Dethatching during periods of rapid shoot growth in the spring and prior to the advent of high summer temperatures should be avoided, however, since serious thinning of the turf can occur without adequate recovery.

The timing of vertical mowing should also be adjusted in relation to the germination of undesirable weedy species. Dethatching is a vertical cutting action that provides openings or avenues for the germination and emergence of weedy species. For example, fall vertical mowing of bermudagrass may cause an increased invasion of annual bluegrass (31). Similarly, vertical mowing of a Kentucky bluegrass turf in early summer is undesirable because of potentially serious crabgrass encroachment problems. If this problem occurs, the appropriate control measures should be practiced to eliminate these weeds before they become permanently established.

References

1. ALDERFER, R. B., and R. R. ROBINSON. 1947. Runoff from pastures in relation to grazing intensity and soil compaction. Journal of American Society of Agronomy. 39: 948–958.

2. ANONYMOUS. 1957. Sponginess in turf. Rhode Island Agriculture. 4(1): 5.

3. BYRNE, T. G., W. B. DAVIS, L. J. BOOKER, and L. F. WERENFELS. 1965. A further evaluation of the vertical mulching method of greens improvement. California Turfgrass Culture. 15(2): 9–11.

4. DAWSON, R. B. 1934. Forking as an aid to turf recovery after drought. Journal of the Board of Greenkeeping Research. 3(11): 233–241.

5. DAWSON, R. B., and R. B. FERRO. 1939. Scarification and turf improvement. Journal of the Board of Greenkeeping Research. 6(20): 11–17.

6. EDMOND, D. B., and S. T. J. COLES. 1958. Some long-term effects of fertilizers on a mown turf of browntop and chewing's fescue. New Zealand Journal of Agricultural Research. 1: 665–674.

7. ENGEL, R. E. 1951. Some preliminary results from turf cultivation. Rutgers University 19th Annual Short Course in Turf Management. pp. 17–22.

8. ———. 1951. Studies of turf cultivation and related subjects. Ph. D. Thesis. Rutgers University. pp. 1–100.

9. ———. 1952. Turf cultivation. Rutgers Short Course in Turf Management. pp. 9–11.

10. ———. 1967. A note on the development of puffiness in ¼-inch bentgrass turf with varied nitrogen fertilization. 1967 Report on Turfgrass Research at Rutgers University. pp. 46–47.

11. ENGEL, R. E., and R. B. ALDERFER. 1967. The effect of cultivation, topdressing, lime, nitrogen and wetting agent on thatch development in ¼-inch bentgrass turf over a ten-year period. 1967 Report on Turfgrass Research at Rutgers University. pp. 32–45.

12. FITTS, O. B. 1925. The vital importance of topdressing in the maintenance of satisfactory creeping bent greens. Bulletin of Green Section of USGA. 5(11): 242–245.

13. HARPER, J. C. 1952. The effect of aerification on phosphorus penetration. Pennsylvania State College 21st Annual Turf Conference. pp. 32–36.

14. HEMSTREET, C. L., and F. DORMAN. 1964. Reviving old putting greens. California Turfgrass Culture. 14(3): 20–23.

15. LAIRD, A. S. 1930. A study of the root systems of some important sod-forming grasses. Florida Agricultural Experiment Station Bulletin 211. pp. 1–27.

16. LEDEBOER, F. B. 1966. Investigations into the nature of lawn thatch and methods for its decomposition. M.S. Thesis. University of Rhode Island. pp. 1–59.

17. LEDEBOER, F. B., and C. R. SKOGLEY. 1967. Investigations into the nature of thatch and methods for its decomposition. Agronomy Journal. 59: 320–323.

18. MADISON, J. H., and R. M. HAGAN. 1962. Extraction of soil moisture by Merion bluegrass (Poa pratensis L. 'Merion') turf, as affected by irrigation frequency, mowing height, and other cultural operations. Agronomy Journal. 54: 157–160.

19. MENDENHALL, C. 1949. Fifty years of progress in aerating established turf. Golfdom. 23(3): 56–58.

20. MERKEL, E. J. 1952. The effect of aerification on water absorption and runoff. Pennsylvania State College 21st Annual Turf Conference. pp. 37–41.

21. MEYERS, H. C. 1964. Effects of nitrogen levels and vertical mowing intensity and frequency on growth and chemical composition of Ormond bermudagrass. Proceedings of the 12th Annual University of Florida Turf-Grass Management Conference. pp. 77–79.

22. MIDGELY, A. R. 1931. The movement and fixation of phosphates in relation to permanent pasture fertilization. Journal of American Society of Agronomy. 23: 788–799.

23. MORGAN, W. C. 1962. Observations on turfgrass aeration and vertical-mowing. California Turfgrass Culture. 12(2): 12–13.

24. MORGAN, W. C., J. LETEY, and L. H. STOLZY. 1965. Turfgrass renovation by deep aerification. Agronomy Journal. 57: 494–496.

25. PERRY, R. L. 1958. Standardized wear index for turfgrasses. Southern California Turfgrass Culture. 8: 30–31.

26. SCHERY, R. W. 1966. Remarkable Kentucky bluegrass. Weeds, Trees and Turf. 5(10): 16–17.

27. THOMPSON, W. R., JR., and C. Y. WARD. 1965. Evaluation of management practices in controlling thatch accumulation on Tifgreen bermudagrass golf green turf. 1965 Agronomy Abstracts. p. 47.

28. ———. 1966. Prevent thatch accumulation on Tifgreen bermudagrass greens. Golf Superintendent. 34(9): 20–38.

29. WELTON, F. A., and J. C. CARROLL. 1940. Lawn experiments. Ohio Agricultural Experiment Station Bulletin 613. pp. 1–43.

30. YOUNGNER, V. B. 1954. Wear resistance of turfgrasses. Proceedings of the Northern California Turfgrass Conference. pp. 1–3.

31. ———. 1967. Vertical mowing—aerification and Poa annua invasion. USGA Green Section Record. 5(3): 6–7.

chapter 16

Establishment

Introduction

Rapid, successful turfgrass establishment is desirable in dust and erosion control. The longer the period required to achieve complete turfgrass establishment, the greater the likelihood of soil loss by wind and water erosion. Thin, open areas resulting from poor establishment practices also provide avenues for the encroachment of undesirable turfgrass weeds. Executing the proper steps in turfgrass establishment minimizes the time and effort subsequently required to achieve a satisfactory level of turfgrass quality and soil stabilization.

Turfs can be established from seed or vegetative plant parts. There are four basic methods of vegetative establishment: (a) sodding, (b) plugging, (c) stolonizing, and (d) sprigging. Each procedure can be used to advantage under certain situations and turfgrass species. Creeping bentgrass, bermudagrass, zoysiagrass, and St. Augustinegrass are the turfgrasses most commonly propagated by the latter three methods. No matter which method of turfgrass establishment or propagation is used, it is imperative that adequate soil preparation is accomplished prior to planting.

Soil Preparation

The primary objectives of soil preparation are (a) a firm, granular soil for rapid establishment and (b) a soil root zone possessing adequate infiltration, aeration, and drainage to maintain a high quality turf with minimal difficulty. Effective sod rooting requires the same degree of soil preparation as planting seed. Certain procedures should be followed in soil preparation when planting by seed or vegetative means. The steps in soil preparation for turfgrass establishment are

1. Control of persistent weeds;
2. Removal of rocks and debris;
3. Rough grading;
4. Surface and subsurface drainage;
5. Partial or complete soil modification;
6. Application of fertilizer and lime, if needed;
7. Final soil preparation.

Advanced planning is essential in the coordination and proper timing of all operations. This may involve the preparation of working drawings concerning the topography, subsurface drainage, irrigation system, design modifications, grass plantings, and ornamental plantings. A set of specifications should also be prepared to use in drawing contracts and purchasing. The scheduling of construction operations is facilitated by a construction planning flow sheet; this is quite important when more than one contractor is involved. A set of drawings, specifications, and construction schedules is more important on extensive turfgrass installations, particularly golf courses and sports fields.

Control of Persistent Weeds

Seeds, rhizomes, or stolons of weeds that are very difficult to control or cannot be selectively removed from desirable turfgrass species should be controlled prior to planting. Perennial grassy weeds possessing a vigorous, creeping habit are particularly difficult problems. Included in this category are quackgrass, creeping bentgrass, and nutsedge. Annual bluegrass is a prolific seed producer that is also objectionable and should be controlled.

An assessment of the potential weed problem and the need for control should be made prior to disturbance of the proposed establishment site. This is important because certain weeds, such as quackgrass and bentgrass, are controlled with a nonselective grass herbicide that must be applied when the weeds have 6 to 10 in. of active shoot growth.

Amitrole or dalapon can be used for the preplant control of undesirable, perennial grassy weeds at a comparatively low cost. Effective control of rhizomatous or stoloniferous grasses depends on absorption and translocation of the herbicide to the underground plant parts responsible for proliferation of the weed. This is accomplished by applying the amitrole or dalapon to actively growing foliage. Control is most effective if combined with plowing and cultivation at two-week intervals following the initial herbicide application. Planting must be delayed for at least 5 to 6 weeks if amitrole is used and for 6 to 8 weeks with dalapon.

A second approach to controlling undesirable, persistent weedy species is soil fumigation (46, 94, 143). It is effective in the control of most weedy species as well

Figure 16-1. Preplant soil fumigation with methyl bromide.

as diseases, insects, and nematodes (Fig. 16-1). The specific procedures and materials are discussed in Chapter 17. The toxic chemical must extend throughout the soil zone containing rhizomes and stolons in order to be effective. A 6 to 8-in. depth is usually sufficient, although somewhat greater depths may be required for nutsedge and quackgrass. Soil fumigation is a relatively expensive and time-consuming procedure that is usually practiced on limited areas. Many of the chemicals used for soil fumigation are quite toxic and dangerous to handle. Thus, it is suggested that fumigation be done by a trained professional rather than the novice homeowner. Soil fumigation can be accomplished after the site has been disturbed by plowing or cultivation. It is not dependent on the presence of a certain quantity of foliage to achieve control.

Removal of Rocks and Debris

Good soil preparation involves the removal of rocks, trash, and other debris. There is a tendency to bury rocks, wood, concrete, and other materials that accumulate during construction in pits. This practice frequently leads to future problems. The root zone soil and turf perched above the rocks or debris are subjected to severe heat and drought stress during midsummer periods. A distinct, dead patch of turf results that requires more frequent irrigation than adjacent turfgrass areas. Usually these materials must be dug out and removed to correct the problem.

Buried wood and stumps eventually decay leaving undesirable depressions that disrupt the surface uniformity. The burial of tree stumps and wood also provides favorable conditions for the proliferation of certain fungi. This results in an objectionable amount of mushroom growth in the turf above the buried stumps and wood debris. The removal or burning of woody materials at the time of construction avoids these problems.

Removal of even relatively small stones in the upper 4 in. of the root zone is desirable, particularly on areas that are to be cultivated by coring, grooving, or slicing. The presence of 1 to 2-in. diameter stones at the soil surface can interfere with cultivation operations and may damage the machine. A stone-free soil should be selected if topsoil is brought in for soil modification. A stone picking machine can be used for stone and rock removal if the existing topsoil on the site is to be utilized. This is a time-consuming process but is definitely needed on those sites where cultivation, particularly coring, is to be practiced.

Rough Grading

The topsoil should be removed from the site prior to the initiation of construction, assuming the existing topsoil is of the desired texture and is present in sufficient quantities to justify saving. It can be stock-piled in an accessible, nonuse area for redistribution over the site when the construction is completed. This can be accomplished with a bulldozer or soil pan.

The desired subgrade should be shaped after construction is completed. The subgrade should conform to the proposed final surface contour but lowered to allow for soil modification or topsoil placement. Surface contours can be used as a specific feature of the landscape or can be designed to blend into the natural or artificial landscape features such as specimen trees, rock gardens, lakes, or ponds. Steep slopes should be avoided because of difficulties in establishment and mowing. Drastic changes in elevation are more prone to scalping during mowing. The use of a retaining wall or a ground cover is an alternate solution to areas having a rapid change in elevation.

Frequently, the soils on construction sites are severely compacted from the use of heavy trucks and other equipment. Deep cultivation of the compacted subsoil is very important under these conditions. The subsoil should be cultivated to a depth of at least 6 to 8 in. by plowing and disking. Even deeper, selective cultivation can be achieved by using a subsoiler.

Surface and Subsurface Drainage

Providing rapid, effective surface and subsurface drainage prior to establishment can eliminate many future problems in turfgrass culture. It is important to allow sufficient time for shaping surface contours since undesirable contours are very difficult to alter once the turf has been established. A properly contoured turfgrass area is both functional and aesthetically pleasing. Functionally, contours should ensure rapid removal of excess surface water from the area. Pockets or depressional areas should be avoided. Contours of home lawns and similar sites should slope gently away from the building at a grade of not less than 1 %.

Installation of a drain tile system to ensure proper subsurface drainage is usually necessary on poorly drained, impermeable soils such as clays, loamy clays, and some loams. The principles for installing drain tile are outlined in Chapter 10. A satisfactory outlet must be available for rapid removal of excess water from the area. A dry well or slit trench drain can be installed to assist in water movement away from depressional areas that cannot be drained by surface contours, waterways, or a subsurface drain tile system.

Soil Modification

The stockpiled topsoil should be redistributed over the site to a depth of 4 in. or more after completing subsoil cultivation. Mixing a portion of the topsoil into the upper 2 to 3 in. of the subsoil avoids a distinct layering effect between the topsoil and subsoil. Topsoil should not be distributed when the subsoil is very dry, excessively wet, or frozen.

Soil modification may be required if desirable topsoil in the loam, sandy loam, or loamy sand textures is not available. This involves introduction of a foreign soil of improved texture and structure in order to provide a more favorable root zone for turfgrass growth. A sandy loam to loamy sand topsoil of 8 to 12 in. in depth is preferred. The specific procedures and alternatives in soil modification were discussed in Chapter 10. Frequently, homeowners add a very thin layer of organic matter, sand, or loam soil to the existing surface of a soil high in clay or sand content in an attempt to achieve a more favorable root zone. This practice is seldom beneficial and may cause soil layering problems. Also, the incorporation of 0.25 to 0.5 in. of soil into the upper 6 in. of the root zone is a waste of time and money. At least 4 to 6 in. of root zone soil mix must be incorporated into the surface of the soil in order to achieve more favorable water retention, cation exchange capacity, aeration, and infiltration characteristics. After soil modification has been accomplished or the stockpiled topsoil has been applied, it is important to allow the soil to settle for a period of time prior to shaping the final contours. Deep cultivation to the depth of the root zone should also be practiced. One other step that should be done at this time, if desired, is the installation of an underground sprinkler irrigation system.

Application of Fertilizer and Lime

A soil test made from a sample representative of the topsoil should be made to determine the soil pH level. If liming is required, it should be done prior to planting since the lime can be incorporated throughout the soil root zone for the maximum rate and effectiveness in correcting acidity. The types of liming materials and the rates and methods of application were outlined in Chapter 10. Higher rates of lime can be applied prior to planting than after establishment. Lime should be applied as long as possible prior to seedbed fertilization to avoid fixation of phosphorus in unavailable forms.

The soil test is also used as a guide in determining the amount of phosphorus and potassium fertilizer to apply. Certain soils can be extremely deficient in phosphorus, potassium, or both. Only a soil test can give an accurate determination of the amounts of these two nutrients that should be applied. An application of 1 to 2 lb each of nitrogen, phosphorus, and potassium per 1000 sq ft is adequate in many situations. A sufficient level of phosphorus is particularly important in the seedbed since it is one of the most critical nutrients in seedling establishment of grasses (105). The fertilizer should be incorporated into the upper 3 to 4 in. of the soil immediately after application. It is preferable to fertilize just prior to seeding rather than several weeks earlier in order not to enhance weed growth (87).

Final Soil Preparation

The ideal soil for planting is moist, granular, firm, and free of clods, stones, and other debris. The final tillage operations should be accomplished within 24 hrs prior to planting, if at all possible. The soil surface should be granular (1 to 5 mm) but not dusty. A firm, granular soil permits (a) more accurate control of the planting depth and (b) better contact between the seed and the moist soil. The use of

a high speed rotary tilling unit for breaking up large clods should be avoided
since it tends to destroy soil structure.
Cultipacking is quite effective in breaking up large clods. The soil should be
firmed by rolling after incorporating the
fertilizer by disking, harrowing, and/or
cultivating. Rolling also assists in determining the levelness of the soil surface.
Rolling and tilling operations should not
be done if the soil is wet in order to avoid
compaction and destruction of soil structure.

A final light raking following rolling
should provide good soil conditions for
planting. A rigid metal garden rake may
be used on small areas. A finger or tiller
rake mounted on the back of a tractor

Figure 16-2. A rear mounted tiller rake used
in the final leveling and soil preparation.
(Photo courtesy of Roseman Mower Corp.,
Glenview, Illinois).

can be effectively utilized in the final leveling and soil preparation of extensive
turfgrass areas (Fig. 16-2).

Seeding

Most cool season turfgrass species are established from seed, whereas most warm
season species are established vegetatively. The primary factors affecting turfgrass
establishment from seed are (a) the planting procedures, (b) mulching, and (c)
postgermination care. This assumes (a) adequate soil preparation; (b) selection of
the appropriate species, mixture, or blend; and (c) a high quality seed lot was
obtained.

Seed Quality

Seed quality is determined by the seed production and processing practices.
The two most important facets of turfgrass seed quality are (a) purity and (b)
germination. These two basic components of quality can be determined through
standard analytical tests conducted by a reputable seed laboratory. The accuracy
of a seed test depends on obtaining a representative sample. This involves the use
of a seed probe sampler. Seed samples are taken from a number of different bags
within the seed lot and combined for the laboratory analysis. The minimum acceptable purity and germination percentages for turfgrass seeds are shown in Table
16-1.
Seed purity. A purity analysis involves a physical determination of the percentage
of (a) pure seed, (b) other crop seed, (c) weed seed, and (d) inert matter present in

Table 16-1

THE APPROXIMATE MINIMUM SEED PURITY AND GERMINATION
FOR TWENTY-THREE TURFGRASS SPECIES. THE UNDERLINED NUMBER
IS THE PREFERRED MINIMUM PERCENTAGE

Turfgrass species	Minimum seed purity, %	Minimum seed germination, %
Bahiagrass	70–75	70–75
Bentgrass:		
colonial	95–98	85–90
creeping	95–98	85–90
velvet	90–95	85–90
Bermudagrass	95–98	80–85
Bluegrass:		
Canada	85–90	80–85
Kentucky	90–95	75–80
rough	90–95	80–85
Buffalograss	85–90	60–70
Carpetgrass	90–95	85–90
Centipedegrass	45–50	65–70
Fescue:		
chewings	95–97	80–85
meadow	95–97	85–90
red	95–97	80–85
sheep	90–95	80–85
tall	95–98	85–90
Grama	40–50	70–75
Redtop	90–95	85–90
Ryegrass:		
Italian	95–98	90–95
perennial	95–98	90–95
Timothy	95–98	90–95
Wheatgrass	85–90	80–85
Zoysiagrass	95–97	45–50

the seed lot. The purity analysis requires trained seed analysts capable of distinguishing between seeds of various species, some of which may resemble one another very closely. Many of the characteristics utilized in distinguishing related species are quite minute. The "other crop" category in the purity analysis is quite important to the consumer since bentgrass, tall fescue, redtop, rough bluegrass, and ryegrass are usually included within this category. Such species are desirable under certain situations but are objectionable weeds when present as contaminates in seed lots of other desirable turfgrass species.

The purity analysis is one of the most important factors to consider in obtaining quality seed for sod production. Kentucky bluegrass sod growers should obtain seed lots free from annual bluegrass, creeping bentgrass, and rough bluegrass. Many sod growers specify to the seed testing laboratory that an analysis be made of a seed sample approximately two and a half times larger than normally analyzed in order to ensure freedom from undesirable seeds that fall within the "other crop" category.

Seed germination. Specific rules and procedures have been established for determining the germination percentages of various turfgrass species. Germination is evaluated under a controlled environment that is usually more favorable than normally occurs in the field. The standard conditions established for turfgrass seed germination tests are found in the Association of Official Seed Analysts Proceedings (3). Poor germination of a given seed lot may be due to (a) mechanical injury during harvesting, threshing, and processing; (b) insect injury; (c) infection by pathogens; (d) immature seeds due to improper harvest time; or (e) declining vitality because of an extended storage period.

Seed labeling. The seed label is provided as a guide to the buyer in purchasing quality seed. Most states and countries have specific laws requiring the labeling of turfgrass seeds that are sold commercially. The label contains an official statement as to the species or cultivar present, germination with the testing date, and purity. The percentages of (a) pure seed, (b) "other crop" seed, (c) weed seed, and (d) inert matter are usually included on the label. The origin of the seed lot may also be listed. Inert matter consists of chaff, dirt, stones, stems, and pieces of seed that are not viable. The term weed seeds, as used on a seed label, includes all species generally recognized as weeds from an agricultural standpoint. Since this category is based primarily on agricultural considerations, certain weed seeds that are a very serious problem in turfs may not be included in the weed seed category. Seeds of such species as timothy, orchardgrass, redtop, tall fescue, bentgrass, and bromegrass are included in the "other crop" category. Under certain conditions, additional information regarding the specific content of certain potentially undesirable seeds such as annual bluegrass, bentgrass, tall fescue, and ryegrass may need to be determined by further seed purity analyses. The price of a seed lot is frequently determined by its germination and purity.

Seed certification. Certification of turfgrass seed is also practiced. Both the seed production field and harvested seed are inspected under a certification program. The presence of a certified tag on a seed lot is a guarantee of its trueness to type from a genetic standpoint but gives no indication of seed quality in terms of germination or purity. Certification is another factor in addition to seed quality that must be considered in purchasing seed. The purchase of certified seed is an important consideration where trueness to type is critical, such as for greens.

Planting Procedures

Specific factors to consider in developing a series of planting procedures include (a) time, (b) rate, (c) placement and covering of the seed, and (d) type of seeder. Each is an important consideration in ensuring rapid, uniform turfgrass establishment.

Seeding time. The proper seeding time is determined by (a) the level and duration of favorable soil temperature and moisture conditions, (b) anticipated degree of weed competition, (c) anticipated incidence of diseases and insects, and (d) the germination requirements of the specific turfgrass species. Seeding is best accomplished just prior to an anticipated period of optimum soil temperatures and moisture. Seedlings grown under favorable conditions have an opportunity to make the greatest amount of growth and are better able to survive the unfavorable environmental conditions that may subsequently occur (115).

Early spring ranks a poor second to late summer for planting cool season turf-grasses (47, 60, 90, 145, 151). The duration of favorable temperature and moisture conditions is shorter. Competition from broadleaf and annual grassy weeds is also much greater from an early spring planting than from a late summer planting (70, 93, 117).

Late spring or early summer plantings are preferred for warm season turfgrass species (79). Planting at this time permits the longest possible period with high soil temperatures necessary for rapid establishment of warm season species.

Midsummer plantings of cool season turfgrass species are generally avoided due to problems with unfavorable moisture conditions, high temperatures, and excessive weed competition (70, 109, 135, 145, 151). Effective establishment can be achieved during midsummer with cool season turfgrass species, however, if (a) irrigation and seedbed weed control are practiced and (b) excessively high surface soil temperatures do not occur (145). Turfgrass seedlings are quite prone to high temperature kill at the time of emergence (91).

Late summer plantings of cool season turfgrasses are usually preferred because of the favorable moisture and temperature conditions (20, 47, 59, 60, 61, 90, 93, 106, 109, 120, 145, 151). The length of time when soil temperatures are in the 55 to 65°F range during the fall and subsequent early spring is much longer than for a spring planting. In addition, weed competition is minimal compared to seedings made in the spring or early summer (70). Late summer seedings are also better able to survive the subsequent midsummer drought than a spring seeding (61). Late summer or fall plantings of warm season turfgrass species should be avoided due to unfavorably low soil temperatures (79).

Late fall planting should be avoided if soil temperatures permit germination but limit seedling growth and rooting (70). Cool season turfgrass species that enter the winter as very small seedlings are extremely prone to winterkill by heaving (21).

Dormant plantings made when soil temperatures have declined to the level where no seed germination occurs are sometimes effective. The potential for suc-cessful establishment of a dormant seeding is quite good if favorable winter condi-tions exist in terms of temperatures low enough to inhibit seed germination, a continuous snow cover, and minimal erosion (109, 145). An unacceptable stand of turf quite commonly occurs if the winter conditions are quite open with severe desiccation and considerable erosion. Thus, successful establishment of dormant seedings depends on the anticipated winter conditions of the particular region. Dormant seedings are most commonly practiced on wet, imperfectly drained soils that are difficult to cultivate or operate equipment on during the early spring period. A dormant seeding germinates and establishes much more rapidly under these conditions.

Seeding rate. The seeding rate depends on the (a) turfgrass species, (b) percent germination and purity of the seed lot, (c) germination conditions, and (d) estab-lishment rate desired. The approximate number of seeds per pound and seeding rates of the commonly used turfgrass species are shown in Table 16-2 (117). The number of seeds per pound of the various turfgrass species is only approximate.

Table 16-2

THE APPROXIMATE NUMBER OF SEEDS PER POUND AND NORMAL SEEDING RATES FOR
TWENTY-THREE TURFGRASS SPECIES

Turfgrass species	Approximate number of seeds per lb	Normal seeding rates	
		lb per 1000 sq ft	Number of seeds per sq in.
Bahiagrass	166,000	6–8	7–9
Bentgrass:			
colonial	8,723,000	0.5–1	30–60
creeping	7,890,000	0.5–1	27–55
velvet	11,800,000	0.5–1	41–81
Bermudagrass, hulled	1,787,000	1–1.5	12–19
Bluegrass:			
Canada	2,495,000	1–1.5	17–26
Kentucky	2,177,000	1–1.5	15–23
rough	2,540,000	1–1.5	18–26
Buffalograss	50,000	3–6	1–2
Carpetgrass	1,123,000	1.5–2.5	12–19
Centipedegrass	400,000	4–6	11–17
Fescue:			
chewings	546,000	3.5–4.5	13–17
meadow	227,000	7–9	11–14
red	546,000	3.5–4.5	13–17
sheep	530,000	3.5–4.5	13–16
tall	227,000	7–9	11–14
Grama, blue	898,000	1.5–2.5	9–16
Redtop	4,990,000	0.5–1	17–34
Ryegrass:			
Italian	227,000	7–9	11–14
perennial	227,000	7–9	11–14
Timothy	1,134,000	1–2	8–16
Wheatgrass, fairway	324,000	3–5	7–11
Zoysiagrass	1,369,000	2–3	19–28

The actual number for a given seed lot varies with the particular cultivar and the size-density relationships of individual seeds. The latter factor is determined primarily by the environmental conditions under which the seed crop was produced. Favorable moisture and temperature conditions produce turfgrass seeds that are somewhat denser with a lower number of seeds per pound.

The optimum planting rate varies greatly with the specific turfgrass species involved, which is a function of the seed size. The commonly seeded turfgrass species range from 227,000 seeds per pound for ryegrass to more than 8 million for bentgrass. Seed lots having an objectionably low germination percentage must be seeded at a higher rate. An additional consideration is whether the species possesses a bunch-type or creeping habit of growth. The planting rate is more critical for bunch-type species than for creeping types. Rhizomatous or stoloniferous turfgrasses such as Kentucky bluegrass and bentgrass are capable of filling in voids between individual plants and forming a tight sod. Bunch-type turfgrass species such as ryegrass and tall fescue have a limited ability to spread into open areas from tillers. Thus, shoot density of bunch-type turfgrasses is achieved primarily by utilizing the optimum seeding rate at the time of establishment.

The most successful planting rates in assuring rapid, uniform turfgrass establishment apply an approximate number of seeds per square inch that is equal to or slightly less than the ultimate number of shoots per square inch of the mature turf (96). The stable shoot density of most cool season turfgrasses cut at 1.5 in. usually ranges from ten to twenty-five shoots per square inch (90). For example, tall fescue has a relatively low shoot density; red fescue and bentgrass, a high shoot density; and ryegrass and Kentucky bluegrass intermediate. Most seeding rates apply approximately fifteen to twenty-five seeds per square inch.

Relatively low seeding rates can be used. As little as 0.25 lb of Kentucky bluegrass seed per 1000 sq ft ultimately produces a shoot density comparable to a seeding rate of 1.0 lb per 1000 sq ft. However, the length of time required to produce a dense, uniform turf is much longer than for a 1.0-lb planting rate. Greater temperature extremes and a more rapid drying rate occurs at the soil surface of thin stands compared to denser seedling stands (35). The result is an increased likelihood of high temperature kill or desiccation. In addition, the thin stand is ideal for the germination and emergence of competitive weedy species (84, 93, 109). This problem is substantially reduced at the more optimum seeding rate.

Excessively high seeding rates can also create problems (20, 151). The number of tillers per plant decreases at higher seeding rates (59, 60). Kentucky bluegrass seeded at 3 lb per 1000 sq ft forms a dense, uniform cover more rapidly than a 0.5- to 1.0-lb planting rate. The excessively high shoot density results in the production of many spindly, weak seedlings, however, which remain in a juvenile state for an extended period of time (96). The weakened seedlings are more prone to disease injury (96, 117, 132). A certain percentage of the plants within the turf-

grass community gradually increase in size and vigor. The more vigorous plants eventually become dominant and crowd out the weaker ones to form a stand with a shoot density comparable to the optimum seeding rate (96). Although a more rapid, dense, uniform cover is produced initially, the establishment of a dense, tight sod actually is slower than when seeding at a more optimum rate of 1.0 lb per 1000 sq ft.

Placement and covering of seeds. Most turfgrass seeds are quite small and should be placed at a fairly shallow depth in the soil (Fig. 16-3). A planting depth of 0.2 to 0.4 in. below the soil surface is usually most favorable for rapid germination (58, 96, 109, 110,

Figure 16-3. The proper placement and firming of Kentucky bluegrass seed at a shallow depth using a cultipacker seeder.

115). Large seeds can be planted somewhat deeper (96). Planting turfgrass seeds too deeply impairs emergence and may result in a poor stand (58, 96, 110, 115). Also, the seedling emergence rate decreases as the planting depth is increased from 0.2 in. (115).

A poor stand is more likely to occur if the seed is placed on the soil surface rather than at a shallow depth of 0.2 to 0.4 in. Successful establishment of seed applied to the soil surface is possible if adequate moisture is constantly available and the seed is not displaced by the erosive action of wind or water (110).

Controlled firming that provides good contact between the seed and soil is also an important consideration in ensuring rapid germination. Rolling improves the seed–soil contact for rapid imbibition of water by the seed. The use of an irregular surface roller is preferred to a flat roller. A ridging effect at the soil surface tends to control the erosive action of wind and water.

Types of seeders. Many different types of seeders are available including (a) gravity, (b) centrifugal, (c) cultipacker, and (d) hydroseeder. Turfgrass seeds are readily blown by the wind because they are quite small and light. Thus, seedings are best made during periods of minimum wind activity such as late afternoon or early morning. The wind influence is minimal for cultipacker seeders, whereas the centrifugal type is most affected. The seeder should be calibrated prior to use so that the desired rate for a specific species or seed mixture is applied. The application rate adjustment of most seeders is only approximate and can vary considerably. For example, the actual planting rate using a cultipacker-type seeder at a standard setting of 1 lb per 1000 sq ft may vary from 0.75 to 1.25 lb depending on the soil moisture, temperature, degree of seedbed preparation, and characteristics of the seed lot.

The cultipacker seeder is the preferred type for rapid turfgrass establishment because it provides (a) uniform seed distribution, (b) seed placement at the proper depth, and (c) firming of the soil around the seed (Fig. 16-4). They can be purchased as pull-type models in widths of 8 and 10 ft or as a rear mounted tractor unit in a 5 ft width. Both deep and shallow ridge placement cultipacker units are available. Cultipacker seeders are most commonly used on extensive turfgrass areas, particularly in Kentucky bluegrass sod production.

A more uniform seed application is usually accomplished with a gravity-type seeder than with a centrifugal type. Mixtures containing seeds of distinctly different size and density are not distributed uniformily with a centrifugal type seeder. Both types should be equipped with an agitator inside the seed hopper. Some type of shallow raking, spike harrowing, cultipacking, or similar covering process must be accomplished as soon as possible after planting with gravity or centrifugal seeders.

A more recent development in turfgrass seeding techniques known as **hydroseeding** involves the use of a water carrier for the application of seed under pressure (147). Specialized equipment is required including a pump, hose, nozzle, and a 500- to 1500-gal tank with both paddle- and liquid-type agitators. After mixing the seed with the water, the seed-water suspension is applied by pumping through a hose–nozzle arrangement onto the site under a pressure of 90 to 150 psi. Fertilizer and

Figure 16-4. A cultipacker seeder which is utilized for turfgrass establishment on extensive turfgrass areas. (Photo courtesy of P. E. Rieke, Michigan State University, East Lansing, Michigan).

pulp fiber mulches can also be placed in the hydroseeder tank for application in combination with the seed. Hydroseeding has been widely used on roadsides and similar extensive turfgrass areas. This involves hauling large quantities of water, which can be costly and objectionable under certain conditions. An adequate water supply should be available within a relatively short distance.

Hydroseeding applies the seed to the surface in contrast to the cultipacker seeder, which places the seed in the soil followed by firming. Thus, effective establishment by hydroseeding can only be achieved if there is adequate surface soil moisture available and no wind or water erosion (88). A poor stand frequently results under conditions of surface moisture stress or erosion. Hydroseeding is most effectively used on moist soils and areas of high rainfall or where irrigation is practiced. It is particularly effective on (a) roadside slopes and similar extensive turfgrass areas where cultipacker seeders cannot be utilized or (b) areas that are rocky and cannot be readily cultivated.

Mulching

Mulching is one of the most important practices in ensuring uniform, rapid turfgrass germination and establishment. The mulch has a twofold function in (a) controlling erosion and (b) providing a favorable microenvironment for seed germination and seedling growth (60, 101, 120, 121, 137). Mulches not only control soil erosion that causes gullies and rills but also protect against sheet erosion caused by the action of splashing raindrops (56). A mulch functions in reducing the rainfall velocity and impact with soils (19). Erosion control involves soil stabilization from the action of wind and water as well as preventing the displacement of seeds and fertilizer (120). Not all mulches provide both erosion control and a favorable microenvironment for germination and establishment. Critical microenvironmental factors enhanced by mulching include improved surface soil moisture and a moderation in surface temperatures (9, 20, 69, 99, 120, 146). Mulches can function in increasing soil water infiltration and reducing surface runoff (6, 55). Surface crusting is also minimized by mulching.

Another characteristic associated with a favorable microenvironment for germination is a minimum tendency to absorb water (69). Mulch materials having a high moisture absorbing tendency actually compete with the germinating seed or seedling for the available surface moisture and may impair successful turfgrass establishment, particularly under droughty conditions. Peat moss or ground corncobs have a high tendency to absorb water compared to straw.

The selection of a mulch is based on the (a) effectiveness in erosion control, (b) microenvironment provided for germination, (c) availability, (d) freedom from toxic compounds and weeds, (e) cost, (f) ease of application, and (g) application cost. Many types of mulching materials are available.

Straw mulch. Straw is a common mulch for turfgrass establishment from seed and has been utilized effectively for many years. It provides a favorable microenvironment for establishment as well as erosion control, when applied at the proper rate (9, 13, 18, 25, 36, 50, 99, 123, 124, 125, 127, 137). A 60% reduction in evaporative water loss from the soil surface has been achieved by using a straw mulch (69). A straw mulch is quite effective in improving the infiltration of water into the soil (6, 55).

The microenvironment is determined primarily by the thickness of the straw (69). The most effective straw mulching rate ranges from 80 to 100 lb per 1000 sq ft (1.5 to 2 tons per acre). One-half the straw cover is usually removed on high quality turfgrass areas when the seedlings reach a height of 1 to 1.5 in.

Straw mulches can be applied (a) manually with a pitchfork or (b) mechanically with a mulch blower. The latter method is commonly used on more extensive turfgrass areas such as roadsides (Fig. 16-5). The mechanical mulch blower is

Figure 16-5. Application of a straw mulch with an asphalt binder to a roadside using a mechanical mulcher.

pulled by a truck on which the baled straw is the carrier. The blower should have the capacity to uniformly apply straw or hay at the desired mulching rate at a distance of not less than 80 ft. Mulching should be started on the windward side of flat areas or on the upper portion of steep slopes.

One problem with straw is its proneness to blowing. Straw stabilization can be achieved by the application of a light mist of asphalt emulsion. The asphalt can be applied directly at the tip of the blower unit on a mechanical mulch blower or after straw application to the site by direct pumping from a tank through a hose-nozzle arrangement. The asphalt binder is usually applied to the straw mulch at a rate of 200 gal per acre. An alternative for straw stabilization involves the use of a woven mulch net or cheesecloth. It is effective on small lawn sites. Straw can also be stabilized by partial disking into the soil surface (16). Rolling followed by frequent watering also reduces blowing.

When obtaining straw for mulching high quality turfgrass areas, it is important to avoid straw containing objectionable, difficult to control weeds (9). The straw should be fumigated prior to distribution on the site if undesirable weed seeds or propagules are present. Straw may be objectionable for use as a mulch in certain locations because of the fire hazard (16).

Other straw-like materials such as hay and grass can also be used as a mulch (53, 54, 137, 138, 139, 151). The degree of moisture retention is somewhat less, however, and they are more difficult to handle and spread (69).

Wood mulches. Excelsior, wood chips, and shredded bark are wood products that are effective in mulching turfgrass seedings. Excelsior is available in uniform, 4 ft wide rolls or in a loose form that is applied through a mechanical mulching machine similar to straw. The wood fibers expand upon exposure to moisture, forming a fairly thick cover comparable to straw. Excelsior is as effective as straw in erosion control and also provides a comparable microenvironment for germination and establishment (53, 54, 65, 137, 138, 151). In addition, it does not contain weedy contaminates and decomposes readily.

Wood chips, wood shavings, and shredded bark used at a high rate are also effective mulches for erosion control and turfgrass establishment (50, 54, 139, 151). The moisture retention properties are not as good as for straw (69). Its use at a reasonable price depends on the availability in a given location. The wood chips and bark should be processed into a size small enough to facilitate rapid, uniform distribution over a site. The decomposition rate of wood chips is considerably slower than straw or wood fiber. Another acceptable wood mulch that can be used if available is pine needles (137).

Woven net mulches. Woven mesh or net mulches are synthetic mulch covers commonly available in widths of 3 to 4 ft and varying warp sizes and densities. Typical examples are jute and woven paper net. Both provide a degree of initial erosion control but frequently fail to provide a favorable microenvironment for successful turfgrass establishment, particularly under conditions of moisture stress (13, 54, 65, 138). The woven paper net is less effective than the jute net in moisture retention (13). The latter is used as a ditch liner for erosion control and turfgrass establishment (22, 50, 54, 119, 138, 139).

The woven net mulches are stabilized by the use of large staples placed at intervals along the net. The net should be heeled into the soil at the upper end of the slope to force surface water over the top of the net. Burlap is sometimes used for erosion control on irrigated slopes. Proper timing in the removal of burlap covers is critical in avoiding excessive shading of turfgrass seedlings.

Fiber mulches. Materials that can be applied as a slurry through a hydroseeder or hydroplanter are included in the fiber or wood cellulose fiber mulch category. This type of mulch has given a degree of soil erosion control when applied uniformly but has been erratic in providing a favorable microenvironment for turfgrass establishment (34, 54). The fiber mulches have been successfully used in regions of medium to high rainfall of uniform distribution (18, 127). These materials are most effective on steep slopes or rocky areas that cannot be tilled and planted by conventional equipment (Fig. 16-6). The fiber mulches have been quite erratic and frequently result in a stand failure unless supplemental irrigation is provided in climates where droughty periods occur.

Other mulches. Glass fiber has been used as a mulch in turfgrass establishment. It is applied with special equipment utilizing compressed air. Results have been erratic in terms of establishment and soil erosion control (50, 65, 100, 127). In addition, the long strands persist at the soil surface creating problems if picked up by a rotary or vertical-type mower. Thus, glass fiber has been unacceptable as a turfgrass mulch.

Figure 16-6. Use of a hydroseeder for applying seed, fertilizer, and a wood cellulose mulch to a roadside slope. (Photo courtesy of the Finn Equipment Co., Cincinnati, Ohio).

Peat, ground corncobs, sawdust, and similar organic matters are sometimes used as mulches in turfgrass establishment (9). These materials provide a degree of erosion control if applied uniformly at the proper rate but may or may not improve turfgrass establishment. They can be utilized successfully if the area is irrigated or frequent precipitation occurs. The peats and similar organic materials having a high water absorption capacity may actually compete with the seed or seedling for surface moisture. Thus, peats may impair turfgrass establishment under droughty conditions. They can also contribute to a soil layering problem.

Soil stabilization can also be achieved by spraying an elastomeric polymer emulsion, latex, or asphalt formulation directly to the soil surface (27, 53, 66, 112, 123, 139). Elastomeric polymers are only effective in erosion control and turfgrass establishment when adequate soil moisture levels are maintained through regular irrigation or rainfall. Although erosion control is achieved, turfgrass establishment may not occur since the thin surface coating provides little moisture retention or temperature moderation (13, 36, 54, 127, 138). When an asphalt layer is used, it is important that the asphalt be free of toxic materials such as kerosene. Turfgrass seedlings grow through an asphalt layer, provided it is not too thick. Stand losses under asphalt sometimes occur due to a lethal heat buildup (52).

Polyethylene covers. A 4 to 6-mil, clear, polyethylene cover can be effectively used to enhance turfgrass establishment under suboptimal temperature conditions where turfgrass growth would not normally occur. The polyethylene cover serves as a barrier to moisture loss and also functions in trapping heat by means of the "green-house effect." It should be removed if excessively high temperatures occur underneath. Polyethylene covers should not be used at optimum or above optimum temperatures because of the potential for a lethal heat buildup (123). This technique is most commonly practiced on small areas such as greens and tees. The cover should be secured from the effects of wind by the use of staples or spiked wooden lath.

Seed mats. Many types of seed mats have been developed over the years for turfgrass establishment. A seed mat is an organic mulch cover with seed and fertilizer applied on one side. The seed mat is designed to be readily unrolled on the site with a minimum of effort by the novice turfgrass user. It may be composed of such materials as cotton fiber, synthetic fiber, or urethane. The mat serves (a) as the vehicle for applying seed and fertilizer, (b) in erosion control, and (c) to provide a more favorable microenvironment for seed germination and establishment.

Certain thicknesses and pore sizes of urethane seed mat have proved satisfactory as a mulch in terms of both erosion control and moisture retention. Other seed mats composed of cotton or synthetic fibers have not been successful, particularly under droughty conditions (148). The seed mat places the seed on the soil surface. Favorable moisture conditions must exist at the soil surface for successful turfgrass establishment. Certain types of seeds mats containing cotton or pulp fiber materials actually compete with the seed for moisture. Other materials, such as the urethane seed mat, are much less competitive. Seed mats have not been widely used in turfgrass establishment.

Seed Germination

The formation of a dense sod from seed consists of two distinct components: (a) germination and (b) establishment. Germination involves the imbibition of water by the seed and initiation of the primary root and shoot system from which the turfgrass plant develops. Establishment involves successful rooting and growth of a mature plant and lateral shoot development from tillers, stolons, or rhizomes that form a dense, tight sod.

The germination process. The vital part of a turfgrass seed is the embryo, which is essentially a young plant that is arrested in development. A number of specific phases are involved in the germination process. Following imbibition and osmosis of water in which the seed is softened, swelling occurs and enzymes are secreted. The primary materials stored in seeds are starch, hemicellulose, fats, and proteins that are in an insoluble form. These materials are digested by the enzymes to soluble carbohydrates that are translocated to the actively growing parts of the germinating seed such as the radicle and plumule.

The radicle and plumule develop normally, providing adequate carbohydrates are available. The radicle emerges from the seed first, followed by the plumule. Chlorophyll-bearing tissue is formed when the plumule reaches light. The young seedling then transfers its carbohydrate source from the reserve supply in the seed to the active photosynthetic process of its young leaves. The seed has successfully germinated at this point and the seedling enters the establishment phase of sod development (Fig. 16-7).

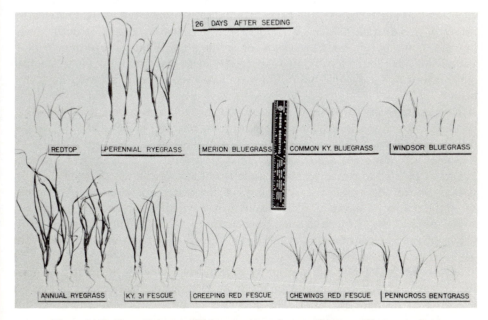

Figure 16-7. The relative establishment rates of ten turfgrasses. (Photo courtesy of P. R. Henderlong, Ohio State University, Columbus, Ohio).

Primary environmental factors influencing germination. Environmental conditions usually necessary for rapid, complete germination of turfgrass seeds include (a) an adequate water supply, (b) favorable temperatures, (c) an adequate oxygen supply, and (d) exposure to light. The water content of mature, fully developed turfgrass seeds is extremely low. Water has a vital role since water absorption by imbibition and osmosis is the first physiological step in seed germination (75). Water functions in softening and swelling of the seed, which facilitates the entrance of oxygen and dilutes the protoplasm. Normal digestion, respiration, and assimilation processes are activated when the protoplasm becomes sufficiently moist. Soluble carbohydrates are transferred from the endosperm to the embryo.

Temperature, oxygen, and light conditions are also major factors controlling germination. Most cool season turfgrass species germinate readily in the 60 to 85°F range, while the warm season species require higher temperatures of 70 to 95°F. Seed germination of certain species is enhanced by exposure to an alternating diurnal temperature cycle (5, 38, 40, 59, 60, 68, 108, 113, 134, 140). Species having this thermoperiodic response include Kentucky bluegrass, bermudagrass, creeping bentgrass, Canada bluegrass, and goosegrass.

An adequate oxygen level is essential for germination since it is needed for respiration. The respiration rate is particularly rapid in germinating seedlings. Adequate soil aeration is important in ensuring an optimum oxygen level for successful seed germination.

Exposure to light is a fourth factor that may or may not be required for the germination of turfgrass seeds. Freshly harvested seed is usually more responsive to light exposure than mature seed (7, 11, 48, 72, 82, 108, 113). Germination of Kentucky bluegrass, Canada bluegrass, and bermudagrass seed is enhanced by light exposure.

Causes of germination failure. Sometimes seed germination does not occur even though a favorable environment exists for germination. Failure to germinate successfully may be subdivided into two aspects: (a) seeds that sprout but fail to emerge and (b) seeds that are viable but fail to germinate even under a favorable environment.

Factors that may cause seeds to sprout but fail to emerge include (a) planting too deeply, (b) crusted soil surface, (c) damage by fertilizer or pesticides, (d) saline soil, (e) direct damage by insects or diseases, and (f) weak or damaged seed.

Some turfgrass seeds are extremely small and must be planted near the soil surface. If planted too deeply, the plumule is not strong enough to reach the soil surface before the carbohydrate supply in the seed is exhausted. A crusted soil surface forms a mechanical barrier that may impair emergence of the germinating seed (61, 94). Surface crusting can be minimized by (a) daily irrigation, (b) a mulch, (c) mechanical breaking of the surface crust, or (d) soil modification with materials that reduce the crusting tendency.

Placement adjacent to the seed of an excessively high concentration of a fertilizer having a high salt index can result in seedling damage or death due to physiological drought (87, 107). The presence of phytotoxic concentrations of insecticides, fungicides, or similar materials in the vicinity of the seed can also cause death of the sensitive seedling (8, 92).

Insect or disease damage may occur if certain pests are present and actively feeding. Damping-off (*Fusarium*, *Pythium*, and *Rhizoctonia* spp.) are common disease problems of turfgrass seedlings. Turfgrass seed treatment with a fungicide such as thiram aids in controlling pre-emergence seedling death caused by damping-off organisms (58, 130, 131). This practice has not been used to any extent. Post-emergence seedling disease development cannot be controlled by seed treatment. Such insects as sod webworm and armyworm are occasional problems that may require chemical control.

Failure to emerge can also be due to infirmities in the seeds caused by damage during harvesting or processing. In addition, older seeds may still be viable but relatively weak and may not survive the germination stage (118).

Failure of viable seeds to germinate even under a favorable moisture, temperature, oxygen, and light environment may be caused by several different factors including (a) the presence of inhibitors and (b) an impermeable seed coat. The presence of an inhibitor in the seed coat or hulls of turfgrass seeds can restrict germination. This is a form of seed dormancy. Many seed inhibitors restrict the activity of the α-amalase enzyme involved in digestion processes during seed germination. Crabgrass and certain Kentucky bluegrass cultivars possess a seed germination inhibitor (64). The dormancy effect is most pronounced on freshly harvested seeds and declines during seed storage (48, 134). The degree of seed dormancy is not affected by the maturity stage at harvest but is affected by the moisture content of the seed (48). The higher the moisture content, the greater the dormancy.

A hard, impermeable seed coat that restricts the entrance of water or gases into the seed can prevent germination even under favorable conditions. This most commonly occurs in legumes such as crownvetch and the clovers (30, 111). Seed germination is enhanced by scarification processes that disrupt the impermeable coat and permit the entrance of moisture and gases. Scarified or dehulled bermudagrass and zoysiagrass seeds germinate more rapidly than seeds that do not have the hulls removed (2, 30, 58, 108). Germination of bahiagrass and centipedegrass seeds is enhanced by acid scarification (30, 74).

Pregerminated seed. The turfgrass establishment rate can be enhanced by using pregerminated seed, especially at cool temperatures. Pregermination involves the partial germination of turfgrass seeds by placing them in a moist environment at favorable temperatures. Once partial germination occurs, the seed must be planted immediately before the plumule and radicle emerge to the extent that they are desiccated or damaged during the seeding operation (33, 41). Planting into a dry seedbed results in severe desiccation damage to pregerminated seeds. Establishment from pregerminated seed is successful only if an adequate seedbed moisture level is maintained after planting.

Post-germination Care

A number of practices can be employed during the establishment phase to ensure the rapid development of a dense, tight sod. A combination of mulching and proper irrigation are the key factors in successful turfgrass establishment (Fig.

Figure 16-8. Successful, uniform establishment of a mixture of cool season turf-grasses resulting from the use of a mulch (left) and the lack of a stand where no mulch was used (right).

16-8). If a straw or hay mulch is used, a portion of it should be removed if the seedlings exhibit chlorosis due to an undesirably low light intensity. This is most likely to occur when the seedlings are 1 to 2 in. tall. The seedlings should also be protected against the effects of traffic by placing warning signs at intervals and even rope, fence, or wooden barricades, if needed.

Mowing. Normal mowing practices should be initiated when the seedlings reach a height one-third greater than the anticipated cutting height (60). For example, if a Kentucky bluegrass turf is to be maintained at a 1.5 in. cutting height, the new seedlings should be mowed when they reach a height of 2 in. Mowing should be continued at the standard frequency and cutting height similar to the practices for established turfs. Permitting the new seedlings to grow to an excessively tall height and then cutting them back to the normal height is not desirable. Although the initial, tall shoot growth results in the development of a somewhat larger root system, the drastic defoliation required to lower the cutting height to the desired level results in a severe setback to the plant. This effect is more drastic than mowing the plant earlier at a height similar to that desired on a long term basis.

Fertility. A light application of nitrogen fertilizer made when the turfgrass seedlings are between 1.5 and 2 in. high enhances the establishment rate substantially. Approximately 0.5 lb of actual nitrogen per 1000 sq ft watered into the soil is quite effective. Excessively high nitrogen levels should be avoided in order not to restrict root, rhizome, and stolon growth (90).

Topdressing. On stoloniferous species such as bentgrass that are to be maintained at a close cutting height, the formation of a dense sod is enhanced by frequent, light topdressings that cover the stolons. This practice encourages rooting of the stolons and the initiation of shoots at the nodes. It is not unusual to topdress new bentgrass plantings at 2- to 3-week intervals.

Irrigation. One of the most critical practices during both the germination and establishment phases is irrigation (9). The surface soil zone where seed germination and seedling growth activity occur should be maintained in a moist condition at all times. Failure to maintain an adequate moisture level is one of the major causes of poor turfgrass establishment. Young turfgrass seedlings have an extremely short root system that depends on a readily available moisture supply at the soil surface. The surface of unmulched soils can dry out very rapidly during periods of high light intensity and high temperature. A light application of water at midday is suggested under these conditions, particularly if the area is not mulched. Less frequent irrigation is required where mulching is practiced. The quantity of water applied during the daily irrigations is small, assuming the upper 6 in. of the seedbed was thoroughly soaked immediately following the completion of the seeding, rolling, and mulching operations. The soil surface should be maintained in a moist condition for at least 3 weeks following planting.

Weed control. One of the most critical decisions in postgermination care of turf-grasses is the timing of weed control practices. Most herbicides are somewhat more toxic to the newly germinated, sensitive turfgrass seedlings than to mature, established turfgrasses (31, 38, 39, 80, 81, 116, 128, 129). The root system is much more prone to injury than the shoots (43). Thus, it is desirable to delay applications of postemergence herbicides as long as possible after seeding. Even then, the herbicide should be applied only if necessary.

A few herbicides can be applied prior to planting for weed control during the establishment phases (80). An example would be the use of siduron for controlling annual grasses during the establishment of Kentucky bluegrass from seed (43). However, most turfgrass weeds must be controlled by postemergence herbicide applications. Postemergence applications should be delayed as long as possible after seeding. Generally, the phenoxy-type herbicides utilized in broadleaf weed control should not be applied for at least 4 weeks following germination (31). The use of postemergence organic arsenicals for the control of certain annual grassy weeds should be delayed even longer if possible (83, 84). If the weed population in the turfgrass stand is so extensive and vigorous that it is competing excessively with the desirable species, steps to control the competing weeds must be taken to avoid partial or complete loss of the desirable turfgrass seedling stand (83, 84).

Sodding

Sodding is the term used to describe the planting or covering of an entire area with pieces of sod. Sodding has been practiced for many centuries. The early use of turf in England dates back to the time when sod from meadows was cut, moved, and transplanted in turfgrass gardens. Starting in the late 1950's, a major commercial sod production industry developed in North America that supplies sod to home, business, institutional, and commercial sites (Fig. 16-9).

Sodding is a method of vegetative establishment where a mature, high quality turf is transplanted. Rooting into the underlying soil is the primary requisite for successful sod establishment. Sodding permits the establishment of a high quality turf in the shortest possible time. Problems with weed competition are minimized compared to establishment from seed, stolons, sprigs, or plugs. Sodded turfs are also available for immediate use. Establishment from sod involves (a) soil preparation, (b) obtaining sod of satisfactory quality, (c) transplanting, and (d) post-transplant care.

Figure 16-9. Transplanting Kentucky bluegrass sod on a home lawn site (Photo courtesy of D. P. Martin, Michigan State University, East Lansing, Michigan).

Sod Harvesting

Sod is ready for harvest when the rhizomes, stolons, and roots have knitted together to the extent that the sod can be harvested and handled without tearing. Characteristics desired in a high quality sod include (a) uniformity; (b) high shoot density; (c) adequate strength for harvesting and handling; (d) freedom from serious weeds, weed seeds, insects, diseases, and nematodes; (e) acceptable color; (f) sufficient maturity in terms of carbohydrate reserves to permit effective rooting; and (g) a minimum thatch layer.

The length of time required to produce a sod crop varies from 6 months to 2 years depending on the particular turfgrass cultivar, soil conditions, climate, and cultural practices. Kentucky bluegrass and bermudagrass sod can be produced in a shorter time than most other turfgrass species. Zoysiagrass, St. Augustinegrass, and red fescue require a relatively long time to produce a sod crop. Centipedegrass and bentgrass rank intermediate between these two groups. The time required to produce a crop is shorter on organic soils where irrigation is practiced and the effective growing season is longer. The **effective growing season** is defined as the number of days that soil temperatures are in the optimum range for sod growth of the particular turfgrass species or cultivar.

Sod cutting. Sod cutting is usually accomplished with powered mechanical sod cutters although manual, push-type units or sod knives can be used where relatively small quantities of sod are involved. The mechanical sod cutter cuts the sod at a specific thickness in widths of 12, 18, or 24 in. A mechanical drop knife attachment can be placed on the sod cutter to cut the sod strips into specific lengths. Kentucky bluegrass and red fescue sod is usually cut in lengths giving a total area of 1.0 or 1.5 sq yd per piece. Warm season turfgrass species are usually cut in 1 by 2 ft pieces having a 2 sq ft area. The sod cutter knife should be kept sharp at all times for the most effective operation. The frequency of knife replacement varies with the cutting conditions and soil type. The sod should be moist at the time of cutting for the most rapid rooting after transplanting. An excessively high moisture content should be avoided, however, since it results in increased shipping costs as well as a greater weight to handle during transplanting.

The thickness at which sod is cut varies with the turfgrass species, uniformity of the soil surface, soil type, and sod strength. The sod cutting thickness varies with the particular turfgrass species as shown in Table 16-3. The extent of rhizome and stolon development is the most critical factor in both sod strength for handling and in the transplant rooting capabilities. The poorer the sod strength and shoot density, the thicker the sod must be cut to permit handling. Sod cut relatively thin roots faster than sod cut thick and it is also lighter in weight for ease in handling (1, 73, 142). Sod cut excessively thick is more tolerant of droughty conditions but is much slower to root. Sod cut excessively thin is very prone to injury from atmospheric drought.

Table 16-3

THE RELATIVE CUTTING DEPTHS FOR SOD HARVESTING
OF ELEVEN TURFGRASS SPECIES

Relative cutting depth	Cutting depth, in.*	Turfgrass species
Shallow	0.3–0.6	Bentgrass
Medium	0.5–0.8	Kentucky bluegrass Bermudagrass Zoysiagrass
Thick	0.7–1.0	Red fescue Tall fescue Meadow fescue
Very thick	0.8–1.3	St. Augustinegrass Bahiagrass Centipedegrass Perennial ryegrass

*Measured from the soil surface.

The sod thickness is also an important factor in the productive life of a sod field. The amount of soil removed during sod production depends on the cutting depth and harvest frequency. The amount of soil removed by an annual sod harvest on organic soil is only slightly more than the loss that occurs by subsidence under an annual row crop of onions (122).

Sod handling. Sod was rolled and loaded manually for many years. More recently a number of different types of mechanical sod rolling, handling, and loading machines has been developed. The sod strips may be handled as flat pieces, rolls, or folded. St. Augustinegrass, centipedegrass, and bahiagrass are usually handled as small flat pieces that are stacked on a pallet holding approximately 500 sq ft of sod. Kentucky bluegrass and red fescue sod have usually been rolled or folded for handling.

There are many different types of sod handling units. Some mechanical units operate independently in sod cutting, rolling, and conveying to trucks. There are also dual purpose sod handling machines that cut and roll the sod within a single unit (Fig. 16-10). Finally, there are mechanical handling units that cut, roll or fold, and convey the sod to pallets or adjacent trucks (Fig. 16-11). Reliability, amount of sod harvested per unit of time, and harvesting cost are the major factors to consider in selecting a method of mechanical harvesting.

Sod shipping. Sometimes the harvested sod remains stacked for an extended period of time (a) during shipment to distant consumer sites or (b) because unfavorable weather conditions do not permit immediate transplanting. Sod can be seriously damaged by sod heating in the stack during this period. The length of time that sod can be stacked before heating damage occurs varies from 10 to 60 hr depend-

Figure 16-10. A dual pupose sod cutter and roller in operation.

Figure 16-11. An automatic sod harvester which cuts, rolls, and conveys the sod for stacking on pallets. (Photo courtesy of Ryan Equipment Co., St. Paul, Minnesota).

ing on the physiological condition of the sod and the temperature of the soil and atmosphere.

The postharvest physiology of sod heating is a complex phenomenon. Kill of stacked sod is caused primarily by high temperature. Lethal heating effects are observed in Kentucky bluegrass sod stacks at 105°F. The killing temperature of warm season turfgrass species is higher than cool season turfgrasses. Disease development (*Fusarium* spp.) in the sod stack occurs as a secondary effect rather than as the primary cause of kill.

A number of environmental and physiological factors contribute to increased sod heating proneness in stacks. Factors included are (a) soil temperature at harvest, (b) amount of leaf area present, (c) nitrogen nutritional level, (d) soil moisture content, (e) amount of clippings present, (f) presence of disease, and (g) developmental state of the sod.

A high soil or sod temperature at the time of harvest results in increased sod heating proneness. This high sod temperature effect can be reduced somewhat by harvesting in early morning when temperatures are normally lowest.

The amount of leaf area present at the time of sod harvest is one of the major cultural factors influencing proneness to sod heating. A low cutting height reduces the amount of respiratory tissue and causes a significant decrease in sod heating problems. A high plant nitrogen nutritional level or the presence of clippings increases the proneness to sod heating under certain conditions.

Increased heating of Kentucky bluegrass sod is also associated with severe disease development, particularly *Helminthosporium* leaf spot. The effect of the developmental stage at the time of sod harvest is quite striking. Sod is much more prone to heating during periods of seed head development.

A ventilation system through the bottom of the sod load may assist in reducing the heating problem when sod is shipped to distant points. A typical ventilation system involves an air scoop mounted at the front top of the load that directs a flow of air through an open, false bottom tunnel located in the center of the truck bed. Artificial cooling of sod loads by means of vacuum cooling has been effective in minimizing sod heating (4). Sod cooling is costly, however, which results in an increased sod market price. Vacuum cooling has not been widely used for this reason.

Sod Transplanting

The primary objective of sod transplanting is to ensure rapid rooting or "knitting" of the sod into the underlying soil. Roots are initiated primarily from the rhizomes of Kentucky bluegrass sod (73, 141). Sod should be stored in a cool, shaded location if there is an unexpected delay in transplanting. Factors that ensure rapid sod rooting include (a) proper soil preparation, (b) adequate moisture in the underlying soil, and (c) transplanting techniques that minimize desiccation.

Soil preparation prior to sodding should be the same as previously discussed for seeding. Transplanting on rocky, improperly tilled soil results in problems. It is important to ensure an adequate soil moisture level prior to transplanting (85). The maximum rooting rate occurs on a moist underlying soil. Sod placed on dry soils experiences a substantial delay in rooting even though the sod and underlying soil are thoroughly soaked following transplanting (Fig. 16-12).

The transplanting procedure should involve staggering the sod pieces in a checkerboard pattern. The ends of individual sod pieces should be joined in good contact but with no overlapping. It is important not to stretch the sod during transplanting since it shrinks upon drying to form objectional void spaces between the sod pieces. It may be necessary to peg sod pieces placed on slopes to prevent slippage.

After the sod is properly placed, the area should be tamped or rolled to ensure good contact between the sod and underlying soil. Rolling should be perpendicular to the direction in which the sod lengths were layed. Air pockets remain between the sod and soil if tamping or rolling is not practiced. This results in drying of sod roots and rhizomes and causes a delay in sod rooting.

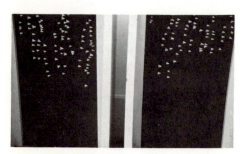

Figure 16-12. The relative depth of rooting of sod transplanted on a dry (left) versus a moist soil (right). White markers indicate the daily rate of growth of individual roots.

Sod can be transplanted at anytime during the year provided an adequate moisture level can be maintained by irrigating (106). Late fall sodding should be early enough to permit rooting into the underlying soil prior to winter dormancy. Dormant sodding, in which temperatures are unfavorable for rooting, may or may not survive the winter. The sod will survive with minimum problems if a snow cover persists throughout the winter. The sod may be seriously damaged, however, if an open winter characterized by severe atmospheric desiccation and soil drought occurs.

The relative transplant rooting ability of sod produced on various soil types is sometimes questioned. Comparisons of the transplant rooting ability of sod grown on organic soil versus mineral soil have generally indicated no significant difference, provided the sods were produced under comparable conditions (85, 86). Much more variability in sod rooting is caused by variations in the cultural practices used in sod production than from the actual soil on which the sod is grown. The physiological condition of the turf at harvest is a factor in the relative rooting rate. Sod that has been forced by high nitrogen fertility levels is immature and possesses a relatively low level of carbohydrate reserves. As a result, the rooting rate is impaired compared to a sod grown under a controlled nitrogen fertility level that has a higher level of carbohydrate reserves.

The presence of a cleavage plane at the interface of the sod piece with the underlying soil is of concern, particularly on sports turfs where play involves the twisting and turning action of cleated shoes. Sod possessing a particle size that is substantially different from the underlying soil has a cleavage plane at which the sod tears more readily when stress is exerted by twisting or turning actions. Sod containing soil of a distinctly different texture can also be a problem on greens and tees because of the disruption in water movement. Thus, the sod used on sports fields, greens, and tees should be grown on a soil having a texture comparable to the underlying soil on the site where the sod is to be transplanted. Certain mineral soils can be just as objectionable as an organic soil under these conditions. Transplanting a sod having a different soil particle size than the underlying soil has not been a significant problem on ornamental and functional turfs such as home lawns.

Post Transplant Care

Irrigation of transplanted sod is quite comparable to that for seeding. Deep irrigation to a 6 to 8 in. soil depth should be accomplished immediately after the sodding is completed. This is best achieved by applying the water at a relatively low rate over an extended period of time. Subsequently, the sodded area should be irrigated lightly every day at noon to maintain an adequate moisture level in the sod. Sod that is cut thick requires less frequent irrigation than thin sod.

The mowing height and frequency on newly sodded areas should be the same as normally practiced on established turfs of the same species. Fertilization of newly sodded areas is generally not required because (a) the sod should have been grown under optimum fertility and (b) fertilizer has usually been incorporated into the soil prior to transplanting. No weed control practices should be necessary, provided quality sod has been purchased. Traffic control may be required.

Plugging

The vegetative propagation of turfgrasses by means of plugs or small sod pieces is called **plugging**. This procedure can be utilized for repairing damaged areas or for the initial establishment of a turf by vegetative propagation. The repair of damaged areas is sometimes referred to as **spot sodding**.

Plugging of damaged areas or spot sodding is most commonly practiced on turfs that were established vegetatively with a specific, uniform cultivar. Typical examples are creeping bentgrass or bermudagrass putting greens. Plugs are usually obtained from a nursery area and carefully plugged into the dead spots on the green. When repairing greens by spot plugging, it is important to obtain the plugs from a nursery having the same cultivar, intensity of culture, and underlying soil texture.

Plugging is also utilized in establishing new turfgrass areas. Plugging is safer than sprigging in terms of surviving drought stress (63). The species most commonly established by plugging are zoysiagrass, manilagrass, St. Augustinegrass, centipedegrass, and buffalograss. The sod plugs are obtained from nurseries or sod fields that are established and maintained in the same way as normal sod production. However, harvesting usually involves a specialized automatic machine that cuts, divides, and separates the sod plugs into 2- to 4-in. diameter circular plugs or 2- to 4-in. squares. The plugs are planted in the prepared soil on 12- to 16-in. centers. The rate of spread from a 4-in. plug is greater than from a 2-in. plug (97). Planting 2- to 4-in. sod squares on 12-in. centers requires 30 to 110 sq ft of sod per 1000 sq ft.

Soil preparation for plug establishment of new turfgrass areas should be the same as for seeding or sodding. After harvest, the plugs are transported to the site and set in a prepared seedbed on 6- to 14-in. centers (Fig. 16-13). The shorter the distance between plugs, the quicker a turf is established (98). Specialized automatic planting machines are available for plug establishment on large areas. Establishment by plugging can be accomplished manually on small areas.

The area should then be rolled to firm the soil around the plug and provide favorable moisture relations. Irrigation during initial rooting of the plug facilitates establishment. On unirrigated areas, the plugs should contain a larger quantity of adhering soil for better drought tolerance. Plug establishment of warm season turfgrass species, such as zoysiagrass, is usually done during late spring or early summer in order to have a maximum period of warm temperatures for rapid turfgrass establishment.

Figure 16-13. A 4-inch diameter zoysiagrass plug set in place with some lateral stolon growth developing. (Photo courtesy of V. B. Youngner, University of California at Riverside, Riverside, California).

A common practice on home lawns is to establish a small area of zoysiagrass turf initially. Then plugs are taken from this area for plugging and spread of the zoysiagrass throughout the remainder of the lawn. The plugs are usually planted directly into the existing turf in this case. The turfgrass establishment rate by plugging into an existing turf is significantly enhanced if the competition from surrounding vegetation is reduced by the use of contact herbicides or by vertical mowing (71, 76). Mowing at 0.75 in. is more favorable than 1.5 in. for zoysiagrass establishment from plugs (71). Applying a growth inhibitor such as maleic hydrazide to the existing vegetation prior to plugging enhances turfgrass establishment from plugs (97, 136). Also, differential fertilizer applications made only into the plug holes stimulate plug establishment (97, 136).

Post plugging care involves mowing at the height and frequency normally required for the specific turfgrass species or cultivar being established. A fertilizer application made 3 to 4 weeks after plugging enhances the establishment rate. Fertilizer applications should be made as required to maintain an adequate rate of outward rhizome and stolon growth into the surrounding area. Irrigation enhances establishment. Topdressing is seldom practiced following plugging except on bentgrass or bermudagrass greens. A more optimum establishment rate is achieved if competitive weeds are controlled as they occur.

Stolonizing and Sprigging

Stolonizing and sprigging are two means of vegetative establishment in which the soil is not transported and planted with the stolons and rhizomes. This contrasts with sodding and plugging where the sod piece including the adhering soil is used as a unit for establishment. **Stolonizing** is the vegetative planting of stolons by broadcasting over a prepared soil and covering by topdressing or press rolling.

It is most commonly practiced in the establishment of creeping bentgrass and bermudagrass. **Topsoil planting** is a modification of stolonizing. This method involves covering the area with a soil containing viable rhizomes and/or stolons for the purpose of producing a turfgrass cover. Topsoil planting is used for the establishment of bermudagrass on highway slopes and airfields.

Sprigging involves the vegetative planting of stolons or rhizomes in furrows or small holes. In this case, the stolons or rhizomes have little or no adhering soil and are called **sprigs**. Bermudagrass, zoysiagrass, and bentgrass are established by sprigging (Fig. 16-14).

Figure 16-14. Sprig row establishment of Meyer zoysiagrass. (Photo courtesy of G. C. Nutter, Jacksonville Beach, Florida).

Stolon and Sprig Production

The production of high quality stolons is a very specialized operation. One of the major concerns in stolon production is to grow plant material that is genetically pure in strain and free of offtype or objectionable weedy species. Soil preparation prior to the establishment of stolon nurseries is the same as for seeding, sodding, or plugging, except for special precautions to ensure that the soil is weed-free. Stolon nurseries are frequently fumigated prior to establishment or are fallowed for an extended period of time. Sprigs for the establishment of stolon nurseries should be obtained from a source that ensures the genetic purity of the cultivar. Sprigs are usually obtained from turfs maintained under greens conditions that do not permit seed formation and the introduction of offtypes.

Stolons are normally propagated in rows spaced 3 to 5 ft apart. The sprigs should be planted in shallow furrows, at a depth of 1.5 to 2 in. in the case of bentgrass. Deeper depths are preferred for bermudagrass and St. Augustinegrass if heat or moisture stress conditions are anticipated. A square foot of sod normally plants 75 to 80 linear ft. Sprigs are covered with soil in such a manner that a portion of the sprigs extends above the soil surface. The soil should be firmed or pressed around the sprigs by a mechanical roller. This is usually accomplished with the feet in the case of small plantings. Stolon planting of cool season species such as bentgrass should be done in late summer, whereas bermudagrass and St. Augustinegrass should be planted in late spring. Planting at these times ensures the longest period of favorable temperature and moisture conditions for growth.

Following planting, the most rapid stolon growth can be achieved through irrigation and regular fertilization to maintain adequate color. Seed heads must be removed immediately to avoid the development of viable seeds that could fall to the soil surface, germinate, and produce offtypes which would contaminate the stolon rows.

Harvesting stolons usually involves a sod cutter for loosening the plant material from the underlying root system. The material is passed through a cutting or shredding unit that chops the stolons into small lengths of 3 to 6 in. The stolons are then bagged and transported to the site for planting.

Sprigs are most commonly harvested from a sod maintained similar to that for commercial sod production. Stolons can also be harvested from sodded turfs. The sod is cut with a regular sod cutter at a very shallow depth. The vegetative material is then passed through a chopping unit or shredder that cuts the sod into relatively small pieces for sprigging. Stolons and sprigs should be planted as soon as possible after being harvested (15).

Fully automatic stolon and sprig harvesting machines are now available that cut, shred, screen, and load the stolons. The stolons or sprigs should be maintained in a cool, moist environment during shipment to the planting site (37). Storage at lower temperatures restricts root and shoot sprout development. Sprouting should be avoided since serious injury to sprouted sprigs or stolons can occur if exposed to heat or desiccation. This would impair the vigor and rate of establishment.

Fungicide applications are sometimes required for the control of certain diseases that develop in the bags during shipment.

High quality, genetically pure stolons or sprigs should be obtained or grown. The purchase of certified vegetative material provides a guarantee as to (a) the genetic purity and (b) the specific cultivar.

Stolonizing

Soil preparation for stolonizing should be the same as for seeding or sodding. The turfgrass species most commonly established by stolonizing are creeping bentgrass, velvet bentgrass, and bermudagrass. The establishment rate for stolonizing is greater than sprigging but the mortality is also greater (63). Late summer is the best time to stolonize the bentgrasses while bermudagrass establishment should be done in late spring. Approximately 8 to 10 bushels of bentgrass stolons are needed to plant 1000 sq ft. This requires approximately 80 to 140 sq ft of nursery sod. Somewhat less area is required per bushel of bermudagrass stolons. Following uniform distribution of the stolons, the area should be rolled or a steel mat placed over the area to press the stolons into the soil (95) (Fig. 16-15).

Figure 16-15. Planting bentgrass stolons using the steel matting technique for pressing the stolons into the soil surface. (Photo courtesy of the Milwaukee Sewerage Commission, Milwaukee, Wisconsin).

When planting bermudagrass on extensive turfgrass areas such as fairways and sports fields, the broadcast stolons are pressed into the upper inch of soil with a series of straight disk-like roller devices. The soil can be firmed around the stolons with a cultipacker. Fully automatic machines are available that broadcast the stolons, press them into the soil, and firm the soil around the stolons (78) (Fig. 16-16).

Stolons can also be applied with a water carrier by means of hydroplanting. This is basically the same procedure as previously described for hydroseeding.

Figure 16-16. A mechanical planter used for stolonizing bermudagrass. (Photo courtesy of Cal-Turf, Camarillo, California).

A hydraulic planter with a gear-type pump should be used. Hydroplanting has the advantage of minimal foot traffic on the area after the final grade is established (126). In addition, considerably less labor and time are involved.

Topdressing is usually applied as a 0.15- to 0.25-in. thick layer that is equivalent to 0.5 to 0.75 cu yd per 1000 sq ft. Rolling the area following topdressing provides good contact between the soil and stolons for more optimum moisture relations during the establishment period. Approximately 15 to 25% of the stolons should extend above the soil upon completion of topdressing. The area should be irrigated immediately after topdressing.

Sprigging

Soil preparation for sprigging should be the same as for seeding, sodding, or stolonizing. Less plant material is required for sprigging than for stolonizing and the mortality is also less. The establishment rate is more rapid than plugging (98). Special mechanical planters are available for sprigging in narrow rows. Sprigging can also be accomplished by hand on smaller areas such as lawns by either row or broadcast sprigging.

Typically, a furrow is made and the sprigs inserted at a depth of 1 to 2 in. A shallower planting depth is preferred provided adequate moisture is available (37, 126). The soil is then pushed over the sprigs and firmed. The rows are normally spaced 10 to 18 in. apart with the sprigs placed approximately 4 to 6 in. apart in the rows. The narrower the spacing between rows and between sprigs within rows, the more rapid the turfgrass establishment rate. After covering and firming into place, 25% of the sprigs should extend above the soil surface. The area should be irrigated deeply as soon as possible after the sprigging operation is completed.

Clonal planting. Beachgrass is established by a modification of sprigging in which clones containing four to six stems each are planted on 12- to 18-in. centers or in rows 18 in. apart with a 3- to 5-in. spacing within a row (25). Clonal planting may be accomplished manually or mechanically (150). Clonal planting of beachgrass should be done in the spring. The clones are collected for planting when in a dormant state. A fertilizer application is particularly important in ensuring establishment on sandy sites.

Post Stolonizing or Sprigging Care

It is imperative to maintain a moist surface during initial establishment from stolons and sprigs. The surface soil layer can become dry quite rapidly causing desiccation of the sprigs or stolons and loss of stand. Thus, the area should be irrigated every day around noon or even more frequently where high rates of evapotranspiration occur.

Topdressing should be practiced at regular intervals during the initial establishment period. It provides more favorable moisture conditions for stolon growth, encourages rooting at the nodes, and enhances the initiation of additional branch stolons that results in increased shoot density. The topdressing material used during stolonizing should be similar to the underlying soil.

The initial mowing of bentgrass and bermudagrass established from stolons or sprigs should be done when the stolons reach a length of 3 to 4 in. The cut stolon pieces should be returned since they may also root and contribute to a more rapid establishment rate. Zoysiagrass establishes more rapidly at a cutting height of 1 in. or less (63).

Weeds should be controlled if they occur. DCPA and bensulide are effective in controlling most weedy annual grasses without significant injury to bermudagrass (44, 51). Mecoprop is commonly used for broadleaf weed control during the establishment of bentgrass greens. Rooting of bentgrass, bermudagrass, St. Augustinegrass, centipedegrass, and carpetgrass stolons is seriously restricted by the application of 2,4-D (31).

Renovation and Reestablishment

Established turfs are sometimes permitted to deteriorate to the extent that they cannot be improved by normal turfgrass cultural practices. The problem can only be corrected by either renovation or reestablishment when this occurs. The primary factors causing severe deterioration of a turf are (a) a predominance of unadapted or undesirable species; (b) undesirable soil conditions such as acidity, salinity, poor drainage, compaction, or the presence of a severe layering problem; (c) an excessive thatch accumulation; (d) severe turfgrass damage caused by diseases, insects, nematodes, or the toxic effect of chemicals; and/or (e) improper cultural practices. The area must be either renovated or reestablished, if the poor turfgrass quality resulting from one or more of these factors cannot be corrected by normal cultural practices such as mowing, fertilization, cultivation, irrigation, and pest control. The most important principle to consider prior to reestablishment or renovation is the correction of the original cause or causes of turfgrass deterioration.

Reestablishment

Reestablishment or rebuilding usually involves (a) complete removal of the turf, (b) deep cultivation, and (c) planting using the procedures previously described in this chapter for establishment of new turfgrass areas. Persistent weedy perennial species such as quackgrass and bentgrass should be eradicated prior to reestablishment. Thatch and soil layering problems can be eliminated and poor soil drainage or compaction corrected by soil modification during reestablishment. The area may also be recontoured. The preferred time for reestablishment is in late summer if cool season turfgrasses are involved and in late spring for warm season turfgrasses.

Renovation

Renovation involves turfgrass improvement beyond the routine cultural practices in which reseeding or replanting is done into the existing live and/or dead vegetation. In this case, the causes of turfgrass deterioration have usually resulted from (a) improper cultural practices or (b) injury caused by turfgrass pests or chemical toxicity. Renovation is generally practiced where the stand of living, desirable turfgrass species is so poor that the area cannot be improved readily through routine fertilization and irrigation but is sufficient to be saved. The basic steps in turfgrass renovation are

1. Eradication of undesirable species.
2. Thatch removal, if present.
3. Cultivation by coring, grooving, slicing, or spiking.
4. Fertilization and possibly liming.
5. Seeding.
6. Irrigation.

A basic principle in successful turfgrass renovation is to ensure good contact between the soil and the seed or vegetative plant parts. The best time to renovate a turf is in late summer where cool season turfgrasses are involved and in late spring if warm season species are used. Sometimes the timing of renovation must be coordinated with periods of minimal use, particularly on sports turfs.

The first renovation step is the control of undesirable species such as broadleaf and grassy weeds (17, 144, 149). Broadleaf weeds can be readily controlled with 2,4-D, silvex, or dicamba. Most annual grasses are controlled by repeat applications of an organic arsenical such as MAMA, while contact kill of the above-ground foliage of weedy perennial grasses such as quackgrass, tall fescue, and bentgrass can be achieved with such materials as sodium arsenite and paraquat (45, 67, 102, 103).

A thorough vertical mowing should be accomplished after killing the undesirable vegetation. This serves to remove the dead, undesirable weedy vegetation and also removes any thatch that may have accumulated. Next, the area should be thoroughly cultivated by coring, grooving, or slicing. From two to five repeat cultivations may be required. These operations are preferred to spiking since some soil is brought to the surface and penetration is to a greater depth. A vertical mowing or matting of the area assists in breaking up the cores if coring is practiced. The area should be fertilized following the cultivation operations. Liming may also be required.

The next step is to apply the seed uniformly over the area. Matting, dragging, or raking following seeding tends to work the seed down into the cultivated areas. Topdressing is frequently practiced following renovation seeding of greens. A light rolling after topdressing assists in rapid germination. The area should be thoroughly soaked immediately following seeding and kept moist throughout the initial germination and establishment period.

Specialized renovation machines have been developed that provide a grooving or slicing operation in combination with the insertion of seed into the slits made within the existing turf (Fig. 16-17). This procedure has been used successfully in renovating extensive turfgrass areas.

A similar procedure is involved in the renovation of turfgrass areas such as greens where stolons or sprigs are to be planted. This is a particularly important practice on greens containing a bentgrass cultivar that can only be established vegetatively. Overseeding with another cultivar could destroy the uniformity of

Figure 16-17. A turfgrass renovation machine for inserting seed into the soil with the aid of a slicing operation that follows a grooving cultivation which brings soil to the surface.

the turfgrass surface. All steps are essentially the same as renovation seeding except that the planting operation involves inserting the sprigs into narrow slits located on 8- to 16-in. centers in parallel rows (57). Renovation establishment by vegetative means can also be achieved by working stolons into holes produced by coring. After inserting the sprigs into slits or after a broadcast stolon application into coring holes, the green is topdressed to ensure good soil contact and favorable moisture relations for rapid establishment.

Winter Overseeding

The seeding of cool season turfgrasses into warm season species to provide a green, actively growing winter cover requires specific planting techniques. Winter overseeding is frequently practiced on bermudagrass greens and also to a certain extent on lawn areas containing winter-dormant warm season species.

Seedbed preparation commonly involves vertical mowing or very close cutting immediately prior to overseeding to reduce competition from the warm season species. This practice also improves seed contact with the soil. Coring approximately 3 to 4 weeks prior to planting is beneficial, particularly on greens planted with small seeded species. The seed is then distributed uniformly over the area. Spiking at the time of seeding also assists in seed–soil contact.

The area should be topdressed immediately following the overseeding operation. The topdressing rate ranges from 0.4 to 0.7 cu yd per 1000 sq ft. The lower rate is used for the smaller-seeded turfgrass species. The soil mix utilized for topdressing should be comparable to the underlying soil. A light matting is sometimes practiced following topdressing. The overseeded area should then be irrigated thoroughly and kept moist throughout the establishment period.

The seedbed preparation procedures required for large seeded turfgrasses such as ryegrass do not have to be as extensive as for the small seeded turfgrasses. In home lawn situations, winter overseeding of ryegrass can be accomplished relatively simply by close mowing, seeding, and topdressing at 0.5 cu yd per 1000 sq ft. Successful winter overseeding is more difficult to achieve on warm season turfs that contain an excessive thatch accumulation.

Proper timing in the overseeding operation is an important consideration. Basically, overseeding should be done late enough in the fall that bermudagrass growth is slowed by the low temperatures and will not compete excessively with the overseeded cool season turfgrass species. On the other hand, overseeding should not be so late that the soil temperatures are below the favorable range for seed germination of the cool season species, thus resulting in an inferior stand.

The seeding rates used for winter overseeding are much higher than typically used in the establishment of cool season turfgrasses in the cool humid region. The combination of a surface seeding and the effects of traffic results in a greater mortality rate that necessitates a higher seeding rate. The seeding rates for greens are 4 to 8 times as great as for lawns, depending on the particular species utilized. Typical seeding rates for lawns by species are 3 to 4 lb per 1000 sq ft for bluegrass, 6 to 8 lb for red fescue, and 12 to 16 lb for ryegrass. This contrasts with greens where 4 to 8 lb of bentgrass, 10 to 14 lb of bluegrass, 30 to 35 lb of red fescue, and 50 to 60 lb of ryegrass per 1000 sq ft are commonly used when seeded alone. These rates are reduced proportionally when a mixture is used, as is more commonly the case.

The use of fungicides for seedling disease control is frequently necessary following winter overseeding with cool season turfgrasses. This is particularly important on greens where higher nitrogen and moisture levels usually exist. A fungicide application should be made immediately after overseeding. Additional fungicide applications may be required depending on the specific conditions.

References

1. ANONYMOUS. 1925. How thick to cut sod for putting greens. Bulletin of Green Section of USGA. 5: 172–173.

2. ANONYMOUS. 1936. Bermuda grass seed. Turf Culture. 1(2): 6.

3. ANONYMOUS. 1964. Rules for testing seeds. Proceedings of the Association of Official Seed Analysts. 54(2): 1–112.

4. ANONYMOUS. 1964. Vacuum cooling of sod aids shipping problems. American Nurseryman. 120(6): 52–54.

5. AKAMINE, E. K. 1944. Germination of Hawaiian range grass seeds. Hawaii Agricultural Experiment Station Technical Bulletin No. 2. pp. 1–60.

6. ALDERFER, R. B., and F. G. MERKLE. 1943. The comparative effects of surface application vs. incorporation of various mulching materials on structure, permeability, run-off and other soil properties. Soil Science Society of America Proceedings. 8: 79–86.

7. ANDERSON, A. M. 1947. Some factors influencing the germination of seed of Poa compressa L. Proceedings of the Association of Official Seed Analysts. 37: 134–143.

8. Atsatt, P. R., and L. C. Bliss. 1963. Some effects of emulsified hexa-octadecanol on germination, establishment, and growth of Kentucky bluegrass. Agronomy Journal. 55: 533–537.

9. Barkley, D. G., R. E. Blaser, and R. E. Schmidt. 1965. Effect of mulches on microclimate and turf establishment. Agronomy Journal. 57: 189–192.

10. Barnett, A. P., E. G. Diseker, and E. C. Richardson. 1967. Evaluation of mulching methods for erosion control on newly prepared and seeded highway backslopes. Agronomy Journal. 59: 83–85.

11. Bass, L. N. 1951. Effect of light intensity and other factors on germination of seeds of Kentucky bluegrass (*Poa pratensis* L.). Proceedings of the Association of Official Seed Analysts. 41: 83–86.

12. ———. 1953. Relationships of temperature, time and moisture content to the viability of seeds of Kentucky bluegrass. Proceedings of the Iowa Academy of Science. 60: 86–88.

13. Beard, J. B. 1966. A comparison of mulches for erosion control and grass establishment on light soil. Michigan Agricultural Experiment Station Quarterly Bulletin. 48(3): 369–376.

14. Beard, J. B., and P. E. Rieke. 1964. Sod production in Michigan. Michigan State University Mimeo. pp. 1–7.

15. Beaty, E. R. 1966. Sprouting of Coastal bermudagrass stolons. Agronomy Journal. 58: 555–556.

16. Bell, R. S., and J. C. F. Tedrow. 1945. The control of wind erosion by the establishment of turf under airport conditions. Rhode Island Agricultural Experiment Station Bulletin 295. pp. 1–22.

17. Bichowsky, E. R., and V. B. Youngner. 1956. Chemicals for lawn renovation. Southern California Turf Culture. 6: 3.

18. Blaser, R. E. 1962. Soil mulches for grassing. Roadside Development 1962, Highway Research Board. pp. 15–20.

19. Borst, H. L., and R. Woodburn. 1942. The effect of mulching and methods of cultivation on run-off and erosion from Muskingum silt loam. Agricultural Engineering. 23: 19–22.

20. Bredakis, E. J., and J. M. Zak. 1965. Timing of seeding throughout the growing season. 24th Short Course on Roadside Development. 24: 65–69.

21. Brink, V. C., J. R. Mackay, S. Freyman, and D. C. Pearce. 1967. Needle ice and seedling establishment in southwestern British Columbia. Canadian Journal of Plant Science. 47: 135–139.

22. Broack, R. V. D., C. S. Slater, M. T. Augustine, and E. E. Evaul. 1959. Jute for waterways. Journal of Soil and Water Conservation. 14: 117–119.

23. Brock, J. R. 1964. Grass under plastic. Golf Course Reporter. 32(10): 62–64.

24. Brown, R. L., and A. L. Hafenrichter. 1948. Factors influencing the production and use of beachgrass and dunegrass clones for erosion control. I. Effect of date of planting. Journal of American Society of Agronomy. 40: 512–521.

25. ———. 1948. Factors influencing the production and use of beachgrass and dunegrass clones for erosion control. II. Influence of density of planting. Journal of American Society of Agronomy. 40: 603–609.

26. ———. 1948. Factors influencing the production and use of beachgrass and dunegrass clones for erosion control. III. Influence of kinds and amounts of fertilizer on production. Journal of American Society of Agronomy. 40: 677–684.

27. Bruto, F. R. 1947. Developments in the use of mulches and grasses for erosion control on highway roadsides. Highway Research Abstracts. 140: 16–19.

28. Bryan, W. E. 1918. Hastening the germination of bermuda grass seed by the sulfuric acid treatment. Journal of American Society of Agronomy. 10: 279–281.

29. Burt, E. O. 1965. Warm season turfgrasses and preemergence herbicides. Golf Course Reporter. 33: 54–56.

30. BURTON, G. W. 1939. Scarification studies on southern grass seeds. Journal of American Society of Agronomy. 31: 179–187.

31. ———. 1947. 2,4-D aids the establishment of southern turf grasses. Timely Turf Topics. 1: 3.

32. BUTTON, E. F. 1959. Effect of gibberellic acids on laboratory germination of creeping red fescue (*Festuca rubra*). Agronomy Journal. 51: 60–61.

33. BUTTON, E. F., and C. F. NOYES. 1964. Effect of seaweed extract upon emergence and survival of seedlings of creeping red fescue. Agronomy Journal. 56: 444–445.

34. BUTTON, E. F., and K. POTHARST. 1962. Comparison of mulch materials for turf establishment. Journal of Soil and Water Conservation. 17(4): 166–169.

35. CHAMPNESS, S. S. 1950. Effect of microclimate on the establishment of timothy grass. Nature. 165: 325.

36. CHEPIL, W. S., N. P. WOODRUFF, F. H. SIDDOWAY, and D. V. ARMBRUST. 1963. Mulches for wind and water erosion control. USDA Agricultural Research Service. 41–84. pp. 1–23.

37. CHILES, R. E., W. W. HUFFINE, and J. Q. LYND. 1966. Differential response of *Cynodon* varieties to type of sprig storage and planting depth. Agronomy Journal. 58: 231–234.

38. CHIPPINDALE, H. G. 1949. Environment and germination in grass seeds. British Grassland Society Journal. 4: 57–61.

39. CORNMAN, J. F., F. M. MADDEN, and N. J. SMITH. 1964. Tolerance of established lawn grasses, putting greens, and turfgrass seeds to pre-emergence crabgrass control chemicals. Proceedings of the 18th Northeastern Weed Control Conference. 18: 519–522.

40. CULLIMAN, B. 1941. Germinating seeds of southern grasses. Proceedings of the Association of Official Seed Analysts. 33: 74–76.

41. DANIEL, W. H. 1958. Why soak grass seed? Proceedings of the 1958 Midwest Regional Turf Conference. pp. 30–31.

42. DAWSON, R. B., and T. W. EVANS. 1931. The establishment of grasses on very acid moorland with a view to turf formation. Journal of the Board of Greenkeeping Research. 2(5): 106–111.

43. DAWSON, R. B., and N. L. FERGUSON. 1938. The effect of lead arsenate on the germination of seed and subsequent development of certain grasses. Journal of the Board of Greenkeeping Research. 5(19): 274–281.

44. DEAL, E. E. 1967. Tufcote bermudagrass establishment using preemergence herbicides. 1967 Agronomy Abstracts. p. 56.

45. DEFRANCE, J. A. 1948. Effect of certain chemicals on the killing of putting-green turf and their influence on various methods of reseeding. Greenkeeper's Reporter. 16: 36–38.

46. DEFRANCE, J. A., R. S. BELL, and T. E. ODLAND. 1947. Killing weed seeds in the grass seedbed by the use of fertilizers and chemicals. Journal of American Society of Agronomy. 39: 530–535.

47. DEFRANCE, J. A., and J. A. SIMMONS. 1951. Relative period of emergence and initial growth of turf grasses and their adaptibility under field conditions. Proceedings of the American Society for Horticultural Science. 57: 439–442.

48. DELOUCHE, J. C. 1958. Germination of Kentucky bluegrass harvested at different stages of maturity. Proceedings of the Association of Official Seed Analysts. 48: 81–84.

49. DEWITT, J. L., C. L. CANODE, and J. K. PATTERSON. 1962. Effects of heating and storage on the viability of grass seed harvested with high moisture content. Agronomy Journal. 54: 126–129.

50. DOLLING, H., and W. D. SHRADER. 1964. Comparison of fiber glass and other mulch materials for erosion control on highway backslopes in Iowa. 23rd Short Course on Roadside Development. 23: 83–88.

51. DUBLE, R. L., and E. C. HOLT. 1965. Preemergence herbicides as they influence bermudagrass establishment. Golf Course Reporter. 33: 50–52.

52. DUDECK, A. E., N. P. SWANSON, and A. R. DEDRICK. 1966. Protecting steep construction slopes against water erosion. II. Effect of selected mulches on seedling stand, soil temperature, and moisture relations. 1966 Agronomy Abstracts. p. 38.

53. ———. 1967. Mulches for grass establishment on fill slopes. 1967 Agronomy Abstracts. p. 51.

54. ———. 1967. Mulch performance on steep construction slopes. Rural and Urban Roads. pp. 59–62.

55. DULEY, F. L. 1939. Surface factors affecting the rate of intake of water by soils. Soil Science Society of America Proceedings. 4: 60–64.

56. ELLISON, W. D. 1944. Studies of raindrop erosion. Agricultural Engineering. 25: 131–136, 181–182.

57. FITTS, O. B. 1925. Converting established turf to creeping bent by broadcasting stolons and topdressing. Bulletin of the Green Section of USGA. 5: 223–224.

58. FORBES, I., and M. H. FERGUSON. 1948. Effects of strain difference, seed treatment, and planting depth on seed germination of *Zoysia* spp. Journal of American Society of Agronomy. 40: 725–732.

59. FRAZIER, S. L. 1960. Turfgrass seedling development under measured environment and management conditions. M.S. Thesis. Purdue University. pp. 1–61.

60. ———. 1960. How grass seedlings make turf. Proceedings of the 1960 Midwest Regional Turf Conference. pp. 43–44.

61. FRISCHKNECHT, N. C. 1951. Seedling emergence and survival of range grasses in central Utah. Agronomy Journal. 43: 177–182.

62. GARMAN, H., and E. C. VAUGHN. 1916. The curing of blue-grass seeds as affecting their viability. Kentucky Agricultural Experiment Station Bulletin No. 198. pp. 25–39.

63. GARY, J. E. 1967. The vegetative establishment of four major turfgrasses and the response of stolonized "Meyer" zoysiagrass (*Zoysia japonica* var. Meyer) to mowing height, nitrogen fertilization and light intensity. M.S. Thesis. Mississippi State University. pp. 1–50.

64. GIANFAGNA, A. J., and A. M. S. PRIDHAM. 1951. Some aspects of dormancy and germination of crabgrass seed, *Digitaria sanguinalis* Scop. Proceedings of the American Society for Horticultural Science. 58: 291–297.

65. GILBERT, W. B., and E. E. DEAL. 1964. Temporary ditch liners for erosion control and sod establishment. 1964 Agronomy Abstracts. p. 101.

66. HAGER, O. B. 1961. Mulch as an aid to erosion control. 20th Short Course on Roadside Development. 20: 76–90.

67. HALLOWELL, C. K., and M. E. FARNHAM. 1948. Sodium arsenite as an agent in renovating turf. Greenkeeper's Reporter. 16(2): 12–14.

68. HARRINGTON, G. T. 1923. Use of alternating temperatures in the germination of seeds. Journal of Agricultural Research. 23(5): 295–332.

69. HARRIS, F. S., and H. H. YAO. 1923. Effectiveness of mulches in preserving soil moisture. Journal of Agricultural Research. 23: 727–742.

70. HARRISON, C. M. 1944. Rough grasses for parks, highway, and recreational areas. Greenkeeper's Reporter. 12: 30–31.

71. HART, S. W., and J. A. DEFRANCE. 1955. Behavior of *Zoysia japonica* Meyer in cool-season turf. USGA Journal and Turf Management. 8: 25–28.

72. HENDRICKS, S. B., V. B. TOOLE, and H. A. BORTHWICK. 1968. Opposing actions of light in seed germination of *Poa pratensis* and *Amaranthus arenicola*. Plant Physiology. 43: 2023–2028.

73. HODGES, T. K. 1958. Cutting sod for rhizome values. Proceedings of the 1958 Midwest Regional Turf Conference. pp. 40–42.

74. HODGSON, H. J. 1949. Effect of heat and acid scarification on germination of seed of bahia grass, *Paspalum notatum*, Flugge. Agronomy Journal. 41: 531–533.

75. HUGHES, T. D., J. F. STONE, W. W. HUFFINE, and J. R. GINGRICH. 1966. Effect of soil bulk density and soil water pressure on emergence of grass seedlings. Agronomy Journal. 58: 549–553.

76. HUME, E. P., and R. H. FREYRE. 1950. Propagation trials with Manila grass, *Zoysia matrella* in Puerto Rico. American Society for Horticultural Science Proceedings. 55: 517–518.

77. JACKSON, N. 1958. Seed dressing trial, 1958. Journal of Sports Turf Research Institute. 9(34): 454–458.

78. JENSEN, R. 1965. Planting large turf areas. Proceedings of the University of Florida Turf-Grass Management Conference. 8: 130–132.

79. JOHNSON, C. M., and W. R. THOMPSON. 1961. Fall and winter seeding of lawns. Mississippi Farm Research. 24(9): 4.

80. JUSKA, F. V. 1961. Pre-emergence herbicides for crabgrass control and their effects on germination of turfgrass species. Weeds. 9: 137–144.

81. JUSKA, F. V., and A. A. HANSON. 1964. Effect of pre-emergence crabgrass herbicides on seedling emergence of turfgrass species. Weeds. 12: 97–101.

82. KEARNS, V., and E. H. TOOLE. 1939. Temperature and other factors affecting the germination of fescue seed. USDA Technical Bulletin No. 638. pp. 1–35.

83. KEMMERER, H. R. 1962. Effects of post-emergence herbicides on establishment of spring seeded Kentucky bluegrass. Illinois Turfgrass Conference Proceedings. pp. 11–13.

84. KEMMERER, H. R., and J. D. BUTLER. 1964. The effect of seeding rates, fertility and weed control on the spring establishment of several lawn grasses. American Society for Horticultural Science Proceedings. 85: 599–604.

85. KING, J. W., and J. B. BEARD. 1967. Soil and management factors affecting the rooting capability of organic and mineral grown sod. 1967 Agronomy Abstracts. p. 53.

86. ———. 1969. Measuring rooting of sodded turfs. Agronomy Journal. 61: 497–498.

87. KOLLETT, J. R., A. J. WISNIEWSKI, and J. A. DEFRANCE. 1958. The effect of ureaform fertilizers in seedbeds for turfgrass. Park Maintenance. 11(5): 12–20.

88. KULFINSKI, F. B. 1957. Establishment of vegetation on highway backslopes in Iowa. Iowa Highway Research Board Bulletin No. 11. pp. 1–135.

89. KURTZ, K. W. 1967. Effect of nitrogen fertilizer on the establishment, density, and strength of Merion Kentucky bluegrass sod grown on a mineral soil. M.S. Thesis. Western Michigan University. pp. 1–45.

90. LAGE, D. P., and E. C. ROBERTS. 1964. Influence of nitrogen and a calcined clay soil conditioner on development of Kentucky bluegrass. American Society for Horticultural Science Proceedings. 84: 630–635.

91. LAUDE, H. M., J. E. SHRUM, JR., and W. E. BIEHLER. 1952. The effect of high soil temperatures on the seedling emergence of perennial grasses. Agronomy Journal. 44: 110–112.

92. LEGG, D. C. 1967. A note on the effect of chlordane on seedling turf. Journal of Sports Turf Research Insitute. 43: 59–60.

93. LONGLEY, L. E. 1941. Date and rate of lawn seeding. Minnesota Agricultural Experiment Station Bulletin 355. pp. 1–8.

94. LUNT, O. R., P. A. MILLER, and C. C. WYCKOFF. 1955. Seedbed preparation for turfgrasses. Golf Course Reporter. 23: 5–9.

95. LYONS, W. E. 1956. How to make quick work of planting a nursery green. Golfdom. 30: 49–52.

96. MADISON, J. H. 1966. Optimum rates of seeding turfgrasses. Agronomy Journal. 58: 441–443.

97. MAHDI, Z., and V. T. STOUTEMEYER. 1953. The change of populations of turf grasses by plugging. Golf Course Reporter. 21: 22–27.

98. MAHDI, Z., and C. G. WYCKOFF. 1953. A comparison of two methods of establishing turf from vegetative division. Golf Course Reporter. 21: 20–21.

99. McCALLA, T. M., and F. L. DULEY. 1946. Effect of crop residues on soil temperature. Journal of American Society of Agronomy. 38: 75–89.

100. McCOOL, D. K., and W. O. REE. 1965. Tests for a glass fiber channel liner. USDA Agricultural Research Service. 41–111. pp. 1–14.

101. McCULLY, W. G., E. L. WHITELY, and W. J. BOWMER. 1964. The influence of mulching and tillage on turf establishment. 1964 Agronomy Abstracts. p. 101.

102. McFAUL, J. A. 1954. Renovating old turf with calcium cyanamid. New York State Turf Association Bulletin 49. pp. 187–188.

103. ———. 1955. Further work with cyanamid in seedbeds. New York State Turf Association Bulletin 50. pp. 191–192.

104. McVEY, G. R. 1967. Response of turfgrass seedlings to various phosphorus sources. 1967 Agronomy Abstracts. p. 53.

105. ———. 1968. How seedlings respond to phosphorus. Weeds, Trees and Turf. 7(6): 18–19.

106. MILLER, R. A. 1962. Seed sod study. Illinois Turfgrass Conference Proceedings. pp. 39–45.

107. ———. 1963. Bluegrass growth response to various formulations, placement, and rates of plant food. Illinois Turfgrass Conference Proceedings. pp. 19–22.

108. MORINAGA, T. 1926. Effect of alternating temperatures upon the germination of seeds. American Journal of Botany. 13: 141–166.

109. MORRISH, R. H., and C. M. HARRISON. 1948. The establishment and comparative wear resistance of various grasses and grass-legume mixtures to vehicular traffic. Journal of American Society of Agronomy. 40: 168–179.

110. MURPHY, R. P., and A. C. ARNY. 1939. The emergence of grass and legume seedlings planted at different depths in five soil types. Journal of American Society of Agronomy. 31: 17–28.

111. MUSSER, H. B., W. L. HOTTENSTEIN, and J. P. STANFORD. 1954. Penngift crown vetch for slope control on Pennsylvania highways. Pennsylvania Agricultural Experiment Station Bulletin 576. pp. 1–21.

112. MYERS, H. E., and R. I. THROCKMORTON. 1941. Some experiences with asphalt in the establishment of grasses and legumes for erosion control. Soil Science Society of America Proceedings. 6: 459–461.

113. NELSON, A. 1927. The germination of *Poa* spp. Annals of Applied Biology. 14(2): 157–174.

114. PIETERS, A. J., and E. BROWN. 1902. Kentucky bluegrass seed: Harvesting, curing, and cleaning. USDA Bureau of Plant Industry Bulletin No. 19. pp. 1–19.

115. PLUMMER, A. P. 1943. The germination and early seedling development of twelve range grasses. Journal of American Society of Agronomy. 35: 19–34.

116. PRIDHAM, A. M. S. 1946. 2,4-dichlorophenoxyacetic acid reduces germination of grass seed. American Society for Horticultural Science Proceedings. 47: 439–445.

117. RABBITT, A. E. 1950. Economics in turf maintenance through seed usage. Proceedings of National Turf Field Days. pp. 45–49.

118. RADKO, A. M. 1955. Zoysia seed storage and germination tests. USGA Journal and Turf Management. 8(6): 25–26.

119. REE, W. O. 1960. The establishment of vegetation lined waterways. 19th Short Course on Roadside Development. 19: 54–65.

120. RICHARDSON, E. C., and E. G. DISEKER. 1961. Control of roadbank erosion in the southern Piedmont. Agronomy Journal. 53: 292–294.

121. ———. 1964. Establishing and maintaining roadside cover in Georgia. 1964 Agronomy Abstracts. p. 102.

122. RIEKE, P. E., J. B. BEARD, and R. E. LUCAS. 1968. Grass sod production on organic soils in Michigan. Proceedings 3rd International Peat Congress. pp. 350–354.

123. SCHMIDT, B. L. 1961. Methods of controlling erosion on newly-seeded highway backslopes in Iowa. Iowa Highway Research Board Bulletin No. 24. pp. 1–47.

124. ———. 1963. Methods of controlling erosion on newly seeded highway backslopes in Iowa. Iowa Highway Research Board Bulletin 23. pp. 60–65.

125. SCHMIDT, B. L., G. S. TAYLOR, and R. W. MILLER. 1965. Roadside stabilization and vegetative establishment with corn steep liquor solutions and straw mulch. 1965 Agronomy Abstracts. p. 49.

126. SCHMIDT, R. E. 1967. Hydraulic vegetative planting of turfgrasses. Golf Superintendent. 35: 10–69.

127. SIMPSON, S. L., and C. Y. WARD. 1964. Manufactured mulches compared with grain straw for mulching roadside seedings. 1964 Agronomy Abstracts. p. 103.

128. SKOGLEY, C. R. 1961. The effect of Zytron on seedling turfgrasses. Proceedings of the Northeastern Weed Control Conference. 15: 258–263.

129. SKOGLEY, C. R., and J. A. JAGSCHITZ. 1964. The effect of various preemergence crabgrass herbicides on turfgrass seed and seedlings. Proceedings of the Northeastern Weed Control Conference. 18: 523–529.

130. SMITH, J. D. 1955. Turf disease notes 1955. Journal of Sports Turf Research Institute. 9(31): 60–75.

131. ———. 1956. Seed dressing trials, 1956. Journal of Sports Turf Research Institute. 9(32): 244–250.

132. ———. 1957. Seed dressing trial. Journal of Sports Turf Research Institute. 9: 369–372.

133. SPENCER, J. T. 1950. Seed production of Kentucky 31 fescue and orchard grass as influenced by rate of planting, nitrogen fertilization and management. Kentucky Agricultural Experiment Station Bulletin 554. pp. 1–18.

134. SPRAGUE, V. G. 1940. Germination of freshly harvested seeds of several *Poa* species and of *Dactylis glomerata*. Journal of American Society of Agronomy. 32: 715–721.

135. ———. 1943. The effects of temperature and day length on seedling emergence and early growth of several pasture species. Soil Science Society of America Proceedings. 8: 287–294.

136. STOUTEMEYER, V. T. 1954. Plugs to change turf. California Agriculture. 8(2): 5–6.

137. STURKIE, D. G. 1964. Mulches and land preparation in establishing roadside vegetation. 1964 Agronomy Abstracts. p. 103.

138. SWANSON, N. P., A. R. DEDRICK, and A. E. DUDECK. 1966. Protecting steep construction slopes against water erosion. I. Effects of selected mulches on seed, fertilizer and soil loss. 1966 Agronomy Abstracts. p. 40.

139. ———. 1967. Protecting steep fill slopes against water erosion. 1967 Agronomy Abstracts. p. 55.

140. TOOLE, E. H., and V. K. TOOLE. 1940. Germination of seed of goosegrass, *Eleusine indica*. Journal of Amercan Society of Agronomy. 32: 320–321.

141. WARREN, B. O. 1955. Merion bluegrass experiences. Midwest Regional Turf Foundation Conference Proceedings. pp. 31–32.

142. WARREN, B., and D. HABENICHT. 1959. Highway sodding tests and results. Proceedings of the 1959 Midwest Regional Turf Conference. pp. 28–29.

143. WELTON, F. A. 1937. A weed-free seedbed for lawns. Ohio Agricultural Experiment Station Bulletin 579. pp. 28–30.

144. WELTON, F. A., and J. C. Carroll. 1934. Renovation of an old lawn. Journal of American Society of Agronomy. 26: 486–491.

145. ———. Lawn experiments. Ohio Agricultural Experiment Station Bulletin 613. pp. 1–43.

146. WHITE, A. W., JR., J. E. GIDDENS, and H. D. MORRIS. 1959. The effect of sawdust on crop growth and physical and biological properties of Cecil soil. Soil Science Society of America Proceedings. 23: 365–368.

147. WRIGHT, J. L. 1949. Seeding slopes along Connecticut's highways. Journal of New York Botanical Garden. 50(593): 113–114.

148. YOUNGNER, V. B. 1958. The turfgrass seed mat. Southern California Turfgrass Culture. 8: 32.

149. YOUNGNER, V. B., and W. B. TOVENER. 1956. Weed free turfgrass seedbeds. Southern California Turfgrass Culture. 6: 27–32.

150. ZAK, J. M., and E. J. BREDAKIS. 1964. Dune stabilization at Provincetown. 1964 Agronomy Abstracts. p. 104.

151. ———. 1967. Establishment and management of roadside vegetative cover in Massachusetts. Massachusetts Agricultural Experiment Station Bulletin 562. pp. 1–28.

chapter 17

Turfgrass Pests

Introduction

One of the most important components of turfgrass quality is uniformity. The presence of weeds in a turfgrass community disrupts the uniformity due to the variability in leaf width, color, and growth habit. Similarly, turfgrass injury caused by diseases, insects, nematodes, and other small animals disrupts the uniformity of the turf and may result in a substantial reduction in shoot density. Because of the objectionable nature of these turfgrass pests, cultural practices should be used to discourage them. Appropriate measures must be taken to achieve control should disease, insect, or nematode infestations become sufficiently severe to cause turfgrass injury. The emphasis in this chapter is on environmental, soil, and cultural conditions that affect the occurrence of turfgrass pests rather than on chemicals for controlling the pests.

Turfgrass Weeds

A **weed** is a plant growing where it is not wanted, or essentially a plant out of place. Human attitude determines which plants are not wanted in turfs and are therefore considered weeds. A plant is usually called a weed when it disrupts the uniformity of a turf due to a substantially different (a) leaf width and/or shape, (b) growth habit, or (c) color. In addition to detracting from turfgrass uniformity, weeds also compete with the desirable turfgrass species for light, soil moisture, soil nutrients, and carbon dioxide.

Leaf width and shape are major factors determining whether a plant is a turf-grass weed. For example, plants of broadleaf plantain or crabgrass disrupt the uniformity of a red fescue turf because of the vast difference in leaf shape and width (Fig. 17-1). Most desirable turfgrass species have a leaf width that does not exceed 4 mm.

Growth habit is another factor determining whether a plant is considered a weed. Creeping bentgrass grows in patches that tend to crowd out the desirable Kentucky bluegrass or red fescue (87, 88, 90, 213, 279). As a result, the bentgrass appears as scattered patches throughout a Kentucky bluegrass–red fescue turf giving an undesirable, weedy appearance. Bermudagrass behaves similarly as a patchy weed in St. Augustinegrass or zoysiagrass turfs.

Color can be a factor in determining which plants are weeds. Rough bluegrass is considered a weed when present in small amounts in a Merion Kentucky blue-grass turf. The distinct, light green color, usually in patches, is quite objectionable in a dark green turf.

A particular grass plant may be considered a weed under certain situations but a desirable species under other cultural conditions. Creeping bentgrass or the improved bermudagrass cultivars are frequently used on close cut greens as the desirable species. These same grasses can be very objectionable weeds, however, when occurring in high cut Kentucky bluegrass or St. Augustinegrass turfs (87, 88, 90, 213, 251, 279).

Figure 17-1. The disruption of turfgrass uniformity caused by the different leaf width, shape, and growth habit of broadleaf plantain (left) and crabgrass (right). (Photo courtesy of O. M. Scott, Marysville, Ohio).

Turfgrass Weed Ecology

The ecology of turfgrass weeds involves the growth characteristics and adaptations that permit weeds to persist and invade turfgrass communities. An understanding of turfgrass weed ecology is necessary in order to develop cultural practices that minimize weed encroachment into turfs.

Dormancy and persistence factors. Evolutionary processes have resulted in the development of specific characteristics that permit the common turfgrass weeds to survive and persist for an extended period of time even under unfavorable conditions. Specific plant organs were developed that enable turfgrass weeds to survive. These include (a) seeds, (b) rhizomes, (c) stolons, (d) bulbs, and (e) tubers. Seeds function as the primary survival mechanism for annual turfgrass weeds, while the rhizomes, stolons, bulbs, and tubers are vegetative structures that enhance the survival of perennial turfgrass weeds. Typically, the annual weeds are prolific seed producers.

Many of the specialized morphological structures of turfgrass weeds possess the additional physiological characteristic of dormancy. The dormancy factor permits weeds to survive in the soil for an extended period of time even though the soil may be frequently disturbed by cultivation. Dormancy is more common in turfgrass weeds adapted to the cool humid regions. The dormancy mechanism may be of several different types including a (a) germination inhibitor in the seed-coat, (b) hard, impermeable seed coat that impairs the entrance of water and gases, (c) hard seed coat that mechanically restricts germination, or (d) physiologically immature embryo. Common turfgrass weeds having a dormancy factor that delays seed germination are annual bluegrass and large crabgrass (151).

The dormancy factor is a remarkably efficient mechanism for weed survival under unfavorable conditions. The importance of dormancy is readily observed when a soil is tilled for turfgrass establishment. Invariably weed seeds and other propagules are brought to the soil surface where growth and development processes are initiated that produce weedy plants capable of competing with the desirable turfgrass species.

The survival mechanism of turfgrass weeds is further facilitated by requiring a specific set of environmental conditions for seed germination or bud growth to occur. These conditions usually involve an extended period of moist soil conditions and favorable temperatures that contribute to the ability of the weed to complete its life cycle. Weed survival may also involve an enforced dormancy where the environmental conditions prevent the seeds from germinating. For example, germination is impaired by plowing or tillage in which the seeds are deeply buried in a habitat that is not favorable for germination (298).

Climatic factors. Major climatic factors influencing the relative competitive ability and incidence of specific turfgrass weeds are (a) temperature, (b) water, and (c) light. Temperature is a very important factor controlling seed germination and bud growth. The temperature requirement varies with the particular weedy species. For example, annual bluegrass seed germination and growth is most active at relatively cool soil temperatures, which result in it being a more serious weed problem during the summer in cool humid climates and during the winter in warm humid climates (118, 178, 201, 339). Annual bluegrass seed germination essentially ceases at soil temperatures above 80°F (118). In contrast, certain other annual grasses such as crabgrass and goosegrass are most active during the high temperature periods of midsummer (142). Growth and development of these latter turfgrass weeds are terminated by exposure to freezing temperatures since they behave as summer annuals.

Water is another significant factor in weed seed germination. For example, germination of annual bluegrass, crabgrass, chickweed, and sedge is favored by very moist conditions. Subsequent shoot growth of these weeds as well as bentgrass is substantially enhanced by moist to wet soil conditions (175, 312, 359). In contrast, weeds such as the cinquefoils and quackgrass can compete more favorably with the desirable turfgrass species under droughty conditions.

Light intensity is a particularly important factor controlling the initial seed germination and emergence of certain common turfgrass weeds. Crabgrass (351) and goosegrass (142, 350) seldom occur in shaded areas or dense, high cut turfs since light is required for seed germination.

Soil factors. A knowledge of the soil characteristics that favor a specific turfgrass weed can assist in developing cultural practices to discourage encroachment of the weed into the turfgrass community. Soil factors influencing weed survival and competitive ability are (a) soil reaction, (b) water content, (c) fertility, (d) aeration, and (e) temperature.

The soil reaction can be an important factor in the encroachment of certain turfgrass weeds (25, 177, 279). Neutral soil pH's favor annual bluegrass (217, 339). Red sorrel and bentgrass grow best on relatively acidic soils (25, 237). Certain other species such as knotweed can grow under a relatively wide range of soil pH's.

The soil water content can also be an important factor influencing the encroachment of certain turfgrass weeds. The sedges, creeping bentgrass, annual bluegrass, crabgrass, and chickweed are favored by waterlogged soil conditions, whereas bermudagrass is adapted to droughty soils (312).

Annual bluegrass, bermudagrass, and creeping bentgrass are turfgrass weeds that respond to a high soil fertility level. The level of specific nutrients can also influence weed encroachment. Clover is favored by a high potassium level, while annual bluegrass responds to high phosphorus levels (217, 339). Soil aeration is also a factor in the encroachment of certain turfgrass weeds. Annual bluegrass and knotweed are adapted to compacted soils having poor aeration.

Cultural factors. Cultural factors such as (a) cutting height, (b) fertility level, (c) irrigation intensity, and (d) cultivation can significantly influence the ability of a turfgrass weed to encroach into the turfgrass community. A close cutting height that weakens the turf frequently results in increased weed problems due to the reduced competitive ability of the turf (16, 87, 89, 116, 143, 144, 176, 365). For example, creeping bentgrass, bermudagrass, and annual bluegrass encroachment into Kentucky bluegrass or St. Augustinegrass is significantly enhanced by close cutting (87, 129, 144, 202, 250, 377) (Fig. 17-2). Close mowing also assists in seed germination and seedling growth of crabgrass and goosegrass since it increases light penetration to the soil surface.

A nutrient deficiency favors the invasion of broadleaf weeds due to the thin turf and resulting lack of competition (144). On the other hand, intensive fertility and irrigation practices encourage the encroachment of certain weeds such as

annual bluegrass, bermudagrass, crabgrass, and creeping bentgrass (144, 175,
312, 359, 360). Frequent irrigation and
a moist soil surface enhance weed seed
germination (250, 366).

Cultivation practices such as coring,
which improve soil aeration and encourage the growth of desirable species,
impair the relative competitive ability of
such turfgrass weeds as annual bluegrass
and knotweed. The timing of cultivation
and vertical mowing practices in relation
to the optimum weed seed germination
periods is particularly important. For
example, vertical mowing in early fall
when annual bluegrass seed germination
is quite intense results in a substantial
increase in annual bluegrass encroachment (379).

Figure 17-2. Invasion of a creeping bentgrass
patch into a closely cut Kentucky bluegrass
turf.

Weed Dissemination

The wide dispersal of most turfgrass weeds is illustrated by the fact that a major
portion of the common turfgrass weeds found in the United States are of European
or Eurasian origin. Turfgrass weeds can be disseminated by specialized morphological structures such as seeds, rhizomes, stolons, bulbs, bulblets, or tubers.
Many common turfgrass weeds are prolific seed producers. For example, one
annual bluegrass plant produced 360 seeds within a single growing season in
western British Columbia, Canada (300).

Weed seeds can be widely disseminated by the action of wind, water, and
animals (including man). Certain weed seeds contain modified structures that
facilitate wind dispersal. Dandelion, thistle, and pigweed are examples of weeds
possessing specialized parachute, plumed, or winged seeds of low density that are
readily carried by the wind. Surface water movement also functions in weed dispersal. Plantain, peppergrass, and sedge have unique structures that permit them
to survive submersion without injury to the embryo and, thus, to be widely dispersed by water. Seeds of the sedges are quite buoyant and readily disseminated by
floating on water.

Weeds are widely spread by animals since many seeds can pass through the
digestive tract without loss of viability. Birds are particularly effective in dissemi-
nating weed seeds since they consume large quantities of seeds and scatter them
in droppings. Certain seeds can also cling to the fur, feathers, and feet of animals
and to shoes or clothing of humans, which introduces them into new areas. Man
also is an important factor in weed dissemination on equipment and in seed,

stolons, and sod that are marketed commercially. Bentgrass and annual bluegrass have been widely disseminated in turfgrass seeds.

The spread of turfgrass weeds by vegetative means such as rhizomes, stolons, bulbs, bulblets, and tubers is generally quite slow. However, the activities of man greatly enhance weed distribution by vegetative means. Rhizomes, stolons, and bulbs can be carried on cultivating and mowing equipment from an infested area and introduced into a high quality, weed-free turf. Such common turfgrass weeds as quackgrass, bermudagrass, nutsedge, and creeping bentgrass are frequently spread by vegetative means. Rhizomes, stolons, and bulbs can also be carried in topsoil to be used in soil modification and topdressing.

Correcting Turfgrass Weed Problems

Weed encroachment into an established turf usually results when the turf has become weak and thin due to (a) environmental stress, (b) damage of turfgrass pests, (c) improper turfgrass cultural practices, or (d) intense wear from vehicles or foot traffic. Weed control practices are not effective on a long term basis unless the original cause of weed encroachment is corrected. Thus, the avoidance or correction of turfgrass weed problems involves a twofold program: weed prevention and weed control.

Weed Prevention

The effort required and success achieved in weed prevention practices vary with the particular species involved. A preventive program is most effective with those turfgrass weeds that are disseminated primarily by vegetative plant parts such as rhizomes or stolons but quite ineffective on turfgrass weeds having reproductive plant parts that are readily disseminated by the wind. The three major methods of turfgrass weed prevention are (a) sanitary, (b) cultural, and (c) preplant control.

Sanitary Practices

Sanitary practices that avoid the introduction and spread of turfgrass weeds are very important. The introduction into a turfgrass area of only a few seeds or vegetative plant parts of certain weedy species such as annual bluegrass, creeping bentgrass, or bermudagrass can result in the development of a major weed problem within 2 to 4 years. Weeds can be introduced (a) in turfgrass seeds, stolons, sprigs, plugs, or sod; (b) in soil or topdressing; (c) on turfgrass equipment; or (d) from adjacent unmowed areas.

Sanitary practices during turfgrass establishment involve the use of turfgrass seeds, stolons, sprigs, plugs, or sod that are free of objectionable weedy species. The seed label should be checked for "weed" and "other crop" content when purchasing seed. A number of very objectionable turfgrass weeds may be included in the "other crop" category. A limited percentage of certain weed seeds can be permitted in a turfgrass seed lot, whereas other more objectionable weeds such as annual bluegrass, creeping bentgrass, and tall fescue should be avoided. It is also important to purchase vegetative stolons, sprigs, plugs, or sod from a reputable grower known for the production of weed-free material. The vegetative material should be inspected for freedom from undesirable weedy species.

Typical examples of the more objectionable turfgrass weeds are annual bluegrass, creeping bentgrass, or rough bluegrass in Kentucky bluegrass seed and sod; annual bluegrass in creeping bentgrass seed, stolons, or sod; bermudagrass in St. Augustinegrass sod; nutsedge in bermudagrass or Kentucky bluegrass sod; tall fescue or velvetgrass in red fescue seed or sod; and quackgrass in Kentucky bluegrass sod.

Many species of weed seeds are present in the upper profile of most soils (313). Rhizomes or stolons of undesirable perennial grassy weeds may also occur in the topsoil. Annual bluegrass seed, quackgrass rhizomes, underground structures of nutsedge, and bermudagrass rhizomes are common contaminants of turfgrass soils used for soil modification or topdressing. The seeds or vegetative plant parts of these weedy species can remain dormant in the soil for an extended period of time. New plants are produced by seed germination or bud development on the nodes of rhizomes and stolons when favorable moisture, temperature, and light conditions occur. Materials used for soil modification or topdressing should be free of undesirable weed seeds or vegetative plant parts or the contaminants controlled by fumigation prior to planting. Either fumigation or heat treatment can be used on smaller quantities of soil being prepared for topdressing.

Turfgrass weed seeds and vegetative plant parts can also be moved from infested areas to noninfested turfs on maintenance equipment. Weed seeds and stolons are readily moved on mowers and cultivation equipment. Seeds or stolons of certain weeds can also be moved from one area to another on shoes, particularly when wet. Rhizomes as well as seeds and stolons of weeds can be readily moved from infested to noninfested areas during soil tillage operations prior to planting. The introduction of weeds on turfgrass equipment can be avoided by proper sanitary practices. Specifically, mowers and cultivating equipment should be thoroughly cleaned prior to moving them from weed infested turfgrass areas to noninfested areas. This can be accomplished by thorough washing or steaming.

A fourth means of introducing objectionable turfgrass weeds is from adjacent, unmowed areas such as ditches or fence rows. If the weeds in these areas are permitted to flower, the seed can be readily spread to adjacent turfgrass areas through the action of wind and water. This can be a serious problem on sod farms having an elaborate system of open ditches. Cultural practices such as timely mowing or the use of growth inhibitors can be used to prevent seed formation of objectionable or difficult to control weeds. It may be necessary to eradicate certain undesirable weedy species from these areas.

Cultural Practices

Certain weeds may not become established and encroach into a turfgrass area because of the dense, vigorous, competitive nature of the turf. Cultural practices that ensure such a turf include the proper (a) turfgrass species; (b) cutting height and frequency; (c) soil fertility and pH level; (d) irrigation frequency and intensity; (e) disease, insect, and nematode controls; and (f) cultivation of compacted areas as needed. A dense, actively growing turf is best able to compete with weeds for light, nutrients, and water. For example, weed encroachment into Kentucky bluegrass-red fescue turfs is less at cutting heights above 1.5 in. Optimum soil fertility and moisture levels should be maintained to ensure a vigorous growing turf of maximum competitive capability and recuperative potential (340). Certain turfgrass weeds such as crabgrass and goosegrass require light for seed germination. Thus, higher cutting heights and adequate fertility levels maintain a high shoot density and an active growth rate that impair weed encroachment due to the limited light penetration to the soil surface. Excessive fertilization that reduces the hardiness of turfgrasses to environmental stress can be detrimental to the turf and result in increased weed encroachment.

Excessive irrigation beyond the requirements of the turfgrass plant is undesirable since it can stimulate weed encroachment. A waterlogged soil condition restricts turfgrass growth and vigor due to the lack of soil aeration. Turfgrass weeds such as annual bluegrass and creeping bentgrass are more tolerant of these conditions. Light, frequent irrigation is particularly favorable for the germination of turfgrass weed seeds such as crabgrass.

Insects, diseases, and nematodes should be controlled whenever necessary to prevent serious thinning or open spots in the turf (173). Such thinned or bare areas provide ideal sites for the introduction of weeds into a turfgrass area.

Cultivation practices can effectively improve soil aeration and water movement particularly on fine textured soils subjected to intense traffic. Turfgrass cultivation or vertical mowing should be scheduled for times when the weed encroachment potential is minimal. For example, cultivation or vertical mowing in late summer or early fall results in increased annual bluegrass invasion into turfs (379).

Preplant Weed Control

Turfs are particularly prone to weed encroachment during establishment from seed, stolons, sprigs, or plugs. The quantity and type of weed seeds and vegetative propagules present in the soil influence the ability of desirable turfgrass seedlings to establish and become the dominant component in the turfgrass community. Improper seedbed preparation and planting techniques frequently result in voids or a thin stand that is an ideal avenue for the invasion and spread of turfgrass weeds. This problem can be minimized by the use of preplant weed control techniques involving mechanical and/or chemical means of killing weed seeds or vegetative plant parts prior to planting (Fig. 17-3). Weed eradication at this time

Figure 17-3. Effective preplant weed control and the resultant minimum weed competition (right) compared to the severe weed competition on the left.

minimizes competition and encroachment during the establishment period, provided weeds are not introduced from other sources.

Fallowing. If sufficient time is available, the soil can be fallowed or mechanically tilled for a period of time to control a major portion of the weeds arising from the existing seeds and vegetative plant parts present in the soil. Control of weeds by fallowing may be achieved by (a) encouraging the germination of existing weed seeds in the soil, (b) depletion of food reserves in vegetative plant parts such as rhizomes or stolons, or (c) exposure of the vegetative plant parts to desiccation or low temperature stress. Fallowing has the disadvantage of being relatively expensive. The area is also left in a barren condition and prone to erosion for an extended period of time. Fallowing generally involves tilling at intervals of 4 to 6 weeks to destroy newly emerging weeds and to bring additional dormant weed seeds and vegetative plant parts to the surface where conditions are favorable for germination and development. Maximum effectiveness of fallowing is achieved where surface soil moisture and temperatures are favorable for weed seed germination or shoot development from vegetative plant parts. Fallowing rarely controls all weeds due to the dormancy factor. It is most effective when combined with chemical weed control applied prior to fallowing. The longer the area is fallowed at the optimum soil moisture content, the greater the degree of weed control achieved.

Soil fumigation. Preplant chemical weed control usually involves the use of a **temporary soil fumigant** that controls viable weed seeds, vegetative propagules, soil-borne diseases, nematodes, and rodents and has a toxic residual period of no more than 4 to 6 weeks (78, 94). Characteristics desired in a temporary soil fumigant include (a) effective control of viable seeds and vegetative plant parts of undesirable weedy species, (b) rapid toxic action, (c) effectiveness to a depth of 6 to 8 in., (d) ability to plant the treated areas promptly, (e) low cost, (f) ease of application, and (g) safe handling and use. Preplant chemical weed control is relatively expensive. It is usually practiced only on turfgrass sites of high economic value. The use of temporary soil fumigants has not been extensive and is most likely to be practiced on (a) topdressing materials, (b) relatively limited areas such as greens and tees, and (c) areas such as stolon nurseries and sod farms where a weed-free soil is required prior to planting.

Certain steps in soil preparation should be followed prior to applying a temporary soil fumigant. Soil fumigants are most effective when soil temperatures at the 2- to 4-in. depth are above 60°F. The soil should be cultivated somewhat deeper than the anticipated treatment depth. The desired penetration of the temporary soil fumigant in a gaseous form depends on the depth to which weed seeds and rhizomes occur. Treatment to a 6- to 9-in. depth is usually adequate. Thorough cultivation and aeration of the soil prior to applying a soil fumigant ensure maximum distribution of the fumigant vapors as well as good penetration of the liquid into the soil. The soil should be thoroughly loosened and all clods broken up during the tillage operations. Never apply soil fumigants to a dry soil. The soil should be moist to a depth of 6 in. at least 1 week prior to treatment and maintained in this condition until the fumigant is applied. The area should be irrigated to ensure an adequate soil moisture level if precipitation is lacking.

Calcium cyanamide, chloropicrin, dazomet, formaldehyde, metham, and methyl bromide are the more commonly used temporary soil fumigants for turfgrass purposes. The characteristics and guidelines for use are presented in Table 17-1. Any of the temporary soil fumigants discussed are toxic to living plants or root systems growing in the vicinity of the treated area. Proper care in handling and use of these materials is important. A gas mask or respirator should be used whenever toxic gaseous forms are used, particularly in enclosed areas. Contact with skin, eyes, clothing, or shoes should be avoided. Exposed clothing or shoes should be removed immediately and thoroughly aerated before wearing again.

Weed Control

Specific weed control practices may be required at intervals on established turfs to maintain an acceptable level of turfgrass quality. Control involves the use of techniques that limit weed infestations. A certain degree of weed control can be achieved by the manipulation of cultural practices and the environment to reduce the competitive capability of a particular weed and increase the competitive ability of the desirable turfgrass species. Turfgrass weed control usually involves a combination of methods including sanitary practices and cultural regulation as well as mechanical and chemical control methods.

There are varying degrees of weed control. **Weed eradication** involves the complete elimination of all living weed plants and dormant seeds or vegetative buds

Table 17-1

CHARACTERISTICS OF SIX SOIL FUMIGANTS PLUS STEAM

Name of fumigant		Minimum exposure time, days	Aeration time,* days	Comments
Common	Chemical			
Calcium cyanamide	Calcium cyanamide	20	0	Must thoroughly incorporate into the upper 2 in. of soil surface
Chloropicrin	Trichloronitromethane	2	10–15	Inject 6 in. deep into the soil, pack, and apply water seal at the soil surface
Dazomet (DMTT)	Tetrahydro-3, 5-dimethyl-2H–1, 3, 5-thia-diazine-2-thione	2	20–30	Apply liquid drench or granular to soil, incorporate into surface 2 in. and apply water seal or plastic cover
Formaldehyde	Formaldehyde	2	15	Apply as a drench and apply a plastic cover; not effective for nematodes
Metham (SMDC)	Sodium methyldithiocarbamate	2	15–20	Apply as a drench, pack, and apply a water seal or plastic cover
Methyl bromide	—	1–2	2–3	Inject under an airtight plastic cover
Steam, 212° F		30‡	0	Use usually limited to small areas or topdressing

* Longer aeration time is required at soil temperatures below 60° F or on poorly aerated, fine textured soils.
‡ Time in minutes.

contained in the soil. This is rarely achieved under turfgrass conditions, except on a temporary basis. Eradication of all living weedy plants is the more common procedure in intensely cultured turfs, particularly greens. Elimination of such common turfgrass weeds as clover, crabgrass, dandelion, and plantain is possible with the available herbicides and cultural practices. Eradication of certain weeds may not be practical due to the lack of an effective herbicide for selective control. Annual bluegrass and creeping bentgrass are included in this category. Partial control can be attempted in this situation through the manipulation of turfgrass cultural practices and the environment. This is done to prevent the weedy species from becoming the dominant component in the turfgrass community.

As weeds appear in a turfgrass community, the first step in control is identification of the particular weedy species. Then an appropriate method of control can be selected. The method of weed control used is determined by such plant characteristics as the (a) method of reproduction, (b) life cycle, and (c) season in which growth occurs. Turfgrass weeds vary from relatively easily controlled species to those that cannot be selectively controlled with available techniques and materials. The more nearly a weed resembles the desirable turfgrass species in terms of ecological requirements and life cycle, the more difficult it is to control the weed without injuring the desirable turfgrass species.

The three primary methods of weed control are (a) physical, (b) chemical, and (c) biological. The first two methods are discussed in detail. The third method involves the use of insects, pathogens, parasites, or predators to reduce the population of weedy species. This method is cheap, permanent, and involves no repetitious treatment when effective natural enemies of weeds are available. Biological control is only effective on a limited number of weedy species. As yet, none of these are of any significance as far as turfgrass culture is concerned.

With both physical and chemical methods, control should be initiated (a) prior to flowering to avoid the production of viable seed and (b) at a time of year favorable for turfgrass growth in order to fill in the voids left within the turfgrass community after the weeds are removed. No matter which method is selected, it should be used in conjunction with the proper cultural practices to ensure a dense, vigorous turf. This provides a high degree of competition to weeds that may encroach into the turf.

Physical Methods of Turfgrass Weed Control

One of the oldest approaches to weed control involves the physical method. The two basic types of mechanical weed control are (a) manual removal and (b) mowing. Manual weed control may involve hand pulling or spudding.

Hand pulling is most effectively used on annual weeds, particularly broadleafs. They are readily destroyed by severing the plant from the root at the soil surface. Hand weeding of annuals is less successful in turfs having a high soil infestation of dormant annual weed seeds. It should be practiced before the plant produces viable seeds. Biennials can be controlled if the root system is removed. Hand pulling can also be used effectively against seedlings of perennial weeds. Established perennial weeds are much more difficult to control by hand pulling since new plants can be initiated from underground plant parts. Repeated removal of the aboveground parts of perennial weedy species tends to exhaust the carbohydrate reserves and weaken the weed to the extent that it may eventually be eliminated from the turfgrass community. However, a great deal of time and effort is required to achieve control of most perennial weedy species through manual removal. Quackgrass, creeping bentgrass, bermudagrass, nutsedge, and nimblewill are perennial weedy species that are particularly difficult to control by hand pulling. Thus, chemical methods are more commonly practiced in controlling perennial weeds.

Spudding is frequently used in combination with hand pulling. A special implement having a metal blade with a forked, sharpened end called a **spud** is used. The spud is pushed into the soil to sever the roots of the turfgrass weed and the plant plus a portion of the root system is pulled by hand from the soil. Hand pulling and spudding are manual methods that are most commonly used (a) for controlling annual broadleaf weeds, (b) on relatively small turfgrass areas, (c) where only a few occasional weeds develop in more extensive turfgrass areas, and (d) in turfs adjacent to valuable ornamental plantings that can be seriously damaged if a herbicide is used.

Mowing is a routine practice in turfgrass culture that is effective in controlling a broad spectrum of weeds. Most large, erect growing weedy species are controlled in turfs simply by frequent, close mowing at a height of less than 2 in. For example, lambsquarters and pigweed are commonly a problem in newly seeded turfs. These species are usually eliminated under regular mowing at 1 to 1.5 in. Mowing cannot be effectively used to control certain prostrate, low growing weedy species. Such weeds are a more serious problem in turfs. Repeated, close mowing of some prostrate, perennial weedy species may result in control by the depletion of their underground carbohydrate reserves.

The proper timing of mowings is important in preventing the development of viable weed seeds (238). The seed production phase in the life cycle of annual weedy species is a particularly vulnerable period. Control can frequently be achieved simply by preventing seed head development or removing the seed heads. Weeds should usually be cut when in the bud stage or earlier.

Heat is another physical method for controlling weeds that has not been used to any extent on turfs. It is quite economical and is widely used in commercial turfgrass seed production operations following harvesting. Special equipment has been developed for the selective control of young weedy species by flaming that involves the controlled burning of petroleum products at very high temperatures. Flaming has not been widely used in turfgrass culture primarily because of the expensive, specialized equipment required and the limited range of weedy species that can be controlled by this method. The advantages of flaming include (a) no

Table 17-2

CHEMICAL NAME AND CHARACTERISTICS OF 25 HERBICIDES COMMONLY USED IN CONTROLLING TURFGRASS WEEDS

Name of herbicide		Primary weed group controlled*	Type of Control‡	Comments
Common	Chemical			
Amitrole	3-amino-1, 2, 4-triazole	G	NS	Systemic; 4 to 6 week residual
Atrazine	2-chloro-4-(ethylamino)-6-(isopropyla-mino)-s-triazine	G	NS	Long residual; selective on certain warm season grasses
Benefin	N-butyl-N-ethyl-α, α, α-trifluoro-2,6-dini-tro-p-toluidine	A	Pre	Apply before weed seed germination; has low solubility
Bensulide	O,O-diisopropyl phosphorodithioate S-ester with N-(2-mercaptoethyl) benzene-sulfonamide	A	Pre	Toxic to turfgrass seedlings; has a long residual period
Bromoxynil	3,5-dibromo-4-hydroxybenzonitrile	B, G	Post	No residual; nonsystemic
Cacodylic acid	Hydroxydimethylarsine acid	B, G	C	No residual; nonsystemic
Calcium arsenate	Calcium arsenate	A	Pre	Toxicity buffered by the soil phosphorus level
Dalapon	2,2-dichloropropionic acid and sodium 2, 2-dichloropropionate	G	NS	Systemic; 6 to 8 week residual
DCPA	Dimethyl 2, 3,5,6-tetrachloroterephthalate	A	Pre	Toxic to fescues
Dicamba	2-methoxy-3,6-dichlorobenzoic acid	B	Post	Systemic, toxic to certain trees and shrubs
DSMA	Disodium methanearsonate	A	Post	Requires 2 to 3 applications
Endothall	7-oxabicyclo(2.2.1)heptane-2,3-dicarbox-ylic acid	A	Post	Marginal selectivity to desirable species
Lead arsenate	Lead arsenate	A	Pre	Residual of up to 4 years; toxicity buffered by phosphorus
MAMA	Monoammonium methanearsonate	A	Post	Requires 2 applications
Mecoprop (MCPP)	2-[(4-chloro-o-tolyl)oxy] propionic acid	B	Post	Nontoxic to bentgrasses; systemic
Monuron	3-(p-chlorophenyl)-1,1-dimethylurea	B, G	NS	Long residual up to 20 months
MSMA	Monosodium methanearsonate	A	Post	Requires 2 applications
Paraquat	1,1' dimethyl 4,4' bipyridinium	B, G	C	No residual
PMA	(acetato) phenylmercury	A(B)	Post	Controls seedlings only; toxic to Merion
Siduron	1-(2-methylcyclohexyl)-3-phenylurea	A	Pre	Nontoxic to most turfgrass seedlings

Table 17-2 (Cont.)

Silvex	2-(2,4,5-trichlorophenoxy)propionic acid	B	Post	Toxic to bentgrass turfs; systemic
Simazine	2-chloro-4,6-bis(ethylamino)-s-triazine	A, B	Pre(NS)	Long residual; selective on certain warm season grasses
Sodium arsenite	Sodium arsenite	B, G	C	No residual; nonsystemic
Terbutol	2,6-di-tert-butyl-p-tolyl-methylcarbamate	A	Pre	Apply before weed seed germination
2,4-D	2,4-dichlorophenoxyacetic acid	B	Post	Systemic; toxic to bentgrass turfs

* A-annual weedy grasses, which may or may not include annual bluegrass.
B-broadleaf types or dicotyledons.
G-most grasses.

‡ C-contact, nonselective, postemergence herbicide.
Pre-preemergence herbicide.
Post-postemergence herbicide.
NS-nonselective.

559

toxic residue, (b) no drift hazard to adjacent, desirable species, and (c) the additional benefit in controlling other turfgrass pests such as insects and diseases.

Chemical Methods of Turfgrass Weed Control

A number of chemical methods for effective turfgrass weed control have evolved since 1950. Chemical weed control involves the use of **herbicides**, which are chemicals that kill weedy plants or inhibit normal growth (Table 17-2). Chemical control is frequently preferable to mechanical methods such as hand pulling or spudding since an acceptable degree of weed control can usually be achieved with considerably less cost and time.

Herbicides are distinguished by the type of kill as either (a) contact or (b) systemic. **Contact herbicides** kill only those plant parts or living cells to which the chemical is applied. There is no significant translocation of a contact herbicide through the vascular system of the plant. Kill occurs quite rapidly after application. The use of contact herbicides is limited primarily to the control of annual weedy species since only the aboveground plant parts are killed. Cacodylic acid, paraquat, and sodium arsenite are contact-type herbicides.

Systemic herbicides are absorbed by either the roots or aboveground plant parts and are then translocated through the plant to tissues or areas where toxicity occurs. They usually kill through a chronic effect resulting from a disruption of plant growth and metabolic processes. Kill may not be evident until 1 to 4 weeks after application. Most selective turfgrass herbicides are of the systemic type. Typical examples are mecoprop, siduron, silvex, and 2,4-D. Selective control of perennial weedy species possessing stolons, rhizomes, or other underground plant parts is most likely to be achieved with a systemic herbicide. When applied at the proper rate, a systemic herbicide is absorbed by the foliage and translocated through the phloem downward to the crowns, rhizomes, stolons, bulbs, or tubers, which are eventually killed. It is important to avoid applying excessive amounts of a systemic herbicide as this might result in immediate or contact toxicity to the leaves. Absorption and translocation of the herbicide is impaired if this occurs and herbicidal effectiveness is lessened.

Herbicides are also distinguished in terms of the time at which the herbicide is applied with respect to turfgrass and/or weed seed germination. **Preplant herbicide** applications are made before seeding or vegetative planting. This commonly involves the use of a soil fumigant such as dazomet, metham, or methyl bromide. **Preemergence herbicides** involve treatments made before the emergence of a specified weed or group of weeds from the soil. Preemergence treatments include those applied just before the turfgrass seed is planted or before the weeds emerge from an established turf. Benefin, bensulide, DCPA, lead arsenate, and siduron are examples of preemergence herbicides used in turfgrass weed control. **Postemergence herbicide** treatments are made after emergence of a specific weed. Dicamba, DSMA, MAMA, MSMA, silvex, and 2,4-D are commonly used as postemergence herbicides.

Mode of Herbicide Action

Kill of turfgrass weeds by a systemic herbicide depends on a complex sequence of events. Included are (a) herbicide absorption by the weed, (b) herbicide movement through the plant to the specific sites or tissues where the herbicide is active, and (c) death of the weed resulting from the disruption of some process that is vital for plant survival. Phytotoxicity is only achieved if a sufficient herbicide concentration occurs at the appropriate sensitive sites within the plant at the desired time. A number of factors can interfere during the three phases in the mode of action of a herbicide.

Absorption. Herbicides enter weeds primarily through the foliage or roots, although penetration through the stem may occur in some woody species. Herbicides are absorbed through both the upper and lower leaf surfaces with entry being either stomatal or cuticular in nature. The rate of herbicide entry is most rapid through open stomata. Oils and aqueous sprays of low surface tension enter the stomata most readily.

Penetration of herbicides through the cuticle occurs by diffusion. Dalapon and 2,4-D enter foliage directly through the cuticle. The waxy cuticle and cellulose of weed foliage is nonpolar. As a result, nonpolar herbicides are absorbed more readily than polar herbicides. This is the reason that the ester of 2,4-D is more effective in controlling wild onion and wild garlic than the amine form. Considerable variation exists in the composition, thickness, and structure of the cuticle of different weed species. Wild onion and wild garlic possess a waxy, relatively impermeable leaf surface. Absorption of herbicides through waxy surfaces is greatly facilitated by the use of surfactants that are incorporated into the foliar sprays. Also, higher cutting heights increase the leaf area available for absorption and the total quantity of herbicide absorbed per plant.

Herbicide absorption through the roots occurs primarily with soil applied herbicides. In turfgrass weed control, this type of absorption is most commonly associated with the preemergence herbicides. Root absorption is most effective with polar herbicides. Herbicide absorption by the root system is similar to the mechanisms for inorganic ions discussed in Chapter 13. A specific herbicide may be absorbed by active or passive mechanisms, or both, depending on the chemical and physical properties of the herbicide molecule. Herbicides applied to the soil may be absorbed by the soil complex and thus are not available for uptake by the root system. Such factors as (a) soil pH, (b) concentration of exchangeable ions, and (c) amount of herbicide applied affect the degree of herbicide absorption on the soil complex relative to the amount of herbicide available in the soil solution for uptake by the root system.

Translocation. The herbicide moves toward the site of phytotoxic action following absorption. Herbicide translocation is imperative for controlling certain weedy grasses such as creeping bentgrass, bermudagrass, nutsedge, and quackgrass that possess underground rhizomes, stolons, bulbs, or tubers. Movement of the herbicide may involve simple diffusion within the leaf or translocation to distant sensitive organs through the xylem or phloem. Contact herbicides such as sodium arsenite move short distances by means of simple diffusion within the leaf. In the case of simazine, movement occurs with the xylem or upward transporation stream of the plant. Translocation through xylem tissue is not affected by the herbicide concentration since these tissues are nonliving. Thus, very toxic chemicals such as the arsenicals can be translocated through the xylem.

Herbicide translocation in the phloem results in movement from carbohydrate-synthesizing leaves to carbohydrate-dependent tissues such as roots, buds, and similar meristematic tissues that are not photosynthetically active. Thus, the downward translocation of herbicides is most effective when large amounts of carbohydrates are being moved from the leaf toward the roots, rhizomes, stolons, and tubers. This usually occurs at or just after full leaf development. Amitrole, dalapon, and 2,4-D are herbicides that are translocated through the phloem. Effective translocation of herbicides through the phloem depends on maintaining live phloem cells. The use of excessively high concentrations of herbicides such as 2,4-D results in toxicity to the phloem cells and impairs 2,4-D translocation to the lower plant parts. Use of the recommended rate combined with repeated applications is more effective in achieving total weed control with phloem translocated herbicides.

Toxicity. Kill occurs by one of several different mechanisms after herbicide movement to the site of sensitivity within the plant occurs. The mechanism of toxicity may involve (a) inhibition of photosynthesis, which is typical of the triazines and substituted ureas; (b) interference with nucleic acid synthesis caused by auxin-like herbicides; (c) protein metabolism disruption caused by auxin-like herbicides; or (d) inhibition of the activity of certain enzymes such as in the case of dalapon. Visible plant symptoms resulting from the toxic action of herbicides range from a stunting or cessation of growth and loss of chlorophyll to very distinct morphological aberrations. This is followed by rapid desiccation and eventual death of the weed. Leaf chlorosis is one of the earliest visual symptoms of herbicide injury since many chemicals affect chlorophyll synthesis and/or destruction. A typical example is amitrole, which causes chlorosis due to an inhibition of normal chloroplast development in young tissues.

Herbicide Selectivity

The use of chemicals to control weeds within a turfgrass community without killing or no more than slightly affecting the desirable turfgrass species is termed **herbicide selectivity** (Fig. 17-4). Selectivity is relative and can be achieved in a number of different ways and at several different times during application, absorption, and translocation of the herbicide. The first requisite in achieving effective herbicide kill is proper application and contact of the herbicide on the plant surface. Differential selectivity can be obtained between the turfgrass species and the weed based on the degree of herbicide contact. Differential contact is achieved by (a) plant morphological differences, (b) plant anatomical differences, (c) selective herbicide placement, or (d) time of application.

Morphologically, the basal location of turfgrass meristems is an important factor in herbicide selectivity. Dicotyledons have an apical-type growth with the meristems located in the terminal portion of the stems. Meristematic areas located at the stem tips are more exposed to herbicide sprays than the meristems of turfgrasses that are located at the base of the leaf near the soil and are protected by the leaf sheath. Similarly, the vegetative buds on rhizomes are protected due to their location below ground.

The degree of pubescence on the leaf surface can also affect the amount of contact and degree of herbicide selectivity. Pubescence can prevent the spray droplets from reaching the leaf epidermis, which in turn impairs absorption. Pubescence can increase herbicide effec-

Figure 17-4. Selective kill of broadleaf weeds (left) without injury to the desirable turfgrass species.

tiveness, however, if the leaf surface becomes saturated with the herbicide. In this case, pubescence actually increases the quantity of herbicide spray held on the leaf surface for absorption.

The chemical composition of the plant surface may also influence herbicide selectivity. Weedy species such as wild garlic, wild onion, and purslane have a very waxy leaf surface that tends to repel herbicide spray droplets. The reduced herbicide contact caused by the waxy leaf surface can be partially alleviated by the use of a surfactant.

Rooting depth is another important factor in achieving selectivity of herbicides that are placed and tend to remain near the soil surface. In this case, the perennial, deep rooted turfgrass species do not absorb as much of the herbicide as shallow rooted weed seedlings because of the location of a major portion of the root absorption surface relative to the location of the herbicide. Partial selectivity can also be achieved through selective placement. This is frequently achieved through spot spraying of weedy clumps in turfs.

The relative degree of absorption of a specific herbicide by different plants can also be a factor in selectivity. In this case it is assumed that proper contact with the plant surface has been achieved. Differential absorption of herbicides from species to species is associated with inherent morphological, chemical, and electrical properties of the plant surface. Morphologically, the number and size of stomata vary among species and can be an important factor in selectivity, particularly with herbicides absorbed through the stomata. Similarly, the thickness and chemical nature of the cuticle varies from species to species as well as with plant age and environmental conditions. A degree of selectivity also results from the differential absorption of specific herbicides by roots. Roots are best adapted to absorb polar herbicides. Even within polar herbicides, differential selectivity and absorption may occur depending on the molecular size of the herbicide and type of root absorption mechanism involved.

Herbicide selectivity can also be achieved through proper timing of herbicide applications in relation to growth of the desirable turfgrass species versus the weed. Maximum absorption and kill of plants occurs during periods of rapid respiration and growth. Weeds that have just emerged into a mature turfgrass community are more easily controlled with greater selectivity at this early growth stage. The winter annual weeds can be controlled selectively during the dormancy period of perennial warm season turfgrasses such as bermudagrass and zoysiagrass.

Kill depends on the amount of herbicide that can be translocated to the sensitive sites within the plant at a particular time. This factor is of most concern in the case of systemic herbicides. Selectivity can result from variable translocation rates of a particular herbicide or from herbicide detoxification during translocation to the sensitive sites within the plant. In addition, the herbicide may be absorbed by nonsensitive tissues during translocation and never reach the sensitive sites. Simazine is an example of a herbicide that is metabolized quite rapidly by the enzymes of certain tolerant species during translocation. Herbicide translocation also varies with the age of the plant species, with the environmental conditions under which the plant is growing, and among species. For example, the translocation rate of 2,4-D in turfgrasses is much slower than in broadleaf weeds and contributes to the selectivity of the herbicide.

Herbicides can be divided into two major groups based on selectivity. A **selective herbicide** kills the weed without injuring the turfgrass plant to the extent that it cannot recover. The effectiveness of certain herbicides may vary depending on the rate of application with higher rates resulting in nonselective kill, whereas selectivity is achieved at lower rates. Most herbicides used for turfgrass weed control are of the selective type including benefin, bensulide, dicamba, DSMA, MAMA, siduron, silvex, and 2,4-D.

The herbicide tolerance of turfgrasses varies among species and cultivars (4, 52, 122, 217, 337). For example, Kentucky bluegrass is generally more tolerant than red fescue. Bentgrass and St. Augustinegrass are quite prone to shoot injury by 2,4-D and silvex (4, 53, 117, 257, 258, 265). It would be ideal for selective turfgrass herbicides to provide control of undesirable weeds without any injury to the desirable species. Selectivity in terms of minimal visible injury to aboveground plant parts has been achieved with a number of the turfgrass herbicides now used. Many are used, however, that cause a certain degree of injury to the root system (31, 53, 120, 312, 337). The roots of turfgrass seedlings are quite sensitive to injury from 2,4-D (319). This type of injury does not affect turfgrass quality directly but can result in reduced drought tolerance, turfgrass vigor, and recuperative potential. Increased proneness to disease development may also occur. A restriction in shoot, rhizome, and tiller numbers and growth may occur with certain less selective herbicides (56, 117, 120, 123, 146, 216).

A **nonselective herbicide** is toxic to all plants within the turfgrass community. Sodium arsenite, paraquat, and cacodylic acid are typical examples of nonselective herbicides that are sometimes used in turfgrass weed control (Fig. 17-5). Other herbicides that are nonselective to grasses include amitrole and dalapon. Nonselective herbicides are utilized primarily where there are no selective herbicides available for control of a particular turfgrass weed or where complete long term control of vegetation is desired such as in fence rows and around posts, road signs, and similar objects that would otherwise require laborious trimming operations.

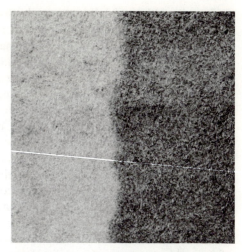

Figure 17-5. Nonselective kill (left) of a creeping bentgrass-annual bluegrass turf by paraquat.

Factors Affecting Herbicide Control

Satisfactory control of a turfgrass weed can only be achieved if the herbicide is applied at the proper rate. The optimum rate of herbicide application varies depending on a number of factors. The plant age and use of wetting agents affect the amount of herbicide required to achieve control. Young, actively growing tissues are more susceptible to herbicides. The use of wetting agents improves surface contact of the herbicide, enhances herbicide absorption, and reduces the amount of herbicide required to achieve an acceptable degree of control. With certain species, however, the use of a wetting agent lowers the surface tension to the extent that a substantial portion of the herbicide spray runs off the leaf surface. As a result, the herbicide is lost and its effectiveness decreased. This negative effect of wetting agents is primarily a problem where large spray volumes are used. The wetting agent may also reduce selectivity if it is based on differential wetting and herbicide retention.

Environmental factors. The influence of the environment on herbicide contact, absorption, and translocation in turfgrass weeds is also a factor to consider. Wind, temperature, rainfall, irrigation, and light intensity can influence the degree of control achieved with a given herbicide.

The effect of wind activity on herbicide contact is twofold in (a) reducing the degree of control achieved due to a lack of contact with the target species and (b) carrying herbicides to sensitive nontarget species, which results in unwanted injury. This movement of herbicide spray droplets away from the target species by atmospheric air currents is termed **spray drift**. The greater the extent of spray drift, the less the amount of herbicide that comes in contact with the target species and the lower the degree of control achieved. The direction and amount of spray drift varies with the (a) direction and velocity of wind, (b) size of the spray droplets, (c) type of application equipment, and (d) type of spray or spray additives used. The use of nozzles or high pressures that produce small spray droplets results in greater proneness to spray drift. The amount of drift varies with the type of applicator. Aerial application results in greater spray drift. Shielded booms can be utilized on ground operated spray equipment to minimize the amount of drift. In addition, spray drift can be reduced considerably by the use of certain types of herbicide carriers and additives. Drift, particularly from aerial applications, can be minimized significantly by the use of invert emulsions, polymeric thickeners, or particulating agents.

Herbicide vapors can also move in the atmosphere and are strongly influenced by wind movement. This is of primary concern with those herbicides whose vapors can be toxic to adjacent, sensitive ornamental plants or crops. This has been a common problem in the use of 2,4-D on turfgrass areas where grapes or tomatoes are grown in the vicinity and on lawns where roses are propagated. The movement of herbicide vapors in the atmosphere can be minimized by (a) using herbicides having low volatility, (b) application at the proper rate, and (c) use under atmospheric conditions where vaporization and vapor movement are minimal. High surface temperatures increase the extent of herbicide vaporization.

Environmental conditions can affect the amount of herbicide control achieved by its influence on the physiological condition of the plant and by influencing the degree of herbicide contact, absorption, and translocation. An environment that enhances growth of the weedy target species results in increased control since actively growing tissues are most susceptible to herbicide toxicity. Low temperatures restrict herbicide absorption and translocation. Warm to moderately high temperatures stimulate herbicide absorption and translocation, which increases the degree of control achieved (315). Excessively high temperatures can reduce herbicide absorption, however, due to rapid drying of spray droplets on the leaf surface.

A high atmospheric water vapor content generally stimulates herbicide absorption and translocation. Under these conditions water stress is minimal, the drying rate of spray droplets on the leaf surface is reduced, stomata are open for a longer period of time, and physiological processes are not disrupted due to periodic internal moisture stresses.

The role of light in influencing weed control with herbicides involves a twofold effect. First of all, light stimulates the opening of stomata, which serve as a major avenue for the absorption of certain herbicides. In addition, light stimulates photosynthesis and the production of carbohydrates that are translocated throughout the plant. Herbicides translocated primarily through the phloem have an increased translocation rate during periods of active photosynthesis due to the higher rate of carbohydrate translocation. A negative aspect is that certain herbicides are subject to photo decomposition by the shorter wavelengths of the visible spectrum.

The occurrence of rain or irrigation relative to the time of herbicide application can affect the amount of weed control achieved. The degree of control is substantially reduced if the herbicide spray is washed from the foliage before a sufficient amount of herbicide is absorbed.

Soil factors. The herbicides utilized for preemergence weed control in established turfs are primarily types that come into contact with the soil and must be absorbed from the soil solution through the root system. The effectiveness of preemergence herbicides depends on the presence of a high herbicide concentration in the upper 0.5 in. of soil. This is the zone in which the seed of most annual weedy species germinate. The rate of preemergence herbicide application for effective weed control varies depending on the texture and chemical properties of the soil.

Herbicide persistence, or the length of time a herbicide remains active in the soil, is important in turfgrass weed control from two standpoints. It is desirable for the preemergence herbicides to remain active in the soil for an extended period of time. In contrast, where a nonselective herbicide is used, it is desirable to have a minimum length of persistence in order for the area to be planted or reestablished as soon as possible. Herbicides vary greatly in length of persistence, from 1 to 3 weeks in the case of 2,4-D and from 3 to 5 weeks for amitrole used on a nonselective basis. Monuron used as a permanent soil sterilant has a persistence of 6 to 20 months.

Herbicide persistence in the soil is influenced by such soil environmental factors as the content of clay and organic matter, pH, moisture level, soil temperature, and degree of microbial activity. The activity of herbicides can be lost through chemical decomposition, microorganism decomposition, adsorption on clay or organic matter colloids, leaching, and volatilization.

Turfgrass Weeds

Many species of weeds can occur in turfgrass areas. Forty-three of the major turfgrass weeds, their characteristics and chemical controls, are presented in Table 17-3. Another thirty-nine less common weeds, their characteristics, and chemical controls, are summarized in Table 17-4. It is not practical to cover all weeds that might occur in a turf in a given region. Thus, only the more widely occurring problem weeds are included. For information and plant descriptions of other turfgrass weeds, the reader is referred to the many books on this subject (8, 11, 51, 187).

Turfgrass Diseases

A **disease** is defined as any disturbance caused by a living organism or environmental factor that interferes with the synthesis, translocation, or utilization of plant metabolites, mineral nutrients, or water to the extent that the affected plant changes in appearance or is substantially altered in growth rate from that of a comparable, healthy plant. Turfgrass diseases can be divided into noninfectious and infectious types. Noninfectious diseases are physiological disorders resulting from (a) temperature, water, or light stresses; (b) essential nutrient deficiencies;

Table 17-3

SCIENTIFIC NAME, CHARACTERISTICS, AND CONTROL FOR 43 OF THE MOST COMMON TURFGRASS WEEDS

| Name of weed | | Weed type* | Length of life† | Primary dissemination‡ | Chemical control§ |
Common	Scientific				
Annual bluegrass	*Poa annua* L.	BG-CG	A (P)	S	PE except siduron
Barnyardgrass	*Echinochloa crusgalli* (L.) Beauv.	BG	A	S	PT
Bentgrass, creeping	*Agrostis palustris* Huds.	CG	P	V (S)	NS
Bermudagrass	*Cynodon* spp. Rich.	CG	P	V	NS
Carpetweed	*Mollugo verticillata* L.	CB	A	S	C
Chickweed, common	*Stellaria media* (L.) Cyrillo	CB	A	S	Dicamba, mecoprop, silvex, or PT
mouse ear	*Cerastium vulgatum* L.	CB	P	S	Dicamba, mecoprop, silvex, or PT
Cinquefoil, common	*Potentilla canadensis* L.	CB	P	S	Dicamba + 2,4-D
Clover, white	*Trifolium repens* L.	CB	P	S (V)	Dicamba, mecoprop, or silvex
Crabgrass, large	*Digitaria sanguinalis* (L.) Scop.	BG	A	S	PE or PT
smooth	*Digitaria ischaemum* (Schreb.) Muhl.	BG	A	S	PE or PT
Dallisgrass	*Paspalum dilatatum* Poir.	BG	P	S	PT
Dandelion, common	*Taraxacum officinale* Weber	BB	P	S	C
Dock, curly	*Rumex crispus* L.	BB	P	S	Dicamba or 2,4-D (r)
Fescue, tall	*Festuca arundinacea* Schreb.	BG	P	S	NS
Garlic, wild	*Allium vineale* L.	BB	P	S (V)	Dicamba or 2,4-D (r)
Goosegrass	*Eleusine indica* (L.) Gaertn.	BG	A	S	PE or PT
Ground ivy	*Galechoma hederacea* L.	CB	P	S	Dicamba, mecoprop, or silvex (r)
Hawkweed, mouse-ear	*Hieracium pilosella* L.	CB	P	S (V)	Dicamba or 2,4-D (r)
Healall	*Prunella vulgaris* L.	CB	P	S (V)	Dicamba + 2,4-D or 2,4-D (r)
Henbit	*Lamium amplexicaule* L.	CB	A	S	Dicamba or silvex
Knotweed, prostrate	*Polygonum aviculare* L.	CB	A	S	Dicamba

Table 17-3 (Cont.)

Common name	Scientific name	Type	Life	Repro.	Control
Lambsquarters, common	*Chenopodium album* L.	BB	A (B)	s	C
Mallow, common	*Malva neglecta* Wallr.	BB	A (B)	s	Dicamba + 2,4-D or silvex (r)
Moneywort	*Lysimachia nummularia* L.	CB	P	V (S)	2,4-D
Nimblewill	*Muhlenbergia schreberi* Gmel.	CG	P	S (V)	NS
Nutsedge, yellow	*Cyperus esculentus* L.	CS	P	V (S)	PT or 2,4-D (r)
purple	*Cyperus rotundus* L.	CS	P	V (S)	PT or 2,4-D (r)
Onion, wild	*Allium canadense* L.	BB	P	S	Dicamba or 2,4-D
Plantain, broadleaf	*Plantago major* L.	BB	P	s	2,4-D or silvex
buckhorn	*Plantago lanceolata* L.	BB	P	s	2,4-D or silvex
Purslane, common	*Portulaca oleracea* L.	CB	A	s	Dicamba or silvex
Quackgrass	*Agropyron repens* (L.) Beauv.	CG	P	V (S)	NS
Sandbur, field	*Cenchrus pauciflorus* Benth.	BG	A	s	PT
Sheperdspurse	*Capsella bursa-pastoris* (L.) Medic.	BB	A	s	C
Sorrel, red	*Rumex acetosella* L.	CB	P	S (V)	Dicamba
Speedwell, thymeleaf	*Veronica serpyllifolia* L.	CB	P	S (V)	Dicamba, silvex, or 2,4-D (m)
Spurge, prostrate	*Euphorbia supina* Raf.	CB	A	s	Dicamba
Velvetgrass	*Holcus lanatus* L.	BG	P	s	NS
Witchgrass	*Panicum capillare* L.	BG	A	s	PT
Woodsorrel, yellow	*Oxalis stricta* L.	CB	P (A)	s	Silvex
Yarrow, common	*Achillea millefolium* L.	CB	P	S (V)	Dicamba or 2,4-D (r)
Yellow rocket	*Barbarea vulgaris* (R.) Br.	BB	B-P	s	Dicamba or 2,4-D

* BB—bunch broadleaf type.
  CB—creeping broadleaf type.
  BG—bunch grass type.
  CG—creeping grass type.
  CS—creeping sedge type.
† A—annual.
  B—biennial.
  P—perennial.
‡ S—seed.

V—vegetative stolons or rhizomes.
§ m—marginal control.
  r—repeat applications usually required.
NS—nonselective kill with dalapon or amitrole and reestablishment.
PE—controlled preemergence with such materials as benefin, bensulide, calcium arsenate, DCPA, or siduron.
PT—controlled postemergence by 2 applications of arsonates such as DSMA, MAMA, or MSMA.
C—controlled with dicamba, mecoprop, silvex, 2,4-D, or a combination of these.

Table 17-4

SCIENTIFIC NAME, CHARACTERISTICS, AND CONTROL FOR 39 OF THE LESS COMMON TURFGRASS WEEDS

Name of weed		Weed type*	Length of life†	Primary dissemination‡	Chemical control§
Common	Scientific				
Bedstraw	Galium spp. L.	CB	A (P)	S	Dicamba or silvex (r)
Bellflower, creeping	Campanula rapunculoides L.	CB	P	S	Dicamba
Buffalobur	Solanum rostratum Dunal.	BB	A	S	NS
Buttercup, creeping	Ranunculus repens L.	CB	P	S	Dicamba or 2,4-D
Catnip	Nepeta cataria L.	CB	P	S (V)	2,4-D (r)
Catsear, potted	Hypochoeris radicata L.	BB	P	S	Silvex or 2,4-D
Chicory	Cichorium intybus L.	BB	P	S	C
Daisy, English	Bellis perennis L.	BB	P	S	Dicamba or silvex
Daisy, oxeye	Chrysanthemum leucanthemum L.	CB	P	S (V)	Dicamba, mecoprop, or silvex (r)
Deadnettle, red	Lamium purpureum L.	CB	A	S	Dicamba
Dichondra	Dichondra spp. Forst.	CB	P	V	2,4-D
Dropseed, annual	Sporobolus neglectus Nash	BG	A	S	PE or PT
Foxtail, green	Setaria viridis (L.) Beauv.	BG	A	S	PE or PT
yellow	Setaria glauca (L.) Beauv.	BG	A	S	PE or PT
Geranium, Carolina	Geranium carolinianum L.	BB	A (B)	S	C
Kikuyugrass	Pennisetum clandestinum Hochst.	CG	P	V (S)	MSMA (r)
Kochia	Kochia scoparia (L.) Schrad.	CB	A	S	2,4-D
Mayweed	Anthemis cotula L.	BB	A	S	Dicamba + 2,4-D
Medic, black	Medicago lupulina L.	BB	A (B)	S	Dicamba or silvex
Mosses	Bryum, Ceratodon, Hypnum, or Polytrichum spp.	—	—	spores	Iron sulfate or mercury compounds (m)
Mustard, wild	Brassica kaber L.	BB	A	S	C
Orchardgrass	Dactylis glomerata L.	BG	P	S	NS
Panicum, fall	Panicum dichotomiflorum Michx.	BG	A	S	PE or PT
Paspalum, fringeleaf	Paspalum ciliatifolium Michx.	BG	P	S	PT
Pearlwort, birdseye	Sagina procumbens L.	CB	P	V	Dicamba, endothall, or mecoprop (m)
Pennycress, field	Thlaspi arvense L.	BB	A	S	C
Pennywort, lawn	Hydrocotyle sibthorpioides Lam.	CB	P	V	Silvex (r)

Table 17-4 (Cont.)

Pigweed	*Amaranthus* spp. L.	CB	A	S	C
Pineappleweed	*Matricaria matricarioides* (Less.) Porter	BB	A (P)	S	Dicamba, mecoprop, or silvex (r)
Puncturevine	*Tribulus terrestris* L.	CB	A	S	2,4-D (r)
Speedwell, creeping	*Veronica filiformis* Sm.	CB	A	S	Endothall (r)
Speedwell, purslane	*Veronica peregrina* L.	BB	A	S	Dicamba + 2,4-D or silvex (r)
Starwort, little	*Stellaria graminea* L.	CB	P	S	Dicamba
Thistle, musk	*Carduus nutans* L.	BB	B	S	Dicamba or 2,4-D
Timothy	*Phleum pratense* L.	BG	P	S	NS
Torpedograss	*Panicum repens* L.	CG	P	V	NS (m)
Vervain, prostrate	*Verbena bracteata* Lag. and Rodr.	CB	A (P)	S	2,4-D or silvex (r)
Violets	*Viola* spp.	CB	P	S	Silvex (r)
Waterleaf	*Ellisia myctelea* L.	BB	A	S	Silvex

* BB—bunch broadleaf type.
  CB—creeping broadleaf type.
  BG—bunch grass type.
  CG—creeping grass type.
  CS—creeping sedge type.
† A—annual.
  B—biennial.
  P—perennial.
‡ S—seed.
  V—vegetative stolons or rhizomes.
§ m—marginal control.
  r—repeat applications usually required.
  NS—nonselective kill with dalapon or amitrole and reestablishment.
  PE—controlled preemergence with such materials as benefin, bensulide, calcium arsenate, DCPA, or siduron.
  PT—controlled postemergence by 2 applications of arsonates such as DSMA, MAMA, or MSMA.
  C—controlled with dicamba, mecoprop, silvex, 2,4-D, or a combination of these.

(c) inorganic and organic toxicities; (d) soil oxygen deficiencies; (e) atmospheric pollution; (f) soil acidity or alkalinity; (g) mechanical injury; or (h) genetic defects. The symptoms, causes, and prevention of the noninfectious diseases have been discussed in previous chapters.

Infectious plant diseases are those caused by fungi, viruses, bacteria, nematodes, or parasitic higher plants. The current knowledge of viruses as causes of turfgrass disease is limited and discussed only briefly. Similarly, information on bacteria as causal organisms of turfgrass diseases is lacking. Viruses and bacteria do not appear to be significant factors in turfgrass diseases, but this may be more apparent than real. Nematodes are discussed in a subsequent section.

The following section deals primarily with the fungi because they cause most of the infectious turfgrass diseases. Fungal pathogens can cause turfgrass diseases by (a) consuming the vital contents of host cells; (b) disrupting the photosynthesis, respiration, and metabolism of host cells through the secretion of toxins, enzymes, or growth regulating substances; or (c) blocking the transport of carbohydrates, mineral nutrients, and water through conductive tissues.

Characteristics of Fungi

Fungi are nucleated, spore-bearing plants that are grouped within the division *Mycota*. They lack chlorophyll and reproduce sexually and asexually. Structurally they are multicellular, filamentous, and usually nonmotile; they possess polysaccharide cell walls.

A fungus is composed of a simple vegetative body termed the **thallus** that consists of microscopic tube-like filaments that branch and grow at either apex in all directions over and within the host. It is not differentiated into true leaves, stems, and roots. Each of the filaments, known as a **hypha**, is composed of a thin, transparent tube bounded by a wall. The hyphae are capable of indefinite growth under favorable conditions of temperature, water, and nutrient availability. A mass of hyphae is referred to as the **mycelium** (Fig. 17-6). A minute fragment from almost any portion of the mycelium is sufficient to initiate a new individual by continued apical growth.

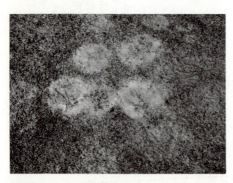

The life cycle of most fungi begins with germination of a spore or sclerotium. **Spores** are minute propagating units from which the fungus mycelium is produced. The spore germinates under favorable conditions by initiating a short germ tube from which the mycelium grows. The mycelium usually grows radially from the central origin of the germinating spore. A compact, hard, resting body called a **sclerotium** is formed

Figure 17-6. Grayish-white mycelium of *Typhula itoana* covering the leaves of creeping bentgrass.

by the mycelium of some fungi. The closely packed cells of the sclerotium differ from the loose, disassociated hyphae that make up the thallus. Because of this organizational structure, the sclerotium is able to survive unfavorable environmental conditions and germinate when favorable growth conditions occur. Germination of sclerotium may involve the production of hyphae or the formation of spore-bearing structures.

Fungi may also reproduce by asexual or sexual means during their life cycle. Asexual reproduction is generally more important in the propagation and dissemination of fungi because of the prolific nature of this process, which can occur several times a year. During asexual reproduction, specialized cells called **conidia** are formed by budding at the tips of mycelial stalks called **conidiophores**. Mature conidia are readily detached from the conidiophore and distributed by means of (a) rain or irrigation water, (b) mowing equipment, (c) wind, or (d) the movement of humans or machines. Conidia movement may range from a few inches in the case of splashing raindrops up to substantial distances through the action of wind, mowers, or human feet.

Conidia are capable of germinating soon after formation. Germination is stimulated by favorable temperatures, water, and nutrient availability. Following hydration, a slender tube called a **germ tube** is formed. It grows apically over the surface of the host until a penetration site is encountered. Common penetration sites for germ tubes include stomata, mower wounds, or areas damaged by the feeding activity of insects or nematodes. At the penetration site the germ tube usually enlarges into a pear-shaped swelling called an **appressorium** that attaches the germ tube to the host surface (100, 112, 231). An infection hypha is formed at the bottom of the appressorium that penetrates the cell wall or stomatal opening. With some fungal species, the germ tube may enter the plant without forming an appressorium (20, 72, 112, 185). A supply of free water is required by most fungi throughout this process. Following penetration, infection may be established with the mycelium proliferating inside the host plant. The mycelium may grow between the host cells, inside the host cells, or both, depending on the fungal species.

Sexual reproduction which involves the mating of two fungi may also occur. The process is important because genetically different fungi are formed and also because sexual spores tend to be thick-walled and, therefore, function as resistant survival structures. Sexual reproduction provides a means for producing new races of a particular fungal species. Races are individuals that differ in their ability to infect different cultivars of a turfgrass species. Races of a number of turfgrass disease-causing fungi have been isolated including stripe smut (92, 149, 171, 348) and dollar spot (28, 135). The fruiting bodies of each fungus species that produces sexual spores have a characteristic appearance that is used to identify the fungus.

Parasitic nature of fungi. Fungi are primitive plants that must obtain their nutrients and carbohydrates from other living organisms and/or nonliving organic materials. Fungi that obtain their food by infecting living organisms are called **parasites**. If they cause a disease in doing this, they are termed **pathogens**. Fungi that can multiply and reproduce only on living protoplasm are **obligate parasites**. Typical examples are the rusts and powdery mildew. The plant that is infected by a parasite is the **host**. Fungi that feed on dead organic matter are called **saprophytes**. The saprophytic fungi have a very beneficial function in contributing to the decomposition of dead plants within the turfgrass community. Fungi that can live on either dead or living tissues are **facultative parasites**. *Fusarium* is a typical example of a facultative parasite that lives most of the time as a saprophyte and is occasionally a parasite. A **facultative saprophyte** is an organism that is usually a parasite but is occasionally a saprophyte. Typical examples are the *Helminthosporium* and *Sclerotinia* organisms.

Both the obligate and facultative parasites can function as the causal organism in producing turfgrass diseases. The facultative parasites are relatively weak pathogens and generally cause damage to seedlings and to turfgrasses that are in a weakened condition due to (a) inadequate nutrition, (b) environmental stress, or (c) injury caused by some other turfgrass pest or man. In contrast, obligate parasites generally develop best on vigorously growing turfgrass plants since they can live only on living plant cells.

Parasitic fungi remove both water and nutrients from the host turfgrass plant. The rust and powdery mildew fungi, which grow intracellularly within a host, obtain their food through specialized absorbing organs called **haustoria**. The hyphae located between the cells can absorb food directly from the host cells. Disease attacks by fungi may result in restricted turfgrass growth and an overall decline in vigor. Turfgrass shoot density may also be reduced under extreme parasitism.

Fungi as pathogens. Parasitism is directly involved with pathogenicity since it is required in order to develop and support the fungi–host interrelationship. In terms of overall damage to the turfgrass plant, however, parasitism may or may not be the major cause. Disease-inciting pathogens may cause damage by (a) direct parasitism or (b) secretion of enzymes, toxins, growth regulators, polysaccharides, or antibiotics that contribute to disease formation. Direct parasitism, or the procuring of food from the host, is not necessarily the major cause of damage but it is usually involved in establishing the intimate relationship between the host and fungus that provides the base from which the disease condition develops.

Enzymes, toxins, and growth regulators are the most common causes of disease. Some fungal enzymes act in the breakdown of the polysaccharide polymers that make up the host cell walls, while others may destroy the protoplasm. Some toxins are capable of disrupting the permeability and function of the protoplast (130). For example, *Sclerotinia homoeocarpa* produces a toxin that is injurious to the root tips of bentgrass (108, 113, 114, 255). Similarly, toxic cyanogenic compounds are produced by certain fairy ring-causing fungi (15, 130, 243) and by a low temperature Basidiomycete (357). The growth regulators influence, either negatively or positively, the rate of cell division and enlargement. Thus, disease incitement in turfgrasses can be the result of several mechanisms that interfere with one or more essential functions of the plant.

The Disease Cycle

The distinct series of events associated with disease development is known as the **disease cycle**. The first event, **inoculation**, involves bringing the pathogen or its reproductive structures into contact with the host. The most common type of fungal inoculum is the spore, particularly conidia. Following germination on the host surface, the germ tube forms as described earlier.

The second event, **penetration**, involves the entrance of the pathogen into the host. Penetration may occur through (a) wounds, (b) natural openings, or (c) the plant surface directly. Stomata are the common means of penetration through natural openings in turfgrasses (86) (Fig. 17-7). Wound penetration sites may be caused by mechanical damage, mowing, cultivation, foot traffic, or the feeding activity of insects and nematodes (231). One of the common means of penetration is through mower wounds (72). Penetration by fungi directly through the intact host plant surface is achieved by (a) mechanical pressure of the infection hyphae on the cuticle and epidermal cell walls of the plant surface and/or (b) the secretion of fungal enzymes that soften or dissolve the host cuticle or epidermal cell walls (277, 348).

The third event in the disease cycle is **infection**. The pathogen contacts the susceptible cells or tissues of the host during this process and establishes a stable parasitic relationship in which the fungus absorbs host nutrients. Plant tissues are not killed by the pathogen during infection.

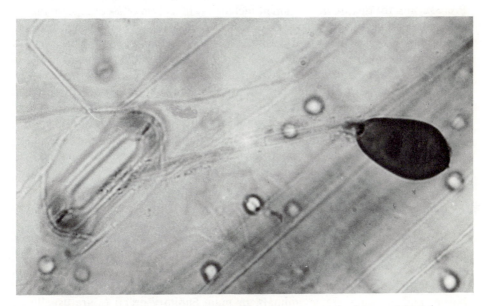

Figure 17-7. Penetration of a *Helminthosporium* germ tube into a stomatal opening in a Kentucky bluegrass leaf. (Photo courtesy of R. M. Endo, University of California at Riverside, Riverside, California).

Following infection there is a fourth event termed the **incubation period**. This is the interval between plant infection and the appearance of disease symptoms. The incubation period may vary from a few days to many weeks, depending on the particular pathogen, host, and environmental conditions. The pathogen invades and spreads further into the tissues of the host plant during this stage of disease development. The degree of spread depends on the resistance of the host cells and the pathogenic potential of the parasite. The extent of invasion varies with the particular pathogen from simple penetration into the epidermal cells, typical of powdery mildew, to a more extensive invasion into the cells and tissues of the host plant, typical of many parasitic fungi. The pathogen may secrete biochemically active substances that disrupt vital physiological processes or the structural integrity of the protoplasm.

The pathogen is generally capable of reproducing rather rapidly after infection has been established. Eventually, spores are produced in extremely large numbers and are disseminated to other turfgrasses. Some fungi that attack turfgrasses exist primarily as hyphae. *Corticium fuciforme*, *Rhizoctonia solani*, and *Sclerotinia homoeocarpa* are examples of fungi that apparently do not form either sexual or asexual spores.

Preventing Turfgrass Diseases

Prerequisites for disease development include (a) inoculum of a pathogenic fungus, (b) a susceptible turfgrass species or cultivar, and (c) favorable environmental conditions. Disease prevention practices can be effective in minimizing turfgrass injury from a number of infectious diseases and is preferable to chemical control whenever it can be utilized. The five major methods of minimizing or preventing turfgrass diseases are (a) sanitary, (b) environmental, (c) cultural, (d) preplant control, and (e) use of resistant cultivars. Manipulation of turfgrass cultural practices and the environment has the twofold objective of (a) improving the vigor of the host so that it is best able to resist, survive, and recover from pathogen infections and (b) providing an environment that is unfavorable for germination, growth, sporulation, and development of the pathogen.

Sanitary Practices

Sanitary practices involve minimizing or avoiding the introduction of pathogens into healthy, disease-free turf. Inoculum of fungi can be introduced by (a) infected turfgrass seeds, rhizomes, stolons, sprigs, plugs, or sod; (b) contaminated equipment; (c) infested soil or topdressing material, particularly if the disease is soil-borne; and (d) wind, water, animals, or man. Sanitary practices are usually not effective for those turfgrass pathogens that are readily spread by spores. In the case of soil-borne diseases, sanitary practices can be effective in avoiding the introduction of parasitic fungi in sod; plugs; soil; topdressing; or on soil attached to cultivation equipment, rhizomes, stolons, or sprigs. Plugs and sod should be inspected for freedom from symptoms of soil-borne diseases before the materials are introduced into a disease-free area. Similarly, soil or topdressing materials that contain potentially serious parasitic turfgrass fungi can be heat-treated or fumigated with such materials as dazomet, metham, or methyl bromide. Turfgrass cultivation equipment should be washed free of soil and thoroughly cleaned, preferably by steam treatment, prior to movement from a disease infested to an uninfested area.

In the case of plant materials such as stolons, sprigs, rhizomes, plugs, or sod, it is important to avoid inoculum and/or disease present at the time of harvest, shipment, and planting. This reduces the introduction of pathogens, particularly new races of pathogens, into disease-free areas. Freedom from pathogens also ensures that the plant materials possess maximum vigor for rapid establishment. Sanitary practices should be recognized as one aspect in an overall turfgrass disease prevention program. The effectiveness of sanitary disease prevention is somewhat less important, however, compared to the other methods available.

Environmental Factors

A favorable environment is one of the critical prerequisites necessary for disease development. The environmental conditions existing after contact of the pathogen with host influences turfgrass disease development and may determine whether the disease will or will not occur. Since disease development depends on a favorable combination of environmental factors, a sudden change in one or more of these factors can determine its severity. The environmental conditions necessary for disease development should be understood so that appropriate measures can be taken to maintain a turfgrass environment that is generally unfavorable for turfgrass pathogens but is favorable for growth of the potential host plant.

Temperature. Each species of fungi has an optimum, suboptimal, and supraoptimal temperature for germination, hyphal growth, infection, disease development, sporulation, and survival (20, 28, 95, 108, 124, 170, 232). The optimum temperature for spore production may be different from the optimum for mycelial growth (232). The optimum temperature for development of a given turfgrass disease can range from as low as 35 to 40°F for *Fusarium* and *Typhula* species to as high as 90 to 95°F for certain *Pythium* species (86, 107, 108, 137, 199, 240, 263, 269, 273, 299). The optimum temperature may also vary with different races of a fungal species such as has been found for *Sclerotinia homoeocarpa* (28, 108, 135) and *Rhizoctonia solani* (84, 107, 254).

Temperature is an important factor influencing the growth rate of the host plant. The optimum temperature for disease development may be distinctly different from the optimum temperature for fungal growth (86, 95, 108). Temperatures that are optimum for growth of the host produce a more vigorous plant that is better able to resist disease attacks and recover from the effects. Temperature extremes may injure the turfgrass host, causing a weakened plant that is more susceptible to infection from fungi, including the relatively weak facultative parasites. Thus, temperature influences (a) the growth rate and relative susceptibility of the host, (b) the multiplication and growth of the pathogen, and (c) the interaction of the host and pathogen.

The most rapid disease development occurs at optimum temperatures for development of the pathogen, which are above or below the optimum for growth of the host turfgrass plant. The frequency and severity of a disease is usually associated with a particular season of the year. Infectious diseases such as *Typhula* blight (299) and *Fusarium* patch (86, 240) are usually associated with cool seasons, while *Fusarium* blight (19), *Pythium* blight (108, 137, 269, 273, 363), and *Rhizoctonia* brown patch (81, 84, 95, 285, 293) are more commonly associated with warm seasons.

Water. Free water influences turfgrass disease development in many ways. One of the most critical periods occurs during inoculation and penetration since free water is required to initiate germination of most fungal spores and to maintain growth of the germ tube. In the case of water mold fungi such as the *Pythium* species, free water is also necessary for the formation and movement of motile zoospores to the host. The need for free water during penetration is associated with fungi that infect the aboveground portions of turfgrass plants. Free water can originate from rain, irrigation, fog, dew, or leaf exudates. Leaf exudates formed at mowing wounds serve as a favorable media for spore germination and mycelial growth because they contain mineral nutrients, carbohydrates, and nitrogenous compounds (109, 112, 115, 182, 183, 184, 185, 372). A proliferation of *Helminthosporium* and *Rhizoctonia* mycelium in the exudates at the tips of turfgrass leaves can be observed

in the early morning (111, 314) (Fig. 17-8). Saprophytic organism such as bacteria that are present on the host surface may utilize the nutrients in the exudate before the turfgrass pathogen.

The occurrence and severity of many warm weather turfgrass diseases such as *Pythium* and *Rhizoctonia* are closely associated with the amount and distribution of rainfall (70, 85, 269). The number of disease cycles per year is closely related to the number of rains occurring within that period of time. The mycelia of many fungi become desiccated during periods of water stress. Germination, mycelial growth, and sporulation may be reinitiated when favorable moisture conditions occur.

Water stresses or excesses that weaken the host turfgrass plant increase

Figure 17-8. A proliferation and bridging of *Sclerotinia homoeocarpa* mycelium in the exudate of Seaside creeping bentgrass leaf tip mower wounds. (Photo courtesy of R. M. Endo, University of California at Riverside, Riverside, California).

the susceptibility to diseases caused by facultative parasites. Infection by certain turfgrass pathogens such as the fungi causing dollar spot, *Fusarium* blight, *Pythium* blight, red thread, and stripe smut is apparently enhanced by exposure to water stress (18, 37, 70, 71, 75, 199, 252, 273, 278). In the case of soil-borne organisms that affect underground plant parts, the severity of the disease usually increases with higher soil water levels (70, 360). Excessively high soil water levels enhance the multiplication and movement of pathogens and impair root growth of the host. High soil water levels tend to increase the succulence of the host plant, which increases its susceptibility to fungal infections.

Moisture is also important in other ways. Sporulation of fungi is generally most active during moist weather. The splashing action of raindrops and irrigation water as well as the movement of surface water can be significant factors in the dissemination of fungal inoculum of such species as *Gloeocercospora* and *Pythium* (70, 269). Excess surface water moving across a turfgrass area can carry spores, sclerotia, and mycelial fragments from an infected turfgrass area to uninfected areas. For example, the spread of *Pythium* over a green is frequently associated with the drainage pattern (269).

Atmospheric water vapor. A high atmospheric water vapor content enhances the activity of certain turfgrass pathogens, particularly *Fusarium*, *Rhizoctonia*, and *Sclerotinia* spp. (20, 70, 81, 159, 285, 293, 329). Dense, high quality turfs generally have a higher atmospheric water vapor content within the stand, which results in a more favorable environment for disease development. The dense vegetation impairs air movement and causes stratification within the vertical profile of the turf. An additional factor associated with intensively cultured turfs is the high soil water level that contributes to a high atmospheric water vapor content within the turfgrass stand.

The atmospheric water vapor content contributes indirectly to disease development by affecting the evaporation rate of free water from the turfgrass plant surfaces (314). The higher the atmospheric water vapor content, the slower the evaporation rate. This increases the time that free water remains on the plant surfaces to stimulate spore germination and mycelial growth. The water vapor content tends to be high under tree canopies, particularly if air movement is restricted by screens of shrubs and trees. Thus, a high water vapor content combined with a lack of incoming radiation results in the shaded host surface remaining wet longer following irrigation, rainfall, or dew formation. The occurrence and severity of diseases can be significantly influenced by the frequency and duration of dews (105, 271, 318, 376).

Light. The influence of light on disease development is much less significant than that of temperature or water. High light intensities reduce sporulation, spore germination, and mycelial growth of most fungal diseases. Host plants grown at low light intensities are frequently more susceptible to fungal diseases. The severity of powdery mildew on Kentucky bluegrass is usually increased under shaded conditions in cool humid climates (22, 150) (Fig. 17-9). Rust infections are usually

Figure 17-9. A severe powdery mildew infection on Merion Kentucky bluegrass growing in the shade.

more severe on shaded turfs, particularly in warm humid climates (168, 230, 234). In contrast, *Fusarium* blight is more serious in full sunlight (19). The responses may be related to altered temperature and moisture conditions under the shade as well as to light effects. The influence of light quality is illustrated by the enhanced sporulation of *Typhula* blight when exposed to the ultraviolet wavelengths (299).

Wind. Wind can be an important factor in the occurrence of diseases. Spores of fungi that cause rust, smut, mildew, and *Helminthosporium* diseases are widely disseminated by wind movement. Wind also influences the rate of free water evaporation from the leaf surfaces of host plants. The longer free water remains on the leaf, the more likely that fungal penetration and infection will occur.

Stratification of higher temperatures and water vapor contents occurs adjacent to turfs when air movement is impaired by a surrounding screen of trees, shrubs, or hills. Turfs growing in such locations are more frequently attacked and damaged by diseases, especially high temperature fungi, than similar turfs growing on exposed sites. Wind provides a mixing action that lowers the atmospheric water vapor content and temperature adjacent to the turf. As a result, the evaporation of free water is stimulated and environmental conditions are created that are less favorable for disease development.

Cultural Factors

Turfgrass cultural practices significantly influence the development of turfgrass diseases. Many of the routine cultural practices utilized in maintaining turfs, particularly those of high quality, tend to favor diseases caused by obligate parasites. This may result from increasing the susceptibility of the host or by providing a more favorable microenvironment for fungal spore germination, penetration, and infection. Cultural practices that result in a weakened turf tend to favor diseases caused by facultative parasites.

Mowing. Mowing generally enhances disease development. Wounds produced at the leaf tips serve as major avenues for the penetration of many fungal pathogens such as *Rhizoctonia* and *Fusarium* (314). Exudations from the wounds form droplets containing nutrients that enhance spore germination and mycelial growth (112, 115, 182, 183, 184, 372). Disease development increases with an increased mowing frequency (252). Spore dissemination is also facilitated by mowing, particularly when greens are mowed in early morning. For example, the pattern of *Pythium* disease development is affected by the mowing direction (269).

Height of cut is another aspect of mowing that may affect the development of turfgrass diseases. A close cutting height frequently results in increased suscepti-bility to disease injury, particularly from *Helminthosporium*, *Puccinia*, *Rhizoctonia*, and *Sclerotinia* species (103, 116, 144, 171, 173, 215, 252, 253, 314). The increased shoot density occurring at short cutting heights contributes to a higher water vapor content within the plant community. Also, individual plants within a closely cut turfgrass community are relatively small, have a restricted rooting depth, and are reduced in overall plant vigor. The net result is increased susceptibility to disease injury. In addition, bridging of mycelium from leaf tip to leaf tip across a turf is generally much more rapid under close, frequent mowing than in comparable, higher cut turfs. In contrast, turfs entering the winter with an excessive amount of shoot growth are more susceptible to *Typhula* blight damage than turfs maintained at a cutting height of 1.5 in. or less.

Defoliation resulting from mowing can be beneficial to turfs having a rapid rate of vertical leaf extension. Some parasitic fungi that penetrate through wounds at the leaf tip have a relatively slow rate of downward infection to the meristematic tissues. In such cases the infected leaf tips may be removed by mowing while the lower, disease-free tissues continue to grow.

Clipping removal. The removal of diseased clippings from a turf can reduce the amount of inoculum available for current and future pathogen activity. This does not imply that the inoculum supply will be eliminated but only that the relative amount of inoculum present can be substantially reduced under certain conditions. This may reduce the likelihood of severe disease development, especially for fungi that are unable to grow in the thatch but can survive on the clippings as dormant structures. A typical example occurs with the *Helminthosporium* species. The removal of disease-free clippings may also be important, especially for soil-inhabit-ing pathogenic fungi that are able to colonize these substrates for readily decom-posable sugar-containing compounds. *Pythium* and *Rhizoctonia* are common examples of the latter situation.

Fertility. Turfgrasses need a minimum level of essential nutrients in the proper balance to resist and recover rapidly from disease (32, 33, 155). This balanced level of nutrition should provide a controlled shoot growth rate, a deep extensive root system, and good recuperative potential (248). The timing of fertilizer appli-cations should also be considered in relation to potential disease development (86). For example, early fall applications of nitrogen fertilizer can cause an increased incidence of *Typhula* blight compared to earlier or later nitrogen applications. The type of nitrogen carrier used can also influence the degree of disease develop-ment (369). Fertilization with activated sewage sludge reduces the severity of dollar spot on bentgrass (63, 64, 256, 310) and *Pythium* blight on ryegrass (362) in com-parison to the use of soluble nitrogen carriers. The mechanisms involved in these responses are not understood.

Nitrogen. Excessively high nitrogen fertility levels may increase the susceptibility of turfgrass to certain pathogens such as *Fusarium* (145, 157, 158, 165, 254), *Helminthosporium* (59, 60, 144, 145, 173, 192, 249), *Ophiobolus* (154, 155), *Piricul-aria* (139), *Rhizoctonia* (7, 33, 69, 93), and *Typhula* (355). *Fusarium* patch and brown patch are particularly destructive at excessive nitrogen fertility levels because the cell walls of the turfgrass host are thinner and more easily penetrated by fungal hyphae. Excessive amounts of nitrogen also increase the succulence of the tissue as well as the glutamine content and frequency of leaf exudates (183).

In contrast, the development of other turfgrass diseases caused by such fungal pathogens as *Corticium* (124, 154, 157, 165, 278, 331), *Puccinia* (36, 227, 309), and *Sclerotinia* (7, 63, 64, 66, 75, 93, 109, 110, 119, 136, 192, 193, 196, 256, 262, 309, 310, 369), are particularly severe at low levels of nitrogen nutrition (Fig. 17-10). It has been suggested that higher nitrogen fertility levels stimulate leaf growth causing the potential host plant to outgrow or avoid severe disease development (75, 309). Regardless of the detrimental or beneficial effects of high nitrogen fertility levels, the recovery rate from disease injury is more rapid at higher levels of nitrogen nutrition.

Other nutrients. The potassium, phosphorus, calcium, and iron nutritional levels are also factors in turfgrass disease development. Adequate potassium nutrition is important in reducing the disease proneness of turfgrasses. For example, high potassium levels apparently reduce the incidence of brown patch (69, 93), dollar spot (256), *Fusarium* patch (156, 269), *Helminthosporium* species (128, 192), *Ophiobolus* patch (155, 156, 157), and red thread (157). Optimum phosphorus levels stimulate root development and reduce the susceptibility of turfgrasses to

Figure 17-10. Increased proneness of red fescue to red thread when grown under a low nitrogen fertility level (0.15 lb per 1,000 sq. ft. per month on the left) compared to a high nitrogen fertility level (0.5 lb per 1,000 sq. ft. per month on the right). (Photo courtesy of N. Jackson, University of Rhode Island, Kingston, R.I.)

seedling damping-off diseases. High calcium levels tend to reduce the susceptibility of turfgrasses to certain diseases including *Pythium* blight (273) and red thread (278). Applications of iron are reported to reduce the severity of *Fusarium* patch.

Soil reaction. The degree of soil acidity can affect the pathogen directly and, in turn, influence turfgrass disease development. Most parasitic soil fungi are favored by an acidic soil pH of less than 6 (33, 124). Neutral and alkaline soil pH's increase the incidence of a few diseases such as *Ophiobolus* patch and *Fusarium* patch (333, 334)

There is also an indirect influence of the soil reaction on the disease susceptibility of turfgrasses. The turfgrass root system is restricted under extremely alkaline or acidic soil conditions and the availability of essential nutrients is reduced to the extent that the plant is weakened and susceptible to fungal infections (273). The soil reaction also influences the activity of saprophytic soil organisms that directly or indirectly affect the pathogen population.

Irrigation. The proper frequency and intensity of irrigation should provide a controlled shoot growth rate that enables the turf to resist and rapidly recover from disease. Applying an excessive amount of water or irrigating too frequently increases turfgrass disease susceptibility in several ways (82, 254, 360). Cells of the leaf and stem tissues have thinner walls and increased succulence, which permit fungal hyphae to penetrate more readily. Also, the atmospheric water vapor content within the turfgrass community tends to be higher and more favorable for fungal activity. Finally, soil waterlogging due to excessive water applications causes (a) a reduced root system, (b) thinning, and (c) decreased turfgrass vigor.

The timing of irrigation is also a factor influencing disease development (254). If possible, turfs should be irrigated when the evaporation of water droplets from the shoots is most rapid. Irrigation during the evening usually permits the water droplets to remain on the leaves for an extended period of time compared to early morning irrigation. As a result, the likelihood of spore germination, growth, and penetration into the host is increased (82).

Excessive water applications that saturate the soil cause an increased frequency and intensity of leaf exudate formation. Early morning syringing is beneficial because it removes the leaf exudates and disperses dew droplets so that the leaf surface dries more rapidly (133, 318). Thus, syringing can be used to minimize conditions favorable for spore germination and mycelial growth. Midday syringing, which lowers turfgrass temperatures, can be used to reduce the likelihood of infection from high temperature pathogens.

Thatch. A thick thatch layer offers a very favorable microenvironment for the saprophytic development of fungal pathogens, particularly if maintained in a moist condition through irrigation or poor drainage. Facultative parasites, such as *Fusarium*, *Pythium* and *Rhizoctonia* species, are capable of multiplying on thatch in competition with other organisms. Certain fungi, such as *Helminthosporium* species, can produce large quantities of spore inoculum while growing on dead thatch material (185). Other fungi, such as *Corticium* and *Sclerotinia*, survive in the thatch by means of dormant structures. Thus, thatch contributes to increased disease development.

The formation of excessive thatch layers can be reduced through topdressing. Topdressing can also provide beneficial bacteria and fungi that (a) stimulate thatch decomposition and (b) are antagonistic to the parasitic fungi. This is true only if the topdressing material has not been fumigated or has been composted for an extended period of time following fumigation to permit development of a favorable fungi and microorganism population.

Drainage, soil modification, and cultivation. Adequate soil drainage is particularly effective in reducing the inoculum production and mycelial growth of water mold fungi, such as *Pythium* species, and may contribute to reduced disease development. Removal of excess water can be achieved through proper soil modification, surface contouring, and use of a subsurface drainage system. Such practices also encourage the development of a deep root system and healthy, vigorous turf. Soil cultivation practices can be utilized as needed under intense traffic to ensure an adequate infiltration rate.

Preplant Disease Control

The usefulness of preplant chemical disease control may be questioned since inoculum for most of the major turfgrass pathogens can be readily reintroduced into established turfs by means of wind, water, equipment, and humans. Preplant disease control can be effectively employed in special situations where a build-up in the soil fungal population would prevent establishment of the turf. Also, certain soil inhabiting fungi, such as the fairy ring causing pathogens are more easily controlled through preplant soil fumigation (161). Preplant disease control practices generally involve fumigation or heat treatment. Chemical fumigation can be achieved with such materials as dazomet, metham, and methyl bromide.

Use of Resistant Cultivars

From the standpoint of cost and effort, the preferred method of disease prevention involves the use of disease-resistant turfgrass species or cultivars. A plant is referred to as immune, resistant, or susceptible to a pathogen. Total **immunity** is a condition where the plant is not susceptible to infection caused by a pathogen even under the most favorable conditions. A completely **susceptible** plant is one that is unable to resist the effects of a pathogen and is severely damaged.

Varying degrees of **resistance** occur between the extremes of immunity and complete susceptibility (Fig. 17-11). Resistance is determined by various heritable internal and external plant characteristics. Resistance may involve (a) the localization and isolation of the pathogen at the point of entry by means of mechanical barriers that the fungus cannot penetrate, (b) detoxification of toxic substances secreted by the pathogen, (c) restricting dissemination of the pathogen by inhibiting its reproductive processes, or (d) an environmental requirement for optimum growth of the host cultivar that is not favorable for pathogen development.

Physical barriers that impair penetration of the pathogen into the host tissues include such structural defenses as the (a) thickness of the waxy cuticle layer on the leaf surface, (b) thickness and toughness of the epidermal walls, and (c) stomatal structure and duration the stomata are open. Internal structural barriers may also be formed by host cells as a result of pathogen penetration.

Most resistance mechanisms associated with turfgrass pathogens involve an internal biochemical defense rather than the structural type. One mechanism involves detoxification of toxins produced by the pathogen. In this case the toxins are inactivated by binding or are rapidly metabolized by the resistant host plant. Another type of resistance involves inhibition of fungal growth and reproduction by the host plant. Germi-

Figure 17-11. *Helminthosporium* leaf spot injury to the susceptible Kenblue Kentucky bluegrass (left) compared to no injury to the resistant Merion Kentucky bluegrass (right). (Photo courtesy of R. Funk, Rutgers University, New Brunswick, New Jersey.)

nating spores may penetrate the host but are not able to infect the host tissues (277). Mechanisms involved in this type of defense include the (a) formation of biochemical inhibitors by the host plant that restrict pathogen growth, (b) potential host lacking certain organic or mineral substances essential for pathogen growth, or (c) production of fungi-toxic substances by the host plant.

Turfgrass species and cultivars vary greatly in resistance to various diseases. Included are brown patch (267, 285, 293), dollar spot (13, 219, 309), *Fusarium* blight (18, 72), *Fusarium* patch (86, 254), *Helminthosporium* species (20, 21, 134, 138, 170, 173, 236, 272, 277), *Ophiobolus* patch (162), *Pythium* blight (141, 272), red thread (132), rust (14, 38, 39, 40, 140, 230, 235), stripe smut (92, 148, 149, 171, 186, 233), and *Typhula* blight (23, 73, 355). The reported resistance of a turfgrass cultivar to a given pathogen may also vary from region to region due to the modifying influence of environmental conditions and the occurrence of different races of a disease-causing organism (149, 171). There is no species or cultivar available that is highly resistant to all the major turfgrass diseases. A primary objective of most active turfgrass breeding programs is the development of an acceptable level of resistance to the major turfgrass diseases.

Controlling Turfgrass Diseases

A key prerequisite in controlling turfgrass diseases is the use of cultural practices that ensure a vigorous, healthy turf having minimal susceptibility to fungal infection, maximum host resistance, and adequate recuperative potential from disease injury. Judicious irrigation and fertilization practices that avoid moisture or nutritional deficiencies are particularly important.

Identification. The disease must be recognized and the causal organism identified before the appropriate disease control practice can be initiated. Both aboveground and belowground turfgrass plant parts should be examined for disease symptoms and the presence of characteristic fungal structures. Use of a hand lens is helpful, particularly when reproductive structures are present. A knowledge of (a) turfgrass cultivar susceptibility, (b) environmental and cultural conditions, and (c) time of year when each disease commonly occurs provides supplemental information in identifying the particular disease problem. Sometimes positive identification of the causal organism can be accomplished only by culturing mycelial strands on nutrient agar in a petri dish. The characteristic reproductive structures formed during growth of each fungal species are then used for microscopic identification of the pathogen.

Disease control can be achieved through physical, biological, or chemical methods. The specific control procedure selected varies with the particular type of causal organism associated with the disease. Chemical methods are the most reasonable and effective means currently available for controlling diseases in established turfs.

Physical Control

Physical treatment may involve exposure to heat, cold, or certain types of radiation. These methods of disease control have not been utilized to any extent on established turfs. Soil sterilization by heat treatment can be practiced for the control of soil-borne pathogenic organisms present in soils prior to turfgrass establishment or in topdressing or soil modification materials.

Biological Control

Biological methods involve the control of pathogenic fungi by using other fungi or viruses that either parasitize or antagonize the causal organism. As yet, this method has not been of any value in controlling turfgrass diseases.

Chemical Control

Chemical methods of turfgrass disease control involve the use of chemical compounds called fungicides that (a) are lethal to the fungi or (b) inhibit pathogen germination, growth, and multiplication (Fig. 17-12). An effective turfgrass fungicide must be toxic to the pathogen without causing turfgrass injury. Certain fungicides are toxic to a wide range of turfgrass pathogens, while others have specific activity for only one or a few pathogens (76). Most turfgrass fungicides are used to prevent infections by pathogens rather than to cure an established disease. The former is referred to as **preventive disease control**, while the latter is termed **curative disease control**. Fungicides can also be effectively utilized to control fac-

Figure 17-12. Fungicidal control of *Sclerotinia* dollar spot (center) without injury to the Toronto creeping bentgrass (Photo courtesy of J. M. Vargas, Michigan State University, East Lansing, Michigan).

ultative parasites present in the thatch and, thus, to reduce or minimize a potential source of inoculum for disease development.

Nonsystemic fungicides achieve disease control by (a) killing the spores, mycelia, and sclerotia present on the turfgrass surface at the time of fungicide application and (b) killing inoculum subsequently deposited on the turfgrass surface during the active residue period of the fungicide. Thus, effective control is only achieved on those plant parts to which the nonsystemic fungicide is applied and only for the period of time that the fungicide remains on the plant surface in an active condition. Certain fungicides can kill the fungi even after they have penetrated the host plant.

In the past, most fungicides utilized for controlling turfgrass diseases that infect aboveground plant parts have been of the nonsystemic type. The residual activity of the nonsystemic fungicide is gradually lost by (a) removal during mowing, (b) degradation of the active ingredient, and (c) being washed from the shoots by irrigation or precipitation. In addition, unprotected shoot growth is constantly being formed within the turfgrass community. As a result, preventive, nonsystemic fungicides must be applied at intervals to maintain adequate protection on the turfgrass leaf surfaces.

Since control with nonsystemics is achieved by a protective coat of a toxic compound on the plant surface, the fungicide must be applied uniformly over the surface of the plant. An effective fungicide must be soluble enough to be absorbed at a toxic level by the pathogen and still have a sufficient degree of insolubility to remain at a toxic level on plant surfaces for an extended period of time. As a result, spreader-stickers are frequently utilized in the fungicide formulation. The interval between applications of a nonsystemic fungicide usually varies from 7 to 14 days depending on the (a) stability and tenacity of the fungicide, (b) particular fungal pathogen involved, (c) time of year, (d) frequency and duration of rain and irrigation, and (e) growth rate of the turf.

The preferred type of turfgrass fungicide is one that is absorbed and translocated throughout the plant. This type, called a **systemic fungicide**, can prevent subsequent pathogen infections and is also capable of eradicating fungi that have already infected the host plant. A systemic fungicide is preferred to a nonsystemic, providing the degree of disease control is comparable and it is not phytotoxic to the turf. Several promising systemic fungicides for turfgrass disease control are finally being developed.

Table 17-5

CHEMICAL NAMES OF 21 COMMONLY USED TURFGRASS FUNGICIDES

Name of fungicide		Nonsystemic, N or systemic, S
Common	Chemical	
Anilazine	2, 4-dichloro-6-*o*-chloro-anilino-*s*-triazine	N
Benomyl	Methyl 1-(butylcarbamoyl)-2-benzimidazole-carbamate	S
Cadmium compounds	Cadmium carbonate Cadmium chloride Cadmium sebacate Cadmium succinate	N
Captan	*N*-(trichloromethylmer-capto)-4-cyclohexene-1,2-dicarboximide	N
Chloroneb	1,4-dichloro-2,5-dimethoxybenzene	N (S)
Cycloheximide	Cycloheximide	N (S)
Daconil	2,4,5,6-tetrachloroisoph-thalonitrile	N
Dexon	*p*-dimethylaminobenzene diazo sodium sulfonate	N
Difolatan	*cis*-*N*-[(1,1,2,2-tetrachloro-ethyl)thio]-4-cyclohexene-1,2-dicarboximide	N
Folpet	*N*-(trichloromethylthio) phthalimide	N
Karathane	Dinitro(1-methyl heptyl) phenyl crotonate	N
Koban	5-ethoxy-3-trichloromethyl-1,2,4-thiadiazole	N
Mancozeb	Coordination product of zinc ion and maneb	N
Maneb	Manganous ethylenebisdi-thiocarbamate	N
Mercury, inorganic	Mercurous chloride + mercuric chloride	N
organic	Cyano (methylmercuril) guanidine	N
PCNB	Pentachloronitrobenzene	N (S)
PMA	(Acetato) phenylmercury	N
Sulfur	Sulfur	N
Thiram	Tetramethylthiuram disulfide	N
Zineb	Zinc ethylenebisdithio-carbamate	N

Applying fungicides. Turfgrass fungicides are available as sprays, granules, and dusts. Dust formulations are usually not used because they are objectionable to apply and coverage is not as good as with sprays. Sprays are preferred, particularly for active fungi, because more uniform coverage of the aboveground shoots can be achieved. The sprays used are primarily wettable powders, although emulsifiable concentrates are available. The volume of water used to apply fungicides should be increased when the objective is also to kill potential sources of inoculum in the thatch layer. Effective application of fungicidal sprays requires the use of a properly adjusted and operated sprayer.

Granular formulations of fungicides used for foliar disease control are generally inferior to sprays since coverage of the aboveground plant parts is poor. Granular materials have been effective on facultative parasites that are active in the thatch layer. Typical examples are *Typhula* and *Fusarium* that grow saprophytically on the thatch and produce spores as inoculum for leaf infection. Fungicides, such as captan, dexon, PCNB, and thiram, are applied to the soil as drenches or granulars for controlling damping-off, seedling blight, and crown or root rot diseases.

Types of fungicides. A number of fungicides are available for turfgrass disease control. Inorganic compounds that exhibit varying degrees of fungicidal activity include cadmium, copper, manganese, mercury, sulfur, and zinc. Groups of organic compounds that are particularly important as fungicides include the organic mercuries, organic sulfur, benzene, dithiocarbamate, and heterocyclic compounds (Table 17-5).

Certain fungicides can be quite toxic to man and animals if improperly used. The use of fungicides containing cadmium or mercury should be minimized if alternate fungicides are available that give comparable disease control and are biodegradable. This practice reduces the pollution problem caused by pesticides having a long residual life. The use of mercury and cadmium fungicides is forbidden by law in certain areas. It is very important to comply with the directions on the label and avoid repeated or prolonged contact with the skin as well as inhalation of dust, mist, or vapors. Fungicides must be used at the recommended rate. Higher rates are usually phytotoxic to the turf, while lower rates frequently result in ineffective disease control.

Turfgrass Diseases

Twenty-five of the more important turfgrass diseases, their characteristics, and chemical controls are presented in Table 17-6. For more detailed descriptions of the causal organisms involved, the reader is referred to the references included at the end of this chapter (70, 160, 199, 271, 341). Other less common diseases that occasionally cause problems include (a) a snow mold (*Sclerotinia borealis* Bub. and Vleug.) (167), (b) *Helminthosporium stenospilum* Drechs. (134, 138), (c) *Cercospora* leaf spot (*Cercospora* spp.), (d) *Septoria* leaf spot (*Septoria* spp.), and (e) frost scorch (*Sclerotium rhizoides* Auersw.).

Table 17-6

CAUSAL ORGANISM, SYMPTOMS, AND CHEMICAL CONTROLS FOR 24 COMMON TURFGRASS DISEASES

Disease	Causal organism	Turfgrass disease symptoms	Conditions favoring disease development	Chemical control
Anthracnose (371)	Colletotrichum graminicola Ces. Wils.	Yellowish to brown leaf lesions; 2–12 in. irregular patches	Low nitrogen (N) levels and moisture stress	None
Brown patch (6, 33, 73, 82, 84, 194, 266, 267, 285, 287, 293, 314, 324, 380)	Rhizoctonia solani Kuhn [Pellicularia filamentosa (Pat.) D. P. Rogers]	Leaves turn light brown; tan, circular patches up to 3 ft in diameter	High N level and irrigation; wet, warm weather	Anilazine, daconil, mancozeb, mercuries, thiram, or zineb
Copper spot	Gloeocercospora sorghi Bain and Edgerton	Reddish leaf lesions; 1–3 in. copper colored patches	Cool, wet weather; acidic soils	Anilazine, cadmiums, daconil, or thiram
Dollar spot (13, 28, 73, 76, 135, 169, 203, 309, 316, 332, 369)	Sclerotinia homoeocarpa F. T. Bennett	Light tan lesions banding the leaf; 1.5–2.5 in., bleached spots	Low N level, dry soil; high humidity and moderate temperatures	Anilazine, benomyl, cadmiums, cycloheximide-thiram, or daconil
Fairy ring (5, 130, 131, 161, 166, 174, 241, 322)	Many†	Dark green rings of 2–15 ft; sometimes dead in center; occasional mushrooms	Decomposing organic matter	Fumigation
Fusarium blight (18, 19, 72)	Fusarium roseum (LK) Snyd. and Hans. or F. tricinctum (Cda.) Snyd. and Hans.	Tan blotches starting at leaf tip; circular "frog-eye" patches up to 2 ft	Moderate, dry weather and thatch	Benomyl
Fusarium patch (pink snow mold) (26, 86, 163, 164, 240, 254, 328)	Fusarium nivale (Fr.) Ces. [Calonectria graminicola) Berk. and Ber.) Wr.]	Pink mycelium on leaves; 1–2 in. tan circular patches; or white mycelial mass on leaves, white to pink, circular patch up to 2 ft	Cool, wet weather or cold, moist conditions under a snow cover	Cadmiums, benomyl, daconil, mercuries, or PMA
Gray leaf spot (139)	Piricularia grisea (Cke.) Sacc.	Oblong gray to ash-brown spots with purple-brown margins	High N level; hot, wet weather	Thiram

Disease	Pathogen	Symptoms	Conditions favoring disease	Chemical control
Helminthosporium blight or netblotch	Helminthosporium dictyoides Drechsl.	Brown leaf lesions each net-like; irregular, brown patches	High N levels and excessive irrigation	Anilazine, captan, daconil, difolatan, folpet, mancozeb, PMA, or zineb
Helminthosporium leaf blotch (134, 138)	Helminthosporium cynodontis Margi.	Brown to tan leaf lesions; straw colored, irregular patches	Lesions develop in cool, moist weather	Anilazine, cycloheximide-thiram, organic mercuries, or thiram
Helminthosporium leaf spot (185)	Helminthosporium sorokinianum Sacc. ex. Sorokin.	Lesions 0.5-1.5 mm with tan to brown center and darker margin; thinning and browning of turf	High N level, close cutting, and excessive irrigation	Anilazine, captan, cycloheximide-thiram, daconil, folpet, PMA, or zineb
Melting out (68, 73, 103, 170, 173, 272, 275, 277, 361, 369)	Helminthosporium vagans Drechsl.	Whitish-tan lesions with purple margins; thinning and browning of turf	High N level, close cutting, and excessive irrigation	Anilazine, captan, cycloheximide-thiram, daconil, folpet, mancozeb, PMA, or zineb
Mushrooms and puffballs (174)	Many genera†	White to tan-brown fruiting structures	Cool, wet weather; excessive irrigation	Fumigate
Ophiobolus patch (162, 333)	Ophiobolus graminis Sacc.	Light-brown rings up to 2 ft; weeds may invade dead center	Moderate temperatures; moist, thatched conditions	Organic mercuries
Powdery mildew (150)	Erysiphe graminis D.C.	Fine, white to grayish, cobweb-like mycelial growth on the leaves	Cool, humid, shaded conditions; high N level	Benomyl, cycloheximide-thiram, karathane, or sulfur
Pythium blight (cottony blight) (137, 141, 231, 263, 269, 272, 273, 363)	Pythium ultimum Trow. or P. aphanidermatum (Edson) Fitzpatrick	Water-soaked leaves; small, light-brown, irregular spots or streaks	Hot, humid, wet, thatched conditions	Chloroneb, dexon, or koban
Red leaf spot	Helminthosporium erythrospilum Drechsl.	Tan to brown lesions with reddish-brown margins; tanning and thinning of turf	Warm, wet weather; high N level	Anilazine, captan, cycloheximide-thiram, or daconil
Red thread (pink patch) (27, 124, 132, 165, 330, 331)	Corticium fuciforme (Berk.) Wakef.	Red mycelial growth on water-soaked lesions; irregular patches up to 2 ft	Deficient N level, humid conditions, and moderate temperatures	Anilazine, cadmiums, cycloheximide-thiram, or daconil
Rusts (40, 235)	Puccinia spp.*	Elongated, reddish leaf pustules; turf reddish brown and thin	Deficient N level; moderate, moist conditions	Cycloheximide-thiram or zineb

591

Table 17-6 (Cont.)

Disease	Causal organism	Turfgrass disease symptoms	Condition favoring disease development	Chemical control
Slime mold	*Physarium cinereum* (Batsch) Pers.	White to grayish, slimy, irregular shaped mass up to 12 in. covers the leaves	Moderate temperatures and wet conditions	Not required; wash and brush mycelium from leaves
Smut, flag	*Urocystis agropyri* (Preuss) Schrot.	Long, narrow, gray to black streaks on the leaves, curling of leaf tip; browning and thinning of turf	Moderate temperatures; thatched, dry soil conditions	Benomyl
stripe (37, 91, 171, 188, 232, 248, 348)	*Ustilago striiformis* (West.) Niessl	"	"	"
St. Augustine decline virus (SADV) (12)	A virus	Chlorotic mottling of leaf blades initially; shortened stolon internodes; thinning that progresses to complete kill after 3 to 4 years	Low nutritional level, moisture stress, or improper culture	None
Typhula blight (gray snow mold) (1, 23, 80, 299, 316, 355, 361)	*Typhula itoana* Imai, *T. idahoenis* Remsberg, or *T. ishikariensis* Imai	Light gray mycelium on leaves; whitish-gray, circular patches up to 2 ft in diameter	Cold, humid conditions especially under a snow cover	Cadmiums, chloroneb, or mercuries
Zonate eyespot (101, 102)	*Helminthosporium giganteum* Drechsl.	Brown to purplish, elongated leaf lesions; browning and thinning of turf	Warm, wet weather	Anilazine, cycloheximide-thiram, organic mercury, or thiram

* Leaf rust—*P. brachypodii* Otth. var, *poae-nemoralis*.
Stem rust—*P. graminis* Pers.
Stripe rust—*P. striiformis* West (39).
Crown rust—*P. coronata* Cda.
Bermudagrass rust—*P. cynodontis* Deam.

Zoysiagrass rust—*P. zoysiae* Diet. (14, 140, 168, 230, 234).
St. Augustinegrass rust—*P. stenotaphri* Cumm.
† *Agaricus, Collybia, Hygrophrous, Lepiota, Lycoperdon, Marasmius. Psalliota, Scleroderma,* and *Tricholoma.*

Other diseases with unknown causes. There are three diseases causing significant damage to turfgrasses for which a specific causal organism has not been ascertained. A disease that has been called **spring dead spot** has become a serious problem on bermudagrass turfs in the transitional zone and cooler portions of the warm humid regions. Most bermudagrass cultivars are affected by this disease. Spring dead spot appears as irregular, circular dead spots up to 3 ft in diameter. Shoots, rhizomes, and roots within the affected area are completely killed. Damage to bermudagrass turfs is not evident until winter dormancy is broken and shoot growth reinitiated. Recovery of the affected area must occur from surrounding, unaffected plants. The affected spots reoccur in the same locations each year and gradually enlarge. Development of an effective control has been impaired by an inability to identify the causal organism.

The second disease is a **winter crown rot** caused by an unidentified low temperature Basidiomycete fungus. The turfgrass injury symptoms are similar to typical snow mold damage. As the turf breaks dormancy in the spring, irregular shaped patches initially appear yellowish and gradually deteriorate to a straw color (65). A light gray mycelial growth may be evident on the leaves (45). Individual patches up to 1 ft in diameter may occur in sufficient density that they coalesce causing extensive damage to large areas of turf. Damage and crown rot of individual turfgrass plants occur while in the dormant stage. Damage is most severe under a snow cover at temperatures slightly below freezing (240).

Yellow tuft is another disease for which a specific causal organism has not been determined (286, 345, 346). Symptoms of the disease appear in close cut turfs as small, yellowish tufts of turf having a more rapid rate of vertical shoot growth. A detailed examination shows many extremely fine shoots and leaves arising from a single node on a stolon. General turfgrass symptoms appear as a yellowish mottling or spotting. Yellow tuft symptoms are most commonly observed during the spring and fall periods. The disease usually occurs on bentgrass greens with scattered reports of a similar problem on bermudagrass greens. Varying degrees of resistance to yellow tuft have been observed among the creeping bentgrass cultivars. Yellow tuft is objectionable from the standpoint that it disrupts the uniformity of turfgrass color and can also create an unevenness that disturbs ball roll on greens. Nematodes, viruses, and bacteria have all been suggested as possible causal organisms of yellow tuft but no positive, reproducible causal agent has been identified.

Turfgrass Nematodes

A **nematode** is a microscopic animal that can live saprophytically in water and soil or as a parasite of plants and animals. Most nematode species are beneficial since they feed on fungi, bacteria, and insects that are potentially damaging to turfgrasses. The discussion in this section is devoted to the small group of nematodes that have been shown to be parasitic or are generally considered as potential pathogens of turfgrasses. The knowledge of parasitic turfgrass nematodes is quite limited compared to most turfgrass pests. Most turfgrass nematode investigations have been conducted since 1960. A determination of the extent and type of injury caused by nematodes has been made on a relatively limited number of turfgrass species, most of which are adapted primarily to warm humid climates.

Characteristics of Parasitic Nematodes

Parasitic plant nematodes are eel-like microscopic roundworms having translucent, unsegmented bodies possessing no legs or other appendages. They generally range from 350 to 2500 μ in length and are round in cross section with a diameter of 15 to 35 μ. Nematodes cannot be observed without the use of a microscope because of the nearly transparent nature and very small size.

Nematodes possess primitive but relatively efficient internal organs. The digestive system is in the form of a hollow tube extending from the mouth through the general length of the body. The nervous and reproductive systems of plant nematodes are relatively well developed. Respiratory and circulatory systems are lacking. Reproduction is through eggs and usually involves sexual means, although some species reproduce by parthenogenesis.

Approximately 3 to 4 weeks is required to complete the life cycle of a nematode under favorable conditions. Egg production and laying usually occur in the soil. The eggs hatch into larvae that possess most of the characteristics and structural features of an adult. As the larvae grow in size, they undergo a series of four larval stages, each of which is terminated by a molt. Nematodes depend on a susceptible host for food during the infective larval stages and as adults. If an acceptable host is not present, the nematode starves to death and does not complete its life cycle.

Most ectoparasitic plant nematodes live in the soil for a major portion of their life cycle. Eggs of certain nematode species may lie dormant in the soil for an extended period of time. In the later larval stages the nematodes are found in association with underground roots and stems upon which they feed. The nematode population is usually concentrated in the upper 6 in. of the soil profile. The greatest nematode population found in the soil is usually in the immediate vicinity of host plant roots or underground stems.

The lateral movement of nematodes in the soil seldom exceeds a fraction of a foot per year. Agents for the dissemination of nematodes include equipment, irrigation water, animals, dust particles suspended in the air, vegetative plant parts, and flood or drainage waters.

Effects of Parasitic Nematodes on Turfgrasses

Parasitic plant nematodes feed on living plant tissues of turfgrasses by puncturing host plant cells with a hollow, needle-like structure called a **stylet**. Once achieved, the stylet functions as the structure through which the extracorporeal enzymes are injected into the cells. The enzymes predigest the cell contents, which facilitates withdrawal of the contents through the stylet into the nematode for ingestion and assimilation. The enzymes also aid in penetrating the plant cell wall.

Each nematode species possesses a characteristic feeding position. Location of the nematode body during feeding may vary from an ectoparasitic to endoparasitic orientation. **Ectoparasitic** nematodes feed only on cells located near the surface of the tissue and have a major portion of their body outside the host (Fig. 17-13). Turfgrass damage is more commonly associated with the ectoparasitic nematodes. **Endoparasitic** species actually penetrate directly into the plant tissue and feed from within the host.

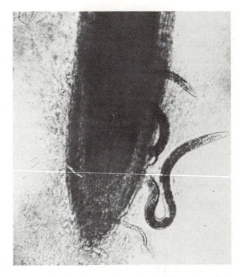

Figure 17-13. A nematode feeding in the ectoparasitic orientation on a grass root. (Photo courtesy of V. G. Perry, University of Florida, Gainesville, Florida).

Direct damage caused by the mechanical feeding of nematodes is seldom of significance. Most nematode damage is associated with the enzymes secreted into the plant cells during the feeding process. The secretions cause the formation of lesions, swelling, galls, or death of meristematic areas. Water and nutrient conducting tissues may also be destroyed and vital plant processes disrupted.

Parasitic nematodes can cause direct damage to plants by functioning as the only active pathogen or as part of a complex with an infectious fungus. The nematode-fungus complex produces a pathological effect that is far greater than either organism can produce alone. Complexes of nematodes with fungi have been reported for such fungal organisms as *Fusarium, Rhizoctonia,* and *Pythium.* The mechanism of this complex involves two aspects. One is mechanical wounding by the nematode that provides an avenue for penetration of the fungus into the host. In addition, it has been suggested that nematodes may lower the host resistance to a particular fungus. Another facet to the pathological activity of nematodes is its role in the transmission of viruses through the soil. The dagger, needle, and stubby root nematodes can function as vectors in virus transmission.

Nematode injury symptoms. Initial symptoms that may indicate nematode activity include stunting of the roots and the formation of root lesions, root knots, or excessive root branching. The specific type of root symptom varies depending on the species of (a) parasitic nematode involved and (b) host plant. Root lesions characterized by discoloration and deterioration frequently result from nematode feeding activity. Digestive secretions may cause a stimulation or modification in cell growth that results in a root enlargement called a **root knot** or gall. The development of numerous lateral roots may be associated with nematode feeding. Also associated with root nematode damage is the secondary invasion of plant pathogenic or saprophytic fungi and bacteria.

Aboveground visual symptoms of nematode damage are generally associated with the effects of a restricted, nonfunctioning root system. Typical shoot symptoms include a light green to yellow chlorosis and a gradual reduction in vigor as well as an inability to withstand such stresses as heat, cold, drought, and disease. Eventually, the leaf blades die back from the tips and a general thinning of the turf results. Chlorosis and thinning occur in a spotty or random nature. Nematode injury to shoots is most likely to be observed (a) during periods of moisture stress, (b) during disease attacks, or (c) on turfs maintained at a low fertility level.

Absolute determination of a nematode problem from visual symptoms or a soil nematode analysis is very difficult and erratic at best. Visual symptoms such as shoot chlorosis, a reduction in vigor and tolerance to stress, and root browning can be caused by many other factors in addition to parasitic nematodes. High populations of a number of different species of parasitic nematodes can be found in turfgrass soils (41, 61, 97, 126, 153, 223, 281, 282, 288, 344, 347, 354). However, significant damage to the turf may or may not be associated with high parasitic nematode populations (125). As a result, one of the best indicators of a parasitic nematode problem on turfs is a positive response to a nematicide application. Previous history of nematode problems associated with a particular soil, climatic condition, or turfgrass species can also be used as an indicator of a potential nematode problem.

Control of Turfgrass Nematodes

The presence of a significant population of parasitic plant nematodes should be confirmed before initiating nematode control practices. The normal procedure involves the collection of a moist soil-root sample representative of the turfgrass area where a nematode problem is suspected. The soil-root sample should be taken from the margins of the affected area to a depth of 4 to 6 in. The sample should be submitted to a reputable laboratory for analysis of the numbers and specific kinds of parasitic plant nematodes present. The analysis procedure involves the separation of nematodes from the soil by a washing and screening technique followed by microscopic examination for parasitic nematodes. Nematode control practices should be initiated if the presence of a significant population of parasitic nematodes is confirmed by the laboratory analysis and symptoms of nematode damage have been observed to the extent that turfgrass quality or vigor is impaired. Both preventive practices and chemical controls are available for solving potential or existing nematode problems.

Nematode Prevention

The first step in nematode prevention is the use of cultural practices that ensure an actively growing, vigorous turf. A healthy turf can frequently support a relatively high population of parasitic nematodes without any foliar symptoms. The type of nitrogen carrier used is reported to affect the nematode population. Activated sewage sludge caused a reduction in the nematode population when compared to inorganic nitrogen carriers (180, 283). It is thought that the organic nitrogen carrier enhances the development of predators that kill or are antagonistic to the parasitic nematodes. The three major methods of turfgrass nematode prevention are (a) sanitary practices, (b) preplant nematode control, and (c) use of resistant turfgrass cultivars.

Sanitary practices. Parasitic turfgrass nematodes can be introduced into an uninfected turfgrass area by (a) rhizomes, stolons, or sod pieces; (b) soil or topdressing; (c) contaminated turfgrass equipment; (d) the feet of humans or animals; and (e) surface drainage water. Sanitary practices involve the use of turfgrass stolons, sprigs, plugs, or sod that are free of parasitic nematodes. The control of nematodes in bermudagrass vegetative propagation material involves soaking sod plugs or cores in hot water at temperatures of 120 to 125°F for 30 to 45 min (181). Similarly, materials used in soil modification or topdressing should be free of parasitic nematodes. Heat treatment, fumigation, or sterilization can be used for controlling parasitic nematodes in soils.

All vegetative plant parts and soil should be removed from maintenance equipment, tires, and shoes before moving from a nematode infested turfgrass area to a noninfested area. This is particularly important for cultivation equipment involving soil tillage, coring, slicing, or spiking. Equipment, tires, and shoes can be cleaned by thorough washing or steaming. Sanitary practices for preventing the introduction of parasitic nematodes into a turfgrass area are particularly important in those regions where there has been a history of nematode injury to turfgrasses.

Preplant control. Preplant nematode control practices should be considered on soils infested with parasitic nematodes. Nematode control can be achieved by fumigation or heat treatment prior to turfgrass establishment. Chemicals utilized for preplant control of parasitic nematodes include chloropicrin, dazomet, metham, and methyl bromide. The application rate and procedures should be designed to achieve control throughout the upper 8 to 10 in. of the soil profile. Complete eradication of the nematode population is seldom achieved but the population can be reduced by 80 to 90%.

Resistant cultivars. The preferred method of minimizing parasitic nematode problems on turfs is through the development and use of resistant turfgrass cultivars. This approach is relatively inexpensive and convenient. Unfortunately, little emphasis has been placed on breeding nematode-resistant cultivars. Turfgrass species vary in susceptibility to the various parasitic nematodes. Varying degrees of resistance to certain species of parasitic nematodes have been reported for certain bahiagrass, bermudagrass, St. Augustinegrass, and zoysiagrass cultivars (147, 259, 306, 327). No turfgrass cultivars are available that are resistant to all species of parasitic nematodes.

Chemical Control

An effective nematicide must provide control against all species of parasitic nematodes without significant injury to the desirable turfgrass species. Nematicides that can be applied to established turfs include (a) DBCP (1,2 dibromo-3-chloropropane) and (b) diazinon [O,O-diethyl O-(2-isopropyl-4-methyl-6-pyrimidinyl) phosphorothioate] (34, 239, 283, 284, 291, 297, 343, 373) (Fig. 17-14). Two other highly toxic organophosphate nematicides for use on established turfs are (a) mocap (O-ethyl S,S-dipropyl phosphorodithioate) and (b) O,O-diethyl O-p(methyl sulfinyl) phenyl phosphorothioate. These latter two are not registered for use on home lawns. The available turfgrass nematicides are not very effective on certain genera of endoparasitic nematodes (126, 343).

Nematicide applications can be made to the surface as an emulsified liquid applied with a low pressure sprayer or in a granular formulation applied through a spreader. The nematicide should be throughly watered in immediately after application to (a) minimize foliar burn to the turf and (b) move the nematicide into the soil-root zone. A minimum of 0.2 in. of water should be applied with up to 1 in. suggested to ensure maximum effectiveness. Movement and uniform distribution of the nematicide by soil drenching is more readily achieved in well-aerated porous soils.

Nematicide applications should be timed so that a high percentage of the nematode population is in the larval or adult stage. The percentage of nematodes persisting in the more resistant egg stage is lower at soil temperatures above 60°F. The available nematicides provide a substantial reduction in the population of parasitic nematodes but eradication is seldom achieved. Thus, repeat applications are necessary whenever nematode injury symptoms become evident on the turf. A minimum of one nematicide application per year is usually required where nematode problems persist.

Figure 17-14. The high quality of a Tifgreen bermudagrass turf treated with a nematicide (above) compared to the untreated, chlorotic turf (below) which is infested with parasitic nematodes. (Photo courtesy of V. G. Perry and G. C. Horn, University of Florida, Gainesville, Florida).

Nematode Pests of Turf

The parasitic plant nematodes belong to the *Nemathalminthes* phylum in the *Nematoda* class. Fifteen genera of nematodes reported or suspected of being parasitic on turfgrasses are included in Table 17-7. Uncertainty as to the parasitic nature of some nematodes is due to the lack of pathogenicity studies needed to definitely implicate a nematode species as a parasite of turfgrasses. Other genera of nematodes that are occassionally found in association with turfgrasses and may be pathogenic include the *Aphelenchoides, Ditylenchus, Eucephalobus, Hemicriconemoides, Longidorus, Panagrolaimus, Peltamigratus, Scutellonema*, and *Tetylenchus*.

Turfgrass Insects

Insects belong to the *Arthropoda* branch of the animal kingdom. Typical characteristics of the arthropods are segmented bodies, jointed appendages, and an exoskeleton. Insects comprise by far the largest phylum within the animal kingdom. The *Insecta* class of the arthropods comprises all the insects and is characterized by three separate body divisions, three pairs of legs, one pair of antennae, and wings usually present.

Several other classes of *Arthropoda* are turfgrass pests or nuisances in turfs. The mites, spiders, scorpions, and ticks are in the *Arachnida* class characterized by two main body divisions, four pairs of jointed legs, wings, and the absence of antennae. The centipedes are in the *Chilopoda* class characterized by one pair of legs on each body segment, a worm-like body, and more than five pairs of legs. The millipedes are within the *Diplopoda* class and have two pairs of legs for each body segment, a worm-like body, and more than five pairs of legs. The sowbugs are classified in the *Crustacea* class and have five or more pairs of legs, a head that is merged with the thorax, and two pairs of antennae. The arthropods not included in the *Insecta* class that are mentioned above are generally a nuisance in turfgrass areas but cause little damage to the turf. An exception to this is the mite. The discussions in this section generally refer to the *Insecta* class.

Characteristics of Turfgrass Insects

Most insects are relatively small with a more or less elongated, cylindrical form that is bilaterally symmetrical. The insect body is surrounded by an exoskeleton that functions as the supporting structure and outer covering. The exoskeleton is composed of a relatively hard, nitrogenous polysaccharide called chitin. The insect body has three distinct segments known as the head, thorax, and abdomen. The **head** is located at the anterior of the body and bears the mouth parts, eyes, and antennae. Typically the eyes are compound. The mouth parts vary in type and structure depending on whether the insects feed by chewing or sucking. The

Table 17-7

A SUMMARY OF THE CHARACTERISTICS AND PARASITIC NATURE OF FOURTEEN TURFGRASS NEMATODE PESTS

| Name of nematode | | Characteristics | Feeding habit* | Turfgrass injury symptoms | | Proved host† |
Common	Genera			Roots	Shoots	
Awl	*Dolichodorus*	2.5–3 mm; long stylet	E	Stunting	Yellowing	T
Burrowing	*Radopholus*	0.5–1 mm; overlapping esophagus	I	Necrosis; rot	Chlorotic and stunted	T
Cyst	*Heterodera*	Female, pear-shaped, and white	I	White cysts (brown in soil)	Chlorotic	St. Augustinegrass (98, 99)
Dagger	*Xiphinema*	1.8 mm; stylet without knobs	E	Brown necrosis	Chlorotic	T
Lance	*Hoplolaimus*	1.5 mm; distinct stylet	E (I)	Browning	Growth restricted	Bermudagrass (98, 308)
Lesion	*Pratylenchus*	0.5–0.8 mm; overlapping esophagus	I	Brown lesions	Chlorotic	Italian ryegrass (352)
Pin	*Paratylenchus*	0.2–0.3 mm	E	Lesions	Stunted	Tall fescue (77)
Ring	*Criconemoides*	0.5 mm; stout; heavy annulation	E (I)	Necrosis and rot	Chlorotic and stunted	Carpetgrass (343)
Root-knot	*Meloidogyne*	Pear-shaped female of 0.5–1.3 mm	I	Knots and stunting	Chlorotic and stunted	Bentgrass, bahiagrass, zoysiagrass, + W (24, 179, 189, 190, 264, 292, 305, 306, 326, 343) T
Sheath	*Hemicycliophora*	1.2–1.3 mm; sheath	E	—	Chlorotic	
Spiral	*Helicotylenchus*	0.6–1 mm; spiral shape	E (I)	Browning and necrosis	Chlorotic and stunted	Bermudagrass, Kentucky bluegrass (290, 291)
Sting	*Belonolaimus*	2–2.5 mm; long stylet	E	Lesions and stunted	Chlorotic and stunted	W + centipedegrass (98, 301, 302, 308, 374)
Stubby root	*Trichodorus*	0.7 mm; curved stylet	E	Stubby tips and stunting	Chlorotic and stunted	St. Augustinegrass (301, 302, 303)
Stunt	*Tylenchorhynchus*	0.8–0.9 mm	I	Stunted	Stunting	T

* E—ectoparasitic.
I—endoparasitic.

† T—thought to be pathogenic and commonly associated with turfgrasses.
W—major warm season turfgrasses (bermudagrass and St. Augustinegrass).

antennae occur in many different sizes and forms and function as sensory organs. The middle region of the body is known as the **thorax**. It bears membranous wings and legs that are jointed. The third body segment, the **abdomen**, typically has eleven segments.

Insects possess extraordinary reproductive powers. Most insects develop from eggs. A typical life cycle involves hatching of the eggs into a **larva** that is worm-like in nature. Frequently, the larva has a form and habitat that is vastly different from the mature adult. The larva grows by periodically shedding its outer skin. Each shedding of the cuticle is known as a **molt** with four to eight molts normally occurring in many of the common turfgrass insect pests. The form of the insect between molts is known as an **instar**.

When insects change in form significantly during molting, it is referred to as **metamorphosis**. Complete metamorphosis occurs when insects change from the worm-like larva into a pupa following the last larval instar. The insect does not feed and is usually inactive during the pupa stage. Many insects also overwinter in the pupa stage. The final molt at the end of the pupa stage results in the insect being transformed into an adult that does not change in size. Some insects do not undergo complete metamorphosis but possess molts in which the young have a form similar to the adult. Young of insects having a simple metamorphosis are called **nymphs** and contain wing and eye structures similar to the adult. The various molting stages involve primarily an increase in size when simple metamorphosis is involved. Intermediate types of metamorphosis exist between the simple and complete types just described. The length of the life cycle varies with the insect species from one generation per year for many to several generations within a year or, in some cases, to more than a year.

Certain insects are distinctly beneficial and others highly detrimental to turfs, while some are considered primarily nuisances to human activities. Beneficial effects of insects include (a) the activity of certain burrowing insects in soil cultivation to enhance air and water movement, (b) functioning in the decomposition of organic materials in thatch and their incorporation into the soil by burrowing insects, (c) the pollination of many important economical and ornamental plants, and (d) predators or parasites that control certain weed and insect pests.

The primary emphasis in this section is placed on insects that behave as turfgrass pests. Turfgrass insects can cause direct damage to the turf by a feeding or chewing action on the roots, rhizomes, stolons, or shoots. Damage is also produced by sucking plant juices from shoot tissues. Insects cause indirect damage by functioning as a carrier of certain virus diseases.

Symptoms of direct damage to turfs caused by insects include (a) weakened plants having a reduced shoot growth rate, (b) defoliation of the leaves and shoots, or (c) lack of a root system. Irregular patches of yellowish or chlorotic turf may appear that eventually turn brown, thin, and die (Fig. 17-15). Turfs damaged or weakened by insects are generally less tolerant of environmental stresses and diseases.

Certain insect pests and other classes of arthropods associated with turfgrass

Figure 17-15. Typical damage symptoms to a Kentucky bluegrass turf caused by white grubs.

areas do not cause direct damage to the turf but are objectionable due to their burrowing or nesting habits. For example, the surface mounding habits of ants on greens or the burrowing of mole crickets can be particularly objectionable. Other nuisance pests are objectionable because they bite or sting humans who use turfgrass areas. Included are centipedes, chiggers, fleas, knats, mosquitos, spiders, scorpions, and cicada killer wasps.

Control of Turfgrass Insects

Because the incidence of turfgrass insect damage is less frequent than many other problems, the damage is sometimes incorrectly attributed to disease, drought, or improper nutrition. As a result, recognition of an insect as the cause of injury is delayed until extensive damage has occurred. Considerable time and money may be wasted in the application of unneeded fungicides and nematicides. Thus, it is important to carefully examine the aboveground and belowground turfgrass plant parts for symptoms of damage and the presence of insects. A periodic examination of the turf is necessary for the presence of harmful turfgrass insects to be detected early and proper control procedures initiated. Turfgrass insect problems usually occur at relatively unpredictable intervals because of the variability in distribution and frequency of occurrence from year to year in a given location.

Insect Prevention

Preventive methods that minimize the occurrence of destructive turfgrass insects include (a) sanitary practices, (b) preplant insect control, and (c) use of resistant turfgrass cultivars. The effectiveness of sanitary practices and preplant insect control prior to turfgrass establishment depends on the mobility of the particular insects involved. In general, such preventive practices are not as effective as for nematodes. However, sanitary practices should be followed to avoid the introduction of turfgrass insects into a noninfested area by means of contaminated sod, soil, or turfgrass equipment. Sod containing harmful turfgrass insects should be treated with an appropriate insecticide prior to harvesting. Similarly, soil contaminated with harmful turfgrass insects should be fumigated or heat-treated prior to use for soil modification or topdressing. Preplant insect control involves the use of heat treatment or such materials as dazomet, metham, or methyl bromide at the appropriate rates.

The use of turfgrass cultivars that are resistant to the particular insect problem in a given area is the preferred method of insect prevention. It is particularly desirable because of the convenience, minimum expense, and avoidance of residue problems associated with certain insecticides. However, most turfgrass breeding programs have not developed to the point where insect resistance is one of the major objectives in cultivar improvement.

A degree of resistance to certain turfgrass insects has been observed. This suggests that genetic material is available for incorporating resistance to insect pests into the improved turfgrass cultivars. For example, a degree of resistance to sod webworm has been observed in cultivars of Kentucky bluegrass (47, 289). Certain bermudagrass cultivars have resistance to the bermudagrass mite (49). It appears unlikely, however, that a turfgrass cultivar resistant to all the common destructive turfgrass insects will be developed in the near future.

Associated with a preventive insect control program is the use of cultural practices that ensure a dense, vigorous, actively growing turf having good recuperative potential. A vigorous, healthy turf can survive a much larger population of destructive insects without serious injury than a turf that is in a thin, weakened condition due to improper mowing, fertility, or irrigation practices.

Chemical Control

The first step in the use of insecticides is identifying the particular insect involved. The type of insect, its feeding habit, and its life cycle determine the particular insecticide to be used as well as the rate and time of application that will achieve satisfactory control.

Most insecticides available for the control of destructive turfgrass insects can be divided into three major groups: (a) organophosphates, (b) carbamates, and (c) chlorinated hydrocarbons. No one insecticide is effective against all destructive turfgrass insects. The organophosphates and carbamates are widely used. They have a more limited period of effective control with the toxic residue dissipating more rapidly than insecticides in the chlorinated hydrocarbon group. The organophosphates as a group are quicker acting and have shorter residual properties than the carbamates. The organophosphates are acutely toxic to man and must be handled with extreme care.

The chlorinated hydrocarbons have been widely used for turfgrass insect control in the past because of the good control achieved over an extended period of time. The extremely long period of residual toxicity of these compounds is objectionable, however, from the standpoint of potential contamination and toxicity to fish, wildlife, and humans. The use of certain chlorinated hydrocarbons such as aldrin, dieldrin, DDT, and heptachlor is no longer permitted by laws passed in certain states and countries. Similarly, the use of lead arsenate is forbidden in certain regions.

Twelve widely used turfgrass insecticides along with their chemical names and particular insecticide grouping are summarized in Table 17-8.

Insecticides can be applied in the form of dusts, granules, or sprays marketed as wettable powders or emulsified concentrates. Sprays are generally preferred to dusts. No matter which form is applied, it is important to use the proper equipment calibrated to uniformly apply the insecticide at the desired rate. Insecticides should not be used indiscriminately on a preventive program, but only as needed to control insects causing serious damage to a turfgrass area.

Most insecticides can cause injury to man and animals if improperly used. Special precautions should be taken to comply with the directions on the label. Repeated or prolonged contact with the skin should be avoided as well as the inhalation of dust, mist, or vapors of the insecticide. Applications should be directed only to those turfgrass areas where insect control is desired. Insecticides applied for the control of soil insects should be thoroughly watered in immediately after

Table 17-8

CHEMICAL NAMES AND CHARACTERISTICS OF 12 INSECTICIDES COMMONLY USED
FOR THE CONTROL OF TURFGRASS INSECT PESTS

Name of insecticide		Chemical grouping*
Common	Chemical	
Carbaryl	1-naphthyl N-methylcarbamate	C
Carbophenothion	S-(p-chlorophenylthiomethyl) O,O-diethyl phosphorodithioate	OP
Chlordane	1,2,4,5,6,7,8,8-octachloro-2,3,3a,4,7,7a-hexahydro-4,7-methanoindene	CH
Diazinon	O,O-diethyl O-(2-isopropyl-4-methyl-6-pyrimidinyl) phosphorodithioate	OP
Dicofol	1,1-bis(p-chlorophenyl)-2, 2,2-trichloroethanol	M
Dursban®	O,O-diethyl O-(3,5,6-trichloro-2-pyridyl) phosphorothioate	OP
Ethion	O,O,O',O'-tetraethyl S,S'-methylene bisphosphorodithioate	OP
Lead arsenate	Lead arsenate	—
Malathion	O,O-dimethyl dithiophosphate of diethyl mercaptosuccinate	OP
Methoxychlor	2,2-bis (p-methoxyphenyl)-1,1,1-trichlorothane	CH
Propoxur	o-isopropoxyphenyl methylcarbamate	C
Trichlorofon	Dimethyl (2,2,2,-trichloro-1-hydroxyethyl) phosphonate	OP

* C—carbamates.
  CH—chlorinated hydrocarbons.
  OP—organophosphates.
  M—a miticide only.

application and the turf permitted to completely dry before humans and animals are permitted on the area.

Insect Pests of Turf

The scientific name, injury symptoms, and chemical controls for the major turfgrass insect pests are summarized in Table 17-9. One noninsect, a mite, is contained in the table. Included are insects that cause damage to turfs by direct feeding injury or by mounding that disrupts turfgrass uniformity. The feeding damage is caused by the larvae in most cases. For detailed descriptions of the insects, the reader is referred to the many entomology texts available.

A number of other insects occasionally cause damage to turfgrass areas but they are not as general a problem as those insects listed in Table 17-9. Potentially destructive insects that are soil inhabitants and feed on underground plant parts include the (a) oriental beetle (*Anomala orientalis* Waterhouse), (b) Asiatic garden beetle (*Maladera castanea* Arrow), (c) rose chafer (*Macrodactylus subspinosus* Fabricius), and (d) white-fringed beetle (*Graphognathus leucoloma* Boheman). Practically all damage caused by these four insects is associated with larval feeding activity. They are soil inhabitants with the characteristic white to grayish body, brownish head, and curled position when resting. Feeding activity is concentrated 1 to 2 in. below the soil surface with the grubs moving deeper during winter low temperature periods. Grubs of these four insects can be chemically controlled with chlordane.

Potentially destructive insects that feed on turfgrass leaves and stems include the (a) field skipper (*Atalopedes campestris* Boisduval) (358), (b) fiery skipper (*Hylephila phyleus* Drury) (35, 209), (c) lucerne moth (*Nomophila noctuella* Denis and Schiffermiiller) (35, 336), (d) striped grass loopers (*Mocis* spp.), and (e) grasshoppers (order *Orthoptera*). With the exception of the grasshoppers, most of the damage caused by these insects is a result of larval feeding activity. These insects feed on a relatively wide range of species including turfgrasses. The lucerne moth, skippers, and striped grass loopers can be controlled with chlordane, while grasshoppers are controlled with carbaryl.

Insects that occasionally damage turfgrasses by sucking juices from the leaf and stem tissues include the (a) meadow spittlebug (*Philaenus spumarius* L.), (b) two-lined spittlebug (*Philaenus leueophthalmas* C.F.), and (c) false chinch bug (*Nysius ericae* Schill) (222). The two-lined spittlebug lives within a mass of spittle in the nymph stage and attacks bermudagrass, St. Augustinegrass, and centipedegrass. The other two species feed on turfgrasses infrequently. Spittlebugs can be controlled with malathion or propoxur.

Insects that are undesirable in turfgrass areas because their soil nesting habits cause unsightly mounds include the (a) green June bug (*Cotinus nitida* L.), (b) mound-building bee (*Nomia heteropoda* Say) (222), (c) rhinoceros beetle (*Xyloryctes satyrus* Fabricius) (222), (d) periodical cicada (*Magicicada septendecim* L.), and (e)

Table 17-9

SCIENTIFIC NAMES, CHARACTERISTICS, AND CONTROLS FOR 28 COMMON INSECT PESTS OF TURFS

Name of insect		Type of feeding*	Turfgrass damage symptoms	Chemical controls§	Comments
Common	Scientific (or family)				
Ants† (35, 321)	Formicidae family	—	Mounds smother turf; tunnels dry out roots	Chlordane or diazinon	A social, nest-building insect
Cicada killer	Sphecius speciosus Drury	—	Mounds smother turf and disrupt uniformity	Chlordane or diazinon	Active in midsummer; wasp nests in ground
Armyworm, (35) common	Noctuidae family Pseudaletia unipuncta Haworth	S	Somewhat circular, defoliated area	Carbaryl, chlordane, diazinon, or methoxychlor	Larvae feed nocturnally
fall	Spodoptera frugiperda J. E. Smith	S	"	"	"
Billbug, hunting (212, 221)	Calendrinae subfamily Calendra venatus vestita Chttn.	R (S)	Puncture of stems and crown; irregular, brown patches	Diazinon or propoxur	Larvae may burrow into stem; prefer moist areas
Phoenix (35, 226)	Sphenophorus phoeniciensis Chittenden	R (S)	"	"	"
Chafer, (307) European	Chrysomeloidea superfamily Amphimallon majalis Razoumowsky	R	Lack of root system; irregular, brown patches	Chlordane	Larvae feed in midsummer
northern masked	Cyclocephala borealis Arrow	R	"	"	"
southern masked	C. immaculata Olivier	R	"	"	"
Chinch bug, (106, 320, 323) common (321)	Blissus spp. B. leucopterus Say	J	Large, irregular yellowish to brown patches	Carbaryl, carbophenothion, diazinon, ethion, or propoxur	Occur in large concentrations
hairy (295)	B. hirtus Montandon	J	"	"	"

Table 17-9 (Cont.)

southern (208, 222, 224, 228, 375)	*B. insularis* Barber	J	"	"	"
Crane fly (leatherjackets) (368)	*Tipulidae* family	R	Irregular, brown patches	M	Larvae prefer moist, cool conditions
Cutworms	*Noctuidae* family	S	Leaves and stems severed at soil surface	Carbaryl, chlordane, diazinon, methoxychlor, or trichlorofon	Larvae feed nocturnally
Frit fly (207, 321)	*Oscinella frit* L.	S	Turf stunted and thin	Carbaryl or diazinon	Larvae burrow beneath leaf sheath and into stem
Leafhopper (35)	*Cicadellidae* family	J	White spots on shoots; browning of turf	Carbaryl, diazinon, or malathion	High mobility; may require repeat treatments
Japanese beetle	*Popillia japonica* Newm.	R	Lack of root system; irregular, brown patches	Chlordane	Most damage is caused by the larvae
June beetle (222, 247, 296, 307)	*Phyllophaga* spp.	R	Lack of root system; large, irregular, brown patches	Chlordane	Damage is caused by the larvae
Mites, (35) bermudagrass (48, 49, 50, 207)	*Acarina* order *Aceria neocynodomis* Keifer	J	Blotching of leaves; chlorosis and tufted, stunting of shoots	Diazinon or dicofol	Prefers warm, humid conditions
clover grass	*Bryobia praetiosa* Koch *Oligonychus stickenyi* McGregor	J J	" "	" "	" "
Mole cricket (220, 222)	*Gryllotalpidae* spp.	R	Lack of root system; burrowing uproots seedlings	Chlordane or diazinon	Active nocturnally; prefers warm, humid conditions
Scale insects, (222, 375) bermudagrass scale	*Coccidea* superfamily *Odonaspis ruthae* Kotinsky	J	Yellowish, weakened turf that may turn brown	Malathion (M)‡	Feeding activity is on the nodes of stems

Table 17-9 (Cont.)

| Name of insect | | Type of feeding* | Turfgrass damage symptoms | Chemical controls§ | Comments |
Common	Scientific (or family)				
ground pearls	Margarodes spp.	J	"	Fumigation	Prefers warm, humid conditions
rhodesgrass scale (57)	Antonina graminis Maskell	J	"	Diazinon or malathion (M)‡	Feeding activity is near the grass crown
Sod webworm|| (35, 210, 336, 349)	Crambinae subfamily	S	Shoots defoliated to the soil; large, irregular, brown patches	Carbaryl, diazinon, ethion, propoxur, or trichlorofon	Larvae feed nocturnally
Turfgrass weevil (55)	Hyperodes spp.	S	Irregular patches of thin, brown turf	Diazinon or Dursban®	Most damage caused by larvae feeding on the stem base
Wireworms (click beetles)	Elateridae family	R	Lack of root system; irregular, brown patches	Chlordane	Larvae may bore into rhizomes; quite damaging to seedlings

* R—root and rhizome feeders.
S—shoot feeders.
J—suck juicies.

† Allegheny mound ant—Formica exsectiodes Forel.
Cornfield ant—Lasius alienus Forester.
Fire ant—Solenopsis germinata Fabricius.
Harvester ant—Pogonomyrmex spp.
Pavement ant—Tetramorium caestipum L.
Red ant—F. palliedefulva Latreille.

‡ M—marginal control.

§ Insecticide and miticide use patterns for turf are under constant revision. Several of the insecticides effective against certain turfgrass pests have been restricted. Before using any pesticide check with a local authoritive federal, provincial, or state information agency for specific recommendations and use restrictions.

|| Bluegrass webworm—C. teterrellus Zincken (3, 280).
Burrowing sod webworm—Acroloplus popeanellus Clemens.
Larger sod webworm—C. trisectus Walker.
Striped sod webworm—C. mutabilis Clemens.
Tropical sod webworm—Herpetogramma phaeopteralis Guenee (222).

desert termite (*Gnathamitermes* spp.). The adults and larvae of these mound and burrowing insects are relatively large, ranging from 1 to 3 in. in length. The green June bug grubs and periodical cicada nymphs cause smothering of turf because of their mound-building habit. In the case of the mound-building bee and rhinoceros beetle, the unsightly mounding habit and potential smothering of turf is associated primarily with the nest-building habits of the adults. The desert termite is objectionable because it constructs thin crusts of soil of about 1 to 4 in. in diameter above the soil surface. Chlordane is the preferred chemical control for these insects.

Other Larger Turfgrass Pests

Larger animals that occasionally become destructive pests or objectionable nuisances in turfgrass areas include moles, rodents, slugs, snails, birds, and house pets. Problems associated with these larger pests are usually not as frequent as some of the other more common turfgrass pests but they can cause serious damage.

Moles (*Scapanus* spp.) are small animals primarily of the *Talpidae* family that feed on insects and usually live underground. They have a velvety fur; very small, concealed eyes; sharp, pointed teeth; a length of from 5 to 7 in. and front paws with stout claws and an outward orientation for burrowing activities. The burrowing or tunneling of moles is objectionable because it disrupts the surface uniformity of the turf and severs the turfgrass root system, which increases proneness to desiccation. Further damage results from smothering of the turf at the entrance to the tunnels due to soil mounding. During tunneling activities, the mole is feeding on insects and other small soil animals such as earthworms that inhabit the upper 3 to 4 in. of the soil. Techniques for eliminating objectionable mole activity in turfs include (a) elimination of the food supply by use of insecticides, (b) trapping and removal, (c) use of poison baits, or (d) fumigation. The first two methods of control are generally more effective than the latter two.

Rodents that can cause damage to turfgrass areas include mice, voles, gophers, and ground squirrels. They belong to the *Rodentia* order of gnawing or nibbling mammals. Turfgrass damage usually results from numerous holes and runways produced by the tunneling and burrowing activity of the rodents. The mice (*Muridae* family) and voles (*Microtus* spp.) form runways in the turf, particularly under a winter snow cover. An elaborate system of runways and nesting areas can be developed by these animals over a period of several months that can extensively damage a turf. Burrowing rodents such as the pocket gopher (*Thomomys* spp.) and ground squirrels (*Spermophilus* spp.) cause damage by burrowing through a turfgrass area and throwing up large piles of soil at tunnel entrances that smother the turf. The elimination of rodent problems in turfs involves the use of (a) traps, (b) poison baits containing zinc phosphide (voles) or strychnine (ground squirrels and pocket gophers), or (c) fumigation. Where the animals are tunneling below the soil, lethal gases can be introduced into the tunnels and runways.

Slugs and **snails** are occasional nuisances in turfgrass areas because they leave a slimy mucus on the plants. Slugs (*Limacidae* family) are slimy, elongated terrestrial gastropods that have no shell, while snails (*Gastropoda* class) are a mollusk having a single shell that is usually spirally coiled. Slugs and snails can be controlled by the use of poison baits. One example is a bait containing 1 oz of methaldehyde mixed with 2 lb of wheat bran or cornmeal and moistened with water. It is best to apply the bait in late afternoon.

Birds, *Aves* class, can cause damage to intensely cultured turfs such as greens cut at less than 0.5 in. In the process of scavenging for grubs and other insects to feed upon, a group of birds can seriously disrupt the surface of a green. This problem is alleviated by eliminating the food source through the use of insecticides.

Larger animals such as dogs (*Canis* spp.) and skunks (*Meptitus* spp.), can also cause damage to turfgrass areas in several ways. Animals that follow the same daily routine of travel can form trails due to the concentrated traffic. Such trails of dead grass are unsightly. Also, animals that urinate on turfs can cause a disruption in turfgrass uniformity ranging from direct foliar kill to a darker green area that has a more rapid shoot growth rate due to the high nitrogen content in the urine. Skunks and armadillos can damage turfs by digging for grubs and other insects.

Earthworms

Earthworms are soil-inhabiting *Oligochaeta* belonging to a large group of organisms classified as *Annelida* (127, 205, 206, 274). A moist environment is necessary since they possess no lungs and must breathe through their skin. Earthworms actively burrow through the soil feeding on a mixture of decaying organic matter and soil. Higher earthworm populations are usually associated with soil having a higher organic matter content or undisturbed soils having a permanent sod cover. Earthworm populations are generally reduced in soils having a pH below 4.5 or under saline conditions. The population of earthworms is also usually lower in sandy soils due to the reduced organic matter content and droughty conditions. Vertically, the highest earthworm population is generally in the upper 12 to 18 in. of soil since this is where their primary food source of partially decomposed organic matter is located.

Earthworms move to the surface to feed on dead leaves and stems as well as through the soil where they feed on dead roots. Earthworm activity is concentrated near the soil surface when there is adequate soil moisture and a favorable temperature. The earthworms move to deeper soil depths as the soil dries out or cools. Horizontal movement of earthworm populations in the soil is not particularly rapid with estimates of approximately 30 to 35 ft per year given for some of the more common species. Thus, wide dispersal of earthworms depends on the activities of man through transportation in sod, soil, and topdressing.

Earthworms contribute a number of beneficial effects to the soil: (a) They enhance organic matter decomposition including thatch. (b) The ingestion of organic material and soil during their upward and downward movement provides a mixing action that carries organic matter to deeper soil depths and brings soil of low organic matter content from the lower profile to the surface. (c) The decomposition of dead earthworm bodies provides additional soil organic matter. (d) The tunneling activity modifies certain soil physical properties. Included are improved soil porosity and the associated enhancement of infiltration and aeration as well as reduced surface runoff of water. Earthworm tunnels can represent as much as 5% of the soil volume under a turfgrass area. The soil structure may also be improved through the activity of earthworms.

Earthworms can be objectionable on certain turfgrass areas such as greens because they leave small mounds or castings that disrupt the uniformity, appearance, and playability of the surface (Fig. 17-16). As much as 0.8 lb per sq ft of castings can be brought to the soil surface in 1 year. The castings contain a mixture of soil and decomposed organic material, are particularly high in nutrients, and have improved stability in water. The castings must be removed from the surface of the green prior to mowing operations or active play. The use of insecticides may be necessary where castings are consistently formed due to earthworm activity on greens. Insecticides that are effective include chlordane and lead arsenate.

Figure 17-16. Earthworm castings deposited on the surface of a creeping bentgrass green.

References

1. ADAMS, P. B., and F. L. HOWARD. 1963. Residual control of gray (*Typhula*) snow mold of bentgrass. Golf Course Reporter. 31(5): 20–22.

2. AHRENS, J. F., R. J. LUKENS, and A. R. OLSON. 1962. Preemergence control of crabgrass in turf with fall and spring treatments. Proceedings of Northeastern Weed Control Conference. 16: 511–518.

3. AINSLIE, G. G. 1930. The bluegrass webworm. USDA Technical Bulletin 173. pp. 1–25.

4. ALBRECHT, H. R. 1947. Strain differences in tolerance to 2,4-D in creeping bent grasses. Journal of American Society of Agronomy. 39(2): 163–165.

5. ALEXANDER, P. M. 1964. Nematode and fairy ring control on Tifgreen bermuda. Golf Course Reporter. 32: 12–14.

6. ALLISON, S. L., H. S. SHERWIN, I. Forbes, and R. E. WAGNER. 1949. *Rhizoctonia solani*, a destructive pathogen of Alta fescue, smooth brome grass, and birdsfoot trefoil. Phytopathology. 39: 1.

7. ANONYMOUS. 1952. Special purpose turf. Pennsylvania Agriculture Experiment Station Bulletin 553. pp. 52–53.

8. ANONYMOUS. 1954. Weeds of the north central states. North Central Regional Publication No. 36. Illinois Agricultural Experiment Station Circular 718. pp. 1–239.

9. Anonymous. 1962. Life history studies as related to weed control in the northeast. 1—Nutgrass. University of Rhode Island Agricultural Experiment Station Bulletin 364. pp. 1–33.

10. Anonymous. 1962. Life history studies as related to weed control in the northeast. 4-Quackgrass. University of Rhode Island Agricultural Experiment Station Bulletin 365. pp. 1–10.

11. Anonymous. 1967. Herbicide handbook of the Weed Society of America, 1st ed. pp. 1–293.

12. Anonymous. 1970. Texas A & M reports research progress on 'SAD.' Weeds, Trees and Turf. 9(1): 41.

13. Bain, D. C. 1962. *Sclerotinia* blight of bahia and Coastal bermuda grasses. Plant Disease Reporter. 46: 55–56.

14. ———. 1966. *Puccinia zoysiae* in Mississippi. Plant Disease Reporter. 50: 770.

15. Bayliss, J. S. 1911. Observations on *Marasmius oreades* and *Clitocybe gigantea* as parasitic fungi causing "fairy rings." Journal of Economic Biology. 6(4): 111–132.

16. Beach, G. A. 1963. Management practices in the care of lawns. Proceedings of the 10th Annual Rocky Mountain Regional Turfgrass Conference. pp. 42–45.

17. Bean, G. A. 1964. The pathogenicity of *Helminthosporium* spp. and *Curvularia* spp. on bluegrass in the Washington, D.C. area. Plant Disease Reporter. 48: 978–979.

18. ———.1966. *Fusarium* blight of turfgrasses. Golf Superintendent. 34(10): 32–34.

19. ———. 1966. Observations on *Fusarium* blight of turfgrasses. Plant Disease Reporter. 50: 942–945.

20. Bean, G. A., and R. D. Wilcoxson. 1964. *Helminthosporium* leaf spot of bluegrass. Phytopathology. 54: 1065–1070.

21. ———. 1964. Pathogenicity of three species of *Helminthosporium* on roots of bluegrass. Phytopathology. 54: 1084–1085.

22. Beard, J. B. 1965. Factors in the adaptation of turfgrasses to shade. Agronomy Journal. 57(5): 457–459.

23. ———. 1966. Fungicide and fertilizer applications as they affect *Typhula* snow mold control on turf. Quarterly Bulletin of the Michigan Agricultural Experiment Station. 49(2): 221–228.

24. Bell, A. A., and L. R. Krusberg. 1964. Occurrence and control of a nematode of the genus *Hypsoperine* on zoysia and bermuda grasses in Maryland. Plant Disease Reporter. 48: 721–722.

25. Bengtson, J. W., and F. F. Davis. 1939. Experiments with fertilizers on bent turf. Turf Culture. 1: 192–213.

26. Bennett, F. T. 1933. *Fusarium* patch disease of bowling and golf greens. Journal of Board of Greenkeeping Research. 3(9): 79–86.

27. ———. 1935. *Corticium* disease of turf. Journal of Board of Greenkeeping Research. 4(12): 32–39.

28. ———. 1937. Dollarspot disease of turf and its causal organism, *Sclerotinia homoeocarpa* N. SP. Annals of Applied Biology. 24: 236–257.

29. Bibby, F. F., and D. M. Tuttle. 1959. Notes on phytophagous and predatory mites of Arizona. Journal of Economic Entomology. 52(2): 186–190.

30. Bingham, S. W. 1965. Wild garlic control in bermudagrass turf. Weeds, Trees and Turf. 4(12): 10.

31. ———. 1967. Influence of herbicides on root development of bermudagrass. Weeds. 15: 363–365.

32. Bloom, J. R., and H. B. Couch. 1958. Influence of pH, nutrition, and soil moisture on the development of large brown patch. Phytopathology. 48(5): 260.

33. ———. 1960. Influence of environment on diseases of turfgrasses. I. Effect of nutrition, pH, and soil moisture on *Rhizoctonia* brown patch. Phytopathology. 50: 532–535.

34. Bloom, J. R., and P. J. Wuest. 1961. Nematodes and turf. Golf Course Reporter. 29(6): 1–4.

35. BOHART, R. M. 1947. Sod webworms and other lawn pests in California. Hilgardia. 17(9): 267–308.

36. BRITTON, M. P. 1957. Bluegrass rust studies. Proceedings of the 1957 Midwest Regional Turf Conference. pp. 78–79.

37. ———. 1963. Stripe smut damage to Kentucky bluegrass lawns. Illinois Turfgrass Conference Proceedings. pp. 1–2.

38. BRITTON, M. P., and J. D. BUTLER. 1965. Resistance of seven Kentucky bluegrass varieties to stem rust. Plant Disease Reporter. 49: 708–710.

39. BRITTON, M. P., and G. B. CUMMINS. 1956. The reaction of species of *Poa* and other grasses to *Puccinia striiformis*. Plant Disease Reporter. 40: 643–645.

40. ———. 1959. Subspecific identity of the stem rust fungus of Merion bluegrass. Phytopathology. 49: 287–289.

41. BRITTON, M. P., and H. C. HECHLER. 1961. The occurrence of nematodes in golf course greens in Illinois. Illinois Turfgrass Conference Proceedings. pp. 23–24.

42. BRITTON, M. P., and D. P. ROGERS. 1963. *Olpidium brassicae* and *Polymyxa graminis* in roots of creeping bent in golf putting greens. Mycologia. 55: 758–763.

43. BROADFOOT, W. C. 1936. Experiments on the chemical control of snowmold of turf in Alberta. Scientific Agriculture. 16: 615–618.

44. ———. 1937. Snow-mould of turf in Alberta. Journal of Board of Greenkeeping Research. 5(18): 182–183.

45. BROADFOOT, W. C., and M. W. CORMACK. 1941. A low-temperature Basidiomycete causing early spring killing of grasses and legumes in Alberta. Phytopathology. 31: 1058–1059.

46. BRODIE, B. B., and G. W. BURTON. 1967. Nematode population reduction and growth response of bermuda turf as influenced by organic pesticide applications. Plant Disease Reporter. 51: 562–566.

47. BUCKNER, R. C., B. C. PASS, P. B. BURRUS, and J. R. TODD. 1969. Reaction of Kentucky bluegrass strains to feeding by the sod webworm. Crop Science. 9: 744–746.

48. BUTLER, G. D., and A. A. BALTENSPERGER. 1963. The bermudagrass eriophyid mite. California Turfgrass Culture. 13(2): 9–11.

49. BUTLER, G. D., and W. R. KNEEBONE. 1965. Variations in response of bermudagrass varieties to bermudagrass mite infestations with and without chemical control. Arizona Turfgrass Research Report. 230: 7–16.

50. BUTLER, G. D., and D. M. TUTTLE. 1961. New mite is damaging to bermudagrass. Progressive Agriculture in Arizona. 13(1): 11.

51. BUTLER, J. D., and F. W. SLIFE. 1967. Lawn weeds: identification and control. Illinois Cooperative Extension Service Circular 873. pp. 1–28.

52. CALLAHAN, L. M. 1966. Select herbicides carefully . . . turfgrass tolerances do differ. Weeds, Trees and Turf. 5(11): 6–17.

53. CALLAHAN, L. M., and R. E. ENGEL. 1965. Tissue abnormalities induced in roots of colonial bentgrass by phenoxyalkylcarboxylic acid herbicides. Weeds. 13(4): 336–338.

54. CALLAHAN, L. M., R. E. ENGEL, and R. D. ILLNICKI. 1968. Environmental influence on bentgrass treated with silvex. Weed Science. 16: 193–196.

55. CAMERON, R. S., H. J. KASTL, and J. F. Cornman. 1968. Hyperodes weevil damages annual bluegrass. New York Turfgrass Association Bulletin 79. pp. 307–308.

56. CANODE, C. L., and W. C. ROBOCKER. 1966. Annual weed control in seedling grasses. Weeds. 14(4): 306–309.

57. CHADA, H. L., and E. A. WOOD, JR. 1960. Biology and control of the rhodesgrass scale. USDA Technical Bulletin 1221. pp. 1–21.

58. CHAPPELL, W. E., and R. E. SCHMITT. 1962. Phytotoxic effects of certain preemergence crabgrass control treatments on seedling turfgrass. Proceedings of Northeastern Weed Control Conference. 16: 474–478.

59. CHEESMAN, J. H., and E. C. ROBERTS. 1964. Effect of nitrogen level and moisture stress on *Helminthosporium* infection in Merion bluegrass. 1964 Agronomy Abstracts. p. 105.

60. CHEESMAN, J. H., E. C. ROBERTS, and L. H. TIFFANY. 1965. Effects of nitrogen level and osmotic pressure of the solution on incidence of *Puccinia graminis* and *Helminthosporium sativum* infection in Merion Kentucky bluegrass. Agronomy Journal. 57: 599–602.

61. CHRISTIE, J. R., J. M. GOOD, JR, and G. C. NUTTER. 1954. Nematodes associated with injury to turf. Proceedings of the Soil and Crop Science Society of Florida. 14: 167–169.

62. COCKERHAM, S. T., and J. W. WHITWORTH. 1967. Germination and control of annual bluegrass. Golf Superintendent. 35: 10–46.

63. COOK, R. N., and R. E. ENGEL. 1959. The effect of nitrogen carriers on the incidence of disease on bentgrass turf. 1959 Agronomy Abstracts. p. 88.

64. COOK, R. N., R. E. ENGEL, and S. BACHELDER. 1964. A study of the effect of nitrogen carriers on turfgrass disease. Plant Disease Reporter. 48(4): 254–255.

65. CORMACK, M. W. 1952. Winter crown rot or snow mold of alfalfa, clovers and grasses in Alberta. II. Field studies on host and varietal resistance and other factors related to control. Canadian Journal of Botany. 30: 537–548.

66. CORNMAN, J. F. 1969. The influence of maintenance practices on thatch formation by Merion Kentucky bluegrass turf. 1969 Agronomy Abstracts. p. 53.

67. CORNMAN, J. F., F. M. MADDEN, and N. J. SMITH. 1964. Tolerance of established lawn grasses, putting greens, and turfgrass seeds to preemergence crabgrass control chemicals. Proceedings of Northeastern Weed Control Conference. 18: 519–522.

68. COUCH, H. B. 1957. Melting-out of Kentucky bluegrass—its cause and control. Golf Course Reporter. 25(7): 5–7.

69. ———. 1958. Nitrogen applications in brown patch control. Golfdom. 31(4): 64.

70. ———. 1962. Diseases of turfgrasses. Reinhold Publishing Corp., New York. pp. 1–289.

71. ———. 1966. Relationship between soil moisture, nutrition and severity of turfgrass diseases. Journal of Sports Turf Research Institute. 11(42): 54–64.

72. COUCH, H. B., and E. R. BEDFORD. 1966. *Fusarium* blight of turfgrasses. Phytopathology. 56: 781–786.

73. COUCH, H. B., E. R. BEDFORD, L. D. MOORE, J. C. ROGOWICZ, and R. R. MUSE. 1963. Results of '61–62 turf-grass fungicide trials. Golf Course Reporter. 31(5): 24–34.

74. COUCH, H. B., and J. R. BLOOM. 1960. Influence of soil moisture stresses on the development of the root knot nematode. Phytopathology. 50: 319–321.

75. ———. 1960. Influence of environment on diseases of turfgrasses. II. Effect of nutrition, pH, and soil moisture on *Sclerotinia* dollar spot. Phytopathology. 50: 761–763.

76. COUCH, H. B., and L. D. MOORE. 1960. Broad spectrum fungicides tested for control of melting-out of Kentucky bluegrass and *Sclerotinia* dollar spot of Seaside bentgrass. Plant Disease Reporter. 44: 506–509.

77. COURSEN, B. W., and W. R. JENKINS. 1958. Host-parasite relationships of the pin nematode, *Paratylenchus projectus*, on tobacco and tall fescue. Phytopathology. 48: 460.

78. CRAFTS, A. S., H. D. BRUCE, and R. N. RAYNOR. 1941. Plot tests with chemical soil sterilants in California. California Agricultural Experiment Station Bulletin 648. pp. 1–25.

79. CRAFTS, A. S., and R. S. ROSENFELS. 1939. Toxicity studies with arsenic in eighty California soils. Hilgardia. 12(3): 177–200.

80. DAHL, A. S. 1929. Results of snowmold work during winter of 1928–1929. Bulletin of USGA Green Section. 9(8): 134–135.

81. ———. 1933. Effect of temperature and moisture on occurrence of brownpatch. Bulletin of USGA Green Section. 13(3): 53–61.

82. ———. 1933. Effect of watering putting greens on occurrence of brownpatch. Bulletin of USGA Green Section. 13: 62–66.

83. ———. 1933. Relationship between fertilizing and drainage in the occurrence of brownpatch. Bulletin of USGA Green Section. 13: 136–139.

84. ———. 1933. Effect of temperature on brown patch of turf. Phytopathology. 23: 8.

85. ———. 1934. The relation between rainfall and injuries to turf—season 1933. Greenkeepers' Reporter. 2(2): 1–4.

86. ———. 1934. Snowmold of turf grasses as caused by *Fusarium nivale*. Phytopathology. 24: 197–214.

87. DAVIS, R. R. 1958. The effect of other species and mowing height on persistence of lawn grasses. Agronomy Journal. 50: 671–673.

88. ———. 1961. Turfgrass mixtures—influence of mowing height and nitrogen. Proceedings of the 1961 Midwest Regional Turf Conference. pp. 27–29.

89. ———. 1961. No matter how you cut it the roots get hurt. Golfdom. 35(5): 38–40.

90. ———. 1961. Turfgrass mixtures—influence of mowing height and nitrogen. Illinois Turfgrass Conference Proceedings. pp. 42–45.

91. DAVIS, W. H. 1924. Spore germination of *Ustilago striaeformis*. Phytopathology. 14(6): 251–267.

92. ———. 1935. Summary of investigations with *Ustilago striaeformis* parasitizing some common grasses. Phytopathology. 25(8): 810–817.

93. DeFRANCE, J. A. 1938. The effect of different fertilizer ratios on colonial, creeping, and velvet bentgrass. Proceedings of the American Society for Horticultural Science. 36: 773–780.

94. ———. 1952. Weed-free compost and seedbeds for turf. Rhode Island Agricultural Experiment Station Miscellaneous Publication No. 31. pp. 1–15.

95. DICKINSON, L. S. 1930. The effect of air temperature on the pathogenicity of *Rhizoctonia solani* parasitizing grasses on putting-green turf. Phytopathology. 20: 597–608.

96. DICKSON, J. G. 1956. Diseases of field crops. McGraw-Hill Book Co., New York. pp. 1–517.

97. DiEDWARDO, A. A. 1962. Distribution of nematodes on turf in Florida. Proceedings of the University of Florida Turfgrass Management Conference. 10: 23–27.

98. ———. 1963. Pathogenicity and host-parasite relationships of nematodes on turf in Florida. Florida Agricultural Experiment Station Annual Report. p. 109.

99. DiEDWARDO, A. A., and V. G. PERRY. 1964. *Heterodera leuceilyma* N.SP. (*Nemata: Heteroderidae*) a severe pathogen of St. Augustinegrass in Florida. Florida Agricultural Experiment Station Bulletin 687. pp. 1–35.

100. DODMAN, R. L., K. R. BARKER, and J. C. WALKER. 1968. Modes of penetration by different isolates of *Rhizoctonia solani*. Phytopathology. 58: 31–32.

101. DRECHSLER, C. 1928. Zonate eyespot of grasses caused by *Helminthosporium giganteum*. Journal of Agricultural Research. 37: 473–492.

102. ———. 1929. Occurrence of the zonate-eyespot fungus *Helminthosporium giganteum* on some additional grasses. Journal of Agricultural Research. 39: 129–135.

103. ———. 1930. Leaf spot and foot rot of Kentucky bluegrass caused by *Helminthosporium vagans*. Journal of Agricultural Research. 40: 447–456.

104. DUICH, J. M., and H. B. MUSSER. 1960. Response of Kentucky bluegrass, creeping red fescue and bentgrass to nitrogen fertilizers. Pennsylvania Agricultural Experiment Station Progress Report 214. pp. 1–20.

105. DUVDEVANI, S., J. REICHERT, and J. PALTI. 1946. The development of downy and powdery mildew of cucumber as related to dew and other environmental factors. Palestine Journal of Botany, Rehovot Series. 5(2): 127–151.

106. EDEN, W. G., and R. L. SELF. 1960. Controlling chinch bugs on St. Augustinegrass lawns. Auburn University Agricultural Experiment Station Progress Report Series No. 79. pp. 1–3.

107. ENDO, R. M. 1961. Turfgrass diseases in southern California. Plant Disease Reporter. 45(11): 869–873.

108. ———. 1963. Influence of temperature on rate of growth of five fungus pathogens of turfgrass and on rate of disease spread. Phytopathology. 53: 857–861.

109. ———. 1966. Control of dollar spot of turfgrass by nitrogen and its probable bases. Phytopathology. 56: 877.

110. ———. 1967. Why nitrogen fertilization controls the dollar spot disease of turfgrass. California Turfgrass Culture. 17(2): 11.

111. ———. 1967. The role of guttation fluid in fungal disease development. California Turfgrass Culture. 17(2): 12–13.

112. ENDO, R. M., and R. H. AMACHER. 1964. Influence of guttation fluid on infection structures of *Helminthosporium sorokinianum*. Phytopathology. 54(11): 1327–1334.

113. ENDO, R. M., and I. MALCA. 1965. Morphological and cytohistological responses of primary roots of bentgrass to *Sclerotinia homoeocarpa* and *D*-galactose. Phytopathology. 55: 781–789.

114. ENDO, R. M., I. MALCA, and E. M. KRAUSMAN. 1964. Degeneration of the apical meristem and apex of bentgrass roots by a fungal toxin. Phytopathology. 54: 1175–1176.

115. ENDO, R. M., and J. J. OERTLI. 1964. Stimulation of fungal infection of bentgrass. Nature. 201(4916): 313.

116. ENGEL, R. E. 1966. A comparison of colonial and creeping bentgrass for ½ and ¾ inch turf. New Jersey Agricultural Experiment Station Bulletin 816. pp. 45–58.

117. ———. 1966. Response of bentgrass turf to dicamba, mecoprop, and silvex herbicides. New Jersey Agricultural Experiment Station Bulletin 816. pp. 85–92.

118. ———. 1967. Temperatures required for germination of annual bluegrass and colonial bentgrass. Golf Superintendent. 35: 20–28.

119. ENGEL, R. E., and R. B. ALDERFER. 1967. The effect of cultivation, topdressing, lime, nitrogen, and wetting agent on thatch development in ¼-inch bentgrass turf over a ten year period. New Jersey Agricultural Experiment Station Bulletin 818. pp. 32–45.

120. ENGEL, R. E., and L. M. CALLAHAN. 1967. Merion Kentucky bluegrass response to soil residue of preemergence herbicides. Weeds. 15(2): 128–130.

121. ENGEL, R. E., J. H. DUNN, and R. D. ILLNICKI. 1967. Preemergence crabgrass herbicide performance as influenced by dry vs. spray treatments and variation of application date of spring treatments on lawn turf. New Jersey Agricultural Experiment Station Bulletin 818. pp. 112–115.

122. ENGEL, R. E., C. R. FUNK, and D. A. KINNEY. 1968. Effect of varied rates of atrazine and simazine on the establishment of several zoysia strains. Agronomy Journal. 60: 261–262.

123. ENGEL, R. E., and R. D. ILLNICKI. 1963. Injury to established turfgrasses from preemergence herbicides. Proceedings of Northeastern Weed Control Conference. 17: 493.

124. ERWIN, L. E. 1941. Pathogenicity and control of *Corticium fuciforme*. Rhode Island Agricultural Experiment Station Bulletin 278. pp. 1–34.

125. ESSER, R. P. 1962. Turf nematode survey. Proceedings of the University of Florida Turf–Grass Management Conference. 10: 38–42.

126. ESSER, R. P., and E. B. SLEDGE. 1962. Root knot nematodes associated with sodgrass in Florida. Proceedings of the University of Florida Turf–Grass Management Conference. 10: 30–37.

127. EVANS, A. C. 1947. Earthworms. Journal of Board of Greenkeeping Research. 7: 49–54.

128. EVANS, E. M., R. D. ROUSE, and R. T. GUDAUSKAS. 1964. Low soil potassium sets up Coastal for leafspot disease. Highlights of Agricultural Research. 11: 2.

129. EVANS, T. W. 1932. The cutting and fertility factors in relation to putting green management. Journal of Board of Greenkeeping Research. 2: 196–200.

130. FILER, T. H. 1965. Damage to turfgrasses caused by cyanogenic compounds produced by *Marasmius oreades*, a fairy ring fungus. Plant Disease Reporter. 49: 571–574.

131. ———. 1965. Parasitic aspects of a fairy ring fungus, *Marasmius oreades*. Phytopathology. 55: 1132–1134.

132. ———. 1966. Red thread found on bermuda grass. Plant Disease Reporter. 50: 525–526

133. FITTS, O. B. 1924. Early morning watering as an aid to brownpatch control. Bulletin of Green Section of USGA. 4(7): 159.

134. FREEMAN, T. E. 1957. A new *Helminthosporium* disease of bermudagrass. Plant Disease Reporter. 41(5): 389–391.

135. ———. 1959. Florida isolates of dollarspot fungus stand hot weather. Florida Agricultural Experiment Station Research Report. 4: 3.

136. ———. 1960. Control dollarspot by fertilization. Florida Agricultural Experiment Station Research Report. 5: 17.

137. ———. 1960. Effects of temperature on cottony blight of ryegrass. Phytopathology. 50: 575.

138. ———. 1964. *Helminthosporium* diseases of bermudagrass. Golf Course Reporter. 32(5): 24–26.

139. ———. 1964. Influence of nitrogen on severity of *Piricularia grisea* infection of St. Augustine grass. Phytopathology. 54: 1187–1189.

140. ———. 1965. Rust of *Zoysia* spp. in Florida. Plant Disease Reporter. 49: 382.

141. FREEMAN, T. E., and G. C. HORN. 1963. Reaction of turfgrasses to attack by *Pythium aphanidermatum* (Edson) Fitzpatrick. Plant Disease Reporter. 47: 425–427.

142. FULWIDER, J. R., and R. E. ENGEL. 1959. The effect of temperature and light on germination of seed of goosegrass (*Eleusine indica*). Weeds. 7: 359–361.

143. FUNK, C. R., and R. E. ENGEL. 1966. Influence of variety, fertility level, and cutting height on weed invasion of Kentucky bluegrass. Rutgers University Short Course. pp. 10–17.

144. FUNK, C. R., R. E. ENGEL, and P. M. HALISKY. 1966. Performance of Kentucky bluegrass varieties as influenced by fertility level and cutting height. New Jersey Agricultural Experiment Station Bulletin 816. pp. 7–21.

145. ———. 1967. Summer survival of turfgrass species as influenced by variety, fertility level and disease incidence. New Jersey Agricultural Experiment Station Bulletin 818. pp. 71–77.

146. GASKIN, T. A. 1964. Effect of pre-emergence crabgrass herbicides on rhizome development in Kentucky bluegrass. Agronomy Journal. 56: 340–342.

147. ———. 1965. Susceptibility of bluegrass to root-knot nematodes. Plant Disease Reporter. 49: 89–90.

148. ———. 1965. Varietal reaction of creeping bentgrass to stripe smut. Plant Disease Reporter. 49: 268.

149. ———. 1966. Evidence for physiologic races of stripe smut (*Ustilago striiformis*) attacking Kentucky bluegrass. Plant Disease Reporter. 50: 430–431.

150. GASKIN, T. A., and M. P. BRITTON. 1962. The effect of powdery mildew on the growth of Kentucky bluegrass. Plant Disease Reporter. 47: 724–725.

151. GIANFAGNA, A. J., and A. M. S. PRIDHAM. 1951. Some aspects of dormancy and germination of crabgrass seed, *Digitaria sanguinalis* Scop. Proceedings of the American Society for Horticultural Science. 58: 291–297.

152. ———. 1951. Experiments with crabgrass seed. New York State Turf Association Bulletin 29. pp. 114.

153. GOOD, J. M., A. E. STEELE, and T. J. RATCLIFFE. 1959. Occurrence of plant parasitic nematodes in Georgia turf nurseries. Plant Disease Reporter. 43: 236–238.

154. GOSS, R. L. 1962. Fertility effects on disease development. Proceedings of the 16th Annual Northwest Turfgrass Conference. pp. 83–85.

155. GOSS, R. L., and C. J. GOULD. 1967. Some interrelationships between fertility levels and *Ophiobolus* patch disease in turfgrasses. Agronomy Journal. 59: 149–151.

156. ———. 1967. The effect of potassium on turfgrass and its relationship to turfgrass diseases. 1967 Agronomy Abstracts. p. 52.

157. ———. 1968. Turfgrass diseases: the relationship of potassium. USGA Green Section Record. 5(5): 10–13.

158. ———. 1968. Some inter-relationships between fertility levels and *Fusarium* patch disease of turfgrasses. Journal of Sports Turf Research Institute. 44: 19–26.

159. GOULD, C. J. 1963. How climate affects our turfgrass diseases. Proceedings of the 17th Annual Northwest Turfgrass Conference. pp. 29–43.

160. ———. 1964. Turf-grass disease problems in North America. Golf Course Reporter. 32(5): 36–54.

161. GOULD, C. J., H. M. AUSTENSON, and V. L. MILLER. 1958. Fairy ring diseases of lawns. Washington Agricultural Experiment Station Circular 330. pp. 1–5.

162. GOULD, C. J., R. L. GOSS, and M. EGLITIS. 1961. *Ophiobolus* patch disease of turf in western Washington. Plant Disease Reporter. 45: 296–297.

163. GOULD, C. J., R. L. GOSS, and V. L. MILLER. 1961. Fungicidal tests for control of *Fusarium* patch disease of turf. Plant Disease Reporter. 45: 112–118.

164. GOULD, C. J., V. L. MILLER, and R. L. GOSS. 1965. New experimental and commercial fungicides for control of *Fusarium* patch disease of bentgrass turf. Plant Disease Reporter. 49: 923–927.

165. ———. 1967. Fungicidal control of red thread disease of turfgrass in western Washington. Plant Disease Reporter. 51: 215–219.

166. GOULD, C. J., V. L. MILLER, and D. POLLEY. 1955. Fairy ring diseases of lawns. Golf Course Reporter. 23: 16–20.

167. GROVES, J. W., and C. A. BOWERMAN. 1955. *Sclerotinia borealis* in Canada. Canadian Journal of Botany. 33: 591–594.

168. GUDAUSKAS, R. T., and S. M. MCCARTER. 1966. Occurrence of rust on *Zoysia* species in Alabama. Plant Disease Reporter. 50: 885.

169. GUDAUSKAS, R. T., and N. E. MCGLOHON. 1964. *Sclerotinia* blight of bahiagrass in Alabama. Plant Disease Reporter. 48: 418.

170. HALISKY, P. M., and C. R. FUNK. 1966. Environmental factors affecting growth and sporulation of *Helminthosporium vagans* and its pathogenicity to *Poa pratensis*. Phytopathology. 56: 1294–1296.

171. HALISKY, P. M., C. R. FUNK, and S. BACHELDER. 1966. Stripe smut of turf and forage grasses —its prevalence, pathogenicity, and response to management practices. Plant Disease Reporter. 50: 294–298.

172. HALISKY, P. M., C. R. FUNK, and R. E. ENGEL. 1966. Occurrence and pathogenicity of *Helminthosporium sativum* in the Rutgers turf plots in 1964 and 1965. New Jersey Agricultural Experiment Station Bulletin 816. pp. 72–78.

173. ———. 1966. Melting-out of Kentucky bluegrass varieties by *Helminthosporium vagans* as influenced by turf management practices. Plant Disease Reporter. 50: 703–706.

174. HALISKY, P. M., and J. L. PETERSON. 1967. Occurrence of mushrooms and puffballs and their association with fairy rings in turfgrass areas of New Jersey. New Jersey Agricultural Experiment Station Bulletin 818. pp. 15–19.

175. HARPER, J. C., and H. B. MUSSER. 1950. The effect of watering and compaction on fairway turf. Pennsylvania State College 19th Annual Turf Conference. pp. 43–52.

176. HART, R. H., and W. S. MCGUIRE. 1964. Effect of clipping on the invasion of pastures by velvetgrass, *Holcus lanatus* L. Agronomy Journal. 56(2): 187–188.

177. HARTWELL, B. L., and S. C. DAMON. 1917. The persistence of lawn and other grasses as influenced especially by the effect of manures on the degree of soil acidity. Rhode Island Agricultural Experiment Station Bulletin 170. pp. 1–24.

178. HAWES, D. T. 1965. Studies of the growth of *Poa annua* L. as affected by soil temperature, and observations of soil temperature under putting green turf. M. S. Thesis. Cornell University. pp. 1–79.

179. HEALD, C. M. 1969. Histopathology of 'Tifdwarf' bermudagrass infected with *Meloidogyne graminis* (Sledge and Golden) Whitehead. Journal of Nematology. 1(1): 9–10.

180. HEALD, C. M., and G. W. BURTON. 1968. Effect of organic and inorganic nitrogen on nematode populations on turf. Plant Disease Reporter. 52: 46–48.

181. HEALD, C. M., and H. D. WELLS. 1967. Control of endo- and ecto-parasitic nematodes in turf by hot-water treatments. Plant Disease Reporter. 51: 905–907.

182. HEALY, M. J. 1965. Factors influencing disease development on putting green turf. 6th Illinois Turfgrass Conference Proceedings. pp. 1–7.

183. ———. 1967. Factors affecting the pathogenicity of selected fungi isolated from putting green turf. Ph.D. Thesis. University of Illinois. pp. 1–59.

184. HEALY, M. J., and M. P. BRITTON. 1967. The importance of guttation fluid on turf diseases. 8th Illinois Turfgrass Conference Proceedings. 8: 40–46.

185. ———. 1968. Infection and development of *Helminthosporium sorokinianum* in *Agrostis palustris*. Phytopathology. 58: 273–276.

186. HEALY, M. J., M. P. BRITTON, and J. D. BUTLER. 1965. Stripe smut damage on 'Pennlu' creeping bentgrass. Plant Disease Reporter. 49: 710.

187. HERRON, J. W. 1961. Lawn weed control. Kentucky Cooperative Extension Service Circular 577. pp. 1–39.

188. HODGES, C. F. 1967. Etiology of stripe smut, *Ustilago striiformis* (West.) Niessl, on Merion bluegrass, *Poa pratensis*. Ph. D. Thesis. University of Illinois. pp. 1–64.

189. HODGES, C. F., and D. P. TAYLOR. 1965. Morphology and pathological histology of bentgrass roots infected by a member of the *Meloidogyne incognita* group. Nematologica. 11: 40.

190. ———. 1966. Host-parasite interactions of a root knot nematode and creeping bentgrass, *Agrostis palustris*. Phytopathology. 56: 88–91.

191. HODGES, C. F., D. P. TAYLOR, and M. P. BRITTON. 1963. Root-knot nematode on creeping bentgrass. Plant Disease Reporter. 47: 1102–1103.

192. HOLBEN, F. J. 1950. Nitrogen-potash relationships as they affect the growth and diseases of turf grasses. Pennsylvania State College 19th Annual Turf Conference Proceedings. pp. 74–93.

193. ———. 1952. Potash-nitrogen fertilization on fescue and bent turf. Pennsylvania State College 21st Annual Turf Conference Proceedings. pp. 52–56.

194. HOLT, E. C. 1963. Control of large brown patch on St. Augustinegrass. Golf Course Reporter. 31(5): 48–50.

195. HOPEN, H. J., W. E. SPLITTSTOESSER, and J. D. BUTLER. 1966. Intraspecies bentgrass selectivity of siduron. University of Illinois Horticulture Science No. 432.

196. HORN, G. C., and W. L. PRITCHETT. 1968. Role of potash in turfgrass fertilization. Florida Turf. 1(2): 1–8.

197. HOVIN, A. W. 1957. Germination of annual bluegrass seed. Southern California Turfgrass Culture. 7: 13.

198. ———. 1957. Variations in annual bluegrass. Golf Course Reporter. 25(7): 18–19.

199. HOWARD, F. L., J. B. ROWELL, and H. L. KEIL. 1951. Fungus diseases of turf grasses. Rhode Island Agricultural Experiment Station Bulletin 308. pp. 1–56.

200. JACKSON, N. 1959. *Ophiobolus* patch disease fungicide trials, 1958. Journal of Sports Turf Research Institute. 9: 459–461.

201. ———. 1959. Evaluation of some grass varieties. Journal of Sports Turf Research Institute. 10(35): 13–28.

202. ———. 1964. Further notes on the evaluation of some grass varieties. Journal of Sports Turf Research Institute. 40: 67–75.

203. ———. 1966. Dollar spot disease and its control. Illinois Turfgrass Conference Proceedings. pp. 21–25.

204. JACKSON, N., and J. D. SMITH. 1965. Fungal diseases of turf grasses. Sports Turf Research Institute Publication. 2nd ed. pp. 1–97.

205. JEFFERSON, P. 1955. Earthworms and turf culture. Journal of Sports Turf Research Institute. 10: 276–289.

206. ———. 1956. Studies on the earthworms of turf. Journal of Sports Turf Research Institute. 9: 166–179.

207. JEFFERSON, R. N., and J. S. MORISHITA. 1962. Progress report on the bermudagrass mite and the frit fly. California Turfgrass Culture. 12(2): 9–10.

208. ———. 1968. Southern chinch bug, a new pest of turfgrass in California. California Turfgrass Culture. 18(2): 10–11.

209. JEFFERSON, R. N., and E. J. SWIFT. 1956. Control of turfgrass pests. Southern California Turfgrass Culture. 6(2): 9–12.

210. JOHNSON, J. P. 1944. Notes on the sod webworm. Connecticut Agricultural Experiment Station Bulletin 488. p. 420.

211. JUSKA, F. V. 1961. Preemergence herbicides for crabgrass control and their effects on germination of turfgrass species. Weeds. 9: 137–144.

212. ———. 1965. Billbug injury in a zoysia turf. Park Maintenance. 18(5): 38–40.

213. JUSKA, F. V., and A. A. HANSON. 1959. Evaluation of cool-season turfgrasses alone or in mixtures. Agronomy Journal. 51: 597–600.

214. ———. 1959. Evaluation of cool season turfgrasses. Park Maintenance. 12(9): 18–20.

215. ———. 1963. The management of Kentucky bluegrass on extensive turfgrass areas. Park Maintenance. 16: 22–32.

216. ———. 1964. Effect of preemergence crabgrass herbicides on seedling emergence of turfgrass species. Weeds. 12: 97–101.

217. ———. 1967. Factors affecting *Poa annua* L. control. Weeds. 15(2): 98–101.

218. ———. 1969. Nutritional requirements of *Poa annua* L. Agronomy Journal. 61: 466–468.

219. KALLIO, A. 1966. Chemical control of snow mold (*Sclerotinia borealis*) on four varieties of bluegrass (*Poa pratensis*) in Alaska. Plant Disease Reporter. 50: 69–72.

220. KELSHEIMER, E. G. 1950. Control of mole crickets. Florida Agricultural Experiment Station Circular S-15. pp. 1–7.

221. ———. 1956. The hunting billbug, a serious pest of zoysia. Proceeding of the Florida State Horticultural Society. 59: 415–418.

222. KELSHEIMER, E. G., and S. H. KERR. 1957. Insects and other pests of lawns and turf. Florida Agricultural Experiment Station Circular S-96. pp. 1–22.

223. KELSHEIMER, E. G., and A. J. OVERMAN. 1953. Notes on some ectoparasitic nematodes found attacking lawns in the Tampa Bay area. Proceeding of the Florida State Horticultural Society. 66: 301–303.

224. KERR, S. H. 1966. Biology of the lawn chinch bug, *Blissus insularis*. Florida Entomology. 49: 9–18.

225. KLOMPARENS, W. 1953. A study of *Helminthosporium sativum* P. K. B. as an unreported parasite of *Agrostis palustris* Huds. Ph.D. Thesis. Michigan State University. pp. 1–77.

226. KLOSTERMEYER, E. C. 1964. Lawn billbugs—a new pest of lawns in eastern Washington. Irrigation Experiment Station (Prosser) E. M. 2349.

227. KOLLETT, J. R., A. J. WISNIEWSKI, and J. A. DeFRANCE. 1958. The effect of ureaform fertilizers in seedbeds for turfgrass. Park Maintenance. 11(5): 12–20.

228. KOMBLAS, K. N. 1962. Biology and control of the lawn chinch bug, *Blissus insularis*. M.S. Thesis. Louisiana State University. pp. 1–82.

229. KOSHY, T. K. 1969. Breeding systems in annual bluegrass, *Poa annua* L. Crop Science. 9: 40–43.

230. KOZELNICKY, G. M., and W. N. GARRETT. 1966. The occurrence of zoysia rust in Georgia. Plant Disease Reporter. 50: 839.

231. KRAFT, J. M., R. M. ENDO, and D. C. ERWIN. 1967. Infection of primary roots of bentgrass by zoospores of *Pythium aphanidermatum*. Phytopathology. 57: 86–90.

232. KREITLOW, K. W. 1943. *Ustilago striaeformis*. II. Temperature as a factor influencing development of smutted plants of *Poa pratensis* L. and germination of fresh chlamydospores. Phytopathology. 33: 1055–1063.

233. KREITLOW, K. W., and F. V. JUSKA. 1959. Susceptibility of Merion and other Kentucky bluegrass varieties to stripe smut (*Ustilago striiformis*). Agronomy Journal. 51: 596–597.

234. KREITLOW, K. W., F. V. JUSKA, and R. T. HAARD. 1965. A rust on *Zoysia japonica* new to North America. Plant Disease Reporter. 49: 185–186.

235. KREITLOW, K. W., and W. M. MYERS. 1947. Resistance to crown rust in *Festuca elatior* and *F. elatior* var. *arundinacea*. Phytopathology. 37(1): 59–63.

236. KREITLOW, K. W., H. SHERWIN, and C. L. LEFEBVRE. 1950. Susceptibility of tall and meadow fescues to *Helminthosporium* infection. Plant Disease Reporter. 34(6): 189–190.

237. LAPP, W. S. 1943. A study of factors affecting the growth of lawn grasses. Pennsylvania Academy of Science. 17: 117–148.

238. LARUE, C. D. 1935. The time to cut dandelions. Science. 82: 350.

239. LAUTZ, W. 1958. Chemical control of nematodes parasitic on turf and sweet corn. Proceedings of the Florida State Horticultural Society. 71: 38–40.

240. LEBEAU, J. B. 1964. Control of snow mold by regulating winter soil temperature. Phytopathology. 54: 693–696.

241. LEBEAU, J. B., and E. J. HAWN. 1961. Fairy rings in Alberta. Canadian Plant Disease Survey. 41: 317–320.

242. ———. 1963. A simple method for control of fairy ring caused by *Marasmius oreades*. Journal of Sports Turf Research Institute. 11(39): 23–26.

243. ———. 1963. Formation of HCN by the mycelial stage of a fairy ring fungus. Phytopathology. 53(12): 1395–1396.

244. LEBEAU, J. B., M. W. CORMACK, and E. W. B. WARD. 1961. Chemical control of snow mold of grasses and alfalfa in Alberta. Canadian Journal of Plant Science. 41: 744–750.

245. LEVY, E. B., and E. A. MADDEN. 1931. Weeds in lawns and greens—competition effects and control by treatment with chemical sprays. New Zealand Journal of Agriculture. 42: 406–421.

246. LUDBROOK, W. V., and J. BROCKWELL. 1952. Control of dollar spot in Canberra. Journal of the Australian Institute of Agriculture Science. 18: 39–40.

247. LUGINBILL, P., and H. R. PAINTER. 1953. May beetles of the United States and Canada. USDA Technical Bulletin 1060. pp. 1–100.

248. LUKENS, R. J. 1965. Urea, an effective treatment for stripe smut on *Poa pratensis*. Plant Disease Reporter. 49: 361.

249. ———. 1967. Chemical control of stripe smut. Plant Disease Reporter. 51(5): 356.

250. MADISON, J. H. 1958. Turf invasion by weedy grasses. California Agriculture. 12(4): 11.

251. ———. 1962. The effect of management practices on invasion of lawn turf by bermudagrass (*Cynodon dactylon* L.). Proceedings of the American Society for Horticultural Science. 80: 559–564.

252. ———. 1966. Brown patch of turfgrass caused by *Rhizoctonia solani* Kühn. California Turfgrass Culture. 16(2): 9–13.

253. MADISON, J. H., and R. M. HAGAN. 1962. Extraction of soil moisture by Merion bluegrass (*Poa pratensis* L. 'Merion') turf, as affected by irrigation frequency, mowing height, and other cultural operations. Agronomy Journal. 54: 157–160.

254. MADISON, J. H., L. J. PETERSEN, and T. K. HODGES. 1960. Pink snowmold on bentgrass as affected by irrigation and fertilizer. Agronomy Journal. 52: 591–592.

255. MALCA, I. and R. M. ENDO. 1965. Identification of galactose in cultures of *Sclerotinia homoeocarpa* as the factor toxic to bentgrass roots. Phytopathology. 55: 775–780.

256. MARKLAND, F. E., E. C. ROBERTS, and L. R. FREDERICK. 1969. Influence of nitrogen fertilizers on Washington creeping bentgrass, *Agrostis palustris* Huds. II. Incidence of dollar spot,
Sclerotinia homoeocarpa, infection. Agronomy Journal. 61: 701–705.

257. MARTH, P. C., and J. W. MITCHELL. 1944. 2,4-Dichlorophenoxyacetic acid as a differential
herbicide. Botanical Gazette. 106: 224–232.

258. ———. 1946. Effects of spray mixtures containing 2,4-D, urea, and fermate on the growth of
grass. Botanical Gazette. 107: 417–424.

259. MCBETH, C. W. 1945. Tests on the susceptibility and resistance of several southern grasses
to the root-knot nematode, *Heterodera marioni*. Proceedings of the Helminthological Society
of Washington. 12(2): 41–44.

260. MCCAIN, A. H. 1967. Dichondra leafspot. California Turfgrass Culture. 17(4): 26.

261. MEINERS, J. P. 1955. Etiology and control of snow mold of turf in the Pacific northwest.
Phytopathology. 45: 59–62.

262. MEYERS, H. C. 1964. Effects of nitrogen levels and vertical mowing intensity and frequency
on growth and chemical composition of Ormond bermudagrass. Proceedings of the 12th
Annual University of Florida Turf-Grass Management Conference. pp. 77–79.

263. MIDDLETON, J. T. 1943. The taxonomy, host range and geographic distribution of the genus
Pythium. Memoirs of the Torrey Botanical Club. 20(1): 1–171.

264. MIRZA, K., and V. G. PERRY. 1967. Histopathology of St. Augustinegrass roots infected by
Hypsoperine graminis Sledge and Golden. Nematologica. 13: 146–147.

265. MITCHELL, J. W., and P. C. MARTH. 1945. Effects of 2,4-dichlorophenoxyacetic acid on the
growth of grass plants. Botanical Gazette. 107: 276–284.

266. MONTEITH, J., JR. 1925. July experiments for control of brownpatch on Arlington experimental turf garden. Bulletin of USGA Green Section. 5: 173–176.

267. ———. 1926. The brown patch disease of turf: its nature and control. Bulletin of USGA
Green Section. 6: 127–142.

268. ———. 1929. Some effects of lime and fertilizer on turf diseases. Bulletin of USGA Green
Section. 9: 82–99.

269. ———. 1933. A *Pythium* disease of turf. Phytopathology. 23(1): 23–24.

270. MONTEITH, J., JR., and A. S. DAHL. 1928. A comparison of some strains of *Rhizoctonia
solani* in culture. Journal of Agricultural Research. 36(10): 897–903.

271. ———. 1932. Turf diseases and their control. Bulletin of USGA Green Section. 12(4): 85–87.

272. MOORE, L. D., and H. B. COUCH. 1961. *Pythium ultimum* and *Helminthosporium vagans*
as foliar pathogens of *Gramineae*. Plant Disease Reporter. 45: 616–619.

273. MOORE, L. D., H. B. COUCH, and J. R. BLOOM. 1963. Influence of environment on diseases
of turfgrasses. III. Effect of nutrition, pH, soil temperature, air temperature, and soil moisture
on *Pythium* blight of Highland bentgrass. Phytopathology. 53: 53–57.

274. MOOTE, D. S. 1954. Earthworms and their effect on fine turf. Golf Course Reporter. 22: 14–16.

275. MOWER, R. G. 1961. Histological studies of suscept-pathogen relationships of *Helmintho-
sporim sativum* P. K. & B., *Helminthosporium vagans* Drechsl., and *Curvularia lunata*
(Wakk.) Boed. on leaves of Merion and common Kentucky bluegrass (*Poa pratensis* L.).
Ph.D. Thesis. Cornell University. pp. 1–150.

276. MOWER, R. G., and J. F. CORNMAN. 1958. Selective control of *Veronica filiformis* in turf.
Northeastern Weed Control Conference Proceedings. 12: 142–150.

277. MOWER, R. G., and R. L. MILLAR. 1963. Histological relationships of *Helminthosporium
vagans*, *H. sativum*, and *Curvularia lunata* in leaves of Merion and common Kentucky bluegrass. Phytopathology. 53: 351.

278. MUSE, R. R., and H. B. COUCH. 1965. Influence of environment on diseases of turfgrasses.
IV. Effect of nutrition and soil moisture on *Corticium* red thread of creeping red fescue.
Phytopathology. 55: 507–510.

279. MUSSER, H. B. 1948. Effects of soil acidity and available phosphorus on population changes

in mixed Kentucky bluegrass-bent turf. Journal of American Society of Agronomy. 40: 614–620.

280. NORTH, H. F. A., and G. A. THOMPSON, JR. 1933. Investigations regarding bluegrass webworms in turf. Journal of Economic Entomology. 26: 1117–1125.

281. NORTON, D. C. 1959. Relationship of nematodes to small grains and native grasses in north and central Texas. Plant Disease Reporter. 43: 227–235.

282. NUTTER, G. C. 1955. Nematode investigations in turf. Florida Agricultural Experiment Station Annual Report. p. 58.

283. NUTTER, G. C., and J. R. CHRISTIE. 1958. Nematode investigations on putting green turf. Proceedings of the Florida State Horticultural Society. 71: 445–449.

284. ———. 1959. Nematode investigations on putting green turf. USGA Journal and Turf Management. 11(7): 24–28.

285. OAKLEY, R. A. 1924. Brown-patch investigation. Bulletin of USGA Green Section. 4: 87–92.

286. ———. 1924. Mottled condition of turf. USGA Green Section Bulletin. 4: 259.

287. ———. 1925. Some things we have learned about brown-patch. Bulletin of USGA Green Section. 5: 75–77.

288. ORR, C. C., and A. M. Golden. 1966. The pseudo-root-knot nematode of turf in Texas. Plant Disease Reporter. 50: 645.

289. PASS, B. C., R. C. BUCKNER, and P. R. BURRUS, II. 1965. Differential reaction of Kentucky bluegrass strains to sod webworms. Agronomy Journal. 57: 510–511.

290. PERRY, V. G. 1958. A disease of Kentucky blue grass incited by certain spiral nematodes. Phytopathology. 48: 397.

291. PERRY, V. G., H. M. DARLING, and G. THORNE. 1959. Anatomy, taxonomy and control of certain spiral nematodes attacking blue grass in Wisconsion. Wisconsin Agricultural Experiment Station Research Bulletin 207. pp. 1–24.

292. PERRY, V. G., and K. M. MAUR. 1968. The pseudo-root-knot nematode of turf grasses. Florida Turf. 1(1): 3–8.

293. PIPER, C. V., and R. A. OAKLEY. 1921. The brown-patch disease of turf. USGA Green Section Bulletin. 1: 112–115.

294. POLIVKA, J. B. 1959. The biology and control of turf grubs. Ohio Agricultural Experiment Station Research Bulletin 829. pp. 1–30.

295. ———. 1963. Control of the hairy chinch bug, *Blissus leucopterus hirtus*, Mont., in Ohio. Ohio Agricultural Experiment Station Research Circular 122. pp. 3–8.

296. ———. 1965. Effectiveness of insecticides for the control of white grubs in turf. Ohio Agricultural Experiment Station Research Circular 140. pp. 3–7.

297. POWELL, W. M. 1964. The occurrence of *Tylenchorhynchus maximus* in Georgia. Plant Disease Reporter. 48: 70.

298. RAMPTON, H. H., and T. M. CHING. 1966. Longevity and dormancy in seeds of several cool-season grasses and legumes buried in soil. Agronomy Journal. 58: 220–222.

299. REMSBURG, R. E. 1940. Studies in the genus *Typhula*. Mycologia. 32: 52–96.

300. RENNEY, A. J. 1964. Preventing *Poa annua* infestations. Proceedings of the 18th Annual Northwest Turfgrass Conference. pp. 3–5.

301. RHOADES, H. L. 1962. Effects of sting and stubby-root nematodes on St. Augustinegrass. Proceedings of the University of Florida Turf-Grass Management Conference. 10: 28–29.

302. ———. 1962. Effects of sting and stubby-root nematodes on St. Augustine grass. Plant Disease Reporter. 46: 424–427.

303. ———. 1965. Parasitism and pathogenicity of *Trichodorus proximus* to St. Augustine grass. Plant Disease Reporter. 49: 259–262.

304. RICE, E. J., and C. R. SKOGLEY. 1962. Seed and seedling tolerance of lawn grasses to crabgrass herbicides. Proceedings of Northeastern Weed Control Conference. 16: 466–473.

305. RIFFLE, J. W. 1964. Root-knot nematode on African bermuda grass in New Mexico. Plant Disease Reporter. 48: 964–965.

306. RIGGS, R. D., J. L. DALE, and M. L. HAMBLEN. 1962. Reaction of bermuda grass varieties and lines to root-knot nematodes. Phytopathology. 52: 587–588.

307. RITCHER, P. O. 1940. Kentucky white grubs. Kentucky Agricultural Experiment Station Bulletin 401. pp. 73–149.

308. RIVERS, J. E., and A. A. DiEDWARDO. 1963. The effect of sting (*Belonolaimus longicaudatus*) and lance (*Hoplolaimus coronatus*) nematodes on bermudagrass. Proceedings of the University of Florida Turf-Grass Management Conference. 11: 150–151.

309. ROBERTS, E. C. 1963. Relationships between mineral nutrition of turf-grass and disease susceptibility. Golf Course Reporter. 31(5): 52–57.

310. ———. 1965. What to expect from a nitrogen fertilizer. Summary of the 16th Annual RCGA Sports Turfgrass Conference. pp. 37–46.

311. ROBERTS, E. C., and D. R. BROCKSHUS. 1966. Kind and extent of injury to greens from pre-emergence herbicides. Golf Superintendent. 34(4): 13–36.

312. ROBERTS, E. C., F. E. MARKLAND, and H. M. PELLETT. 1966. Effects of bluegrass stand and watering regime on control of crabgrass with preemergence herbicides. Weeds. 14(2): 157–161.

313. ROBINSON, R. J. 1949. Annual weeds and their viable seed population in the soil. Agronomy Journal. 41: 513–518.

314. ROWELL, J. B. 1951. Observations on the pathogenicity of *Rhizoctonia solani* on bent grass. Plant Disease Reporter. 35: 240–242.

315. RUMBURG, C. B., R. E. ENGEL, and W. F. MEGGITT. 1960. Effect of temperature on the herbicidal activity and translocation of arsenicals. Weeds. 8: 582–588.

316. RUNNELS, H. A. 1963. Tests on dollar spot and snow mold. Golf Course Reporter. 31(5): 42.

317. SAMPSON, K., and J. H. WESTERN. 1941. Diseases of British grasses and herbage legumes. Cambridge University Press. London, England. pp. 1–85.

318. SCHARDT, A. 1925. Brown-patch control resulting from early-morning work on greens. Bulletin of Green Section of USGA. 5(11): 254–255.

319. SCHMIDT, R. E., and H. B. MUSSER. 1958. Some effects of 2,4-D on turfgrass seedlings. USGA Journal and Turf Management. 11: 28–32.

320. SCHREAD, J. C. 1963. The chinch bug and its control. Connecticut Agricultural Experiment Station Circular 223. pp. 1–3.

321. ———. 1964. Insect pests of Connecticut lawns. Connecticut Agricultural Experiment Station Circular 212. pp. 3–12.

322. SHANTZ, H. L., and R. L. PIEMEISEL. 1917. Fungus fairy rings in eastern Colorado and their effect on vegetation. Journal of Agricultural Research. 11: 191–245.

323. SHELFORD, V. E. 1932. An experimental and observational study of the chinch bug in relation to climate and weather. Illinois Natural History Survey Bulletin. 19: 487–547.

324. SHURTLEFF, M. C. 1954. New look at brown patch. Rhode Island Agricultural Experimental Research. 4: 42–43.

325. SKOGLEY, C. R., and J. A. JAGSCHITZ. 1964. The effect of various crabgrass herbicides on turfgrass seed and seedlings. Proceedings of Northeastern Weed Control Conference. 18: 523–529.

326. SLEDGE, E. B. 1960. Studies on *Meloidogyne* sp. on grass. 23rd Biennial Report, State Plant Board of Florida. 14: 108–110.

327. ———. 1962. Preliminary report on a *Meloidogyne* sp. parasite of grass in Florida. Plant Disease Reporter. 46: 52–54.

328. SMITH, J. D. 1953. *Fusarium* patch disease. Journal of Sports Turf Research Institute. 8: 1–23.

329. ———. 1953. Fungi and turf diseases. Journal of the Sports Turf Research Institute. 8(29): 230–252.

330. ———. 1953. *Corticium* disease. Journal of Sports Turf Research Institute. 8: 252–258.

331. ———. 1954. Fungi and turf diseases. 4. *Corticium* disease. Journal of Sports Turf Research Institute. 8(30): 365–377.

332. ———. 1955. Fungi and turf diseases. 5. Dollar spot disease. Journal of Sports Turf Research Institute. 8(31): 35–39.

333. ———. 1956. Fungi and turf diseases. 6. *Ophiobolus* patch disease. Journal of Sports Turf Research Institute. 9: 180–202.

334. ———. 1958. The effect of lime application on the occurrence of *Fusarium* patch disease on a forced *Poa annua* turf. Journal of Sports Turf Research Institute. 9(34): 467–470.

335. ———. 1959. The effect of lime application on the occurrence of *Fusarium* patch disease on a forced *Poa annua* turf. Journal of Sports Turf Research Institute. 9: 467–470.

336. SMITH, R. C. 1942. *Nomophila noctuella* as a grass and alfalfa pest in Kansas (*Lepidoptera, Pyralididae*). Journal of Kansas Entomological Society. 15(1): 25–34.

337. SPLITTSTOESSER, W. E., and H. J. HOPEN. 1967. Response of bentgrass to siduron. Weeds. 15: 82–83.

338. SPRAGUE, H. B. 1933. Root development of perennial grasses and its relation to soil conditions. Soil Science. 36: 189–209.

339. SPRAGUE, H. B., and G. W. BURTON. 1937. Annual bluegrass (*Poa annua* L.) and its requirements for growth. New Jersey Agricultural Experiment Station Bulletin 630. pp. 1–24.

340. SPRAGUE, H. B., and E. E. EVAUL. 1930. Experiments with turfgrasses. New Jersey Agricultural Experiment Station Bulletin 497. pp. 1–55.

341. SPRAGUE, R. 1950. Diseases of cereals and grasses in North America. Ronald Press Co., New York. pp. 1–538.

342. STURKIE, D. G. 1937. Control of weeds in lawns with calcium cyanamide. Journal of American Society of Agronomy. 29(10): 803–808.

343. TARJAN, A. C. 1964. Rejuvenation of nematized centipedegrass turf with chemical drenches. Proceedings of the Florida State Horticultural Society. 77: 456–461.

344. TARJAN, A. C., and P. C. CLEO. 1955. Reduction of root-parasitic nematode populations in established bent grass turf by use of chemical drenches. Phytopathology. 45: 350.

345. TARJAN, A. C., and M. H. FERGUSON. 1951. Observations of nematodes in yellow tuft of bentgrass. USGA Journal and Turf Management. 4(2): 28–30.

346. TARJAN, A. C., and S. W. HART. 1955. Occurrence of yellow tuft of bentgrass in Rhode Island. Plant Disease Reporter. 39: 185.

347. TAYLOR, D. P., M. P. BRITTON, and H. C. HECHLER. 1963. Occurrence of plant parasitic nematodes in Illinois golf greens. Plant Disease Reporter. 47: 134–135.

348. THIRUMALACHAR, M. J., and J. G. DICKSON. 1953. Spore germination, cultural characters, and cytology of *Ustilago striiformis* and the reaction of hosts. Phytopathology. 43: 527–535.

349. THOMPSON, H. E. 1967. Buffalograss sod webworm. Turf-Grass Times. 2(5): 7.

350. TOOLE, E. H., and V. K. TOOLE. 1940. Germination of seed of goosegrass, *Eleusine indica*. Journal of American Society of Agronomy. 32(4): 320–321.

351. ———. 1941. Progress of germination of seed of *Digitaria* as influenced by germination temperature and other factors. Journal of Agricultural Research. 63(2): 65–90.

352. TROLL, J., and R. A. ROHDE. 1965. Pathogenicity of the nematodes *Pratylenchus penetrans* and *Tylenchorynchus claytoni* on turfgrasses. Phytopathology. 56: 1285.

353. ———. 1966. The effects of nematicides on turfgrass growth. Plant Disease Reporter. 50: 489–492.

354. TROLL, J., and A. C. TARJAN. 1954. Widespread occurrence of root parasitic nematodes in golf course greens in Rhode Island. Plant Disease Reporter. 38(5): 342–344.

355. TYSON, J. 1936. Snowmold injury to bent grasses. Quarterly Bulletin of the Michigan Agricultural Experiment Station. 19(2): 87–92.

356. VAN WEERDT, L. G., and W. BIRCHFIELD, and R. P. ESSER. 1959. Observations on some subtropical plant parasitic nematodes in Florida. Proceedings of the Soil and Crop Science Society of Florida. 19: 443–451.

357. WARD, E. W. B., and G. D. THORN. 1965. Evidence for the formation of HCN from glycine by a snow mold fungus. Phytopathology. 55: 1081.

358. WARREN, L. O., and J. E. Roberts. 1956. A hesperid, *Atalopedes campestris* (BDV), as a pest of bermuda grass pastures. Journal of Kansas Entomological Society. 29(4): 139–141.

359. WATSON, J. R. 1949. Compaction and irrigation studies on established fairway turf. Pennsylvania State College 18th Annual Turf Conference. pp. 88–90.

360. ———. 1950. Irrigation and compaction on established fairway turf. Ph.D. Thesis. Pennsylvania State University. pp. 1–69.

361. WATSON, J. R., and J. L. KOLB. 1956. Snowmold control. Golf Course Reporter. 24(7): 5–10.

362. WELLS, H. D. 1957. Southern turfgrass disease control. Proceedings of the 11th Annual Southeastern Turfgrass Conference. pp. 16–28.

363. WELLS, H. D., and B. P. ROBINSON. 1954. Cottony blight of ryegrass caused by *Pythium aphanidermatum*. Phytopathology. 44(9): 509–510.

364. WELTON, F. A., and J. C. CARROLL. 1938. Crabgrass in relation to arsenicals. Journal of American Society of Agronomy. 30: 816.

365. ———. 1940. Lawn experiments. Ohio Agricultural Experiment Station Bulletin 613. pp. 1–43.

366. ———. 1941. Control of lawn weeds and the renovation of lawns. Ohio Agricultural Experiment Station Bulletin 619. pp. 1–85.

367. ———. 1947. Lead arsenate for the control of crabgrass. Journal of American Society of Agronomy. 39: 513–521.

368. WILKINSON, A. T. 1968. Leatherjackets: new pest for golf greens. USGA Green Section Record. 6(1): 20–21.

369. WILLIAMS, A. S., and R. E. SCHMIDT. 1963. Studies on dollar spot and melting out. Golf Course Reporter. 31(5): 36–38.

370. ———. 1964. Studies on dollar spot and melting out. Golf Course Reporter. 32: 36–40.

371. WILSON, G. W. 1914. The identity of the anthracnoses of grasses in the United States. Phytopathology. 4(2): 106–113.

372. WILSON, J. K. 1923. The nature and reaction of water from hydathodes. Cornell University Experiment Station Memoir 65. pp. 1–11.

373. WINCHESTER, J. A. 1964. Nematode research: south Florida. Proceedings of the 12th Annual University of Florida Turf-Grass Management Conference. 12: 148–151.

374. WINCHESTER, J. A., and E. O. BURT. 1964. The effect and control of sting nematodes on Ormond bermuda grass. Plant Disease Reporter. 48: 625–628.

375. WOLFENBARGER, D. O. 1953. Insects and mite control problems on lawn and golf grasses. Florida Entomologist. 36(1): 9–12.

376. YARWOOD, C. E. 1939. Relation of moisture to infection with some downy mildews and rusts. Phytopathology. 29: 933–945.

377. YOUNGNER, V. B. 1959. Ecological studies on *Poa annua* in turfgrasses. Journal of British Grassland Society. 14: 233–237.

378. ———. 1961. Observations on the ecology and morphology of *Pennisetum clandestinum*. Orton. 16(1): 77–84.

379. YOUNGNER, V. B., and E. J. NUDGE. 1968. Chemical control of annual bluegrass as related to vertical mowing. California Turfgrass Culture. 18: 17–18.

380. ZUMMO, N., and A. G. PLAKIDAS. 1958. Brown patch of St. Augustine grass. Plant Disease Reporter. 42: 1141–1147.

Index

* Boldface numbers indicate pages on which index entries are discussed in greatest detail.